Advances in Experimental Medicine and Biology

More information about this series at http://www.springer.com/series/5584

Advances in Experimental Medicine and Biology

Gene W. Yeo
Editor

Systems Biology of RNA Binding Proteins

Springer

Editor
Gene W. Yeo
UCSD Moores Cancer Center
Institute for Genomic Medicine UCSD Stem Cell Program
La Jolla, CA, USA

ISSN 0065-2598 ISSN 2214-8019 (electronic)
ISBN 978-1-4939-1220-9 ISBN 978-1-4939-1221-6 (eBook)
DOI 10.1007/978-1-4939-1221-6
Springer New York Heidelberg Dordrecht London

Library of Congress Control Number: 2014947870

Printed on acid-free paper

Springer is part of Springer Science+Business Media (www.springer.com)

Preface

Ribonucleic acid (RNA)-binding proteins are rapidly being recognized as a major class of approximately a thousand proteins that are widely conserved throughout eukaryotes and play key roles in almost every aspect of RNA metabolism. RNA-binding proteins interact with hundreds to thousands of RNA substrates, including coding transcripts and long and short noncoding RNAs via *cis*-regulatory sequences or guided by associated small RNAs. Defects in the regulation of RNA targets by mutations within the RNA regulatory proteins, RNA *cis*-elements, or changes in protein availability or expression lead to numerous diseases.

With the surge of high-throughput, massively parallel microarray and sequencing technologies in the past few years, there have been tremendous advances in genomics and systems-level approaches and computational methods to probe how RNA-binding proteins affect various aspects of the RNA processing life cycle and uncover key RNA substrates. Advances in this area have increased recognition of the relevance of RNA-binding proteins to neurological disease, heart and muscle abnormalities, germ-line defects, and other genetic diseases. Additional developments include new molecular engineering approaches that utilize RNA-binding proteins to control gene expression, computational models of splicing regulation, and successful therapeutic strategies to modify RNA-binding protein-RNA interactions.

Given the exciting onset of intersection between systems approaches and RNA processing, I felt that it was timely to assemble the first book focused on genome-wide and systems-level perspectives on the diverse roles that RNA-binding proteins play in development and disease. In particular, I wanted to bring to the forefront some of the major questions that are currently under intense investigation by the experts contributing to this book.

The content of this book surveys a wide range of genome-wide and systems approaches to studying RNA-binding proteins in a myriad of organisms, cells, and tissues. More importantly, many chapters illustrate major steps in the processing of RNA and development and diseases caused by defects in these steps. Lastly, many authors discuss open questions, anticipated answers, and potential new areas of research on posttranscriptional gene regulation.

Tuschl and colleagues present a comprehensive analysis of RNA-binding proteins in humans, their evolutionary conservation, structural domains and expression, and a survey of various classes of RNA-binding proteins implicated in human disease.

Lecuyer and Bergalet opine on the importance of and regulatory principles involved in RNA localization and review emerging genomic and imaging technologies that have provided insights into RNA localization and diseases associated with localization defects.

Tian and Zheng review the *cis* elements and *trans* factors involved in cleavage and polyadenylation, emphasizing the importance of alternative cleavage and polyadenylation in gene regulation and the contribution of transcriptome-wide technologies in identifying the diversity of alternative 3′ ends.

Zhao and Chen provide an overview of the exciting new field of long noncoding RNAs and describe the methodologies used to study protein and DNA interactions with long noncoding RNAs during development and in disease states.

Lau and Clark recount the systems-wide approaches that have contributed to our understanding of Piwi proteins and Piwi-interacting RNAs on germ-line genome regulation.

Wang and Choudhury discuss current advances in the molecular engineering of RNA-binding proteins, emphasizing design principles and their applications as therapeutic agents and basic tools in biology.

Fairbrother and colleagues present current practices in medical genetics, the principles behind biochemical binding and functional assays, and advances in scaling up assays in assessing how genetic variation impacts RNA-binding protein interaction.

Carstens and colleagues feature the impact of RNA-binding proteins on key posttranscriptional changes during the epithelial to mesenchymal transition.

Bennett and colleagues provide an inside view of antisense oligonucleotide-based therapies for diseases caused by defects in premature messenger RNA processing.

Swanson and Goodwin emphasize the importance of altering RNA-binding protein functions in microsatellite expansion diseases.

Cooper and Giudice discuss how activities of RNA-binding proteins affect heart development and cardiomyopathy and how these RNA-binding proteins modulate these processes.

Pasquinelli and Azoubel Lima highlight genome-wide methodologies utilized to find microRNAs and their mRNA targets in *C. elegans*.

Barash and Vaquero-Garcia survey the current state of developing regulatory models for splicing using machine learning-based approaches and an introduction to the AVISPA web tool.

The contributors of this book are internationally recognized leaders in the arena of RNA processing, and we envision that this book will serve as a valuable resource for both experts and non-experts. Advanced undergraduates, entering graduate students in biology, chemistry, molecular engineering, computer science, and bioinformatics, and medical students and postdoctoral fellows who are new to the arena of posttranscriptional gene regulation should find this book accessible. We hope that the chapters in this volume will stimulate interest and appreciation of the complexity and importance of posttranscriptional gene regulation to its readers and even lead them to pose new solutions to the many challenges that lie ahead in understanding how RNA-binding proteins affect gene regulation.

I sincerely express my greatest gratitude to the contributors to this book, Yoseph Barash, Jorge Vaquero-Garcia, Thomas Bebee, Benjamin Cieply, Russ Carstens, Jimena Guidice, Thomas Cooper, Rachel Soemedi, Hugo Vega, Judson M Belmont, Sohini Ramachandran, William Fairbrother, Josef Clark, Nelson Lau, Julie Bergalet, Eric Lecuyer, Sarah Azoubel Lima, Amy Pasquinelli, Punit Seth, Frank Rigo, Frank Bennett, Marianne Goodwin, Maurice Swanson, Dinghai Zheng, Bin Tian, Stefanie Gerstberger, Markus Hafner, Manuel Ascano, Thomas Tuschl, Rajarshi Choudhury, Zefeng Wang, Ling-ling Chen, and Jing Crystal Zhao.

La Jolla, CA, USA Gene W. Yeo

Contents

Chapter 1
Evolutionary Conservation and Expression of Human RNA-Binding Proteins and Their Role in Human Genetic Disease

Stefanie Gerstberger, Markus Hafner, Manuel Ascano, and Thomas Tuschl

Abstract RNA-binding proteins (RBPs) are effectors and regulators of posttranscriptional gene regulation (PTGR). RBPs regulate stability, maturation, and turnover of all RNAs, often binding thousands of targets at many sites. The importance of RBPs is underscored by their dysregulation or mutations causing a variety of developmental and neurological diseases. This chapter globally discusses human RBPs and provides a brief introduction to their identification and RNA targets. We review RBPs based on common structural RNA-binding domains, study their evolutionary conservation and expression, and summarize disease associations of different RBP classes.

Keywords RNA-binding domains, overview • RNA-binding proteins, tissue specificity • RNA-binding proteins, abundance • RNA-binding proteins, genetic diseases

1 Principles of Posttranscriptional Gene Regulation

RNA is an essential constituent of all living organisms and central to decoding the genetic information of every cell. Recent advances in RNA sequencing technologies have facilitated the discovery of novel transcripts and we will soon know the precise composition of most cellular transcriptomes. While functional annotation for many RNAs is still in progress, the major classes of RNAs have now been described (Table 1.1). The most abundant RNAs, constituting 90 % of cellular RNAs by copy number, are shared by all organisms and required for protein synthesis: rRNAs, tRNAs, and mRNAs (Table 1.1). The remaining 10 % are noncoding RNAs (ncRNAs) that mainly serve as guides or molecular scaffolds in a variety of processes including RNA splicing, RNA modification, and RNA silencing. The structure, length, and composition of these RNAs and their

S. Gerstberger • M. Hafner • M. Ascano • T. Tuschl (✉)
e-mail: ttuschl@mail.rockefeller.edu

G.W. Yeo (ed.), *Systems Biology of RNA Binding Proteins*, Advances in Experimental Medicine and Biology 825, DOI 10.1007/978-1-4939-1221-6_1

Table 1.1 Functional description of the main RNA classes in humans and their length distribution

RNA class	Size (nt)	Biological role (additional reviews on function and biogenesis)
Messenger RNA (mRNA)	~200–100,000	Encodes the information for protein-coding genes, translated by ribosomes (Dreyfuss et al. 2002; Glisovic et al. 2008; Müller-McNicoll and Neugebauer 2013)
Transfer RNA (tRNA)	~70–95	RNA adaptor molecule, transports amino acids to ribosome and recognizes specific triplet codons on mRNA (Suzuki et al. 2011; Maraia and Lamichhane 2011; Simos and Hurt 1999)
Ribosomal RNA (rRNA)	121–5,072	Structural component of ribosomes (Boisvert et al. 2007; Ciganda and Williams 2011; Granneman and Baserga 2004)
Small nuclear RNA (snRNA)	~70–190	snRNAs U1, U2, U4, U5, U6, U11, U12, U4atac, and U6atac are core components of the spliceosome; U7 snRNA functions in 3′end maturation of histone RNAs (Kiss 2004; Matera et al. 2007)
Small nucleolar RNA (snoRNA) and small Cajal-body-specific RNA (scaRNA)	~50–450	Guide chemical modifications (methylation and pseudouridylation) of rRNAs, snRNAs, and snoRNAs (Filipowicz and Pogaćić 2002; Kiss et al. 2006; Matera et al. 2007)
microRNA (miRNA) and small interfering RNA (siRNA)	21–22	Associate with AGO proteins, guide them to target sequences predominantly in the 3′UTRs of mRNAs, induce degradation and translational repression (Bartel 2009; Kim et al. 2009)
piwi-interacting RNA (piRNA)	~28–32	Associates with PIWI proteins; PIWI RNP complexes induce ribonucleolytic cleavage and epigenetic silencing of transposable elements (Kim et al. 2009; Siomi et al. 2011)
Long intervening noncoding RNA (lincRNA), 7SK RNA	>200	Recruits chromatin modifiers and remodeling complexes, modulates transcription by recruitment of protein cofactors to transcription starts sites and enhancers, functions as molecular scaffolds for nuclear RBPs (Batista and Chang 2013; Ulitsky and Bartel 2013); 7SK RNA regulates transcription elongation (Peterlin et al. 2011)
Ribonuclease P/(RNase P) and mitochondrial RNA-processing endonuclease (MRP RNase)	~260–340	Ribonucleolytic RNP complexes that carry out processing of precursor tRNAs, rRNAs, snRNAs, and other noncoding RNAs (Xiao et al. 2002; Jarrous 2002; Ellis and Brown 2009; Esakova and Krasilnikov 2010)
Y RNA	~80–110	Small noncoding RNAs that form an RNP complex with TROVE2 (Ro60) protein and act as RNA chaperones, have a role in DNA replication and immune response (Hall et al. 2013; Köhn et al. 2013)

(continued)

Table 1.1 (continued)

RNA class	Size (nt)	Biological role (additional reviews on function and biogenesis)
Signal recognition particle RNA (7SL/SRP RNA)	~300	RNA of the signal recognition particle; the complex recognizes signal sequences of newly synthesized peptides and targets them to the endoplasmatic reticulum (Akopian et al. 2013)
Vault-associated RNA (vtRNA)	~80–120	Small noncoding RNAs, part of the vault RNP complex, involved in drug resistance, downregulate mRNA targets through posttranscriptional gene silencing (Berger et al. 2008)
Telomerase RNA (telRNA)	~450	RNA component of the telomerase complex TERC, which acts as reverse transcriptase and elongates telomerase repeats, TERC is structurally related to box H/ACA snoRNAs (Egan and Collins 2012)

Additional reviews on biogenesis pathways and RBP components interacting with each class of RNA are referenced

ribonucleoprotein particles (RNP) are distinct and allow their integration into diverse functions and layers of regulation to control target RNAs and their many functions.

Posttranscriptional gene regulation (PTGR) is a term that refers to the cellular processes that control gene expression at the level of RNA; it encompasses RNA maturation, modification, transport, and degradation. Consequently, every RNA molecule independent of its ultimate function is at some level subject to PTGR. RNA-binding proteins (RBPs) are central players of PTGR, as they directly bind to RNAs to form RNPs. In many cases, the RNP is the most basic unit, comprising a complex of obligate RNA and protein partners (e.g., snRNPs, snoRNPs, RNase P, ribosome subunits), which elicits its respective function. However, many other types of RNAs, particularly mRNAs and tRNAs, only transiently associate with RBPs, whose functions are necessary for their proper maturation, localization, and turnover (Dreyfuss et al. 2002; Granneman and Baserga 2004; Phizicky and Hopper 2010; Müller-McNicoll and Neugebauer 2013). Indeed most mRNA-binding proteins (mRBPs) have thousands of targets they regulate (Ascano et al. 2011). Hence the proper assembly and function of RNA-protein complexes are critical for development and maintenance of all cells and organisms. For a large fraction of RBPs, we are only starting to understand the complexity of their basic molecular roles, modes of recognition, and global targets.

In this chapter we review the current state of knowledge of the protein components involved in PTGR in humans. We discuss common patterns found among RBPs, based on targets, evolutionary conservation, shared structural domains, and cell-type-specific or ubiquitous expression. We then examine various classes of RBPs commonly implicated in human disease.

2 Human RBPs

2.1 Experimental and Bioinformatic Approaches Leading Towards a Census of RBPs

A complete catalogue of the proteins involved in PTGR is an important goal. Historically, different strategies have been employed towards the identification of RBPs (Ascano et al. 2013). Common approaches used RNA pull-down assays to recover associated proteins in cell lysates, followed by their mass spectrometric identification, or candidate proteins were recombinantly expressed and interrogated for their RNA-binding properties in vitro. These RNA-centric approaches identified the interactome of subsets of RNAs but did not capture the whole RBP proteome, and were not suitably of high throughput.

The first genome-wide approaches for the identification of proteins involved in PTGR utilized predictive methods and searched for the presence of protein domains conferring RNA binding. Early studies on the protein components of heteronuclear RNPs (hnRNPs) led to the identification of the first conserved, canonical RNA-binding domain (RBD) within RBPs (Burd and Dreyfuss 1994). Following these initial discoveries, and facilitated by advances in genome sequencing and the acquisition of protein structures, more precise classification of structural and functional protein domains followed rapidly (Henikoff et al. 1997). Computational prediction algorithms that use probability matrices from multiple sequence alignments enabled the detection of structural domains in uncharacterized protein sequences across organisms. The results of these predictions are publically available in a number of databases such as Interpro, Pfam, SCOP, SMART, or CDD (Murzin et al. 1995; Apweiler et al. 2001; Marchler-Bauer et al. 2003; Letunic et al. 2009; Finn et al. 2010). Among these domain classifications, at least 600 can be found with annotation referring to involvement in RNA-related processes.

Predicting the number of RBPs encoded in various genomes has remained a challenge. RBPs were defined by the presence of one or more canonical RBDs, such as RRM, KH, CSD, zinc fingers, and PUF domains (Lunde et al. 2007). Selecting these predominantly mRNA-binding RBDs, the number of RBPs was initially estimated to ~400–500 in human and mouse (McKee et al. 2005; Galante et al. 2009; Cook et al. 2011), ~300 in *D. melanogaster* (Lasko 2000; Gamberi et al. 2006), ~250–500 in *C. elegans* (Lasko 2000; Lee and Schedl 2006; Tamburino et al. 2013), and ~500 RBPs in *S. cerevisiae* (Hogan et al. 2008). Inclusion of RNA-processing domains involved in RNA metabolism of every known type of RNA leads to numbers near ~700 RBPs in humans (Anantharaman et al. 2002).

Other predictive approaches such as the Kyoto Encyclopedia of Genes and Genomes database (KEGG) (Kanehisa and Goto 2000) and the Gene Ontology project (GO) (Ashburner et al. 2000) integrate domain annotations, protein homologies, and searches of scientific literature statements. These estimate the number of human proteins with RNA-related functions to ~1,800 proteins (Fig. 1.1). However, these methods are often not reliable due to false classifications of proteins, leading to a large number false positives and false negatives.

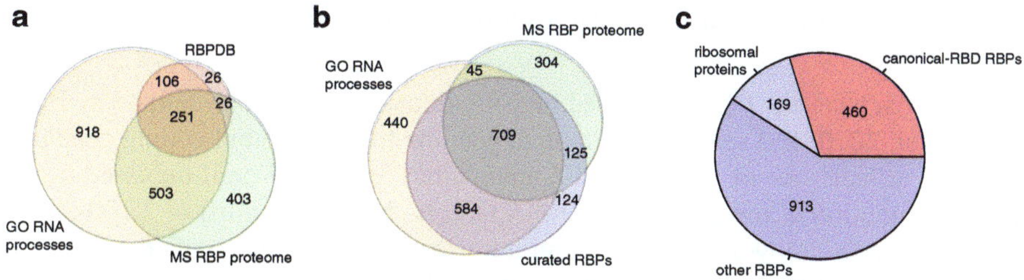

Fig. 1.1 Different approaches to define the catalogue of human RBPs. (**a**) Venn diagram showing the overlap of proteins with RNA-related Gene Ontology (GO) categories (Ashburner et al. 2000) (*orange*), the human RNA-binding proteome identified by RNA-cross-linking and mass spectrometry studies (MS RBP proteome, *green*) (Baltz et al. 2012; Castello et al. 2012; Kwon et al. 2013), and the RBPDB database of human RBPs with canonical RBDs (Cook et al. 2011) (*red*). (**b**) Venn diagram showing the overlaps of GO RBPs (*orange*), MS RBP proteome (*green*), and the curated RBP list based on analysis of RNA-binding domains and experimental evidence of RNA binding found in the literature (*violet*). (**c**) Composition of RBPs in the curated RBP list: Canonical-RBD RBPs (containing canonical RBDs (Lunde et al. 2007; Cook et al. 2011), *red*), ribosomal proteins (*bright violet*), other RBPs (*dark violet*)

In parallel, experimental proteome-wide methods were employed to identify the number of known and novel RBPs such as the development of protein microarrays, which allowed increased throughput for probing the RNA-binding capabilities of a fraction of the proteome in vitro, using RNA probes of defined sequence (Scherrer et al. 2010; Tsvetanova et al. 2010; Siprashvili et al. 2012). In an attempt to comprehensively identify existing and novel RBPs in human at large scale with a singular approach, cross-linking-based methods were recently introduced. In these methods, RBPs were covalently cross-linked to endogenous RNAs using in vivo UV cross-linking, followed by polyA selection of mRNAs, and subsequent identification of interacting proteins by mass spectrometry. These approaches identified ~800 mRBPs in human HEK293 and HeLa cell lines, respectively (Baltz et al. 2012; Castello et al. 2012), 555 in mouse embryonic stem cells (mESCs) (Kwon et al. 2013), and 200 mRBPs in yeast (Mitchell et al. 2013). Together, 1,100 of known and putative human mRBPs were experimentally defined and, assuming homologous function between mouse and human proteins, an additional ~80 proteins may be added (Fig. 1.1a, b). A significant portion of these (64 %) overlapped with known GO-classified RBPs (Fig. 1.1a, b). Many of the residual mRBP candidates did not contain previously described RBDs and require further experimental validation, while other known and expressed RBPs were missed due to the sensitivity of the experiments. However, in comparison to earlier predictive counts of the number of mRBPs (Cook et al. 2011), this approach expanded the mRBP proteome from ~400 to ~1,200 proteins and may, with increasing sensitivity, represent the most suitable method to identify novel RBPs in proteome-wide experiments in different cell types.

Here, we describe our attempt in generation of a curated and comprehensive list of RBPs involved in PTGR processes to guide us in their study of molecular and cellular function and definition of all RNA-related processes.

2.2 Generation of a Curated List of Human RBPs

Our approach selects RBDs involved in RNA-related processes as defined by Pfam (Finn et al. 2010) and searches the human genome for any protein-coding gene that contains at least one of the selected domains, but remains overall unbiased to the putative function of the gene and its RNA targets. Arriving at a list of 2,130 candidates, we added known RBPs from literature searches with unclassified RBDs and additionally screened proteins defined as RBPs by GO (Ashburner et al. 2000) and proteome-wide mass spectrometry datasets (Baltz et al. 2012; Castello et al. 2012; Kwon et al. 2013) for literature-based evidence of their involvement in PTGR.

The list of RBPs was finalized according to the following main criteria: (1) the proteins possessed defined RNA-binding or RNA-enzymatic domains, (2) the proteins were experimentally shown to be part of RNP complexes and thereby involved in RNA metabolic pathways, or (3) they possessed high sequence identity to homologs and paralogs involved in PTGR. Some of the candidate RBPs identified in the recent cross-linking-mass spectrometry studies were not considered as RBPs, if their RNA-binding activity could not be confirmed independently in other published datasets or their domain structure, family members, and homologs were not indicative of an RBP. We furthermore disregarded proteins containing putative RBDs if they showed strong evidence for exclusive roles in RNA-unrelated pathways, such as the majority of C2H2 zinc finger transcription factors of which only a small subset are RNA-binding, e.g., TF3A binding to 5S rRNA (Brown 2005). We included proteins, which are components of well-defined, large RBP complexes, such as the ribosome or the spliceosome, as it is difficult to establish with certainty which proteins interact with RNA directly or indirectly in these large RNPs. This approach is supported by recent proteome-wide cross-linking studies, and the RNA-binding properties of scaffold proteins CNOT1 and TDRD3 have for example emerged through this process (Thomson and Lasko 2005; Siomi et al. 2010; Baltz et al. 2012; Castello et al. 2012; Kwon et al. 2013). Through this curation process we reduced the union of ~3,700 proteins, derived from domain annotation, mass spectrometry datasets, literature search, and GO annotation, to arrive at a final of 1,542 proteins (Fig. 1.1b). The resulting curated list of RBPs contains proteins interacting with all RNA classes. A comparison to the conventionally named "canonical RBPs" (Lunde et al. 2007; Cook et al. 2011) shows that canonical RBPs only represent one-third of RBPs in this set and the majority of RBPs in our set would not be considered using currently available datasets (Fig. 1.1c). The following sections discuss abundance, evolution, expression, and RBPs in human disease based on this curated set of RBPs.

3 Quantitative Aspects of Proteins in RNA Metabolism

The dynamics of complex assembly and composition of RNPs, their targets, and protein cofactors are extremely sensitive to the quantitative relationship between the abundance of RBPs and their targets (Dreyfuss et al. 2002; Müller-McNicoll and

Neugebauer 2013). RBPs are in constant competition for binding to frequently occurring short and degenerated RNA sequence elements and thus the cellular compartment concentration of RNA and RBPs will affect the equilibrium of dynamic RNP formation and disassembly. Processes such as pre-mRNA splicing and alternative polyadenylation, where the choice of alternative splice sites or 3′UTR lengths is dependent on the abundance of splicing enhancers, silencers, or U1 snRNP (Smith and Valcarcel 2000; Kaida et al. 2010; Berg et al. 2012; Kornblihtt et al. 2013), emphasize the importance of determining precise RBP levels.

Approximately 7 % of all protein-coding genes are committed to PTGR, but their contribution to the pool of expressed proteins in cells is much higher. We analyzed expression level of ubiquitous RBPs based on RNA-seq data in HEK293 cells (Teplova et al. 2013) (Fig. 1.2). In this cell line, RBPs represented 9 % (1,364 genes) of the ~16,300 expressed genes (expressed with RPKM > 1), but their corresponding transcripts represent more than 25 % of total cellular mRNA, including 7 % mRBPs and 14 % ribosomal proteins (RBP categorization discussed in next section), stressing the abundance of mRNA metabolism and the central role of protein translation (Fig. 1.2). In contrast, transcription factors and cytoskeletal proteins were not

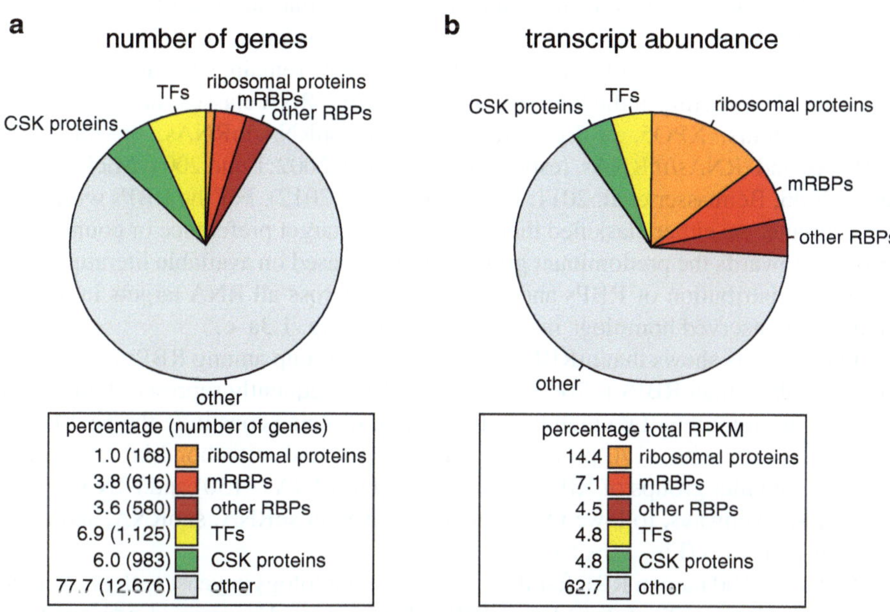

Fig. 1.2 Analysis of RBP abundance. Curated RBPs are subclassified into ribosomal proteins (*orange*), mRBPs (*bright red*), and other RBPs (*dark red*). Percentages of RBPs are compared to a set of GO-defined transcription factors (TFs, *yellow*), a set of GO-defined cytoskeletal proteins (CSK proteins, *green*) (Ashburner et al. 2000), and all other expressed genes (*grey*). (**a**) Count of expressed genes with RPKM > 1 in RNA-seq data from HEK293 cells (Teplova et al. 2013). (**b**) Relative abundance of each gene group is given by the summation of expression levels of genes in each category

overrepresented in the transcriptome of HEK293 cells. In summary, about a quarter of the transcriptome is committed to RNA metabolism, highlighting its fundamental role in the cell.

4 Paralogous RBP Families and their Targets

Determining the evolutionary relationships and the conservation of gene families has been critical for understanding gene function and emphasized the utility of model organisms for the study of fundamental biological processes (Henikoff et al. 1997). To account for redundancies among RBPs it is therefore beneficial to consider the RBP family as the smallest functional unit. Grouping the 1,542 RBPs into paralogous gene families with at least 20 % homology gives 1,113 RBP families with one or two members on average. The large number of families reflects a high diversity of RBPs in human.

Here, we categorized RBPs and RBP families based on their reported natural targets and examined their distribution and evolutionary relationships among different classes. Although RBPs often show some degree of interaction with a range of target RNAs in vivo, most of them are committed to one subtype of RNA (Hafner et al. 2010; Wang et al. 2012; Ascano et al. 2012; Hussain et al. 2013; Lovci et al. 2013; Wang et al. 2013) (Table 1.1). Some exceptions remain, such as RNA nucleases, and RBPs acting at the interface of two different RNA classes, such as spliceosomal proteins, XPO5, or EEF1A, recognizing snRNA/mRNAs, pre-miRNAs/mRNAs, and tRNAs/mRNAs, respectively (Liu et al. 2002; Lund 2004; Mickleburgh et al. 2006; Bennasser et al. 2011; Dever and Green 2012). For the RBPs with multiple targets, we either classified them as diverse in target preference or counted the proteins towards the predominant group of targets based on available literature. The resulting distribution of RBPs and RBP families across all RNA targets in human and their conserved homologs in yeast is shown in Fig. 1.3a–c.

Our analysis shows that mRBPs form the largest group among RBPs comprising 45 % of all human RBPs (~700 proteins). mRBPs frequently represent families of RBPs with more than two members. Ribosomal proteins constitute the next larger group of RBPs with ~170 proteins of the cytosolic and mitochondrial ribosomes. The next smaller groups of RBPs are committed to tRNA (~150) and rRNA (~120) biogenesis pathways, followed by proteins involved in snRNA, snoRNA, and other ncRNA pathways (Fig. 1.3a, b).

Of the ~1,100 human RBP families, ~550 have homologs in yeast with on average 30 % homology. Different clades of RBPs display varied degrees of conservation. Cytosolic ribosomal proteins are the most conserved with ~57 % homology, while proteins associating with mRNAs or ncRNAs are least conserved, 27 % and 20 %, respectively, and also have the least number of conserved homologs, 45 % and 21 %, respectively (Fig. 1.3c). Nevertheless, despite the gene expansions within protein

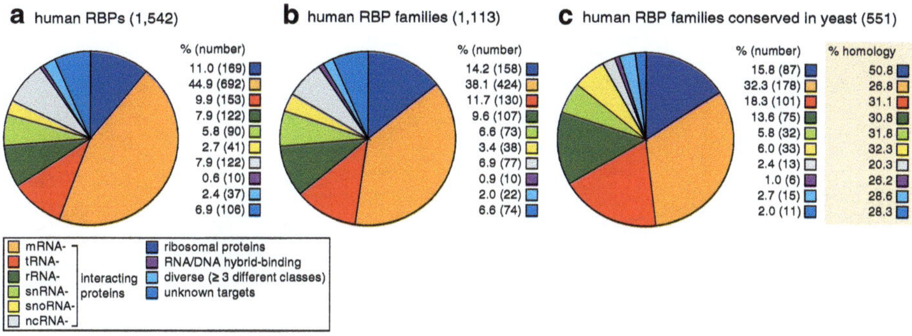

Fig. 1.3 Number of RBPs involved in different RNA pathways. Curated RBPs are categorized into the following groups: (*1*) Ribosomal proteins and RBP-interacting proteins (e.g., TUDOR proteins, RBP transport proteins) (*dark blue*), (*2*) mRNA-binding proteins (*orange*), tRNA-binding proteins (*red*), rRNA-binding proteins (*dark green*), snRNA-binding proteins (*bright green*), snoRNA-binding proteins (*yellow*), ncRNA-binding proteins (ncRNAs defined as miRNA, piRNA, MRP, 7SL, XIST, lincRNAs, telRNA, etc.) (*light grey*), RNA/DNA-hybrid-interacting proteins (*violet*), RBPs interacting unselectively with a range of RNA targets (*light blue*), RBPs with unknown RNA targets (*marine blue*). Distribution into the listed categories of the (**a**) 1,542 curated human RBPs, (**b**) 1,113 human paralogous RBP families, and (**c**) conserved paralogous RBP families in *S. cerevisiae* and their average conservation score (*orange box*)

families at later evolutionary stages (Venter 2001; Van de Peer et al. 2009), the relative ratios of paralogous RBPs families invested in the different RNA pathways remain approximately the same across evolution as seen for the distribution between human and yeast RBP families (Fig. 1.3b, c). This breadth of PTGR factors agrees with an earlier analysis of 32 RBP domain classes of canonical RBDs (including RRM, KH, dsrm, DEAD, PUF, Piwi, PAZ, zinc finger, LSM) showing that the large diversity of RBPs found in contemporary metazoans was already established in the last common ancestor (LCA) of animals, and which possessed an estimated total number of 88 RRM, 15 KH, 49 DEAD box, 9 dsrm, and 38 other RBD proteins (Kerner et al. 2011). Thus the complexity of PTGR was present at the earliest stages of evolution, reflecting that RNA metabolism lies at the heart of eukaryotic gene regulation.

Visualization of the evolutionary relationships of RBP families facilitates systems biology approaches to dissect their regulatory roles. Phylogenetic trees give an intuitive graphic representation of the conservation of proteins, highlight closely related homologs, and thereby provide a glimpse into function of uncharacterized RBPs if function has been already established for a relative. Phylogenetic comparison of the predominantly mRNA-binding KH-domain-containing proteins and the proteins of the small subunit of the cytosolic ribosome illustrates the differences in their evolutionary trajectory (Fig. 1.4).

KH proteins experienced multiple gene expansions, as noted earlier for mRBPs, and evolved new RBP families at the later metazoan stages, thereby

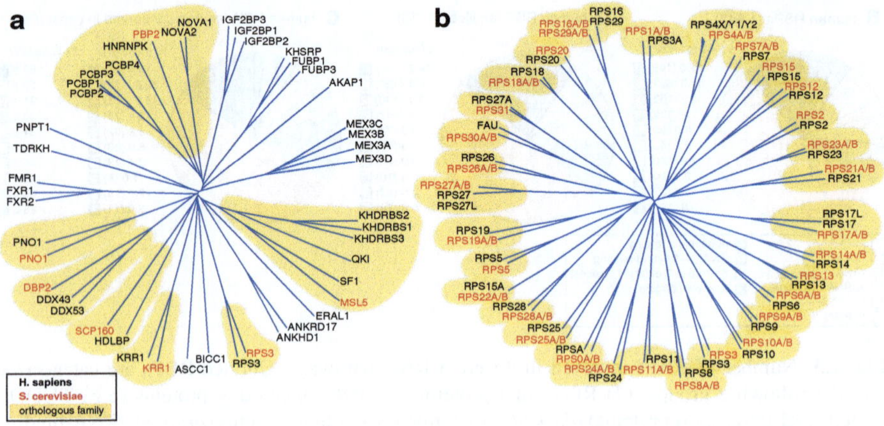

Fig. 1.4 Phylogenetic trees of (**a**) KH-containing proteins and (**b**) ribosomal proteins of the small subunit. Branch lengths are scaled to the sequence identity of the proteins. *S. cerevisiae* proteins are marked in *red*, human proteins in *black*, homologous families with conserved members in *S. cerevisiae* highlighted in *yellow*

expanding and diversifying components involved in various regulatory pathways, such as mRNA splicing, translational regulation, and transport. KH protein families contain between one and four members in human, and possess generally one distantly related homolog in yeast (Fig. 1.4a). Multiple family members often have redundant biological functions and RNA target spectra. For example, members of the FMR1 family (FMR1, FXR1, FXR2) or the IGF2BP1 family (IGF2BP1, 2, and 3) show >90 % identical RNA-binding specificities (Hafner et al. 2010; Ascano et al. 2012).

In contrast, cytosolic ribosomal proteins display an unusually high conservation, not too surprising, given that the process of protein translation is conserved to such a high degree between prokaryotes and all clades of eukaryotes that functional details of translation determined in bacteria are almost identical to higher systems (Wool et al. 1995; Melnikov et al. 2012; Dever and Green 2012). The ~90 human ribosomal cytosolic proteins are highly similar in structure and function between yeast and human and show late divergence in evolution, as illustrated for the phylogenetic tree of small ribosomal subunit proteins (Fig. 1.4b). With on average 57 % protein identity, all human cytosolic ribosomal proteins have direct one-to-one, or due to a whole-genome duplication in yeast, one-to-two or two-two matching homologs (Wool 1979; Wool et al. 1995; Anger et al. 2013). In contrast, the majority (80 %) of the ~80 human mitochondrial ribosomal proteins (Matthews et al. 1982) have no homologs in yeast, and the few that are conserved have comparatively low homology (22 % identity), reflecting that mitochondrial ribosomes, acquired through eubacterial endosymbiosis, rapidly evolved independently across species and that major remodeling events happened later in evolution (Cavdar Koc et al. 2001; O'Brien 2003).

5 Structural Analysis of RNA-Binding Domains

Analysis of structural features in proteins and the grouping of proteins into domain classes can help to understand their biological function (Henikoff et al. 1997; Anantharaman et al. 2002). Structural domains can predict how RBPs recognize and bind RNAs. It can also uncover redundancies to other RBPs in target recognition, as well as highlight families of RBPs that remain to be characterized. The size of a domain class mirrors its diversity and evolutionary adaptation to biological pathways.

Structure-guided searches can be valuable to place proteins into biological pathways and, for instance, DICER1 and DROSHA were identified as the endonucleases responsible for double-stranded RNA (dsRNA) processing in microRNA (miRNA) maturation based on known structure and substrate preferences of the dsRNA-processing bacterial and yeast RNase III enzymes (Hammond et al. 2000) (Bernstein et al. 2001; Lee et al. 2003). Similarly, the structural similarity of AGO proteins and the germline-specific PIWI proteins sparked the search for PIWI-interacting small RNAs with similar features as miRNAs, now known as piRNAs (Girard et al. 2006; Grivna 2006; Aravin et al. 2006).

For a review of the structural features of RBPs, we analyzed characteristic domain combinations of RBD classes (Fig. 1.5a) and will give here a brief overview over the abundant RBD classes and their modes of RNA-binding, natural targets, and the processes they are involved in. For excluded classes, we refer to a number of excellent review articles (Burd and Dreyfuss 1994; Sommerville 1999; Aravind and Koonin 2001a; Aravind and Koonin 2001b; Anantharaman et al. 2002; Arcus 2002; Szymczyna et al. 2003; Kim and Bowie 2003; Maraia and Bayfield 2006; Lunde et al. 2007; Rajkowitsch et al. 2007; Glisovic et al. 2008; Curry et al. 2009; Mihailovich et al. 2010; Zhang et al. 2010). To give additional insight into the structural properties of RBPs, we distinguished between RBDs with only RNA-binding properties (nonenzymatic RBDs), and RBDs that also contain enzymatic functions (enzymatic RBDs), such as RNA helicases, ATPases, polymerases, editing enzymes, and nucleases.

5.1 Modes of RNA Interaction by RBPs and their Domain Organization

Prototypical single-stranded RNA (ssRNA)-binding domains interact with their targets in a nucleobase-sequence-specific manner typically binding between 4 and 8 nucleotides (Singh and Valcarcel 2005; Lunde et al. 2007; Glisovic et al. 2008). Specificity is introduced mainly by hydrogen bonding and van der Waals interactions of the nucleobases with the protein side chains or the carbonyl and amide groups of the main chain (Auweter et al. 2006), often leaving the RNA phosphate backbone exposed to the solvent. Additional base stacking interactions with aromatic amino acids or positively charged residues in cationic π interactions serve to

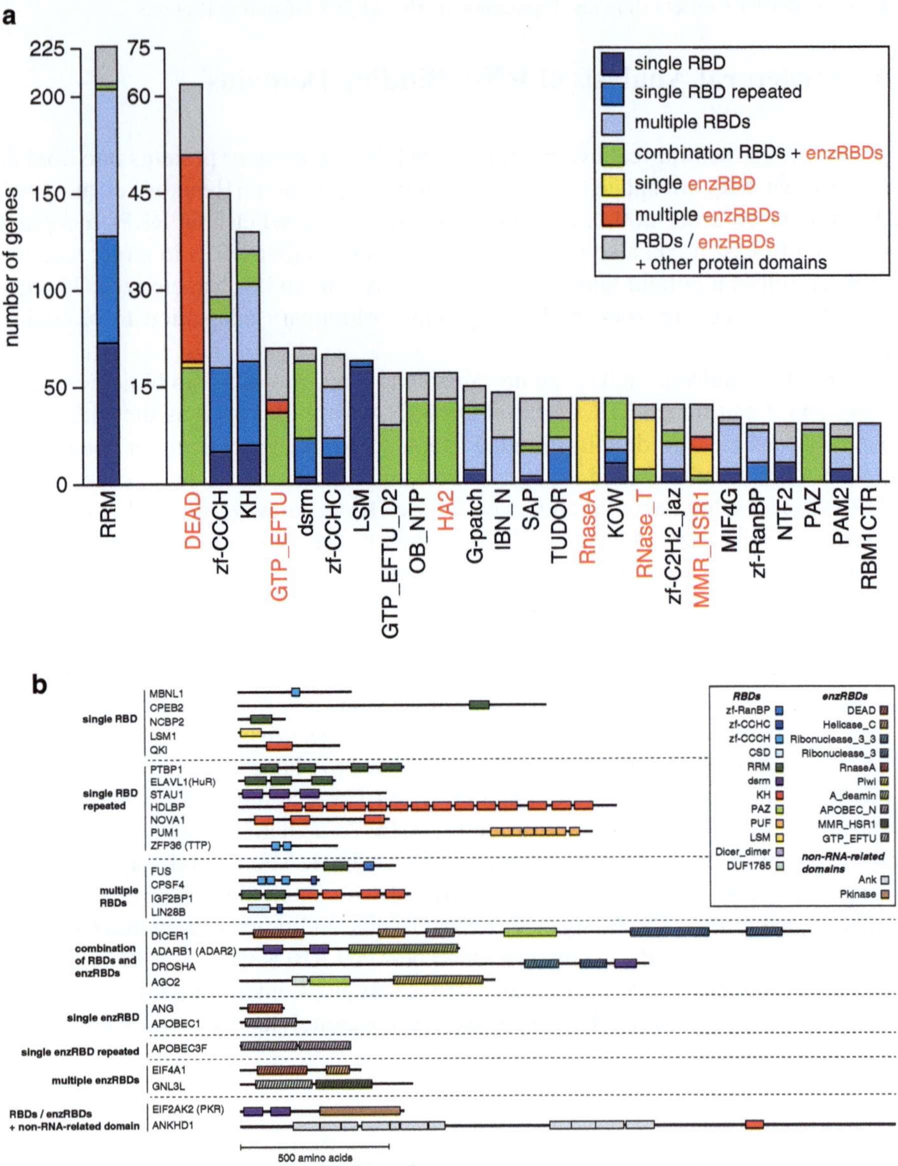

Fig. 1.5 (**a**) Analysis of structural patterns of the most abundant RNA-binding domains (with ≥ 9 members) in humans. Shown are the counts for the number of genes containing the listed RNA-binding domains (with ≥ 9 members) named by its Pfam abbreviation (Finn et al. 2010). RNA-binding domains are categorized into those binding RNA without additional enzymatic activity (RBD) (*black*) and those with additional enzymatic activity (enzRBD) (*red*). The RBD category was broadly defined to include protein-protein interacting domains known to interact with RBPs, such as those found in TUDOR family proteins (TUDOR) or ribosomal proteins. The following structural patterns are counted: (*1*) singular occurrence of an RNA-binding domain (RBD—*dark blue*, enzRBD—*yellow*), (*2*) single RBD repeated (RBD—*marine blue*), (*3*) multiple RBDs (RBD—*light blue*, enzRBD—*red*), (*4*) combinations of RBDs and enzRBDs (*green*), and (*5*) combination of at least one RBD/enzRBD with at least one other, non-RNA-related protein domain (*grey*). (**b**) Scheme of domain structure organization of representative RBPs, categorized into the domain combination classes listed in (**a**)

increase affinity. Double-stranded RNA (dsRNA)-binding proteins achieve specificity through recognition of shape of RNA secondary structure, such as stem–loops (Masliah et al. 2013). Non-sequence-specific RBDs generally interact with the negatively charged phosphate backbone, leaving the bases exposed to the solvent. To achieve specificity, these RBPs can interact with cofactors recruiting them to specific targets, as has been observed for many RNA helicases (Rocak and Linder 2004; Auweter et al. 2006).

Many RBDs (but also DNA-binding domains) derive from a few common superfamily folds, such as the oligonucleotidyl transferase fold and the oligosaccharide-binding fold (OB-fold). Oligonucleotidyl transferase fold proteins include enzymatic RBPs such as TUTases, polyA polymerases, RNA ligases, tRNA CCA-adding enzymes, and immune-stimulatory 2′,5′-oligoadenylate synthases (Kuchta et al. 2009). RBDs of the OB-fold superfamily are the S1, PAZ, and CSD domains (Murzin 1993; Arcus 2002; Lunde et al. 2007). However, RBDs largely diversified throughout evolution and most RBD classes have only one member, while only 4 % of all RBD classes found in human have more than eight members. Members of the 26 most abundant RBD classes (with 9 and more members) constitute a third of the 1,542 curated RBPs (Fig. 1.5a); most of them are mRBPs. Some of the highly studied RBDs (Lunde et al. 2007), such as the PUF (two proteins), S1, CSD, and PIWI domains (eight members each), define smaller RBD classes in humans. Particularly, ribosomal structural components and proteins involved in processes related to ribosome maturation are unique and thus cannot be classified into large families of related structural organization (Korobeinikova et al. 2012).

More than half of all RBPs contain only one RBD; mRBPs, however, form a notable exception and often have multiple RBDs, either one repeated RBD or multiple RBDs in combination. This modular design allows flexibility and versatility for target recognition and, as RBDs usually recognize relatively short stretches of RNA, increases the affinity and specificity for RNA targets by extending the RNA recognition element (RRE) of the protein (Lunde et al. 2007). A few RBDs are exclusively found in combination with other conserved protein domains, such as RBM1CTR in hnRNPs or the PAZ domain found in the Argonaute proteins and DICER1.

5.2 Abundant Nonenzymatic RBDs

5.2.1 RNA-Recognition Motif

The ssRNA-recognition motif (RRM) is the most frequently found RNA-binding domain in eukaryotes, and has 226 members in humans. The ~90 amino-acid-long domain adopts a βαββαβ topology and is composed of two RNP consensus motifs that recognize 4–6 nucleotides (nts) by stacking interactions of the bases with three conserved aromatic amino acids in the β-sheets (Auweter et al. 2006; Lunde et al. 2007; Cléry et al. 2008). Its small size and modular organization yield flexibility to adaptive change and allowed the RRM domain to vastly expand during evolution (Anantharaman et al. 2002).

Deviations from this canonical binding mode, including N- and C-terminal extensions of the domain, as well as usage of the linker regions and other regions outside of the β-sheet, have been characterized and allow for recognition of up to 8 nt (Maris et al. 2005; Auweter et al. 2006; Lunde et al. 2007; Cléry et al. 2008; Muto and Yokoyama 2012).

Sixty-one RRM proteins only comprise a single isolated RRM domain; examples include the polyA-binding protein CPEB family and the nuclear cap-binding protein NCBP2 (CBP20) (Fig. 1.5b). Sixty RBPs occur with several repeated RRMs; among them are the PTBP and the ELAVL families, regulators of mRNA splicing, stability, and localization (Sawicka et al. 2008; Simone and Keene 2013). Another 68 RRM proteins are found in combination with other RBDs; prominent examples in this group include the IGF2BP1 and FUS families, involved in translational regulation, mRNA transport, and splicing (Tan and Manley 2009; Bell et al. 2013) (Fig. 1.5b). Hence, RRM proteins are found in a variety of biological pathways, the majority of which involve mRNA-related processes, such as regulation of mRNA stability, splicing, translation, and transport. RRM domains are generally a diagnostic indicator for ssRNA binding. In a few cases, however, protein-binding partners have been shown to occlude interaction of the RRM domain with RNA, as seen for the RRM domain of the exon junction complex protein RBM8A (Y14), which binds to the protein cofactor MAGOH (Maris et al. 2005; Glisovic et al. 2008).

5.2.2 K-Homology Domain

The heterogeneous nuclear RNP K-homology (KH) domain binds to ssRNA and ssDNA, and has 39 members in humans. KH domains are ~70 amino acids long and characterized by a hydrophobic core domain with an (I/L/V)IGXXGXX(I/L/V) consensus sequence in the center. Structurally, all KH domains form a three-stranded β-sheet packed against three α-helices and belong either to the eukaryotic type I, of βααβα topology, or the prokaryotic type II, of αββααβ topology (Grishin 2001; Lunde et al. 2007; Valverde et al. 2008). KH domains typically recognize 4-nt ssRNA sequences through electrostatic interactions. Signal transduction and activation of RNA (STAR) proteins, such as SAM68 and QKI, contain just a single KH domain sandwiched between two short signaling motifs, which modulate the protein activity through posttranslational modifications in response to intracellular signaling pathways (Lasko 2003; Chénard and Richard 2008). However, most KH proteins contain combinations of RBDs including the IGF2BP1 family, with four KH and two RRM motifs (Bell et al. 2013), and the brain-specific NOVA splicing family with three repeated KH domains (Li et al. 2007) (Fig. 1.5b). The most extreme example of multiplication of RBDs is found in HDLBP, conserved from yeast to humans, which has 14 repeated KH domains in human (Fig. 1.5b). Analogous to RRM proteins, KH domain proteins predominantly interact with mRNAs and are found in posttranscriptional processes, such as mRNA splicing (PCBP and NOVA family), transport, and translation

(IGF2BP1, FMR1, and the MEX family). The two, highly conserved, KH domain proteins PNO1 and RPS3 represent an exception in their target specificity and interact with rRNA during ribosome biogenesis (Vanrobays et al. 2004; Anger et al. 2013).

5.2.3 Double-Stranded RNA-Binding Motif

The dsRNA-binding motif (dsrm) has 21 members and 4 dsrm-like members in humans. Dsrm domains are ~70 amino acids long and adopt an αβββα topology, in which the two α-helices are packed against the three β-sheets. This facilitates nonspecific, shape-dependent contacts with the RNA backbone along the minor and major grooves of A-form dsRNA helix, as well as base contacts along the minor groove and the apical loop (Chang and Ramos 2005; Lunde et al. 2007; Masliah et al. 2013). Dsrm domains are rarely found alone; 24 of the 25 of human dsrm proteins contain multiple dsrm domains or other enzymatic and nonenzymatic RBDs that modulate their function. While the best known dsrm-containing proteins are the Staufen family (STAU1, STAU2) of mRNA stability and transport regulators (Miki et al. 2005; Park and Maquat 2013), this domain type is not confined to mRBPs; instead, most dsrm proteins interact with a range of RNA substrates and are commonly found in RNA enzymes. Members include the adenine-to-inosine RNA-editing ADAR family, processing stem–loops or double strands in mRNAs, viral RNAs, and miRNA precursors (Savva et al. 2012), the two miRNA-processing endonucleases DROSHA and DICER1 (Kim et al. 2009; Wilson and Doudna 2013), as well as the interferon-inducible protein kinase EIF2AK2 (PKR), which, upon binding dsRNA, activates its kinase domain (Saunders and Barber 2003; Raven and Koromilas 2008) (Fig. 1.5b).

5.2.4 CCCH and CCHC Zinc Fingers

The two ssRNA-binding zinc fingers (zf), zf-CCCH and zf-CCHC, form rigid structures by coordination of a Zn^{2+} ion with three cysteine (C) and one histidine (H) residues. In humans, 45 genes contain the zf-CCCH (C-×8-C-×5-C-×3-H type) and 21 contain the zf-CCHC (C-×2-C-×4-H-×4-C) motif, also known as zinc knuckle. Zf proteins form sequence-specific interaction with RNAs through hydrogen bonding and van der Waals interactions of the protein backbone (Lunde et al. 2007; Kaymak et al. 2010), and use stacking interactions of aromatic side chains with the bases to increase RNA-binding affinity. In contrast to other ssRNA-binding domains, the rigidity and shape of the protein structure are the key determinant for specificity of zinc-finger proteins to their target RNAs. The domains generally occur in repeats or in combination with other RBDs. While for most of the CCHC and CCCH zf proteins the molecular function remains unclear, characterized zf proteins are predominantly involved in regulation of mRNA-related processes. Classic examples of zf-CCCH proteins are the AU-rich-binding ZFP36 (TTP) proteins, which participate in rapid degradation of mRNAs transcribed after immune stimulation (Sandler and Stoecklin 2008; Brooks and Blackshear 2013), and the muscleblind (MBNL1,2,3) family, which regulates alternative splicing during muscle differentiation (Pascual et al. 2006; Cooper et al. 2009). Characterized zf-CCHC proteins include the CPSF4 mRNA polyadenylation and cleavage factor (Colgan and Manley 1997; Shatkin and Manley 2000; Proudfoot and O'Sullivan 2002; Proudfoot 2004) (Fig. 1.5b), and the ZCCHC7 (AIR1) protein, member of the nuclear polyadenylation TRAMP complex, required for the degradation of aberrant nuclear ncRNAs (Anderson and Wang 2009). Another prominent member in this class is the LIN28 family (Fig. 1.5b), which posttranscriptionally maintains pluripotency in early embryonic development by inhibiting maturation of miRNA let-7 family precursors and increasing stability and translation of mRNA targets (Thornton and Gregory 2012; Wilbert et al. 2012; Cho et al. 2012; Hafner et al. 2013).

5.2.5 LSM Domain

The LSM domain is found in 19 proteins in humans. First discovered in Sm proteins, it later was re-named LSM ("like-Sm") to include proteins outside of its founding members. The LSM fold is a bipartite domain that stretches along a region of ~65 amino acids and folds into an N-terminal α-helix followed by a twisted five-stranded β-strand (Wilusz and Wilusz 2005; Tharun 2009). The two motifs, motif I, ~22 residues long, and motif II, ~16 residues long, are connected by a variable linker region. LSM proteins form hexa- and heptameric-ring-shaped complexes around RNA. Binding to short, internal polyU- and polyA-rich stretches, they generally associate with snRNAs and some snoRNAs to form stable snRNP complexes (Achsel et al. 2001; Khusial et al. 2005). These RNP complexes commonly act as RNA-RNA and RNA-protein chaperones (Wilusz and Wilusz 2005; Tharun 2009).

The originally characterized Sm proteins (SNRPB, SNRBPD1, SNRBPD2, SNRBPD3, SNRBPE, SNRBPF, SNRBPG) form the core protein complex around the snRNAs of the major and minor spliceosome (U1, U2, U4, U5, U11, U12, U4atac) (Tharun 2009). Other combinations of LSM-fold proteins and their RNA components lead to functional complexes (Beggs 2005; Wilusz and Wilusz 2005; Tharun 2009), such as the nuclear LSM2-8 protein complex (LSM2, LSM3, LSM4, LSM5, LSM6, LSM7, LSM8), which forms around U6 and U6atac snRNA and participates in mRNA splicing. Association of the same LSM proteins around the C/D box U8 snoRNA results in an RNP complex functioning in rRNA maturation (Pannone et al. 2001; Tomasevic and Peculis 2002). In addition, binding of LSM2-8 to nuclear polyadenylated mRNAs promotes mRNA decapping (Kufel et al. 2004). Other LSM complexes include the U7 snRNP complex (SNRPB, LSM10, SNRPD3, LSM11, SNRPE, SNRPF, SNRPG), which is essential for histone 3′ end processing (Pillai et al. 2001), and the cytoplasmic LSM1-7 complex (LSM1, LSM2, LSM3, LSM4, LSM5, LSM6, LSM7), which localizes to P bodies and facilitates mRNA decapping after deadenylation (Tharun et al. 2000; Coller and Parker 2004; Parker and Song 2004; Parker and Sheth 2007). We can expect the identification of novel functions and target specificities in PTGR by as yet uncharacterized variations in the composition of LSM complexes (Wilusz and Wilusz 2005).

5.2.6 PIWI and PAZ Domain

The combination of PIWI, PAZ, and MID domains characterizes the Argonaute RBP family, a clade wirh four AGO and four PIWI protein members in humans (Peters and Meister 2007; Kim et al. 2009). Proteins of this clade bind miRNAs, siRNAs, and piRNAs by anchoring the 5′phosphate in the MID-domain pocket and the 3′end in the PAZ domain, while the PIWI domain interacts with the RNA backbone (Song 2004; Song and Joshua-Tor 2006; Wang et al. 2008b; Tian et al. 2011; Simon et al. 2011; Schirle and MacRae 2012). The PAZ RBD is also a structural component of the miRNA-processing endonuclease DICER1 (Zhang et al. 2004).

The 110 amino-acid-long PAZ domain consists of a β-barrel followed by an αβ-domain, and is structurally related to OB folds, S1, and LSM domains (Lunde et al. 2007). Forming a clamp-like structure, the PAZ domain selectively binds the two-nucleotide overhangs of small RNA duplexes at the 3′end, thereby acting as an anchor to position small RNAs for cleavage (Jinek and Doudna 2009). The PIWI domain is structurally similar to the RNase H endonuclease domain; however, in mammals, only PIWI proteins (Siomi et al. 2011) and AGO2 (Meister et al. 2004; Liu 2004) display nuclease activity, while in other AGO proteins subtle changes in the active site or the N-terminal regulatory domain prevent catalytic activity (Hauptmann et al. 2013; Faehnle et al. 2013; Nakanishi et al. 2013). AGO proteins initiate, guided by miRNAs and siRNAs, posttranscriptional silencing of mRNAs (Hutvagner and Simard 2008), and PIWI proteins, guided by piRNAs, silence transposons at the posttranscriptional and epigenetic levels (Kim et al. 2009; Siomi et al. 2011). Given the variability of possible guide RNA sequences, Argonaute proteins are tremendously versatile and by using different endogenously expressed guide RNA sequences they can form hundreds of distinct RNP complexes in vivo. Capable of targeting virtually any given cytosolic RNA sequence in a specific manner, they are used extensively as a tool in biotechnological applications (Dorsett and Tuschl 2004).

5.2.7 PUF Repeat

In humans, PUF repeats are only found in the two members of the Pumilio family; however, the structure and RNA-recognition mechanism of this domain are highly conserved and probably the best understood among all RBDs. PUF domains are ~40 amino acids long and consist of three α-helices that pack together into a half-ring structure. Each PUF domain recognizes only one nucleotide, but multiple repeats additively increase the number of bases recognized, and Pumilio proteins contain multiple PUF repeats that recognize highly sequence-specifically stretches within mRNAs (Wang et al. 2001). The extremely high specificity is achieved by hydrogen bonding interactions of two residues per repeat, while aromatic side chains wrap the bases into a tight fit. Human Pumilio proteins (PUM1, PUM2) contain eight PUF repeats, which together recognize the sequence UGUANAUA frequently located within the 3′UTRs of its targets to regulate mRNA stability and translation (Wickens et al. 2002; Wang et al. 2002; Hafner et al. 2010). Compared to other RBDs recognizing short and often degenerated RNA sequences, the RRE of Pumilio repeats is highly predictive for identifying Pumilio protein targets. Indeed, predictions of conserved RREs within 3′UTRs of mRNAs mainly identified, next to miRNA, Pumilio protein-binding sites (Xie et al. 2005), highlighting the exceptionally high information content of the Pumilio RRE. The high molecular specificity of the interaction has allowed engineering of RNA-binding specificity of Pumilio proteins to recognize different sequences (Cheong and Hall 2006).

5.3 Predominant Enzymatic RNA-Binding Domains

5.3.1 DExD/H helicases

DExD/H helicases, comprising DEAD and DEAH box helicases, are ATP-dependent enzymes that are involved in RNA-protein remodeling in the cell. They form the second largest class of RBPs comprising 73 members in humans, of which 62 interact specifically with RNA, and the remaining with DNA. The majority of the human RNA-binding DexD/H helicases, 42 members, belong to the DEAD box class, while the others are DEAH and DExH Ski-like helicases (named after its founding member Ski2p) (la Cruz et al. 1999). DExD/H RNA helicases belong to the SF2 helicase superfamily and contain NTPase characteristic Walker A and B motifs; their seven helicase signature motifs extend over ~400 amino acids (Tanner and Linder 2001; Rocak and Linder 2004; Pyle 2008; Jankowsky and Fairman-Williams 2010; Fairman-Williams et al. 2010). The helicases are differentiated by their catalytic core residues Asp-Glu-Ala-Asp for DEAD box helicases, and Asp-Glu-Ala/x-His for the related DEAH box and Ski2-like helicases. The enzymatic core arranges into two discrete domains connected by a linker that forms a cleft, in which an ATP can bind (Tanner and Linder 2001), whose hydrolysis provides the energy for unwinding RNA secondary structures or reorganizing RNPs in either a directional (DEAH helicases) or a bidirectional (DEAD helicases) manner. DExD/H RNA helicases generally lack substrate specificity, or even affinity, towards RNA and DNA. This allows them to promiscuously unwind and remodel a broad range of targets, but also requires their association with cofactors that give specificity and affinity for their targets (Rocak and Linder 2004; Jankowsky and Fairman-Williams 2010). While most members of DExD/H helicases are involved in mRNA-related processes, in particular splicing, they play essential roles in diverse PTGR pathways such as transcriptional regulation, rRNA and tRNA maturation, viral defense, miRNA RISC loading, translation initiation, RNA export, and degradation (Rocak and Linder 2004; Fukuda et al. 2007; Pyle 2008; Jankowsky 2011; Linder and Jankowsky 2011; Martin et al. 2013; Schmidt and Butler 2013; Fullam and Schröder 2013). Next to RNA-RNA and RNA-protein remodeling, DExD/H helicases are also important in RNA-protein complex disassembly and facilitate removal of protein interactors from their targets during RNA export (Linder and Jankowsky 2011).

Paralogs within one RBP family can function in highly diverse roles and pathways, but even one helicase can assume a variety of different biological functions depending on its associated cofactors. For instance, EIF4A1, the first DEAD box helicase for which remodeling and unwinding was mechanistically characterized, forms the EIF4F translation initiation complex, together with the cap-binding protein EIF4E and the scaffolding protein EIF4G (Gingras et al. 1999; Andreou and Klostermeier 2013). Complexed with EIF4H or EIF4B cofactors, EIF4A1 unwinds secondary structures in the 5′UTR, allowing binding of the 43S ribosome complex for AUG start codon scanning. In contrast, although structurally very similar (65 %), the family member EIF4A3 is a core component of the exon junction complex (EJC), in which it acts as an RNA clamp to assist correct positioning of the EJC 20–24 nt upstream of mRNA exon-exon junctions (Linder and Fuller-Pace 2013).

5.3.2 EF-Tu GTP-Binding Domain

The EF-Tu GTP-binding domain (GTP_EFTU), named after its prokaryotic founding member EF-Tu, is a highly conserved domain across all kingdoms of life, and shared by 21 genes in humans. The domain is typically found in GTP-binding translation elongation factors, which are composed of three structural domains, the GTP-binding domain, and two β-barrel nucleotide-binding domains, D2 and D3, which bind to aminoacylated tRNAs (Nissen et al. 1995; Wang et al. 1997; Negrutskii and El'skaya 1998). Eukaryotic EF-1α (human ortholog EEF1A1) has also been shown to interact with higher molecular weight G/U-rich RNAs and rRNAs at a tRNA-independent binding site (Negrutskii and El'skaya 1998). Translation elongation factors are essential for protein synthesis; they bind aminoacyl-tRNAs in a GTP-dependent manner and direct them to the A-site of the ribosome where, upon codon recognition by the tRNA, GTP is hydrolyzed and the factor released (Dever and Green 2012). Furthermore, the GTP_EFTU domains are not only found in combination with D2 and D3 in various translation initiation and release factors, but also alone in GTPases involved in mRNA splicing (EFTUD2) (Fabrizio et al. 1997) and rRNA biogenesis (GNL3 family) (Du et al. 2006; Kressler et al. 2010). The biological role for a majority of the human GTP_EFTU-containing proteins has not been characterized. Particularly, the translation-independent roles of EF1-α and family members, which have been found to directly interact with stem–loops in β-actin (ACTB) and metallothionein 1 (human orthologs MT1 family) mRNAs, are not well understood (Liu et al. 2002; Mickleburgh et al. 2006). More than half of the GTP_EFTU-containing proteins were isolated from direct mRNA-cross-linking experiments by Castello et al. (Castello et al. 2012), supporting a wider role for these proteins in PTGR.

6 Tissue Specificity of RNA-Binding Proteins

Given the evolutionary conservation of RBPs and their involvement in general RNA metabolism, one anticipates predominantly ubiquitous expression for the majority of RBPs. Most tissue-specific proteins evolved recently and have ancestral proteins with low conservation, or no homologs in lower eukaryotes (Winter et al. 2004). Hence, tissue-specific RBPs evolved at later stages of metazoan evolution, the most extreme example of which is represented by the vertebrate-specific secreted RNA ribonucleases A (RNase A) family comprising eight nucleolytic and five inactive pseudo-nuclease members (Rosenberg 2011). Among the active members of this class are angiogenin (ANG), which promotes blood vessel growth, cleaves tRNAs under oxidative stress and is predominantly expressed and secreted from the liver, as well as the EDN and ECP RNases (RNASE2 and RNASE3), which play a role in innate immune response and are expressed in bone marrow eosinophiles (Yamasaki et al. 2009; Rosenberg 2011; Ivanov et al. 2011) (Fig. 1.6a).

In order to obtain an overview of the number of tissue-specific RBPs, we examined mRNA expression levels across tissues, which provide a reliable estimate of protein abundance in the cell (Guo et al. 2010). Current RNA-seq resources do not provide a comprehensive collection of tissues profiled in one study and thus we used a publically available microarray tissue atlas comprising 31 different human tissues to examine the tissue specificity of ~17,000 profiled genes (Dezso et al. 2008). We devised a metric measuring tissue specificity by calculating the deviation for each gene from a uniform expression across all tissues. Our analysis likely underestimates tissue specificity, particularly of low-expressed genes, due to tissue heterogeneity, which adds additional noise to the true expression profiles. In addition, mRNA splicing or alternative polyadenylation events, which can give rise to cell-type-specific protein isoforms with different biological functions (Black 2000; Kornblihtt et al. 2013) (Matlin et al. 2005; Sandberg et al. 2008; Mayr and Bartel 2009; Hogg and Goff 2010), cannot be captured by microarray analysis. Nevertheless, a few general themes contrasting RBPs from other protein-coding genes emerge from the present analysis.

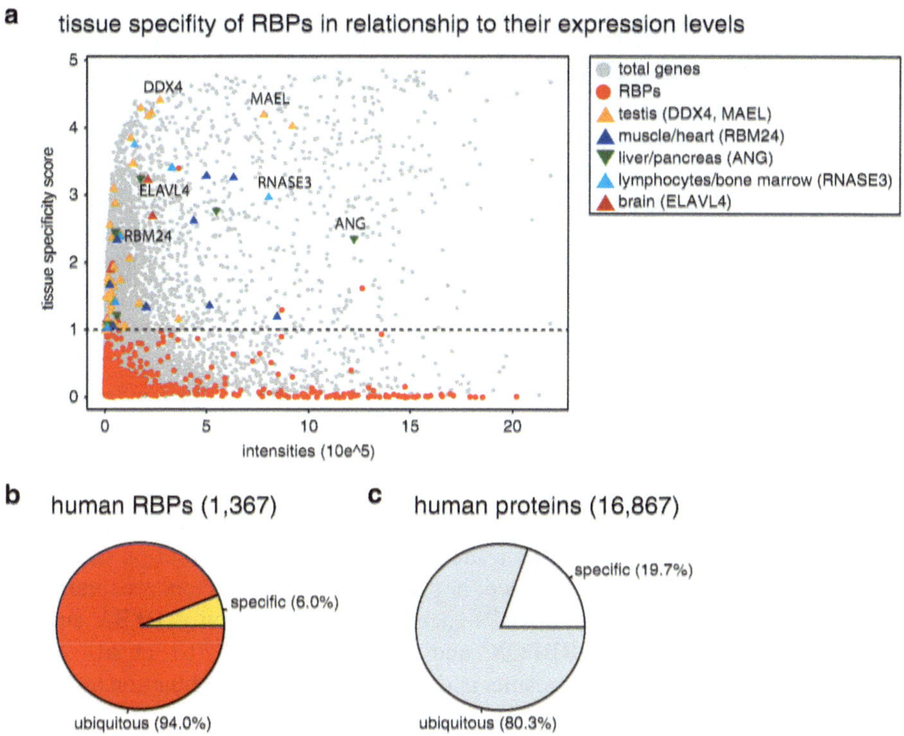

Fig. 1.6 Tissue specificity of RBPs based on their mRNA expression profiles in a human microarray tissue atlas (Dezso et al. 2008). (**a**) Plot of maximum expression intensity versus their tissue specificity score. The threshold for tissue specificity is set at 1 (*dashed line*), to include the brain-specific ELAVL and NOVA splicing factor families and the germline-specific PIWI family. Protein-coding genes are marked as *grey-filled circles* and RBPs as *red-filled circles*. Tissue-specific RBPs are marked based on the tissue with their highest expression: testis (*orange triangle*), muscle/heart (*dark blue triangle*), liver/pancreas (*dark green inverted triangle*), lymphocytes/bone marrow (*light blue triangle*), brain (*dark red triangle*). In (**b**) the number of tissue-specific RBPs is compared to all expressed RBPs and in (**c**) the number of tissue-specific proteins is compared to total expressed proteins

About 90 % of the curated RBPs (~1,400) were profiled in the study. While RBPs showed similar average expression levels as compared to other protein-coding genes, they displayed markedly lower variation of expression across tissues. Only ~2 % of RBPs showed very high, tissue-specific expression on the mRNA level compared to 8 % among other protein-coding genes (Dezso et al. 2008). Even when setting the criteria for tissue-specific expression more lenient to include known RBPs with enrichment in selected tissues, only about ~6 % of RBPs were classified as tissue specific, while at the same threshold 20 % of non-RNA-binding proteins displayed tissue-specific expression (Fig. 1.6b, c). Tissue-specific RBPs were predominantly located to the germline, and to a lesser degree to brain, bone marrow, liver, and muscle, where they play distinct biological roles in RNA metabolic pathways (Fig. 1.6a).

RNA metabolism in the germline is uniquely specialized and many of the exclusively expressed RBPs are involved in the piRNA-induced silencing pathway, comprising piRNA biogenesis as well as their effector components such as PIWI (PIWIL1,2,3,4) and TUDOR proteins, MOV10L1, Maelstrom, and PLD6 and DDX4 (VASA) (Siomi et al. 2010; Siomi et al. 2011). Other germline-specific RBPs are the DAZL family (DAZ1,2,3, DAZL, BOLL) and the nanos zinc fingers (NANOS2, NANOS3), both of which are regulators of mRNA translation (Parisi and Lin 2000; Brook et al. 2009; Barckmann and Simonelig 2013). Loss of function of these genes commonly results in arrest of germline development and leads to infertility (Yen 2004; Klattenhoff and Theurkauf 2008; Kim et al. 2009; Brook et al. 2009; Thomson and Lin 2009; Siomi et al. 2011; Massart et al. 2012; Ishizu et al. 2012; Lasko 2013).

The brain shows higher levels of alternative splicing, polyadenylation, and editing compared to other tissues, which are required for neuronal development, plasticity, and memory functions (Xu et al. 2002; Yeo et al. 2005; Wang et al. 2008a; Grabowski 2011; Miura et al. 2013). The higher level of posttranscriptional regulation points towards the brain's demand to rapidly adapt to evolutionary pressure and PTGR can be considered a means for accelerating evolution of protein diversity (Xing and Lee 2006). Hence, in particular splicing factors are overrepresented among the brain-specific RBPs such as the NOVA family, PTB2, and the neuronal members of the RBFOX and ELAVL families (Li et al. 2007). Dysregulation of these proteins results in defects in neurological function and brain development.

We examined paralogous RBP protein families in further detail and defined three categories for tissue specificity in RBP families: (1) highly tissue-specific RBP families, (2) ubiquitous RBP families containing tissue-specific members that evolved independently to meet cell-type-specific needs, and (3) ubiquitous families wherein the individual members display a gradient of expression in different tissues and cell types.

About 2 % of RBP families fall into the first category and display a high degree of tissue specificity in all of their members, expressed in only one or two tissues. Among them are the DAZL, NOVA, RNase A, and PIWI family.

RBP families in the second category (~3 %) have ubiquitously expressed members and one or two tissue-specific paralogs. Members generally have highly overlapping functions, with the tissue-specific paralogs carrying out similar functions to the ubiquitous paralog. A number of germline-restricted helicases, essential for gametogenesis, belong to this category, including DDX4 and DDX3Y, the RNA export factor DDX25, and the piRNA pathway MOV10L1 helicase, all of which have ubiquitously expressed paralogs (DDX3X paralog of DDX4 and DDX3Y, DDX19A and DDX19B paralog of DDX25, MOV10 paralog of MOV10L) (Abdelhaleem 2005; Tsai-Morris et al. 2010; Frost et al. 2010; Zheng et al. 2010; Lasko 2013). Other RBPs include the AU-rich element-binding, nucleocytoplasmic shuttling ELAVL mRBP family (ELAVL1,2,3,4, also known as HuR,B,C,D) (Brennan and Steitz 2001; Simone and Keene 2013).

Tissue-specific paralogs likely fulfill a specific demand in PTGR by potentially regulating an increased abundance or number of RNA targets in this tissue. Several mRBP families, such as the ELAVL family, possess identical RNA-binding specificities (Ascano et al. 2011). Whether ubiquitous paralogs are able to functionally compensate for loss of function of their tissue-specific counterparts has not been evaluated systematically in vivo. However, despite their close homology, overexpression of ELAVL4 but not ELAVL1 in cultured neural crest cells is able to induce neurite outgrowth (Wakamatsu and Weston 1997), suggesting that, differences in expression levels alone may not fully account for the different biological functions of paralogs.

About 25 % of paralogous families are ubiquitously expressed but their members display some tissue-specific variation in expression. For these proteins, different expression levels in different tissues can reflect a higher demand or sensitivity for the expressed RBP due to different metabolic activities in different cell types. As expected, dysregulation of the RBPs shows the strongest phenotype in the tissues with highest expression. For instance loss of FMR1, expressed ubiquitously in every tissue but found elevated in brain and gonads (Wang et al. 2008a), leads to mental retardation and autism, macroorchidism, and ovarian insufficiency (Ascano et al. 2012). The variations in RBP expression leading to tissue-specific phenotypes may be considered in the selection of suitable model systems for the investigation of their biological function.

The residual 70 % of paralogous RBP families are expressed ubiquitously with marginal variation in expression. Dysregulation of ubiquitous RBPs impacts PTGR in multiple tissues; however, the severity of the phenotype in different tissues varies. Thus, counterintuitively, loss of abundantly and ubiquitously expressed SMN1 leads to spinal muscular atrophy (SMA), and mutation or loss of TARDBP/TDP43 is associated with amyotrophic lateral sclerosis (ALS) (Ule 2008; Lagier-Tourenne et al. 2010). These examples indicate that tissue-specific defects caused by dysregulation of an RBP do not strictly correlate with levels of expression and that differences in expression of RNA targets, protein partners, or additional biological roles of the protein must also be taken into consideration. To understand tissue-specific phenotypes of ubiquitous RBPs it may be necessary to also consider the tissue-specific expression levels of RNA targets or protein cofactors.

7 RBPs in Human Disease

Efforts to understand the role and the pathomechanisms of RBPs in disease have focused predominantly on mRBPs (Wang and Cooper 2007; Lukong et al. 2008; Ule 2008; Cooper et al. 2009; Hanson et al. 2011; Kapeli and Yeo 2012; Castello et al. 2013; Ramaswami et al. 2013). Disease associations of RBPs targeting other classes of RNA such as tRNAs, rRNAs, snoRNAs, snRNAs, and others have been less extensively covered (Scheper et al. 2007a, b; Antonellis and Green 2008; Perron

Table 1.2 Overview of RBPs with identified genetic disease-causing mutations collected in the OMIM database (Hamosh et al. 2005), categorized into their main RNA target groups Mitochondrially localized proteins are indicated with (mt). Proteins within the same RBP family are written in one line; if family members are also involved in disease they are highlighted in bold; if family members are involved in diseases other than the listed category they are highlighted in bold brown and found elsewhere in a separate category in the table

RBP class	disease category	RBP family	disease	genetic mutation	reference
mRNA-binding	Cancer	**EWSR1**, **FUS**, **TAF15**	Ewing sarcoma, soft tissue tumors	Gene fusion	(May et al. 1993; Ichikawa et al. 1994; Panagopoulos et al. 1994; Gill et al. 1995; Panagopoulos et al. 1999)
		TPR	Gastric, thyroid carcinoma, sarcoma	Gene fusion	(Dean et al. 1987; Gonzatti-Haces et al. 1988)
	Muscular/ cardiac disease	**CNBP**, ZCCHC13	Myotonic dystrophy	RNA repeat expansion sequesters RBPs	(Liquori et al. 2001)
		MBNL1, **MBNL2**, **MBNL3**	Myotonic dystrophy	Sequestered RBP in repeat expansion	(Miller 2000; Mankodi et al. 2001; Fardaei 2002)
		CELF1, CELF2-6	Myotonic dystrophy	Sequestered RBP in repeat expansion	(Timchenko et al. 1996; Roberts et al. 1997)
		MATR3, **RBM20**	Cardio-/distal myopathy	Missense mutation	(Senderek et al. 2009; Brauch et al. 2009)
		NOL3	Myoclonus	Missense mutation	(Russell et al. 2012)
		PABPN1, PABN1L	Muscular dystrophy	Polyalanine expansion leading to protein aggregation	(Brais et al. 1998)
	Neurological disease	AFF1, **AFF2**, AFF3, AFF4	Mental retardation	Deletion, loss-of-function through repeat expansion in mRNA	(Knight et al. 1993; Stettner et al. 2011)
		ATXN1, ATXN1L	Spinocerebellar ataxia	Polyglutamine expansion leading to protein aggregation	(Orr et al. 1993; Banfi et al. 1994; Servadio et al. 1995)
		ATXN2, ATXN2L	Spinocerebellar ataxia, susceptibility to late-onset Parkinson disease, susceptibility to amyotrophic lateral sclerosis (ALS)	Polyglutamine expansion leading to protein aggregation	(Pulst et al. 1996; Cancel et al. 1997; Gwinn-Hardy et al. 2000; Elden et al. 2010)
		DYNC1H1, DNAH1-11, DNAH17, DYNC2H1	Charcot-Marie-Tooth disease, mental retardation, spinal muscular atrophy (SMA)	Missense mutation	(Vissers et al. 2010; Weedon et al. 2011; Harms et al. 2012)
		UBA1, UBA6-7	Spinal muscular atrophy (SMA)	Missense mutation	(Ramser et al. 2008)
		EIF2B1	Leukoencephalopathy with vanishing white matter	Missense mutation	(van der Knaap et al. 2002)
		EIF2B2	Leukoencephalopathy with vanishing white matter	Missense, nonsense mutation	(Leegwater et al. 2001)
		EIF2B3	Leukoencephalopathy with vanishing white matter	Missense mutation	(van der Knaap et al. 2002)
		EIF2B4	Leukoencephalopathy with vanishing white matter	Missense mutation	(van der Knaap et al. 2002)
		EIF2B5	Leukoencephalopathy with vanishing white matter	Missense mutation	(Leegwater et al. 2001; van der Knaap et al. 2002; Fogli et al. 2002)
		EIF4G1, EIF4G2, EIF4G3	Parkinson disease	Missense mutation	(Chartier-Harlin et al. 2011)
		FMR1, FXR1, FXR2	Fragile X mental retardation syndrome (FXS), fragile X tremor/ataxia syndrome (FXTAS), premature ovarian failure	Deletion, repeat expansion leading to protein loss-of-function (FXS) or RNA-gain-of-function (FXTAS) sequesters RBPs	(Kremer et al. 1991; Devys et al. 1992; Gedeon et al. 1992; Wöhrle et al. 1992; Murray et al. 1998; Hagerman et al. 2001)
		EWSR1, **FUS**, **TAF15**	Amyotrophic lateral sclerosis (ALS)	Missense mutation leading to prion-like protein aggregation	(Kwiatkowski et al. 2009; Vance et al. 2009)
		HNRNPA2B1, HNRNPA0, HNRNPAB, HNRNPA1L2, **HNRNPA1**, HNRNPA3, HNRNPD, HNRNPDL	Amyotrophic lateral sclerosis (ALS)	Missense mutation leading to prion-like protein aggregation	(Kim et al. 2013)
		SETX	Amyotrophic lateral sclerosis (ALS)	Missense, nonsense mutation, deletion	(Moreira et al. 2004)
		TARDBP	Amyotrophic lateral sclerosis (ALS)	Missense mutation leading to protein aggregation	(Sreedharan et al. 2008)
		IGHMBP2	Distal spinal muscular atrophy (DSMA1)	Missense mutation	(Grohmann et al. 2001)
		LRPPRC	Leigh syndrome	Missense mutation	(Mootha et al. 2003)
		MECP2	Rett syndrome, X-linked mental retardation	Missense, nonsense mutation, frameshift, deletion	(Amir et al. 1999; Wan et al. 1999; Cheadle et al. 2000; Huppke et al. 2000)
		MTPAP, TUT1	Spastic ataxia	Missense mutation	(Crosby et al. 2010)
		PARK7	Parkinson disease	Missense mutation	(Bonifati et al. 2003)
		PQBP1	Renpenning syndrome 1	Frameshift	(Kalscheuer et al. 2003)
		PRKRA, **TARBP2**	Dystonia	Frameshift, missense mutation	(Camargos et al. 2008; Seibler et al. 2008)
		RANBP2, RGPD1-6, RGPD8	Acute, infection-induced susceptibility to encephalopathy	Missense mutation	(Neilson et al. 2009)
		RBM28	Alopecia, neurologic defects, endocrinopathy syndrome	Missense mutation	(Nousbeck et al. 2008)
		UPF3A, **UPF3B**	Mental retardation	Frameshift, missense, nonsense mutation	(Tarpey et al. 2007)
		TIA1, TIAL1	Welander distal myopathy	Missense mutation	(Hackman et al. 2012)
		RBFOX1, RBFOX2, RBFOX3	Mental retardation, epilepsy	Deletion, breakpoint	(Bhalla et al. 2004; Martin et al. 2007)
	Neurological/ developmental disease	**GLE1**	Lethal congenital contracture syndrome	Splice site mutation, missense mutation	(Nousiainen et al. 2008)
	Developmental disease	**BICC1**	Susceptibility to renal dysplasia	Missense, nonsense mutation	(Kraus et al. 2012)
		EIF2AK1, EIF2AK2, **EIF2AK3**	Wolcott-Rallison syndrome, multiple epiphyseal dysplasia	Missense, nonsense mutation, splice site mutation	(Delépine et al. 2000; Brickwood et al. 2003; Durocher et al. 2006)
		FTO	Growth, developmental delay	Missense mutation	(Boissel et al. 2009)
		MKRN1, MKRN2, **MKRN3**	Central precocious puberty 2	Missense mutation, insertion	(Abreu et al. 2013)
		NR0B1, NR0B2	Congenital adrenal hypoplasia	Deletion, missense mutation	(Muscatelli et al. 1994; Yanase et al. 1996)
		RBM5, RBM6, **RBM10**	TARP syndrome	Frameshift, missense mutation	(Johnston et al. 2010)
		SF3B4	Acrofacial dysostosis	Missense, nonsense mutation, frameshift	(Bernier et al. 2012; Czeschik et al. 2013)
		SKIV2L	Trichohepatonenteric syndrome 2	Missense, nonsense mutation	(Fabre et al. 2012)
	Infertility	BOLL, **DAZ1-4**, **DAZL**	Azoospermia	Deletion	(Reijo et al. 1995)
	Metabolic disease	**AUH**, ECH1, ECHS1, ECHDC2, ECHDC3	3-methylglutaconic aciduria	Nonsense mutation, frameshift, splice site mutation	(IJlst et al. 2002; Ly et al. 2003)
		C12ORF65 (mt)	Combined oxidative phosphorylation deficiency, spastic paraplegia-55 (SPG55)	Frameshift, missense, nonsense mutation	(Antonicka et al. 2010; Shimazaki et al. 2012)

(continued)

Table 1.2 (continued)

		GFM1 (mt)	Combined oxidative phosphorylation deficiency	Missense, nonsense mutation	(Coenen et al. 2004; Valente et al. 2007)
		TSFM (mt)	Combined oxidative phosphorylation deficiency	Missense mutation	(Smeitink et al. 2006)
		EEFSEC, **TUFM** (mt)	Combined oxidative phosphorylation deficiency	Missense mutation	(Valente et al. 2007)
		SECISBP2, SECISBP2L	Abnormal thyroid metabolism	Missense mutation	(Dumitrescu et al. 2005)
	Hematologic disease	**FIP1L1**	Spontaneous hypereosinophilic syndrome	Deletion leading to gene fusion	(Cools et al. 2003; Griffin et al. 2003)
		U2AF1, U2AF1L4	Myelodysplastic syndrome	Missense mutation	(Graubert et al. 2012)
	Ophthalmologic disease	**TDRD7**	Cataract	Frameshift	(Lachke et al. 2011)
	Immunological/ skin disease	ADAD1, ADAD2, ADAT, **ADAR**, ADARB1, ADARB2	Aicardi-Goutieres syndrome (AGS), dyschromatosis symmetrica hereditaria 1 (DSH1)	Missense, nonsense mutation	(Miyamura et al. 2003; Rice et al. 2012)
tRNA-binding	Cancer/ metabolic disease	**ELAC2**	Prostate cancer, combined oxidative phosphorylation deficiency	Missense, nonsense mutation, frameshift	(Tavtigian et al. 2001; Haack et al. 2013)
	Muscular/ metabolic/ hematologic disease	**PUS1**	Myopathy, lactic acidosis and sideroblastic anemia 1	Missense, nonsense mutation	(Bykhovskaya et al. 2004; Fernandez-Vizarra et al. 2007)
		YARS2 (mt)	Myopathy, lactic acidosis and sideroblastic anemia 2	Missense mutation	(Riley et al. 2010)
	Neurological disease	**AARS**, AARS2 (mt)	Charcot-Marie Tooth disease	Missense mutation	(Latour et al. 2010; Lin et al. 2011)
		ADAT3	Mental retardation	Missense mutation	(Alazami et al. 2013)
		AIMP1	Hypomyelinating leukodystrophy	Frameshift	(Feinstein et al. 2010)
		YARS	Charcot-Marie Tooth disease	Missense mutation, frameshift	(Jordanova et al. 2006)
		KARS	Charcot-Marie Tooth disease, deafness	Missense mutation, frameshift	(McLaughlin et al. 2010; Santos-Cortez et al. 2013)
		GARS	Charcot-Marie Tooth disease	Missense mutation	(Antonellis et al. 2003)
		CLP1	Microencephalopathy, motor sensory defects	Missense mutation	(Karaca et al. 2014)
		DARS	Hypomyelination with brainstem and spinal cord involvement	Missense mutation	(Taft et al. 2013)
		DARS2 (mt)	Leukoencephalopathy	Frameshift, missense, nonsense mutation, splice site mutation	(Scheper et al. 2007)
		NSUN2	Mental retardation	Nonsense, missense mutation, splice site mutation	(Abbasi-Moheb et al. 2012; Khan et al. 2012)
		FTSJ1	Mental retardation	Frameshift, splice site mutation	(Freude et al. 2004; Ramser et al. 2004)
		RARS, **RARS2** (mt)	Pontocerebellar hypoplasia	Splice site mutation, missense mutation	(Edvardson et al. 2007; Rankin et al. 2010)
		SEPSECS	Pontocerebellar hypoplasia	Missense mutation	(Agamy et al. 2010)
		TSEN2	Pontocerebellar hypoplasia	Missense mutation	(Budde et al. 2008)
		TSEN34	Pontocerebellar hypoplasia	Missense mutation	(Budde et al. 2008)
		TSEN54	Pontocerebellar hypoplasia	Missense, nonsense mutation, deletion	(Budde et al. 2008; Cassandrini et al. 2010)
	Developmental disease	EEF2, **EFTUD2**	Mandibulofacial dysostosis with microcephaly	Splice site mutation, nonsense, missense mutation, frameshift	(Lines et al. 2012; Bernier et al. 2012; Gordon et al. 2012)
	Metabolic disease	**EEF2**, EFTUD2	Spinocerebellar ataxia	Missense mutation	(Hekman et al. 2012)
		AARS, **AARS2** (mt)	Combined oxidative phosphorylation deficiency	Missense mutation	(Götz et al. 2011)
		MTFMT (mt)	Combined oxidative phosphorylation deficiency	Missense, nonsense mutation, deletion	(Neeve et al. 2013; Haack et al. 2014)
		MTO1 (mt)	Combined oxidative phosphorylation deficiency	Frameshift, missense mutation	(Ghezzi et al. 2012)
		EARS2 (mt)	Combined oxidative phosphorylation deficiency	Missense mutation, insertion	(Steenweg et al. 2012; Talim et al. 2013)
		FARS2 (mt)	Combined oxidative phosphorylation deficiency	Missense mutation	(Shamseldin et al. 2012; Elo et al. 2012)
		SARS2 (mt)	Hyperuricemia, pulmonary hypertension, renal failure, and alkalosis	Missense mutation	(Belostotsky et al. 2011)
		TRMU (mt)	Liver failure, deafness	Missense mutation	(Guan et al. 2006; Zeharia et al. 2009)
	Opthalmologic disease/ hearing loss	**HARS**, **HARS2** (mt)	Usher syndrome, Perrault syndrome	Missense mutation	(Pierce et al. 2011; Puffenberger et al. 2012)
		LARS2 (mt)	Perrault syndrome	Deletion	(Pierce et al. 2013)
rRNA-binding	Developmental disease	**EMG1**	Bowen-Conradi syndrome	Missense mutation	(Armistead et al. 2009)
		MURC, **PTRF**, PRKCDBP, SDPR	Lipodystrophy, muscular dystrophy	Frameshift	(Hayashi et al. 2009; Shastry et al. 2010)
		TCOF1	Treacher Collins syndrome 1	Missense, nonsense mutation, deletion, insertion	(Caluseriu et al. 2013; Marszalek et al. 2003; Splendore et al. 2002; Wise et al. 1997)
	Developmental/ hematologic disease	**SBDS**	Shwachman-Diamond syndrome	Frameshift, missense, nonsense mutation	(Boocock et al. 2003; Nakashima et al. 2004)
	Opthalmologic disease	**WDR36**	Open angle glaucoma	Missense mutation	(Monemi et al. 2005)
snRNA-binding	Neurological disease	**SMN1**, **SMN2**	Spinal muscular atrophy (SMA)	Missense, nonsense mutation, frameshift, deletion	(Lefebvre et al. 1995; Cobben et al. 1995; Parsons et al. 1996; Hahnen et al. 1997; Gambardella et al. 1998; Sossi et al. 2001)
	Skin disease	**SART3**	Porokeratosis	Missense mutation	(Zhang et al. 2005)
		SNRPE	Hypotrichosis	Missense mutation	(Pasternack et al. 2013)
		USB1	Poikiloderma with neutropenia	Deletion, splice site mutation, frameshift	(Volpi et al. 2010; Tanaka et al. 2010)
	Opthalmologic disease	**SNRNP200**, ASCC3	Retinitis pigmentosa	Missense mutation	(Zhao et al. 2009)
		PRPF3	Retinitis pigmentosa	Missense mutation	(Chakarova et al. 2002)
		PRPF31	Retinitis pigmentosa	Splice site mutation/deletion, missense mutation	(Vithana et al. 2001)
		PRPF6	Retinitis pigmentosa	Missense mutation	(Tanackovic et al. 2011)
		PRPF8	Retinitis pigmentosa	Missense mutation	(McKie et al. 2001)
		RP9	retinitis pigmentosa	Missense mutation	(Keen et al. 2002)
snoRNA-binding	Neurological disease	**NOP56**	Spinocerebellar ataxia	RNA repeat expansion sequesters RBPs	(Kobayashi et al. 2011)
	Hematologic/ neurodevelopmental/ developmental disorder	**DKC1**	Dyskeratosis congenita	Missense mutation, deletion, intron insertion, splice site mutation	(Heiss et al. 1998; Knight et al. 1999; Vulliamy et al. 1999; Knight et al. 2001; Kanegane et al. 2005; Pearson et al. 2008)
		NHP2	Dyskeratosis congenita	Missense mutation	(Vulliamy et al. 2008)

(continued)

Table 1.2 (continued)

		NOP10	Dyskeratosis congenita	Missense mutation	(Walne et al. 2007)
cytosolic ribosomal proteins	Neurological disease	**RPL10**, RPL10L	Autism	Missense mutation	(Klauck et al. 2006)
	Hematologic disease	**RPL11**	Diamond-Blackfan anemia	Frameshift, deletion, splice site mutation, nonsense mutation	(Gazda et al. 2008)
		RPL35A	Diamond-Blackfan anemia	Missense, nonsense mutation, deletion	(Farrar et al. 2008)
		RPL5	Diamond-Blackfan anemia	Missense, nonsense mutation, frameshift, splice site mutation	(Gazda et al. 2008)
		RPS10	Diamond-Blackfan anemia	Missense, nonsense mutation, frameshift	(Doherty et al. 2010)
		RPS17, RPS17L	Diamond-Blackfan anemia	Missense, frameshift	(Cmejla et al. 2007; Gazda et al. 2008)
		RPS19	Diamond-Blackfan anemia	Missense, nonsense mutation, frameshift	(Draptchinskaia et al. 1999; Matsson et al. 1999)
		RPS24	Diamond-Blackfan anemia	Nonsense mutation, frameshift	(Gazda et al. 2006)
		RPS26	Diamond-Blackfan anemia	Missense mutation, splice site mutation, frameshift	(Doherty et al. 2010)
		RPS7	Diamond-Blackfan anemia	Splice site mutation	(Gazda et al. 2008)
mitochondrial (mt) ribosomal proteins	Metabolic disease	**MRPL3**	Combined oxidative phosphorylation deficiency	Missense mutation	(Galmiche et al. 2011)
		MRPS16	Combined oxidative phosphorylation deficiency	Nonsense mutation	(Miller et al. 2004)
		MRPS22	Combined oxidative phosphorylation deficiency	Missense mutation	(Saada et al. 2007)
lincRNA-binding	Cancer	**BRCA1**	Breast, ovarian, pancreatic cancer	Missense, nonsense mutation, deletion, frameshift	(Castilla et al. 1994; Simard et al. 1994; Al-Sukhni et al. 2008)
	Neurological disease	**DNMT1**	Cerebellar ataxia, neuropathy	Missense mutation	(Klein et al. 2011; Winkelmann et al. 2012)
	Developmental disorder	EZH1, **EZH2**	Weaver syndrome 2	Missense mutation, frameshift	(Gibson et al. 2012)
miRNA-binding	Cancer	**DICER1**	Pleuropulmonary blastoma, goiter with testicular tumours, embryonal habdomyosarcoma	Missense, nonsense mutation, frameshift	(Hill et al. 2009; Rio Frio et al. 2011; Foulkes et al. 2011)
		XPO5	Colecteral cancer	Frameshift, insertion	(Melo et al. 2010)
	Cancer/ developmental disorder	SMAD1, SMAD2, **SMAD3**, **SMAD4**, SMAD5, SMAD6, **SMAD7**, SMAD9	Colorectal cancer, Loeys-Dietz syndrome, juvenile polyposis, pancreatic cancer	Missense, nonsense mutation, frameshift	(Schutte et al. 1996; Howe et al. 1998; Broderick et al. 2007; van de Laar et al. 2011; Regalado et al. 2011)
		PRKRA, **TARPBP2**	Colecteral cancer	Frameshift	(Melo et al. 2010)
	Neurological disease	**SNIP1**	Psychomotor retardation, epilepsy, craniofacial dymorphism	Missense mutation	(Puffenberger et al. 2012)
	Pulmonary disease	**SMAD1**, SMAD2, SMAD3, SMAD4, SMAD5, **SMAD6**, SMAD7, **SMAD9**	Pulmonary hypertension, aortic valve disease (AOVD2)	Missense, nonsense mutation	(Nasim et al. 2011; Drake et al. 2011; Tan et al. 2012)
telRNA-binding	Developmental/ cardiovascular/ pulmonary disease	**TERT**	Coronary artery disease, dyskeratosis congenita, aplastic anemia	Missense mutation, frameshift	(Yamaguchi et al. 2005; Armanios et al. 2007; Tsakiri et al. 2007; Marrone et al. 2007)
	Developmental disorder	**WRAP53**	Dyskeratosis congenita	Missense mutation	(Zhong et al. 2011)
7SL-RNA-binding	Hematologic disease	**SRP72**	Bone marow failure	Missense mutation	(Kirwan et al. 2012)
	Developmental disorder	**LARP7**, SSB	Alazami syndrome	Base insertion	(Alazami et al. 2012)
immune-stimulatory RNA-binding	Autoimmune disease	**IFIH1**, DDX58, DHX58	Aicardi-Goutieres syndrome (AGS)	Missense mutation	(Rice et al. 2014)
	Metabolic disease	**OAS1**, OAS2, OAS3, OASL	Susceptibility to diabetes mellitus	Missense mutation	(Tessier et al. 2006)
	Cancer	**RNASEL**	Prostate cancer	Nonsense, missense mutation	(Carpten et al. 2002; Casey et al. 2002)
RNA/DNA-hybrid-binding	Autoimmune/ neurological disease	**RNASEH2A**	Aicardi-Goutieres syndrome (AGS)	Missense mutation	(Crow et al. 2006; Rice et al. 2013)
		RNASEH2B	Aicardi-Goutieres syndrome (AGS)	Missense mutation	(Crow et al. 2006)
		RNASEH2C	Aicardi-Goutieres syndrome (AGS)	Missense mutation	(Crow et al. 2006)
		SAMHD1	Aicardi-Goutieres syndrome (AGS), Chilbain lupus 2	Missense mutation	(Rice et al. 2009; Ravenscroft et al. 2011)
		DNMT3A, **DNMT3B**, DNMT3L	Immunodeficiency-centromeric instability-facial anomalies syndrome 1	Missense mutation, deletion	(Xu et al. 1999)
	Neurological disease	**APTX**	Ataxia	Deletion, splice site mutation, missense mutation	(Amouri et al. 2004; Criscuolo et al. 2005)
diverse targets	Neurological disease	**ANG**, RNASE1-4, RNASE6-8	Amyotrophic lateral sclerosis (ALS)	Missense mutation	(Greenway et al. 2006)
		EXOSC3	Pontocerebellar hypoplasia, spinal motor neuron degeneration	Missense mutation, deletion	(Wan et al. 2012)
		RNASET2	Leukoencephalopathy, cancer	Missense mutation, deletion, splice site mutation	(Henneke et al. 2009)
	Developmental disorder	DIS3, DIS3L, **DIS3L2**	Perlman syndrome	Deletion, splice site mutation, missense mutation	(Astuti et al. 2012)
	Metabolic disease/ deafness	**PNPT1**	Combined oxidative phosphorylation deficiency	Missense mutation	(Ameln et al. 2012; Vedrenne et al. 2012)
unknown targets	Cardiac disease	**CALR3**, CALR, CANX, CLGN	Cardiomyopathy	Missense mutation	(Chiu et al. 2007)
	Developmental disorder	**ASCC1**	Barrett esophagus	Missense mutation	(Orloff et al. 2011)
	Pulmonary disease	**DNAAF2**	Ciliary dyskinesia	Insertion, nonsense mutation	(Omran et al. 2008)

and Provost 2009; Narla and Ebert 2010; Wapinski and Chang 2011; Esteller 2011; Batista and Chang 2013). Here we will give a brief overview of RBPs identified as Mendelian disease factors, categorized by targets, and will focus on their characteristic disease phenotypes (Table 1.2).

Despite the interdependence of RNA regulatory pathways, many of the observed disease phenotypes are specific for RBPs binding to particular classes of RNAs, and symptoms of diseases with unknown molecular mechanisms could point towards putative dysregulated RNA metabolic pathways. The majority of RBPs have not been associated with diseases yet, but the recurring patterns occurring for the known proteins may support, facilitated by the growing genomic and transcriptomic data from patients, the interpretation of the dysregulated pathways involving new candidates in disease.

7.1 Diseases Involving mRNA-Binding Proteins

Most of the ~150 RBPs currently listed in the OMIM database (Hamosh et al. 2005) are mRNA-binding proteins with phenotypes characteristically showing neurological and neuromuscular dysfunctions due to dysregulation of splicing, translation, localization, or protein aggregation (Lukong et al. 2008; Ule 2008; Cooper et al. 2009; Hanson et al. 2011; Kapeli and Yeo 2012). Family members tend to have overlapping phenotypes reflecting functional redundancies; for instance the paralogs RBM20 and MATR3 are both involved in myopathies due to dysregulated splicing of their targets (Senderek et al. 2009; Guo et al. 2012).

RNA gain of function leading to altered mRBP-binding patterns is a common pathomechanism (Cooper et al. 2009; Echeverria and Cooper 2012; Nelson et al. 2013). In these disorders, repeat expansions in introns or UTRs of mRNAs often lead to sequestration of mRBPs in the nucleus, thus causing dysregulation of their respective targets. Such diseases comprise FXTAS (fragile X-associated tremor/ataxia syndrome), caused by trinucleotide repeat expansions in the 5′UTR of the FMR1 mRNA (Hagerman 2013), and the myotonic dystrophies DM1 and DM2, caused by repeat expansions in the 3′UTR of DMPK and the intron of CNBP (ZNF9), sequestering the mRBPs CELF1 and MBNL1 and their paralogs (Echeverria and Cooper 2012).

Mutations leading to prion-like aggregation of cytoplasmic or shuttling mRBPs and the dysregulation in assembly and disassembly processes of cytoplasmic RNP granules have been found to be the underlying cause in a range of neurodegenerative disorders. Cytoplasmic inclusion of TARDBP/TDP43 and FUS has been found in amyotrophic lateral sclerosis, a motor neuron disease leading to muscle atrophy (Anthony and Gallo 2010; Lagier-Tourenne et al. 2010; Lee et al. 2012; Ling et al. 2013). Accumulation and inefficient removal of these RNA-protein granules lead to cellular stress predominantly affecting neuronal cells (Liu-Yesucevitz et al. 2011; Li et al. 2013; Buchan et al. 2013; Ramaswami et al. 2013).

7.2 Mitochondrial RBPs in Disease

mRBPs involved in mitochondrial metabolism, such as translation elongation factors GFM1 and TSFM, typically cause deficiencies in oxidative phosphorylation that manifest themselves on a physiological level as neurological and muscular myopathies (Smeitink et al. 2006). Analogously, defects in mitochondrially localized tRNA- and rRNA-binding proteins and subunits of the mitochondrial ribosome also cause deficiencies in oxidative phosphorylation (Smits et al. 2010; Yao and Fox 2013) (Table 1.2).

7.3 Diseases Involving snRNA-Binding Proteins

Mutation or loss of function of snRBPs involved in snRNP complex formation or snRNA biogenesis causes defects in proper mRNA splicing and affects mRNA splicing patterns. Overall, only a few snRBPs have been linked to disease.

The most studied snRBP is the SMN protein, the loss of function of which causes spinal muscular atrophy (SMA). SMN proteins form a multimeric complex with Gemin proteins, which carries out the assembly of snRNPs and other RNP complexes in the cytoplasm (Paushkin et al. 2002; Battle et al. 2006). Autosomal recessive loss of function of the SMN1 gene is the molecular cause for SMA, affecting 1 in 6,000 births (Gubitz et al. 2004). The pathomechanism of SMA exemplifies a striking example of how a single point mutation within a cryptic splice-enhancer site can lead to dramatically altered protein levels that result in global defects in mRNA splicing. Two copies, SMN1 and SMN2, are found in the human genome encoding identical forms of the mature protein. Transcription from the SMN1 locus gives full-length mRNA. The SMN2 gene, however, harbors a single, silent nucleotide mutation in exon 7 leading to frequent exon skipping and resulting in only a small fraction of functional protein being produced, with the majority of transcripts giving rise to truncated, inactive protein. In the disease, deletion of the SMN1 gene results in the SMN2 locus being the only source for full-length mRNAs and protein. The very low amounts of SMN protein produced lead to highly skewed ratios of snRNPs, resulting in global aberrant splicing patterns (Zhang et al. 2008; Cooper et al. 2009). While snRNP assembly defects are detected in all tissues upon SMN1 deletion, the physiological phenotype manifests itself mainly in motor neurons (Glisovic et al. 2008; Cooper et al. 2009; Liu-Yesucevitz et al. 2011).

Mutations in other snRNP complexes have also been shown to cause disease. Loss of function of snRBPs of the U2, U12, and U4/U6-U5 snRNP complexes specifically cause retinitis pigmentosa, a retinal degeneration leading to blindness caused by incorrect splicing of mRNAs encoding for photoreceptors (Daiger et al. 2013). Why mutations in components of the general splicing machinery display highly tissue-specific phenotypes in the eye remains unclear (Wang and Cooper 2007; Singh and Cooper 2012).

7.4 Diseases Involving tRNA Metabolism of tRNA-Binding Proteins

Disease-causing mutations in tRNA-binding proteins are found in the tRNA maturation and aminoacylation pathways and show predominantly neurological phenotypes (Scheper et al. 2007a, b) (Table 1.2). Mutations in a number of cytoplasmic tRNA synthetases cause Charcot-Marie-Tooth disease, affecting the peripheral nervous system and leading to muscular atrophy (Antonellis and Green 2008; Yao and Fox 2013). Mutations in TSEN tRNA-splicing endonucleases lead to pontocerebellar hypoplasia, a sometimes fatal underdevelopment of the cerebellum, causing intellectual disability and impairing muscle control and motor skills (Budde et al. 2008). Dysfunctions of cytoplasmic tRNA aminoacyl synthetases lead to inefficient translation and molecularly link tRNAs with mRNA-binding proteins (Scheper et al. 2007a, b), explaining the overlapping range of symptoms observed for these classes of RBPs in neurological diseases.

7.5 Diseases Involving rRNA-Biogenesis Factors and Ribosomal Proteins

Loss of function of rRNA biogenesis factors and ribosomal proteins is generally embryonically lethal and only few diseases, classified as ribosomopathies, are known for these RBPs (Ruggero and Pandolfi 2003; Narla and Ebert 2010). Ribosomopathies commonly show growth retardation, organ malformation, and frequently bone marrow failure (Liu 2006). Examples of ribosomopathies include mutations in the SBDS rRNA biogenesis gene that cause Shwachman-Bodian syndrome, displaying a deficit in neutrophils and other blood cell types (Boocock et al. 2003), as well as a number of ribosomal proteins predominantly belonging to the small ribosomal subunit that cause Diamond-Blackfan anemia, an impairment of red blood cell formation (Narla and Ebert 2010).

7.6 Diseases Involving snoRNA-Binding Proteins

snoRNA-binding proteins are involved in maturation pathways of rRNAs, snRNAs, and the H/ACA-snRNA-like telomerase RNA. Dysregulation of snoRNPs leads to deficiencies in nucleotide modifications, pseudouridylation, or methylation of their targets (Bachellerie et al. 2002; Filipowicz and Pogacić 2002). As a consequence, snoRNP disease phenotypes overlap with those observed for defects in rRNA biogenesis, ribosomal proteins, as well as proteins involved in the telomerase assembly pathway, such as the protein components TERT and WRAP53 of the telomerase complex. Defects in snoRNA biogenesis manifest themselves in the severe developmental disorder dyskeratosis congenita, leading to bone marrow failure, growth retardation, neurological defects, and premature aging (Filipowicz and Pogacić 2002; Smogorzewska and de Lange 2004).

7.7 Diseases Involving microRNA Pathway Components

Mutations in miRNA-binding proteins are involved in different cancers and developmental disorders and have been well characterized (Merritt et al. 2008; Perron and Provost 2009; Kaneko et al. 2011). For example, mutations and loss of function or reduced levels of DICER1, TARBP2, and XPO5 have been found in pleuropulmonary blastomas and ovarian and other types of cancers (Zhang et al. 2006; Melo et al. 2009; Hill et al. 2009; Melo et al. 2010). In addition, changes in miRNA profiles are observed in many more human diseases due to changes in the cellular metabolism, and may prove valuable diagnostic tools, analogous to mRNA expression profiles (Esquela-Kerscher and Slack 2006; Kloosterman and Plasterk 2006; Calin and Croce 2006; Li and Kowdley 2012).

7.8 Immune Stimulatory and Stress-Related Diseases Caused by RBPs

In recent years it has become evident that nucleic acids play a central role in autoimmune and cellular stress-related diseases. Mutations or loss of function in three RNA/DNA nucleases, SAMHD1, the RNase H2 complex, and TREX1, leads to development of the autoimmune disease Aicardi-Goutieres syndrome (AGS), a neurodevelopmental disorder causing white matter abnormalities and cerebral atrophy. Symptomatically, AGS overlaps with the autoimmune disorder systemic lupus erythematosus (SLE) (Crow et al. 2006b). In both cases, a failure to remove accumulating nucleic acids is central to disease development and activates the innate immune system by type I interferon signaling (Atianand and Fitzgerald 2013; Rabe 2013).

The triphosphatase SAMHD1 possesses 3′–5′ exonuclease activity for ssRNA, ssDNA, and DNA/RNA hybrids. Its antiviral and autoimmune-suppressive function has been attributed to its role in the removal of nucleotides and nucleic acids in the cell (Beloglazova et al. 2013). The heterotrimeric RNase H2 complex endonucleolytically cleaves DNA/RNA hybrids and is thought to be required for the removal of Okazaki fragments during DNA replication (Cerritelli and Crouch 2009; Rabe 2013). Mutations in all three subunits of the RNase H2 complex (RNASEH2A, RNASEH2B, RNASEH2C) have been found to cause AGS (Crow et al. 2006b; Rabe 2013). The 3′–5′ DNA exonuclease TREX1 is implicated in degradation of ssDNA fragments during replication and antiviral defense (Rabe 2013). Mutations in TREX1 are found in AGS patients (Crow et al. 2006a) and TREX1 knockout mice accumulate endogenous retroelements; the accumulating nucleic acids are thought to trigger a subsequent interferon response (Stetson et al. 2008). Interestingly, loss of function of the dsRNA-editing ADAR enzyme has also been shown to cause AGS by an as yet unknown mechanism (Rice et al. 2012).

Many autoantibodies against other RBPs and even RNA have been detected in paraneoplastic syndromes and autoimmune diseases (DeHoratius et al. 1975; Hendrick et al. 1981; Pettersson et al. 1984; Gold et al. 1988; Gelpi et al. 1992; Sakai et al. 1994; Buckanovich et al. 1996). It is thought that dysregulation of RNA clearance mechanisms triggers innate immune responses and leads to apoptosis and release of RBP-RNA complexes into circulation. There these granules mobilize the immune system to develop autoantibodies against self-RNA-protein complexes

(Gaipl et al. 2005; Muñoz et al. 2010). The Ro60 complex, consisting of the TROVE2 (Ro60) protein and Y RNAs, was among the first identified targets of autoimmune antibodies in SLE patients was the Ro-RNP particle (Lerner et al. 1981; Hendrick et al. 1981). The Ro60 complex plays a regulatory role in DNA replication and stress response, removing misfolded RNAs, and mice lacking Ro60 develop lupus-like syndromes (Chen and Wolin 2004; Sim and Wolin 2011; Hall et al. 2013). Cleavage of tRNAs and Y RNAs accompanies cellular stress response and apoptosis (Phizicky and Hopper 2010; Nawrot et al. 2011; Hall et al. 2013; Köhn et al. 2013) and these stress-induced small RNA fragments may also act as immune-stimulatory RNAs. Autoantibodies against RBPs associating with these RNAs have been found in serum of SLE patients, stressing the importance of efficient clearance of circulating RNP granules in autoimmune diseases.

Dysfunctional nucleic acid clearance has not only been associated with autoimmune diseases but also with neurological defects of the peripheral nervous system, as seen in loss of function of the RNA exosome component EXOSC3 and RNASET2 (Henneke et al. 2009; Wan et al. 2012), which also acts as a tumor suppressor (Monti et al. 2008). Given the autoimmune-stimulatory role of nucleic acids it is surprising that loss of these general RNA turnover factors more closely resembles loss of mRBPs or tRNA-binding proteins. It will be interesting to understand in mechanistic detail whether the defects caused by these general factors are due to accumulation of RNA or protein aggregates inside the cell, or due to impaired RNA biogenesis pathways.

8 Conclusion

The central role of PTGR in cellular metabolism can be appreciated by considering the large number of proteins interacting with RNA. Over 1,500 of the 21,000 unique human proteins are directly contributing to PTGR. RBPs form many distinct families with few members and human RBPs can be grouped into ~1,100 paralogous families related by 20 % identity.

The complexity of PTGR was established early evolutionary time scales. The lowest common ancestor of metazoans had a set of ~200 RBPs (Kerner et al. 2011), and of the ~1,100 human RBP families ~600 families have homologs in yeast. As a comparison, a census of human and *S. cerevisiae* transcription factors containing transcription factor domains estimated ~1,400 and ~150 genes in human and yeast, respectively (Costanzo et al. 2000; Lee et al. 2002; Vaquerizas et al. 2009). The difference reflects that transcription factors underwent large expansions in the eumetazoan lineage (Larroux et al. 2008; Degnan et al. 2009), suggesting that the rapidly evolved complexity of transcriptional regulation was required for the development of multiple cell types in eukaryotes.

Consistent with their high degree of conservation, most RBPs (98 %) do not display highly tissue-specific expression, but they are abundant and make up to 25 % of the total cellular protein content. Interestingly, dysregulation of ubiquitous and general components in PTGR often shows highly tissue-specific phenotypes; for instance defects involving mRNA- and tRNA-binding proteins are most frequently associated with neurological diseases, especially of the peripheral nervous system.

Given that the common RBD folds have been characterized and the majority of RBPs do not fall into large families, novel RBPs are most probably singular or have recently evolved RNA-binding activity independent of their family. This makes RBP prediction highly challenging and leaves experimental approaches as the most suitable strategy for their identification. Novel genome-wide experimental methods such as covalent RNA-protein cross-linking coupled with high-throughput sequencing to identify RNA target sites, or combined with mass spectrometric approaches to identify proteome-wide RBPs cross-linked to RNAs, have substantially advanced the efforts towards the elucidation of posttranscriptional regulatory networks. With the increasing sensitivity of experimental approaches, the number of RBPs is likely to grow. Even at present though, the precise biological functions and RNA targets for the majority of known human RBPs have not been characterized and some processes, such as noncoding RNA maturation pathways, control of RNA transport, sensing of intracellular RNA, and mechanisms of RNA/RNP clearance, remain poorly understood. Thus, the main challenges in the field are the characterization of these processes and the mechanisms leading to human disease.

References

Abbasi-Moheb L, Mertel S, Gonsior M et al (2012) Mutations in NSUN2 cause autosomal-recessive intellectual disability. Am J Hum Genet 90:847–855. doi:10.1016/j.ajhg.2012.03.021

Abreu AP, Dauber A, Macedo DB et al (2013) Central precocious puberty caused by mutations in the imprinted gene MKRN3. N Engl J Med 368:2467–2475. doi:10.1056/NEJMoa1302160

Abdelhaleem M (2005) RNA, helicases: regulators of differentiation. Clinical Biochem 38:499–503. doi:10.1016/j.clinbiochem.2005.01.010

Achsel T, Stark H, Luhrmann R (2001) The Sm domain is an ancient RNA-binding motif with oligo(U) specificity. Proc Natl Acad Sci U S A 98:3685–3689. doi:10.1073/pnas.071033998

Agamy O, Ben Zeev B, Lev D et al (2010) Mutations disrupting selenocysteine formation cause progressive cerebello-cerebral atrophy. Am J Hum Genet 87:538–544. doi:10.1016/j.ajhg.2010.09.007

Akopian D, Shen K, Zhang X, Shan S-O (2013) Signal recognition particle: an essential protein-targeting machine. Annu Rev Biochem 82:693–721. doi:10.1146/annurev-biochem-072711-164732

Alazami AM, Al-Owain M, Alzahrani F et al (2012) Loss of function mutation in LARP7, chaperone of 7SK ncRNA, causes a syndrome of facial dysmorphism, intellectual disability, and primordial dwarfism. Hum Mutat 33:1429–1434. doi:10.1002/humu.22175

Alazami AM, Hijazi H, Al-Dosari MS et al (2013) Mutation in ADAT3, encoding adenosine deaminase acting on transfer RNA, causes intellectual disability and strabismus. J Med Genet 50:425–430. doi:10.1136/jmedgenet-2012-101378

Al-Sukhni W, Rothenmund H, Borgida AE et al (2008) Germline BRCA1 mutations predispose to pancreatic adenocarcinoma. Hum Genet 124:271–278. doi:10.1007/s00439-008-0554-0

Amir RE, Van den Veyver IB, Wan M et al (1999) Rett syndrome is caused by mutations in X-linked MECP2, encoding methyl-CpG-binding protein 2. Nat Genet 23:185–188. doi:10.1038/13810

Amouri R, Moreira M-C, Zouari M et al (2004) Aprataxin gene mutations in Tunisian families. Neurology 63:928–929

Anantharaman V, Koonin EV, Aravind L (2002) Comparative genomics and evolution of proteins involved in RNA metabolism. Nucleic Acids Res 30:1427–1464

Anderson JT, Wang X (2009) Nuclear RNA surveillance: no sign of substrates tailing off. Crit Rev Biochem Mol Biol 44:16–24. doi:10.1080/10409230802640218

Andreou AZ, Klostermeier D (2013) The DEAD-box helicase eIF4A: paradigm or the odd one out? RNA Biol 10:19–32. doi:10.4161/rna.21966

Anger AM, Armache J-P, Berninghausen O et al (2013) Structures of the human and Drosophila 80S ribosome. Nature 497:80–85. doi:10.1038/nature12104

Anthony K, Gallo JM (2010) Aberrant RNA processing events in neurological disorders. Brain Res 1338:67–77. doi:10.1016/j.brainres.2010.03.008

Antonellis A, Green ED (2008) The role of aminoacyl-tRNA synthetases in genetic diseases. Annu Rev Genomics Hum Genet 9:87–107. doi:10.1146/annurev.genom.9.081307.164204

Antonellis A, Ellsworth RE, Sambuughin N et al (2003) Glycyl tRNA synthetase mutations in Charcot-Marie-Tooth disease type 2D and distal spinal muscular atrophy type V. Am J Hum Genet 72:1293–1299. doi:10.1086/375039

Antonicka H, Ostergaard E, Sasarman F et al (2010) Mutations in C12orf65 in patients with encephalomyopathy and a mitochondrial translation defect. Am J Hum Genet 87:115–122. doi:10.1016/j.ajhg.2010.06.004

Apweiler R, Attwood TK, Bairoch A et al (2001) The InterPro database, an integrated documentation resource for protein families, domains and functional sites. Nucleic Acids Res 29:37–40

Aravin A, Gaidatzis D, Pfeffer S et al (2006) A novel class of small RNAs bind to MILI protein in mouse testes. Nature 442:203–207. doi:10.1038/nature04916

Aravind L, Koonin EV (2001a) A natural classification of ribonucleases. Meth Enzymol 341:3–28

Aravind L, Koonin EV (2001b) THUMP–a predicted RNA-binding domain shared by 4-thiouridine, pseudouridine synthases and RNA methylases. Trends Biochem Sci 26:215–217

Arcus V (2002) OB-fold domains: a snapshot of the evolution of sequence, structure and function. Curr Opin Struct Biol 12:794–801. doi:10.1016/S0959-440X(02)00392-5

Armanios MY, Chen JJ-L, Cogan JD et al (2007) Telomerase mutations in families with idiopathic pulmonary fibrosis. N Engl J Med 356:1317–1326. doi:10.1056/NEJMoa066157

Armistead J, Khatkar S, Meyer B et al (2009) Mutation of a gene essential for ribosome biogenesis, EMG1, causes Bowen-Conradi syndrome. Am J Hum Genet 84:728–739. doi:10.1016/j.ajhg.2009.04.017

Ascano M, Hafner M, Cekan P et al (2011) Identification of RNA-protein interaction networks using PAR-CLIP. WIREs RNA 3:159–177. doi:10.1002/wrna.1103

Ascano M, Mukherjee N, Bandaru P et al (2012) FMRP targets distinct mRNA sequence elements to regulate protein expression. Nature 492:382–386. doi:10.1038/nature11737

Ascano M, Gerstberger S, Tuschl T (2013) Multi-disciplinary methods to define RNA-protein interactions and regulatory networks. Curr Opin Genet Dev 23:20–28. doi:10.1016/j.gde.2013.01.003

Ashburner M, Ball CA, Blake JA et al (2000) Gene ontology: tool for the unification of biology. The Gene Ontology Consortium. Nat Genet 25:25–29. doi:10.1038/75556

Astuti D, Morris MR, Cooper WN et al (2012) Germline mutations in DIS3L2 cause the Perlman syndrome of overgrowth and Wilms tumor susceptibility. Nat Genet 44:277–284. doi:10.1038/ng.1071

Atianand MK, Fitzgerald KA (2013) Molecular basis of DNA recognition in the immune system. J Immunol 190:1911–1918. doi:10.4049/jimmunol.1203162

Auweter SD, Oberstrass FC, Allain FHT (2006) Sequence-specific binding of single-stranded RNA: is there a code for recognition? Nucleic Acids Res 34:4943–4959. doi:10.1093/nar/gkl620

Bachellerie JP, Cavaille J, Huttenhofer A (2002) The expanding snoRNA world. Biochimie 84:775–790

Baltz AG, Munschauer M, Schwanhausser B et al (2012) The mRNA-bound proteome and its global occupancy profile on protein-coding transcripts. Mol Cell 46:674–690. doi:10.1016/j.molcel.2012.05.021

Banfi S, Servadio A, Chung MY et al (1994) Identification and characterization of the gene causing type 1 spinocerebellar ataxia. Nat Genet 7:513–520. doi:10.1038/ng0894-513

Barckmann B, Simonelig M (2013) Control of maternal mRNA stability in germ cells and early embryos. Biochim Biophys Acta 1829:714–724. doi:10.1016/j.bbagrm.2012.12.011

Bartel DP (2009) MicroRNAs: target recognition and regulatory functions. Cell 136:215–233. doi:10.1016/j.cell.2009.01.002

Batista PJ, Chang HY (2013) Long noncoding RNAs: cellular address codes in development and disease. Cell 152:1298–1307. doi:10.1016/j.cell.2013.02.012

Battle DJ, KASIM M, Yong J et al (2006) The SMN complex: an assembly machine for RNPs. Cold Spring Harbor Symp Quant Biol 71:313–320. doi:10.1101/sqb.2006.71.001

Beggs JD (2005) Lsm proteins and RNA processing. Biochem Soc Trans 33:433–438. doi:10.1042/BST0330433

Bell JL, Wächter K, Mühleck B et al (2013) Insulin-like growth factor 2 mRNA-binding proteins (IGF2BPs): post-transcriptional drivers of cancer progression? Cell Mol Life Sci 70:2657–2675. doi:10.1007/s00018-012-1186-z

Beloglazova N, Flick R, Tchigvintsev A et al (2013) Nuclease activity of the human SAMHD1 protein implicated in the Aicardi-Goutieres syndrome and HIV-1 restriction. J Biol Chem 288:8101–8110. doi:10.1074/jbc.M112.431148

Belostotsky R, Ben-Shalom E, Rinat C et al (2011) Mutations in the mitochondrial seryl-tRNA synthetase cause hyperuricemia, pulmonary hypertension, renal failure in infancy and alkalosis, HUPRA syndrome. Am J Hum Genet 88:193–200. doi:10.1016/j.ajhg.2010.12.010

Bennasser Y, Chable-Bessia C, Triboulet R et al (2011) Competition for XPO5 binding between Dicer mRNA, pre-miRNA and viral RNA regulates human Dicer levels. Nat Struct Mol Biol 18:323–327. doi:10.1038/nsmb.1987

Berg MG, Singh LN, Younis I et al (2012) U1 snRNP determines mRNA length and regulates isoform expression. Cell 150:53–64. doi:10.1016/j.cell.2012.05.029

Berger W, Steiner E, Grusch M et al (2008) Vaults and the major vault protein: Novel roles in signal pathway regulation and immunity. Cell Mol Life Sci 66:43–61. doi:10.1007/s00018-008-8364-z

Bernier FP, Caluseriu O, Ng S et al (2012) Haploinsufficiency of SF3B4, a component of the pre-mRNA spliceosomal complex, causes Nager syndrome. Am J Hum Genet 90:925–933. doi:10.1016/j.ajhg.2012.04.004

Bernstein E, Caudy AA, Hammond SM, Hannon GJ (2001) Role for a bidentate ribonuclease in the initiation step of RNA interference. Nature 409:363–366. doi:10.1038/35053110

Bhalla K, Phillips HA, Crawford J et al (2004) The de novo chromosome 16 translocations of two patients with abnormal phenotypes (mental retardation and epilepsy) disrupt the A2BP1 gene. J Hum Genet 49:308–311. doi:10.1007/s10038-004-0145-4

Black DL (2000) Protein diversity from alternative splicing: a challenge for bioinformatics and post-genome biology. Cell 103:367–370

Boissel S, Reish O, Proulx K et al (2009) Loss-of-function mutation in the dioxygenase-encoding FTO gene causes severe growth retardation and multiple malformations. Am J Hum Genet 85:106–111. doi:10.1016/j.ajhg.2009.06.002

Boisvert F-M, van Koningsbruggen S, Navascués J, Lamond AI (2007) The multifunctional nucleolus. Nat Rev Mol Cell Biol 8:574–585. doi:10.1038/nrm2184

Bonifati V, Rizzu P, van Baren MJ et al (2003) Mutations in the DJ-1 gene associated with autosomal recessive early-onset parkinsonism. Science 299:256–259. doi:10.1126/science.1077209

Boocock GRB, Morrison JA, Popovic M et al (2003) Mutations in SBDS are associated with Shwachman-Diamond syndrome. Nat Genet 33:97–101. doi:10.1038/ng1062

Brais B, Bouchard J-P, Xie Y-G et al (1998) Short GCG expansions in the PABP2 gene cause oculopharyngeal muscular dystrophy. Nat Genet 18:164–167. doi:10.1038/ng0298-164

Brauch KM, Karst ML, Herron KJ et al (2009) Mutations in ribonucleic acid binding protein gene cause familial dilated cardiomyopathy. J Am Coll Cardiol 54:930–941. doi:10.1016/j.jacc.2009.05.038

Brennan CM, Steitz JA (2001) HuR and mRNA stability. Cell Mol Life Sci 58:266–277

Brickwood S, Bonthron DT, Al-Gazali LI et al (2003) Wolcott-Rallison syndrome: pathogenic insights into neonatal diabetes from new mutation and expression studies of EIF2AK3. J Med Genet 40:685–689

Broderick P, Carvajal-Carmona L, Pittman AM et al (2007) A genome-wide association study shows that common alleles of SMAD7 influence colorectal cancer risk. Nat Genet 39:1315–1317. doi:10.1038/ng.2007.18

Brook M, Smith JWS, Gray NK (2009) The DAZL and PABP families: RNA-binding proteins with interrelated roles in translational control in oocytes. Reproduction 137:595–617. doi:10.1530/REP-08-0524

Brooks SA, Blackshear PJ (2013) Tristetraprolin (TTP): Interactions with mRNA and proteins, and current thoughts on mechanisms of action. Biochim Biophys Acta 1829:666–679. doi:10.1016/j.bbagrm.2013.02.003

Brown RS (2005) Zinc finger proteins: getting a grip on RNA. Curr Opin Struct Biol 15:94–98. doi:10.1016/j.sbi.2005.01.006

Buchan JR, Kolaitis R-M, Taylor JP, Parker R (2013) Eukaryotic stress granules are cleared by autophagy and Cdc48/VCP function. Cell 153:1461–1474. doi:10.1016/j.cell.2013.05.037

Buckanovich RJ, Yang YY, Darnell RB (1996) The onconeural antigen Nova-1 is a neuron-specific RNA-binding protein, the activity of which is inhibited by paraneoplastic antibodies. J Neurosci 16:1114–1122

Budde BS, Namavar Y, Barth PG et al (2008) tRNA splicing endonuclease mutations cause pontocerebellar hypoplasia. Nat Genet 40:1113–1118. doi:10.1038/ng.204

Burd CG, Dreyfuss G (1994) Conserved structures and diversity of functions of RNA-binding proteins. Science 265:615–621

Bykhovskaya Y, Casas K, Mengesha E et al (2004) Missense mutation in pseudouridine synthase 1 (PUS1) causes mitochondrial myopathy and sideroblastic anemia (MLASA). Am J Hum Genet 74:1303–1308. doi:10.1086/421530

Calin GA, Croce CM (2006) MicroRNA signatures in human cancers. Nat Rev Cancer 6:857–866. doi:10.1038/nrc1997

Caluseriu O, Lowry BR, McLeod R et al (2013) The hutterite variant of Treacher Collins syndrome: a 28-year-old story solved. Am J Med Genet 161A:2855–2859. doi:10.1002/ajmg.a.36172

Camargos S, Scholz S, Simón-Sánchez J et al (2008) DYT16, a novel young-onset dystonia-parkinsonism disorder: identification of a segregating mutation in the stress-response protein PRKRA. Lancet Neurol 7:207–215. doi:10.1016/S1474-4422(08)70022-X

Cancel G, Dürr A, Didierjean O et al (1997) Molecular and clinical correlations in spinocerebellar ataxia 2: a study of 32 families. Hum Mol Genet 6:709–715

Carpten J, Nupponen N, Isaacs S et al (2002) Germline mutations in the ribonuclease L gene in families showing linkage with HPC1. Nat Genet 30:181–184. doi:10.1038/ng823

Casey G, Neville PJ, Plummer SJ et al (2002) RNASEL Arg462Gln variant is implicated in up to 13 % of prostate cancer cases. Nat Genet 32:581–583. doi:10.1038/ng1021

Cassandrini D, Biancheri R, Tessa A et al (2010) Pontocerebellar hypoplasia: clinical, pathologic, and genetic studies. Neurology 75:1459–1464. doi:10.1212/WNL.0b013e3181f88173

Castello A, Fischer B, Eichelbaum K et al (2012) Insights into RNA biology from an atlas of mammalian mRNA-binding proteins. Cell 149:1393–1406. doi:10.1016/j.cell.2012.04.031

Castello A, Fischer B, Hentze MW, Preiss T (2013) RNA-binding proteins in Mendelian disease. Trends Genet 29:318–327. doi:10.1016/j.tig.2013.01.004

Castilla LH, Couch FJ, Erdos MR et al (1994) Mutations in the BRCA1 gene in families with early-onset breast and ovarian cancer. Nat Genet 8:387–391. doi:10.1038/ng1294-387

Cavdar Koc E, Burkhart W, Blackburn K et al (2001) The small subunit of the mammalian mitochondrial ribosome. Identification of the full complement of ribosomal proteins present. J Biol Chem 276:19363–19374. doi:10.1074/jbc.M100727200

Cerritelli SM, Crouch RJ (2009) Ribonuclease H: the enzymes in eukaryotes. FEBS J 276: 1494–1505. doi:10.1111/j.1742-4658.2009.06908.x

Chakarova CF, Hims MM, Bolz H et al (2002) Mutations in HPRP3, a third member of pre-mRNA splicing factor genes, implicated in autosomal dominant retinitis pigmentosa. Hum Mol Genet 11:87–92

Chang K-Y, Ramos A (2005) The double-stranded RNA-binding motif, a versatile macromolecular docking platform. FEBS J 272:2109–2117. doi:10.1111/j.1742-4658.2005.04652.x

Chartier-Harlin M-C, Dachsel JC, Vilariño-Güell C et al (2011) Translation initiator EIF4G1 mutations in familial Parkinson disease. Am J Hum Genet 89:398–406. doi:10.1016/j.ajhg.2011.08.009

Cheadle JP, Gill H, Fleming N et al (2000) Long-read sequence analysis of the MECP2 gene in Rett syndrome patients: correlation of disease severity with mutation type and location. Hum Mol Genet 9:1119–1129

Chen X, Wolin SL (2004) The Ro 60 kDa autoantigen: insights into cellular function and role in autoimmunity. J Mol Med 82:232–239

Chénard CA, Richard S (2008) New implications for the QUAKING RNA binding protein in human disease. J Neurosci Res 86:233–242. doi:10.1002/jnr.21485

Cheong C-G, Hall TMT (2006) Engineering RNA sequence specificity of Pumilio repeats. Proc Natl Acad Sci U S A 103:13635–13639. doi:10.1073/pnas.0606294103

Chiu C, Tebo M, Ingles J et al (2007) Genetic screening of calcium regulation genes in familial hypertrophic cardiomyopathy. J Mol Cell Cardiol 43:337–343. doi:10.1016/j.yjmcc.2007.06.009

Cho J, Chang H, Kwon SC et al (2012) LIN28A is a suppressor of ER-associated translation in embryonic stem cells. Cell 151:765–777. doi:10.1016/j.cell.2012.10.019

Ciganda M, Williams N (2011) Eukaryotic 5S rRNA biogenesis. WIREs RNA 2:523–533. doi:10.1002/wrna.74

Cléry A, Blatter M, Allain FHT (2008) RNA recognition motifs: boring? Not quite. Curr Opin Struct Biol 18:290–298. doi:10.1016/j.sbi.2008.04.002

Cmejla R, Cmejlova J, Handrkova H et al (2007) Ribosomal protein S17 gene (RPS17) is mutated in Diamond-Blackfan anemia. Hum Mutat 28:1178–1182. doi:10.1002/humu.20608

Cobben JM, van der Steege G, Grootscholten P et al (1995) Deletions of the survival motor neuron gene in unaffected siblings of patients with spinal muscular atrophy. Am J Hum Genet 57: 805–808

Coenen MJH, Antonicka H, Ugalde C et al (2004) Mutant mitochondrial elongation factor G1 and combined oxidative phosphorylation deficiency. N Engl J Med 351:2080–2086. doi:10.1056/NEJMoa041878

Colgan DF, Manley JL (1997) Mechanism and regulation of mRNA polyadenylation. Genes Dev 11:2755–2766. doi:10.1101/gad.11.21.2755

Coller J, Parker R (2004) Eukaryotic mRNA decapping. Annu Rev Biochem 73:861–890. doi:10.1146/annurev.biochem.73.011303.074032

Cook KB, Kazan H, Zuberi K et al (2011) RBPDB: a database of RNA-binding specificities. Nucleic Acids Res 39:D301–D308. doi:10.1093/nar/gkq1069

Cools J, DeAngelo DJ, Gotlib J et al (2003) A tyrosine kinase created by fusion of the PDGFRA and FIP1L1 genes as a therapeutic target of imatinib in idiopathic hypereosinophilic syndrome. N Engl J Med 348:1201–1214. doi:10.1056/NEJMoa025217

Cooper TA, Wan L, Dreyfuss G (2009) RNA and disease. Cell 136:777–793. doi:10.1016/j.cell.2009.02.011

Costanzo MC, Hogan JD, Cusick ME et al (2000) The yeast proteome database (YPD) and Caenorhabditis elegans proteome database (WormPD): comprehensive resources for the organization and comparison of model organism protein information. Nucleic Acids Res 28:73–76

Criscuolo C, Mancini P, Menchise V et al (2005) Very late onset in ataxia oculomotor apraxia type I. Ann Neurol 57:777. doi:10.1002/ana.20463

Crosby AH, Patel H, Chioza BA et al (2010) Defective mitochondrial mRNA maturation is associated with spastic ataxia. Am J Hum Genet 87:655–660. doi:10.1016/j.ajhg.2010.09.013

Crow YJ, Hayward BE, Parmar R et al (2006a) Mutations in the gene encoding the 3 "-5" DNA exonuclease TREX1 cause Aicardi-Goutieres syndrome at the AGS1 locus. Nat Genet 38: 917–920. doi:10.1038/ng1845

Crow YJ, Leitch A, Hayward BE et al (2006b) Mutations in genes encoding ribonuclease H2 subunits cause Aicardi-Goutieres syndrome and mimic congenital viral brain infection. Nat Genet 38:910–916. doi:10.1038/ng1842

Curry S, Kotik-Kogan O, Conte MR, Brick P (2009) Getting to the end of RNA: structural analysis of protein recognition of 5′ and 3′ termini. Biochim Biophys Acta 1789:653–666. doi:10.1016/j.bbagrm.2009.07.003

Czeschik JC, Voigt C, Alanay Y et al (2013) Clinical and mutation data in 12 patients with the clinical diagnosis of Nager syndrome. Hum Genet 132:885–898. doi:10.1007/s00439-013-1295-2

Daiger SP, Sullivan LS, Bowne SJ (2013) Genes and mutations causing retinitis pigmentosa. Clin Genet 84:132–141. doi:10.1111/cge.12203

de la Cruz J, Kressler D, Linder P (1999) Unwinding RNA in Saccharomyces cerevisiae: DEAD-box proteins and related families. Trends Biochem Sci 24:192–198

Dean M, Park M, Vande Woude GF (1987) Characterization of the rearranged tpr-met oncogene breakpoint. Mol Cell Biol 7:921–924

Degnan BM, Vervoort M, Larroux C, Richards GS (2009) Early evolution of metazoan transcription factors. Curr Opin Genet Dev 19:591–599. doi:10.1016/j.gde.2009.09.008

DeHoratius RJ, Pillarisetty R, Messner RP, Talal N (1975) Anti-nucleic acid antibodies in systemic lupus erythematosus patients and their families. Incidence and correlation with lymphocytotoxic antibodies. J Clin Invest 56:1149–1154. doi:10.1172/JCI108190

Delépine M, Nicolino M, Barrett T et al (2000) EIF2AK3, encoding translation initiation factor 2-alpha kinase 3, is mutated in patients with Wolcott-Rallison syndrome. Nat Genet 25:406–409. doi:10.1038/78085

Dever TE, Green R (2012) The elongation, termination, and recycling phases of translation in eukaryotes. Cold Spring Harbor Pers Biol 4:a013706. doi:10.1101/cshperspect.a013706

Devys D, Biancalana V, Rousseau F et al (1992) Analysis of full fragile X mutations in fetal tissues and monozygotic twins indicate that abnormal methylation and somatic heterogeneity are established early in development. Am J Med Genet 43:208–216

Dezso Z, Nikolsky Y, Sviridov E et al (2008) A comprehensive functional analysis of tissue specificity of human gene expression. BMC Biol 6:49. doi:10.1186/1741-7007-6-49

Doherty L, Sheen MR, Vlachos A et al (2010) Ribosomal protein genes RPS10 and RPS26 are commonly mutated in Diamond-Blackfan anemia. Am J Hum Genet 86:222–228. doi:10.1016/j.ajhg.2009.12.015

Dorsett Y, Tuschl T (2004) siRNAs: applications in functional genomics and potential as therapeutics. Nat Rev Drug Discov 3:318–329. doi:10.1038/nrd1345

Drake KM, Zygmunt D, Mavrakis L et al (2011) Altered MicroRNA processing in heritable pulmonary arterial hypertension: an important role for Smad-8. Am J Respir Crit Care Med 184:1400–1408. doi:10.1164/rccm.201106-1130OC

Draptchinskaia N, Gustavsson P, Andersson B et al (1999) The gene encoding ribosomal protein S19 is mutated in Diamond-Blackfan anaemia. Nat Genet 21:169–175. doi:10.1038/5951

Dreyfuss G, Kim VN, Kataoka N (2002) Messenger-RNA-binding proteins and the messages they carry. Nat Rev Mol Cell Biol 3:195–205. doi:10.1038/nrm760

Du X, Rao MRKS, Chen XQ et al (2006) The homologous putative GTPases Grn1p from fission yeast and the human GNL3L are required for growth and play a role in processing of nucleolar pre-rRNA. Mol Biol Cell 17:460–474. doi:10.1091/mbc.E05-09-0848

Dumitrescu AM, Liao X-H, Abdullah MSY et al (2005) Mutations in SECISBP2 result in abnormal thyroid hormone metabolism. Nat Genet 37:1247–1252. doi:10.1038/ng1654

Durocher F, Faure R, Labrie Y et al (2006) A novel mutation in the EIF2AK3 gene with variable expressivity in two patients with Wolcott-Rallison syndrome. Clin Genet 70:34–38. doi:10.1111/j.1399-0004.2006.00632.x

Echeverria GV, Cooper TA (2012) RNA-binding proteins in microsatellite expansion disorders: Mediators of RNA toxicity. Brain Res 1462:100–111. doi:10.1016/j.brainres.2012.02.030
Edvardson S, Shaag A, Kolesnikova O et al (2007) Deleterious mutation in the mitochondrial arginyl-transfer RNA synthetase gene is associated with pontocerebellar hypoplasia. Am J Hum Genet 81:857–862. doi:10.1086/521227
Egan ED, Collins K (2012) Biogenesis of telomerase ribonucleoproteins. RNA 18:1747–1759. doi:10.1261/rna.034629.112
Elden AC, Kim H-J, Hart MP et al (2010) Ataxin-2 intermediate-length polyglutamine expansions are associated with increased risk for ALS. Nature 466:1069–1075. doi:10.1038/nature09320
Ellis JC, Brown JW (2009) The RNase P family. RNA Biol 6:362–369
Elo JM, Yadavalli SS, Euro L et al (2012) Mitochondrial phenylalanyl-tRNA synthetase mutations underlie fatal infantile Alpers encephalopathy. Hum Mol Genet 21:4521–4529. doi:10.1093/hmg/dds294
Esakova O, Krasilnikov AS (2010) Of proteins and RNA: The RNase P/MRP family. RNA 16:1725–1747. doi:10.1261/rna.2214510
Esquela-Kerscher A, Slack FJ (2006) Oncomirs - microRNAs with a role in cancer. Nat Rev Cancer 6:259–269. doi:10.1038/nrc1840
Esteller M (2011) Non-coding RNAs in human disease. Nat Rev Genet 12:861–874. doi:10.1038/nrg3074
Fabre A, Charroux B, Martinez-Vinson C et al (2012) SKIV2L mutations cause syndromic diarrhea, or trichohepatoenteric syndrome. Am J Hum Genet 90:689–692. doi:10.1016/j.ajhg.2012.02.009
Fabrizio P, Laggerbauer B, Lauber J et al (1997) An evolutionarily conserved U5 snRNP-specific protein is a GTP-binding factor closely related to the ribosomal translocase EF-2. EMBO J 16:4092–4106. doi:10.1093/emboj/16.13.4092
Faehnle CR, Elkayam E, Haase AD et al (2013) The making of a slicer: activation of human argonaute-1. Cell Rep 3:1901–1909. doi:10.1016/j.celrep.2013.05.033
Fairman-Williams ME, Guenther U-P, Jankowsky E (2010) SF1 and SF2 helicases: family matters. Curr Opin Struct Biol 20:313–324. doi:10.1016/j.sbi.2010.03.011
Fardaei M (2002) Three proteins, MBNL, MBLL and MBXL, co-localize in vivo with nuclear foci of expanded-repeat transcripts in DM1 and DM2 cells. Hum Mol Genet 11:805–814. doi:10.1093/hmg/11.7.805
Farrar JE, Nater M, Caywood E et al (2008) Abnormalities of the large ribosomal subunit protein, Rpl35a, in Diamond-Blackfan anemia. Blood 112:1582–1592. doi:10.1182/blood-2008-02-140012
Feinstein M, Markus B, Noyman I et al (2010) Pelizaeus-Merzbacher-like disease caused by AIMP1/p43 homozygous mutation. Am J Hum Genet 87:820–828. doi:10.1016/j.ajhg.2010.10.016
Fernandez-Vizarra E, Berardinelli A, Valente L et al (2007) Nonsense mutation in pseudouridylate synthase 1 (PUS1) in two brothers affected by myopathy, lactic acidosis and sideroblastic anaemia (MLASA). J Med Genet 44:173–180. doi:10.1136/jmg.2006.045252
Filipowicz W, Pogacić V (2002) Biogenesis of small nucleolar ribonucleoproteins. Curr Opin Cell Biol 14:319–327. doi:10.1016/s0955-0674(02)00334-4
Finn RD, Mistry J, Tate J et al (2010) The Pfam protein families database. Nucleic Acids Res 38:D211–D222. doi:10.1093/nar/gkp985
Fogli A, Wong K, Eymard-Pierre E et al (2002) Cree leukoencephalopathy and CACH/VWM disease are allelic at the EIF2B5 locus. Ann Neurol 52:506–510. doi:10.1002/ana.10339
Foulkes WD, Bahubeshi A, Hamel N et al (2011) Extending the phenotypes associated with DICER1 mutations. Hum Mutat 32:1381–1384. doi:10.1002/humu.21600
Freude K, Hoffmann K, Jensen L-R et al (2004) Mutations in the FTSJ1 gene coding for a novel S-adenosylmethionine-binding protein cause nonsyndromic X-linked mental retardation. Am J Hum Genet 75:305–309. doi:10.1086/422507
Frost RJA, Hamra FK, Richardson JA et al (2010) MOV10L1 is necessary for protection of spermatocytes against retrotransposons by Piwi-interacting RNAs. Proc Natl Acad Sci U S A 107:11847–11852. doi:10.1073/pnas.1007158107

Fukuda T, Yamagata K, Fujiyama S et al (2007) DEAD-box RNA helicase subunits of the Drosha complex are required for processing of rRNA and a subset of microRNAs. Nat Cell Biol 9: 604–611. doi:10.1038/ncb1577

Fullam A, Schröder M (2013) DExD/H-box RNA helicases as mediators of anti-viral innate immunity and essential host factors for viral replication. Biochim Biophys Acta 1829:854–865. doi:10.1016/j.bbagrm.2013.03.012

Gaipl US, Voll RE, Sheriff A et al (2005) Impaired clearance of dying cells in systemic lupus erythematosus. Autoimmun Rev 4:189–194. doi:10.1016/j.autrev.2004.10.007

Galante PAF, Sandhu D, de Sousa Abreu R et al (2009) A comprehensive in silico expression analysis of RNA binding proteins in normal and tumor tissue: Identification of potential players in tumor formation. RNA Biol 6:426–433. doi:10.4161/rna.6.4.8841

Galmiche L, Serre V, Beinat M et al (2011) Exome sequencing identifies MRPL3 mutation in mitochondrial cardiomyopathy. Hum Mutat 32:1225–1231. doi:10.1002/humu.21562

Gambardella A, Mazzei R, Toscano A et al (1998) Spinal muscular atrophy due to an isolated deletion of exon 8 of the telomeric survival motor neuron gene. Ann Neurol 44:836–839. doi:10.1002/ana.410440522

Gamberi C, Johnstone O, Lasko P (2006) Drosophila RNA binding proteins. Int Rev Cytol 248:43–139. doi:10.1016/S0074-7696(06)48002-5

Gazda HT, Grabowska A, Merida-Long LB et al (2006) Ribosomal protein S24 gene is mutated in Diamond-Blackfan anemia. Am J Hum Genet 79:1110–1118. doi:10.1086/510020

Gazda HT, Sheen MR, Vlachos A et al (2008) Ribosomal protein L5 and L11 mutations are associated with cleft palate and abnormal thumbs in Diamond-Blackfan anemia patients. Am J Hum Genet 83:769–780. doi:10.1016/j.ajhg.2008.11.004

Gedeon AK, Baker E, Robinson H et al (1992) Fragile X syndrome without CCG amplification has an FMR1 deletion. Nat Genet 1:341–344. doi:10.1038/ng0892-341

Gelpi C, Sontheimer EJ, Rodriguez-Sanchez JL (1992) Autoantibodies against a serine tRNA-protein complex implicated in cotranslational selenocysteine insertion. Proc Natl Acad Sci U S A 89:9739–9743

Ghezzi D, Baruffini E, Haack TB et al (2012) Mutations of the mitochondrial-tRNA modifier MTO1 cause hypertrophic cardiomyopathy and lactic acidosis. Am J Hum Genet 90: 1079–1087. doi:10.1016/j.ajhg.2012.04.011

Gibson WT, Hood RL, Zhan SH et al (2012) Mutations in EZH2 cause Weaver syndrome. Am J Hum Genet 90:110–118. doi:10.1016/j.ajhg.2011.11.018

Gill S, McManus AP, Crew AJ et al (1995) Fusion of the EWS gene to a DNA segment from 9q22-31 in a human myxoid chondrosarcoma. Genes Chromosome Cancer 12:307–310

Gingras AC, Raught B, Sonenberg N (1999) eIF4 initiation factors: Effectors of mRNA recruitment to ribosomes and regulators of translation. Annu Rev Biochem 68:913–963

Girard A, Sachidanandam R, Hannon GJ, Carmell MA (2006) A germline-specific class of small RNAs binds mammalian Piwi proteins. Nature 442:199–202. doi:10.1038/nature04917

Glisovic T, Bachorik JL, Yong J, Dreyfuss G (2008) RNA-binding proteins and post-transcriptional gene regulation. FEBS Lett 582:1977–1986

Gold HA, Craft J, Hardin JA et al (1988) Antibodies in human serum that precipitate ribonuclease P. Proc Natl Acad Sci U S A 85:5483–5487

Gonzatti-Haces M, Seth A, Park M et al (1988) Characterization of the TPR-MET oncogene p65 and the MET protooncogene p140 protein-tyrosine kinases. Proc Natl Acad Sci U S A 85:21–25

Gordon CT, Petit F, Oufadem M et al (2012) EFTUD2 haploinsufficiency leads to syndromic oesophageal atresia. J Med Genet 49:737–746. doi:10.1136/jmedgenet-2012-101173

Götz A, Tyynismaa H, Euro L et al (2011) Exome sequencing identifies mitochondrial alanyl-tRNA synthetase mutations in infantile mitochondrial cardiomyopathy. Am J Hum Genet 88:635–642. doi:10.1016/j.ajhg.2011.04.006

Grabowski P (2011) Alternative splicing takes shape during neuronal development. Curr Opin Genet Dev 21:388–394. doi:10.1016/j.gde.2011.03.005

Granneman S, Baserga SJ (2004) Ribosome biogenesis: of knobs and RNA processing. Exp Cell Res 296:43–50. doi:10.1016/j.yexcr.2004.03.016

Graubert TA, Shen D, Ding L et al (2012) Recurrent mutations in the U2AF1 splicing factor in myelodysplastic syndromes. Nat Genet 44:53–57. doi:10.1038/ng.1031

Greenway MJ, Andersen PM, Russ C et al (2006) ANG mutations segregate with familial and "sporadic" amyotrophic lateral sclerosis. Nat Genet 38:411–413. doi:10.1038/ng1742

Griffin JH, Leung J, Bruner RJ et al (2003) Discovery of a fusion kinase in EOL-1 cells and idiopathic hypereosinophilic syndrome. Proc Natl Acad Sci U S A 100:7830–7835. doi:10.1073/pnas.0932698100

Grishin NV (2001) KH domain: one motif, two folds. Nucleic Acids Res 29:638–643

Grivna ST (2006) A novel class of small RNAs in mouse spermatogenic cells. Genes Dev 20: 1709–1714. doi:10.1101/gad.1434406

Grohmann K, Schuelke M, Diers A et al (2001) Mutations in the gene encoding immunoglobulin mu-binding protein 2 cause spinal muscular atrophy with respiratory distress type 1. Nat Genet 29:75–77. doi:10.1038/ng703

Guan M-X, Yan Q, Li X et al (2006) Mutation in TRMU related to transfer RNA modification modulates the phenotypic expression of the deafness-associated mitochondrial 12S ribosomal RNA mutations. Am J Hum Genet 79:291–302. doi:10.1086/506389

Gubitz AK, Feng W, Dreyfuss G (2004) The SMN complex. Exp Cell Res 296:51–56. doi:10.1016/j.yexcr.2004.03.022

Guo H, Ingolia NT, Weissman JS, Bartel DP (2010) Mammalian microRNAs predominantly act to decrease target mRNA levels. Nature 466:835–840. doi:10.1038/nature09267

Guo W, Schafer S, Greaser ML et al (2012) RBM20, a gene for hereditary cardiomyopathy, regulates titin splicing. Nat Med 18:766–773. doi:10.1038/nm.2693

Gwinn-Hardy K, Chen JY, Liu HC et al (2000) Spinocerebellar ataxia type 2 with parkinsonism in ethnic Chinese. Neurology 55:800–805

Haack TB, Gorza M, Danhauser K et al (2014) Phenotypic spectrum of eleven patients and five novel MTFMT mutations identified by exome sequencing and candidate gene screening. Mol Genet Metab 111:342–352. doi:10.1016/j.ymgme.2013.12.010

Haack TB, Kopajtich R, Freisinger P et al (2013) ELAC2 Mutations Cause a Mitochondrial RNA Processing Defect Associated with Hypertrophic Cardiomyopathy. Am J Hum Genet 93:211–223. doi:10.1016/j.ajhg.2013.06.006

Hackman P, Sarparanta J, Lehtinen S et al (2012) Welander distal myopathy is caused by a mutation in the RNA-binding protein TIA1. Ann Neurol 73:500–509. doi:10.1002/ana.23831

Hafner M, Landthaler M, Burger L et al (2010) Transcriptome-wide identification of RNA-binding protein and MicroRNA target sites by PAR-CLIP. Cell 141:129–141. doi:10.1016/j.cell.2010.03.009

Hafner M, Max KEA, Bandaru P et al (2013) Identification of mRNAs bound and regulated by human LIN28 proteins and molecular requirements for RNA recognition. RNA 19:613–626. doi:10.1261/rna.036491.112

Hagerman P (2013) Fragile X-associated tremor/ataxia syndrome (FXTAS): pathology and mechanisms. Acta Neuropathol 126:1–19. doi:10.1007/s00401-013-1138-1

Hagerman RJ, Leehey M, Heinrichs W et al (2001) Intention tremor, parkinsonism, and generalized brain atrophy in male carriers of fragile X. Neurology 57:127–130

Hahnen E, Schönling J, Rudnik-Schöneborn S et al (1997) Missense mutations in exon 6 of the survival motor neuron gene in patients with spinal muscular atrophy (SMA). Hum Mol Genet 6:821–825

Hall AE, Turnbull C, Dalmay T (2013) Y RNAs: recent developments. BioMol Concepts 4:103–110

Hammond SM, Bernstein E, Beach D, Hannon GJ (2000) An RNA-directed nuclease mediates post-transcriptional gene silencing in Drosophila cells. Nature 404:293–296. doi:10.1038/35005107

Hamosh A, Scott AF, Amberger JS et al (2005) Online Mendelian Inheritance in Man (OMIM), a knowledgebase of human genes and genetic disorders. Nucleic Acids Res 33:D514–D517. doi:10.1093/nar/gki033

Hanson KA, Kim SH, Tibbetts RS (2011) RNA-binding proteins in neurodegenerative disease: TDP-43 and beyond. WIREs RNA 3:265–285. doi:10.1002/wrna.111

Harms MB, Ori-McKenney KM, Scoto M et al (2012) Mutations in the tail domain of DYNC1H1 cause dominant spinal muscular atrophy. Neurology 78:1714–1720. doi:10.1212/WNL.0b013e3182556c05

Hauptmann J, Dueck A, Harlander S et al (2013) Turning catalytically inactive human Argonaute proteins into active slicer enzymes. Nat Struct Mol Biol 20:814–817. doi:10.1038/nsmb.2577

Hayashi YK, Matsuda C, Ogawa M et al (2009) Human PTRF mutations cause secondary deficiency of caveolins resulting in muscular dystrophy with generalized lipodystrophy. J Clin Invest 119:2623–2633. doi:10.1172/JCI38660

Heiss NS, Knight SW, Vulliamy TJ et al (1998) X-linked dyskeratosis congenita is caused by mutations in a highly conserved gene with putative nucleolar functions. Nat Genet 19:32–38. doi:10.1038/ng0598-32

Hekman KE, Yu G-Y, Brown CD et al (2012) A conserved eEF2 coding variant in SCA26 leads to loss of translational fidelity and increased susceptibility to proteostatic insult. Hum Mol Genet 21:5472–5483. doi:10.1093/hmg/dds392

Hendrick JP, Wolin SL, Rinke J et al (1981) Ro small cytoplasmic ribonucleoproteins are a subclass of La ribonucleoproteins: further characterization of the Ro and La small ribonucleoproteins from uninfected mammalian cells. Mol Cell Biol 1:1138–1149

Henikoff S, Greene EA, Pietrokovski S et al (1997) Gene families: the taxonomy of protein paralogs and chimeras. Science 278:609–614

Henneke M, Diekmann S, Ohlenbusch A et al (2009) RNASET2-deficient cystic leukoencephalopathy resembles congenital cytomegalovirus brain infection. Nat Genet 41:773–775. doi:10.1038/ng.398

Hill DA, Ivanovich J, Priest JR et al (2009) DICER1 mutations in familial pleuropulmonary blastoma. Science 325:965. doi:10.1126/science.1174334

Hogan DJ, Riordan DP, Gerber AP et al (2008) Diverse RNA-binding proteins interact with functionally related sets of RNAs, suggesting an extensive regulatory system. PLoS Biol 6:e255. doi:10.1371/journal.pbio.0060255

Hogg JR, Goff SP (2010) Upf1 senses 3′UTR length to potentiate mRNA decay. Cell 143:379–389. doi:10.1016/j.cell.2010.10.005

Howe JR, Roth S, Ringold JC et al (1998) Mutations in the SMAD4/DPC4 gene in juvenile polyposis. Science 280:1086–1088

Huppke P, Laccone F, Krämer N et al (2000) Rett syndrome: analysis of MECP2 and clinical characterization of 31 patients. Hum Mol Genet 9:1369–1375

Hussain S, Sajini AA, Blanco S et al (2013) NSun2-mediated cytosine-5 methylation of vault noncoding RNA determines its processing into regulatory small RNAs. Cell Rep 4:255–261. doi:10.1016/j.celrep.2013.06.029

Hutvagner G, Simard MJ (2008) Argonaute proteins: key players in RNA silencing. Nat Rev Mol Cell Biol 9:22–32. doi:10.1038/nrm2321

Ichikawa H, Shimizu K, Hayashi Y, Ohki M (1994) An RNA-binding protein gene, TLS/FUS, is fused to ERG in human myeloid leukemia with t(16;21) chromosomal translocation. Cancer Res 54:2865–2868

IJlst L, Loupatty FJ, IRuiter JPN et al (2002) 3-Methylglutaconic aciduria type I is caused by mutations in AUH. Am J Hum Genet 71:1463–1466. doi:10.1086/344712

Ishizu H, Siomi H, Siomi MC (2012) Biology of PIWI-interacting RNAs: new insights into biogenesis and function inside and outside of germlines. Genes Dev 26:2361–2373. doi:10.1101/gad.203786.112

Ivanov P, Emara MM, Villen J et al (2011) Angiogenin-induced tRNA fragments inhibit translation initiation. Mol Cell 43:613–623. doi:10.1016/j.molcel.2011.06.022

Jankowsky E (2011) RNA helicases at work: binding and rearranging. Trends Biochem Sci 36:19–29. doi:10.1016/j.tibs.2010.07.008

Jankowsky E, Fairman-Williams ME (2010) An introduction to RNA helicases: superfamilies, families, and major themes. In: Jankowsky E (ed) RNA helicases, vol 19. RSC Biomolecular Sciences, London, pp 1–31

Jarrous N (2002) Human ribonuclease P: subunits, function, and intranuclear localization. RNA 8:1–7

Jinek M, Doudna JA (2009) A three-dimensional view of the molecular machinery of RNA interference. Nature 457:405–412. doi:10.1038/nature07755

Johnston JJ, Teer JK, Cherukuri PF et al (2010) Massively parallel sequencing of exons on the X chromosome identifies RBM10 as the gene that causes a syndromic form of cleft palate. Am J Hum Genet 86:743–748. doi:10.1016/j.ajhg.2010.04.007

Jordanova A, Irobi J, Thomas FP et al (2006) Disrupted function and axonal distribution of mutant tyrosyl-tRNA synthetase in dominant intermediate Charcot-Marie-Tooth neuropathy. Nat Genet 38:197–202. doi:10.1038/ng1727

Kaida D, Berg MG, Younis I et al (2010) U1 snRNP protects pre-mRNAs from premature cleavage and polyadenylation. Nature 468:664–668. doi:10.1038/nature09479

Kalscheuer VM, Freude K, Musante L et al (2003) Mutations in the polyglutamine binding protein 1 gene cause X-linked mental retardation. Nat Genet 35:313–315. doi:10.1038/ng1264

Kanegane H, Kasahara Y, Okamura J et al (2005) Identification of DKC1 gene mutations in Japanese patients with X-linked dyskeratosis congenita. Br J Haematol 129:432–434. doi:10.1111/j.1365-2141.2005.05473.x

Kanehisa M, Goto S (2000) KEGG: kyoto encyclopedia of genes and genomes. Nucleic Acids Res 28:27–30

Kaneko H, Dridi S, Tarallo V et al (2011) DICER1 deficit induces Alu RNA toxicity in age-related macular degeneration. Nature 471:325–330. doi:10.1038/nature09830

Kapeli K, Yeo GW (2012) Genome-wide approaches to dissect the roles of RNA binding proteins in translational control: implications for neurological diseases. Front Neurosci 6:144. doi:10.3389/fnins.2012.00144

Karaca E, Weitzer S, Pehlivan D et al (2014) Human CLP1 mutations alter tRNA biogenesis, affecting both peripheral and central nervous system function. Cell 157:636–650. doi:10.1016/j.cell.2014.02.058

Kaymak E, Wee LM, Ryder SP (2010) Structure and function of nematode RNA-binding proteins. Curr Opin Struct Biol 20:305–312. doi:10.1016/j.sbi.2010.03.010

Keen TJ, Hims MM, McKie AB et al (2002) Mutations in a protein target of the Pim-1 kinase associated with the RP9 form of autosomal dominant retinitis pigmentosa. Eur J Hum Genet 10:245–249. doi:10.1038/sj.ejhg.5200797

Kerner P, Degnan SM, Marchand L et al (2011) Evolution of RNA-binding proteins in animals: insights from genome-wide analysis in the sponge Amphimedon queenslandica. Mol Biol Evol 28:2289–2303. doi:10.1093/molbev/msr046

Khan MA, Rafiq MA, Noor A et al (2012) Mutation in NSUN2, which encodes an RNA methyltransferase, causes autosomal-recessive intellectual disability. Am J Hum Genet 90:856–863. doi:10.1016/j.ajhg.2012.03.023

Khusial P, Plaag R, Zieve GW (2005) LSm proteins form heptameric rings that bind to RNA via repeating motifs. Trends Biochem Sci 30:522–528. doi:10.1016/j.tibs.2005.07.006

Kim CA, Bowie JU (2003) SAM domains: uniform structure, diversity of function. Trends Biochem Sci 28:625–628. doi:10.1016/j.tibs.2003.11.001

Kim VN, Han J, Siomi MC (2009) Biogenesis of small RNAs in animals. Nat Rev Mol Cell Biol 10:126–139. doi:10.1038/nrm2632

Kim HJ, Kim NC, Wang Y-D et al (2013) Mutations in prion-like domains in hnRNPA2B1 and hnRNPA1 cause multisystem proteinopathy and ALS. Nature 495:467–473. doi:10.1038/nature11922

Kirwan M, Walne AJ, Plagnol V et al (2012) Exome sequencing identifies autosomal-dominant SRP72 mutations associated with familial aplasia and myelodysplasia. Am J Hum Genet 90:888–892. doi:10.1016/j.ajhg.2012.03.020

Kiss T (2004) Biogenesis of small nuclear RNPs. J Cell Sci 117:5949–5951. doi:10.1242/jcs.01487

Kiss T, Fayet E, Jády BE et al (2006) Biogenesis and intranuclear trafficking of human box C/D and H/ACA RNPs. Cold Spring Harbor Symp Quant Biol 71:407–417. doi:10.1101/sqb.2006.71.025

Klattenhoff C, Theurkauf W (2008) Biogenesis and germline functions of piRNAs. Development 135:3–9. doi:10.1242/dev.006486

Klauck SM, Felder B, Kolb-Kokocinski A et al (2006) Mutations in the ribosomal protein gene RPL10 suggest a novel modulating disease mechanism for autism. Mol Psychiatry 11:1073–1084. doi:10.1038/sj.mp.4001883

Klein CJ, Botuyan M-V, Wu Y et al (2011) Mutations in DNMT1 cause hereditary sensory neuropathy with dementia and hearing loss. Nat Genet 43:595–600. doi:10.1038/ng.830

Kloosterman WP, Plasterk RHA (2006) The diverse functions of microRNAs in animal development and disease. Dev Cell 11:441–450. doi:10.1016/j.devcel.2006.09.009

Knight SJL, Flannery AV, Hirst MC et al (1993) Trinucleotide repeat amplification and hypermethylation of a CpG island in FRAXE mental retardation. Cell 74:127–134. doi:10.1016/0092-8674(93)90300-F

Knight SW, Heiss NS, Vulliamy TJ et al (1999) X-linked dyskeratosis congenita is predominantly caused by missense mutations in the DKC1 gene. Am J Hum Genet 65:50–58. doi:10.1086/302446

Knight SW, Vulliamy TJ, Morgan B et al (2001) Identification of novel DKC1 mutations in patients with dyskeratosis congenita: implications for pathophysiology and diagnosis. Hum Genet 108:299–303

Kobayashi H, Abe K, Matsuura T et al (2011) Expansion of intronic GGCCTG hexanucleotide repeat in NOP56 causes SCA36, a type of spinocerebellar ataxia accompanied by motor neuron involvement. Am J Hum Genet 89:121–130. doi:10.1016/j.ajhg.2011.05.015

Köhn M, Pazaitis N, Hüttelmaier S (2013) Why YRNAs? About versatile RNAs and their functions. Biomolecules 3:143–156. doi:10.3390/biom3010143

Kornblihtt AR, Schor IE, Allo M et al (2013) Alternative splicing: a pivotal step between eukaryotic transcription and translation. Nat Rev Mol Cell Biol 14:153–165. doi:10.1038/nrm3525

Korobeinikova AV, Garber MB, Gongadze GM (2012) Ribosomal proteins: structure, function, and evolution. Biochem (Mosc) 77:562–574. doi:10.1134/S0006297912060028

Kraus MR-C, Clauin S, Pfister Y et al (2012) Two mutations in human BICC1 resulting in Wnt pathway hyperactivity associated with cystic renal dysplasia. Hum Mutat 33:86–90. doi:10.1002/humu.21610

Kremer EJ, Pritchard M, Lynch M et al (1991) Mapping of DNA instability at the fragile X to a trinucleotide repeat sequence p(CCG)n. Science 252:1711–1714

Kressler D, Hurt E, Bassler J (2010) Driving ribosome assembly. Biochim Biophys Acta 1803:-673–683. doi:10.1016/j.bbamcr.2009.10.009

Kuchta K, Knizewski L, Wyrwicz LS et al (2009) Comprehensive classification of nucleotidyltransferase fold proteins: identification of novel families and their representatives in human. Nucleic Acids Res 37:7701–7714. doi:10.1093/nar/gkp854

Kufel J, Bousquet-Antonelli C, Beggs JD, Tollervey D (2004) Nuclear pre-mRNA decapping and 5′ degradation in yeast require the Lsm2-8p complex. Mol Cell Biol 24:9646–9657. doi:10.1128/MCB.24.21.9646-9657.2004

Kwiatkowski TJ, Bosco DA, Leclerc AL et al (2009) Mutations in the FUS/TLS gene on chromosome 16 cause familial amyotrophic lateral sclerosis. Science 323:1205–1208. doi:10.1126/science.1166066

Kwon SC, Yi H, Eichelbaum K et al (2013) The RNA-binding protein repertoire of embryonic stem cells. Nat Struct Mol Biol 20:1122–1130. doi:10.1038/nsmb.2638

Lachke SA, Alkuraya FS, Kneeland SC et al (2011) Mutations in the RNA granule component TDRD7 cause cataract and glaucoma. Science 331:1571–1576. doi:10.1126/science.1195970

Lagier-Tourenne C, Polymenidou M, Cleveland DW (2010) TDP-43 and FUS/TLS: emerging roles in RNA processing and neurodegeneration. Hum Mol Genet 19:R46–R64. doi:10.1093/hmg/ddq137

Larroux C, Luke GN, Koopman P et al (2008) Genesis and expansion of metazoan transcription factor gene classes. Mol Biol Evol 25:980–996. doi:10.1093/molbev/msn047

Lasko P (2000) The Drosophila melanogaster genome: translation factors and RNA binding proteins. J Cell Biol 150:F51–F56
Lasko P (2003) Gene regulation at the RNA layer: RNA binding proteins in intercellular signaling networks. Sci STKE 2003:RE6. doi:10.1126/stke.2003.179.re6
Lasko P (2013) The DEAD-box helicase Vasa: evidence for a multiplicity of functions in RNA processes and developmental biology. Biochim Biophys Acta 1829:810–816. doi:10.1016/j.bbagrm.2013.04.005
Latour P, Thauvin-Robinet C, Baudelet-Méry C et al (2010) A major determinant for binding and aminoacylation of tRNA(Ala) in cytoplasmic Alanyl-tRNA synthetase is mutated in dominant axonal Charcot-Marie-Tooth disease. Am J Hum Genet 86:77–82. doi:10.1016/j.ajhg.2009.12.005
Lee M-H, Schedl T (2006) RNA-binding proteins. WormBook 1–13. doi: 10.1895/wormbook.1.79.1
Lee TI, Rinaldi NJ, Robert F et al (2002) Transcriptional regulatory networks in Saccharomyces cerevisiae. Science 298:799–804. doi:10.1126/science.1075090
Lee Y, Ahn C, Han JJ et al (2003) The nuclear RNase III Drosha initiates microRNA processing. Nature 425:415–419. doi:10.1038/nature01957
Lee EB, Lee VMY, Trojanowski JQ (2012) Gains or losses: molecular mechanisms of TDP43-mediated neurodegeneration. Nat Rev Neurosci 13:38–50. doi:10.1038/nrn3121
Leegwater PA, Vermeulen G, Könst AA et al (2001) Subunits of the translation initiation factor eIF2B are mutant in leukoencephalopathy with vanishing white matter. Nat Genet 29:383–388. doi:10.1038/ng764
Lefebvre S, Bürglen L, Reboullet S et al (1995) Identification and characterization of a spinal muscular atrophy-determining gene. Cell 80:155–165
Lerner MR, Boyle JA, Hardin JA, Steitz JA (1981) Two novel classes of small ribonucleoproteins detected by antibodies associated with lupus erythematosus. Science 211:400–402
Letunic I, Doerks T, Bork P (2009) SMART 6: recent updates and new developments. Nucleic Acids Res 37:D229–D232. doi:10.1093/nar/gkn808
Li Y, Kowdley KV (2012) MicroRNAs in common human diseases. Genom Proteom Bioinform 10:246–253. doi:10.1016/j.gpb.2012.07.005
Li Q, Lee J-A, Black DL (2007) Neuronal regulation of alternative pre-mRNA splicing. Nat Rev Neurosci 8:819–831. doi:10.1038/nrn2237
Li YR, King OD, Shorter J, Gitler AD (2013) Stress granules as crucibles of ALS pathogenesis. J Cell Biol 201:361–372. doi:10.1083/jcb.201302044
Lin K-P, Soong B-W, Yang C-C et al (2011) The mutational spectrum in a cohort of Charcot-Marie-Tooth disease type 2 among the Han Chinese in Taiwan. PLoS One 6:e29393. doi:10.1371/journal.pone.0029393
Linder P, Fuller-Pace FV (2013) Looking back on the birth of DEAD-box RNA helicases. Biochim Biophys Acta 1829:750–755. doi:10.1016/j.bbagrm.2013.03.007
Linder P, Jankowsky E (2011) From unwinding to clamping - the DEAD box RNA helicase family. Nat Rev Mol Cell Biol 12:505–516. doi:10.1038/nrm3154
Lines MA, Huang L, Schwartzentruber J et al (2012) Haploinsufficiency of a spliceosomal GTPase encoded by EFTUD2 causes mandibulofacial dysostosis with microcephaly. Am J Hum Genet 90:369–377. doi:10.1016/j.ajhg.2011.12.023
Ling S-C, Polymenidou M, Cleveland DW (2013) Converging mechanisms in ALS and FTD: disrupted RNA and protein homeostasis. Neuron 79:416–438. doi:10.1016/j.neuron.2013.07.033
Liquori CL, Ricker K, Moseley ML et al (2001) Myotonic dystrophy type 2 caused by a CCTG expansion in intron 1 of ZNF9. Science 293:864–867. doi:10.1126/science.1062125
Liu J (2004) Argonaute2 Is the catalytic engine of Mammalian RNAi. Science 305:1437–1441. doi:10.1126/science.1102513
Liu JM (2006) Ribosomes and marrow failure: coincidental association or molecular paradigm? Blood 107:4583–4588. doi:10.1182/blood-2005-12-4831
Liu G, Grant WM, Persky D et al (2002) Interactions of elongation factor 1alpha with F-actin and beta-actin mRNA: implications for anchoring mRNA in cell protrusions. Mol Biol Cell 13:579–592. doi:10.1091/mbc.01-03-0140

Liu-Yesucevitz L, Bassell GJ, Gitler AD et al (2011) Local RNA translation at the synapse and in disease. J Neurosci 31:16086–16093. doi:10.1523/JNEUROSCI.4105-11.2011
Lovci MT, Ghanem D, Marr H et al (2013) Rbfox proteins regulate alternative mRNA splicing through evolutionarily conserved RNA bridges. Nat Struct Mol Biol 20:1434–1442. doi:10.1038/nsmb.2699
Lukong KE, Chang KW, Khandjian EW, Richard S (2008) RNA-binding proteins in human genetic disease. Trends Genet 24:416–425. doi:10.1016/j.tig.2008.05.004
Lund E (2004) Nuclear export of MicroRNA precursors. Science 303:95–98. doi:10.1126/science.1090599
Lunde BM, Moore C, Varani G (2007) RNA-binding proteins: modular design for efficient function. Nat Rev Mol Cell Biol 8:479–490. doi:10.1038/nrm2178
Ly TBN, Peters V, Gibson KM et al (2003) Mutations in the AUH gene cause 3-methylglutaconic aciduria type I. Hum Mutat 21:401–407. doi:10.1002/humu.10202
Mankodi A, Urbinati CR, Yuan QP et al (2001) Muscleblind localizes to nuclear foci of aberrant RNA in myotonic dystrophy types 1 and 2. Hum Mol Genet 10:2165–2170
Maraia RJ, Bayfield MA (2006) The La protein-RNA complex surfaces. Mol Cell 21:149–152. doi:10.1016/j.molcel.2006.01.004
Maraia RJ, Lamichhane TN (2011) 3′ processing of eukaryotic precursor tRNAs. WIREs RNA 2:362–375. doi:10.1002/wrna.64
Marchler-Bauer A, Anderson JB, DeWeese-Scott C et al (2003) CDD: a curated Entrez database of conserved domain alignments. Nucleic Acids Res 31:383–387
Maris C, Dominguez C, Allain FHT (2005) The RNA recognition motif, a plastic RNA-binding platform to regulate post-transcriptional gene expression. FEBS J 272:2118–2131. doi:10.1111/j.1742-4658.2005.04653.x
Marrone A, Walne A, Tamary H et al (2007) Telomerase reverse-transcriptase homozygous mutations in autosomal recessive dyskeratosis congenita and Hoyeraal-Hreidarsson syndrome. Blood 110:4198–4205. doi:10.1182/blood-2006-12-062851
Marszalek B, Wisniewski SA, Wojcicki P et al (2003) Novel mutation in the 5' splice site of exon 4 of the TCOF1 gene in the patient with Treacher Collins syndrome. Am J Med Genet 123A:169–171. doi:10.1002/ajmg.a.20312
Martin CL, Duvall JA, Ilkin Y et al (2007) Cytogenetic and molecular characterization of A2BP1/FOX1 as a candidate gene for autism. Am J Med Genet B Neuropsychiatr Genet 144B:869–876. doi:10.1002/ajmg.b.30530
Martin R, Straub AU, Doebele C, Bohnsack MT (2013) DExD/H-box RNA helicases in ribosome biogenesis. RNA Biol 10:4–18. doi:10.4161/rna.21879
Masliah G, Barraud P, Allain FHT (2013) RNA recognition by double-stranded RNA binding domains: a matter of shape and sequence. Cell Mol Life Sci 70:1875–1895. doi:10.1007/s00018-012-1119-x
Massart A, Lissens W, Tournaye H, Stouffs K (2012) Genetic causes of spermatogenic failure. Asian J Androl 14:40–48. doi:10.1038/aja.2011.67
Matera AG, Terns RM, Terns MP (2007) Non-coding RNAs: lessons from the small nuclear and small nucleolar RNAs. Nat Rev Mol Cell Biol 8:209–220. doi:10.1038/nrm2124
Matlin AJ, Clark F, Smith CWJ (2005) Understanding alternative splicing: towards a cellular code. Nat Rev Mol Cell Biol 6:386–398. doi:10.1038/nrm1645
Matsson H, Klar J, Draptchinskaia N et al (1999) Truncating ribosomal protein S19 mutations and variable clinical expression in Diamond-Blackfan anemia. Hum Genet 105:496–500
Matthews DE, Hessler RA, Denslow ND et al (1982) Protein composition of the bovine mitochondrial ribosome. J Biol Chem 257:8788–8794
May WA, Gishizky ML, Lessnick SL et al (1993) Ewing sarcoma 11;22 translocation produces a chimeric transcription factor that requires the DNA-binding domain encoded by FLI1 for transformation. Proc Natl Acad Sci U S A 90:5752–5756
Mayr C, Bartel DP (2009) Widespread shortening of 3′UTRs by alternative cleavage and polyadenylation activates oncogenes in cancer cells. Cell 138:673–684. doi:10.1016/j.cell.2009.06.016

McKee AE, Minet E, Stern C et al (2005) A genome-wide in situ hybridization map of RNA-binding proteins reveals anatomically restricted expression in the developing mouse brain. BMC Dev Biol 5:14. doi:10.1186/1471-213X-5-14
McKie AB, McHale JC, Keen TJ et al (2001) Mutations in the pre-mRNA splicing factor gene PRPC8 in autosomal dominant retinitis pigmentosa (RP13). Hum Mol Genet 10:1555–1562
McLaughlin HM, Sakaguchi R, Liu C et al (2010) Compound heterozygosity for loss-of-function lysyl-tRNA synthetase mutations in a patient with peripheral neuropathy. Am J Hum Genet 87:560–566. doi:10.1016/j.ajhg.2010.09.008
Meister G, Landthaler M, Patkaniowska A et al (2004) Human Argonaute2 mediates RNA cleavage targeted by miRNAs and siRNAs. Mol Cell 15:185–197. doi:10.1016/j.molcel.2004.07.007
Melnikov S, Ben-Shem A, Garreau de Loubresse N et al (2012) One core, two shells: bacterial and eukaryotic ribosomes. Nat Struct Mol Biol 19:560–567. doi:10.1038/nsmb.2313
Melo SA, Ropero S, Moutinho C et al (2009) A TARBP2 mutation in human cancer impairs microRNA processing and DICER1 function. Nat Genet 41:365–370. doi:10.1038/ng.317
Melo SA, Moutinho C, Ropero S et al (2010) A genetic defect in exportin-5 traps precursor micrornas in the nucleus of cancer cells. Cancer Cell 18:303–315. doi:10.1016/j.ccr.2010.09.007
Merritt WM, Lin YG, Han LY et al (2008) Dicer, Drosha, and outcomes in patients with ovarian cancer. N Engl J Med 359:2641–2650. doi:10.1056/NEJMoa0803785
Mickleburgh I, Chabanon H, Nury D et al (2006) Elongation factor 1alpha binds to the region of the metallothionein-1 mRNA implicated in perinuclear localization–importance of an internal stem-loop. RNA 12:1397–1407. doi:10.1261/rna.2730106
Mihailovich M, Militti C, Gabald6n T, Gebauer F (2010) Eukaryotic cold shock domain proteins: highly versatile regulators of gene expression. Bioessays 32:109–118. doi:10.1002/bies.200900122
Miki T, Takano K, Yoneda Y (2005) The role of mammalian Staufen on mRNA traffic: a view from its nucleocytoplasmic shuttling function. Cell Struct Funct 30:51–56
Miller JW (2000) Recruitment of human muscleblind proteins to (CUG)n expansions associated with myotonic dystrophy. EMBO J 19:4439–4448. doi:10.1093/emboj/19.17.4439
Miller C, Saada A, Shaul N et al (2004) Defective mitochondrial translation caused by a ribosomal protein (MRPS16) mutation. Ann Neurol 56:734–738. doi:10.1002/ana.20282
Mitchell SF, Jain S, She M, Parker R (2013) Global analysis of yeast mRNPs. Nat Struct Mol Biol 20:127–133. doi:10.1038/nsmb.2468
Miura P, Shenker S, Andreu-Agullo C et al (2013) Widespread and extensive lengthening of 3′ UTRs in the mammalian brain. Genome Res 23:812–825. doi:10.1101/gr.146886.112
Miyamura Y, Suzuki T, Kono M et al (2003) Mutations of the RNA-specific adenosine deaminase gene (DSRAD) are involved in dyschromatosis symmetrica hereditaria. Am J Hum Genet 73:693–699. doi:10.1086/378209
Monemi S, Spaeth G, DaSilva A et al (2005) Identification of a novel adult-onset primary open-angle glaucoma (POAG) gene on 5q22.1. Hum Mol Genet 14:725–733. doi:10.1093/hmg/ddi068
Monti L, Rodolfo M, Rudolfo M et al (2008) RNASET2 as a tumor antagonizing gene in a melanoma cancer model. Oncol Res 17:69–74
Mootha VK, Lepage P, Miller K et al (2003) Identification of a gene causing human cytochrome c oxidase deficiency by integrative genomics. Proc Natl Acad Sci U S A 100:605–610. doi:10.1073/pnas.242716699
Moreira M-C, Klur S, Watanabe M et al (2004) Senataxin, the ortholog of a yeast RNA helicase, is mutant in ataxia-ocular apraxia 2. Nat Genet 36:225–227. doi:10.1038/ng1303
Müller-McNicoll M, Neugebauer KM (2013) How cells get the message: dynamic assembly and function of mRNA-protein complexes. Nat Rev Genet 14:275–287. doi:10.1038/nrg3434
Muñoz LE, Lauber K, Schiller M et al (2010) The role of defective clearance of apoptotic cells in systemic autoimmunity. Nat Rev Rheumatol 6:280–289. doi:10.1038/nrrheum.2010.46
Murray A, Webb J, Grimley S et al (1998) Studies of FRAXA and FRAXE in women with premature ovarian failure. J Med Genet 35:637–640

Murzin AG (1993) OB(oligonucleotide/oligosaccharide binding)-fold: common structural and functional solution for non-homologous sequences. EMBO J 12:861–867

Murzin AG, Brenner SE, Hubbard T, Chothia C (1995) SCOP: a structural classification of proteins database for the investigation of sequences and structures. J Mol Biol 247:536–540. doi:10.1006/jmbi.1995.0159

Muscatelli F, Strom TM, Walker AP et al (1994) Mutations in the DAX-1 gene give rise to both X-linked adrenal hypoplasia congenita and hypogonadotropic hypogonadism. Nature 372:672–676. doi:10.1038/372672a0

Muto Y, Yokoyama S (2012) Structural insight into RNA recognition motifs: versatile molecular Lego building blocks for biological systems. WIREs RNA 3:229–246. doi:10.1002/wrna.1107

Nakanishi K, Ascano M, Gogakos T et al (2013) Eukaryote-specific insertion elements control human ARGONAUTE slicer activity. Cell Rep 3:1893–1900. doi:10.1016/j.celrep.2013.06.010

Nakashima E, Mabuchi A, Makita Y et al (2004) Novel SBDS mutations caused by gene conversion in Japanese patients with Shwachman-Diamond syndrome. Hum Genet 114:345–348. doi:10.1007/s00439-004-1081-2

Narla A, Ebert BL (2010) Ribosomopathies: human disorders of ribosome dysfunction. Blood 115:3196–3205. doi:10.1182/blood-2009-10-178129

Nasim MT, Ogo T, Ahmed M et al (2011) Molecular genetic characterization of SMAD signaling molecules in pulmonary arterial hypertension. Hum Mutat 32:1385–1389. doi:10.1002/humu.21605

Nawrot B, Sochacka E, Düchler M (2011) tRNA structural and functional changes induced by oxidative stress. Cell Mol Life Sci 68:4023–4032. doi:10.1007/s00018-011-0773-8

Neeve VCM, Pyle A, Boczonadi V et al (2013) Clinical and functional characterisation of the combined respiratory chain defect in two sisters due to autosomal recessive mutations in MTFMT. Mitochondrion 13:743–748. doi:10.1016/j.mito.2013.03.002

Negrutskii BS, El'skaya AV (1998) Eukaryotic translation elongation factor 1 alpha: structure, expression, functions, and possible role in aminoacyl-tRNA channeling. Prog Nucleic Acid Res Mol Biol 60:47–78

Neilson DE, Adams MD, Orr CMD et al (2009) Infection-triggered familial or recurrent cases of acute necrotizing encephalopathy caused by mutations in a component of the nuclear pore, RANBP2. Am J Hum Genet 84:44–51. doi:10.1016/j.ajhg.2008.12.009

Nelson DL, Orr HT, Warren ST (2013) The unstable repeats—three evolving faces of neurological disease. Neuron 77:825–843. doi:10.1016/j.neuron.2013.02.022

Nissen P, Kjeldgaard M, Thirup S et al (1995) Crystal structure of the ternary complex of Phe-tRNAPhe, EF-Tu, and a GTP analog. Science 270:1464–1472

Nousbeck J, Spiegel R, Ishida-Yamamoto A et al (2008) Alopecia, neurological defects, and endocrinopathy syndrome caused by decreased expression of RBM28, a nucleolar protein associated with ribosome biogenesis. Am J Hum Genet 82:1114–1121. doi:10.1016/j.ajhg.2008.03.014

Nousiainen HO, Kestila M, Pakkasjarvi N et al (2008) Mutations in mRNA export mediator GLE1 result in a fetal motoneuron disease. Nat Genet 40:155–157. doi:10.1038/ng.2007.65

O'Brien TW (2003) Properties of human mitochondrial ribosomes. IUBMB Life 55:505–513. doi:10.1080/1521654031000162661 0

Omran H, Kobayashi D, Olbrich H et al (2008) Ktu/PF13 is required for cytoplasmic pre-assembly of axonemal dyneins. Nature 456:611–616. doi:10.1038/nature07471

Orloff M, Peterson C, He X et al (2011) Germline mutations in MSR1, ASCC1, and CTHRC1 in patients with Barrett esophagus and esophageal adenocarcinoma. JAMA 306:410–419. doi:10.1001/jama.2011.1029

Orr HT, Chung MY, Banfi S et al (1993) Expansion of an unstable trinucleotide CAG repeat in spinocerebellar ataxia type 1. Nat Genet 4:221–226. doi:10.1038/ng0793-221

Panagopoulos I, Aman P, Fioretos T et al (1994) Fusion of the FUS gene with ERG in acute myeloid leukemia with t(16;21)(p11;q22). Genes Chromosome Cancer 11:256–262

Panagopoulos I, Mencinger M, Dietrich CU et al (1999) Fusion of the RBP56 and CHN genes in extraskeletal myxoid chondrosarcomas with translocation t(9;17)(q22;q11). Oncogene 18:7594–7598. doi:10.1038/sj.onc.1203155

Pannone BK, Kim SD, Noe DA, Wolin SL (2001) Multiple functional interactions between components of the Lsm2-Lsm8 complex, U6 snRNA, and the yeast La protein. Genetics 158:187–196

Parisi M, Lin HF (2000) Translational repression: A duet of Nanos and Pumilio. Curr Biol 10: R81–R83. doi:10.1016/s0960-9822(00)00283-9

Park E, Maquat LE (2013) Staufen-mediated mRNA decay. WIREs RNA 4:423–435. doi:10.1002/wrna.1168

Parker R, Sheth U (2007) P bodies and the control of mRNA translation and degradation. Mol Cell 25:635–646. doi:10.1016/j.molcel.2007.02.011

Parker R, Song H (2004) The enzymes and control of eukaryotic mRNA turnover. Nat Struct Mol Biol 11:121–127. doi:10.1038/nsmb724

Parsons DW, McAndrew PE, Monani UR et al (1996) An 11 base pair duplication in exon 6 of the SMN gene produces a type I spinal muscular atrophy (SMA) phenotype: further evidence for SMN as the primary SMA-determining gene. Hum Mol Genet 5:1727–1732

Pascual M, Vicente M, Monferrer L, Artero R (2006) The Muscleblind family of proteins: an emerging class of regulators of developmentally programmed alternative splicing. Differentiation 74:65–80. doi:10.1111/j.1432-0436.2006.00060.x

Pasternack SM, Refke M, Paknia E et al (2013) Mutations in SNRPE, which encodes a core protein of the spliceosome, cause autosomal-dominant hypotrichosis simplex. Am J Hum Genet 92:81–87. doi:10.1016/j.ajhg.2012.10.022

Paushkin S, Gubitz AK, Massenet S, Dreyfuss G (2002) The SMN complex, an assemblyosome of ribonucleoproteins. Curr Opin Cell Biol 14:305–312

Pearson T, Curtis F, Al-Eyadhy A et al (2008) An intronic mutation in DKC1 in an infant with Høyeraal-Hreidarsson syndrome. Am J Med Genet 146A:2159–2161. doi:10.1002/ajmg.a.32412

Perron MP, Provost P (2009) Protein components of the microRNA pathway and human diseases. Methods Mol Biol 487:369–385. doi:10.1007/978-1-60327-547-7_18

Peterlin BM, Brogie JE, Price DH (2011) 7SK snRNA: a noncoding RNA that plays a major role in regulating eukaryotic transcription. WIREs RNA 3:92–103. doi:10.1002/wrna.106

Peters L, Meister G (2007) Argonaute proteins: mediators of RNA silencing. Mol Cell 26: 611–623. doi:10.1016/j.molcel.2007.05.001

Pettersson I, Hinterberger M, Mimori T et al (1984) The structure of mammalian small nuclear ribonucleoproteins. Identification of multiple protein components reactive with anti-(U1)ribonucleoprotein and anti-Sm autoantibodies. J Biol Chem 259:5907–5914

Phizicky EM, Hopper AK (2010) tRNA biology charges to the front. Genes Dev 24:1832–1860. doi:10.1101/gad.1956510

Pierce SB, Chisholm KM, Lynch ED et al (2011) Mutations in mitochondrial histidyl tRNA synthetase HARS2 cause ovarian dysgenesis and sensorineural hearing loss of Perrault syndrome. Proc Natl Acad Sci U S A 108:6543–6548. doi:10.1073/pnas.1103471108

Pierce SB, Gersak K, Michaelson-Cohen R et al (2013) Mutations in LARS2, encoding mitochondrial leucyl-tRNA synthetase, lead to premature ovarian failure and hearing loss in Perrault syndrome. Am J Hum Genet 92:614–620. doi:10.1016/j.ajhg.2013.03.007

Pillai RS, Will CL, Luhrmann R et al (2001) Purified U7 snRNPs lack the Sm proteins D1 and D2 but contain Lsm10, a new 14 kDa Sm D1-like protein. EMBO J 20:5470–5479. doi:10.1093/emboj/20.19.5470

Proudfoot N (2004) New perspectives on connecting messenger RNA 3′ end formation to transcription. Curr Opin Cell Biol 16:272–278. doi:10.1016/j.ceb.2004.03.007

Proudfoot N, O'Sullivan J (2002) Polyadenylation: a tail of two complexes. Curr Biol 12:R855–R857

Puffenberger EG, Jinks RN, Sougnez C et al (2012) Genetic mapping and exome sequencing identify variants associated with five novel diseases. PLoS One 7:e28936. doi:10.1371/journal.pone.0028936

Pulst SM, Nechiporuk A, Nechiporuk T et al (1996) Moderate expansion of a normally biallelic trinucleotide repeat in spinocerebellar ataxia type 2. Nat Genet 14:269–276. doi:10.1038/ng1196-269

Pyle AM (2008) Translocation and unwinding mechanisms of RNA and DNA helicases. Annu Rev Biophys 37:317–336. doi:10.1146/annurev.biophys.37.032807.125908

Rabe B (2013) Aicardi-Goutières syndrome: clues from the RNase H2 knock-out mouse. J Mol Med 91:1235–1240. doi:10.1007/s00109-013-1061-x

Rajkowitsch L, Chen D, Stampfl S et al (2007) RNA chaperones, RNA annealers and RNA helicases. RNA Biol 4:118–130

Ramaswami M, Taylor JP, Parker R (2013) Altered ribostasis: RNA-protein granules in degenerative disorders. Cell 154:727–736. doi:10.1016/j.cell.2013.07.038

Ramser J, Ahearn ME, Lenski C et al (2008) Rare missense and synonymous variants in UBE1 are associated with X-linked infantile spinal muscular atrophy. Am J Hum Genet 82:188–193. doi:10.1016/j.ajhg.2007.09.009

Ramser J, Winnepenninckx B, Lenski C et al (2004) A splice site mutation in the methyltransferase gene FTSJ1 in Xp11.23 is associated with non-syndromic mental retardation in a large Belgian family (MRX9). J Med Genet 41:679–683. doi:10.1136/jmg.2004.019000

Rankin J, Brown R, Dobyns WB et al (2010) Pontocerebellar hypoplasia type 6: A British case with PEHO-like features. Am J Med Genet 152A:2079–2084. doi:10.1002/ajmg.a.33531

Raven JF, Koromilas AE (2008) PERK and PKR: old kinases learn new tricks. Cell Cycle 7:1146–1150

Ravenscroft JC, Suri M, Rice GI et al (2011) Autosomal dominant inheritance of a heterozygous mutation in SAMHD1 causing familial chilblain lupus. Am J Med Genet 155A:235–237. doi:10.1002/ajmg.a.33778

Regalado ES, Guo D-C, Villamizar C et al (2011) Exome sequencing identifies SMAD3 mutations as a cause of familial thoracic aortic aneurysm and dissection with intracranial and other arterial aneurysms. Circ Res 109:680–686. doi:10.1161/CIRCRESAHA.111.248161

Reijo R, Lee TY, Salo P et al (1995) Diverse spermatogenic defects in humans caused by Y chromosome deletions encompassing a novel RNA-binding protein gene. Nat Genet 10:383–393. doi:10.1038/ng0895-383

Rice GI, Bond J, Asipu A et al (2009) Mutations involved in Aicardi-Goutières syndrome implicate SAMHD1 as regulator of the innate immune response. Nat Genet 41:829–832. doi:10.1038/ng.373

Rice GI, del Toro DY, Jenkinson EM et al (2014) Gain-of-function mutations in IFIH1 cause a spectrum of human disease phenotypes associated with upregulated type I interferon signaling. Nat Genet 46:503–509. doi:10.1038/ng.2933

Rice GI, Kasher PR, Forte GMA et al (2012) Mutations in ADAR1 cause Aicardi-Goutières syndrome associated with a type I interferon signature. Nat Genet 44:1243–1248. doi:10.1038/ng.2414

Rice GI, Reijns MAM, Coffin SR et al (2013) Synonymous mutations in RNASEH2A create cryptic splice sites impairing RNase H2 enzyme function in Aicardi-Goutières syndrome. Hum Mutat 34:1066–1070. doi:10.1002/humu.22336

Riley LG, Cooper S, Hickey P et al (2010) Mutation of the mitochondrial tyrosyl-tRNA synthetase gene, YARS2, causes myopathy, lactic acidosis, and sideroblastic anemia–MLASA syndrome. Am J Hum Genet 87:52–59. doi:10.1016/j.ajhg.2010.06.001

Rio Frio T, Bahubeshi A, Kanellopoulou C et al (2011) DICER1 mutations in familial multinodular goiter with and without ovarian Sertoli-Leydig cell tumors. JAMA 305:68–77. doi:10.1001/jama.2010.1910

Roberts R, Timchenko NA, Miller JW et al (1997) Altered phosphorylation and intracellular distribution of a (CUG)n triplet repeat RNA-binding protein in patients with myotonic dystrophy and in myotonin protein kinase knockout mice. Proc Natl Acad Sci U S A 94:13221–13226

Rocak S, Linder P (2004) DEAD-box proteins: the driving forces behind RNA metabolism. Nat Rev Mol Cell Biol 5:232–241. doi:10.1038/nrm1335

Rosenberg HF (2011) Vertebrate secretory (RNase A) ribonucleases and host defense. Nucleic Acids Mol Biol 26:35–53
Ruggero D, Pandolfi PP (2003) Does the ribosome translate cancer? Nat Rev Cancer 3:179–192. doi:10.1038/nrc1015
Russell JF, Steckley JL, Coppola G et al (2012) Familial cortical myoclonus with a mutation in NOL3. Ann Neurol 72:175–183. doi:10.1002/ana.23666
Saada A, Shaag A, Arnon S et al (2007) Antenatal mitochondrial disease caused by mitochondrial ribosomal protein (MRPS22) mutation. J Med Genet 44:784–786. doi:10.1136/jmg.2007.053116
Sakai K, Gofuku M, Kitagawa Y et al (1994) A hippocampal protein associated with paraneoplastic neurologic syndrome and small cell lung carcinoma. Biochem Biophys Res Commun 199:1200–1208. doi:10.1006/bbrc.1994.1358
Sandberg R, Neilson JR, Sarma A et al (2008) Proliferating cells express mRNAs with shortened 3′ untranslated regions and fewer MicroRNA target sites. Science 320:1643–1647. doi:10.1126/science.1155390
Sandler H, Stoecklin G (2008) Control of mRNA decay by phosphorylation of tristetraprolin. Biochem Soc Trans 36:491. doi:10.1042/BST0360491
Santos-Cortez RLP, Lee K, Azeem Z et al (2013) Mutations in KARS, encoding lysyl-tRNA synthetase, cause autosomal-recessive nonsyndromic hearing impairment DFNB89. Am J Hum Genet 93:132–140. doi:10.1016/j.ajhg.2013.05.018
Saunders LR, Barber GN (2003) The dsRNA binding protein family: critical roles, diverse cellular functions. FASEB J 17:961–983
Savva YA, Rieder LE, Reenan RA (2012) The ADAR protein family. Genome Biol 13:252. doi:10.1186/gb-2012-13-12-252
Sawicka K, Bushell M, Spriggs KA, Willis AE (2008) Polypyrimidine-tract-binding protein: a multifunctional RNA-binding protein. Biochem Soc Trans 36:641–647. doi:10.1042/BST0360641
Scheper GC, van der Knaap MS, Proud CG (2007a) Translation matters: protein synthesis defects in inherited disease. Nat Rev Genet 8:711–723. doi:10.1038/nrg2142
Scheper GC, van der Klok T, van Andel RJ et al (2007b) Mitochondrial aspartyl-tRNA synthetase deficiency causes leukoencephalopathy with brain stem and spinal cord involvement and lactate elevation. Nat Genet 39:534–539. doi:10.1038/ng2013
Scherrer T, Mittal N, Janga SC, Gerber AP (2010) A screen for RNA-binding proteins in yeast indicates dual functions for many enzymes. PLoS One 5:e15499. doi:10.1371/journal.pone.0015499
Schirle NT, MacRae IJ (2012) The crystal structure of human argonaute2. Science 336: 1037–1040. doi:10.1126/science.1221551
Schmidt K, Butler JS (2013) Nuclear RNA surveillance: role of TRAMP in controlling exosome specificity. WIREs RNA 4:217–231. doi:10.1002/wrna.1155
Schutte M, Hruban RH, Hedrick L et al (1996) DPC4 gene in various tumor types. Cancer Res 56:2527–2530
Seibler P, Djarmati A, Langpap B et al (2008) A heterozygous frameshift mutation in PRKRA (DYT16) associated with generalised dystonia in a German patient. Lancet Neurol 7:380–381. doi:10.1016/S1474-4422(08)70075-9
Senderek J, Garvey SM, Krieger M et al (2009) Autosomal-dominant distal myopathy associated with a recurrent missense mutation in the gene encoding the nuclear matrix protein, matrin 3. Am J Hum Genet 84:511–518. doi:10.1016/j.ajhg.2009.03.006
Servadio A, Koshy B, Armstrong D et al (1995) Expression analysis of the ataxin-1 protein in tissues from normal and spinocerebellar ataxia type 1 individuals. Nat Genet 10:94–98. doi:10.1038/ng0595-94
Shamseldin HE, Alshammari M, Al-Sheddi T et al (2012) Genomic analysis of mitochondrial diseases in a consanguineous population reveals novel candidate disease genes. J Med Genet 49:234–241. doi:10.1136/jmedgenet-2012-100836
Shastry S, Delgado MR, Dirik E et al (2010) Congenital generalized lipodystrophy, type 4 (CGL4) associated with myopathy due to novel PTRF mutations. Am J Med Genet 152A:2245–2253. doi:10.1002/ajmg.a.33578
Shatkin AJ, Manley JL (2000) The ends of the affair: capping and polyadenylation. Nat Struct Biol 7:838–842. doi:10.1038/79583

Shimazaki H, Takiyama Y, Ishiura H et al (2012) A homozygous mutation of C12orf65 causes spastic paraplegia with optic atrophy and neuropathy (SPG55). J Med Genet 49:777–784. doi:10.1136/jmedgenet-2012-101212

Sim S, Wolin SL (2011) Emerging roles for the Ro 60-kDa autoantigen in noncoding RNA metabolism. WIREs RNA 2:686–699. doi:10.1002/wrna.85

Simard J, Tonin P, Durocher F et al (1994) Common origins of BRCA1 mutations in Canadian breast and ovarian cancer families. Nat Genet 8:392–398. doi:10.1038/ng1294-392

Simon B, Kirkpatrick JP, Eckhardt S et al (2011) Recognition of 2"-O-methylated 3-"end of piRNA by the PAZ domain of a Piwi protein. Structure 19:172–180. doi:10.1016/j.str.2010.11.015

Simone LE, Keene JD (2013) Mechanisms coordinating ELAV/Hu mRNA regulons. Curr Opin Genet Dev 23:35–43. doi:10.1016/j.gde.2012.12.006

Simos G, Hurt E (1999) Transfer RNA biogenesis: A visa to leave the nucleus. Curr Biol 9:R238–R241

Singh RK, Cooper TA (2012) Pre-mRNA splicing in disease and therapeutics. Trends Mol Med 18:472–482. doi:10.1016/j.molmed.2012.06.006

Singh R, Valcarcel J (2005) Building specificity with nonspecific RNA-binding proteins. Nat Struct Mol Biol 12:645–653. doi:10.1038/nsmb961

Siomi MC, Mannen T, Siomi H (2010) How does the royal family of tudor rule the PIWI-interacting RNA pathway? Genes Dev 24:636–646. doi:10.1101/gad.1899210

Siomi MC, Sato K, Pezic D, Aravin AA (2011) PIWI-interacting small RNAs: the vanguard of genome defence. Nat Rev Mol Cell Biol 12:246–258. doi:10.1038/nrm3089

Siprashvili Z, Webster DE, Kretz M et al (2012) Identification of proteins binding coding and non-coding human RNAs using protein microarrays. BMC Genomics 13:633. doi:10.1186/1471-2164-13-633

Smeitink JAM, Elpeleg O, Antonicka H et al (2006) Distinct clinical phenotypes associated with a mutation in the mitochondrial translation elongation factor EFTs. Am J Hum Genet 79: 869–877. doi:10.1086/508434

Smith CW, Valcarcel J (2000) Alternative pre-mRNA splicing: the logic of combinatorial control. Trends Biochem Sci 25:381–388

Smits P, Smeitink J, van den Heuvel L (2010) Mitochondrial translation and beyond: processes implicated in combined oxidative phosphorylation deficiencies. J Biomed Biotech 2010:737385. doi:10.1155/2010/737385

Smogorzewska A, de Lange T (2004) Regulation of telomerase by telomeric proteins. Annu Rev Biochem 73:177–208. doi:10.1146/annurev.biochem.73.071403.160049

Sommerville J (1999) Activities of cold-shock domain proteins in translation control. Bioessays 21:319–325. doi:10.1002/(SICI)1521-1878(199904)21:4<319::AID-BIES8>3.0.CO;2-3

Song JJ (2004) Crystal structure of argonaute and its implications for RISC slicer activity. Science 305:1434–1437. doi:10.1126/science.1102514

Song JJ, Joshua-Tor L (2006) Argonaute and RNA–getting into the groove. Curr Opin Struct Biol 16:5–11. doi:10.1016/j.sbi.2006.01.010

Sossi V, Giuli A, Vitali T et al (2001) Premature termination mutations in exon 3 of the SMN1 gene are associated with exon skipping and a relatively mild SMA phenotype. Eur J Hum Genet 9:113–120. doi:10.1038/sj.ejhg.5200599

Splendore A, Passos-Bueno MR, Jabs EW et al (2002) TCOF1 mutations excluded from a role in other first and second branchial arch-related disorders. Am J Med Genet 111:324–327. doi:10.1002/ajmg.10567

Sreedharan J, Blair IP, Tripathi VB et al (2008) TDP-43 mutations in familial and sporadic amyotrophic lateral sclerosis. Science 319:1668–1672. doi:10.1126/science.1154584

Steenweg ME, Ghezzi D, Haack T et al (2012) Leukoencephalopathy with thalamus and brainstem involvement and high lactate "LTBL" caused by EARS2 mutations. Brain 135:1387–1394. doi:10.1093/brain/aws070

Stetson DB, Ko JS, Heidmann T, Medzhitov R (2008) Trex1 prevents cell-intrinsic initiation of autoimmunity. Cell 134:587–598. doi:10.1016/j.cell.2008.06.032

Stettner GM, Shoukier M, Höger C et al (2011) Familial intellectual disability and autistic behavior caused by a small FMR2 gene deletion. Am J Med Genet 155:2003–2007. doi:10.1002/ajmg.a.34122

Suzuki T, Nagao A, Suzuki T (2011) Human mitochondrial tRNAs: biogenesis, function, structural aspects, and diseases. Annu Rev Genet 45:299–329. doi:10.1146/annurev-genet-110410-132531

Szymczyna BR, Bowman J, McCracken S et al (2003) Structure and function of the PWI motif: a novel nucleic acid-binding domain that facilitates pre-mRNA processing. Genes Dev 17:461–475. doi:10.1101/gad.1060403

Talim B, Pyle A, Griffin H et al (2013) Multisystem fatal infantile disease caused by a novel homozygous EARS2 mutation. Brain 136:e228. doi:10.1093/brain/aws197

Tamburino AM, Ryder SP, Walhout AJM (2013) A compendium of caenorhabditis elegans RNA binding proteins predicts extensive regulation at multiple levels. G3 (Bethesda) 3:297–304. doi:10.1534/g3.112.004390

Tan AY, Manley JL (2009) The TET family of proteins: functions and roles in disease. J Mol Cell Biol 1:82–92. doi:10.1093/jmcb/mjp025

Tan HL, Glen E, Töpf A et al (2012) Nonsynonymous variants in the SMAD6 gene predispose to congenital cardiovascular malformation. Hum Mutat 33:720–727. doi:10.1002/humu.22030

Tanackovic G, Ransijn A, Ayuso C et al (2011) A missense mutation in PRPF6 causes impairment of pre-mRNA splicing and autosomal-dominant retinitis pigmentosa. Am J Hum Genet 88:643–649. doi:10.1016/j.ajhg.2011.04.008

Tanaka A, Morice-Picard F, Lacombe D et al (2010) Identification of a homozygous deletion mutation in C16orf57 in a family with Clericuzio-type poikiloderma with neutropenia. Am J Med Genet 152A:1347–1348. doi:10.1002/ajmg.a.33455

Tanner NK, Linder P (2001) DExD/H box RNA helicases: from generic motors to specific dissociation functions. Mol Cell 8:251–262

Tarpey PS, Raymond FL, Nguyen LS et al (2007) Mutations in UPF3B, a member of the nonsense-mediated mRNA decay complex, cause syndromic and nonsyndromic mental retardation. Nat Genet 39:1127–1133. doi:10.1038/ng2100

Tavtigian SV, Simard J, Teng DH et al (2001) A candidate prostate cancer susceptibility gene at chromosome 17p. Nat Genet 27:172–180. doi:10.1038/84808

Teplova M, Hafner M, Teplov D et al (2013) Structure-function studies of STAR family Quaking proteins bound to their in vivo RNA target sites. Genes Dev 27:928–940. doi:10.1101/gad.216531.113

Tessier M-C, Qu H-Q, Fréchette R et al (2006) Type 1 diabetes and the OAS gene cluster: association with splicing polymorphism or haplotype? J Med Genet 43:129–132. doi:10.1136/jmg.2005.035212

Tharun S (2009) Roles of eukaryotic Lsm proteins in the regulation of mRNA function. Int Rev Cell Mol Biol 272:149–189. doi:10.1016/S1937-6448(08)01604-3

Tharun S, He W, Mayes AE et al (2000) Yeast Sm-like proteins function in mRNA decapping and decay. Nature 404:515–518. doi:10.1038/35006676

Thomson T, Lasko P (2005) Tudor and its domains: germ cell formation from a Tudor perspective. Cell Res 15:281–291. doi:10.1038/sj.cr.7290297

Thomson T, Lin H (2009) The biogenesis and function of PIWI proteins and piRNAs: progress and prospect. Annu Rev Cell Dev Biol 25:355–376. doi:10.1146/annurev.cellbio.24.110707.175327

Thornton JE, Gregory RI (2012) How does Lin28 let-7 control development and disease? Trends Cell Biol 22:474–482. doi:10.1016/j.tcb.2012.06.001

Tian Y, Simanshu DK, Ma J-B, Patel DJ (2011) Structural basis for piRNA 2"-O-methylated 3-"end recognition by Piwi PAZ (Piwi/Argonaute/Zwille) domains. Proc Natl Acad Sci U S A 108:903–910. doi:10.1073/pnas.1017762108

Timchenko LT, Timchenko NA, Caskey CT, Roberts R (1996) Novel proteins with binding specificity for DNA CTG repeats and RNA CUG repeats: implications for myotonic dystrophy. Hum Mol Genet 5:115–121

Tomasevic N, Peculis BA (2002) Xenopus LSm proteins bind U8 snoRNA via an internal evolutionarily conserved octamer sequence. Mol Cell Biol 22:4101–4112

Taft RJ, Vanderver A, Leventer RJ et al (2013) Mutations in DARS cause hypomyelination with brain stem and spinal cord involvement and leg spasticity. Am J Hum Genet 92:774–780. doi:10.1016/j.ajhg.2013.04.006

Tsai-Morris CH, Sheng Y, Gutti RK et al (2010) Gonadotropin-regulated testicular RNA helicase (GRTH/DDX25): a multifunctional protein essential for spermatogenesis. J Androl 31:45–52. doi:10.2164/jandrol.109.008219

Tsakiri KD, Cronkhite JT, Kuan PJ et al (2007) Adult-onset pulmonary fibrosis caused by mutations in telomerase. Proc Natl Acad Sci U S A 104:7552–7557. doi:10.1073/pnas.0701009104

Tsvetanova NG, Klass DM, Salzman J, Brown PO (2010) Proteome-wide search reveals unexpected RNA-binding proteins in Saccharomyces cerevisiae. PLoS One 5(9). doi: 10.1371/journal.pone.0012671

Ule J (2008) Ribonucleoprotein complexes in neurologic diseases. Curr Opin Neurobiol 18: 516–523. doi:10.1016/j.conb.2008.09.018

Ulitsky I, Bartel DP (2013) lincRNAs: genomics, evolution, and mechanisms. Cell 154:26–46. doi:10.1016/j.cell.2013.06.020

Valente L, Tiranti V, Marsano RM et al (2007) Infantile encephalopathy and defective mitochondrial DNA translation in patients with mutations of mitochondrial elongation factors EFG1 and EFTu. Am J Hum Genet 80:44–58. doi:10.1086/510559

Valverde R, Edwards L, Regan L (2008) Structure and function of KH domains. FEBS J 275: 2712–2726. doi:10.1111/j.1742-4658.2008.06411.x

van de Laar IMBH, Oldenburg RA, Pals G et al (2011) Mutations in SMAD3 cause a syndromic form of aortic aneurysms and dissections with early-onset osteoarthritis. Nat Genet 43: 121–126. doi:10.1038/ng.744

Van de Peer Y, Maere S, Meyer A (2009) The evolutionary significance of ancient genome duplications. Nat Rev Genet 10:725–732. doi:10.1038/nrg2600

van der Knaap MS, Leegwater PAJ, Könst AAM et al (2002) Mutations in each of the five subunits of translation initiation factor eIF2B can cause leukoencephalopathy with vanishing white matter. Ann Neurol 51:264–270

Vance C, Rogelj B, Hortobagyi T et al (2009) Mutations in FUS, an RNA processing protein, cause familial amyotrophic lateral sclerosis type 6. Science 323:1208–1211. doi:10.1126/science.1165942

Vanrobays E, Gélugne J-P, Caizergues-Ferrer M, Lafontaine DLJ (2004) Dim2p, a KH-domain protein required for small ribosomal subunit synthesis. RNA 10:645–656

Vaquerizas JM, Kummerfeld SK, Teichmann SA, Luscombe NM (2009) A census of human transcription factors: function, expression and evolution. Nat Rev Genet 10:252–263. doi:10.1038/nrg2538

Vedrenne V, Gowher A, de Lonlay P et al (2012) Mutation in PNPT1, which encodes a polyribonucleotide nucleotidyltransferase, impairs RNA import into mitochondria and causes respiratory-chain deficiency. Am J Hum Genet 91:912–918. doi:10.1016/j.ajhg.2012.09.001

Venter JC (2001) The sequence of the human genome. Science 291:1304–1351. doi:10.1126/science.1058040

Vissers LELM, de Ligt J, Gilissen C et al (2010) A de novo paradigm for mental retardation. Nat Genet 42:1109–1112. doi:10.1038/ng.712

Vithana EN, Abu-Safieh L, Allen MJ et al (2001) A human homolog of yeast pre-mRNA splicing gene, PRP31, underlies autosomal dominant retinitis pigmentosa on chromosome 19q13.4 (RP11). Mol Cell 8:375–381

Volpi L, Roversi G, Colombo EA et al (2010) Targeted next-generation sequencing appoints c16orf57 as clericuzio-type poikiloderma with neutropenia gene. Am J Hum Genet 86:72–76. doi:10.1016/j.ajhg.2009.11.014

von Ameln S, Wang G, Boulouiz R et al (2012) A mutation in PNPT1, encoding mitochondrial-RNA-import protein PNPase, causes hereditary hearing loss. Am J Hum Genet 91:919–927. doi:10.1016/j.ajhg.2012.09.002

Vulliamy TJ, Knight SW, Heiss NS et al (1999) Dyskeratosis congenita caused by a 3′ deletion: germline and somatic mosaicism in a female carrier. Blood 94:1254–1260

Vulliamy T, Beswick R, Kirwan M et al (2008) Mutations in the telomerase component NHP2 cause the premature ageing syndrome dyskeratosis congenita. Proc Natl Acad Sci U S A 105:8073–8078. doi:10.1073/pnas.0800042105
Wakamatsu Y, Weston JA (1997) Sequential expression and role of Hu RNA-binding proteins during neurogenesis. Development 124:3449–3460
Walne AJ, Vulliamy T, Marrone A et al (2007) Genetic heterogeneity in autosomal recessive dyskeratosis congenita with one subtype due to mutations in the telomerase-associated protein NOP10. Hum Mol Genet 16:1619–1629. doi:10.1093/hmg/ddm111
Wan M, Lee SS, Zhang X et al (1999) Rett syndrome and beyond: recurrent spontaneous and familial MECP2 mutations at CpG hotspots. Am J Hum Genet 65:1520–1529. doi:10.1086/302690
Wan J, Yourshaw M, Mamsa H et al (2012) Mutations in the RNA exosome component gene EXOSC3 cause pontocerebellar hypoplasia and spinal motor neuron degeneration. Nat Genet 44:704–708. doi:10.1038/ng.2254
Wang G-S, Cooper TA (2007) Splicing in disease: disruption of the splicing code and the decoding machinery. Nat Rev Genet 8:749–761. doi:10.1038/nrg2164
Wang Y, Jiang Y, Meyering-Voss M et al (1997) Crystal structure of the EF-Tu.EF-Ts complex from Thermus thermophilus. Nat Struct Biol 4:650–656
Wang XQ, Zamore PD, Hall TMT (2001) Crystal structure of a Pumilio homology domain. Mol Cell 7:855–865. doi:10.1016/s1097-2765(01)00229-5
Wang X, McLachlan J, Zamore PD, Hall TMT (2002) Modular recognition of RNA by a human pumilio-homology domain. Cell 110:501–512
Wang ET, Sandberg R, Luo S et al (2008a) Alternative isoform regulation in human tissue transcriptomes. Nature 456:470–476. doi:10.1038/nature07509
Wang Y, Sheng G, Juranek S et al (2008b) Structure of the guide-strand-containing argonaute silencing complex. Nature 456:209–213. doi:10.1038/nature07315
Wang ET, Cody NAL, Jog S et al (2012) Transcriptome-wide regulation of pre-mRNA splicing and mRNA localization by muscleblind proteins. Cell 150:710–724. doi:10.1016/j.cell.2012.06.041
Wang X, Lu Z, Gomez A et al (2013) N(6)-methyladenosine-dependent regulation of messenger RNA stability. Nature 505:117–120. doi:10.1038/nature12730
Wapinski O, Chang HY (2011) Long noncoding RNAs and human disease. Trends Cell Biol 21:354–361. doi:10.1016/j.tcb.2011.04.001
Weedon MN, Hastings R, Caswell R et al (2011) Exome sequencing identifies a DYNC1H1 mutation in a large pedigree with dominant axonal Charcot-Marie-Tooth disease. Am J Hum Genet 89:308–312. doi:10.1016/j.ajhg.2011.07.002
Wickens M, Bernstein DS, Kimble J, Parker R (2002) A PUF family portrait: 3′UTR regulation as a way of life. Trends Genet 18:150–157
Wilbert ML, Huelga SC, Kapeli K et al (2012) LIN28 binds messenger RNAs at GGAGA motifs and regulates splicing factor abundance. Mol Cell 48:195–206. doi:10.1016/j.molcel.2012.08.004
Wilson RC, Doudna JA (2013) Molecular Mechanisms of RNA Interference. Annu Rev Biophys 42:217–239. doi:10.1146/annurev-biophys-083012-130404
Wilusz CJ, Wilusz J (2005) Eukaryotic Lsm proteins: lessons from bacteria. Nat Struct Mol Biol 12:1031–1036. doi:10.1038/nsmb1037
Winkelmann J, Lin L, Schormair B et al (2012) Mutations in DNMT1 cause autosomal dominant cerebellar ataxia, deafness and narcolepsy. Hum Mol Genet 21:2205–2210. doi:10.1093/hmg/dds035
Winter EE, Goodstadt L, Ponting CP (2004) Elevated rates of protein secretion, evolution, and disease among tissue-specific genes. Genome Res 14:54–61. doi:10.1101/gr.1924004
Wise CA, Chiang LC, Paznekas WA et al (1997) TCOF1 gene encodes a putative nucleolar phosphoprotein that exhibits mutations in Treacher Collins Syndrome throughout its coding region. Proc Natl Acad Sci U S A 94:3110–3115
Wöhrle D, Kotzot D, Hirst MC et al (1992) A microdeletion of less than 250 kb, including the proximal part of the FMR-I gene and the fragile-X site, in a male with the clinical phenotype of fragile-X syndrome. Am J Hum Genet 51:299–306

Wool IG (1979) The structure and function of eukaryotic ribosomes. Annu Rev Biochem 48: 719–754. doi:10.1146/annurev.bi.48.070179.003443

Wool IG, Chan YL, Glück A (1995) Structure and evolution of mammalian ribosomal proteins. Biochem Cell Biol 73:933–947

Xiao S, Scott F, Fierke CA, Engelke DR (2002) Eukaryotic ribonuclease P: a plurality of ribonucleoprotein enzymes. Annu Rev Biochem 71:165–189. doi:10.1146/annurev.biochem.71.110601.135352

Xie X, Lu J, Kulbokas EJ et al (2005) Systematic discovery of regulatory motifs in human promoters and 3′ UTRs by comparison of several mammals. Nature 434:338–345. doi:10.1038/nature03441

Xing Y, Lee C (2006) Alternative splicing and RNA selection pressure–evolutionary consequences for eukaryotic genomes. Nat Rev Genet 7:499–509. doi:10.1038/nrg1896

Xu GL, Bestor TH, Bourc'his D et al (1999) Chromosome instability and immunodeficiency syndrome caused by mutations in a DNA methyltransferase gene. Nature 402:187–191. doi:10.1038/46052

Xu Q, Modrek B, Lee C (2002) Genome-wide detection of tissue-specific alternative splicing in the human transcriptome. Nucleic Acids Res 30:3754–3766

Yamaguchi H, Calado RT, Ly H et al (2005) Mutations in TERT, the gene for telomerase reverse transcriptase, in aplastic anemia. N Engl J Med 352:1413–1424. doi:10.1056/NEJMoa042980

Yamasaki S, Ivanov P, Hu G-F, Anderson P (2009) Angiogenin cleaves tRNA and promotes stress-induced translational repression. J Cell Biol 185:35–42. doi:10.1083/jcb.200811106

Yanase T, Takayanagi R, Oba K et al (1996) New mutations of DAX-1 genes in two Japanese patients with X-linked congenital adrenal hypoplasia and hypogonadotropic hypogonadism. J Clin Endocrinol Metabol 81:530–535

Yao P, Fox PL (2013) Aminoacyl-tRNA synthetases in medicine and disease. EMBO Mol Med 5:332–343. doi:10.1002/emmm.201100626

Yen PH (2004) Putative biological functions of the DAZ family. Int J Androl 27:125–129. doi:10.1111/j.1365-2605.2004.00469.x

Yeo GW, Van Nostrand E, Holste D et al (2005) Identification and analysis of alternative splicing events conserved in human and mouse. Proc Natl Acad Sci U S A 102:2850–2855. doi:10.1073/pnas.0409742102

Zeharia A, Shaag A, Pappo O et al (2009) Acute infantile liver failure due to mutations in the TRMU gene. Am J Hum Genet 85:401–407. doi:10.1016/j.ajhg.2009.08.004

Zhang H, Kolb FA, Jaskiewicz L et al (2004) Single processing center models for human dicer and bacterial RNase III. Cell 118:57–68. doi:10.1016/j.cell.2004.06.017

Zhang ZH, Niu ZM, Yuan WT et al (2005) A mutation in SART3 gene in a Chinese pedigree with disseminated superficial actinic porokeratosis. Br J Dermatol 152:658–663. doi:10.1111/j.1365-2133.2005.06443.x

Zhang L, Huang J, Yang N et al (2006) microRNAs exhibit high frequency genomic alterations in human cancer. Proc Natl Acad Sci U S A 103:9136–9141. doi:10.1073/pnas.0508889103

Zhang Z, Lotti F, Dittmar K et al (2008) SMN deficiency causes tissue-specific perturbations in the repertoire of snRNAs and widespread defects in splicing. Cell 133:585–600. doi:10.1016/j.cell.2008.03.031

Zhang Z, Theler D, Kaminska KH et al (2010) The YTH domain is a novel RNA binding domain. J Biol Chem 285:14701–14710. doi:10.1074/jbc.M110.104711

Zhao C, Bellur DL, Lu S et al (2009) Autosomal-dominant retinitis pigmentosa caused by a mutation in SNRNP200, a gene required for unwinding of U4/U6 snRNAs. Am J Hum Genet 85:617–627. doi:10.1016/j.ajhg.2009.09.020

Zheng K, Xiol J, Reuter M et al (2010) Mouse MOV10L1 associates with Piwi proteins and is an essential component of the Piwi-interacting RNA (piRNA) pathway. Proc Natl Acad Sci U S A 107:11841–11846. doi:10.1073/pnas.1003953107

Zhong F, Savage SA, Shkreli M et al (2011) Disruption of telomerase trafficking by TCAB1 mutation causes dyskeratosis congenita. Genes Dev 25:11–16. doi:10.1101/gad.2006411

Wool IG (1979) The structure and function of eukaryotic ribosomes. Annu Rev Biochem 48:719–754. doi:10.1146/annurev.bi.48.070179.003443

Wool IG, Chan YL, Glück A (1995) Structure and evolution of mammalian ribosomal proteins. Biochem Cell Biol 73:933–947

Xiang S, Cooke A, Tong L (2003) Eukaryotic mRNA decapping. Pi a ... and ... Biochem Sci 28:...–...

Xie X, Lu J, Kulbokas EJ et al (2005) Systematic discovery of regulatory motifs in human promoters and 3′ UTRs by comparison of several mammals. Nature 434:338–345. doi:10.1038/nature03441

Xing Y, Lee C (2006) Alternative splicing and RNA selection pressure—evolutionary consequences for eukaryotic genomes. Nat Rev Genet 7:499–509. doi:10.1038/nrg1896

Xu GL, Bestor TH, Bourc'his D et al (1999) Chromosome instability and immunodeficiency syndrome caused by mutations in a DNA methyltransferase gene. Nature 402:187–191

[illegible]

Chapter 2
The Functions and Regulatory Principles of mRNA Intracellular Trafficking

Julie Bergalet and Eric Lécuyer

Abstract The subcellular localization of RNA molecules is a key step in the control of gene expression that impacts a broad array of biological processes in different organisms and cell types. Like other aspects of posttranscriptional gene regulation discussed in this collection of reviews, the intracellular trafficking of mRNAs is modulated by a complex regulatory code implicating specific *cis-regulatory* elements, RNA-binding proteins, and cofactors that function combinatorially to dictate precise localization mechanisms. In this review, we first discuss the functional benefits of transcript localization, the regulatory principles involved, and specific molecular mechanisms that have been described for a few well-characterized mRNAs. We also overview some of the emerging genomic and imaging technologies that have provided significant insights into this layer of gene regulation. Finally, we highlight examples of human diseases where defective transcript localization has been documented.

Keywords RNA localization • Localized translation • Cytoskeletal transport • Diffusion and entrapment • Degradation protection • RNA visualization techniques • Neuro- and musculo-degenerative disorders Cancer

---

J. Bergalet
Institut de Recherches Cliniques de Montréal (IRCM), Montréal, QC, Canada

E. Lécuyer (✉)
Institut de Recherches Cliniques de Montréal (IRCM), Montréal, QC, Canada

Département de Biochimie, Université de Montréal, Montréal, QC, Canada

Division of Experimental Medicine, McGill University, Montréal, QC, Canada

RNA Biology Laboratory, IRCM, 110 Avenue des Pins, Ouest, Montréal, QC, Canada H2W 1R7
e-mail: Eric.Lecuyer@ircm.qc.ca

G.W. Yeo (ed.), *Systems Biology of RNA Binding Proteins*, Advances in Experimental Medicine and Biology 825, DOI 10.1007/978-1-4939-1221-6_2

1 Introduction

In 1983, Jeffery et al. were the first to report the asymmetric localization of *β-actin* mRNA in ascidian oocytes and embryos (Jeffery et al. 1983). At the time of this discovery, it was not clear whether other transcripts could localize to specific subcellular sites in other organisms. However, over the last few decades many examples of localized mRNAs have been characterized in a wide range of organisms, including prokaryotes (Keiler 2011), yeast (Heym and Niessing 2012), algae (Serikawa et al. 2001), fungi (Zarnack and Feldbrugge 2007, 2010), plants (Crofts et al. 2004), invertebrates (Henry et al. 2010), ascidians (Prodon et al. 2007), insects (Lasko 2012; Gaspar 2011), amphibians, fish, and mammals (Medioni et al. 2012; King et al. 2005; Wang et al. 2012). Also, thanks to technical advances in microscopy and genomics approaches, thousands of mRNAs are now known to be regulated at the level of their subcellular trafficking (Medioni et al. 2012; Blower et al. 2007; Gumy et al. 2011; Mili et al. 2008; Lecuyer et al. 2007, 2009).

In this chapter, we first discuss the biological functions and known molecular mechanisms of RNA intracellular localization, focusing on well-characterized examples of RNA-binding proteins (RBPs) that have been implicated in this process in different organisms and cell types. We then describe the established and emerging techniques to study mRNA localization, including live-cell RNA imaging methodologies and systematic genome-wide approaches. Finally, we detail a few examples of disorders for which defective RNA localization appears to underlie human disease pathogenesis.

2 Advantages and Biological Functions of RNA Localization

The intracellular trafficking of mRNA molecules is a prevalent posttranscriptional step of gene regulation that plays an important role in modulating the functions of both coding and noncoding RNA molecules. For example, many recently identified long noncoding RNAs exhibit precise targeting to specific subregions of the nucleus, where they appear to control key aspects of gene expression and chromatin structure (Batista and Chang 2013). In the case of mRNAs, the main focus of this review, localization regulation typically occurs in the cytoplasm and is generally thought to result in targeted translation, thus enriching the encoded protein product in the corresponding region of the cell (Fig. 2.1, steps 1–4). Indeed, ribosomes have been found in most subcellular territories where mRNAs are targeted and multiple studies have shown that mRNA localization is often correlated with the known distribution patterns and biological functions of the encoded proteins (Lecuyer et al. 2007, 2009). As a single mRNA can be simultaneously scanned by several ribosomes, a large quantity of protein can be generated from a few mRNA molecules, thus allowing the rapid enrichment of the encoded protein product at the corresponding site. This process thus represents an attractive strategy for the cell to conserve energy, by

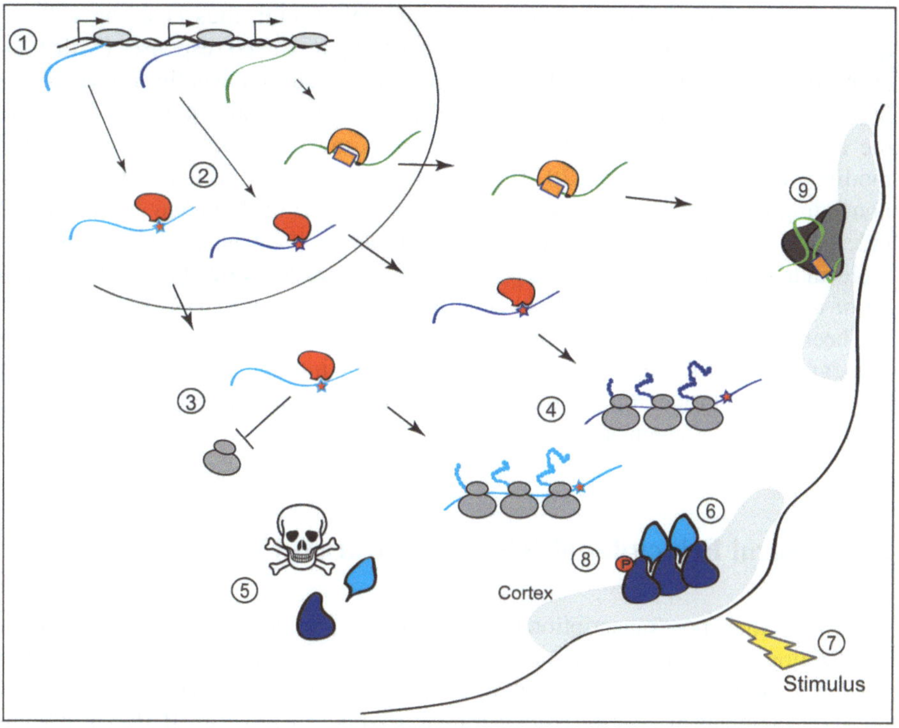

Fig. 2.1 Functional benefits of mRNA localization. Following transcription (*1*), mRNAs are bound in the nucleus by RBPs that function to ensure their maturation and nuclear export (*2*). Once in the cytoplasm, mRNPs travel in a silent state (*3*) toward their target destination where they are integrated into polysomes to be translated (*4*). Targeted translation can help prevent proteins from becoming localized to regions where they may go to waste or have deleterious effects (*5*). Furthermore, targeted translation could serve to nuclear the formation of localized protein complexes (*6*). The process of localized translation can be impacted by extracellular stimuli (*7*), for instance to allow production of protein species with distinct posttranslational modifications (*8*). Finally, localized mRNAs may also fulfill non-coding functions, for instance by acting as molecular scaffolds within RNP complexes (*9*)

simply transporting a few mRNA transcripts rather than trafficking many individual protein molecules to the target destination. Since mRNAs that code for functionally related proteins often share similar localization profiles, their simultaneous transport and translation can also presumably facilitate the co-translational assembly of localized protein complexes (Mingle et al. 2005). In several cases, localized mRNAs have been shown to be translationally repressed during their transit, a mechanism that would seem necessary in order to achieve a tight coupling of mRNA and protein distribution patterns (Jeffery 1988; Long et al. 1997) (Fig. 2.1, step 5). The process of localized mRNA translation thus offers an exquisite mechanism to spatially control the distribution and assembly of protein complexes in the cell while also preventing proteins from entering into the wrong subcellular compartment, which could have wasteful or deleterious consequences for the cell (Fig. 2.1, step 6). Moreover, this process is also temporally controlled since, in some cases, localized translation was shown to be induced by external stimuli, thus enabling cells to quickly adapt to external cues by mounting spatially restricted responses (Mili et al. 2008; Jung et al. 2012) (Fig. 2.1, step 7). As external stimuli usually activate localized signaling proteins such as kinases, mRNA targeting is also a way to modulate the activity of the encoded proteins by controlling their local posttranslational modifications (Medioni et al. 2012) (Fig. 2.1, step 8). Finally, as in the case of many noncoding RNAs, recent studies have also suggested that mRNAs may fulfill coding-independent functions, for example by acting as scaffolds in the assembly of macromolecular structures (Fig. 2.1, Step 9) (Jenny et al. 2006; Kloc et al. 2002). Given all of these possible functional benefits, it is not surprising that RNA localization has been implicated in a wide spectrum of biological processes, both during development and in differentiated cell types, and a growing number of studies suggest that the aberrant regulation of transcript localization can contribute significantly to human disease pathogenesis.

3 General Principles of RNA Localization Control

As with other posttranscriptional steps of gene expression regulation, RNA localization is coordinated, on one hand, by *cis*-regulatory elements residing within the RNA molecule and, on the other, by *trans*-acting factors. The recognition of these *cis*-elements by specific *trans*-acting RNA-binding proteins (RBPs) is thought to result in the formation of ribonucleoprotein complexes (RNPs) that dictate the targeting of the RNA to specific regions of a cell. RNA localization in the cytoplasm is most frequently thought to involve directed transport on cytoskeletal networks driven by the association of RNPs with specific molecular motors. However, alternative mechanisms have been characterized, such as random diffusion and entrapment or localized protection from degradation (Lipshitz and Smibert 2000). Once they reach their final destination, RNPs are thought to be anchored at the site of translation, while different mechanisms may exist to keep RNAs in a translationally repressed state during their transport. In this section, we review some of the general regulatory principles of RNA intracellular trafficking control.

3.1 Cis-Regulatory Motifs Involved in RNA Trafficking

The identification of *cis*-regulatory motifs (CRMs) or "zipcode" elements responsible for RNA localization represents an important facet for understanding this post-transcriptional regulatory process (reviewed in Jambhekar and Derisi (2007)). The most common experimental strategy to identify CRMs is to track the subcellular localization of fragments derived from the localized RNA, either on their own or fused to a reporter transcript. These approaches are generally feasible since these elements are often capable of conferring localization independently of the adjacent

RNA sequence. Ultimately, this allows one to identify regions of the RNA-harboring CRMs that are necessary and/or sufficient for proper RNA targeting. While CRMs modulating mRNA localization have most often been characterized in the 3′ UTR of transcripts, these regulatory elements can also reside in the 5′ UTR or coding region of the mRNA (Van de Bor and Davis 2004). For example, the minimal CRM involved in targeting *sensorin* mRNA to the synapse consists in a 66-nucleotide (nt) element located just upstream the translation start site in the 5′ UTR (Meer et al. 2012). In contrast, the *Drosophila bitesize* mRNA contains a 2.2-kb sequence within its ORF that is necessary and sufficient for its apical targeting in polarized epithelial cells (Serano and Rubin 2003). These examples illustrate how zipcodes can vary in their relative size and positioning within the mRNA. However, in some cases this variability may reflect differences in the resolution of the structure-function studies that have been conducted over the years, as well as the experimental complexity in pinning down minimal regions that mediate localization.

As with other aspects of gene regulation, efforts to define the molecular features of consensus RNA localization CRMs can be greatly aided by the use of bioinformatics approaches (Hamilton and Davis 2011). In contrast to DNA regulatory elements, which in many cases can be identified via simple primary motif search algorithms, RNA-level consensus CRMs can be much more challenging to predict computationally. Indeed, the capacity of RNA molecules to assume dynamic structural folds, which can either hinder or promote CRM formation, implies that bioinformatics approaches to define RNA *cis*-elements need to take into consideration both primary sequence and secondary/tertiary structure information. Indeed, there exists a diversity of RBPs with distinct binding preferences for single versus structured elements in their targets, which underlines the need to understand these various levels of organization (Kazan and Morris 2013; Lunde et al. 2007). Hamilton and Davis provided an extensive survey of available computational algorithms and web resources that can be harnessed to identify putative localization CRMs, which includes tools to define consensus primary sequences within sets of co-localized mRNAs or to predict their structural properties based on principles of free-energy minimization, evolutionary sequence covariation, or local base pairing probabilities (Hamilton and Davis 2011).

As mentioned above, the molecular features of localization CRMs can be quite variable, either being defined via their primary sequence, secondary/tertiary structure features, or a combination of both. Indeed, some zipcodes consist in simple primary sequences, such as the hnRNPA2 responsive element (A2RE), an 11-nt element present in the *myelin basic protein* (*MBP*) transcript that controls the localization of this mRNA to the dendritic compartment of oligodendrocytes (Munro et al. 1999). By contrast, some display secondary or tertiary structures comprising simple stem–loop elements or more complex structural features. For example, Subramanian et al. recently described a G-quadruplex structure in *CAMKIIα* mRNA that is sufficient for neurite targeting in cortical neurons (Subramanian et al. 2011). Interestingly, 30 % of mRNAs found in neurites appear to contain a consensus G-quadruplex in their 3′ UTR, suggesting that this structural CRM could be a common neurite localization signal (Subramanian et al. 2011). However, several other zipcodes have been described for *CAMKIIα* mRNA. First, a 94-nt-long element in the 3′ UTR was shown to be sufficient to target a GFP reporter transcript to neurites in vitro (Mori et al. 2000). However, experiments done in transgenic mice that express a *CAMKIIα* mRNA lacking a part of its 3′ UTR, but still carrying the 94-nt element, revealed that this element alone was not sufficient to localize this mRNA in vivo (Miller et al. 2002). Using the same reporter assay, another group identified a distinct CRM in the middle of the 3′ UTR (Blichenberg et al. 2001). In addition, Huang et al. described two cytoplasmic polyadenylation element (CPE) motifs located in close proximity to the G-quadruplex in the 3′ UTR, which were shown to be important for dendritic localization in hippocampal neurons (Huang et al. 2003). Thus, *CAMKIIα* mRNA is a good example that conveys how difficult it can be to interpret the results of mutational analyses and overexpression studies to delineate specific zipcodes. Moreover, it underlines the concept that CRMs may be employed alternatively depending on the cell type, e.g., G-quadruplexes in cortical neurons versus CPEs in hippocampal neurons.

The presence of multiple localization CRMs in a given mRNA also suggests that these elements may function in a combinatorial fashion, with multiple primary motifs and/or structural elements acting collaboratively to efficiently localize mRNAs to specific sites in the cell (Subramanian et al. 2011). A good example of this is *VG1* mRNA targeting to the vegetal pole of *Xenopus* oocytes, which is mediated by a 340-nt-long vegetal localization element (VLE), composed of four repeated sequences (termed E1 to E4) present in multiple copies, each of which is required for proper localization of the transcript (Deshler et al. 1997). In some cases different CRMs may be required to carry out distinct phases of a multi-step localization mechanism. Indeed, *MBP* mRNA localization in oligodendrocytes involves two types of CRMs responsible for distinct steps of the targeting process: the 11-nt A2RE motif described above mediates the initial transport of the mRNA, while a second more complex element of >1 kb in length is necessary for anchoring the mRNA within the myelin compartment (Ainger et al. 1997).

Multi-step targeting mechanisms can also involve events that take place in different compartments of the cell. Indeed, while experiments using direct injection of RNA fragments into cells/oocytes have indicated that nuclear processing is not always required for precise localization, some examples have highlighted the importance of nuclear history for proper cytoplasmic targeting. For example, Hachet et al. showed that *Oskar* mRNA splicing at the first exon-exon junction is essential for proper mRNA localization (Hachet and Ephrussi 2004). More recent work has shown that this splicing resulted in the formation of a stem–loop motif, designated as the spliced oskar localization element (SOLE), which is required for posterior localization of *oskar* mRNA in *Drosophila* oocytes (Ghosh et al. 2012). In parallel, a recent study revealed a novel role for nuclear RNA editing that functions in association with alternative splicing to influence the dendritic localization of *AMPAR* mRNA in mammals (La Via et al. 2013).

Although many CRMs are expected to function constitutively, provided that the RBPs that recognize them are available in the cell, their functionality may also be regulated intrinsically. Indeed, in the case of structured zipcodes, the conformation of these elements is not frozen, but may be modulated by intracellular/extracellular environmental conditions. For example, the *BC1* noncoding RNA is localized to distal dendrites in neurons thanks to a noncanonical GA kink-turn motif, an RNA element that features both single- and double-stranded components in which the GA base pair introduces a sharp kink in the stem (Muslimov et al. 2006). Interestingly, these types of motifs exist in a dynamic equilibrium between an open structure and a highly ordered tightly kinked conformation (Dennis and Omer 2005). Hence, Muslinov et al. showed that, under conditions of high Mg^{2+} or Ca^{2+} concentrations, the kink-turn motif of *BC1* assumes a kinked conformation that increases the binding affinity of hnRNPA2 on this motif (Muslimov et al. 2006, 2011). These results therefore suggest that changes in the intracellular milieu, such as an increase in divalent cations, may induce a switch in the conformation of structured CRMs, thus modulating the recruitment of trans-acting factors.

3.2 Trans-*Acting Factors or RNA-Binding Proteins Implicated in RNA Localization*

Over the last few decades, a large collection of RBPs have been identified and classified according to the types of RNA-binding domains (RBDs) that they contain and their RNA-binding specificities (Kazan and Morris 2013). These proteins are often highly modular in nature, containing an assortment of RBDs, as well as protein interaction motifs that allow their association with accessory factors (Lunde et al. 2007). These characteristics enable RBPs to coordinately associate with specific RNA targets and recruit important cofactors to allow the precise coordination of posttranscriptional gene regulatory steps, including RNA maturation, nuclear export, stability, localization, and translation. RBPs also often belong to large and well-conserved protein families, for which the highest degree of conservation usually resides within the RBDs (Yisraeli 2005; Zivraj et al. 2013), suggesting that their RNA-binding specificities and functional properties are also likely to be highly conserved.

In Table 2.1, we have listed many of the RBPs and prominent cofactors that have been implicated in the control of RNA subcellular localization in different model systems. Not surprisingly, many of these proteins have also been implicated in other aspects of posttranscriptional gene regulation, suggesting that these factors may fulfill multiple functions during the localization process. Indeed, while in transit, mRNAs have to be protected from degradation and kept in a silent state until they reach their final destination. Therefore, it would seem functionally beneficial for localized mRNAs to recruit such multipurpose RBPs (Lecuyer et al. 2009; Gagnon and Mowry 2011; Jansen and Niessing 2012).

Based on these properties of RBPs and localization CRMs described above, a general model of RNA trafficking emerges. The association of RBPs and interacting cofactors on target RNAs leads to the assembly of large RNPs, sometimes characterized as RNA transport particles or granules based upon microscopic visualization

Table 2.1 RBPs and protein components of trafficking granules

RBP/proteins	Molecular functions	Species	Biological roles	Example mRNAs	Family	Ref.
She2p	Adaptator protein	Yeast	Asymmetric division	*ASH1*		Bohl et al. (2000 #669)
She3p	Adaptator protein	Yeast	Asymmetric division	*ASH1*		Estrada et al. (2003 #670)
VgRBP60	Adaptator protein	Xenopus	Embryonic axis formation	*Vg1/VgT*	HnRNP I	Kroll et al. (2009 #191)
40 LoVe	Adaptator protein	Xenopus	Embryonic axis formation	*Vg1/VgT*	HnRNP D	Kroll et al. (2009 #191)
HnRNP A2	Adaptator protein	Mammals	Synaptic plasticity	*CAMKIIα, neurogranin, Arc, MBP*	HnRNP A/B	Hoek et al. (1998 #142)
Egalitarian	Adaptator protein	Drosophila	Embryonic axis formation	*Gurken, bicoid, oskar*		Dienstbier et al. (2009 #230)
Bicaudal D	Adaptator protein	Drosophila	Embryonic axis formation	*Gurken, bicoid, oskar*		Dienstbier et al. (2009 #229)
Swallow	Adaptator protein	Drosophila	Embryonic axis formation	*Bicoid*		Stephenson et al. (2004 #921)
Squid/Hrp40	Adaptator protein	Drosophila	Embryonic axis formation	*Gurken, bicoid, oskar*		Norvell et al. (2005 #870)
Hrp48	Adaptator protein	Drosophila	Embryonic axis formation	*Gurken, bicoid, oskar*	HnRNP A/B	Yano et al. (2004 #871)
CBF-A	Adaptator protein	Mammals	Neuromorphogenesis	*MBP*	HnRNP A/B	Raju et al. (2008 #916)
Huntingtin	Adaptator protein	Mammals	Neuromorphogenesis	*β-Actin*		Ma et al. (2011 #918)
Miranda	Adaptator protein	Drosophila	Embryonic axis formation neurogenesis	*Oskar* *Prospero*		Irion et al. (2006 #923), Schuldt et al. (1998 #929)

Barentz	Adaptator protein	Drosophila	Embryonic axis formation	*Oskar*		van Eeden et al. (2001 #924), Macchi et al. (2003 #931)
		Mammals	neuromorphogenesis	*BC1*		
ZBP2	Adaptator protein	Chicken	Cell migration	*β-Actin*		Pan et al. (2007 #146)
		Mammals	growth cone turning			
Sm	Nuclear export	Xenopus	Embryonic axis formation	*Xcat2*		Bilinski et al. (2004 #905)
				Xdaz1		
Xstau/Staufen	Adaptator protein	Xenopus	Embryonic axis formation	*Vg1, oskar, CAMKIIα*		Yoon and Mowry (2004 #777), Micklem et al. (2000 #925)
	Translational regulator	Drosophila	Synaptic plasticity			
		Mammals				
ZBP1	Adaptator protein, translational regulator	Chicken	Cell migration	*β-Actin*	VICKZ	Oleynikov and Singer (2003 #145), Deshler et al. (1997 #113)
IMP		Mammals	Growth cone turning	*Vg1, VgT*		
VgRBP1		Xenopus	Embryonic axis formation			
FMRP	Adaptator protein, translational regulator	Drosophila	Synaptic plasticity neuromorphogenesis	*CAMKIIα*		Antar et al. (2005 #459), Bianco et al. (2010 #462)
		Mammals				
Purα	Adaptator protein, translational regulator	Mammals	Synaptic plasticity	*CAMKIIα*		Kanai et al. (2004 #233), Aumiller et al. (2012 #474)
		Drosophila	Embryogenesis			
SMN	Adaptator protein, translational regulator	Mammals	Axon growth	*Cpg15*		Rossoll et al. (2003 #375), Akten et al. (2011 #379)
XVLG1	Adaptator protein	Xenopus	Embryonic axis formation	*Xcat2*	DEAD box	Bilinski et al. (2004 #905)
Vasa	Translational regulator			*Xdaz1*		

(continued)

Table 2.1 (continued)

RBP/proteins	Molecular functions	Species	Biological roles	Example mRNAs	Family	Ref.
Marta1/2	Adaptator protein, stabilization/degradation, translational regulator	Mammals	Synaptic plasticity	*MAP2, Vg1, VgT*		Zivraj et al. (2013 #919), Rehbein et al. (2002 #934), Kolev and Huber (2003 #209)
VgRBP71		Xenopus	Embryonic axis formation			
KSRP						
HuD	Stabilization/degradation	Mammals	Axonal outgrowth	*Neuritin, GAP-43, CAMKIIα*		Yoo (2013 #920), Tiruchinapalli et al. (2008 #937)
Mago-nashi	Splicing	Drosophila	Embryonic axis formation	*Oskar*		Mohr et al. (2001 #927)
Y14 (tsunagi)	Splicing	Drosophila	Embryonic axis formation	*Oskar*		Mohr et al. (2001 #927)
Mbnl 1/2	Splicing	Mammals	Cellular polarity	*Integrin a3*		Adereth et al. (2005 #938), Wang et al. (2012 #310)
		Drosophila				
Rumpelstiltskin	Stabilization/degradation	Drosophila	Embryonic axis formation	*Nanos*	hnRNP M	Jain and Gavis (2008 #902)
Smaug	Stabilization/degradation	Drosophila	Embryonic axis formation	*Nanos, Hsp83*		Semotok et al. (2005 #176)
loc1p	Translational regulator	Yeast	Asymmetric division	*ASH1*		Shen et al. (2009 #685)
puF6p	Translational regulator	Yeast	Asymmetric division	*ASH1*	Pumillio and FBF	Shen et al. (2009 #685)
Khd1p	Translational regulator	Yeast	Asymmetric division	*ASH1*		Paquin et al. (2007 #213)
CPEB/orb	Translational regulator	Mammals	Synaptic plasticity	*CAMKIIα, ZO-1, gurken*		Wu et al. (1998 #208), Nagaoka et al. (2012 #644), Chang et al. (2001 #941)
		Drosophila	cell polarity			
			Embryonic axis formation			

ElrB	Translational regulator	Xenopus	Embryonic axis formation	*Vg1, VgT*	elaV	Colegrove-Otero et al. (2005 #784)
PTB	Translational regulator	Drosophila	Embryonic axis formation	*Oskar*	hnRNP I	Besse et al. (2009 #877)
Bruno	Translational regulator	Drosophila	Embryonic axis formation	*Oskar*	CELF	Kim-Ha et al. (1995 #879)
				Gurken		
Me31B	Translational regulator	Drosophila	Embryonic axis formation/ neuromorphogenesis	*Oskar, CAMKIIα*		Nakamura et al. (2001 #928), Hillebrand et al. (2010 #943)
Rumpelstiltskin	Translational regulator	Drosophila	Embryonic axis formation	*Nanos*	HnRNP M	Jain and Gavis (2008 #902)
Aubergine	Translational regulator	Drosophila	Embryonic axis formation	*Nanos*		Becalska et al. (2011 #904)
Sh5p/Bn1p	Anchorage	yeast	Asymmetric division	*ASH1*		Beach et al. (1999 #680)
Bud6p/Aip3p	Anchorage	yeast	Asymmetric division	*ASH1*		Beach et al. (1999 #680)
eiF1α	Anchorage	Mammals	Cell migration	*β-Actin*		Liu et al. (2002 #760)
			Growth cone turning			
Spectrin	Anchorage	Xenopus	Embryonic axis formation	*Xcat2, Xdaz1*		Vaccaro et al. (2010 #909)
XNOA36	Anchorage	Xenopus	Embryonic axis formation	*Xcat2, Xdaz1*		Vaccaro et al. (2010 #909)

(Antar et al. 2005; Bertrand et al. 1998). The formation and composition of these granules are thought to be spatiotemporally regulated, evolving as RNAs are first synthesized in the nucleus and subsequently exported to the cytoplasm (Oleynikov and Singer 2003; Shen et al. 2009; Kroll et al. 2009; Yoon and Mowry 2004). Once they reach the cytoplasm, these particles can incorporate adaptor proteins that will dictate the downstream localization process (Dienstbier et al. 2009; Dienstbier and Li 2009; Macchi et al. 2003; Takizawa et al. 2000). To prevent mRNAs from being translated during transport, translational repressors may also be recruited as RNA granule components (Shen et al. 2009; Besse et al. 2009; Nagaoka et al. 2012). Once they reach the target destination, RBPs or associated cofactors will allow anchorage of the mRNA to the target destination (Krauss et al. 2009; Delanoue and Davis 2005), where translational repression can be reversed in order to allow localized translation of the mRNA. This could occur via the recruitment of activators, or through the release of translational repression, which can be induced notably by posttranslational modifications in response to local signaling cues (Huttelmaier et al. 2005; Kolev and Huber 2003).

4 Molecular Mechanisms of mRNA Trafficking

Three primary mechanisms of intracellular RNA trafficking have been documented over the years: (1) directed transport by cytoskeletal networks, (2) diffusion coupled to local entrapment, and (3) localized protection from degradation (Fig. 2.2). Below, we discuss in more detail a few examples where these mechanisms have been precisely characterized in different organisms and cell types.

4.1 Cytoskeletal Transport

Transport along polarized cytoskeletal networks is the most predominant mechanism for mRNA localization characterized to date. Once they reach the cytoplasm, RNP complexes are thought to recruit specific adaptor proteins allowing interactions with molecular motors such as kinesins, dyneins, or actinomyosins (Gagnon and

Fig. 2.2 (continued) mRNAs in *Drosophila* oocytes, for which the localization to the anterior and posterior poles is essential for establishment patterning the egg along the anteroposterior axis. The other pathways involved in mRNA trafficking are cytoskeleton independent, involving either transcript diffusion or local entrapment (**e**), an example being the METRO pathway in *Xenopus* oocytes, in which *Xcat2* mRNA is localized to the vegetal cortex via association with the mitochondrial cloud. The last scenario involves a degradation protection mechanism, (**f**) an example of which is *nanos* mRNA in *Drosophila*, which is degraded in the yolk cytoplasm while being stabilized and translated in the posterior pole of the developing embryo

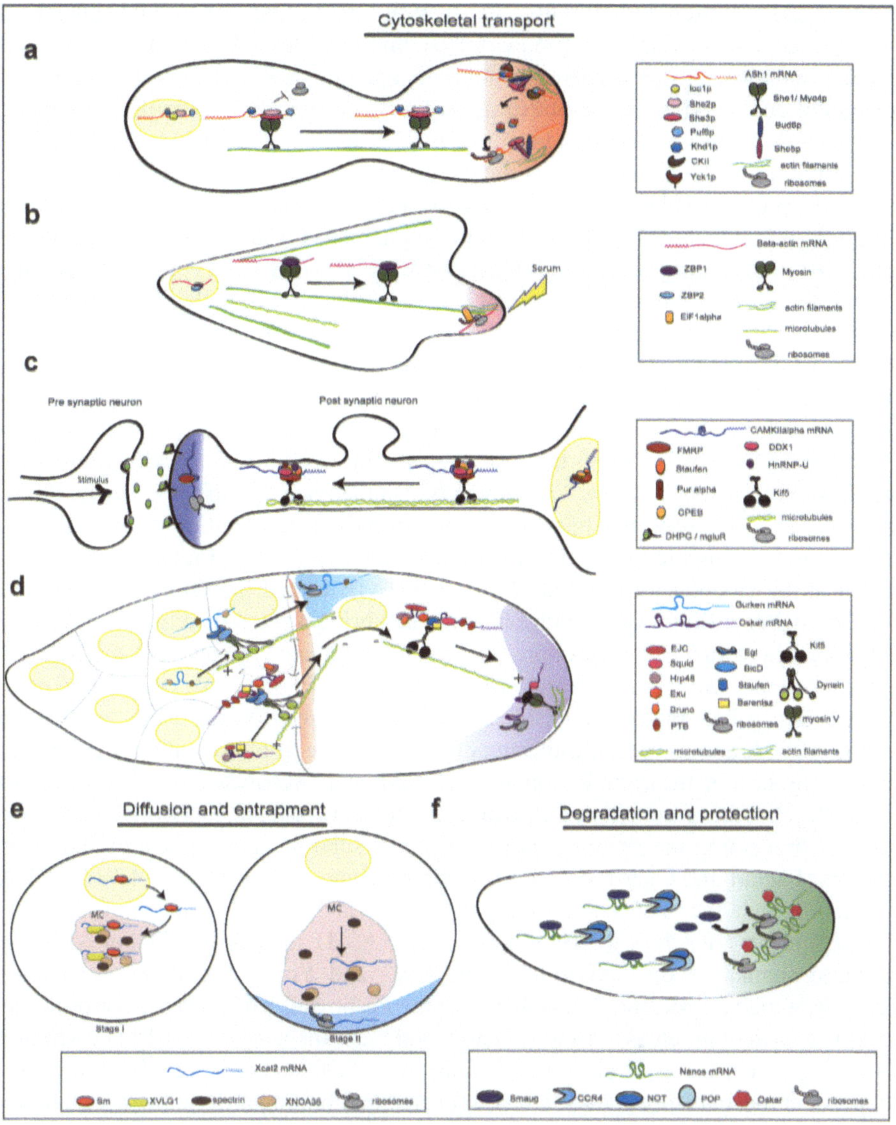

Fig. 2.2 Examples of mRNA localization mechanisms in different organisms and cell types. RNA localization is crucial for the regulation of several biological processes in different cells and organisms. By precisely studying the molecular mechanism involved for the localization of specific mRNAs, three predominant trafficking mechanisms have been elucidated. In the first pathway, RNPs travel along cytoskeletal networks. Some can move on actin microfilaments, such as (**a**) *ASH1* mRNAs in the yeast *S. cerevisiae*, an mRNA that is transported to the bud tip and serves to control mating type switching. (**b**) The *β-actin* mRNA, whose localization to the leading edge of cells is important for cell migration. Other mRNAs can travel along microtubules via associationwith dynein or kinesin motors, such as (**c**) *CAMKIIα* mRNA whose targeting to dendritic synapses of neurons is crucial for synaptic plasticity and memory, or (**d**) *gurken* and *oskar*

Mowry 2011; Jansen and Niessing 2012). While microfilaments are considered as the cytoskeletal network used for short trips, microtubules (MTs) and their molecular motors appear to be involved in long-distance mRNA transport. However, the examples described below demonstrate that this concept is still a work in progress.

Perhaps the best understood example of RNA targeting mediated through transport on actin filament networks is the asymmetric localization of *ASH1* mRNA to the bud tip of the baker's yeast *Saccharomyces cerevisiae*. This process has been implicated in the determination of mother-daughter cell fates through differential modulation of mating type switching (Long et al. 1997; Bobola et al. 1996; Sil and Herskowitz 1996; Takizawa et al. 1997) (Fig. 2.2a). During this process, *ASH1* mRNA is bound by several RBPs, such as She2p and the translational repressors Puf6p, Khd1p, and Loc1p (Jansen and Niessing 2012; Muller et al. 2011) (Fig. 2.2a). Indeed, She2p is loaded onto the mRNA in the nucleus and travels into the nucleolus to recruit Loc1p. While this factor is known to be strictly nuclear, its association with the *ASH1* RNP is important for downstream cytoplasmic localization (Shen et al. 2009; Long et al. 2001). *ASH1* mRNA is then joined by Puf6p (Gu et al. 2004); however even if Puf6p has been shown to strongly accumulate in the nucleus, its association in this compartment has not been clearly demonstrated. Once translocated in the cytoplasm (Du et al. 2008), Khd1p (Irie et al. 2002) is then recruited into the *ASH1* mRNPs along with She3p, which bridges the interaction with She1p/Myo4p, a myosin V motor protein, thus allowing directed transport of the RNP on actin filaments (Bohl et al. 2000; Estrada et al. 2003; Jansen et al. 1996; Long et al. 2000) (Fig. 2.2a). Interestingly, She2p has been shown to bind as a tetramer on *ASH1* mRNA and to recruit multimers of Myo4p-She3p motors allowing the processive motion of the complex on actin (Chung and Takizawa 2010; Krementsova et al. 2011). During transport, Puf6p and Khd1p act to repress *ASH1* mRNA translation by preventing the recruitment of ribosomal subunits and translation initiation factors (Paquin and Chartrand 2008). Upon reaching its destination, *ASH1* mRNA is anchored via the She5p, and Bud6p proteins, which act to bundle actin filaments in the bud tip (Beach et al. 1999) (Fig. 2.2a). ASH1 mRNA translation is finally induced thanks to the activities of the CK2 and Yck1p membrane-associated kinases, which respectively phosphorylate the Puf6p and Khd1p repressor proteins, thus triggering the release of these factors and the translation of Ash1p (Jansen and Niessing 2012; Deng et al. 2008; Paquin et al. 2007) (Fig. 2.2a). This myosin-dependent mechanism appears to be implicated in directing the localized translation of several other transcripts in *S. cerevisiae* (Casolari et al. 2012), and many mechanistic aspects of this regulation also appear to be at work in higher eukaryotes (Krauss et al. 2009; Condeelis and Singer 2005; Latham et al. 2001).

Another well-studied example is *β-actin* mRNA trafficking, for which the localization to lamellipodia in fibroblasts and other migratory cell types or structures, such as axonal projections in neurons, is crucial for directed cell migration (Fig. 2.2b). A key player in this process is the zipcode-binding protein 1 (ZBP1, also known as IMP1), an RBP that associates with *β-actin* mRNA in the nucleus, at the transcription site, and subsequently accompanies this transcript during its cytoplasmic transport (Oleynikov and Singer 2003; Gu et al. 2002; Ross et al. 1997;

Pan et al. 2007) (Fig. 2.2b). It was recently shown that ZBP1 recognizes *β-actin* mRNA through a bipartite *cis* element, composed of two primary sequence motifs separated by a spacer region, a CRM that appears to be conserved in a large collection of putative ZBP1 targets (Chao et al. 2010; Patel et al. 2012). In addition to its role in mRNA trafficking, ZBP1 also acts as a translational repressor of *β-actin* mRNA during its transport (Huttelmaier et al. 2005). The mechanism of cytoskeletal transport mediated by ZBP-1 is still unclear, since this factor has been reported to colocalize and to move along both microfilaments (Oleynikov and Singer 2003; Sundell and Singer 1991; Yamagishi et al. 2009a, b; Eliscovich et al. 2013) and MTs in fibroblasts (Yoshimura et al. 2012; Fusco et al. 2003; Yamada et al. 2011; Lifland et al. 2011). Since its localization is more sensitive to cytochalasin D treatment compared to colchicine, respective inhibitors of actin and MT polymerization, it is thought that actin filaments likely play a predominant role in *β-actin* mRNA localization in fibroblasts (Oleynikov and Singer 2003) (Fig. 2.2b). Interestingly, the trafficking of ZBP1/*β-actin* mRNA is responsive to extracellular stimuli and is activated following serum stimulation, a treatment that is well known to induce the formation of lamellipodia and cell migration (Pan et al. 2007; Oleynikov and Singer 2003). This transport is preceded by the recruitment of the translational elongation factor EiF1α at the leading edge of the fibroblasts, which binds both to *β-actin* mRNA and F-actin, suggesting that this factor may play a role in the transcript anchorage and localized translation (Liu et al. 2002) (Fig. 2.2b).

In parallel to its role in cell migration in fibroblasts, trafficking of *β-actin* RNA granules has also been characterized in axonal growth cones of developing neurons (Bassell et al. 1998) (Fig. 2.2c). As in fibroblasts, *β-actin* mRNA localization in axons can be induced by extracellular neurotrophic stimuli, such as neurotrophin-3 (Zhang et al. 2001) or brain-derived neurotrophic factor (BDNF) (Huttelmaier et al. 2005). Indeed, Huttelmaier et al. reported that after BDNF stimulation, the binding of ZBP1 to *β-actin* mRNA was induced, leading to its increased targeting to growth cones. In addition to stimulating mRNA transport, BDNF also induces the activation of Src family kinases in the growth cone, leading to the phosphorylation of ZBP1 at as it enters this subcellular compartment. This phosphorylation event reduces ZBP1-binding affinity toward *β-actin* mRNA, thus allowing the release and localized translation of β-actin, which supports growth cone turning during the axon guidance process (Huttelmaier et al. 2005; Sasaki et al. 2010) (Fig. 2.2c). In neurons, ZBP1-associated RNA granules are preferentially transported through the MT networks (Bassell et al. 1998) (Fig. 2.2c). In this process, RNA trafficking is ensured either by the kinesin motor proteins that transport cargos to the MT plus ends, which are generally oriented toward the cell periphery, or by the dyneins that transport cargos to MT minus ends positioned within the cell body (Gagnon and Mowry 2011). The case of *β-actin* mRNA thus illustrates how specific RBPs and CRMs can be re-utilized in different cell types to achieve a common functional output, in this case modulating cell migration behavior, although some aspects of the regulation are cell type specific, reflecting the differences in morphological organization of these cells.

The involvement of specific motor proteins in RNA intracellular transport has also been well established. For example, the kinesin plus-end-directed MT motor plays a key role the asymmetric localization of mRNAs along the animal/vegetal axis of *Xenopus* oocytes. Indeed, two different pathways of mRNA localization to the vegetal cortex of frog eggs have been described: the early pathway or *me*ssage *tr*ansport *o*rganizer (designated the METRO pathway), involving RNA diffusion and local entrapment during early stages of oogenesis (stages I–II, described below), and the late pathway, which occurs during stages III–IV of oogenesis and involves MT network trafficking (Kloc and Etkin 1995). The molecular mechanism of *VG1* mRNA localization to the vegetal pole during the late pathway has been well studied. As mentioned above, this transcript is localized thanks to the VLE, a 340-nt CRM located in the 3′ UTR of this mRNA (Mowry and Melton 1992). The VLE is bound in the nucleus by hnRNP1/VgRBP60, which interacts with 40LoVe and Vera/Vg1RBP (Kroll et al. 2009). Once exported from the nucleus, this RNP complex is remodeled, notably via the recruitment of the *Xenopus* Staufen (XStau) protein and Prrp, a factor that can also bind to the VLE (Kroll et al. 2009; Yoon and Mowry 2004). These remodeled cytoplasmic *Vg1* mRNA granules are then transported to the vegetal cortex along MTs by the kinesin 1 and kinesin 2 motor proteins, both of which are required for proper transport (Betley et al. 2004; Messitt et al. 2008). During transit, *Vg1* mRNA translation is repressed via a 252-nt AU-rich sequence, termed the Vg1 translation element (VTE), positioned downstream of the VLE (Otero et al. 2001; Wilhelm et al. 2000a). This VTE is bound by the ElrB protein, the *Xenopus* homologue of HuB, an important translational regulatory RBP (Colegrove-Otero et al. 2005). Once *VG1* mRNA reaches the vegetal pole at stage IV of oogenesis, the RBP VgRBP71/KSRP binds to the 3′ end of the VLE and induces a cleavage at an adjacent polyadenylation signal, thus removing the VTE element and allowing polyadenylation and translation of the mRNA (Kolev and Huber 2003).

In parallel to kinesin-dependent transport, the dynein minus-end-directed MT motors have also been implicated in mRNA trafficking. Notably, this has been shown for a subset of *Drosophila* mRNAs, such as *bicoid* and *gurken*, which are localized to the anterior pole of the oocyte during mid-oogenesis (stages 8–10) (Berleth et al. 1988; Kugler and Lasko 2009; Ephrussi et al. 1991; Kim-Ha et al. 1991) and play a central role in anteroposterior axis specification (Fig. 2.3d). In the egg chamber at stages 8–10, the minus end of microtubules is oriented toward the oocyte while the plus end is turned to the nurse cells. *gurken* and *bicoid* mRNAs, which are produced in the nurse cells, are transported into the oocyte via dyneins on this MT network (Clark et al. 2007) (Fig. 2.3d). This process involves recruitment of the RBP Egalitarian (Egl) and its partner Bicaudal D (BicD) to these mRNAs in the nurse cells, which then associate with dynein motors to ensure RNP transport into the oocytes via ring canals (Clark et al. 2007). In this RNP complex, Egl recognizes these mRNAs via stem–loop CRMs, while BicD acts as an adaptor that interacts with the dynein complex and thus provides linkage to the MT cytoskeleton (Dienstbier et al. 2009; Dienstbier and Li 2009; Bullock and Ish-Horowicz 2001) (Fig. 2.3d).

Dynein, Egl, and BicD also participate in the posterior localization of *oskar* mRNA during oogenesis (Krauss et al. 2009). However, the travels of this mRNA are more complex and divided into different steps, first being transported from the nurse cells to the anterior pole of the oocyte by dyneins (Clark et al. 2007), which is quickly followed by its trafficking from the anterior to the posterior poles of the oocyte thanks to the kinesin 1 motor protein (Mhlanga et al. 2009; Zimyanin et al. 2008) (Fig. 2.3d). The first crucial event is the splicing of *oskar* mRNA at the first exon-exon junction, which allows the recruitment of the Mago-Nashi and Tsunagi exon junction complex (EJC) components and also leads to formation of the SOLE stem–loop element (Hachet and Ephrussi 2004; Ghosh et al. 2012). In addition to EJC factors, the Hrp48, Squid/Hrp40, Cup, and Barentsz proteins are also recruited onto *oskar* mRNA while it is still in the nurse cell nuclei (van Eeden et al. 2001). Before being transported to the oocyte by dyneins, this RNP complex is exported from the nucleus of nurse cells where it is bound by the protein Exuperantia (Exu) (Wilhelm et al. 2000b) and two translational repressors, PTB/hnRNPI (Besse et al. 2009) and Bruno (Kim-Ha et al. 1995) (Fig. 2.3d). Within the oskar mRNA particles, the Barentsz protein is thought to allow the coupling of oskar mRNA to kinesin 1 motors (Wilhelm et al. 2003) (Fig. 2.3d). Once these RNPs reach the posterior pole, the myosin V motor protein allows the entrapment of the mRNA (Krauss et al. 2009), suggesting that an interplay between MT- and microfilament-based motors is important for oskar mRNA targeting (Gagnon and Mowry 2011; Jansen and Niessing 2012; Krauss et al. 2009). Notably, it is known that even if MTs seem to direct polarized transport, many RNA granules can be transported bidirectionally, suggesting a more complex coordination between the motor proteins, RNPs, and cytoskeleton. This was recently shown by Nayak et al., who employed an in vitro reconstitution assay with single-molecule resolution to show that dynein-associated RNP granules could move bidirectionally on microtubules. Also, they showed that mRNA localization to the minus end of microtubules was more efficient for RNAs containing zipcodes compared to those without zipcodes due to the recruitment of a larger number of dyneins to the localizing RNPs (Amrute-Nayak and Bullock 2012). Moreover, while some granules have shown to contain several mRNAs either due to the capacity of RBPs to bind multiple mRNAs or due to RNA-RNA interactions, as described in this last example, those in vitro studies highlight that mRNPs can also transport a single mRNA (Amrute-Nayak and Bullock 2012). In addition to their involvement in embryonic axis formation, Egl and BicD also participate in cell polarity in *Drosophila* syncytial blastoderm embryo by apically localizing many pair rule mRNAs which define the pattern of segmentation including *fushi tarazu* (*ftz*), *hairy* (*h*), *even skipped* (*eve*), and *runt* (*run*) demonstrating that the same machineries can be recycled and used in the localization of different mRNAs in different contexts (Amrute-Nayak and Bullock 2012; Vendra et al. 2007).

4.2 Diffusion Coupled to Local Entrapment

This mechanism has been described for few mRNAs in insects, fish, and amphibians. The first example was *nanos* mRNA which localizes to the posterior pole of *Drosophila* oocytes through a cytoplasmic movement induced by nurse cell dumping and ooplasmic streaming revealed by live time-lapse imaging of RNP particle movements using the MS2 system (Forrest and Gavis 2003), which will be described below. At the posterior pole of the oocyte, the entrapment of *nanos* to the cortex then occurs through its association with the actin cytoskeleton (Forrest and Gavis 2003). Interestingly, the RBPs Rumpelstiltskin (Rump) and Aubergine (Aub) have also been shown to participate in the posterior localization of *nanos* mRNA (Jain and Gavis 2008; Becalska et al. 2011). Since Rump, an hnRNP M homolog, is expressed in the nurse cell nuclei and is found in the cytoplasm of the oocytes, but not in the posterior pole, it has been proposed to participate either in the destabilization or in the translational repression of *nanos* mRNA in the cytoplasm of the oocyte (Jain and Gavis 2008).

Another good example resides in the METRO pathways of RNA transport in early-stage Xenopus oocytes (described above), which enhances mRNA localization to the vegetal cortex by mRNA diffusion and entrapment (Kloc and Etkin 1995) (Fig. 2.2e). This pathway occurs in two main steps. First, in stage I oocytes, mRNAs such as *Xcat2* and *Xdaz1* are localized to the mitochondrial cloud (MC), which lies close to the oocyte nucleus. During stage II, the mitochondrial cloud breaks loose from the nucleus and merges into a nearby cortical region, thereby co-transporting mRNAs to the vegetal cortex (Chang et al. 2004). Bilinski et al. demonstrated that Sm proteins, components of the spliceosome, and the Vasa-like protein XVLG1, a DEAD box RNA helicase, are present in the MC. The exact functional role of those RBPs is not well understood, although it has been suggested that they could assist the export of mRNAs from the nucleus to the MC (Bilinski et al. 2004) (Fig. 2.3e). Once at the MC, mRNAs are immobilized thanks to the aggregation of the endoplasmic reticulum, mitochondria, and proteins such as spectrin or XNOA36 (Vaccaro et al. 2010, 2012), which constitute a type of matrix for RNA entrapment (Fig. 2.2e).

4.3 Local Protection Against Degradation

This mechanism consists in the full degradation of an mRNA except in a particular site in the cell. In *Drosophila* embryos, two different degradation pathways involved in maternal RNA turnover have been proposed to involve this type of mechanism: the first occurs before fertilization thanks to maternal factors (Tadros et al. 2003), while the second occurs 2 h after fertilization and involves zygotic factors (Bashirullah et al. 1999, 2001). The best known maternal mRNAs that are subjected to this targeting process are *hsp83* (Ding et al. 1993) and *nanos*

(Bashirullah et al. 1999) (Fig. 2.2f). These mRNAs are degraded before fertilization in the bulk cytoplasm and protected from degradation solely in the posterior pole plasm of the embryo. Concerning the specific mechanism of cytoplasmic degradation, this appears to be initiated by the Smaug RBP, which recognizes RNAs via a specific zipcode designated as the Smaug response element (SRE), consisting in stem–loop motifs that are present in the 3′ UTR and the coding region of *nanos* and *hsp83* mRNAs, respectively (Zaessinger et al. 2006; Semotok et al. 2008). Indeed, once bound on the mRNA, Smaug recruits the CCR4/POP2/NOT deadenylase complex, which then induces deadenylation and degradation of the mRNA (Semotok et al. 2005). While mRNAs are degraded in the bulk cytoplasm, the posterior set of mRNAs is stabilized. Zaessinger et al. demonstrated that Oskar proteins expressed in the posterior pole prevent the binding of Smaug on *nanos* mRNA, thus blocking its deadenylation and allowing localized translation (Zaessinger et al. 2006) (Fig. 2.2f).

5 Harnessing Genomic and Imaging Technologies for the Study of RNA Localization

RNA localization was initially viewed as a peculiar mechanism limited to a specialized subset of transcripts. However, genomic and high-throughput microscopy approaches to systematically interrogate this layer of gene regulation have revealed that large populations of mRNAs undergo regulation at the level of their subcellular trafficking (Lecuyer et al. 2009). These approaches are strongly complemented by live-cell imaging methods that allow the tracking of RNA intracellular movements with ever-improving spatiotemporal resolution. We devote this next section to reviewing some of the established and emerging techniques that hold great promise for advancing our understanding of RNA localization pathways and regulatory mechanisms.

5.1 Global Studies of mRNA Localization

The advent of high-throughput approaches to globally survey transcriptome localization dynamics has had an important impact on this field of study. These strategies include high-throughput microscopy surveys, as well as biochemical fractionation combined with microarray/RNA sequencing analysis to identify transcripts enriched within specific subcellular compartments or associated with particular RBPs. Indeed, several in situ hybridization-based studies conducted in different model organisms, tissues, and cell types have characterized a wide variety of examples of intracellular RNA localization phenotypes, both for coding and noncoding RNAs (Lecuyer et al. 2007; Dubowy and Macdonald 1998; Makabe et al. 2001; Tomancak et al. 2002; Kingsley et al. 2007; Mercer et al. 2008; Lein et al. 2007). For example, in a large-scale fluorescent in situ hybridization survey of over 3,000 distinct mRNA species during *Drosophila* embryogenesis, approximately 70 % of transcripts were found to exhibit asymmetric subcellular distribution patterns (Lecuyer et al. 2007). This study further identified many novel varieties of mRNAs targeted to precise organelles and cellular structures (e.g., cytoskeleton, cell junctions, membrane territories, mitotic apparatus, nuclei and chromatin) while also uncovering broad correlations between mRNA distribution patterns and the functions/localization of the encoded proteins, suggesting that mRNA localization is a key determinant of localized protein complex assembly and function (Lecuyer et al. 2007, 2009).

In addition to microscopy, cell fractionation using well-established biochemical procedures has been adapted successfully to characterize the transcript populations that are enriched within specific cellular compartments. In this way, mRNAs associated with different organelles, such as mitochondria (Marc et al. 2002; Garcia et al. 2007), the endoplasmic reticulum (Chen et al. 2011; Jagannathan et al. 2011; Kopczynski et al. 1998; Lerner et al. 2003), membranes (Wang et al. 2012), the mitotic spindle (Blower et al. 2007; Sharp et al. 2011), and insoluble and cytosol cell fractions (Wang et al. 2012; Chen et al. 2011) have been identified. Functionally coherent populations of RNAs have also been isolated from cell protrusions induced in response to exogenous stimuli such as in neuronal dendrites (Eberwine et al. 2001; Buckley et al. 2011), axons (Taylor et al. 2009; Willis and Twiss 2011), and pseudopodia (Mili et al. 2008). Interestingly, this general strategy is also a great way to characterize variations in global mRNA localization profiles that can occur during dynamic processes, such as development and in response to specific genetic manipulations (Wang et al. 2012; Gumy et al. 2011; Zivraj et al. 2010). A good example is the work of Saint-Georges et al. who compared mitochondrial transcriptomes from cells expressing or lacking the Puf3 protein, which leads to the characterization of two different mRNA clusters requiring or not Puf3 for their proper localization (Saint-Georges et al. 2008). Also, purification of axons from adult versus embryonic tissues has allowed the characterization of different population of RNAs depending on developmental timing (Gumy et al. 2011; Zivraj et al. 2010).

Finally, as will be thoroughly detailed in other chapters of this book, an alternative strategy that is frequently applied to the study of posttranscriptional gene regulatory pathways consists in biochemically purifying a specific RBP from a cell extract and identifying its associated RNA targets using microarrays or RNA deep sequencing, a general strategy designated as RIP-Chip/Seq, CLIP-Seq, or HITS-Clip (Ule et al. 2005; Darnell 2011; Jensen and Darnell 2008; Licatalosi et al. 2008; Tenenbaum et al. 2002). Using such approaches, it has become clear that RBPs often associate with functionally related RNA targets, some of which have clear links to subcellular localization properties. This is perhaps best characterized by the example of the Puf family of RBPs in yeast, for which the different members interact with mRNAs linked to specific organelles such as mitochondria or membrane-linked fractions (Gerber et al. 2004).

5.2 Microscopy Methods to Visualize RNA Movements in Live Cells

Over the years, several strategies have been developed to follow mRNA localization in living cells or tissues, approaches that have proven instrumental in the characterization of RNA trafficking mechanisms. In general, these methods make use of fluorescently labeled exogenous RNAs, fluorescent probes, or indirect RNA labeling via fluorescent RBPs. The first method to visualize mRNA localization in live cells or tissues consists in the injection of fluorescently labeled exogenous mRNAs followed by time-lapse microscopy in order to visualize the movement of RNP particles. As this assay is quite rapid, it has been useful for the identification of CRMs and RBPs involved in RNA localization (Bullock and Ish-Horowicz 2001; Cha et al. 2001; Glotzer et al. 1997; Yisraeli and Melton 1988). However, the microinjection step can be deleterious for the cells and the introduction of a large amount of labeled mRNA could saturate the endogenous trafficking machineries leading to experimental artifacts.

An alternative set of methods involves the transfection fluorescently labeled oligonucleotides capable of hybridizing to a cellular RNA target of interest. Since the main disadvantage of early versions of this method was the high background fluorescence signal produced by unbound probe, several strategies have been used to circumvent this drawback. Peptide nucleic acid-forced intercalation (PNA-FIT) probes, for example, contain a cyanine base for which the fluorescence is only induced upon binding of the probe to its target mRNA (Kohler et al. 2005; Kummer et al. 2011) (Fig. 2.3a). Santangelo et al. designed multiply labeled tetravalent RNA probes (MTRIPs), consisting in conjugates of labeled oligo probes linked to streptavidin molecule, which dramatically increase the intensity and signal to background ratio of the procedure (Santangelo et al. 2009) (Fig. 2.3b). Another option, described in Fig. 2.3c, involves the use of two chemically reactive probes for which the simultaneous binding to the target RNA results in fluorescence resonance energy transfer (FRET). An improvement to this latter strategy, aimed at minimizing the FRET signal caused by unbound probes, consists in the use of a quencher molecule. Removal of the quencher and restoration of the fluorescence signal only occur when the probes are bound to the target RNA and in close proximity to each other (Abe and Kool 2006) (Fig. 2.3d). In contrast to linear oligo probes, molecular beacon consists of 25-nt stem–loop probes that carry a quencher and fluorophore moieties at their 5′ and 3′ extremities. In the stem–loop conformation, the fluorophore is quenched, a property that is reversed upon binding to the RNA target, resulting in a strong fluorescent signal (Tyagi and Kramer 1996) (Fig. 2.3e). To decrease background signal due to unintended unfolding, an improved method was developed using two different beacons that bind neighboring sites of the target RNA in order to generate a FRET reaction (Santangelo et al. 2004) (Fig. 2.3f). Combined, these methods can prove advantageous for visualizing mRNAs in their endogenous context of regulation while also allowing interesting multiplexing capability. Their disadvantages include their propensity to exhibit high nuclear background and their

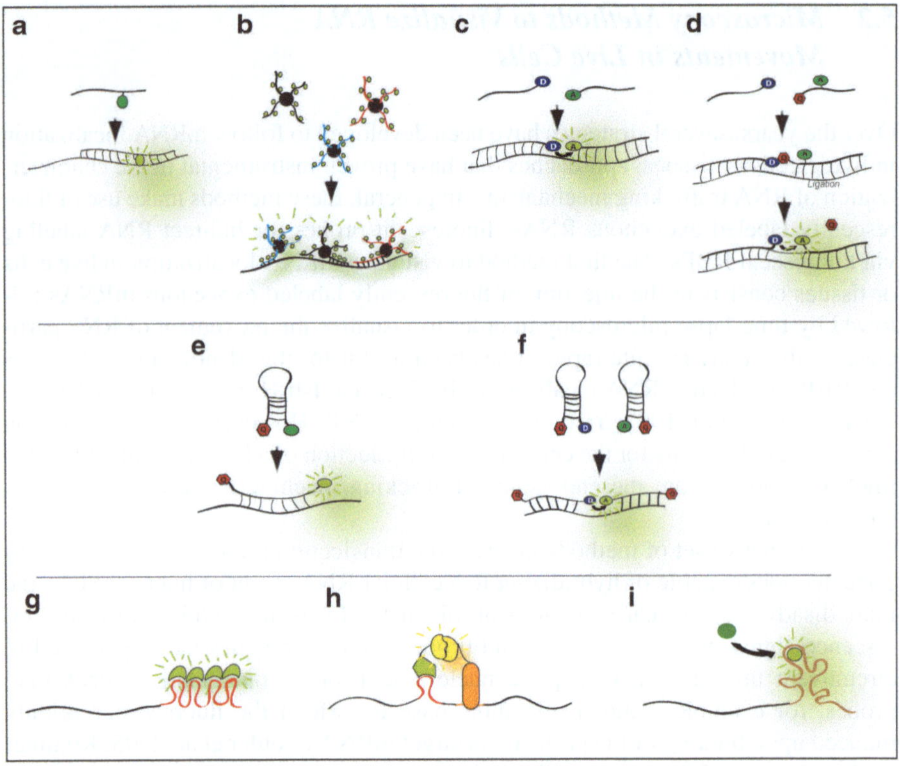

Fig. 2.3 Live imaging methods. Several approaches have been developed to visualize mRNA localization in living cells and organisms. These can involve the use of fluorescent molecules, such as linear probes and molecular beacons, or of fluorescent-tagged RBPs. (**a**) PNA-FIT probes containing a fluorogenic base will induce fluorescence only when hybridized to the targeted mRNA. (**b**) MTRIPs consist in the use of multiple fluorescent probes coated on streptavidin beads. (**c**) Chemically reactive probes coupled to donor and acceptor molecules result in fluorescence resonance energy transfer (FRET), which (**d**) can also be coupled to a quencher in order to avoid nonspecific transfer of energy due to inappropriate association. (**e**) Molecular beacons consist in stem–loop-forming oligonucleotides that contain fluorophore and a quencher molecules at their extremities, which allow the emission of fluorescence only when the molecular beacon is linearized and bound to its target mRNA. (**f**) The FRET strategy can also be applied to molecular beacons, by coupling them to acceptor and donor fluorochromes, which reduces nonspecific signal. (**g**) The most common method to visualize mRNAs in live cells is the MS2 detection system that involves the addition of a multimerized MS2 RNA hairpin within an RNA of interest, which is bound by a co-expressed MS2 coat protein (MCP)-GFP fusion protein. (**h**) Since unbound proteins can emit fluorescence, an alternative strategy is the TriFC system, which consists in the expression of two non-fluorescent moieties of a fluorescent protein fused to an MS2 protein or another RNA-binding domain, for which the simultaneous binding to the RNA target results in the restoration of the fluorescence signal. (**i**) Recently, a novel strategy has emerged involving the use of an RNA aptamer, designated as spinach, which is inserted in the targeted mRNA and can associate with fluorogenic small molecules

binding to the target may by hindered by bound RBPs or RNA structural constraints, although strategies have been implemented to address these drawbacks (Kummer et al. 2011; Santangelo et al. 2009; Nitin et al. 2004; Chen et al. 2009).

Fifteen years ago, a powerful alternative approach, the MS2 system, was invented for the study of *ASH1* mRNA transport in living yeast cells (Bertrand et al. 1998). This method involves two components; a fusion of GFP to the MS2 bacteriophage coat protein, an RBP that binds with high affinity to the 19-nt MS2 stem–loop element (Bertrand et al. 1998), as well as a chimeric RNA molecule containing multimerized copies of the MS2 hairpin element. When both components are co-expressed, the MS2-GFP protein interacts with the MS2 hairpins in the target RNA, thus allowing the visualization of the RNA movements in live cells (Fig. 2.3g). A similar approach is the λN-GFP system, which involves the λN bacteriophage coat protein that recognizes a 15-nt stem–loop designated as the BoxB hairpin (Daigle and Ellenberg 2007). Interestingly, both the MS2 and λN systems were successfully combined to follow the trafficking of individual pairs of transcripts in yeast (Lange et al. 2008). The MS2 system has found widespread usage in the community through the generation of transgenic models expressing MS2-GFP fusion proteins in vivo in a range of organisms, including yeast, *Drosophila*, and mice (Forrest and Gavis 2003; Lionnet et al. 2011). For example, multiple models of living yeast *S. cerevisiae* obtained by homologous recombination have been used to follow the localization of several mRNAs including *ASH1*, *SRO7* to the tip of the bud, *PEX3* to the RE, and *OXA1* to the mitochondria (Haim et al. 2007). Nevertheless, a key drawback of the procedure is that the insertion of multimerized MS2 stem–loops may interfere with the normal regulation of the target RNA.

A limitation of these methods is the high background fluorescence that can be generated due to unbound GFP protein. This has been addressed via the development of the "split GFP system"-type methods, which consist in the expression of two non-fluorescent moieties of the GFP protein fused to distinct RBDs. Hence, the simultaneous recruitment of these proteins to the RNA target results in the restoration of the fluorescence signal (Ozawa et al. 2007; Rackham and Brown 2004) (Fig. 2.3h). A good example of this is the TriFC method described by Rackam et al., involving the co-recruitment of non-fluorescent moieties of the Venus fluorescent protein to a target RNA (Rackham and Brown 2004) (Fig. 2.3h). An alternative strategy to allow the visualization of endogenous RNA without inserting additional sequences emerged from the use of the Pumilio protein. Indeed, the characteristic of this protein is to have two RNA-binding domains (PUM-HD) that can be engineered to recognize different specific sequences (Lu et al. 2009). Ozawa et al. designed two PUM-HD domains, each connected to a split fluorescent protein, to specifically recognize a 16-base sequence of the endogenous mitochondrial RNA encoding the NADH dehydrogenase subunit 6 (ND6) (Ozawa et al. 2007).

Finally, an alternative to using fluorescent protein was recently developed involving a fluorogenic RNA aptamer system with GFP-like fluorescence properties. Indeed, Paige et al. devised the "spinach" system, which consists in a series of RNA aptamers that, upon binding to specific fluorogenic small molecules, will emit fluorescence signals that cover a wide spectrum of emission properties

(Paige et al. 2012) (Fig. 2.3i). Those aptamers are particularly interesting since the small molecules that they associate with are cell permeable and nontoxic. Also, they are resistant to photobleaching and the fluorescence signal is rapidly observed following transcription of the aptamer-containing RNA in the cells. This system was tested to visualize the relocalization of spinach-labeled 5S rRNA into stress granules, suggesting that the use of genetically encoded RNA-fluorophore complexes could emerge as a powerful strategy for tracking single or multiple RNA molecules in living cells (Paige et al. 2012).

6 Implications of mRNA Localization in Human Disease Pathogenesis

As described in the previous sections, mRNA localization has emerged as a prevalent and important layer of gene expression. Not surprisingly, there are several examples for which mRNA localization defects are thought to contribute to disease etiology. This includes functions in several neuro and musculo degenerative disorders, and emerging links to cancer (Table 2.2).

6.1 RNA Localization Defects in Neuro and Musculo Degenerative Disorders

Impaired mRNA trafficking has mainly been reported in a specific group of neuromuscular disorders known as triplet repeat expansion diseases (TREDs). TREDs are genetic disorders caused by abnormal and recurrent amplications of trinucleotide repeats either in the coding or the untranslated regions of certain genes. Different expanded triplet repeats, such as CGG, CAG, CTG, or GAA, have been mapped to specific genes, thus allowing the general classification of >15 TREDs characterized to data (Orr and Zoghbi 2007). Interestingly, the molecular pathogenesis of some of these TREDs is associated with mRNA metabolism defects, in particular with aberrant mRNA localization in neurons or muscle cells. Indeed, some trinucleotide repeats lead to transcriptional silencing of genes encoding important RBP, such as FMRP, implicated in neuronal RNA trafficking. Alternatively, repeat expansions can result in the production of "toxic" RNA molecules that can interfere with the function of important cellular RBPs involved in RNA localization control (Galka-Marciniak et al. 2012).

Perhaps the two best characterized examples of TREDs that result in the loss of expression of specific RBPs are fragile X syndrome (FXS) and spinal muscular atrophy (SMA) (Table 2.2). FXS is characterized by severe intellectual disabilities and is the most frequent form of inherited mental retardation linked to autism. This syndrome occurs as a result of mutations in the *fragile X mental retardation 1* gene

Table 2.2 RBPs involved in RNA localization and disease

Disease	Affected RBP	Genetic defect	Biological roles	Affected mRNAs	Ref.
RBPs expression defects					
Fragile X syndrome	FMRP	FMR1 gene with CGG repeats >200 nt	Found in neuronal RNA granules, translational repressor	*SAPA4*	Dictenberg et al. (2008 #460)
				GABAR-d	
				MAP1b	
Spinal muscular atrophy	SMN	Loss or mutations	Found in neuronal RNA granules, precise role unknown	*β-Actin*	Rossoll et al. (2003 #375)
				PolyA mRNAs	
RBPs sequestration					
Fragile X-associated tremor/ataxia syndrome	Purα	FMR1 gene with CGG repeats between 50 and 200 nt	Found in neuronal RNA granules, precise role unknown	?	Kanai et al. (2004 #233)
Myotonic dystrophy	Mbnl	Dmpk gene with CGG repeats	Found in neuronal RNA granules	?	Wang et al. (2012 #310)
			RNA localization to cytoplasmic membranes		
RBPs overexpression					
Epithelial cancers (colon, breast, pancreas, lung)	IMP/ZBP1	IMP/ZBP1 gene amplification	Found in epithelial cells RNA granules	*β-Actin*	Tessier et al. (2004 #484), Gu et al. (2012 #494)
			RNA localization and translational repressor	*E-cadherin*	
				ARP2/3	

(*FMR1*) located on the X chromosome, which prevents the expression of the encoded FMRP protein. FMRP is an RBP present in many tissues, but predominantly expressed in brain and testis (Siomi et al. 1993; Ashley et al. 1993; Devys et al. 1993; Hinds et al. 1993). While in normal individuals the FMR1 gene contains between 6 and 44 CGG repeats in its 5′ UTR, these triplets show dramatic expansion (>200) in symptomatic individuals suffering from FXS. These 5′ UTR expansions lead to inappropriate promoter methylation and transcriptional silencing of the *FMR1* gene (Bagni et al. 2012). Studies examining the brains of deceased FXS patients (Hinton et al. 1991; Rudelli et al. 1985), or neuronal tissues of *FMR1* knockout mice (reviewed in Portera-Cailliau (2012)), have revealed that loss of *FMR1* function results in neurons with dendritic spine morphological defects (Gibson et al. 1993). Also, in *Drosophila*, mutation of the orthologous *dfmr1* gene is associated with a decrease in branched neuronal dendrites (Bianco et al. 2010). It is now well established that FMRP, by associating with polyribosomes, regulates the translation of many proteins involved in synaptic plasticity (Sidorov et al. 2013). In addition, FMRP plays an important role in dendritic RNA localization, a process which is crucial for neuronal morphogenesis (Bianco et al. 2010) and synaptic function (Dictenberg et al. 2008). Indeed, using a mouse knockout model of *FMR1*, Dictenberg et al. nicely demonstrated that loss of FMRP function impairs the dendritic localization of an important set of mRNAs (e.g., *SAPA4*, *GABAR-d*, *MAP1b*) encoding proteins with established functions in synapse structure and signaling regulation (Dictenberg et al. 2008). Thus, disruption of FMRP function appears to be the primary underlying cause of FXS pathogenesis.

Like FXS, SMA is induced in ~90 % of cases by loss-of-function mutations in the *SMN1* gene, which encodes the survival of motor neuron (SMN) protein (Burghes and Beattie 2009). SMA is the most common genetic neurodegenerative disorder observed in children and is characterized by muscle weakness and atrophy associated with a loss of reflexes affecting many vital motor functions (Crawford and Pardo 1996). The complete loss of function of SMN proteins is lethal in all the organisms previously tested (Burghes 1997; Schrank et al. 1997). In SMA, the disruption of *SMN1* function is partly rescued by the *SMN2* gene, which encodes a paralogue of SMN1 (Lefebvre et al. 1997; Coovert et al. 1997). The reduced level of SMN proteins in SMA patients leads to a specific degeneration of motor neurons in the spinal cord (Burghes and Beattie 2009). However, as those proteins are ubiquitously expressed, it is not understood why spinal motor neurons are more sensitive than other cells. SMN proteins are found both in the cytoplasm and the nuclear Cajal bodies or "gems," where they interact with Geminin proteins implicated in the assembly of snRNPs and mRNA splicing. Cells derived from SMA patients, or from SMA mouse models, have been reported to exhibit a decrease in the assembly of specific nuclear snRNPs and aberrant splicing activity (Gabanella et al. 2007; Wan et al. 2005; Zhang et al. 2008). However, as this defect is observed in different tissues, the link between the collapse of spliceosome function and a specific deficiency of motor neurons is still unclear. Interestingly, growing evidence supports a role for SMN proteins in mRNA transport and localized translation, specifically in axons and growth cones of motor neurons (Bechade et al. 1999; Fallini et al. 2010; Zhang et al. 2003). In this context, SMN proteins have been shown to interact with cytoskeletal elements allowing their antero- and retrograde movement along axonal microtubules (Fallini et al. 2010; Pagliardini et al. 2000). It has subsequently been shown that SMN1 is transported within granular structures in axons of motor neurons, a process mediated by its interaction with several neuronal RBPs involved in RNA transport processes, such as hnRNPR, hnRNPQ, HuD, and KSRP (Zhang et al. 2003; Rossoll et al. 2002; Tadesse et al. 2008; Fallini et al. 2011; Akten et al. 2011). Interestingly, those interactions are mediated by the Tudor domain of SMN1, a region that is frequently mutated in SMA patients (Hubers et al. 2011; Piazzon et al. 2008). Thus, SMN has been proposed to participate in the assembly of neuronal mRNA granules, which is supported by the observation of defective mRNA localization in SMN1-deficient motor neurons (Fallini et al. 2011). Indeed, Fallini et al. showed that *SMN1* knockdown in primary motor neurons leads to a decrease in poly(A) mRNA levels in the axonal compartment (Fallini et al. 2011). Also, SMA mouse models exhibit a deficiency in axonal growth, which is accompanied by a reduction in the localization of *β-actin* mRNA and protein in axons (Rossoll et al. 2003). Finally, a new role for SMN proteins in translational repression has recently been reported, which might be an important feature for efficient transport of translationally silent RNA granules in motor neurons (Sanchez et al. 2013).

In contrast to FXS and SMA, other TREDs, such as fragile X-associated tremor/ataxia syndrome (FXTAS) and myotonic dystrophy type I (DM1), appear to impact RNA localization pathways through a distinct mechanism, via the production of toxic RNAs that interfere with the function of certain RBPs normally implicated in RNA trafficking (Table 2.2). As its name suggests, FXTAS is a progressive neurodegenerative disorder characterized by the induction of intense tremors, ataxia, and Parkinsonism (Leehey 2009). This disease manifests itself in older males that harbor intermediate expansions of CGG repeats (between 50 and 200) in the *FMR1* gene. Repeat-containing RNAs have been shown to accumulate within nuclear inclusions in cells of FXTAS patients (Tassone et al. 2004). These RNA aggregates are thought to cause the nuclear sequestration of rCGG-binding proteins, such as Pur-alpha and hnRNP A2/B1, thus interfering with the normal cellular functions of these RBPs (Li and Jin 2012). In FXTAS patients, neuronal inclusions containing many of these RBPs have been observed leading to the hypothesis that neuronal degeneration could be the result of the sequestration of important RBPs (Greco et al. 2006; Jin et al. 2007). Interestingly, Pur-alpha is well known to bind RNA and modulate the translation of dendritically localized transcripts such as *CAMKIIα* and *MAP2* (Johnson et al. 2006). It was recently shown that the *Drosophila* orthologue of Pur-alpha is implicated in the transport of RNAs during oogenesis, suggesting that its function in RNA transport is evolutionarily conserved (Aumiller et al. 2012). Interestingly, *Drosophila* models of FXTAS have also demonstrated that over-expression of either Pur-alpha or hnRNP A2/B1 can counteract toxicity and neurodegeneration phenotypes induced by CGG repeat RNA (Jin et al. 2007; Sofola et al. 2007).

A similar mechanism is thought to operate in the case of DM1, a multisystemic degenerative disorder that leads to progressive muscle wasting, ataxia, and neurological defects (Udd and Krahe 2012). In this disease, the *DMPK* gene exhibits an aberrant expansion of -CUG- repeats in its 3′ UTR (Brook et al. 1992). Similar to the situation in FXTAS, the *DMPK* CUG repeat containing RNA exhibits a striking accumulation within large nuclear aggregates in the cells of DM1 patients. Several rCUG repeat interacting RBPs become aberrantly sequestered within these nuclear foci, including members of the Muscleblind-like (MBNL) family of RBPs, in particular MBNL1 and -2 (Kanadia et al. 2003; Jiang et al. 2004). While these proteins have mainly been implicated in the regulation of mRNA alternative splicing, it was recently found that they also play an evolutionarily conserved role in the control of mRNA intracellular localization (Wang et al. 2012). These findings suggest that defective mRNA intracellular trafficking may represent a major mechanism of DM1 pathogenesis.

6.2 mRNA Localization and Cancers

A crucial step of tumorigenesis is the epithelial to mesenchymal transition (EMT) that often occurs, leading to a scattering of tumor cells and metastases (Tse and Kalluri 2007). This step is typically characterized by the loss of internal cell polarity and cell-cell adhesion properties of tumor cells, leading to the acquisition of invasive properties that allow the migration of the cells into the vessels following a chemoattractive gradient. One of the invasive properties is the capability to form cellular protrusions or invadopodia (Stylli et al. 2008). In light of the key functions that they play in modulation of cell migration and cytoskeletal dynamics, many studies sought to define possible roles played by members of the VICKZ family of proteins, which includes IMP (CRD-BP), IGF2BP, or ZBP1 in cancer etiology (Table 2.2).

VICKZ proteins are over-expressed in many human breast, colon, pancreas, and lung cancers (Ross et al. 2001; Wang et al. 2004; Yantiss et al. 2005). Interestingly, Ross et al. reported a gene amplification of ZBP1 in primary tumors of one-third of breast cancer patients (Ross et al. 2001). Also, numerous studies have shown a differential expression of ZBP1 in tumors and metastasis. Indeed, while this protein is highly expressed in primary tumor cells in vitro and in vivo, it is not expressed in invasive or metastatic cells (Wang et al. 2004; Yantiss et al. 2005; Gu et al. 2009). This difference is explained on one hand by the methylated profile of ZBP1 promoter observed in metastatic cells preventing its expression (Gu et al. 2009; Noubissi et al. 2006), and on the other hand by the activation of its transcription by the Wnt/β-catenin pathway, which often occurs in tumor cells. To understand the role of ZBP1 in cancer, transgenic mice expressing ZBP1 in epithelial mammary cells have been generated. Since these mice spontaneously develop mammary tumors and facilitate the formation of metastases, ZBP1 has been considered as a proto-oncogene (Tessier et al. 2004). The implication of ZBP1 in tumorigenesis has thus been thought to involve the regulation of mRNAs encoding important proteins involved in cell proliferation, invasion, and metastasis (Liao et al. 2004; Stohr and Huttelmaier 2012; Stohr et al. 2012). For example, Vikesaa et al. have shown using

Hela cells that the expression of IMP proteins was crucial for the proper expression of some mRNAs encoding extracellular matrix and cell adhesion proteins. In this case, loss of IMP function affected invadopodia formation and cell spreading, consistent with an important role in cell migration (Vikesaa et al. 2006). Also, Stohr et al. have recently confirmed the involvement of IMP proteins in the promotion of directional movement of U2OS tumor cells by controlling the translation and the stability of mRNAs encoding important signaling proteins (Stohr et al. 2012). However, contradictory studies suggest that the involvement of ZBP1 in cancer is more complex. To date, Gu et al. have suggested that the variable activities of these proteins could be due to the variable transformed status of the broad diversity of tumor cell types studied (Gu et al. 2012). Indeed, Shestakova et al. were the first to find, in rat carcinoma cell lines, that *β-actin* mRNA was spontaneously mislocalized in metastatic versus non-metastatic cells (Shestakova et al. 1999). This mislocalization was associated with a change in the behavior of the cells from a polarized movement to a "random walk" phenotype often seen in invasive cells (Shestakova et al. 2001). In addition, other studies have shown that the ZBP1 RNA localization function is crucial for the establishment of cell-cell adhesion between the tumors cells, thus serving to prevent their invasive properties (Lapidus et al. 2007). By localizing different mRNAs such as *β-actin*, *E-cadherin*, *ARP2/3* complex, or *α-actinin*, ZBP1 may serve an important function in maintaining cell polarity in order to prevent uncontrolled cell migration (Gu et al. 2012; Shestakova et al. 2001; Lapidus et al. 2007).

7 Conclusions

As described in this review, mRNA localization control has emerged as a key layer of gene regulation that impacts a wide variety of biological process. The regulation of this process involves a complex interplay between diverse types of *cis*-acting elements, RNA-binding proteins, and accessory factors. These interactions can evolve dynamically in different cellular compartments and in response to signals from the environment, thus serving to impose regulatory coherence in the form of distinctive RNA trafficking mechanisms and pathways. Technical innovations in the fields of genomics and imaging should lead to important new insights into the mechanisms of RNA localization control in normal cells and their aberrant behavior in different human diseases.

References

Abe H, Kool ET (2006) Flow cytometric detection of specific RNAs in native human cells with quenched autoligating FRET probes. Proc Natl Acad Sci U S A 103(2):263–268

Adereth Y et al (2005) RNA-dependent integrin alpha3 protein localization regulated by the Muscleblind-like protein MLP1. Nat Cell Biol 7(12):1240–1247

Ainger K et al (1997) Transport and localization elements in myelin basic protein mRNA. J Cell Biol 138(5):1077–1087

Akten B et al (2011) Interaction of survival of motor neuron (SMN) and HuD proteins with mRNA cpg15 rescues motor neuron axonal deficits. Proc Natl Acad Sci U S A 108(25):10337–10342

Amrute-Nayak M, Bullock SL (2012) Single-molecule assays reveal that RNA localization signals regulate dynein-dynactin copy number on individual transcript cargoes. Nat Cell Biol 14(4):416–423

Antar LN, Dictenberg JB, Plociniak M, Afroz R, Bassell GJ (2005) Localization of FMRP-associated mRNA granules and requirement of microtubules for activity-dependent trafficking in hippocampal neurons. Genes Brain Behav 4(6):350–359

Ashley CT Jr, Wilkinson KD, Reines D, Warren ST (1993) FMR1 protein: conserved RNP family domains and selective RNA binding. Science 262(5133):563–566

Aumiller V, Graebsch A, Kremmer E, Niessing D, Forstemann K (2012) Drosophila Pur-alpha binds to trinucleotide-repeat containing cellular RNAs and translocates to the early oocyte. RNA Biol 9(5):633–643

Bagni C, Tassone F, Neri G, Hagerman R (2012) Fragile X syndrome: causes, diagnosis, mechanisms, and therapeutics. J Clin Invest 122(12):4314–4322

Bashirullah A et al (1999) Joint action of two RNA degradation pathways controls the timing of maternal transcript elimination at the midblastula transition in Drosophila melanogaster. EMBO J 18(9):2610–2620

Bashirullah A, Cooperstock RL, Lipshitz HD (2001) Spatial and temporal control of RNA stability. Proc Natl Acad Sci U S A 98(13):7025–7028

Bassell GJ et al (1998) Sorting of beta-actin mRNA and protein to neurites and growth cones in culture. J Neurosci 18(1):251–265

Batista PJ, Chang HY (2013) Cytotopic localization by long noncoding RNAs. Curr Opin Cell Biol 25(2):195–199

Beach DL, Salmon ED, Bloom K (1999) Localization and anchoring of mRNA in budding yeast. Curr Biol 9(11):569–578

Becalska AN et al (2011) Aubergine is a component of a nanos mRNA localization complex. Dev Biol 349(1):46–52

Bechade C et al (1999) Subcellular distribution of survival motor neuron (SMN) protein: possible involvement in nucleocytoplasmic and dendritic transport. Eur J Neurosci 11(1):293–304

Berleth T et al (1988) The role of localization of bicoid RNA in organizing the anterior pattern of the Drosophila embryo. EMBO J 7(6):1749–1756

Bertrand E et al (1998) Localization of ASH1 mRNA particles in living yeast. Mol Cell 2(4):437–445

Besse F, Lopez de Quinto S, Marchand V, Trucco A, Ephrussi A (2009) Drosophila PTB promotes formation of high-order RNP particles and represses oskar translation. Genes Dev 23(2): 195–207

Betley JN et al (2004) Kinesin II mediates Vg1 mRNA transport in Xenopus oocytes. Curr Biol 14(3):219–224

Bianco A, Dienstbier M, Salter HK, Gatto G, Bullock SL (2010) Bicaudal-D regulates fragile X mental retardation protein levels, motility, and function during neuronal morphogenesis. Curr Biol 20(16):1487–1492

Bilinski SM, Jaglarz MK, Szymanska B, Etkin LD, Kloc M (2004) Sm proteins, the constituents of the spliceosome, are components of nuage and mitochondrial cement in Xenopus oocytes. Exp Cell Res 299(1):171–178

Blichenberg A et al (2001) Identification of a cis-acting dendritic targeting element in the mRNA encoding the alpha subunit of Ca2+/calmodulin-dependent protein kinase II. Eur J Neurosci 13(10):1881–1888

Blower MD, Feric E, Weis K, Heald R (2007) Genome-wide analysis demonstrates conserved localization of messenger RNAs to mitotic microtubules. J Cell Biol 179(7):1365–1373

Bobola N, Jansen RP, Shin TH, Nasmyth K (1996) Asymmetric accumulation of Ash1p in postanaphase nuclei depends on a myosin and restricts yeast mating-type switching to mother cells. Cell 84(5):699–709

Bohl F, Kruse C, Frank A, Ferring D, Jansen RP (2000) She2p, a novel RNA-binding protein tethers ASH1 mRNA to the Myo4p myosin motor via She3p. EMBO J 19(20):5514–5524
Brook JD et al (1992) Molecular basis of myotonic dystrophy: expansion of a trinucleotide (CTG) repeat at the 3′ end of a transcript encoding a protein kinase family member. Cell 68(4): 799–808
Buckley PT et al (2011) Cytoplasmic intron sequence-retaining transcripts can be dendritically targeted via ID element retrotransposons. Neuron 69(5):877–884
Bullock SL, Ish-Horowicz D (2001) Conserved signals and machinery for RNA transport in Drosophila oogenesis and embryogenesis. Nature 414(6864):611–616
Burghes AH (1997) When is a deletion not a deletion? When it is converted. Am J Hum Genet 61(1):9–15
Burghes AH, Beattie CE (2009) Spinal muscular atrophy: why do low levels of survival motor neuron protein make motor neurons sick? Nat Rev Neurosci 10(8):597–609
Casolari JM et al (2012) Widespread mRNA association with cytoskeletal motor proteins and identification and dynamics of myosin-associated mRNAs in S. cerevisiae. PLoS One 7(2):e31912
Cha BJ, Koppetsch BS, Theurkauf WE (2001) In vivo analysis of Drosophila bicoid mRNA localization reveals a novel microtubule-dependent axis specification pathway. Cell 106(1):35–46
Chang JS et al (2001) Functioning of the Drosophila orb gene in gurken mRNA localization and translation. Development 128(16):3169–3177
Chang P et al (2004) Localization of RNAs to the mitochondrial cloud in Xenopus oocytes through entrapment and association with endoplasmic reticulum. Mol Biol Cell 15(10):4669–4681
Chao JA et al (2010) ZBP1 recognition of beta-actin zipcode induces RNA looping. Genes Dev 24(2):148–158
Chen J, Liu TW, Lo PC, Wilson BC, Zheng G (2009) "Zipper" molecular beacons: a generalized strategy to optimize the performance of activatable protease probes. Bioconjug Chem 20(10): 1836–1842
Chen Q, Jagannathan S, Reid DW, Zheng T, Nicchitta CV (2011) Hierarchical regulation of mRNA partitioning between the cytoplasm and the endoplasmic reticulum of mammalian cells. Mol Biol Cell 22(14):2646–2658
Chung S, Takizawa PA (2010) Multiple Myo4 motors enhance ASH1 mRNA transport in Saccharomyces cerevisiae. J Cell Biol 189(4):755–767
Clark A, Meignin C, Davis I (2007) A Dynein-dependent shortcut rapidly delivers axis determination transcripts into the Drosophila oocyte. Development 134(10):1955–1965
Colegrove-Otero LJ, Devaux A, Standart N (2005) The Xenopus ELAV protein ElrB represses Vg1 mRNA translation during oogenesis. Mol Cell Biol 25(20):9028–9039
Condeelis J, Singer RH (2005) How and why does beta-actin mRNA target? Biol Cell 97(1): 97–110
Coovert DD et al (1997) The survival motor neuron protein in spinal muscular atrophy. Hum Mol Genet 6(8):1205–1214
Crawford TO, Pardo CA (1996) The neurobiology of childhood spinal muscular atrophy. Neurobiol Dis 3(2):97–110
Crofts AJ et al (2004) Targeting of proteins to endoplasmic reticulum-derived compartments in plants. The importance of RNA localization. Plant Physiol 136(3):3414–3419
Daigle N, Ellenberg J (2007) LambdaN-GFP: an RNA reporter system for live-cell imaging. Nat Methods 4(8):633–636
Darnell RB (2011) HITS-CLIP: panoramic views of protein-RNA regulation in living cells (Translated from eng). Wiley Interdiscip Rev RNA 1(2):266–286 (in eng)
Delanoue R, Davis I (2005) Dynein anchors its mRNA cargo after apical transport in the Drosophila blastoderm embryo. Cell 122(1):97–106
Deng Y, Singer RH, Gu W (2008) Translation of ASH1 mRNA is repressed by Puf6p-Fun12p/ eIF5B interaction and released by CK2 phosphorylation. Genes Dev 22(8):1037–1050
Dennis PP, Omer A (2005) Small non-coding RNAs in Archaea. Curr Opin Microbiol 8(6): 685–694

Deshler JO, Highett MI, Schnapp BJ (1997) Localization of Xenopus Vg1 mRNA by Vera protein and the endoplasmic reticulum. Science 276(5315):1128–1131

Devys D, Lutz Y, Rouyer N, Bellocq JP, Mandel JL (1993) The FMR-1 protein is cytoplasmic, most abundant in neurons and appears normal in carriers of a fragile X premutation. Nat Genet 4(4):335–340

Dictenberg JB, Swanger SA, Antar LN, Singer RH, Bassell GJ (2008) A direct role for FMRP in activity-dependent dendritic mRNA transport links filopodial-spine morphogenesis to fragile X syndrome. Dev Cell 14(6):926–939

Dienstbier M, Li X (2009) Bicaudal-D and its role in cargo sorting by microtubule-based motors. Biochem Soc Trans 37(Pt 5):1066–1071

Dienstbier M, Boehl F, Li X, Bullock SL (2009) Egalitarian is a selective RNA-binding protein linking mRNA localization signals to the dynein motor. Genes Dev 23(13):1546–1558

Ding D, Parkhurst SM, Halsell SR, Lipshitz HD (1993) Dynamic Hsp83 RNA localization during Drosophila oogenesis and embryogenesis. Mol Cell Biol 13(6):3773–3781

Du TG et al (2008) Nuclear transit of the RNA-binding protein She2 is required for translational control of localized ASH1 mRNA. EMBO Rep 9(8):781–787

Dubowy J, Macdonald PM (1998) Localization of mRNAs to the oocyte is common in Drosophila ovaries. (Translated from eng). Mech Dev 70(1–2):193–195 (in eng)

Eberwine J, Miyashiro K, Kacharmina JE, Job C (2001) Local translation of classes of mRNAs that are targeted to neuronal dendrites. Proc Natl Acad Sci U S A 98(13):7080–7085

Eliscovich C, Buxbaum AR, Katz ZB, Singer RH (2013) mRNA on the move: the road to its biological destiny. J Biol Chem 288(28):20361–20368

Ephrussi A, Dickinson LK, Lehmann R (1991) Oskar organizes the germ plasm and directs localization of the posterior determinant nanos. Cell 66(1):37–50

Estrada P et al (2003) Myo4p and She3p are required for cortical ER inheritance in Saccharomyces cerevisiae. J Cell Biol 163(6):1255–1266

Fallini C, Bassell GJ, Rossoll W (2010) High-efficiency transfection of cultured primary motor neurons to study protein localization, trafficking, and function. Mol Neurodegener 5:17

Fallini C et al (2011) The survival of motor neuron (SMN) protein interacts with the mRNA-binding protein HuD and regulates localization of poly(A) mRNA in primary motor neuron axons. J Neurosci 31(10):3914–3925

Forrest KM, Gavis ER (2003) Live imaging of endogenous RNA reveals a diffusion and entrapment mechanism for nanos mRNA localization in Drosophila. Curr Biol 13(14):1159–1168

Fusco D et al (2003) Single mRNA molecules demonstrate probabilistic movement in living mammalian cells. Curr Biol 13(2):161–167

Gabanella F et al (2007) Ribonucleoprotein assembly defects correlate with spinal muscular atrophy severity and preferentially affect a subset of spliceosomal snRNPs. PLoS One 2(9):e921

Gagnon JA, Mowry KL (2011) Molecular motors: directing traffic during RNA localization. Crit Rev Biochem Mol Biol 46(3):229–239

Galka-Marciniak P, Urbanek MO, Krzyzosiak WJ (2012) Triplet repeats in transcripts: structural insights into RNA toxicity. Biol Chem 393(11):1299–1315

Garcia M, Darzacq X, Devaux F, Singer RH, Jacq C (2007) Yeast mitochondrial transcriptomics. Methods Mol Biol 372:505–528

Gaspar I (2011) Microtubule-based motor-mediated mRNA localization in Drosophila oocytes and embryos. Biochem Soc Trans 39(5):1197–1201

Gerber AP, Herschlag D, Brown PO (2004) Extensive association of functionally and cytotopically related mRNAs with Puf family RNA-binding proteins in yeast. (Translated from eng). PLoS Biol 2(3):E79

Ghosh S, Marchand V, Gaspar I, Ephrussi A (2012) Control of RNP motility and localization by a splicing-dependent structure in oskar mRNA. Nat Struct Mol Biol 19(4):441–449

Gibson TJ, Rice PM, Thompson JD, Heringa J (1993) KH domains within the FMR1 sequence suggest that fragile X syndrome stems from a defect in RNA metabolism. Trends Biochem Sci 18(9):331–333

Glotzer JB, Saffrich R, Glotzer M, Ephrussi A (1997) Cytoplasmic flows localize injected oskar RNA in Drosophila oocytes. Curr Biol 7(5):326–337

Greco CM et al (2006) Neuropathology of fragile X-associated tremor/ataxia syndrome (FXTAS). Brain 129(Pt 1):243–255

Gu W, Pan F, Zhang H, Bassell GJ, Singer RH (2002) A predominantly nuclear protein affecting cytoplasmic localization of beta-actin mRNA in fibroblasts and neurons. J Cell Biol 156(1):41–51

Gu W, Deng Y, Zenklusen D, Singer RH (2004) A new yeast PUF family protein, Puf6p, represses ASH1 mRNA translation and is required for its localization. Genes Dev 18(12):1452–1465

Gu W, Pan F, Singer RH (2009) Blocking beta-catenin binding to the ZBP1 promoter represses ZBP1 expression, leading to increased proliferation and migration of metastatic breast-cancer cells. J Cell Sci 122(Pt 11):1895–1905

Gu W et al (2012) Regulation of local expression of cell adhesion and motility-related mRNAs in breast cancer cells by IMP1/ZBP1. J Cell Sci 125(Pt 1):81–91

Gumy LF et al (2011) Transcriptome analysis of embryonic and adult sensory axons reveals changes in mRNA repertoire localization. RNA 17(1):85–98

Hachet O, Ephrussi A (2004) Splicing of oskar RNA in the nucleus is coupled to its cytoplasmic localization. Nature 428(6986):959–963

Haim L, Zipor G, Aronov S, Gerst JE (2007) A genomic integration method to visualize localization of endogenous mRNAs in living yeast. Nat Methods 4(5):409–412

Hamilton RS, Davis I (2011) Identifying and searching for conserved RNA localisation signals. Methods Mol Biol 714:447–466

Henry JJ, Perry KJ, Fukui L, Alvi N (2010) Differential localization of mRNAs during early development in the mollusc, Crepidula fornicata. Integr comp Biol 50(5):720–733

Heym RG, Niessing D (2012) Principles of mRNA transport in yeast. Cell Mol Life Sci 69(11):1843–1853

Hillebrand J et al (2010) The Me31B DEAD-box helicase localizes to postsynaptic foci and regulates expression of a CaMKII reporter mRNA in dendrites of Drosophila olfactory projection neurons. Front Neural Circ 4:121

Hinds HL et al (1993) Tissue specific expression of FMR-1 provides evidence for a functional role in fragile X syndrome. Nat Genet 3(1):36–43

Hinton VJ, Brown WT, Wisniewski K, Rudelli RD (1991) Analysis of neocortex in three males with the fragile X syndrome. Am J Med Genet 41(3):289–294

Hoek KS et al (1998) hnRNP A2 selectively binds the cytoplasmic transport sequence of myelin basic protein mRNA. Biochemistry 37(19):7021–7029

Huang YS, Carson JH, Barbarese E, Richter JD (2003) Facilitation of dendritic mRNA transport by CPEB. Genes Dev 17(5):638–653

Hubers L et al (2011) HuD interacts with survival motor neuron protein and can rescue spinal muscular atrophy-like neuronal defects. Hum Mol Genet 20(3):553–579

Huttelmaier S et al (2005) Spatial regulation of beta-actin translation by Src-dependent phosphorylation of ZBP1. Nature 438(7067):512–515

Irie K et al (2002) The Khd1 protein, which has three KH RNA-binding motifs, is required for proper localization of ASH1 mRNA in yeast. EMBO J 21(5):1158–1167

Irion U et al (2006) Miranda couples oskar mRNA/Staufen complexes to the bicoid mRNA localization pathway. Dev Biol 297(2):522–533

Jagannathan S, Nwosu C, Nicchitta CV (2011) Analyzing mRNA localization to the endoplasmic reticulum via cell fractionation. Methods Mol Biol 714:301–321

Jain RA, Gavis ER (2008) The Drosophila hnRNP M homolog Rumpelstiltskin regulates nanos mRNA localization. Development 135(5):973–982

Jambhekar A, Derisi JL (2007) Cis-acting determinants of asymmetric, cytoplasmic RNA transport. RNA 13(5):625–642

Jansen RP, Niessing D (2012) Assembly of mRNA-protein complexes for directional mRNA transport in eukaryotes—an overview. Curr Protein Pept Sci 13(4):284–293

Jansen RP, Dowzer C, Michaelis C, Galova M, Nasmyth K (1996) Mother cell-specific HO expression in budding yeast depends on the unconventional myosin myo4p and other cytoplasmic proteins. Cell 84(5):687–697
Jeffery WR (1988) The role of cytoplasmic determinants in embryonic development. Dev Biol (N Y 1985) 5:3–56
Jeffery WR, Tomlinson CR, Brodeur RD (1983) Localization of actin messenger RNA during early ascidian development. Dev Biol 99(2):408–417
Jenny A et al (2006) A translation-independent role of oskar RNA in early Drosophila oogenesis. Development 133(15):2827–2833
Jensen KB, Darnell RB (2008) CLIP: crosslinking and immunoprecipitation of in vivo RNA targets of RNA-binding proteins. (translated from eng). Methods Mol Biol 488:85–98
Jiang H, Mankodi A, Swanson MS, Moxley RT, Thornton CA (2004) Myotonic dystrophy type 1 is associated with nuclear foci of mutant RNA, sequestration of muscleblind proteins and deregulated alternative splicing in neurons. Hum Mol Genet 13(24):3079–3088
Jin P et al (2007) Pur alpha binds to rCGG repeats and modulates repeat-mediated neurodegeneration in a Drosophila model of fragile X tremor/ataxia syndrome. Neuron 55(4):556–564
Johnson EM et al (2006) Role of Pur alpha in targeting mRNA to sites of translation in hippocampal neuronal dendrites. J Neurosci Res 83(6):929–943
Jung H, Yoon BC, Holt CE (2012) Axonal mRNA localization and local protein synthesis in nervous system assembly, maintenance and repair. Nat Rev Neurosci 13(5):308–324
Kanai Y et al (2004) Kinesin transports RNA: isolation and characterization of an RNA-transporting granule. Neuron 43(4):513–525
Kanadia RN et al (2003) A muscleblind knockout model for myotonic dystrophy. Science 302(5652):1978–1980
Kazan H, Morris Q (2013) RBPmotif: a web server for the discovery of sequence and structure preferences of RNA-binding proteins. Nucleic Acids Res 41:W180–186
Keiler KC (2011) RNA localization in bacteria. Curr Opin Microbiol 14(2):155–159
Kim-Ha J, Smith JL, Macdonald PM (1991) oskar mRNA is localized to the posterior pole of the Drosophila oocyte. Cell 66(1):23–35
Kim-Ha J, Kerr K, Macdonald PM (1995) Translational regulation of oskar mRNA by bruno, an ovarian RNA-binding protein, is essential. Cell 81(3):403–412
King ML, Messitt TJ, Mowry KL (2005) Putting RNAs in the right place at the right time: RNA localization in the frog oocyte. Biol Cell 97(1):19–33
Kingsley EP, Chan XY, Duan Y, Lambert JD (2007) Widespread RNA segregation in a spiralian embryo. Evol Dev 9(6):527–539
Kloc M, Etkin LD (1995) Two distinct pathways for the localization of RNAs at the vegetal cortex in Xenopus oocytes. Development 121(2):287–297
Kloc M et al (2002) Three-dimensional ultrastructural analysis of RNA distribution within germinal granules of Xenopus. Dev Biol 241(1):79–93
Kohler O, Jarikote DV, Seitz O (2005) Forced intercalation probes (FIT Probes): thiazole orange as a fluorescent base in peptide nucleic acids for homogeneous single-nucleotide-polymorphism detection. Chembiochem 6(1):69–77
Kolev NG, Huber PW (2003) VgRBP71 stimulates cleavage at a polyadenylation signal in Vg1 mRNA, resulting in the removal of a cis-acting element that represses translation. Mol Cell 11(3):745–755
Kopczynski CC et al (1998) A high throughput screen to identify secreted and transmembrane proteins involved in Drosophila embryogenesis. Proc Natl Acad Sci U S A 95(17):9973–9978
Krauss J, Lopez de Quinto S, Nusslein-Volhard C, Ephrussi A (2009) Myosin-V regulates oskar mRNA localization in the Drosophila oocyte. Curr Biol 19(12):1058–1063
Krementsova EB et al (2011) Two single-headed myosin V motors bound to a tetrameric adapter protein form a processive complex. J Cell Biol 195(4):631–641
Kroll TT et al (2009) Interactions of 40LoVe within the ribonucleoprotein complex that forms on the localization element of Xenopus Vg1 mRNA. Mech Dev 126(7):523–538

Kugler JM, Lasko P (2009) Localization, anchoring and translational control of oskar, gurken, bicoid and nanos mRNA during Drosophila oogenesis. Fly 3(1):15–28

Kummer S et al (2011) Fluorescence imaging of influenza H1N1 mRNA in living infected cells using single-chromophore FIT-PNA. Angew Chem 50(8):1931–1934

La Via L et al (2013) Modulation of dendritic AMPA receptor mRNA trafficking by RNA splicing and editing. Nucleic Acids Res 41(1):617–631

Lange S et al (2008) Simultaneous transport of different localized mRNA species revealed by live-cell imaging. Traffic 9(8):1256–1267

Lapidus K et al (2007) ZBP1 enhances cell polarity and reduces chemotaxis. J Cell Sci 120(Pt 18):3173–3178

Lasko P (2012) mRNA localization and translational control in Drosophila oogenesis. Cold Spring Harb Perspect Biol 4(10)

Latham VM, Yu EH, Tullio AN, Adelstein RS, Singer RH (2001) A Rho-dependent signaling pathway operating through myosin localizes beta-actin mRNA in fibroblasts. Curr Biol 11(13):1010–1016

Lecuyer E et al (2007) Global analysis of mRNA localization reveals a prominent role in organizing cellular architecture and function. Cell 131(1):174–187

Lecuyer E, Yoshida H, Krause HM (2009) Global implications of mRNA localization pathways in cellular organization. Curr Opin Cell Biol 21(3):409–415

Leehey MA (2009) Fragile X-associated tremor/ataxia syndrome: clinical phenotype, diagnosis, and treatment. J Investig Med 57(8):830–836

Lefebvre S et al (1997) Correlation between severity and SMN protein level in spinal muscular atrophy. Nat Genet 16(3):265–269

Lein ES et al (2007) Genome-wide atlas of gene expression in the adult mouse brain. Nature 445(7124):168–176

Lerner RS et al (2003) Partitioning and translation of mRNAs encoding soluble proteins on membrane-bound ribosomes. RNA 9(9):1123–1137

Li Y, Jin P (2012) RNA-mediated neurodegeneration in fragile X-associated tremor/ataxia syndrome (translated from eng). Brain Res 1462:112–117

Liao B et al (2004) Targeted knockdown of the RNA-binding protein CRD-BP promotes cell proliferation via an insulin-like growth factor II-dependent pathway in human K562 leukemia cells. J Biol Chem 279(47):48716–48724

Licatalosi DD et al (2008) HITS-CLIP yields genome-wide insights into brain alternative RNA processing (translated from eng). Nature 456(7221):464–469 (in eng)

Lifland AW, Zurla C, Yu J, Santangelo PJ (2011) Dynamics of native beta-actin mRNA transport in the cytoplasm. Traffic 12(8):1000–1011

Lionnet T et al (2011) A transgenic mouse for in vivo detection of endogenous labeled mRNA. Nat Methods 8(2):165–170

Lipshitz HD, Smibert CA (2000) Mechanisms of RNA localization and translational regulation. Curr Opin Genet Dev 10(5):476–488

Liu G et al (2002) Interactions of elongation factor 1 alpha with F-actin and beta-actin mRNA: implications for anchoring mRNA in cell protrusions. Mol Biol Cell 13(2):579–592

Long RM et al (1997) Mating type switching in yeast controlled by asymmetric localization of ASH1 mRNA. Science 277(5324):383–387

Long RM, Gu W, Lorimer E, Singer RH, Chartrand P (2000) She2p is a novel RNA-binding protein that recruits the Myo4p-She3p complex to ASH1 mRNA. EMBO J 19(23):6592–6601

Long RM et al (2001) An exclusively nuclear RNA-binding protein affects asymmetric localization of ASH1 mRNA and Ash1p in yeast. J Cell Biol 153(2):307–318

Lu G, Dolgner SJ, Hall TM (2009) Understanding and engineering RNA sequence specificity of PUF proteins. Curr Opin Struct Biol 19(1):110–115

Lunde BM, Moore C, Varani G (2007) RNA-binding proteins: modular design for efficient function. Nat Rev Mol Cell Biol 8(6):479–490

Ma B et al (2011) Huntingtin mediates dendritic transport of beta-actin mRNA in rat neurons. Sci Rep 1:140

Macchi P et al (2003) Barentsz, a new component of the Staufen-containing ribonucleoprotein particles in mammalian cells, interacts with Staufen in an RNA-dependent manner. J Neurosci 23(13):5778–5788
Makabe KW et al (2001) Large-scale cDNA analysis of the maternal genetic information in the egg of Halocynthia roretzi for a gene expression catalog of ascidian development (translated from eng). Development 128(13):2555–2567 (in eng)
Marc P et al (2002) Genome-wide analysis of mRNAs targeted to yeast mitochondria. EMBO Rep 3(2):159–164
Medioni C, Mowry K, Besse F (2012) Principles and roles of mRNA localization in animal development. Development 139(18):3263–3276
Meer EJ et al (2012) Identification of a cis-acting element that localizes mRNA to synapses. Proc Natl Acad Sci U S A 109(12):4639–4644
Mercer TR, Dinger ME, Sunkin SM, Mehler MF, Mattick JS (2008) Specific expression of long noncoding RNAs in the mouse brain. Proc Natl Acad Sci U S A 105(2):716–721
Messitt TJ et al (2008) Multiple kinesin motors coordinate cytoplasmic RNA transport on a subpopulation of microtubules in Xenopus oocytes. Dev Cell 15(3):426–436
Mhlanga MM et al (2009) In vivo colocalisation of oskar mRNA and trans-acting proteins revealed by quantitative imaging of the Drosophila oocyte. PLoS One 4(7):e6241
Micklem DR (2000) Distinct roles of two conserved Staufen domains in oskar mRNA localization and translation. EMBO J 19(6):1366–1377
Mili S, Moissoglu K, Macara IG (2008) Genome-wide screen reveals APC-associated RNAs enriched in cell protrusions. Nature 453(7191):115–119
Miller S et al (2002) Disruption of dendritic translation of CaMKIIalpha impairs stabilization of synaptic plasticity and memory consolidation. Neuron 36(3):507–519
Mingle LA et al (2005) Localization of all seven messenger RNAs for the actin-polymerization nucleator Arp2/3 complex in the protrusions of fibroblasts. J Cell Sci 118(Pt 11):2425–2433
Mohr SE et al (2001) The RNA-binding protein Tsunagi interacts with Mago Nashi to establish polarity and localize oskar mRNA during Drosophila oogenesis. Genes Dev 15(21):2886–2899
Mori Y, Imaizumi K, Katayama T, Yoneda T, Tohyama M (2000) Two cis-acting elements in the 3′ untranslated region of alpha-CaMKII regulate its dendritic targeting. Nat Neurosci 3(11):1079–1084
Mowry KL, Melton DA (1992) Vegetal messenger RNA localization directed by a 340-nt RNA sequence element in Xenopus oocytes. Science 255(5047):991–994
Muller M et al (2011) A cytoplasmic complex mediates specific mRNA recognition and localization in yeast. PLoS Biol 9(4):e1000611
Munro TP et al (1999) Mutational analysis of a heterogeneous nuclear ribonucleoprotein A2 response element for RNA trafficking. J Biol Chem 274(48):34389–34395
Muslimov IA, Iacoangeli A, Brosius J, Tiedge H (2006) Spatial codes in dendritic BC1 RNA. J Cell Biol 175(3):427–439
Muslimov IA, Patel MV, Rose A, Tiedge H (2011) Spatial code recognition in neuronal RNA targeting: role of RNA-hnRNP A2 interactions. J Cell Biol 194(3):441–457
Nagaoka K, Udagawa T, Richter JD (2012) CPEB-mediated ZO-1 mRNA localization is required for epithelial tight-junction assembly and cell polarity. Nat Commun 3:675
Nakamura A et al (2001) Me31B silences translation of oocyte-localizing RNAs through the formation of cytoplasmic RNP complex during Drosophila oogenesis. Development 128(17):3233–3242
Nitin N, Santangelo PJ, Kim G, Nie S, Bao G (2004) Peptide-linked molecular beacons for efficient delivery and rapid mRNA detection in living cells. Nucleic Acids Res 32(6):e58
Norvell A et al (2005) Squid is required for efficient posterior localization of oskar mRNA during Drosophila oogenesis. Dev Genes Evol 215(7):340–349
Noubissi FK et al (2006) CRD-BP mediates stabilization of betaTrCP1 and c-myc mRNA in response to beta-catenin signalling. Nature 441(7095):898–901

Oleynikov Y, Singer RH (2003) Real-time visualization of ZBP1 association with beta-actin mRNA during transcription and localization. Curr Biol 13(3):199–207

Orr HT, Zoghbi HY (2007) Trinucleotide repeat disorders. Annu Rev Neurosci 30:575–621

Otero LJ, Devaux A, Standart N (2001) A 250-nucleotide UA-rich element in the 3′ untranslated region of Xenopus laevis Vg1 mRNA represses translation both in vivo and in vitro. RNA 7(12):1753–1767

Ozawa T, Natori Y, Sato M, Umezawa Y (2007) Imaging dynamics of endogenous mitochondrial RNA in single living cells. Nat Methods 4(5):413–419

Pagliardini S et al (2000) Subcellular localization and axonal transport of the survival motor neuron (SMN) protein in the developing rat spinal cord. Hum Mol Genet 9(1):47–56

Paige JS, Nguyen-Duc T, Song W, Jaffrey SR (2012) Fluorescence imaging of cellular metabolites with RNA. Science 335(6073):1194

Pan F, Huttelmaier S, Singer RH, Gu W (2007) ZBP2 facilitates binding of ZBP1 to beta-actin mRNA during transcription. Mol Cell Biol 27(23):8340–8351

Paquin N, Chartrand P (2008) Local regulation of mRNA translation: new insights from the bud. Trends Cell Biol 18(3):105–111

Paquin N et al (2007) Local activation of yeast ASH1 mRNA translation through phosphorylation of Khd1p by the casein kinase Yck1p. Mol Cell 26(6):795–809

Patel VL et al (2012) Spatial arrangement of an RNA zipcode identifies mRNAs under post-transcriptional control. Genes Dev 26(1):43–53

Piazzon N et al (2008) In vitro and in cellulo evidences for association of the survival of motor neuron complex with the fragile X mental retardation protein. J Biol Chem 283(9):5598–5610

Portera-Cailliau C (2012) Which comes first in fragile X syndrome, dendritic spine dysgenesis or defects in circuit plasticity? Neuroscientist 18(1):28–44

Prodon F, Yamada L, Shirae-Kurabayashi M, Nakamura Y, Sasakura Y (2007) Postplasmic/PEM RNAs: a class of localized maternal mRNAs with multiple roles in cell polarity and development in ascidian embryos. Dev Dyn 236(7):1698–1715

Rackham O, Brown CM (2004) Visualization of RNA-protein interactions in living cells: FMRP and IMP1 interact on mRNAs. EMBO J 23(16):3346–3355

Raju CS et al (2008) In cultured oligodendrocytes the A/B-type hnRNP CBF-A accompanies MBP mRNA bound to mRNA trafficking sequences. Mol Biol Cell 19(7):3008–3019

Rehbein M et al (2002) Molecular characterization of MARTA1, a protein interacting with the dendritic targeting element of MAP2 mRNAs. J Neurochem 82(5):1039–1046

Ross AF, Oleynikov Y, Kislauskis EH, Taneja KL, Singer RH (1997) Characterization of a beta-actin mRNA zipcode-binding protein. Mol Cell Biol 17(4):2158–2165

Ross J, Lemm I, Berberet B (2001) Overexpression of an mRNA-binding protein in human colorectal cancer. Oncogene 20(45):6544–6550

Rossoll W et al (2002) Specific interaction of Smn, the spinal muscular atrophy determining gene product, with hnRNP-R and gry-rbp/hnRNP-Q: a role for Smn in RNA processing in motor axons? Hum Mol Genet 11(1):93–105

Rossoll W et al (2003) Smn, the spinal muscular atrophy-determining gene product, modulates axon growth and localization of beta-actin mRNA in growth cones of motoneurons. J Cell Biol 163(4):801–812

Rudelli RD et al (1985) Adult fragile X syndrome. Clinico-neuropathologic findings. Acta Neuropathol 67(3–4):289–295

Saint-Georges Y et al (2008) Yeast mitochondrial biogenesis: a role for the PUF RNA-binding protein Puf3p in mRNA localization. PLoS One 3(6):e2293

Sanchez G et al (2013) A novel function for the survival motoneuron protein as a translational regulator. Hum Mol Genet 22(4):668–684

Santangelo PJ, Nix B, Tsourkas A, Bao G (2004) Dual FRET molecular beacons for mRNA detection in living cells. Nucleic Acids Res 32(6):e57

Santangelo PJ et al (2009) Single molecule-sensitive probes for imaging RNA in live cells. Nat Methods 6(5):347–349

Sasaki Y et al (2010) Phosphorylation of zipcode binding protein 1 is required for brain-derived neurotrophic factor signaling of local beta-actin synthesis and growth cone turning. J Neurosci 30(28):9349–9358
Schrank B et al (1997) Inactivation of the survival motor neuron gene, a candidate gene for human spinal muscular atrophy, leads to massive cell death in early mouse embryos. Proc Natl Acad Sci U S A 94(18):9920–9925
Schuldt AJ et al (1998) Miranda mediates asymmetric protein and RNA localization in the developing nervous system. Genes Dev 12(12):1847–1857
Semotok JL et al (2005) Smaug recruits the CCR4/POP2/NOT deadenylase complex to trigger maternal transcript localization in the early Drosophila embryo. Curr Biol 15(4):284–294
Semotok JL et al (2008) Drosophila maternal Hsp83 mRNA destabilization is directed by multiple SMAUG recognition elements in the open reading frame. Mol Cell Biol 28(22):6757–6772
Serano J, Rubin GM (2003) The Drosophila synaptotagmin-like protein bitesize is required for growth and has mRNA localization sequences within its open reading frame. Proc Natl Acad Sci U S A 100(23):13368–13373
Serikawa KA, Porterfield DM, Mandoli DF (2001) Asymmetric subcellular mRNA distribution correlates with carbonic anhydrase activity in Acetabularia acetabulum. Plant Physiol 125(2): 900–911
Sharp JA, Plant JJ, Ohsumi TK, Borowsky M, Blower MD (2011) Functional analysis of the microtubule-interacting transcriptome. Mol Biol Cell 22(22):4312–4323
Shen Z, Paquin N, Forget A, Chartrand P (2009) Nuclear shuttling of She2p couples ASH1 mRNA localization to its translational repression by recruiting Loc1p and Puf6p. Mol Biol Cell 20(8):2265–2275
Shestakova EA, Wyckoff J, Jones J, Singer RH, Condeelis J (1999) Correlation of beta-actin messenger RNA localization with metastatic potential in rat adenocarcinoma cell lines. Cancer Res 59(6):1202–1205
Shestakova EA, Singer RH, Condeelis J (2001) The physiological significance of beta -actin mRNA localization in determining cell polarity and directional motility. Proc Natl Acad Sci U S A 98(13):7045–7050
Sidorov MS, Auerbach BD, Bear MF (2013) Fragile X mental retardation protein and synaptic plasticity. Mol Brain 6:15
Sil A, Herskowitz I (1996) Identification of asymmetrically localized determinant, Ash1p, required for lineage-specific transcription of the yeast HO gene. Cell 84(5):711–722
Siomi H, Siomi MC, Nussbaum RL, Dreyfuss G (1993) The protein product of the fragile X gene, FMR1, has characteristics of an RNA-binding protein. Cell 74(2):291–298
Sofola OA et al (2007) RNA-binding proteins hnRNP A2/B1 and CUGBP1 suppress fragile X CGG premutation repeat-induced neurodegeneration in a Drosophila model of FXTAS. Neuron 55(4):565–571
Stephenson EC (2004) Localization of swallow-green fluorescent protein in Drosophila oogenesis and implications for the role of swallow in RNA. Genesis 39(4):280–287
Stohr N, Huttelmaier S (2012) IGF2BP1: a post-transcriptional "driver" of tumor cell migration. Cell Adh Migr 6(4):312–318
Stohr N et al (2012) IGF2BP1 promotes cell migration by regulating MK5 and PTEN signaling. Genes Dev 26(2):176–189
Stylli SS, Kaye AH, Lock P (2008) Invadopodia: at the cutting edge of tumour invasion. J Clin Neurosci 15(7):725–737
Subramanian M et al (2011) G-quadruplex RNA structure as a signal for neurite mRNA targeting. EMBO Rep 12(7):697–704
Sundell CL, Singer RH (1991) Requirement of microfilaments in sorting of actin messenger RNA. Science 253(5025):1275–1277
Tadesse H, Deschenes-Furry J, Boisvenue S, Cote J (2008) KH-type splicing regulatory protein interacts with survival motor neuron protein and is misregulated in spinal muscular atrophy. Hum Mol Genet 17(4):506–524

Tadros W et al (2003) Regulation of maternal transcript destabilization during egg activation in Drosophila. Genetics 164(3):989–1001
Takizawa PA, Sil A, Swedlow JR, Herskowitz I, Vale RD (1997) Actin-dependent localization of an RNA encoding a cell-fate determinant in yeast. Nature 389(6646):90–93
Takizawa PA, DeRisi JL, Wilhelm JE, Vale RD (2000) Plasma membrane compartmentalization in yeast by messenger RNA transport and a septin diffusion barrier. Science 290(5490):341–344
Tassone F, Iwahashi C, Hagerman PJ (2004) FMR1 RNA within the intranuclear inclusions of fragile X-associated tremor/ataxia syndrome (FXTAS). RNA Biol 1(2):103–105
Taylor AM et al (2009) Axonal mRNA in uninjured and regenerating cortical mammalian axons. J Neurosci 29(15):4697–4707
Tenenbaum SA, Lager PJ, Carson CC, Keene JD (2002) Ribonomics: identifying mRNA subsets in mRNP complexes using antibodies to RNA-binding proteins and genomic arrays (translated from eng). Methods 26(2):191–198 (in eng)
Tessier CR, Doyle GA, Clark BA, Pitot HC, Ross J (2004) Mammary tumor induction in transgenic mice expressing an RNA-binding protein. Cancer Res 64(1):209–214
Tiruchinapalli DM et al (2008) Activity-dependent expression of RNA binding protein HuD and its association with mRNAs in neurons. RNA Biol 5(3):157–168
Tomancak P et al (2002) Systematic determination of patterns of gene expression during Drosophila embryogenesis (translated from eng). Genome Biol 3(12):RESEARCH0088 (in eng)
Tse JC, Kalluri R (2007) Mechanisms of metastasis: epithelial-to-mesenchymal transition and contribution of tumor microenvironment. J Cell Biochem 101(4):816–829
Tyagi S, Kramer FR (1996) Molecular beacons: probes that fluoresce upon hybridization. Nat Biotechnol 14(3):303–308
Udd B, Krahe R (2012) The myotonic dystrophies: molecular, clinical, and therapeutic challenges. Lancet Neurol 11(10):891–905
Ule J, Jensen K, Mele A, Darnell RB (2005) CLIP: a method for identifying protein-RNA interaction sites in living cells (translated from eng). Methods 37(4):376–386 (in eng)
Vaccaro MC et al (2010) A transient asymmetric distribution of XNOA 36 mRNA and the associated spectrin network bisects Xenopus laevis stage I oocytes along the future A/V axis. Eur J Cell Biol 89(7):525–536
Vaccaro MC, Wilding M, Dale B, Campanella C, Carotenuto R (2012) Expression of XNOA 36 in the mitochondrial cloud of Xenopus laevis oocytes. Zygote 20(3):237–242
Van de Bor V, Davis I (2004) mRNA localisation gets more complex. Curr Opin Cell Biol 16(3):300–307
van Eeden FJ, Palacios IM, Petronczki M, Weston MJ, St Johnston D (2001) Barentsz is essential for the posterior localization of oskar mRNA and colocalizes with it to the posterior pole. J Cell Biol 154(3):511–523
Vendra G, Hamilton RS, Davis I (2007) Dynactin suppresses the retrograde movement of apically localized mRNA in Drosophila blastoderm embryos. RNA 13(11):1860–1867
Vikesaa J et al (2006) RNA-binding IMPs promote cell adhesion and invadopodia formation. EMBO J 25(7):1456–1468
Wan L et al (2005) The survival of motor neurons protein determines the capacity for snRNP assembly: biochemical deficiency in spinal muscular atrophy. Mol Cell Biol 25(13):5543–5551
Wang W et al (2004) Identification and testing of a gene expression signature of invasive carcinoma cells within primary mammary tumors. Cancer Res 64(23):8585–8594
Wang ET et al (2012) Transcriptome-wide regulation of pre-mRNA splicing and mRNA localization by muscleblind proteins. Cell 150(4):710–724
Wilhelm JE, Vale RD, Hegde RS (2000a) Coordinate control of translation and localization of Vg1 mRNA in Xenopus oocytes. Proc Natl Acad Sci U S A 97(24):13132–13137
Wilhelm JE et al (2000b) Isolation of a ribonucleoprotein complex involved in mRNA localization in Drosophila oocytes. J Cell Biol 148(3):427–440
Wilhelm JE, Hilton M, Amos Q, Henzel WJ (2003) Cup is an eIF4E binding protein required for both the translational repression of oskar and the recruitment of Barentsz. J Cell Biol 163(6):1197–1204

Willis DE, Twiss JL (2011) Profiling axonal mRNA transport. Methods Mol Biol 714:335–352
Wu L et al (1998) CPEB-mediated cytoplasmic polyadenylation and the regulation of experience-dependent translation of alpha-CaMKII mRNA at synapses. Neuron 21(5):1129–1139
Yamada T, Yoshimura H, Inaguma A, Ozawa T (2011) Visualization of nonengineered single mRNAs in living cells using genetically encoded fluorescent probes. Anal Chem 83(14): 5708–5714
Yamagishi M, Shirasaki Y, Funatsu T (2009a) Size-dependent accumulation of mRNA at the leading edge of chicken embryo fibroblasts. Biochem Biophys Res Commun 390(3):750–754
Yamagishi M, Ishihama Y, Shirasaki Y, Kurama H, Funatsu T (2009b) Single-molecule imaging of beta-actin mRNAs in the cytoplasm of a living cell. Exp Cell Res 315(7):1142–1147
Yano T et al (2004) Hrp48, a Drosophila hnRNPA/B homolog, binds and regulates translation of oskar mRNA. Dev Cell 6(5):637–648
Yantiss RK et al (2005) KOC (K homology domain containing protein overexpressed in cancer): a novel molecular marker that distinguishes between benign and malignant lesions of the pancreas. Am J Surg Pathol 29(2):188–195
Yisraeli JK (2005) VICKZ proteins: a multi-talented family of regulatory RNA-binding proteins. Biol Cell 97(1):87–96
Yisraeli JK, Melton DA (1988) The material mRNA Vg1 is correctly localized following injection into Xenopus oocytes. Nature 336(6199):592–595
Yoon YJ, Mowry KL (2004) Xenopus Staufen is a component of a ribonucleoprotein complex containing Vg1 RNA and kinesin. Development 131(13):3035–3045
Yoshimura H, Inaguma A, Yamada T, Ozawa T (2012) Fluorescent probes for imaging endogenous beta-actin mRNA in living cells using fluorescent protein-tagged pumilio. ACS Chem Biol 7(6):999–1005
Zaessinger S, Busseau I, Simonelig M (2006) Oskar allows nanos mRNA translation in Drosophila embryos by preventing its deadenylation by Smaug/CCR4. Development 133(22):4573–4583
Zarnack K, Feldbrugge M (2007) mRNA trafficking in fungi. Mol Genet Genomics 278(4):347–359
Zarnack K, Feldbrugge M (2010) Microtubule-dependent mRNA transport in fungi. Eukaryot Cell 9(7):982–990
Zhang HL et al (2001) Neurotrophin-induced transport of a beta-actin mRNP complex increases beta-actin levels and stimulates growth cone motility. Neuron 31(2):261–275
Zhang HL et al (2003) Active transport of the survival motor neuron protein and the role of exon-7 in cytoplasmic localization. J Neurosci 23(16):6627–6637
Zhang Z et al (2008) SMN deficiency causes tissue-specific perturbations in the repertoire of snRNAs and widespread defects in splicing. Cell 133(4):585–600
Zimyanin VL et al (2008) In vivo imaging of oskar mRNA transport reveals the mechanism of posterior localization. Cell 134(5):843–853
Zivraj KH et al (2010) Subcellular profiling reveals distinct and developmentally regulated repertoire of growth cone mRNAs. J Neurosci 30(46):15464–15478
Zivraj KH et al (2013) The RNA-binding protein MARTA2 regulates dendritic targeting of MAP2 mRNAs in rat neurons. J Neurochem 124(5):670–684

Chapter 3
RNA-Binding Proteins in Regulation of Alternative Cleavage and Polyadenylation

Dinghai Zheng and Bin Tian

Abstract Almost all eukaryotic pre-mRNAs are processed at the 3′ end by the cleavage and polyadenylation (C/P) reaction, which preludes termination of transcription and gives rise to the poly(A) tail of mature mRNA. Genomic studies in recent years have indicated that most eukaryotic mRNA genes have multiple cleavage and polyadenylation sites (pAs), leading to alternative cleavage and polyadenylation (APA) products. APA isoforms generally differ in their 3′ untranslated regions (3′ UTRs), but can also have different coding sequences (CDSs). APA expands the repertoire of transcripts expressed from the genome, and is highly regulated under various physiological and pathological conditions. Growing lines of evidence have shown that RNA-binding proteins (RBPs) play important roles in regulation of APA. Some RBPs are part of the machinery for C/P; others influence pA choice through binding to adjacent regions. In this chapter, we review *cis* elements and *trans* factors involved in C/P, the significance of APA, and increasingly elucidated roles of RBPs in APA regulation. We also discuss analysis of APA using transcriptome-wide techniques as well as molecular biology approaches.

Keywords RNA processing • Alternative cleavage and polyadenylation • 3′ UTR • Post-transcriptional regulation • Genomics • Bioinformatics • Deep sequencing

---

D. Zheng • B. Tian (✉)
Department of Biochemistry and Molecular Biology, University of Medicine and Dentistry of New Jersey (UMDNJ)-New Jersey Medical School, 185 South Orange Ave., Newark, NJ 07103, USA
e-mail: btian@umdnj.edu; btian@njms.rutgers.edu

G.W. Yeo (ed.), *Systems Biology of RNA Binding Proteins*, Advances in Experimental Medicine and Biology 825, DOI 10.1007/978-1-4939-1221-6_3

1 Alternative Cleavage and Polyadenylation

1.1 Cleavage and Polyadenylation

Except for some histone mRNAs in metazoans and some protozoans (Marzluff et al. 2008), all eukaryotic pre-mRNAs are processed at the 3′ end by cleavage and polyadenylation (C/P), which involves endonucleolytic cleavage of the nascent RNA and synthesis of a poly(A) tail (Colgan and Manley 1997) and preludes termination of transcription. The C/P site, commonly called polyA site or pA, is defined both by upstream elements (USEs) and downstream elements (DSEs) (Fig. 3.1). For metazoan pAs, USEs include the U-rich element, UGUA element, and pA signal (PAS), comprising AAUAAA, AUUAAA, and their variants; DSEs include the U-rich element and GU-rich element. This *cis* element organization is largely conserved in yeast and plants, except that in these species the GU-rich DSE is not present and the PAS sequence is more variable (Tian and Graber 2012).

The C/P machinery, which carries out the C/P reaction, consists of over 20 core proteins in metazoan cells (Fig. 3.1). Some of the C/P factors form sub-complexes in the machinery (Chan et al. 2011), including the cleavage and polyadenylation stimulation factor (CPSF), which consists of CPSF-100, CPSF-30, CPSF-160, CPSF-73, Fip1, and WDR33; the cleavage stimulation factor (CstF), which includes CstF-77, CstF-50, and CstF-64; the cleavage factor I (CFI), which contains CFI-25 and either CFI-59 or CFI-68 subunits, and the cleavage factor II (CFII), composed of Pcf11 and Clp1. The core C/P complex also includes other factors such as Symplekin, nuclear poly(A) binding protein 1 (PABPN1), Poly(A) polymerase (PAP), and RNA polymerase II (RNAP II).

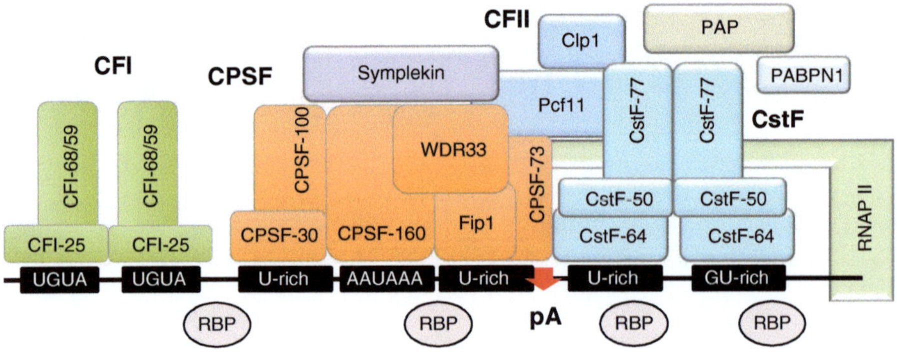

Fig. 3.1 Schematic of the cleavage and polyadenylation (C/P) machinery. The *black line* represents a pre-RNA (5′–3′) made by RNA Polymerase II (RNAP II). *Black boxes* are *cis*-elements. AAUAAA represents the poly(A) signal (PAS), which can also be its variants. The *red arrow* indicates the C/P site (pA). RNA binding proteins (RBPs) are shown to indicate regulation of pA usage. *CFI* cleavage factor I, *CFII* cleavage factor II, *CPSF* cleavage and polyadenylation stimulation factor, *CstF* cleavage stimulation factor, *PAP* poly(A) polymerase, *PABPN1* nuclear poly(A) binding protein. Subunits of the same complex are in the same color. Note that the two CFI-25 subunits bind two UGUA motifs in an antiparallel orientation

PAS, which usually resides within 40 nucleotides (nt) upstream of the pA, is the first *cis* element identified for C/P (Proudfoot and Brownlee 1976). About 53–58 % and 15–17 % of human pAs contain the AAUAAA and AUUAAA PAS sequences, respectively. Single-base variants of A(A/U)UAAA are also widespread, being associated with 10–20 % of the pAs; and 8–12 % of human pAs do not have any of these PAS sequences in the −40 to −1 nt region (Beaudoing et al. 2000; Hoque et al. 2013; Tian et al. 2005). PAS is recognized by CPSF-160 in the CPSF complex (Murthy and Manley 1995). However, a recent PAR-CLIP (Photoactivatable-Ribonucleoside-Enhanced Cross-Linking and Immunoprecipitation) study showed a rather distributive binding of CPSF-160 upstream of the pA (Martin et al. 2012). This binding pattern may indicate that CPSF-160 frequently interacts with pre-mRNA at all stages of the transcriptional cycle, a notion consistent with the observations that the CPSF complex can associate with the transcription factor TFIID during transcription initiation and with RNAP II during transcription elongation (Dantonel et al. 1997; Glover-Cutter et al. 2008; Nag et al. 2007).

The CPSF subunit CPSF-73 is the endonuclease that cleaves pre-mRNA at the pA (Mandel et al. 2006). In mammalian cells, there is a nucleotide preference at the site of cleavage in the order A>U>C≫G (Chen et al. 1995). The cleavage may occur at different positions within a small sequence window (Pauws et al. 2001), which may be due to the imprecise nature of cleavage by CPSF-73 or fuzzy definition of the pA by its surrounding *cis* elements. This micro-heterogeneity of C/P should not be confused with the APA event which involves alternative usage of pAs separated by larger distances (see below).

Several CPSF factors, including CPSF-30 and Fip1, bind U-rich elements (Barabino et al. 1997; Kaufmann et al. 2004). In addition, while not definitively shown, CPSF-100 may also have such binding activity like its yeast homolog Ydh1p/Cft2p (Dichtl and Keller 2001). PAR-CLIP analysis showed that CPSF-30, CPSF-100, and Fip1 binding profiles peak at ~60 nt, ~30 nt, and ~25 nt upstream of the pA, respectively (Martin et al. 2012).

The CstF complex recognizes U- and GU-rich DSEs via the RNA-binding subunit CstF-64 (Perez Canadillas and Varani 2003). CstF-64τ, a paralog of CstF-64, has a slightly different activity in binding poly(U)18 despite extensive sequence similarities (Monarez et al. 2007). Several transcriptome-wide mappings of binding sites using PAR-CLIP, iCLIP-seq (individual nucleotide resolution UV cross-linking and immunoprecipitation sequencing) and CLIP-seq (cross-linking immunoprecipitation high throughput sequencing) have shown that CstF-64 and CstF-64τ binding sites are enriched within ~25 nt downstream of the pA (Hoque et al. 2013; Martin et al. 2012; Yao et al. 2012). Structural studies have indicated that CstF functions as a dimer, owing to the capabilities of both CstF-77 and CstF-50 to form homodimers (Bai et al. 2007; Legrand et al. 2007; Moreno-Morcillo et al. 2011).

The UGUA USE is the binding site for CFI, which facilitates binding of CPSF to pre-mRNA and stimulates both cleavage and poly(A) addition steps (Brown and Gilmartin 2003; Ruegsegger et al. 1996). Structural studies have indicated that the CFI complex functions as a heterotetramer with two CFI-25 subunits forming a homodimer that binds two UGUA elements in an antiparallel conformation

(Li et al. 2011; Yang et al. 2011; Yang et al. 2010). The CFI-25 dimer is embraced on opposite sides by the RRM domains of CFI-68, which enhance RNA binding and facilitate RNA looping (Li et al. 2011; Yang et al. 2011; Yang et al. 2010). PAR-CLIP analysis showed that all CFI subunits are enriched at 40–50 nt upstream of the pA (Martin et al. 2012). Interestingly, half of the strong CFI binding sites lack the UGUA motif within 100 nt upstream from the pA, suggesting that other CFI-interacting factors, such as the splicing factor U2AF65 (Millevoi et al. 2006), may contribute to CFI binding at these sites.

In addition to CPSF, CstF, and CFI, which recognize the USEs and DSEs, other factors in the C/P machinery also play important roles in 3′ end processing. The CFII complex, composed of Clp1 and Pcf11, is critical for cleavage of pre-mRNA (de Vries et al. 2000) and degradation of the 3′ product after cleavage and RNAP II termination (West and Proudfoot 2008). However, the association between CFII and the C/P machinery seems weak and/or transient (Shi et al. 2009). Symplekin interacts with both CstF and CPSF (Hofmann et al. 2002; Takagaki and Manley 2000), thereby bridging the subunits. PAP catalyzes the formation of poly(A) tail on the cleaved nascent RNA, while PABPN1 limits the poly(A) tail length to ~250 nt by regulating the interaction between CPSF and PAP (Kerwitz et al. 2003; Kuhn et al. 2009).

The strength of a pA depends on the combinatorial potency of its *cis*-elements. For example, the pA of human poly(A) polymerase γ (PAPOLG) gene does contain a recognizable PAS, but it has seven UGUA USEs upstream of the pA (Venkataraman et al. 2005). These UGUA elements were shown to interact with the CFI complex, which in turn recruits the CPSF subunit Fip1 and PAP to allow efficient C/P (Venkataraman et al. 2005). In addition, pAs with weak or noncanonical PASs, such as A-rich sequences, tend to have strong downstream U-rich features (Jan et al. 2011; Nunes et al. 2010). Conversely, strong USEs may also compensate weak DSEs (Cheng et al. 2006).

1.2 Prevalent Alternative Cleavage and Polyadenylation in Eukaryotic Genomes

Recent genome-wide studies of pAs using deep sequencing technologies have revealed that APA of pre-mRNA is widespread in all eukaryotes: about 72 % in yeast, 70 % in Arabidopsis, 30 % in *C. elegans*, 43 % in zebrafish, 79 % in mouse, and 69 % in human (Derti et al. 2012; Hoque et al. 2013; Jan et al. 2011; Li et al. 2012; Ozsolak et al. 2010; Shepard et al. 2011; Wu et al. 2011). In addition, 66 % of mouse long noncoding pre-RNAs have also been found to display APA (Hoque et al. 2013).

Alternative pAs can reside in the 3′-most exon or upstream regions (Fig. 3.2). pAs in upstream regions are classified as the "intronic" or "exonic," depending on whether the pA could be removed by splicing. Intronic pAs contain two subgroups: skipped terminal exon pAs or composite terminal exons pA. Skipped terminal

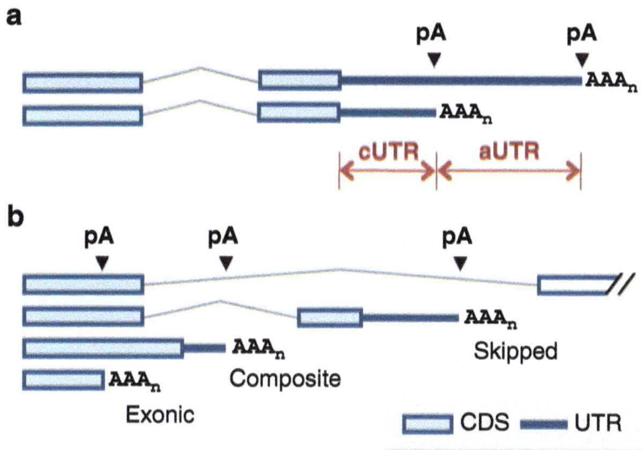

Fig. 3.2 Schematic of different types of pA. (**a**) 3′-most exonic pAs. (**b**) pAs in skipped terminal exons, composite terminal exons, and upstream exons, as indicated by skipped, composite, and exonic, respectively. *Light blue boxes* represent coding sequences (CDSs), *dark blue lines* denote 3′ untranslated regions (3′ UTRs), and *grey lines* indicate splicing. Constitutive 3′ UTR (cUTR) and alternative 3′ UTR (aUTR) are shown in (**a**) to indicate regulation of 3′ UTRs

exons are terminal exons that can be completely skipped by splicing, whereas composite terminal exons are terminal exons that can only be partially removed by splicing (Hoque et al. 2013; Yan and Marr 2005). Weak 5′ splice site (5′ SS) and large intron size are often associated with composite terminal exon pAs, whereas skipped terminal exon pAs tend to have strong PASs (Tian et al. 2007). Among the pAs in the 3′-most exon, the last pAs are usually more conserved and harbor stronger PASs than upstream ones (Hoque et al. 2013; Jan et al. 2011; Tian et al. 2007). In addition, pAs in 3′-most exons are generally more conserved than intronic pAs (Tian et al. 2007).

1.3 Impact of APA on Gene Expression

Like alternative splicing, APA enables a single gene to encode multiple mRNA transcripts. APA in the 3′ UTR can alter the availability of functional motifs important for mRNA metabolism, such as stability, translation, and subcellular localization (Andreassi and Riccio 2009; Di Giammartino et al. 2011; Lutz and Moreira 2011; Tian and Manley 2013). The regions upstream and downstream of a proximal site are named constitutive UTR (cUTR) and alternative UTR (aUTR), respectively. For mammalian genes expressing alternative 3′ UTR isoforms, more than half of the conserved microRNA target sites are located in aUTRs (Ji et al. 2009b; Sandberg et al. 2008). Thus, APA can play significant roles in mRNA metabolism.

Alternative pAs in upstream intronic or coding regions can also lead to isoforms with different coding sequences (CDSs). A cDNA/EST (expressed sequence tags)-based analysis revealed that about 20 % of human genes have at least one intronic

C/P event (Tian et al. 2007). A more recent analysis indicated that about one-third of all alternative pAs in multi-exon protein-coding genes in mouse are in internal exons or introns, over 97 % of which correspond to mRNA isoforms with altered CDSs (Hoque et al. 2013).

1.4 Tissue Specificity, Development, and Disease

The APA isoform expression pattern of a gene varies across tissue types. An analysis of ESTs from 42 distinct tissue types revealed that promoter-proximal pAs are preferentially used in placenta, retina, and blood tissues while bone marrow, uterus, ear, brain, and pancreatic islet have the opposite trend (Zhang et al. 2005). Additionally, proximal pAs were found to be more used in testis (Sandberg et al. 2008), and testis-specific pAs are associated with unique USEs and DSEs (Ji et al. 2009a; Liu et al. 2007). A large scale analysis of five mammals showed that pA usage is more conserved in the same tissue across species than in different tissues within the same species, highlighting an essential role of APA in establishing tissue-specific gene expression profiles (Derti et al. 2012).

APA regulation is prevalent during cell proliferation/differentiation and development. Microarray-based APA analysis (see below) by Sanberg et al. first reported the connection between APA and cell proliferation (Sandberg et al. 2008). Progressive 3′ UTR lengthening coupled with weakening of mRNA polyadenylation activity was observed during mouse embryonic development (Ji et al. 2009a). Conversely, 3′ UTRs become shorter and the expression of mRNAs encoding C/P factors increases during reprogramming of somatic cells from several different cell types (Ji and Tian 2009). These early observations have recently been confirmed by studies using more advanced, deep sequencing-based technologies (Hoque et al. 2013; Li et al. 2012; Shepard et al. 2011; Smibert et al. 2012; Ulitsky et al. 2012). Consistent with the connection with proliferation, expression of mRNA isoforms with shorter 3′ UTRs is generally preferred in cancer cells (Elkon et al. 2012; Fu et al. 2011; Mayr and Bartel 2009; Morris et al. 2012; Singh et al. 2009). Notably, APA isoforms affecting CDSs have been found to be regulated in neuronal cell activation (Flavcll ct al. 2008), cancer development (Elkon et al. 2012), and cell differentiation and development (Hoque et al. 2013).

2 Genome-Wide Approaches to Examine APA

2.1 cDNA/EST-Based Analysis

Early genome-wide studies of APA were based on cDNA sequences and expressed sequence tags (ESTs). While a cDNA sequence typically corresponds to the whole region of an mRNA, an EST is a fragment of cDNA sequence (Okubo et al. 1992).

A large fraction of cDNA/EST sequences are derived from the 3′ end of mRNAs, making them usable to map pAs and study APA (Beaudoing and Gautheret 2001; Gautheret et al. 1998; Tian et al. 2005).

To identify pAs, cDNA/EST sequences from GenBank or dbEST databases are aligned to genomic sequences. Unaligned consecutive As at the 3′ end or Ts at the 5′ end are considered to be from the poly(A) tail, provided that they are longer than a size cutoff (typically 6). Accordingly, the pA is generally considered to be the position right after the 3′-most position of the alignment. Because of the heterogeneity of cleavage, pAs located near each other are clustered to each other. After clustering, the pA with the most cDNA/EST supporting sequences is used to represent all clustered pAs. Using this approach, Beudoing et al. first reported 29 % of human genes have multiple pAs (Beaudoing et al. 2000) and Tian et al. found over half of the human genes and about 30 % of mouse genes have APA (Tian et al. 2005).

The number of cDNA/EST sequences associated with a pA can also reflect the expression level of the APA isoform using the site. Because the tissue source for EST library is usually known, it is possible to examine the expression pattern of APA isoforms across tissues (Gautheret et al. 1998). Using a method named global study of poly(A) site usage by gene-based EST vote (GAUGE), Zhang et al. found bias in APA isoform expression across human tissues (see above) (Zhang et al. 2005).

cDNA/EST databases are valuable but insufficient for in-depth identification of pAs and quantification of APA isoforms. First, the number of cDNA/EST sequences is rather limited, which causes a large difference in APA frequency between human and mouse genomes. Second, only a small fraction of EST sequences have the poly(A) tail information, e.g., less than 10 % for human ESTs, and the expression level of most genes is represented by a small number of ESTs (Gilat et al. 2006). Consequently, only a small number of genes can be quantitatively studied for APA, for example, ~1,000 genes studied in Ji and Tian (2009). Another important issue in pA mapping using cDNA/EST data is false identification of pAs. If an upstream A-rich region of mRNA is targeted by the oligo(dT) primer for cDNA synthesis, the resultant cDNA sequence will contain an oligo(A/T) sequence that can be mistakenly considered to originate from the poly(A) tail. Strikingly, it was estimated that 12 % of cDNA/EST sequences are derived from priming of internal A-rich sequences (Nam et al. 2002). This problem, commonly known as the "internal priming issue" can be partially addressed by examining the genomic sequence surrounding the pA. For example, our lab typically uses the −10 to +10 nt region, and considers a pA to be an internal priming candidate if there are 6 continuous As or more than 7 As in a 10 nt window within this region.

2.2 Serial Analysis of Gene Expression (SAGE)-Based Methods

Serial analysis of gene expression (SAGE) permits quantitative analysis of the transcriptome using short sequences/tags that are linked to the 3′ end of transcripts (Velculescu et al. 1995). The SAGE method involves digestion of cDNA with a frequent cutter restriction enzyme (called anchoring enzyme), such as Nla III (which cuts at CATG), and capture of the 3′-most fragment of the digested products. After adapter ligation and digestion with another enzyme (called tagging enzyme), a short sequence downstream of the cutting site of anchoring enzyme (10 or 17 depending on the type of tagging enzyme used) is cloned and sequenced. In theory, since SAGE tags correspond to the sequence immediately after the 3′-most cutting site of the anchoring enzyme, the pA isoform from which the tags are derived can be inferred (Pauws et al. 2001). Using this approach, we reported 3′ UTR lengthening in mouse embryonic development (Ji et al. 2009a). However, despite that SAGE tags are more quantitative than ESTs, incomplete digestion of cDNA by the anchoring enzyme can lead to false assignment of SAGE tags to isoforms. In addition, due to the high cost associated with cloning and Sanger sequencing, SAGE experiments typically do not generate enough tags for analysis of the whole transcriptome.

2.3 Microarray-Based Methods

Microarrays enable quantitative analysis of gene expression genome-wide. D'Mello et al. reported that a large fraction of Affymetrix 3′ microarray probes target regions in aUTRs (D'Mello et al. 2006), suggesting that APA can affect gene expression analysis using those microarrays. Sandberg et al. took advantage of this unique design of Affymetrix probes and developed a method called PLATA (probe level alternative transcript analysis) to specifically examine APA (Sandberg et al. 2008). A modified *t*-test was used to evaluate differences in mean intensity between of cUTR probes and aUTR probes across samples (Sandberg et al. 2008). A tandem UTR length index (TLI) was calculated to reflect the overall length of all tandem UTR isoforms in a sample, with higher TLI values indicating increased relative expression of distal pA isoforms. With this method, they found that T cells express mRNAs with shortened 3′ UTRs after activation to proliferate, and cell lines in general express shorter 3′ UTRs than their corresponding tissues (Sandberg et al. 2008). Using a similar method to examine the difference between cUTR and aUTR probe signals, we found progressive lengthening of 3′ UTRs in cell differentiation and development (Ji et al. 2009a) and during cell reprogramming (Ji and Tian 2009). A score named Relative Usage of Distal poly(A) site, or RUD, was calculated to represent the relative 3′ UTR length for genes in a sample, with higher RUD indicating longer 3′ UTR (Ji et al. 2009a; Ji and Tian 2009). A more elaborate approach was developed by the Graber lab, which evaluates all probes, instead of separating probes into cUTR and aUTR groups, to identify segmentation points where the sample-to-sample signal ratios on either side of the points are different. This method was implemented into an R program package named Rmodel. Using this approach they observed both shortening and lengthening of 3′ UTRs in multiple tumor types, although the former was more common (Singh et al. 2009).

Although microarray-based APA analysis has greatly advanced our study of APA, it has several limitations. First, only a small fraction of genes have probes targeting both cUTR and aUTR for most types of microarrays. As such, most studies using microarrays analyze fewer than 2,000 genes for APA regulation. Second, microarray data cannot be used to identify pAs de novo. Instead, databases of known pAs are needed to define cUTRs and aUTRs. This latter problem can be a major issue when a gene has more than two pAs or some pAs are located close to each other and the intervening region does not have probes.

2.4 RNA-seq

Deep sequencing of RNA (RNA-seq) is increasingly used to study the transcriptome. Most RNA-seq protocols involve sequencing all regions of the mRNA molecule. Thus, only a small fraction of the sequencing reads correspond to the 3′ region. Nevertheless, poly(A)-containing reads were used to map pAs in the yeast and *Trypanosoma brucei* genomes (Siegel et al. 2010; Wilhelm et al. 2008). To enrich poly(A)-containing reads, Smibert et al. re-amplified RNA-seq libraries by PCR using a primer complementary to the 3′ adaptor plus six T residues at the 3′ end (Smibert et al. 2012). Using this protocol, they reported 3′ UTR shortening in the testis and lengthening in the central nervous system of *Drosophila melanogaster* (Smibert et al. 2012).

Alternatively, RNA-seq reads, including non poly(A)-containing ones, can be treated like microarray probe signals to examine the relative expression of isoforms to which they are mapped (Ji et al. 2011b; Wang et al. 2008). Like in the microarray data analysis, prior annotation of pAs is needed to assign reads to isoforms. This approach has been implemented in the MISO (mixture-of-isoforms) program (Katz et al. 2010). Additionally, RNA-seq can be used to extend annotated 3′ UTRs, thereby supporting newly identified pAs by connecting them to upstream annotated regions. This approach was used to extend the 3′ end for genes in *C. elegans* (Jan et al. 2011), zebrafish (Ulitsky et al. 2012), and mouse (Hoque et al. 2013). In another study, 3′ UTR extension using RNA-seq was performed to identify APA isoforms bearing exceptionally long 3′ UTRs (many >10 kb and some >18 kb), which were also confirmed by northern blotting analysis (Miura et al. 2013).

2.5 pA-Based Deep Sequencing

A number of deep sequencing methods have been developed in the last 2 years to specially interrogate APA isoform expression. Reads generated from these methods have inherent information about the pA identity. They are discussed below and their major features are summarized in Table 3.1.

Table 3.1 Comparison of deep sequencing methods for pA analysis

Method	PAS-seq/SAPAS	MAPS/PolyA-seq	3P-seq	3′ READS	DRS
Fragmentation	Fragmentation buffer	Not needed	RNase T1	Fragmentation buffer	Not needed
Strategy to derive poly(A) tail information	Oligo(dT) priming	Oligo(dT) priming	Splint ligation to the end of poly(A) tail	CU_5T_{45} oligo binding and digestion by RNase H, protection of some As by Us	oligo(dT) capture
Second strand synthesis strategy	Strand switching during RT	Random hexamer priming	Ligation of RNA adapter	Ligation of RNA adapter	Not needed
Internal priming	Yes	Yes	No	No	Little
Sequencing error	Low	Low	Low	Low	High
Simplicity	+++	+++	+	++	++++

2.5.1 Methods Based on Oligo(dT) Priming

Poly(A) Site Sequencing (PAS-Seq) (Shepard et al. 2011) and sequencing APA sites (SAPAS) (Fu et al. 2011) exploit the terminal transferase activity of M-MLV (Moloney Murine Leukemia Virus) reverse transcriptase to add a few untemplated deoxycytodines (dCs) to the 3′ end of the first strand cDNA, which is reverse-transcribed using an oligo(dT)-containing primer (Fig. 3.3a, left). The added dC

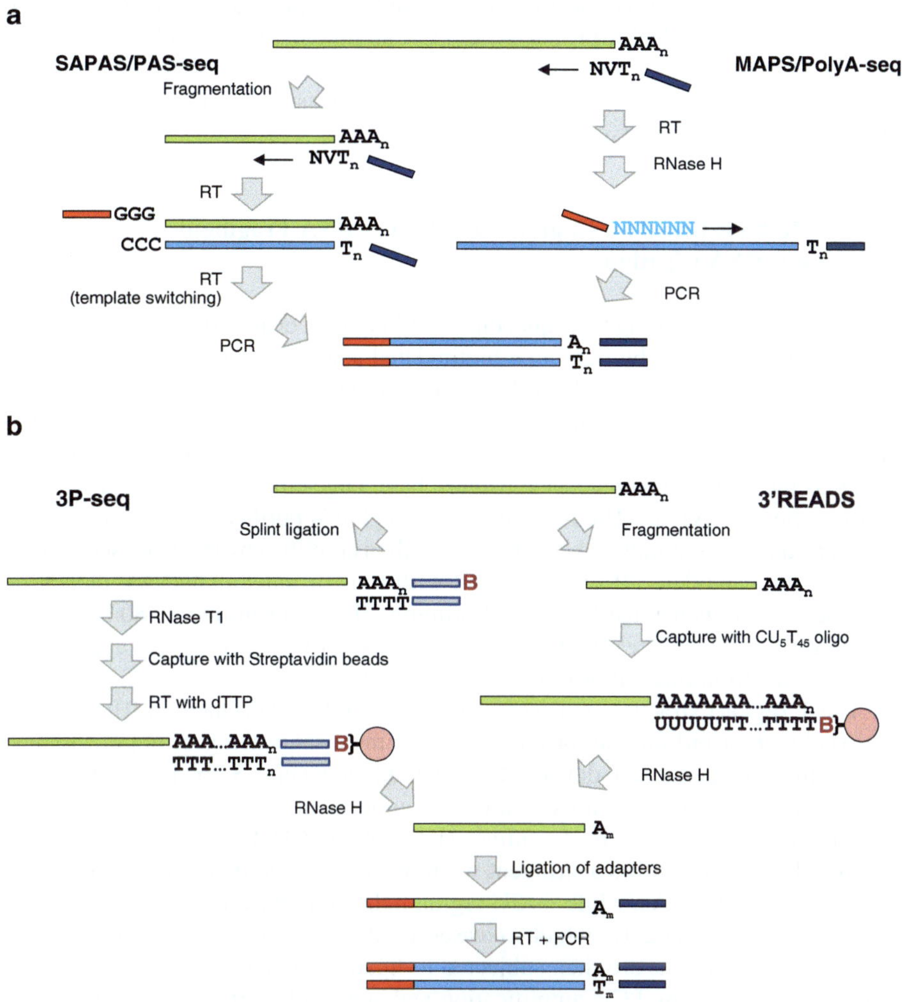

Fig. 3.3 Schematics of deep sequencing methods for pA mapping. (**a**) *Left*, SAPAS/PAS-seq; *right*, MAPS/PolyA-seq. RNA is shown as *green box*, DNA as *blue box*, 5′ adapter as *red box*, and 3′ adapter as *dark blue box*. *RT* reverse transcription, *N* random nucleotide, *V* nucleotide C, G, or T. (**b**) *Left*, 3P-seq; *right*, 3′ READS. Splint adapter is shown as *gray box*. The letter "B" represents biotin and the *pink circle* represents magnetic bead. The letter "m" indicates a short stretch of nucleotides (2–10)

residues anneal with an adaptor containing three terminal guanine (G) residues, allowing the reverse transcriptase to switch template and complete the second strand of cDNA after the three Gs. Double-stranded cDNAs are then amplified by PCR and used for sequencing. Multiplex Analysis of PolyA-linked Sequences (MAPS) (Fox-Walsh et al. 2011) and PolyA-seq (Derti et al. 2012) use random priming for second-strand synthesis (Fig. 3.3a, right), enabling simple construction of cDNA libraries. Reverse transcription is primed with an oligonucleotide consisting of a universal sequence (which is used for PCR amplification) followed by T(10)VN (V: A or C or G; N: A or T or C or G). The second-strand synthesis is primed with random hexamers linked to a second PCR primer. After PCR amplification of the library, a primer ending with 10 Ts is used for sequencing. PAS-Seq/SAPAS/MAPS/PolyA-seq not only identifies pAs and quantifies their isoform expression but also provides information about gene expression.

2.5.2 Methods Based on Capture of Poly(A)+RNA Fragments and RNA Ligation

Poly(A)-position profiling by sequencing (3P-Seq) (Jan et al. 2011) was designed to avoid internal priming by ligating a biotinylated adapter to the end of poly(A) tail via splint-ligation (Fig. 3.3b, left). After partial digestion of the non-poly(A) region of mRNA with RNase T1 (which cuts after the G residue), the 3′ end region is captured on streptavidin-coated beads. The poly(A) tail is reverse-transcribed using dTTP as the only deoxynucleoside triphosphate. Digestion with RNase H removes most of the poly(A) tail. The resultant 3′ end region is purified, ligated to 3′ and 5′ adapters, and reverse-transcribed. cDNAs are then amplified by PCR and sequenced. Most of the candidate 3P-seq reads end with runs of untemplated As, indicating that they are not generated by internal priming. 3P-Seq identified 8,580 additional *C. elegans* UTRs while excluding thousands of isoform sequences that appeared to be generated from internal priming (Jan et al. 2011).

The 3′ region extraction and deep sequencing (3′ READS) method (Hoque et al. 2013) aims to avoid internal priming and minimize sequencing of oligo(A)-tailed RNAs that are generated during RNA degradation (Fig. 3.3b, right). Poly(A)-containing RNA fragments are captured by chimeric oligonucleotide containing 45 thymidines (Ts) followed by 5 uridines (Us)(dubbed CU_5T_{45}) on magnetic beads. Its long T sequence allows a stringent washing condition to be used to enrich RNA fragments with long As, while its 5U region helps to prevent over-digestion of the poly(A) tail by RNase H, which releases RNA fragments from the beads. The eluted RNA fragments are then subjected to ligation of adapters and reverse transcription, followed by PCR amplification and sequencing. Because reads lacking additional untemplated As after alignment are not used for pA identification in 3′ READS, the "internal priming" problem does not exist. Using this method, 5,392 pAs surrounded by genomic A-rich sequences (7.6 % of total) were identified, which would have been filtered out as internal priming candidates if a method employing oligo(dT) in reverse transcription had been used (Hoque et al. 2013).

Compared to PolyA-seq and PAS-seq, 3′ READS generated over ten-times fewer reads aligned to rRNAs, snoRNAs or snRNAs, indicating that 3′ READS can substantially mitigate false positives caused by internal A-rich sequences and oligo(A) tails. Compared with 3P-Seq, 3′ READS produced 54 % more usable reads for pA mapping, presumably due to the stringent washing condition and fewer processing steps (Hoque et al. 2013).

It is noteworthy that 3P-seq and 3′ READS both require adapter ligation to RNA, which may suffer from bias caused by structural features within RNA fragments or adapters (Hafner et al. 2011). Strategies have been developed to mitigate this issue including using random sequences in the adapter near the ligation end (Sorefan et al. 2012; Zhuang et al. 2012).

2.5.3 Direct RNA Sequencing

All the aforementioned methods rely on PCR amplification to generate cDNA libraries, which can introduce bias caused by uneven amplification of different templates and over-amplification (Kozarewa et al. 2009). Direct RNA sequencing (DRS) by Helicos BioSciences avoids PCR amplification. DRS begins with hybridizing 3′-blocked, poly(A)+RNA to oligo(dT)-coated surfaces. The poly(A) region is then filled by reverse-transcription with dTTP, followed by sequencing from the 3′-most non-A position (Ozsolak et al. 2010; Ozsolak et al. 2009). In theory, internal priming can still happen because the oligo(dT) used for capturing mRNA may anneal to internal A-rich regions. However, low internal priming rates have been reported in two studies (Ozsolak et al. 2010; Sherstnev et al. 2012). While DRS needs a lower amount of poly(A)+RNA as input, its raw base error rate is quite high (4–5 %) (Ozsolak and Milos 2011), as compared to the Illumina platform (<0.4–0.5 %) (Luo et al. 2012; Quail et al. 2012).

3 Molecular Analysis of C/P

While genome-wide approaches can globally identify pAs and examine APA events, it is often desirable to confirm the result by molecular assays focusing on individual genes.

3.1 Methods to Examine APA Isoforms

Several molecular methods can be used to analyze expression of APA isoforms (Fig. 3.4), including northern blotting, RNase protection analysis (RPA), RT-qPCR, and 3′ Rapid Amplification of cDNA Ends (3′ RACE).

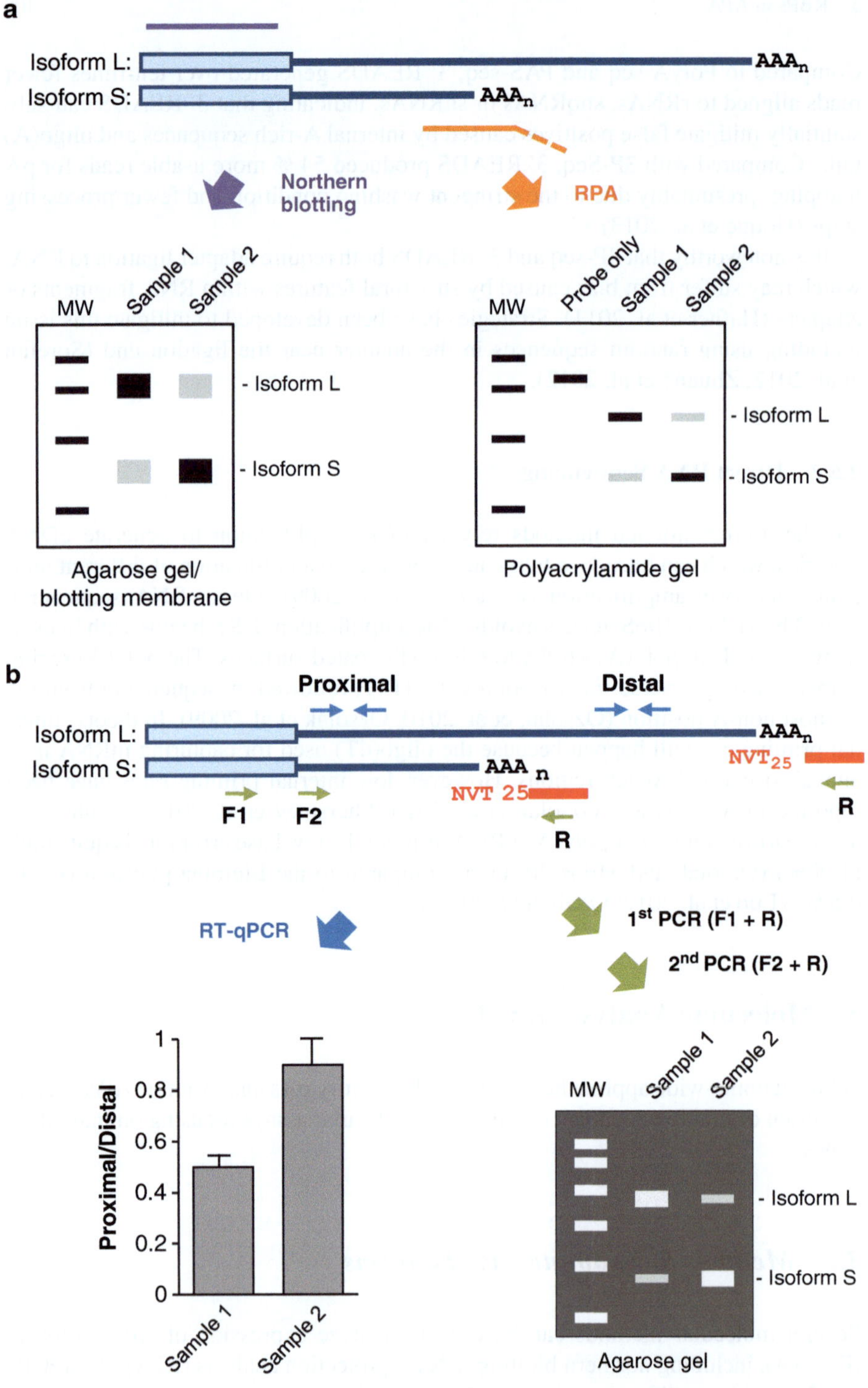

Fig. 3.4 Molecular methods for validation of APA isoforms. (**a**) Northern blotting (*left*) and RNase protection analysis (RPA, *right*). Two hypothetical APA isoforms are shown: long isoform (L) and short isoform (S), differing in the 3′ UTR length. CDS is shown as *light blue box*, and 3′ UTR

In northern blotting analysis (Fig. 3.4a), APA isoforms with different sizes are separated on an agarose gel and transferred to a blotting membrane. A radioactive probe targeting a common region of the isoforms is used for hybridization. The hybridized radioactive signals represent the amounts of isoforms. Northern blotting is widely used to demonstrate the existence of APA isoforms and quantify their expression (Kubo et al. 2006; Mansfield and Keene 2012; Smibert et al. 2012). Although straightforward, northern blotting analysis can be complicated by alternative splicing or alternative initiation.

RPA (Fig. 3.4a) is a sensitive and accurate method to quantify RNA levels (Newman and Krieg 1999). When applied to APA analysis, a radioactive probe that can be annealed to the region containing the pA of interest is used (Ji et al. 2011b; Nagaike et al. 2011). Transcripts using the pA lead to digestion of a longer region of the probe whereas those not using the pA lead to digestion of a shorter region. Digestion products are separated from the input probe on a polyacrylamide gel.

In RT-qPCR assays (Fig. 3.4b), primers are designed to target common and alternative regions of APA isoforms, such as cUTRs and aUTRs, respectively. The ratio of aUTR signal to the cUTR signal for one condition can be compared to that from another condition to infer APA regulation (Licatalosi et al. 2008). Because of different amplification efficiencies for cUTR and aUTR primer pairs, the RT-qPCR method can only be used to detect the relative change of aUTR to cUTR ratio between conditions, not the absolute ratio between isoforms in each condition.

3′ RACE begins with reverse transcription using a primer containing (dT)25VN and an adapter sequence for later amplification (Fig. 3.4b). The RT product can be amplified by nested PCR (Scotto-Lavino et al. 2006), which uses two gene-specific forward primers and a universal reverse primer. Since the two forward primers are designed for common regions, different APA isoforms give rise to PCR products of different sizes. Because the efficiency of reverse transcription can be different for long or short APA isoforms and the PCR reaction is carried out until the saturation phase, the 3′ RACE result is generally considered semiquantitative.

Fig. 3.4 (continued) as *dark blue line*. The *purple line* represents a northern blotting probe which can hybridize to both APA isoforms. The *orange line* denotes an RPA probe. The *dotted part* is the region that cannot anneal to either one of the isoforms. Hypothetical northern blotting and RPA results are shown, which indicates that sample 1 contains more of the long isoform than sample 2 does. (**b**) RT-qPCR (*left*) and 3′ RACE (Rapid Amplification of cDNA Ends). Two hypothetical isoforms are shown as in (**a**). *Small arrows* indicate primers. For RT-qPCR, RNAs are reverse-transcribed with oligo(dT), followed by PCR with two pairs of primers (proximal and distal) targeting different regions of the cDNAs. As indicated, the region targeted by the proximal pair is common to both APA isoforms and the region by the distal pair is unique to the long isoform. qPCR signals from the proximal and distal pairs of primers are compared to indicate the relative expression of the two isoforms. For 3′ RACE, an oligo(dT)25VN primer is used for reverse transcription, and the cDNAs are amplified with two rounds of PCR intended to increase specificity. The first round uses a gene-specific forward primer F1 and a universal reverse primer R. The PCR product is diluted and used as template for the second round of PCR, which uses another gene-specific forward primer F2 and the universal primer R. As in (**a**), hypothetical results are shown to indicate that sample 1 contains more of the long isoform than sample 2 does

3.2 Methods to Examine pA Usage

The pA usage has traditionally been examined by in vitro cleavage and/or polyadenylation assays (Fig. 3.5a), in which a substrate synthesized by in vitro transcription is incubated with nuclear extract. During the coupled cleavage and polyadenylation assay, the pre-mRNA substrate with a pA site is both cleaved and polyadenylated by factors in the nuclear extract (Laishram and Anderson 2010). Because EDTA inhibits the polyadenylation step, the cleavage reaction can be specifically analyzed (Di Giammartino et al. 2013; Shi et al. 2009). In addition, it is possible to examine the polyadenylation step only with an in vitro synthesized substrate without pA (Di Giammartino et al. 2013; Roca and Karginov 2012).

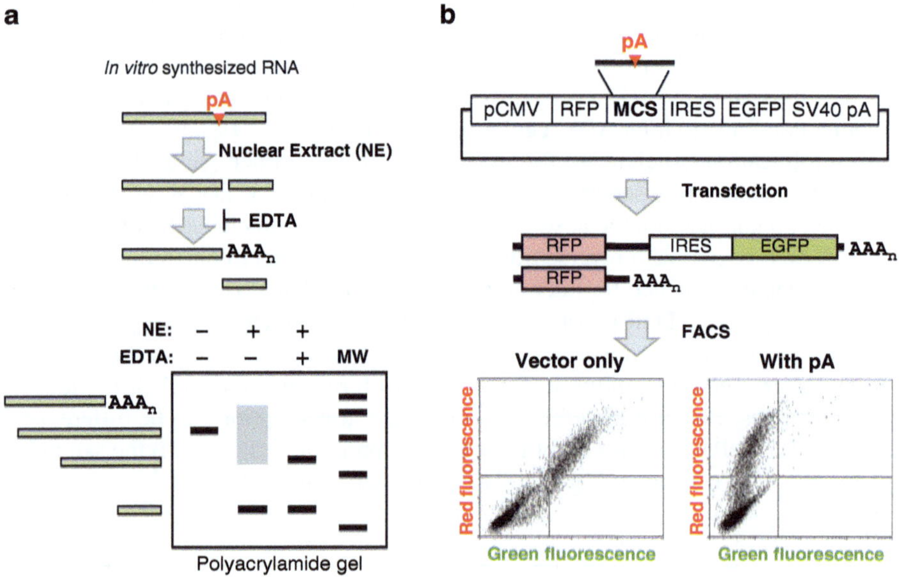

Fig. 3.5 Analysis of pA activity. (**a**) Schematic of in vitro cleavage and polyadenylation (C/P) assay. Nuclear extract (NE) from mammalian cells, such as HeLa cells, is incubated with in vitro synthesized, radioactive RNA substrate containing a pA. The C/P products are shown at the *bottom*. The polyadenylated RNAs have different poly(A) tail lengths, leading to a diffusive band on gel. Only the cleavage reaction is analyzed in the presence of EDTA, as it inhibits the polyadenylation step. (**b**) Reporter assaying using the pRiG vector. pRiG contains, from upstream to downstream, pCMV (Cytomegalovirus immediate early promoter), RFP (red fluorescent protein) coding sequence, multiple cloning sites (MCS), IRES (internal ribosome entry site), EGFP (enhanced green fluorescent protein) coding sequence, and SV40 pA. A sequence containing a pA can be inserted into the MCS region for analysis of its C/P activity. After transfection, cells are subject to fluorescence-activated cell sorting (FACS) analysis. The pRiG containing a stronger pA will express relatively more RFP. The results shown at *bottom* are from our lab. *Left*, pRiG without pA; *right*, pRiG with a strong pA. The expression of isoforms can also be analyzed by the methods shown in Fig. 3.4

Reporter assays provide a simpler way to examine pA usage. For example, our lab used a vector called pRiG to validate pA usage (Fig. 3.5b) (Pan et al. 2006). The vector contains, sequentially, a CMV promoter, a red florescent protein (RFP), multiple cloning sites (MCS), an internal ribosomal entry site (IRES), an enhanced green fluorescent protein (EGFP), and an SV40 pA (Ji et al. 2009a; Pan et al. 2006). A sequence is inserted into MCS for analysis of its activity for C/P. If C/P takes place in the sequence, a reporter transcript containing only RFP is expressed and otherwise a transcript containing both RFP and EGFP is expressed. Therefore the relative amount of two reporter transcripts can reflect how much C/P happens in the sequence. A simple way to examine expression of transcripts is quantification of red and green florescent signals by fluorescence-activated cell sorting (FACS) analysis. However, since the inserted sequence may affect translation of RFP or EGFP, for example, interfering with the function of IRES, RNA analysis using the ones mentioned above sometimes is desirable to confirm the FACS result.

4 The Impact of RBPs on APA

4.1 Core RBPs for Cleavage and Polyadenylation

While the core factors in the 3′ processing machinery are considered essential players in the C/P reaction, regulation of their expression has been implicated in APA control. In perhaps the first study showing that a core C/P factor can regulate APA, Takagaki et al. reported enhanced usage of the upstream intronic pA in IgM heavy chain pre-mRNA upon overexpression of CstF-64 (Takagaki et al. 1996). This result recapitulates the scenario of primary mouse B cell differentiation, during which CstF-64 is upregulated and the intronic pA isoform encoding a secretory form of protein is more expressed compared to the 3′-most pA isoform encoding a membrane-bound form. However, increased CstF-64 level was found to be neither necessary nor sufficient for secretory IgM isoform production in human primary splenic B cells development (Martincic et al. 1998). Recently, Yao et al. examined the global function of CstF-64 on APA in HeLa cells. Using DRS (see above), they found that depletion of CstF-64 alone had little effect, but co-depletion of CstF-64 and its paralog CstF-64τ globally increased the relative usage of distal pAs (Yao et al. 2012). It was proposed that when CstF-64 is abundant, it promotes usage of proximal pAs, which are generally weaker than distal ones, and when CstF-64 is limited, recognition of proximal pAs becomes less efficient, allowing more usage of distal, stronger pAs (Yao et al. 2012). The result also indicates that CstF-64τ and CstF-64 have complementary roles in C/P.

The observation that CFI-25 knockdown enhanced usage of proximal pAs of several genes (Kubo et al. 2006) indicates that CFI can inhibit usage of proximal pAs. Indeed, knockdown of CFI-25 and CFI-68, but not CFI-59, resulted in a global shift to proximal pA usage (Gruber et al. 2012; Martin et al. 2012). While the underlying mechanism for CFI-mediated inhibition of proximal pA usage is unclear, structural analysis of the CFI suggested that the CFI heterotetramer may bind to two UGUA elements flanking the proximal pA and loops out the proximal pA, resulting in inhibition of the usage of proximal site (Yang et al. 2011). Notably, CFI-25 and CFI-68 are highly enriched in mouse male germ cells, and CFI binding sites were identified in many male germ cell transcripts utilizing non-A(A/U)UAAA PASs (Sartini et al. 2008). Whether CFI also plays a role in pA selection in the germ cells, in which proximal sites are generally favored, is to be elucidated in the future.

PABPN1 has recently been identified as a regulator of APA in a reporter-based RNAi screen (Jenal et al. 2012). Transcriptome-wide analysis showed that loss of PABPN1 de-repressed usage of weak upstream pA sites, resulting in widespread 3′ UTR shortening. Consistently, triplet-repeat expansion in PABPN1 (trePABPN1), which causes autosomal-dominant oculopharyngeal muscular dystrophy (OPMD), elicited extensive proximal pA usage in both a mouse model of OPMD and human cells (de Klerk et al. 2012; Jenal et al. 2012). How a factor whose primary role is believed to be in the polyadenylation step can regulate cleavage choice is to be further elucidated in the future.

4.2 Other RBPs Regulating APA

U1 snRNP and its components. The RNA-protein complex U1 snRNP (small nuclear ribonucleoprotein) binds to the 5′ splice site of intron during splicing. U1 snRNP has been shown to suppress C/P, a function attributable to the inhibition of PAP by the U1 snRNP subunit U1-70K (Gunderson et al. 1998). As such, when the 5′ end of U1 snRNA, which is normally used for 5′ splice site recognition, forms base pairs with pre-mRNA near the pA, polyadenylation of pre-mRNA is inhibited (Abad et al. 2008; Fortes et al. 2003). Kaida et al. reported that inhibition of U1 snRNP in HeLa cells by antisense morpholino oligonucleotide not only inhibited splicing but also caused premature C/P, frequently in introns close to (<5 kilobases, kb) the transcription start site (Kaida et al. 2010). In contrast, inhibition of U2 snRNP, another essential component of the splicing machinery (see below), did not induce such effect, suggesting a splicing-independent function for U1 snRNP in protecting pre-mRNAs from premature C/P at cryptic intronic pAs (Kaida et al. 2010). Moreover, the same group reported that shortage of U1 snRNP created by rapid transcriptional upregulation upon neuronal activation results in activation of proximal pAs of pre-mRNAs (Berg et al. 2012).

The U1 snRNP subunit U1A also inhibits polyadenylation of pre-mRNA (Gunderson et al. 1994). This function is employed to autoregulate its own expression because of the presence of its binding site, called Polyadenylation Inhibitory Element (PIE), in its own 3′ UTR (Gunderson et al. 1994). Additionally, U1A has been shown to inhibit the usage of intronic pA of IgM heavy-chain gene.

The mechanism, however, involves inhibition the recruitment of CstF-64 to two GU-rich DSEs due to binding of U1A to an intervening site (Phillips et al. 2004).

U2 snRNP and U2AF. The U2 snRNP, known for recognizing the branch point of the intron, can physically interact with CPSF and promote cleavage of pre-mRNA in vitro (Kyburz et al. 2006). U2 RNA auxiliary factor (U2AF) 65 kDa subunit (U2AF65) binds to the pyrimidine tract in intron near the 3′ splice site, and helps spliceosome assembly during splicing. U2AF65 can also recruit the CFI complex via interaction with the CFI-59 subunit, which leads to stimulation of C/P at a nearby pA (Millevoi et al. 2006). This function may help define the 3′ terminal exon. Moreover, U2AF65, together with U2AF35, another subunit of U2AF, and PTB (see below) bind to the upstream region of the pAs of *Prothrombin*, *Bcl2l2*, *Ivns*, and *Actr* pre-mRNAs to promote their C/P (Danckwardt et al. 2007).

mRNA-protein complex (mRNP) proteins. Nuclear RBPs that bind mRNAs, collectively called mRNP proteins, play important roles in posttranscriptional gene regulation, including pre-mRNA processing, nuclear export, localization, mRNA stability, and translation (Dreyfuss et al. 2002). A large fraction of them are traditionally called hnRNPs (heterogeneous nuclear ribonucleoproteins). Many of the mRNPs have been found to regulate C/P.

The RNA binding protein PTB (hnRNP I) is well known for its function in alternative splicing (Xue et al. 2009). It is also one of the first splicing RBPs that are found to regulate C/P. PTB was first shown to activate the C/P of C2 complement pre-mRNA through its binding site upstream of the pA (Moreira et al. 1998). However, PTB was later shown to compete with CstF-64 for binding to DSEs of the pAs in human *α-globin* and *β-globin* pre-mRNAs, resulting in inhibition of their C/P (Castelo-Branco et al. 2004). This context-dependent regulation of C/P has also been observed later for several other RBPs.

hnRNP F, hnRNP H1 (also known as hnRNP H), and hnRNP H2 (also known as hnRNP H′ or DSEF-1) bind G-rich sequences (Alkan et al. 2006; Bagga et al. 1998; Decorsiere et al. 2011), which are often enriched in the downstream region of pAs (Hu et al. 2005). Indeed, all these factors have been shown to regulate APA. hnRNP F was first found to bind a downstream G-rich sequence and reduce the binding of CstF-64 to the same region in a UV-cross-linking assay (Alkan et al. 2006). Consistently, during B cell development from the plasma stage to the memory stage, increased expression of hnRNP F competes with CstF-64 and inhibits the processing of the intronic pA of IgM heavy-chain gene (Veraldi et al. 2001). A global analysis showed that APA isoforms with more hnRNP H1 binding to the upstream region of pAs, as indicated by CLIP-seq reads, tend to be more expressed in the presence of hnRNP H1, suggesting that hnRNP H1 can promote pA usage (Katz et al. 2010). Interestingly, binding of PTB to the upstream region of pA was found to help recruit hnRNP H1 to G-rich DSEs, and the recruited hnRNP H1 interacts with PAP and stimulates polyadenylation (Millevoi et al. 2009). Similarly, the hnRNP F/H1 dimer binds to the G-rich DSE of the pA of p53 pre-mRNA and stimulates its C/P during DNA damage, when 3′ end processing is generally inhibited

(Decorsiere et al. 2011). hnRNPH 2 is highly similar to hnRNP H1 in sequence (96.2 % identity), and has also been shown to stimulate C/P by enhancing CstF-64 binding (Bagga et al. 1998). Moreover, knockdown of hnRNP H2 reduced the expression of a reporter gene containing G-rich DSEs (Alkan et al. 2006).

hnRNP K was recently found to regulate APA of a noncoding RNA NEAT1 (nuclear paraspeckle assembly transcript 1), which is essential for the formation of paraspeckle (Naganuma et al. 2012). Paraspeckles are subnuclear structures involved in suppression of the expression of hyper-edited transcripts through nuclear retention (Prasanth et al. 2005). Among the two APA isoforms of NEAT1 (the 3.7-kb NEAT1_1 and the 23-kb NEAT1_2), only the long isoform is essential for paraspeckle formation. hnRNP K competes with CFI-68 for binding to CFI-25, thereby inhibiting the binding of CFI complex near the proximal pA (used for NEAT1_1 expression) and suppressing its usage (Naganuma et al. 2012).

Knockdown of hnRNP L in Hela cells increased the usage of an intronic pA of the A*sah1* gene, which harbors an hRNP L binding motif downstream of the pA (Hung et al. 2008). A more recent global analysis using microarrays indicated that hRNP L knockdown elicits hundreds of APA events (Huang et al. 2012). The molecular mechanism for hRNP L-mediated APA, however, still awaits further investigation.

The poly(C)-binding proteins (PCBPs, also known as αCPs) are a family of KH-domain-containing RBPs, including four members: αCP1, αCP2, αCP3, and αCP4 (Chkheidze and Liebhaber 2003). αCPs have been shown to enhance mRNA stability in the cytoplasm (Chkheidze et al. 1999; Kiledjian et al. 1995; Weiss and Liebhaber 1995). In the nucleus, αCP2 associates with CPSF and promotes C/P of the α-globin pre-mRNA through an upstream C-rich motif (Ji et al. 2011a). A recent global study identified 357 significantly regulated APA events after depletion of major αCPs, indicating a general role for αCP in APA (Ji et al. 2013). In the same study, pAs with C-rich motifs within a window 30–40 nt upstream of the PAS were generally repressed upon αCP depletion, consistent with a general role of αCPs as enhancers of C/P (Ji et al. 2013).

Elav (embryonic-lethal abnormal visual system) is a *Drosophila* neuron-specific RBP essential for neuronal development (Campos et al. 1985). Elav binds to AU-rich sequences and has been shown to regulate neuron-specific alternative splicing (Koushika et al. 1996; Lisbin et al. 2001; Soller and White 2005). A recent study indicated that it can play a global role in mediating 3′ UTR extension in the Drosophila nervous system (Hilgers et al. 2012). Forced expression of Elav in non-neuronal ectodermal cells resulted in 3′ UTR extension of the *Brat* mRNA. RNA immunoprecipitation (RIP) confirmed that Elav binds to the downstream region of proximal pAs of seven putative target genes. Tethered Elav, but not tethered Ago1, inhibited usage of the proximal pA of a reporter gene (Hilgers et al. 2012). Mammalian homologs of Elav, Elav-like (Elavl) 1–4 proteins (also known as HuR, HuB, HuC, and HuD, respectively) have also been found to regulate C/P. Using HeLa cell nuclear extract Zhu et al. showed that recombinant HuB can bind to

U-rich USEs of SV40 late pA and calcitonin exon 4 pA, thereby blocking CstF-64 binding and inhibiting C/P (Zhu et al. 2007). During neuronal differentiation, all four Hu proteins can regulate APA of HuR pre-mRNA, inhibiting the expression of a 2.4 kb isoform and enhancing the expression of a 6.0 kb isoform which is both translationally repressed and less stable (Mansfield and Keene 2012). This regulation balances the pro-differentiation activity of HuB/C/D and the pro-proliferation activity of HuR. The mechanism of regulation involves binding of Hu proteins to the downstream region of pA, thereby blocking the recruitment of CstF-64 (Dai et al. 2012).

Nova (neuro-oncological ventral antigen) proteins are neuron-specific RBPs with binding specificity to YCAY motifs (Ule et al. 2006). HITS-CLIP (high-throughput sequencing of RNA isolated by cross-linking immunoprecipitation) experiments using the mouse brain revealed that Nova can bind not only intronic regions, which is consistent with its role in alternative splicing, but also sequences in 3′ UTRs (Licatalosi et al. 2008). Many CLIP read clusters are within a few hundred nt from the pAs. Microarray analysis comparing brains of Nova2 knockout (KO) mice with those of wild type mice identified 297 genes exhibiting regulated APA and, importantly, CLIP read clusters were enriched near pAs of 43 of the genes. APA regulations of *Cugbp2* and *Slc8a1* pre-mRNAs, both of which have Nova2 binding near the regulated pAs, were further confirmed by RPA using the *Nova2* KO brain. Importantly, in both cases, the increased proximal pA usage was associated with reciprocal decreased utilization of distal pAs, indicating that the observed APA isoform regulation is not due to difference in mRNA stability (Licatalosi et al. 2008). Interestingly, using qRT-PCR to quantify the relative abundance of APA isoforms of 29 additional Nova2 target candidates, the authors found that Nova2 suppressed and enhanced the usage of nine and three pAs, respectively (Licatalosi et al. 2008), indicating position- or context-dependent regulation.

ESRP1 (epithelial splicing regulatory protein 1) is an epithelial cell-specific splicing factor with binding specificity to UG-rich motifs (Dittmar et al. 2012; Warzecha et al. 2010; Warzecha et al. 2009). Using DRS (see above), 335 regulated APA events were identified between control and ESRP1-expressing cells. Additional analysis of RNA-seq data further revealed 160 potential regulated APA events. RT-PCR analysis using a gene-specific forward primer and isoform-specific reverse primers confirmed six cases (Dittmar et al. 2012). However, the mechanism for ESRP-mediated APA remains unknown.

SRm160 (serine/arginine repeat-related nuclear matrix protein of 160 kDa) is a component of the splicing-dependent exon-junction complex (EJC) (Blencowe et al. 1998; Le Hir et al. 2001). It was found to associate with CPSF and stimulate C/P of spliced substrates in a manner independent of other EJC subunits (McCracken et al. 2002; McCracken et al. 2003). SRm160 belongs to a group of RBPs called SR proteins, which all contain the arginine/serine (RS) domain and, like hnRNPs, play global roles in alternative splicing. How they are involved in APA is yet to be explored.

5 Discussion

With the advent of deep sequencing technologies, many methods have been developed for global analysis of APA (summarized in Table 3.1), enabling identification of APA events, and study of RBP impacts on APA. It is notable that almost all methods for identification of APA events are based on analysis of APA isoforms at the steady-state level. Because isoforms can have different half-lives owing to stabilizing or destabilizing elements in aUTRs, the readout of APA analysis experiments may not faithfully reflect the regulation of C/P. One solution is to examine the stability of APA isoforms at the same time (Licatalosi et al. 2008). Validation by biochemical assays to examine C/P in vitro or reporter assays to analyze the pA usage (see above) can also mitigate this problem. However, a deep sequencing-based method is still needed to globally examine the C/P activity for all pAs. Recently, it was shown that 4-thiouridine (4sU) added into cell culture can be incorporated into nascent transcripts within 5–15 min (Dolken et al. 2008; Windhager et al. 2012). Assuming that the decay of polyadenylated RNA would be insignificant within such a short window, 4sU labeling coupled with pA-based deep sequencing may help interrogate APA regulation globally.

RBPs in the C/P machinery play critical roles in recognizing the USEs and DSEs of pA. Not surprisingly, regulation of these RBPs impacts pA site choice. However, it is striking that different C/P RBPs have quite different effects on APA. For example, CFI-25/CFI-68 and PABPN1 appear to promote distal pAs, whereas CstF-64/CstF-64 τ seems to have the opposite effect. A more systematic approach to examine all RBPs in the C/P machinery is needed to unravel the detailed mechanisms by which the RBPs regulate pAs. In this vein, it is worth noting that some C/P factors are regulated by extensive posttranslational modifications such as phosphorylation, sumoylation, acetylation, and methylation (Ryan and Bauer 2008). How these modifications impact APA is completely unknown.

A growing number of RBPs previously known to regulate other aspects of RNA metabolism such as splicing, mRNA stability, and translation have been shown to regulate C/P (summarized in Table 3.2). Some RBPs inhibit the usage of target pA by blocking the assembly of the C/P complex or inhibiting the PAP activity, and some enhance the usage of pA by help recruit C/P factors. It is becoming a common theme that, like RBPs involved in splicing, RBP regulate C/P in a context-dependent fashion. Inhibitors of C/P may become enhancers when they associate with different proteins (Table 3.2). However, despite many findings on regulation of APA by RBPs, many questions still remain. How RBPs are involved in setting the tissue-specific APA pattern is to be fully elucidated; how RBPs interact with one another in a combinatorial way to regulate pA is largely unknown; how RNA structures play a role in APA regulation and its relevance to RBPs needs to be addressed; and how RBPs regulate the metabolism of APA isoforms thereby modulating gene expression awaits more analyses. Systems approaches to unravel the binding and functions of RBPs are expected to shed important light on these outstanding questions.

Table 3.2 Effects of RBPs on pA usage

RBP	Effect	Mechanism	Reference
Elav and Elavl (Hu) proteins	Sup	Binds to the U-rich sequences near pAs and blocks the binding of CstF-64.	Hilgers et al. (2012), Soller and White (2003), Dai et al. (2012), Mansfield and Keene (2012), Zhu et al. (2007)
ESRP	Act/Sup	Unknown.	Dittmar et al. (2012)
hnRNP F	Sup	Binds to G-rich sequences and competes with CstF-64 for binding to pAs.	Alkan et al. (2006), Veraldi et al. (2001)
hnRNP F/H1	Act	Unclear.	Decorsiere et al. (2011)
hnRNP H1 (hnRNPH)	Act	Unclear.	Katz et al. (2010)
hnRNP H2 (hnRNPH′ or DSEF-1)	Act	Binds to G-rich sequences near pAs and enhances CstF-64 binding to the pAs	Bagga et al. (1998)
hnRNP I (PTB)	Act/Sup	Competes with CstF-64 Recruits hnRNPH 1, which stimulates polyadenylation.	Castelo-Branco et al. (2004), Millevoi et al. (2009), Moreira et al. (1998)
hnRNP K	Sup	Competes with CFI-68 for binding to CFI-25, thereby inhibiting the binding of CFI complex to pA.	Naganuma et al. (2012)
hnRNP L	Act/Sup	Unknown.	Huang et al. (2012), Hung et al. (2008)
Nova	Act/Sup	Unknown.	Licatalosi et al. (2008)
PCBP (αCP2)	Act	Binds to C-rich motifs and recruits CPSF.	Ji et al. (2011a)
SRm160	Act	Associates with CPSF and stimulates C/P of spliced substrates.	McCracken et al. (2002, 2003)
U1 snRNP	Sup	U1 70K in the U1 snRNP interacts with PAP and inhibits its polyadenylation activity.	Gunderson et al. (1998), Abad et al. (2008), Fortes et al. (2003), Kaida et al. (2010), Berg et al. (2012)
U1A	Sup	Inhibits the recruitment of CstF-64; interacts with PAP and inhibits its polyadenylation activity	Phillips et al. (2004), Gunderson et al. (1994)
U2 snRNP	Act	Interacts with CPSF and promotes cleavage	Kyburz et al. (2006)
U2AF65	Act	Binds to the pyrimidine tract and recruits the CFI complex	Millevoi et al. (2006)

Sup suppress, *Act* activate

References

Abad X, Vera M, Jung SP, Oswald E, Romero I, Amin V, Fortes P, Gunderson SI (2008) Requirements for gene silencing mediated by U1 snRNA binding to a target sequence. Nucleic Acids Res 36:2338–2352

Alkan SA, Martincic K, Milcarek C (2006) The hnRNPs F and H2 bind to similar sequences to influence gene expression. Biochem J 393:361–371

Andreassi C, Riccio A (2009) To localize or not to localize: mRNA fate is in 3′ UTR ends. Trends Cell Biol 19:465–474

Bagga PS, Arhin GK, Wilusz J (1998) DSEF-1 is a member of the hnRNP H family of RNA-binding proteins and stimulates pre-mRNA cleavage and polyadenylation in vitro. Nucleic Acids Res 26:5343–5350

Bai Y, Auperin TC, Chou CY, Chang GG, Manley JL, Tong L (2007) Crystal structure of murine CstF-77: dimeric association and implications for polyadenylation of mRNA precursors. Mol Cell 25:863–875

Barabino SM, Hubner W, Jenny A, Minvielle-Sebastia L, Keller W (1997) The 30-kDa subunit of mammalian cleavage and polyadenylation specificity factor and its yeast homolog are RNA-binding zinc finger proteins. Genes Dev 11:1703–1716

Beaudoing E, Freier S, Wyatt JR, Claverie JM, Gautheret D (2000) Patterns of variant polyadenylation signal usage in human genes. Genome Res 10:1001–1010

Beaudoing E, Gautheret D (2001) Identification of alternate polyadenylation sites and analysis of their tissue distribution using EST data. Genome Res 11:1520–1526

Berg MG, Singh LN, Younis I, Liu Q, Pinto AM, Kaida D, Zhang Z, Cho S, Sherrill-Mix S, Wan L et al (2012) U1 snRNP determines mRNA length and regulates isoform expression. Cell 150:53–64

Blencowe BJ, Issner R, Nickerson JA, Sharp PA (1998) A coactivator of pre-mRNA splicing. Genes Dev 12:996–1009

Brown KM, Gilmartin GM (2003) A mechanism for the regulation of pre-mRNA 3′ processing by human cleavage factor Im. Mol Cell 12:1467–1476

Campos AR, Grossman D, White K (1985) Mutant alleles at the locus elav in Drosophila melanogaster lead to nervous system defects. A developmental-genetic analysis. J Neurogenet 2:197–218

Castelo-Branco P, Furger A, Wollerton M, Smith C, Moreira A, Proudfoot N (2004) Polypyrimidine tract binding protein modulates efficiency of polyadenylation. Mol Cell Biol 24:4174–4183

Chan S, Choi EA, Shi Y (2011) Pre-mRNA 3′-end processing complex assembly and function. Wiley Interdiscip Rev RNA 2:321–335

Chen F, MacDonald CC, Wilusz J (1995) Cleavage site determinants in the mammalian polyadenylation signal. Nucleic Acids Res 23:2614–2620

Cheng Y, Miura RM, Tian B (2006) Prediction of mRNA polyadenylation sites by support vector machine. Bioinformatics 22:2320–2325

Chkheidze AN, Liebhaber SA (2003) A novel set of nuclear localization signals determine distributions of the alphaCP RNA-binding proteins. Mol Cell Biol 23:8405–8415

Chkheidze AN, Lyakhov DL, Makeyev AV, Morales J, Kong J, Liebhaber SA (1999) Assembly of the alpha-globin mRNA stability complex reflects binary interaction between the pyrimidine-rich 3′ untranslated region determinant and poly(C) binding protein alphaCP. Mol Cell Biol 19:4572–4581

Colgan DF, Manley JL (1997) Mechanism and regulation of mRNA polyadenylation. Genes Dev 11:2755–2766

D'Mello V, Lee JY, MacDonald CC, Tian B (2006) Alternative mRNA polyadenylation can potentially affect detection of gene expression by affymetrix genechip arrays. Appl Bioinformatics 5:249–253

Dai W, Zhang G, Makeyev EV (2012) RNA-binding protein HuR autoregulates its expression by promoting alternative polyadenylation site usage. Nucleic Acids Res 40:787–800

Danckwardt S, Kaufmann I, Gentzel M, Foerstner KU, Gantzert AS, Gehring NH, Neu-Yilik G, Bork P, Keller W, Wilm M et al (2007) Splicing factors stimulate polyadenylation via USEs at non-canonical 3′ end formation signals. EMBO J 26:2658–2669

Dantonel JC, Murthy KG, Manley JL, Tora L (1997) Transcription factor TFIID recruits factor CPSF for formation of 3′ end of mRNA. Nature 389:399–402

de Klerk E, Venema A, Anvar SY, Goeman JJ, Hu O, Trollet C, Dickson G, den Dunnen JT, van der Maarel SM, Raz V et al (2012) Poly(A) binding protein nuclear 1 levels affect alternative polyadenylation. Nucleic Acids Res 40:9089–9101

de Vries H, Ruegsegger U, Hubner W, Friedlein A, Langen H, Keller W (2000) Human pre-mRNA cleavage factor II(m) contains homologs of yeast proteins and bridges two other cleavage factors. EMBO J 19:5895–5904

Decorsiere A, Cayrel A, Vagner S, Millevoi S (2011) Essential role for the interaction between hnRNP H/F and a G quadruplex in maintaining p53 pre-mRNA 3′-end processing and function during DNA damage. Genes Dev 25:220–225

Derti A, Garrett-Engele P, Macisaac KD, Stevens RC, Sriram S, Chen R, Rohl CA, Johnson JM, Babak T (2012) A quantitative atlas of polyadenylation in five mammals. Genome Res 22: 1173–1183

Di Giammartino DC, Nishida K, Manley JL (2011) Mechanisms and consequences of alternative polyadenylation. Mol Cell 43:853–866

Di Giammartino DC, Shi Y, Manley JL (2013) PARP1 represses PAP and inhibits polyadenylation during heat shock. Mol Cell 49:7–17

Dichtl B, Keller W (2001) Recognition of polyadenylation sites in yeast pre-mRNAs by cleavage and polyadenylation factor. EMBO J 20:3197–3209

Dittmar KA, Jiang P, Park JW, Amirikian K, Wan J, Shen S, Xing Y, Carstens RP (2012) Genome-wide determination of a broad ESRP-regulated posttranscriptional network by high-throughput sequencing. Mol Cell Biol 32:1468–1482

Dolken L, Ruzsics Z, Radle B, Friedel CC, Zimmer R, Mages J, Hoffmann R, Dickinson P, Forster T, Ghazal P et al (2008) High-resolution gene expression profiling for simultaneous kinetic parameter analysis of RNA synthesis and decay. RNA 14:1959–1972

Dreyfuss G, Kim VN, Kataoka N (2002) Messenger-RNA-binding proteins and the messages they carry. Nat Rev Mol Cell Biol 3:195–205

Elkon R, Drost J, van Haaften G, Jenal M, Schrier M, Vrielink JA, Agami R (2012) E2F mediates enhanced alternative polyadenylation in proliferation. Genome Biol 13:R59

Flavell SW, Kim TK, Gray JM, Harmin DA, Hemberg M, Hong EJ, Markenscoff-Papadimitriou E, Bear DM, Greenberg ME (2008) Genome-wide analysis of MEF2 transcriptional program reveals synaptic target genes and neuronal activity-dependent polyadenylation site selection. Neuron 60:1022–1038

Fortes P, Cuevas Y, Guan F, Liu P, Pentlicky S, Jung SP, Martinez-Chantar ML, Prieto J, Rowe D, Gunderson SI (2003) Inhibiting expression of specific genes in mammalian cells with 5′ end-mutated U1 small nuclear RNAs targeted to terminal exons of pre-mRNA. Proc Natl Acad Sci U S A 100:8264–8269

Fox-Walsh K, Davis-Turak J, Zhou Y, Li H, Fu XD (2011) A multiplex RNA-seq strategy to profile poly(A+) RNA: application to analysis of transcription response and 3′ end formation. Genomics 98:266–271

Fu Y, Sun Y, Li Y, Li J, Rao X, Chen C, Xu A (2011) Differential genome-wide profiling of tandem 3′ UTRs among human breast cancer and normal cells by high-throughput sequencing. Genome Res 21:741–747

Gautheret D, Poirot O, Lopez F, Audic S, Claverie JM (1998) Alternate polyadenylation in human mRNAs: a large-scale analysis by EST clustering. Genome Res 8:524–530

Gilat R, Goncharov S, Esterman N, Shweiki D (2006) Under-representation of PolyA/PolyT tailed ESTs in human ESTdb: an obstacle to alternative polyadenylation inference. Bioinformation 1:220–224

Glover-Cutter K, Kim S, Espinosa J, Bentley DL (2008) RNA polymerase II pauses and associates with pre-mRNA processing factors at both ends of genes. Nat Struct Mol Biol 15:71–78

Gruber AR, Martin G, Keller W, Zavolan M (2012) Cleavage factor Im is a key regulator of 3′ UTR length. RNA Biol 9:1405–1412

Gunderson SI, Beyer K, Martin G, Keller W, Boelens WC, Mattaj LW (1994) The human U1A snRNP protein regulates polyadenylation via a direct interaction with poly(A) polymerase. Cell 76:531–541

Gunderson SI, Polycarpou-Schwarz M, Mattaj IW (1998) U1 snRNP inhibits pre-mRNA polyadenylation through a direct interaction between U1 70K and poly(A) polymerase. Mol Cell 1:255–264

Hafner M, Renwick N, Brown M, Mihailovic A, Holoch D, Lin C, Pena JT, Nusbaum JD, Morozov P, Ludwig J et al (2011) RNA-ligase-dependent biases in miRNA representation in deep-sequenced small RNA cDNA libraries. RNA 17:1697–1712

Hilgers V, Lemke SB, Levine M (2012) ELAV mediates 3′ UTR extension in the Drosophila nervous system. Genes Dev 26:2259–2264

Hofmann I, Schnolzer M, Kaufmann I, Franke WW (2002) Symplekin, a constitutive protein of karyo- and cytoplasmic particles involved in mRNA biogenesis in Xenopus laevis oocytes. Mol Biol Cell 13:1665–1676

Hoque M, Ji Z, Zheng D, Luo W, Li W, You B, Park JY, Yehia G, Tian B (2013) Analysis of alternative cleavage and polyadenylation by 3′ region extraction and deep sequencing. Nat Methods 10:133–139

Hu J, Lutz CS, Wilusz J, Tian B (2005) Bioinformatic identification of candidate cis-regulatory elements involved in human mRNA polyadenylation. RNA 11:1485–1493

Huang Y, Li W, Yao X, Lin QJ, Yin JW, Liang Y, Heiner M, Tian B, Hui J, Wang G (2012) Mediator complex regulates alternative mRNA processing via the MED23 subunit. Mol Cell 45: 459–469

Hung LH, Heiner M, Hui J, Schreiner S, Benes V, Bindereif A (2008) Diverse roles of hnRNP L in mammalian mRNA processing: a combined microarray and RNAi analysis. RNA 14:284–296

Jan CH, Friedman RC, Ruby JG, Bartel DP (2011) Formation, regulation and evolution of Caenorhabditis elegans 3′ UTRs. Nature 469:97–101

Jenal M, Elkon R, Loayza-Puch F, van Haaften G, Kuhn U, Menzies FM, Oude Vrielink JA, Bos AJ, Drost J, Rooijers K et al (2012) The poly(A)-binding protein nuclear 1 suppresses alternative cleavage and polyadenylation sites. Cell 149:538–553

Ji X, Kong J, Liebhaber SA (2011a) An RNA-protein complex links enhanced nuclear 3′ processing with cytoplasmic mRNA stabilization. EMBO J 30:2622–2633

Ji X, Wan J, Vishnu M, Xing Y, Liebhaber SA (2013) The poly-C binding proteins, alphaCPs, act as global regulators of alternative polyadenylation. Mol Cell Biol 33(13):2560–2573

Ji Z, Lee JY, Pan Z, Jiang B, Tian B (2009a) Progressive lengthening of 3′ untranslated regions of mRNAs by alternative polyadenylation during mouse embryonic development. Proc Natl Acad Sci U S A 106:7028–7033

Ji Z, Lee JY, Pan Z, Jiang B, Tian B (2009b) Progressive lengthening of 3′ untranslated regions of mRNAs by alternative polyadenylation during mouse embryonic development. Proc Natl Acad Sci U S A 106:7028–7033

Ji Z, Luo W, Li W, Hoque M, Pan Z, Zhao Y, Tian B (2011b) Transcriptional activity regulates alternative cleavage and polyadenylation. Mol Syst Biol 7:534

Ji Z, Tian B (2009) Reprogramming of 3′ untranslated regions of mRNAs by alternative polyadenylation in generation of pluripotent stem cells from different cell types. PLoS One 4:e8419

Kaida D, Berg MG, Younis I, Kasim M, Singh LN, Wan L, Dreyfuss G (2010) U1 snRNP protects pre-mRNAs from premature cleavage and polyadenylation. Nature 468:664–668

Katz Y, Wang ET, Airoldi EM, Burge CB (2010) Analysis and design of RNA sequencing experiments for identifying isoform regulation. Nat Methods 7:1009–1015

Kaufmann I, Martin G, Friedlein A, Langen H, Keller W (2004) Human Fip1 is a subunit of CPSF that binds to U-rich RNA elements and stimulates poly(A) polymerase. EMBO J 23:616–626

Kerwitz Y, Kuhn U, Lilie H, Knoth A, Scheuermann T, Friedrich H, Schwarz E, Wahle E (2003) Stimulation of poly(A) polymerase through a direct interaction with the nuclear poly(A) binding protein allosterically regulated by RNA. EMBO J 22:3705–3714

Kiledjian M, Wang X, Liebhaber SA (1995) Identification of two KH domain proteins in the alpha-globin mRNP stability complex. EMBO J 14:4357–4364

Koushika SP, Lisbin MJ, White K (1996) ELAV, a Drosophila neuron-specific protein, mediates the generation of an alternatively spliced neural protein isoform. Curr Biol 6:1634–1641

Kozarewa I, Ning Z, Quail MA, Sanders MJ, Berriman M, Turner DJ (2009) Amplification-free Illumina sequencing-library preparation facilitates improved mapping and assembly of (G+C)-biased genomes. Nat Methods 6:291–295

Kubo T, Wada T, Yamaguchi Y, Shimizu A, Handa H (2006) Knock-down of 25 kDa subunit of cleavage factor Im in Hela cells alters alternative polyadenylation within 3′-UTRs. Nucleic Acids Res 34:6264–6271

Kuhn U, Gundel M, Knoth A, Kerwitz Y, Rudel S, Wahle E (2009) Poly(A) tail length is controlled by the nuclear poly(A)-binding protein regulating the interaction between poly(A) polymerase and the cleavage and polyadenylation specificity factor. J Biol Chem 284:22803–22814

Kyburz A, Friedlein A, Langen H, Keller W (2006) Direct interactions between subunits of CPSF and the U2 snRNP contribute to the coupling of pre-mRNA 3′ end processing and splicing. Mol Cell 23:195–205

Laishram RS, Anderson RA (2010) The poly A polymerase Star-PAP controls 3′-end cleavage by promoting CPSF interaction and specificity toward the pre-mRNA. EMBO J 29:4132–4145

Le Hir H, Gatfield D, Izaurralde E, Moore MJ (2001) The exon-exon junction complex provides a binding platform for factors involved in mRNA export and nonsense-mediated mRNA decay. EMBO J 20:4987–4997

Legrand P, Pinaud N, Minvielle-Sebastia L, Fribourg S (2007) The structure of the CstF-77 homodimer provides insights into CstF assembly. Nucleic Acids Res 35:4515–4522

Li H, Tong S, Li X, Shi H, Ying Z, Gao Y, Ge H, Niu L, Teng M (2011) Structural basis of pre-mRNA recognition by the human cleavage factor Im complex. Cell Res 21:1039–1051

Li Y, Sun Y, Fu Y, Li M, Huang G, Zhang C, Liang J, Huang S, Shen G, Yuan S et al (2012) Dynamic landscape of tandem 3′ UTRs during zebrafish development. Genome Res 22:1899–1906

Licatalosi DD, Mele A, Fak JJ, Ule J, Kayikci M, Chi SW, Clark TA, Schweitzer AC, Blume JE, Wang X et al (2008) HITS-CLIP yields genome-wide insights into brain alternative RNA processing. Nature 456:464–469

Lisbin MJ, Qiu J, White K (2001) The neuron-specific RNA-binding protein ELAV regulates neuroglian alternative splicing in neurons and binds directly to its pre-mRNA. Genes Dev 15:2546–2561

Liu D, Brockman JM, Dass B, Hutchins LN, Singh P, McCarrey JR, MacDonald CC, Graber JH (2007) Systematic variation in mRNA 3′-processing signals during mouse spermatogenesis. Nucleic Acids Res 35:234–246

Luo C, Tsementzi D, Kyrpides N, Read T, Konstantinidis KT (2012) Direct comparisons of Illumina vs. Roche 454 sequencing technologies on the same microbial community DNA sample. PLoS One 7:e30087

Lutz CS, Moreira A (2011) Alternative mRNA polyadenylation in eukaryotes: an effective regulator of gene expression. WIREs RNA 2:23–31

Mandel CR, Kaneko S, Zhang H, Gebauer D, Vethantham V, Manley JL, Tong L (2006) Polyadenylation factor CPSF-73 is the pre-mRNA 3′-end-processing endonuclease. Nature 444:953–956

Mansfield KD, Keene JD (2012) Neuron-specific ELAV/Hu proteins suppress HuR mRNA during neuronal differentiation by alternative polyadenylation. Nucleic Acids Res 40:2734–2746

Martin G, Gruber AR, Keller W, Zavolan M (2012) Genome-wide analysis of pre-mRNA 3′ end processing reveals a decisive role of human cleavage factor I in the regulation of 3′ UTR length. Cell Rep 1:753–763

Martincic K, Campbell R, Edwalds-Gilbert G, Souan L, Lotze MT, Milcarek C (1998) Increase in the 64-kDa subunit of the polyadenylation/cleavage stimulatory factor during the G0 to S phase transition. Proc Natl Acad Sci U S A 95:11095–11100

Marzluff WF, Wagner EJ, Duronio RJ (2008) Metabolism and regulation of canonical histone mRNAs: life without a poly(A) tail. Nat Rev Genet 9:843–854
Mayr C, Bartel DP (2009) Widespread shortening of 3′ UTRs by alternative cleavage and polyadenylation activates oncogenes in cancer cells. Cell 138:673–684
McCracken S, Lambermon M, Blencowe BJ (2002) SRm160 splicing coactivator promotes transcript 3′-end cleavage. Mol Cell Biol 22:148–160
McCracken S, Longman D, Johnstone IL, Caceres JF, Blencowe BJ (2003) An evolutionarily conserved role for SRm160 in 3′-end processing that functions independently of exon junction complex formation. J Biol Chem 278:44153–44160
Millevoi S, Decorsiere A, Loulergue C, Iacovoni J, Bernat S, Antoniou M, Vagner S (2009) A physical and functional link between splicing factors promotes pre-mRNA 3′ end processing. Nucleic Acids Res 37:4672–4683
Millevoi S, Loulergue C, Dettwiler S, Karaa SZ, Keller W, Antoniou M, Vagner S (2006) An interaction between U2AF 65 and CF I(m) links the splicing and 3′ end processing machineries. EMBO J 25:4854–4864
Miura P, Shenker S, Andreu-Agullo C, Westholm JO, Lai EC (2013) Widespread and extensive lengthening of 3′ UTRs in the mammalian brain. Genome Res 23(5):812–825
Monarez RR, MacDonald CC, Dass B (2007) Polyadenylation proteins CstF-64 and tauCstF-64 exhibit differential binding affinities for RNA polymers. Biochem J 401:651–658
Moreira A, Takagaki Y, Brackenridge S, Wollerton M, Manley JL, Proudfoot NJ (1998) The upstream sequence element of the C2 complement poly(A) signal activates mRNA 3′ end formation by two distinct mechanisms. Genes Dev 12:2522–2534
Moreno-Morcillo M, Minvielle-Sebastia L, Mackereth C, Fribourg S (2011) Hexameric architecture of CstF supported by CstF-50 homodimerization domain structure. RNA 17:412–418
Morris AR, Bos A, Diosdado B, Rooijers K, Elkon R, Bolijn AS, Carvalho B, Meijer GA, Agami R (2012) Alternative cleavage and polyadenylation during colorectal cancer development. Clin Cancer Res 18:5256–5266
Murthy KG, Manley JL (1995) The 160-kD subunit of human cleavage-polyadenylation specificity factor coordinates pre-mRNA 3′-end formation. Genes Dev 9:2672–2683
Nag A, Narsinh K, Martinson HG (2007) The poly(A)-dependent transcriptional pause is mediated by CPSF acting on the body of the polymerase. Nat Struct Mol Biol 14:662–669
Nagaike T, Logan C, Hotta I, Rozenblatt-Rosen O, Meyerson M, Manley JL (2011) Transcriptional activators enhance polyadenylation of mRNA precursors. Mol Cell 41:409–418
Naganuma T, Nakagawa S, Tanigawa A, Sasaki YF, Goshima N, Hirose T (2012) Alternative 3′-end processing of long noncoding RNA initiates construction of nuclear paraspeckles. EMBO J 31:4020–4034
Nam DK, Lee S, Zhou G, Cao X, Wang C, Clark T, Chen J, Rowley JD, Wang SM (2002) Oligo(dT) primer generates a high frequency of truncated cDNAs through internal poly(A) priming during reverse transcription. Proc Natl Acad Sci U S A 99:6152–6156
Newman CS, Krieg PA (1999) Ribonuclease protection analysis of gene expression in Xenopus. Methods Mol Biol 127:29–40
Nunes NM, Li W, Tian B, Furger A (2010) A functional human Poly(A) site requires only a potent DSE and an A-rich upstream sequence. EMBO J 29:1523–1536
Okubo K, Hori N, Matoba R, Niiyama T, Fukushima A, Kojima Y, Matsubara K (1992) Large scale cDNA sequencing for analysis of quantitative and qualitative aspects of gene expression. Nat Genet 2:173–179
Ozsolak F, Kapranov P, Foissac S, Kim SW, Fishilevich E, Monaghan AP, John B, Milos PM (2010) Comprehensive polyadenylation site maps in yeast and human reveal pervasive alternative polyadenylation. Cell 143:1018–1029
Ozsolak F, Milos PM (2011) Transcriptome profiling using single-molecule direct RNA sequencing. Methods Mol Biol 733:51–61
Ozsolak F, Platt AR, Jones DR, Reifenberger JG, Sass LE, McInerney P, Thompson JF, Bowers J, Jarosz M, Milos PM (2009) Direct RNA sequencing. Nature 461:814–818

Pan Z, Zhang H, Hague LK, Lee JY, Lutz CS, Tian B (2006) An intronic polyadenylation site in human and mouse CstF-77 genes suggests an evolutionarily conserved regulatory mechanism. Gene 366:325–334

Pauws E, van Kampen AH, van de Graaf SA, de Vijlder JJ, Ris-Stalpers C (2001) Heterogeneity in polyadenylation cleavage sites in mammalian mRNA sequences: implications for SAGE analysis. Nucleic Acids Res 29:1690–1694

Perez Canadillas JM, Varani G (2003) Recognition of GU-rich polyadenylation regulatory elements by human CstF-64 protein. EMBO J 22:2821–2830

Phillips C, Pachikara N, Gunderson SI (2004) U1A inhibits cleavage at the immunoglobulin M heavy-chain secretory poly(A) site by binding between the two downstream GU-rich regions. Mol Cell Biol 24:6162–6171

Prasanth KV, Prasanth SG, Xuan Z, Hearn S, Freier SM, Bennett CF, Zhang MQ, Spector DL (2005) Regulating gene expression through RNA nuclear retention. Cell 123:249–263

Proudfoot NJ, Brownlee GG (1976) 3′ non-coding region sequences in eukaryotic messenger RNA. Nature 263:211–214

Quail MA, Smith M, Coupland P, Otto TD, Harris SR, Connor TR, Bertoni A, Swerdlow HP, Gu Y (2012) A tale of three next generation sequencing platforms: comparison of Ion Torrent, Pacific Biosciences and Illumina MiSeq sequencers. BMC Genomics 13:341

Roca X, Karginov FV (2012) RNA biology in a test tube—an overview of in vitro systems/assays. Wiley Interdiscip Rev RNA 3:509–527

Ruegsegger U, Beyer K, Keller W (1996) Purification and characterization of human cleavage factor Im involved in the 3′ end processing of messenger RNA precursors. J Biol Chem 271:6107–6113

Ryan K, Bauer DL (2008) Finishing touches: post-translational modification of protein factors involved in mammalian pre-mRNA 3′ end formation. Int J Biochem Cell Biol 40:2384–2396

Sandberg R, Neilson JR, Sarma A, Sharp PA, Burge CB (2008) Proliferating cells express mRNAs with shortened 3′ untranslated regions and fewer microRNA target sites. Science 320:1643–1647

Sartini BL, Wang H, Wang W, Millette CF, Kilpatrick DL (2008) Pre-messenger RNA cleavage factor I (CFIm): potential role in alternative polyadenylation during spermatogenesis. Biol Reprod 78:472–482

Scotto-Lavino E, Du G, Frohman MA (2006) 3′ end cDNA amplification using classic RACE. Nat Protoc 1:2742–2745

Shepard PJ, Choi EA, Lu J, Flanagan LA, Hertel KJ, Shi Y (2011) Complex and dynamic landscape of RNA polyadenylation revealed by PAS-Seq. RNA 17:761–772

Sherstnev A, Duc C, Cole C, Zacharaki V, Hornyik C, Ozsolak F, Milos PM, Barton GJ, Simpson GG (2012) Direct sequencing of Arabidopsis thaliana RNA reveals patterns of cleavage and polyadenylation. Nat Struct Mol Biol 19:845–852

Shi Y, Di Giammartino DC, Taylor D, Sarkeshik A, Rice WJ, Yates JR 3rd, Frank J, Manley JL (2009) Molecular architecture of the human pre-mRNA 3′ processing complex. Mol Cell 33:365–376

Siegel TN, Hekstra DR, Wang X, Dewell S, Cross GA (2010) Genome-wide analysis of mRNA abundance in two life-cycle stages of Trypanosoma brucei and identification of splicing and polyadenylation sites. Nucleic Acids Res 38:4946–4957

Singh P, Alley TL, Wright SM, Kamdar S, Schott W, Wilpan RY, Mills KD, Graber JH (2009) Global changes in processing of mRNA 3′ untranslated regions characterize clinically distinct cancer subtypes. Cancer Res 69:9422–9430

Smibert P, Miura P, Westholm JO, Shenker S, May G, Duff MO, Zhang D, Eads BD, Carlson J, Brown JB et al (2012) Global patterns of tissue-specific alternative polyadenylation in Drosophila. Cell Rep 1:277–289

Soller M, White K (2003) ELAV inhibits 3′-end processing to promote neural splicing of ewg pre-mRNA. Genes Dev 17:2526–2538

Soller M, White K (2005) ELAV multimerizes on conserved AU4-6 motifs important for ewg splicing regulation. Mol Cell Biol 25:7580–7591

Sorefan K, Pais H, Hall AE, Kozomara A, Griffiths-Jones S, Moulton V, Dalmay T (2012) Reducing ligation bias of small RNAs in libraries for next generation sequencing. Silence 3:4

Takagaki Y, Manley JL (2000) Complex protein interactions within the human polyadenylation machinery identify a novel component. Mol Cell Biol 20:1515–1525

Takagaki Y, Seipelt RL, Peterson ML, Manley JL (1996) The polyadenylation factor CstF-64 regulates alternative processing of IgM heavy chain pre-mRNA during B cell differentiation. Cell 87:941–952

Tian B, Graber JH (2012) Signals for pre-mRNA cleavage and polyadenylation. Wiley Interdiscip Rev RNA 3:385–396

Tian B, Hu J, Zhang H, Lutz CS (2005) A large-scale analysis of mRNA polyadenylation of human and mouse genes. Nucleic Acids Res 33:201–212

Tian B, Manley JL (2013) Alternative cleavage and polyadenylation: the long and short of it. Trends Biochem Sci 38:312–320

Tian B, Pan Z, Lee JY (2007) Widespread mRNA polyadenylation events in introns indicate dynamic interplay between polyadenylation and splicing. Genome Res 17:156–165

Ule J, Stefani G, Mele A, Ruggiu M, Wang X, Taneri B, Gaasterland T, Blencowe BJ, Darnell RB (2006) An RNA map predicting Nova-dependent splicing regulation. Nature 444:580–586

Ulitsky I, Shkumatava A, Jan CH, Subtelny AO, Koppstein D, Bell GW, Sive H, Bartel DP (2012) Extensive alternative polyadenylation during zebrafish development. Genome Res 22: 2054–2066

Velculescu VE, Zhang L, Vogelstein B, Kinzler KW (1995) Serial analysis of gene expression. Science 270:484–487

Venkataraman K, Brown KM, Gilmartin GM (2005) Analysis of a noncanonical poly(A) site reveals a tripartite mechanism for vertebrate poly(A) site recognition. Genes Dev 19:1315–1327

Veraldi KL, Arhin GK, Martincic K, Chung-Ganster LH, Wilusz J, Milcarek C (2001) hnRNP F influences binding of a 64-kilodalton subunit of cleavage stimulation factor to mRNA precursors in mouse B cells. Mol Cell Biol 21:1228–1238

Wang ET, Sandberg R, Luo S, Khrebtukova I, Zhang L, Mayr C, Kingsmore SF, Schroth GP, Burge CB (2008) Alternative isoform regulation in human tissue transcriptomes. Nature 456: 470–476

Warzecha CC, Jiang P, Amirikian K, Dittmar KA, Lu H, Shen S, Guo W, Xing Y, Carstens RP (2010) An ESRP-regulated splicing programme is abrogated during the epithelial-mesenchymal transition. EMBO J 29:3286–3300

Warzecha CC, Shen S, Xing Y, Carstens RP (2009) The epithelial splicing factors ESRP1 and ESRP2 positively and negatively regulate diverse types of alternative splicing events. RNA Biol 6:546–562

Weiss IM, Liebhaber SA (1995) Erythroid cell-specific mRNA stability elements in the alpha 2-globin 3′ nontranslated region. Mol Cell Biol 15:2457–2465

West S, Proudfoot NJ (2008) Human Pcf11 enhances degradation of RNA polymerase II-associated nascent RNA and transcriptional termination. Nucleic Acids Res 36:905–914

Wilhelm BT, Marguerat S, Watt S, Schubert F, Wood V, Goodhead I, Penkett CJ, Rogers J, Bahler J (2008) Dynamic repertoire of a eukaryotic transcriptome surveyed at single-nucleotide resolution. Nature 453:1239–1243

Windhager L, Bonfert T, Burger K, Ruzsics Z, Krebs S, Kaufmann S, Malterer G, L'Hernault A, Schilhabel M, Schreiber S et al (2012) Ultrashort and progressive 4sU-tagging reveals key characteristics of RNA processing at nucleotide resolution. Genome Res 22:2031–2042

Wu X, Liu M, Downie B, Liang C, Ji G, Li QQ, Hunt AG (2011) Genome-wide landscape of polyadenylation in Arabidopsis provides evidence for extensive alternative polyadenylation. Proc Natl Acad Sci U S A 108:12533–12538

Xue Y, Zhou Y, Wu T, Zhu T, Ji X, Kwon YS, Zhang C, Yeo G, Black DL, Sun H et al (2009) Genome-wide analysis of PTB-RNA interactions reveals a strategy used by the general splicing repressor to modulate exon inclusion or skipping. Mol Cell 36:996–1006

Yan J, Marr TG (2005) Computational analysis of 3′-ends of ESTs shows four classes of alternative polyadenylation in human, mouse, and rat. Genome Res 15:369–375

Yang Q, Coseno M, Gilmartin GM, Doublie S (2011) Crystal structure of a human cleavage factor CFI(m)25/CFI(m)68/RNA complex provides an insight into poly(A) site recognition and RNA looping. Structure 19:368–377

Yang Q, Gilmartin GM, Doublie S (2010) Structural basis of UGUA recognition by the Nudix protein CFI(m)25 and implications for a regulatory role in mRNA 3′ processing. Proc Natl Acad Sci U S A 107:10062–10067

Yano M, Hayakawa-Yano Y, Mele A, Darnell RB (2010) Nova2 regulates neuronal migration through an RNA switch in disabled-1 signaling. Neuron 66:848–858

Yao C, Biesinger J, Wan J, Weng L, Xing Y, Xie X, Shi Y (2012) Transcriptome-wide analyses of CstF64-RNA interactions in global regulation of mRNA alternative polyadenylation. Proc Natl Acad Sci U S A 109:18773–18778

Zhang H, Lee JY, Tian B (2005) Biased alternative polyadenylation in human tissues. Genome Biol 6:R100

Zhu H, Zhou HL, Hasman RA, Lou H (2007) Hu proteins regulate polyadenylation by blocking sites containing U-rich sequences. J Biol Chem 282:2203–2210

Zhuang F, Fuchs RT, Sun Z, Zheng Y, Robb GB (2012) Structural bias in T4 RNA ligase-mediated 3′-adapter ligation. Nucleic Acids Res 40:e54

Chapter 4
Functional Analysis of Long Noncoding RNAs in Development and Disease

Ling-Ling Chen and Jing Crystal Zhao

Abstract Once viewed as part of the "dark matter" of genome, long noncoding RNAs (lncRNAs), which are mRNA-like but lack open reading frames, have emerged as an integral part of the mammalian transcriptome. Recent work demonstrated that lncRNAs play multiple structural and functional roles, and their analysis has become a new frontier in biomedical research. In this chapter, we provide an overview of different lncRNA families, describe methodologies available to study lncRNA–protein and lncRNA–DNA interactions systematically, and use well-studied lncRNAs as examples to illustrate their functional importance during normal development and in disease states.

Keywords Epigenetics • Long noncoding RNA • Xist • sno-lncRNA • Cancer • RIP-seq • CLIP-seq • Chart • ChIRP

1 Introduction

In 2002, following high throughput sequencing of mouse cDNA libraries, Okazaki et al. revealed that a vast proportion of the mammalian transcriptome does not code for proteins and defined long noncoding RNAs (lncRNAs) as a significant transcript class (Okazaki et al. 2002). Ten years later, the ENCODE (Encyclopedia of DNA Elements) study reported the existence of over 9,640 lncRNA loci in the human genome, roughly half the number of protein-coding genes (Djebali et al. 2012). These studies have changed our view of the mammalian genome and highlighted the importance of understanding lncRNA function.

L.-L. Chen (✉)
State Key Laboratory of Molecular Biology, Institute of Biochemistry and Cell Biology, Shanghai Institutes for Biological Sciences, Chinese Academy of Sciences, Shanghai 200031, China
e-mail: linglingchen@sibcb.ac.cn

J.C. Zhao
RNA Biology Program, Sanford Burnham Medical Research Institute, La Jolla, CA 92037, USA
e-mail: czhao@sanfordburnham.org

G.W. Yeo (ed.), *Systems Biology of RNA Binding Proteins*, Advances in Experimental Medicine and Biology 825, DOI 10.1007/978-1-4939-1221-6_4

The biogenesis of lncRNAs is similar to that of mRNA in that many lncRNAs are transcribed by RNA polymerase II and have a 5′ cap and 3′ polyadenylation signal. Their size ranges widely from 200 bp to over 100 kb. lncRNAs can reside in either nuclear or cytoplasmic compartments and be soluble or chromatin-bound. Unlike small noncoding RNAs, such as microRNAs or piRNAs, lncRNAs function via diverse mechanisms and comprise different classes.

2 Systematic Discovery of Different lncRNA Classes

The completion of sequencing of the human and mouse genomes together with recent analyses of their transcriptional outputs have revealed that about 80 % of the mammalian genome is transcribed in a cell-specific manner, leading to a new understanding of transcriptional regulation, particularly of noncoding regions (Djebali et al. 2012; Dunham et al. 2012). Earlier studies exploiting the increased sensitivity of genome tiling arrays (Katayama et al. 2005; Kapranov et al. 2007; Preker et al. 2008), together with asymmetric strand-specific analysis of gene expression (ASSAGE) (He et al. 2008) and global run-on sequencing (GRO-seq) (Core et al. 2008), have revealed widespread antisense transcription (natural antisense transcripts, NATs) and promoter-associated transcripts (such as promoter-associated long RNAs, PALRs; promoter-upstream transcripts, PROMPTs) in mammalian cells (Fig. 4.1). More significant progress has been made in discovery of novel lncRNAs following improvement in RNA sequencing (RNA-seq) and application of integrated methodologies. Several studies have systematically identified lncRNAs in a variety of organisms by monitoring patterns of "K4-K36" chromatin modification (Khalil et al. 2009; Cabili et al. 2011; Ulitsky et al. 2011; Guttman and Rinn 2012). For example, it was determined that an intergenic transcript could be defined

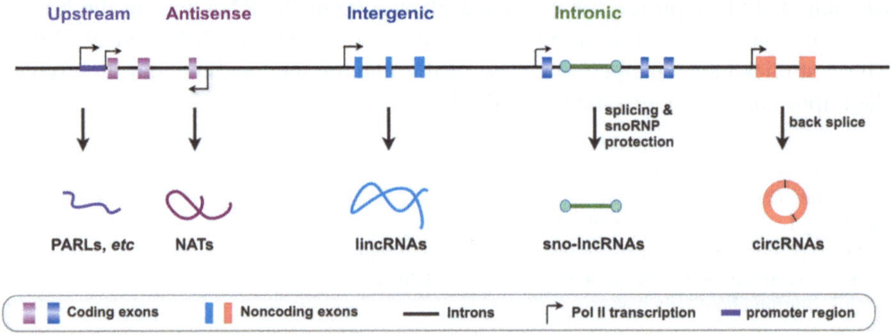

Fig. 4.1 Pervasive transcription of various classes of lncRNA in mammalian genomes. *PARLs* promoter-associated long RNAs, *NATs* natural antisense transcripts, *lincRNAs* large intergenic noncoding RNAs, *sno-lncRNAs* snoRNA-related lncRNAs, *circRNAs* circular RNAs. See text for details

as a potential lncRNA if it exhibited histone 3 Lys 4 trimethylation (H3K4me3) in its promoter region followed by histone 3 Lys 36 trimethylation (H3K36me3) across its actively transcribed region. Other criteria have also been applied to evaluate whether transcripts are true lncRNAs, including the presence of a transcription initiation signal and polyA site, potential coding capacity, orthology features, and analysis of its expression pattern among tissue samples. Since many lncRNAs are transcribed from discrete loci in previously unannotated intergenic regions between protein-coding genes, they were termed large intergenic noncoding RNAs (lincRNAs) (Fig. 4.1). A survey of the entire mouse or human genome using these chromatin marks in numerous cell types or tissues revealed that 5,000–8,000 K4-K36 domains represented lincRNAs (Guttman and Rinn 2012; Cabili et al. 2011).

While a large proportion of the thousands of "K4-K36" lincRNAs identified by chromatin signature are polyadenylated (Cabili et al. 2011; Derrien et al. 2012), recent studies indicate that a number of Pol II-transcribed long noncoding RNAs are processed in alternative ways (Wilusz et al. 2008; Sunwoo et al. 2009; Burd et al. 2010; Yap et al. 2010; Hansen et al. 2011, 2013; Salzman et al. 2012; Yin et al. 2012; Jeck et al. 2013; Memczak et al. 2013). *MALAT1* (also called *NEAT2*) and *Menβ* (also called *NEAT1_2*), which are both nuclear, are processed at their 3′ ends by RNase P (which processes the 5′ ends of tRNAs) (Wilusz et al. 2008; Sunwoo et al. 2009). RNase P cleavage leads to formation of the mature 3′ end of the lncRNA, which is protected by a highly conserved triple helical structures (Brown et al. 2012; Wilusz et al. 2012). Investigators using RNA-seq to define the population of non-polyadenylated "poly(A)-" transcripts in the human transcriptome report that both *MALAT1* and *Menβ* and many previously unannotated intronic transcripts are enriched in poly(A)-transcriptomes (Yang et al. 2011a). Some of these intronic RNAs were further demonstrated to constitute a new family of lncRNAs derived from introns and capped by small nucleolar RNAs (snoRNAs) at both ends (*sno-lncRNAs*) (Yin et al. 2012) (Fig. 4.1). Additional non-polyadenylated lncRNAs from intronic regions were also found in various human cell lines (Derrien et al. 2012) and in *Xenopus tropicalis* (Gardner et al. 2012).

Circular RNAs (circRNAs) are a type of lncRNA that is protected from degradation by head-to-tail circularization. Several recent reports suggest that non-linearized RNAs are largely generated from back-spliced exons, in which splice junctions formed by an acceptor splice site at the 5′ end of an exon and a donor site at a downstream 3′ end (Burd et al. 2010; Yap et al. 2010; Hansen et al. 2011; Salzman et al. 2012). For example, the *INK4a/ARF* locus-associated lncRNA *ANRIL* participates directly in epigenetic transcriptional repression (Yap et al. 2010) (see Part VI for details). This locus also encodes heterogeneous species of RNA transcripts including a circular form of *ANRIL* (*cANRIL*) whose expression is correlated with *INK4/ARF* transcription (Burd et al. 2010). Moreover, sequencing of rRNA-depleted RNAs from human fibroblasts that had been digested with RNase R exonuclease identified numerous circRNAs containing non-colinear exons (Fig. 4.1). These RNAs are proposed to compete with endogenous RNAs in the cytoplasm (Jeck et al. 2013). Very recent studies demonstrate that many circRNAs form by back-spliced exons in animal cells (Memczak et al. 2013) and some of the most abundant ones function as efficient "sponges" to sequester microRNAs to regulate gene expression (Hansen et al. 2013; Memczak et al. 2013).

Pervasively transcribed lncRNAs exhibit several interesting features. First, although most are transcribed by RNA Pol II, many lncRNAs undergo maturation in ways different from mRNAs. Second, their expression is strikingly cell- or tissue-specific compared with coding genes and such RNAs are often coexpressed with neighboring genes (Cabili et al. 2011; Yin et al. 2012). Third, most show low primary sequence similarity with coding sequences (Cabili et al. 2011; Derrien et al. 2012; Zhu et al. 2013). Interestingly, however, loss- and gain-of-function studies of several lncRNAs in zebrafish demonstrate that they play key roles during embryonic development (Ulitsky et al. 2011), indicating the functional conservation despite their limited sequence conservation. Although detailed functions of lncRNAs are only beginning to be defined, several lines of evidence show that they participate in critical processes, such as X chromosome inactivation, genomic imprinting, maintenance of nuclear architecture, pluripotency, and developmental patterning (for reviews, see Wilusz et al. (2009), Chen and Carmichael (2010), Guttman and Rinn (2012), Rinn and Chang (2012)).

3 Novel High Throughput Approaches Enable Analysis of lncRNA–Protein Interactions

As a regulatory mechanism some lncRNAs partner with chromatin modifiers to either silence or activate genes epigenetically. A breakthrough in this field came from pioneering studies that identified the lncRNAs *HOTAIR* and *Xist*, which are transcribed from human *HOX* locus and mouse X-chromosome, respectively, and target chromatin repressor Polycomb proteins onto specific mammalian loci (Rinn et al. 2007; Zhao et al. 2008). Since then, lncRNAs from various genomic locations, including imprinted or cancer gene loci, have been shown to interact not only with chromatin repressors (Nagano et al. 2008; Pandey et al. 2008; Khalil et al. 2009; Yap et al. 2010; Zhao et al. 2010; Cabianca et al. 2012) and activators (Orom et al. 2010; Wang et al. 2011; Gomez et al. 2013; Lai et al. 2013) but with other types of proteins, including DNA methyltransferases (Schmitz et al. 2010), transcription factors (Yao et al. 2010; Jeon and Lee 2011), and splicing factors (Tripathi et al. 2010; Gong and Maquat 2011; Yin et al. 2012). Some lncRNAs, like HOTAIR and *Xist*, function to "guide" proteins to their targets (Rinn et al. 2007; Zhao et al. 2008). Others act as a scaffold to bridge protein complexes at specific genomic loci (Tsai et al. 2010), modify proteins allosterically to alter protein function (Shamovsky et al. 2006; Wang et al. 2008), or serve as a sponge to titrate away proteins in cells (Zhao et al. 2008; Tripathi et al. 2010; Yin et al. 2012).

Several technologies have been developed to allow unbiased identification of protein-interacting lncRNAs genome-wide by coupling RNA immunoprecipitation (RIP) with high throughput sequencing. Conceptually, RIP-seq is analogous to widely used technology ChIP-seq (chromatin IP coupled with high throughput sequencing), which was designed to identify transcription factor binding sites or histone modification patterns genome-wide. Both protocols rely on use of an antibody against the protein of interest to specifically pull-down either RNA or DNA associated with the protein. Major RIP-seq methods include native RIP-seq (nRIP-seq), cross-linked RIP-seq (CLIP-seq), photoactivatable-ribonucleoside-enhanced *CLIP-seq* (PAR-CLIP-seq), and individual-nucleotide resolution CLIP-seq (iCLIP-seq). The similarities and differences among these methods are enlisted in Table 4.1 and discussed below.

3.1 Native RNA Immunoprecipitation Coupled with High Throughput Sequencing (nRIP-seq)

Following our discovery that lncRNA *Xist* acts *in cis* to target Polycomb protein complex 2 (PRC2) onto X-chromosomes to establish the chromosome-wide heterochromatic mark trimethylated histone H3 residue lysine 27 (H3K27-3me) (Zhao et al. 2008), we predicted that other PRC2-interacting lncRNAs or mRNAs likely exist, as PRC2 occupies over 2,000 mammalian DNA loci through unknown mechanisms (Ku et al. 2008). To capture the PRC2 transcriptome in mouse embryonic stem cells (mESCs) we developed nRIP-seq (Zhao et al. 2010), a modification of previous RIP-ChIP strategies (Keene et al. 2006). Briefly, since PRC2 protein is primarily nuclear, an mESC nuclear extract is prepared, and then an antibody targeting Enhancer of zeste homolog 2 (Ezh2), the methyltransferase subunit of PRC2, was added for immunoprecipitation, followed by washing, RNA extraction, library construction, high throughput-sequencing and bioinformatics analysis. One advantage of the method is that it captures protein–RNA interaction in the native state since cross-linking reagents are generally not employed. Using nRIP-seq, we discovered over 9,000 Ezh2-interacting transcripts. Binding specificity was validated by generating a library under the same experimental conditions but made from an Ezh2-null cell line. Compared to the wild type library, we detected tenfold less RNA in the control library, suggesting that the Ezh2 transcriptome is highly enriched. Therefore, nRIP-seq is an excellent tool to study interactions between RNA binding proteins and their targets.

3.2 Cross-Linking and Immunoprecipitation (CLIP)-Seq

One limitation to nRIP-seq is that it cannot distinguish direct from indirect interactions. To do so requires application of techniques that utilize ultraviolet (UV) cross-linking of RNA to protein to precisely map protein binding sites in RNA. One of those methods, called cross-linking and immunoprecipitation (CLIP) (Ule et al. 2003), uses a short wave (254 nm) UV light to create a covalent bond between RNA and interacting proteins in living cells and allows stringent

Table 4.1 Overview of various RIP methods presented in this chapter

	nRIP (Zhao et al. 2010)	CLIP (http://lab.rockefeller.edu/darnell/assets/file/Clip.pdf)	PAR-CLIP (Hafner et al. 2010)	iCLIP (Konig et al. 2010)
Amount of cells needed	1 × T75 flask of mESCs	1 × 15 cm dish	10–20 15 cm dish of HEK 293 cells	1 × 10 cm dish
Cross-linking condition	No cross-linking	254 nM	365 nM	254 nM
Addition of nucleotide analogs	No	No	4-SU or 6-SG	No
RNAse digestion	No	Rnase T1	Rnase T1	RNAse I
SDS-Pageand RNA isolation	No	Yes	Yes	Yes
Linker and adaptor ligation	Yes	Yes	Yes	Yes
cDNA circulization/relinearization	No	No	No	Yes
RNA–protein interaction	Indirect/direct	Direct	Direct; T/C or G/A mutation at binding sites	Direct; binding site to "barcode" sequence

experimental manipulation in order to minimize capture of nonspecific lncRNAs. Following cross-linking, an antibody against a protein of interest is used to immunoprecipitate RNA, and then RNAses are used to digest unbound RNA fragments, leaving a 50–100 bp protein-interacting RNA fragment, which is then radiolabeled and then size-fractionated by SDS-PAGE. The cross-linked complex, usually slightly larger than the protein, is extracted and treated with proteinase K to remove RNA-bound protein. Recovered RNA is then ligated to adapters for reverse transcription and PCR amplification. Amplified cDNA libraries are then sequenced using multiple platforms, such as 454, Illumina, or SOLID, followed by bioinformatics data analysis.

CLIP and CLIP-seq have been used in diverse biological systems, including to identify splicing factor recognition sites, such as the RNA networks of splicing factor NOVA in mouse brain tissue (Ule et al. 2003; Licatalosi et al. 2008; Zhang et al. 2010), FOX2 binding sites in stem cells (Yeo et al. 2009), and SFRS1 interaction sites (Sanford et al. 2009) in human embryonic kidney cells. Multiple laboratories have also mapped mammalian microRNA–mRNA interaction sites through Argonaute CLIP-seq (Chi et al. 2009; Leung et al. 2011) [or Zisoulis et al. 2010 if including nematodes]. In addition, using this method, Guil et al. found nuclear protein hnRNPA1 binds a microRNA precursor and is required for microRNA-mediated repression (Guil and Caceres 2007), while Xu et al. showed that the germ cell-specific DNA/RNA-binding protein MSY2 binds small RNAs (Xu et al. 2009). These studies prove that this method is a powerful tool for identifying protein–RNA interaction in vivo. However, caveats include low cross-linking efficiency at short UV wavelengths (typically 1–5 % of RNA–protein complexes are cross-linked) and the inability to identify the precise nucleotide that binds to protein. Recently, a new method named PAR-CLIP-seq was developed based on CLIP-seq to provide solutions to these problems.

3.3 Photoactivatable-Ribonucleoside-Enhanced (PAR)-CLIP-Seq

The PAR-CLIP-seq method was developed in 2010 by Thomas Tuschl and colleagues (Hafner et al. 2010). Rather than cross-linking cells at 254 nm, Hafner et al. metabolically labeled cells with photoreactive nucleoside analogs, such as 4-thiouridine (4-SU) or 6-thioguanosine (6-SG), allowing more efficient cross-linking at 365 nm. Both nucleoside analogs are readily taken up by mammalian cells following their addition to cell culture medium and are relatively nontoxic, at least in human embryonic kidney (HEK) 293 cancer cells. Like CLIP-seq, cross-linked RNA–protein complexes are digested with *RNases*, followed by fractionation, isolation, proteinase K treatment, adaptor ligation, cDNA library construction, and high throughput sequencing. The advantage of this method is that cross-linking of 4-SU or 6-SG to proteins results in respective thymidine to cytidine and guanosine to adenosine transitions in cDNAs at 4-SU and 6-SG incorporation sites, making it possible to map interacting RNA nucleotides and distinguish true protein-binding RNA species from background.

Using this method, Tuschl and colleagues identified RNA interacting sites of several RNA- or microRNA-binding proteins, including PUM2, QK1, IGF2BP1-3, AGO/EIF2C1-4, and TNRC6A-C (Hafner et al. 2010). Another recent study reported ~26,000 HuR/ELAVL1 binding sites in HeLa cells (Lebedeva et al. 2011). Following comparison of the HuR/ELAVL1 and Argonaute 2 transcriptomes using CLIP-seq and PAR-CLIP-seq methods, Kishore et al reported small differences in accuracies of these methods in identifying binding sites of HuR and Ago2 proteins (Kishore et al. 2011). They also suggested that optimizing conditions used for *RNases* treatment are critical step for library bias (Kishore et al. 2011).

3.4 Individual-Nucleotide Resolution CLIP-Seq (iCLIP-Seq)

Both CLIP-seq and PAR-CLIP-seq require reverse transcription to pass through the amino acid covalently bound to RNA at the cross-linking site. Often, cDNAs are prematurely truncated immediately before that nucleotide (Urlaub et al. 2002). To resolve this problem König et al. developed iCLIP (Konig et al. 2010), in which cleavable adaptors are ligated after reverse transcription allowing RT products to be circularized. This step allows quantification of truncation sites and discrimination between unique cDNA products and PCR duplicates. The group has successfully applied quantitative iCLIP to predict dual splicing effects of T-cell intracellular antigen (TIA)-RNA interactions (Wang et al. 2010) in order to characterize RNA targets of the splicing factor TDP-43 in brain (Tollervey et al. 2011). Work from the same group suggests that direct competition between hnRNP C and U2AF65 protects the transcriptome from exonization of *Alu* elements (Zarnack et al. 2013). Two other groups have also used the method to define landscapes of the RNA splicing factors SRSF3 and SRSF4 (Anko et al. 2012), and U2AF65 (Schor et al. 2012).

Which method to choose for transcriptome analysis largely depends on the nature of RNA–protein interaction, and factors such as protein abundance, cellular location of protein–RNA complex, and specificity of the protein–RNA interaction. An important issue to be considered for any immunoprecipitation-based method is availability of a highly specific and sensitive antibody. Therefore, optimization for the immunoprecipitation should be carried out using different antibodies. Some studies also use epitope-tagged rather than endogenous proteins to isolate binding RNAs when a high quality antibody is not available. However, since overexpression of tagged proteins may alter protein function and protein–RNA interaction, this approach should be taken with additional validation experiments. Furthermore, a proper control dataset, including libraries produced using a nonspecific antibody such as IgG and/or control cells lacking the protein of interest, should be generated side-by-side with the target library in order to separate true sequences from background. With proper controls and experimental optimization, RIP-seq is a powerful tool for global analysis of subsets of mRNAs or lncRNAs bound to various RNA-binding proteins.

4 Use of lncRNA as Bait to Study lncRNA–Protein/DNA Interactions

LncRNAs are implicated in many important biological processes (for reviews, see Wilusz et al. (2009), Chen and Carmichael (2010), Guttman and Rinn (2012), Rinn and Chang (2012)). Although only a handful have been characterized mechanistically, evidence suggests that lncRNAs often function by recruiting, assembling, modifying, or scaffolding other cofactors, including proteins (Tripathi et al. 2010; Tsai et al. 2010; Yap et al. 2010; Yin et al. 2012), DNA (Martianov et al. 2007; Schmitz et al. 2010), and other factors (Cesana et al. 2011; Salmena et al. 2011; Hansen et al. 2013; Memczak et al. 2013). Clearly, identifying these cofactors is of key importance for understanding the function of a specific lncRNA.

4.1 Using lncRNA as a Bait to Investigate an lncRNA–Protein Complex

Although approaches exist to identify RNA binding proteins (RBPs), progress in this area has been hampered due to the instability of lncRNAs and the flexibility of their structures. On the other hand, like proteins, nucleic acids can be affinity-tagged, allowing one to decipher an lncRNA–protein complex using the RNA of interest as bait. A variety of tags, including RNA aptamers and chemical labels, have been developed to allow affinity purification of nucleic acids. Proteins captured using such tags are then systematically identified by mass spectrometry.

RNA aptamers or RNA affinity tags that bind to specific RNA molecules can also be used to identify binding proteins by affinity chromatography or visualize RNA trafficking in living cells (Zhou et al. 2002; Janicki et al. 2004; Mao et al. 2011; Vasudevan and Steitz 2007; Maenner et al. 2010a). Some RNA aptamers are naturally occurring, such as the MS2 coat protein-binding sequence, which is an RNA-hairpin structure that specifically binds to bacteriophage MS2 coat protein (Graveley and Maniatis 1998). Multiple copies of MS2 hairpins can be fused with a bait RNA using recombinant techniques, and then resultant tagged RNAs are obtained by in vitro or in vivo transcription. Affinity purification of the RNA–protein complex is then achieved by incubation of tagged RNAs with a fusion protein containing the MS2 coat protein and maltose-binding protein (MS2-MBP), followed by affinity selection by binding to amylose resin (Zhou et al. 2002; Vasudevan and Steitz 2007). This method has been used to isolate human spliceosomes assembled on a well-characterized model pre-mRNA (Zhou et al. 2002) and protein complexes associated with regulatory lncRNAs (Vasudevan and Steitz 2007; Maenner et al. 2010a). For example, it is known that the lncRNA *Xist* is required to maintain mammalian female X-chromosome inactivation (XCI) by recruiting silencing chromatin remodeling complexes (Penny et al. 1996; Zhao et al. 2008). Interestingly, the most conserved *Xist* RNA regions correspond to repeat elements, among which the A region is the most highly conserved and is critical for XCI initiation (Hoki et al. 2009). To identify proteins interacting with the mouse A Region, the entire A region as well as several fragments from it were fused with the MS2 hairpin followed by affinity purification and mass spectrometry (Maenner et al. 2010a). These analyses revealed components of PRC2 that directly bind A region, a discovery critical to our understanding of X inactivation. This MS2 hairpin and RNA fusion system together with the MS2 coat protein tagged with a fluorescent protein have also been used to track RNA localization in living cells (Janicki et al. 2004; Mao et al. 2011).

Other aptamers, such as streptavidin or Sephadex aptamers, which can bind small molecules as well as complex macromolecules, have been identified by screening synthetic libraries (Srisawat et al. 2001; Walker et al. 2008). Such aptamers have been fused to regulatory RNAs, usually short and abundant ncRNAs, to successfully pull down RNP complexes, including *RPP1* RNA, the large RNA subunit of RNase P (Srisawat and Engelke 2001; Li and Altman 2002). However, as yet there is no generalized rule used to design bait RNAs or aptamers that can fold correctly and maintain a stable conformation; thus fusion of an aptamer to an lncRNA of interest may not always result in an ideal outcome, and synthetic aptamers are not yet widely used to study lncRNA–protein complexes. Recently, however, a streptavidin aptamer has been scaffolded to a tRNA backbone that can stabilize the aptamer RNA conformation (Iioka et al. 2011). This scaffold strategy achieved about a tenfold increase in affinity to efficiently pull down RNA–protein complexes from cell lysates (Iioka et al. 2011). Thus use of scaffolded aptamers remains a promising approach to study lncRNA–protein complexes.

Another alternative to RNA aptamers is chemical labeling. Incoporation of modified ribonucleotide triphosphates (rNTPs) containing compounds such as biotin into RNA during in vitro transcription has been used to isolate RNP complexes and applied to identify proteins specifically associated with lncRNAs (Huarte et al. 2010; Tsai et al. 2010; Yang et al. 2011b; Klattenhoff et al. 2013). In such assays, the biotin-incorporated lncRNA or a fragment of that sequence is first denatured/renatured with RNA structure buffer to maintain correct conformation. RNAs are then incubated with cellular/nuclear extracts and affinity beads, followed by washing and elution to collect associated proteins. Biotin labeling of different fragments of the PRC2-interacting lncRNA *HOTAIR* followed by affinity purification of associated RNP complexes surprisingly revealed that the *HOTAIR* 5′ or 3′ domain could retrieve either PRC2 or the LSD1/CoREST/REST complex (a histone demethylase complex that mediates H3K4me2 demethylation). This approach provided direct evidence that *HOTAIR* acts as a modular scaffold for at least two distinct histone modification complexes to coordinate specific combinations of histone modifications on target gene chromatin (Tsai et al. 2010). Recent studies have applied a similar strategy to reveal that distinct sets of proteins are associated with individual lncRNAs to modulate their function (Huarte et al. 2010; Yang et al. 2011b; Klattenhoff et al. 2013).

There are some limitations to the use of tagged RNAs to identify associated RNP complexes. First, proper folding is required for aptamer/RNA function as well as for RNA–protein interactions. Second, chemical modifications and fusion of affinity tags to RNA may lead to structural perturbations that inhibit RNA–protein complex formation. Third, chemically modified or aptamer-tagged RNAs made from in vitro transcription may not reflect the true nature of complexes formed in vivo. Finally, since cross-linking is rarely used (Huarte et al. 2010; Tsai et al. 2010; Yang et al. 2011b; Klattenhoff et al. 2013) in in vitro studies, transient lncRNA/protein interactions may not be captured, while nonspecific interactions may be unintentionally retrieved, leading to misinterpretation (Riley and Steitz 2013). However, it is worth noting that a recent study aimed at defining the mRNA interactome in human cells developed two complementary protocols for covalent UV cross-linking of RBPs to RNA. This study identified over 800 proteins associated with mRNAs by pull-down with oligo d(T) magnetic beads (Castello et al. 2012). In addition, two recent studies developed genome-wide assays to study both genomic binding sites of an lncRNA and lncRNA–protein complexes by incubating nuclear extracts with biotinylated antisense oligos targeting the lncRNA of interest (Chu et al. 2011; Simon et al. 2011) (see below for details). Thus, despite potential pitfalls inherent in these methods, accumulating evidence suggests that tagging an RNA bait with a reagent in order to "fish out" associated proteins is a reliable way to analyze lncRNA function.

4.2 Global Approaches to Study Genomic Binding Sites of an lncRNA

Many lncRNAs function at the level of chromatin by interacting with chromatin-modifying machinery or acting as scaffolds for multiple complexes (for reviews, see Guttman and Rinn (2012), Rinn and Chang (2012)). While some lncRNAs, such as *Xist* and *Air*, work *in cis* on neighboring genes (Nagano et al. 2008; Zhao et al. 2008), others, such as *roX2* (in *Drosophila*) and *HOTAIR*, work *in trans* to regulate distant genes (Gelbart and Kuroda 2009; Rinn et al. 2007). In these cases, an lncRNA can interact directly or indirectly with a chromatin DNA through a specific RNA or protein. Thus, uncovering binding sites of these lncRNAs genome-wide is essential to understand their function. Recently, a combination of chemical tagging of RNAs and deep sequencing technology have allowed one to systematically identify those sites (Chu et al. 2011; Simon et al. 2011).

Two such approaches, named CHART (Capture Hybridization Analysis of RNA Targets) (Simon et al. 2011) and ChIRP (Chromatin Isolation by RNA Purification) (Chu et al. 2011), were independently developed by the Kingston and Chang laboratories, respectively. These approaches are similar in principle to ChIP-seq methodology. Briefly, chemically labeled oligos complementary to an lncRNA are incubated with cross-linked nuclear extracts, followed by affinity purification and elution of the precipitated materials, including DNAs and proteins. To obtain the genomic map of lncRNA occupancy, the reverse cross-linked DNAs are then subjected to deep sequencing and analyses (Fig. 4.2).

As lncRNAs are known to be highly structured, the key feature of these approaches is to design appropriate chemically labeled oligos that are specific and accessible to a structured lncRNA in cross-linked cells. Two principles are applied to design such oligos. In CHART, a mapping assay using RNase H, which hybrolyzes the RNA strand of a DNA–RNA hybrid, was adapted to probe sites on an lncRNA available to hybridization in extracts of cross-linked chromatin. The most sensitive or "accessible" regions in the lncRNA are then identified by incubating individual 20-mer complementary DNA oligos one at a time with the extracts. Identified regions are then used to design 24–25-mer desthiobiotin-conjugated C-oligos, which are 3′-modified by a single desthiobiotin and four oligoethyleneglycol spacers (Dejardin and Kingston 2009)), to allow biotin elution to achieve a lower background (Fig. 4.2) (Simon et al. 2011). In ChIRP, dozens of 20-mer biotinylated complementary DNA oligo probes that tile the entire length of an lncRNA are synthesized. These probes target all regions of an lncRNA equally, as there is no prior knowledge of its secondary structure or functional domains (Fig. 4.2) (Chu et al. 2011). Such unbiased design of oligonucleotides has been used successfully in single-molecule RNA fluorescent in situ hybridization and yielded highly specific signals (Raj et al. 2008).

To analyze global genomic binding sites of an lncRNA, principles used to design labeled DNA oligos can be slightly varied. In CHART, RNaseH mapping allows one to design labeled DNA oligos accessible to specific regions of an lncRNA (Simon et al. 2011). However, there are limitations to the method with lncRNAs such as *HOTAIR*, which harbors functional domains separate from each other (Tsai et al. 2010). Thus use of selected probes may result in potential loss of information due to failure to identify all functional domains of an lncRNA. This disadvantage potentially can be overcome by ChIRP, which uses unbiased biotinylated tiling oligos. However, ChIRP precipitation of nonspecific DNA fragments may occur due to off-target hybridization of pooled oligonucleotide probes. One way to eliminate this artifact is to split tiling oligos into "even" and "odd" probes based on their relative positions along the target RNA and then pooling them into two groups. Independent experiments run with "even" or "odd" probe groups and analyses then focus only on overlapping signals (Chu et al. 2011). As the two probe sets share no overlapping sequences, they target only the RNA of interest and its associated chromatin.

Both CHART and ChIRP are powerful in that they not only identify genomic maps showing lncRNA occupancy but provide a new insight into RNA–chromatin interaction at almost single nucleotide resolution. The lncRNA *roX2* is known to localize to the X chromosome, where it acts together with the MSL complex to regulate X chromosome dosage compensation in *Drosophila* (Gelbart and Kuroda 2009). CHART has been successfully used to identify *roX2* genomic binding (Simon et al. 2011). CHART-seq analysis of *roX2* yielded the same preference for chromatin entry sites, namely, a GA-rich polypurine motif, as the MSL complex, consistent

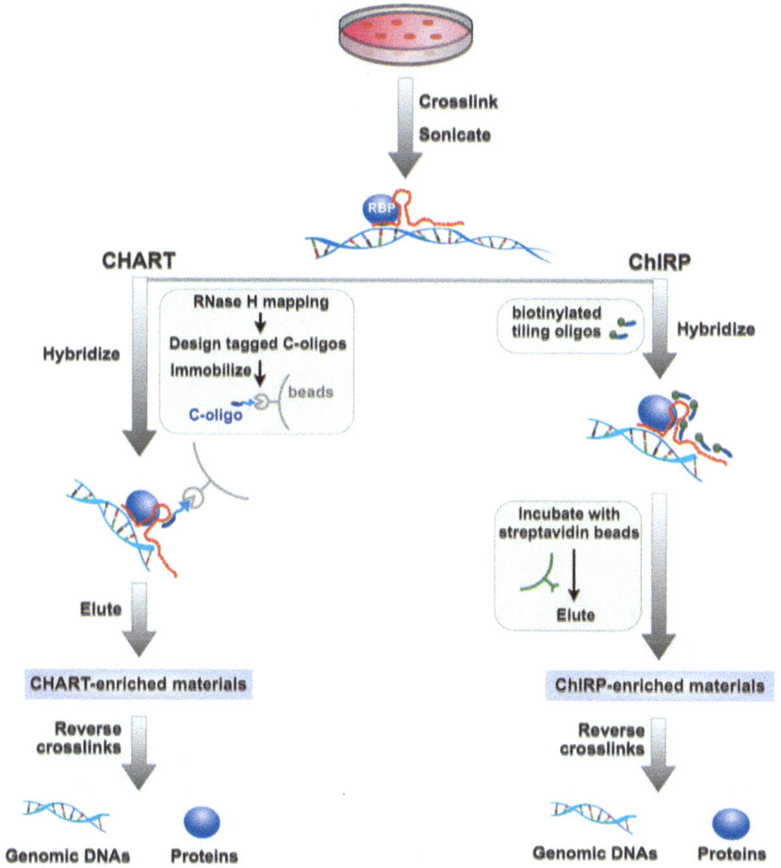

Fig. 4.2 Schematic overview of approaches used to analyze genomic binding sites of an lncRNA. See text for details

with the notion that *roX2* is an integral subunit of the MSL complex bound to chromatin. ChIRP has also been successfully applied to several lncRNAs including *TERC*, *roX2*, and *HOTAIR* (Chu et al. 2011). Interestingly, motif analysis of *HOTAIR* ChIRP-seq-enriched sites revealed a novel GA-rich polypurine motif, suggesting that many lncRNAs function similarly to guide chromatin–lncRNA complexes such as PRC2-*HOTAIR* and MSL-*roX* to appropriate genomic loci. Finally, as both CHART and ChIRP utilize reversible cross-linking, enriched materials can be used to analyze proteins and even RNAs associated with the lncRNA of interest.

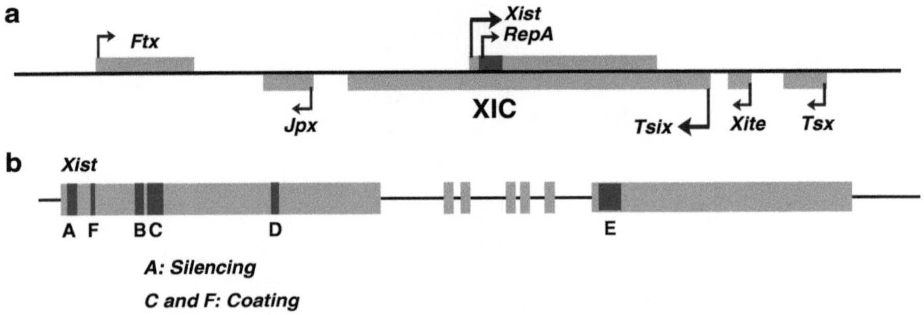

Fig. 4.3 Schematic drawing of X-inactivation center (XIC) and *Xist* RNA. (**a**) A cluster of lncRNAs transcribed from the XIC is required for mammalian dosage compensation. lncRNAs include *Xist* (X-inactivation specific transcript), *Tsix* (antisense of *Xist*), *Tsx* (testis-specific X-linked gene), *Xite* (X-inactivation intergenic transcription element), *RepA* RNA, *Jpx* (also known as *Enox* (Expressed Neighbor of *Xist*)), and *Ftx* (Five prime to *Xist*). *Ftx, JPX,* and *RepA* lncRNAs promote *Xist* transcription, while *Tsixn Xite* and *Tsx* inhibit it. (**b**) Gene structure and repeat regions of *Xist* RNA. Exons are represented as boxes

5 lncRNAs Function in X-Inactivation During Mammalian Development

The most extensively studied lncRNAs come from a region named the X-chromosome inactivation center (XIC) located on the X-chromosome (Rastan 1983; Rastan and Robertson 1985; Brown et al. 1991b). At least six different lncRNAs have been identified from this locus, and they function together to regulate the epigenetic process of X-chromosome inactivation (XCI) (Fig. 4.3a) (Brockdorff et al. 1991; Brown et al. 1991a; Lee et al. 1999; Lee and Lu 1999; Ogawa and Lee 2003; Tian et al. 2010; Anguera et al. 2011; Chureau et al. 2011). XCI failure results in female embryonic lethality; thus these lncRNAs play an essential role in female development. Their mechanisms and regulation are discussed below.

5.1 X-Chromosome Inactivation and Xist lncRNA

Indispensable for XCI is the 17Kb lncRNA named *X-inactivation specific transcript (Xist)* (Borsani et al. 1991; Brockdorff et al. 1991; Brown et al. 1991a). Other lncRNAs relevant to the locus likely function through regulating *Xist* expression. *Xist* is expressed at low levels from two active X-chromosomes in female embryos or female embryo-derived embryonic stem cells before XCI initiation and then is upregulated and, extraordinarily, starts to "coat" almost the entire X-chromosome. *Xist* spreading along the X transforms that allele into heterochromatin. Eventually, that in active allele (Xi) undergoes DNA methylation, repressing over a thousand genes. Xi remains silent during subsequent cell divisions, and *Xist* then becomes dispensable for chromosome-wide silencing, despite the observation that its expression remains high (Brown and Willard 1994; Csankovszki et al. 1999).

5.2 *Xist* Silencing Mechanisms

Xist is the only RNA reported so far capable of "spreading" along almost an entire chromosome and converting an active euchromatin to a condensed heterochromatin. An intensive area of investigation in the field of epigenetics is to understand how *Xist* achieves such silencing. Questions remain as to how it is tethered onto the inactive X and what silencing factors it interacts with.

5.2.1 Loading and Spreading of *XIST* on the Inactive X-Chromosome (Xi)

Currently, evidence suggests that specific DNA or RNA sequences facilitate *Xist* spreading. For example, naturally occurring or induced X:autosome translocations indicate that *Xist* spreads differently along autosomal DNA segments (Surralles and Natarajan 1998; Popova et al. 2006). A search for sequence differences between the X-chromosome and autosomes revealed that LINE (long interspersed elements) elements showed greater density on the X (Lyon 1998). Mary Lyon proposed that interspersed repetitive LINE elements act as booster elements to promote *Xist* spread. Indeed, a transgenic study in mouse embryonic stem cells (mESCs) showed that chromosome regions with greater LINE density are inactivated more efficiently by a *Xist* transgene (Popova et al. 2006). Recently, another study suggested that specific LINEs may participate in local propagation of XCI into regions that would otherwise escape it (Chow et al. 2010). However, the exact function of LINE elements in establishment of XCI remains to be investigated.

Trans-acting factors functioning in this process have been identified only recently. A landmark paper from Jeannie Lee's group showed that the RNA/DNA binding protein Yin-Yang 1 (YY1) is required to tether *Xist* RNA to the inactive X (Jeon and Lee 2011). In YY1 knockdown cells, *Xist* RNA remained highly expressed but exhibited a diffuse localization pattern rather than forming a "cloud" on top of the Xi. However, since YY1 binding sites are widespread throughout the mammalian genome, it remains unclear why YY1 does not guide *Xist* RNA onto an entire autosome in the transgene studies. In addition to YY1, a screening study showed that the nuclear scaffold protein hnRNPU is essential for *Xist* loading and spreading (Hasegawa et al. 2010). Currently, it is not known whether YY1 and hnRNPU form a complex with *Xist*.

5.2.2 *XIST* Silences Genes Through Repressive Polycomb Proteins

Upon *Xist* spreading, the X-chromosome loses active chromatin marks, such as trimethylated histone H3 lysine 4 (H3K4-3me), and gains the heterochromatic marks H3K27-3me and mono ubiquinated H2A (H2AK119u1) (Plath et al. 2003; Silva et al. 2003; de Napoles et al. 2004; Fang et al. 2004; Okamoto et al. 2004). These observations suggest that *Xist* interacts with chromatin modifiers. Indeed, our work and that of others proved *Xist* to be a cofactor to target Polycomb group proteins onto X-chromosome during XCI.

First identified in *Drosophila* as repressing *HOX* loci (Alkema et al. 1995; Harding et al. 1995; Yu et al. 1995), Polycomb proteins are known to play critical roles in epigenetic gene regulation in almost all organisms, including mammals. Polycomb proteins comprise two major complexes, Polycomb Repressive Complex 1 (PRC1) and Polycomb Repressive Complex 2 (PRC2) (Simon and Kingston 2013). PRC2 bears histone methyltransferase activity through its catalytic subunit Ezh2 (Cao et al. 2002; Czermin et al. 2002; Kuzmichev et al. 2002). Once PRC2 establishes the H3K27-3me heterochromatin mark, protein with chromo-domain from PRC1 complex recognize that mark and guide PRC1 to specific genomic loci to establish ubiquinated histone H2A (H2AK119u1) (Fischle et al. 2003; Min et al. 2003; Wang et al. 2004). H2AK119u1 deposition is proposed to repress gene expression by facilitating chromatin compaction or inhibiting RNA polymerase II-dependent transcriptional elongation (Francis et al. 2004; Zhou et al. 2008). Both PRC1 and PRC2 complexes are highly enriched on the inactive X-chromosome during XCI establishment (Silva et al. 2003; de Napoles et al. 2004; Plath et al. 2003; Okamoto et al. 2004). Deletion of some Polycomb proteins, such as Eed from the PRC2 complex, reverses Xi in extra-embryonic tissues, suggesting an essential role for Polycomb proteins in maintaining XCI (Kalantry et al. 2006). To determine how PRC2 is recruited onto the inactive X-chromosome, we undertook studies showing that *Xist* directly binds to the Ezh2 subunit (Zhao et al. 2008). Interestingly, that interaction occurred at the *Xist* silencing domain.

Xist RNA contains six different repeat regions (A to F) (Fig. 4.3b), which likely represent distinct functional domains. The most well studied is the 5′ A region (Wutz et al. 2002), whose sequence is highly conserved among mouse and humans (Hong et al. 2000). A *Xist* transgene lacking the A region cannot initiate silencing (Wutz et al. 2002). Two independent silencing mechanisms are proposed for this region. One study suggests that the A region directly binds the splicing factor ASF/SF2 to ensure proper *Xist* splicing (Royce-Tolland et al. 2010). In another study, we assessed mESCs and discovered that a shorter transcript, named *RepA* RNA, is transcribed through the A region (Zhao et al. 2008). Both *Xist* and *RepA* RNA directly bind Ezh2 through A region repeats. Support for this observation came from the Reinberg group when they showed that Ezh2 phosphorylation enhances interaction between PRC2 and *Xist*/*RepA* RNAs (Kaneko et al. 2010). A different PRC2 subunit, Suz12, reportedly binds *RepA* RNA (Kanhere et al. 2010), although mechanisms underlying complex formation among lncRNAs and PRC2 subunits are not yet understood. We also found that after *RepA* RNA knockdown, *Xist* could not be upregulated; hence the H3K27-3me mark was not established and XCI was not initiated (Zhao et al. 2008). A similar phenotype was observed by Hoki et al. following genetic deletion of the A region in mouse (Hoki et al. 2009). These results suggest that the A region is required for *Xist* upregulation and such regulation requires the *RepA* transcript. Notably, a *Xist* mutant lacking A-repeats can target PRC2 and PRC1 onto Xi (Plath et al. 2003; Schoeftner et al. 2006), indicating that other *Xist* sequences also recruit Polycomb proteins. Such regions remain to be identified.

Several laboratories have studied the structure of the A region in order to understand the molecular basis for *Xist* and Polycomb interaction. Computational analysis revealed that the region contains 7.5–8.5 tandem repeats (variable among mouse and human) of a conserved ~26-mer sequence and predict a double stem loop structure within each repeat (Hendrich et al. 1993; Wutz et al. 2002; Duszczyk et al. 2011). That structure appears to be important for Rep A/PRC2 interaction (Zhao et al. 2008; Kanhere et al. 2010). Interestingly, NMR studies of an in vitro transcribed 26-mer showed that only the first predicted hairpin is formed internally, while the second mediates duplex formation among different repeats (Duszczyk et al. 2008; Duszczyk and Sattler 2012). FRET analysis using chemical and enzymatic probes to examine structure of the whole domain indicated that the entire A region contains two long stem-loop structures, each including four repeats (Maenner et al. 2010b). These studies highlight the importance of RNA structure for function.

Less is known about the structure and function of the other *Xist* repeats. Nevertheless, Repeat regions C and F have been shown to regulate *Xist* spreading, potentially through interaction with YY1 (Jeon and Lee 2011; Wutz et al. 2002; Beletskii et al. 2001; Sarma et al. 2010).

5.3 *Xist Regulatory Mechanisms*

5.3.1 Regulation by Other lncRNAs

Xist upregulation is controlled by *cis*-elements and *trans*-factors. *Cis*-elements include lncRNAs from the XIC, including the *Xist* antisense partner *Tsix* (Lee et al. 1999; Lee and Lu 1999), *Xite (X-inactivation intergenic transcription elements)* (Ogawa and Lee 2003), *Jpx* (also known as *Enox* (*Expressed Neighbor of Xist*) (Tian et al. 2010), *Ftx (Five prime to Xist)* (Chureau et al. 2011), and *Tsx (Testes-specific X-linked)* (Anguera et al. 2011) (Fig. 4.1). *Tsix* was identified as a *Xist* antagonist in 1999 in studies showing that its depletion promotes *Xist* transcriptional activation exclusively from the *Tsix*-disrupted allele (Lee et al. 1999; Lee and Lu 1999). During early development, *Tsix* and *Xist* display opposite expression patterns (Lee et al. 1999; Lee and Lu 1999). *Tsix* is expressed at ~10- to 100-fold higher levels than *Xist* and suppresses *Xist* expression before XCI (Shibata and Lee 2003). *Tsix* levels decrease during XCI initiation, allowing *Xist* upregulation (Lee et al. 1999; Lee and Lu 1999; Shibata and Lee 2003).

Several epigenetic mechanisms are proposed to explain *Tsix* activity. One is that *Tsix* modulates the chromatin state of the *Xist* promoter. This hypothesis is supported by two studies describing a gain of euchromatic marks at the *Xist* promoter following *Tsix* truncation in mESCs (Navarro et al. 2005; Sado et al. 2005), and another indicating an initial gain of heterochromatic marks at the *Xist* promoter followed by switching into euchromatic markers (Sun et al. 2006). These apparent differences in these results stem from use of male (Navarro et al. 2005; Sado et al. 2005) versus female mESCs (Sun et al. 2006). Others propose that *Tsix* forms an RNA duplex with *Xist* and regulates it through the RNAi pathway (Ogawa et al. 2008). Although this presents an attractive model for understanding sense and antisense

RNA interactions, no microRNA has been identified from the *Xist/Tsix* locus, although Dicer and *Tsix* reportedly regulate *Xist* synergistically in mESCs (Ogawa et al. 2008). Recently, we reported that *Tsix* competes with *Xist* for PRC2 binding and interferes with PRC2 loading onto the inactive X at the early stage of XCI (Zhao et al. 2008).

Other lncRNAs functioning in XCI have not been extensively studied. However, it has been shown that deleting *Xite* from one of the Xs in female mESCs downregulates *Tsix in cis* and skews X-inactivation, suggesting that *Xite* promotes *Tsix* expression on the active X (Ogawa and Lee 2003). There are also reports that *Jpx* acts in *trans* to activate *Xist* and that *Ftx* is also a positive regulator of *Xist* (Tian et al. 2010; Chureau et al. 2011). X-inactivation is also mildly affected by loss of *Tsix* by mechanisms as yet uncharacterized (Anguera et al. 2011).

5.3.2 Regulation by Pluripotency Factors

It has long been postulated that XCI is coupled to the mESC pluripotency state, as XCI initiation is tightly associated with mESC differentiation. Furthermore, during reprogramming of mouse somatic cells, Xi reactivation accompanies the establishment of pluripotency (Maherali et al. 2007). Recent studies suggest pluripotent factors play a repressive role in *Xist* regulation. *Xist* upregulation occurs in Oct4 or Nanog conditional knockout male mESCs, and *Xist* coats both X chromosomes in Oct4 knockdown female mESCs as they undergo differentiation (Navarro et al. 2008; Donohoe et al. 2009). The *Xist* first intron is occupied by Oct4, Nanog, and Sox2 in undifferentiated mESCs (Navarro et al. 2008), suggesting that pluripotency factors directly repress *Xist*. However, intron1 deletion results in only a small increase in *Xist* expression, indicating the existence of unidentified *cis*-elements that interact with pluripotency factors (Nesterova et al. 2011). In addition to core pluripotency factors, other pluripotency genes, such as Prdm14, also inhibit *Xist* RNA expression (Ma et al. 2011). Interestingly, while one set of pluripotency factors represses *Xist*, a different set activates *Tsix*. Klf4, c-Myc, and Rex1 reportedly promote *Tsix* expression (Navarro et al. 2010), and Rex1 is required for efficient *Tsix* elongation. Factors such as RNF12 (Ring finger protein 12), an X-linked E3 ubiquitin ligase that targets Rex1 for degradation, are essential to initiate XCI and can activate *Xist* expression, even in male mESCs (Jonkers et al. 2009; Shin et al. 2010). Therefore, pluripotency factors are thought to block *Xist* expression directly or indirectly through *Tsix* activation.

6 ncRNA Function in Disease

In addition to their activity in normal physiological processes, just as other molecules, lncRNAs are also linked to human diseases, including a variety of human cancers and human genetic disorders. We will simply illustrate several lncRNAs and their involvement in human diseases below.

6.1 Xist, XCI and Cancer

Some female cancers, such as breast and ovarian cancers, exhibit loss of Xi and gain of an active X (Pageau et al. 2007), suggesting that XCI mis-regulation functions in tumorigenesis. Some X-linked oncogenes and tumors suppressors that are subject to XCI have been identified. Among them, Zuo et al. found that *FOXP3* is an X-linked breast cancer tumor suppressor gene (Zuo et al. 2007). Additional evidence that *Xist* lncRNA drives cancer development came from a recent study reporting the occurrence of hematological cancer in *Xist* conditional knockout mice (Yildirim et al. 2013). Those authors deleted *Xist* in hematopoietic cells and found that only female mice developed myeloproliferative neoplasm and myelodysplastic syndrome, which is characterized by bone marrow fibrosis and chronic myelomonocytic leukemia. The study suggests that X reactivation perturbs maturation and longevity of hematopoietic stem cells, suggesting a novel role for *Xist* and XCI in adult stem cell and cancer biology.

6.2 Abnormal Expression of lncRNAs in Cancers

Abnormal expression of imprinted lncRNAs is seen in different types of cancer. For example, disruption of the lncRNA-related genomic imprinted *Kcnq1* cluster is associated with human Beckwith–Wiedemann syndrome and Wilms' tumor (O'Neill 2005). In addition, the paternally expressed antisense *PEG8/IGF2AS*, which is transcribed from the *Igf2* locus, is significantly overexpressed in Wilms' tumor samples compared to adjacent normal kidney tissue (Okutsu et al. 2000). Interestingly, although the lncRNA *H19* has no obvious function in imprinted expression of *Igf2* (Jones et al. 1998), it is essential for tumorigenesis (Matouk et al. 2007) and c-Myc directly induces *H19* transcription during tumorigenesis (Barsyte-Lovejoy et al. 2006).

Many tumor suppressor genes are also associated with nearby antisense lncRNA transcripts. For example, the tumor suppressor gene *p15*, a cyclin-dependent kinase inhibitor implicated in leukemia, is regulated by its antisense transcript p15 ncRNA (*p15AS*) (Yu et al. 2008). *p15AS* and *p15* sense expression is inversely correlated in leukemia progression, and *p15* silencing is induced via both *cis* and *trans* mechanisms through heterochromatin formation, as shown by reduced H3K4 methylation and increased H3K9 methylation following introduction of *p15AS* into mammalian cells (Yu et al. 2008). Importantly, *p15* silencing persists even after *p15AS* is downregulated (Yu et al. 2008). Another mode of such regulation is illustrated by *ANRIL*, the long antisense transcript of the *INK4b/ARF/INK4a* tumor suppressor locus (Yap et al. 2010). First, expression of *ANRIL* and its interacting protein chromobox 7 (CBX7), a subunit of the polycomb repressive complex 1(PRC1), is upregulated in prostate cancer tissues. Furthermore, *ANRIL* is associated with PRC1 complex recruitment to the *INK4b/ARF/INK4a* locus by specifically binding to CBX7.

Interestingly, CBX7 contains structurally distinct modes to bind not only *ANRIL*, but also H3K37me, which is methylated by EZH2 of PRC2. Moreover, transfection of *ANRIL* antisense transcripts or expression vectors harboring CBX7 mutants that disrupt H3K27me modification or RNA binding promotes premature growth arrest (Yap et al. 2010). This study suggests that an lncRNA participates directly in *cis-recruitment* of PRC complexes to silence gene expression from a tumor suppressor locus.

Interestingly, the lncRNA *HOTAIR* functions in *trans*-recruitment of PRC2 complexes at non-*HOX* loci not only during developmental patterning (Rinn et al. 2007) but in cancer metastasis (Gupta et al. 2010). *HOTAIR* shows increased expression in primary breast tumor and metastases (Gupta et al. 2010) and in colorectal cancer (Kogo et al. 2011). *HOTAIR* loss decreases cancer invasiveness, while increased *HOTAIR* expression in epithelial cancer cells promotes metastasis by globally altering the chromatin state. Global changes include induction of genome-wide retargeting of PRC2 to an occupancy pattern, leading to PRC2-dependent altered histone H3 lysine 27 methylation and gene expression (Gupta et al. 2010). As noted, *HOTAIR* likely acts as a scaffold for at least two distinct complexes that mediate histone modification at targeted chromatin sites (Tsai et al. 2010), and it binds to many genomic sites through a GA-rich motif (Chu et al. 2011). Taken together, these studies strongly suggest that lncRNAs serve as important regulators in tumorigenesis, probably through direct targeting and recruitment of chromatin-modifying machinery at specific loci.

6.3 Abnormal Expression of lncRNAs in Human Genetic Disorders

lncRNAs also likely function in human genetic disorders, such as facioscapulohumeral muscular dystrophy (FSHD) (Cabianca et al. 2012) and Prader–Willi Syndrome (PWS) (Yin et al. 2012). Facioscapulohumeral muscular dystrophy (FSHD) is an autosomal-dominant disease associated with reduced copy number of a D4Z4 repeat sequences mapping to 4q35. D4Z4 deletion causes an epigenetic switch leading to de-repression of 4q35 genes (Cabianca and Gabellini 2010). The FSHD locus is normally a Polycomb repressive target; however, FSHD patients display loss of Polycomb silencing and a gain in Trithorax-dependent activation (a de-repression process). *DBE-T* is a chromatin-associated lncRNA originating from the FSHD-associated repetitive elements and is produced in FSHD patients. In FSHD patient, the expression of *DBE-T* is to coordinate the de-repression of 4q35 genes (Cabianca et al. 2012). *DBE-T* is thought to recruit the Trithorax group protein Ash1L to the FSHD locus, driving histone H3 lysine 36 dimethylation, chromatin remodeling, and 4q35 gene transcription (Cabianca et al. 2012).

While gain of *DBE-T* expression is likely involved in FSHD (Cabianca et al. 2012), loss of expression of *sno-lncRNAs* likely contributes to the pathogenesis of Prader–Willi Syndrome (PWS) (Yin et al. 2012). PWS is a multiple system human disorder characterized by global developmental delay, mental retardation, and morbid obesity is due to the absence of paternally expressed imprinted genes at 15q11.2-q13 (Cassidy et al. 2012). There are several coding and noncoding genes, including a cluster of 29 box C/D small nucleolar RNAs (snoRNAs) named SNORD116 expressed from this region (Cassidy et al. 2012). Although the role of each transcript in the pathogenesis of PWS remains largely unknown, lines of evidence in human patients suggest the SNORD116 snoRNA gene cluster plays a key role (Sahoo et al. 2008; de Smith et al. 2009; Duker et al. 2010). It is known that Box C/D snoRNAs usually function as guides for ribose methylation of an RNA target (Kiss 2001), however, no targets of SNORD116 snoRNAs has been validated so far (Cassidy et al. 2012). Thus, the function of these SNORD116 snoRNAs and the precise molecular cause of PWS still remain unknown.

Recently, investigators sequenced the repertoire of non-polyadenylated RNAs isolated from human embryonic stem cells (hESCs) (Yang et al. 2011a) and discovered a class of intron-derived and snoRNA-related lncRNAs (*sno-lncRNAs*). Interstingly, five *sno-lncRNAs* are produced from SNORD116 snoRNA gene cluster and are specifically deleted from the minimal deletion in PWS patients (Yin et al. 2012). In wild type hESCs, PWS region *sno-lncRNAs* are expressed at extremely high levels (similar in abundance to some housekeeping mRNAs) and accumulate near their sites of synthesis. Although these lncRNAs have little effect on local gene expression, surprisingly, each of them contains multiple specific binding sites for the alternative splicing regulator Fox2. Thus, PWS region *sno-lncRNAs* act as "molecular sponges" to strongly associate with Fox2 in the nucleus and alter patterns of Fox2-regulated alternative splicing in *sno-lncRNA* depleted cells (Yin et al. 2012). While in PWS patients, these PWS region *sno-lncRNAs* are not expressed, subsequently leading to altered nuclear distribution of Fox proteins and dysregulating splicing embryonically and in adults. These results implicate a new class of lncRNAs in PWS pathogenesis. Overall, all of these studies suggest that lncRNAs could serve as potential targets in treatment of diseases.

7 Conclusions

The number of non protein-coding transcripts identified over the past decade has increased exponentially, causing a dramatic shift in our perception of the mammalian genome from a focus on protein-coding genes to long noncoding elements. Emerging evidence suggests that lncRNAs play essential roles in a wide range of biological functions and various technologies have been developed to study their molecular mechanisms in vivo. These studies pave the way for exciting future discoveries relevant to lncRNAs.

Acknowledgements We are grateful to Li Yang for critical reading of the manuscript. This work is supported by an AACR-Aflac, Inc., Career Development Award for Pediatric Cancer Research (J.C.Z.) and grants XDA01010206, 2012SSTP01 from CAS and grant 31271376 from NSFC (L.-L. Chen).

References

Alkema MJ, van der Lugt NM, Bobeldijk RC, Berns A, van Lohuizen M (1995) Transformation of axial skeleton due to overexpression of bmi-1 in transgenic mice. Nature 374(6524):724–727

Anguera MC, Ma W, Clift D, Namekawa S, Kelleher RJ III, Lee JT (2011) Tsx produces a long noncoding RNA and has general functions in the germline, stem cells, and brain. PLoS Genet 7(9):e1002248

Anko ML, Muller-McNicoll M, Brandl H, Curk T, Gorup C, Henry I, Ule J, Neugebauer KM (2012) The RNA-binding landscapes of two SR proteins reveal unique functions and binding to diverse RNA classes. Genome Biol 13(3):R17

Barsyte-Lovejoy D, Lau SK, Boutros PC, Khosravi F, Jurisica I, Andrulis IL, Tsao MS, Penn LZ (2006) The c-Myc oncogene directly induces the H19 noncoding RNA by allele-specific binding to potentiate tumorigenesis. Cancer Res 66(10):5330–5337

Beletskii A, Hong YK, Pehrson J, Egholm M, Strauss WM (2001) PNA interference mapping demonstrates functional domains in the noncoding RNA Xist. Proc Natl Acad Sci U S A 98(16):9215–9220

Borsani G, Tonlorenzi R, Simmler MC, Dandolo L, Arnaud D, Capra V, Grompe M, Pizzuti A, Muzny D, Lawrence C et al (1991) Characterization of a murine gene expressed from the inactive X chromosome. Nature 351(6324):325–329

Brockdorff N, Ashworth A, Kay GF, Cooper P, Smith S, McCabe VM, Norris DP, Penny GD, Patel D, Rastan S (1991) Conservation of position and exclusive expression of mouse Xist from the inactive X chromosome. Nature 351(6324):329–331

Brown CJ, Ballabio A, Rupert JL, Lafreniere RG, Grompe M, Tonlorenzi R, Willard HF (1991a) A gene from the region of the human X inactivation centre is expressed exclusively from the inactive X chromosome. Nature 349(6304):38–44

Brown CJ, Lafreniere RG, Powers VE, Sebastio G, Ballabio A, Pettigrew AL, Ledbetter DH, Levy E, Craig IW, Willard HF (1991b) Localization of the X inactivation centre on the human X chromosome in Xq13. Nature 349(6304):82–84

Brown CJ, Willard HF (1994) The human X-inactivation centre is not required for maintenance of X-chromosome inactivation. Nature 368(6467):154–156

Brown JA, Valenstein ML, Yario TA, Tycowski KT, Steitz JA (2012) Formation of triple-helical structures by the 3′-end sequences of MALAT1 and MENbeta noncoding RNAs. Proc Natl Acad Sci U S A 109(47):19202–19207

Burd CE, Jeck WR, Liu Y, Sanoff HK, Wang Z, Sharpless NE (2010) Expression of linear and novel circular forms of an INK4/ARF-associated non-coding RNA correlates with atherosclerosis risk. PLoS Genet 6(12):e1001233

Cabianca DS, Casa V, Bodega B, Xynos A, Ginelli E, Tanaka Y, Gabellini D (2012) A long ncRNA links copy number variation to a polycomb/trithorax epigenetic switch in FSHD muscular dystrophy. Cell 149(4):819–831

Cabianca DS, Gabellini D (2010) The cell biology of disease: FSHD: copy number variations on the theme of muscular dystrophy. J Cell Biol 191(6):1049–1060

Cabili MN, Trapnell C, Goff L, Koziol M, Tazon-Vega B, Regev A, Rinn JL (2011) Integrative annotation of human large intergenic noncoding RNAs reveals global properties and specific subclasses. Genes Dev 25(18):1915–1927

Cao R, Wang L, Wang H, Xia L, Erdjument-Bromage H, Tempst P, Jones RS, Zhang Y (2002) Role of histone H3 lysine 27 methylation in Polycomb-group silencing. Science 298(5595): 1039–1043

Cassidy SB, Schwartz S, Miller JL, Driscoll DJ (2012) Prader-Willi syndrome. Genet Med 14(1):10–26

Castello A, Fischer B, Eichelbaum K, Horos R, Beckmann BM, Strein C, Davey NE, Humphreys DT, Preiss T, Steinmetz LM et al (2012) Insights into RNA biology from an atlas of mammalian mRNA-binding proteins. Cell 149(6):1393–1406

Cesana M, Cacchiarelli D, Legnini I, Santini T, Sthandier O, Chinappi M, Tramontano A, Bozzoni I (2011) A long noncoding RNA controls muscle differentiation by functioning as a competing endogenous RNA. Cell 147(2):358–369

Chen LL, Carmichael GG (2010) Long noncoding RNAs in mammalian cells: what, where, and why? Wiley Interdiscip Rev RNA 1(1):2–21

Chi SW, Zang JB, Mele A, Darnell RB (2009) Argonaute HITS-CLIP decodes microRNA-mRNA interaction maps. Nature 460(7254):479–486

Chow JC, Ciaudo C, Fazzari MJ, Mise N, Servant N, Glass JL, Attreed M, Avner P, Wutz A, Barillot E et al (2010) LINE-1 activity in facultative heterochromatin formation during X chromosome inactivation. Cell 141(6):956–969

Chu C, Qu K, Zhong FL, Artandi SE, Chang HY (2011) Genomic maps of long noncoding RNA occupancy reveal principles of RNA-chromatin interactions. Mol Cell 44(4):667–678

Chureau C, Chantalat S, Romito A, Galvani A, Duret L, Avner P, Rougeulle C (2011) Ftx is a non-coding RNA which affects Xist expression and chromatin structure within the X-inactivation center region. Hum Mol Genet 20(4):705–718

Core LJ, Waterfall JJ, Lis JT (2008) Nascent RNA sequencing reveals widespread pausing and divergent initiation at human promoters. Science 322(5909):1845–1848

Csankovszki G, Panning B, Bates B, Pehrson JR, Jaenisch R (1999) Conditional deletion of Xist disrupts histone macroH2A localization but not maintenance of X inactivation. Nat Genet 22(4):323–324

Czermin B, Melfi R, McCabe D, Seitz V, Imhof A, Pirrotta V (2002) Drosophila enhancer of Zeste/ESC complexes have a histone H3 methyltransferase activity that marks chromosomal Polycomb sites. Cell 111(2):185–196

de Napoles M, Mermoud JE, Wakao R, Tang YA, Endoh M, Appanah R, Nesterova TB, Silva J, Otte AP, Vidal M et al (2004) Polycomb group proteins Ring1A/B link ubiquitylation of histone H2A to heritable gene silencing and X inactivation. Dev Cell 7(5):663–676

de Smith AJ, Purmann C, Walters RG, Ellis RJ, Holder SE, Van Haelst MM, Brady AF, Fairbrother UL, Dattani M, Keogh JM et al (2009) A deletion of the HBII-85 class of small nucleolar RNAs (snoRNAs) is associated with hyperphagia, obesity and hypogonadism. Hum Mol Genet 18(17):3257–3265

Dejardin J, Kingston RE (2009) Purification of proteins associated with specific genomic Loci. Cell 136(1):175–186

Derrien T, Johnson R, Bussotti G, Tanzer A, Djebali S, Tilgner H, Guernec G, Martin D, Merkel A, Knowles DG et al (2012) The GENCODE v7 catalog of human long noncoding RNAs: analysis of their gene structure, evolution, and expression. Genome Res 22(9):1775–1789

Djebali S, Davis CA, Merkel A, Dobin A, Lassmann T, Mortazavi A, Tanzer A, Lagarde J, Lin W, Schlesinger F et al (2012) Landscape of transcription in human cells. Nature 489(7414):101–108

Donohoe ME, Silva SS, Pinter SF, Xu N, Lee JT (2009) The pluripotency factor Oct4 interacts with Ctcf and also controls X-chromosome pairing and counting. Nature 460(7251):128–132

Duker AL, Ballif BC, Bawle EV, Person RE, Mahadevan S, Alliman S, Thompson R, Traylor R, Bejjani BA, Shaffer LG et al (2010) Paternally inherited microdeletion at 15q11.2 confirms a significant role for the SNORD116 C/D box snoRNA cluster in Prader-Willi syndrome. Eur J Hum Genet 18(11):1196–1201

Dunham I, Kundaje A, Aldred SF, Collins PJ, Davis CA, Doyle F, Epstein CB, Frietze S, Harrow J, Kaul R et al (2012) An integrated encyclopedia of DNA elements in the human genome. Nature 489(7414):57–74

Duszczyk MM, Sattler M (2012) (1)H, (1)(3)C, (1)(5)N and (3)(1)P chemical shift assignments of a human Xist RNA A-repeat tetraloop hairpin essential for X-chromosome inactivation. Biomol NMR Assign 6(1):75–77

Duszczyk MM, Wutz A, Rybin V, Sattler M (2011) The Xist RNA A-repeat comprises a novel AUCG tetraloop fold and a platform for multimerization. RNA 17(11):1973–1982

Duszczyk MM, Zanier K, Sattler M (2008) A NMR strategy to unambiguously distinguish nucleic acid hairpin and duplex conformations applied to a Xist RNA A-repeat. Nucleic Acids Res 36(22):7068–7077

Fang J, Chen T, Chadwick B, Li E, Zhang Y (2004) Ring1b-mediated H2A ubiquitination associates with inactive X chromosomes and is involved in initiation of X inactivation. J Biol Chem 279(51):52812–52815

Fischle W, Wang Y, Jacobs SA, Kim Y, Allis CD, Khorasanizadeh S (2003) Molecular basis for the discrimination of repressive methyl-lysine marks in histone H3 by Polycomb and HP1 chromodomains. Genes Dev 17(15):1870–1881

Francis NJ, Kingston RE, Woodcock CL (2004) Chromatin compaction by a polycomb group protein complex. Science 306(5701):1574–1577

Gardner EJ, Nizami ZF, Talbot CC Jr, Gall JG (2012) Stable intronic sequence RNA (sisRNA), a new class of noncoding RNA from the oocyte nucleus of Xenopus tropicalis. Genes Dev 26(22):2550–2559

Gelbart ME, Kuroda MI (2009) Drosophila dosage compensation: a complex voyage to the X chromosome. Development 136(9):1399–1410

Gomez JA, Wapinski OL, Yang YW, Bureau JF, Gopinath S, Monack DM, Chang HY, Brahic M, Kirkegaard K (2013) The NeST long ncRNA controls microbial susceptibility and epigenetic activation of the interferon-gamma locus. Cell 152(4):743–754

Gong C, Maquat LE (2011) lncRNAs transactivate STAU1-mediated mRNA decay by duplexing with 3′ UTRs via Alu elements. Nature 470(7333):284–288

Graveley BR, Maniatis T (1998) Arginine/serine-rich domains of SR proteins can function as activators of pre-mRNA splicing. Mol Cell 1(5):765–771

Guil S, Caceres JF (2007) The multifunctional RNA-binding protein hnRNP A1 is required for processing of miR-18a. Nat Struct Mol Biol 14(7):591–596

Gupta RA, Shah N, Wang KC, Kim J, Horlings HM, Wong DJ, Tsai MC, Hung T, Argani P, Rinn JL et al (2010) Long non-coding RNA HOTAIR reprograms chromatin state to promote cancer metastasis. Nature 464(7291):1071–1076

Guttman M, Rinn JL (2012) Modular regulatory principles of large non-coding RNAs. Nature 482(7385):339–346

Hafner M, Landthaler M, Burger L, Khorshid M, Hausser J, Berninger P, Rothballer A, Ascano M Jr, Jungkamp AC, Munschauer M et al (2010) Transcriptome-wide identification of RNA-binding protein and microRNA target sites by PAR-CLIP. Cell 141(1):129–141

Hansen TB, Jensen TI, Clausen BH, Bramsen JB, Finsen B, Damgaard CK, Kjems J (2013) Natural RNA circles function as efficient microRNA sponges. Nature 495(7441):384–388

Hansen TB, Wiklund ED, Bramsen JB, Villadsen SB, Statham AL, Clark SJ, Kjems J (2011) miRNA-dependent gene silencing involving Ago2-mediated cleavage of a circular antisense RNA. EMBO J 30(21):4414–4422

Harding KW, Gellon G, McGinnis N, McGinnis W (1995) A screen for modifiers of deformed function in Drosophila. Genetics 140(4):1339–1352

Hasegawa Y, Brockdorff N, Kawano S, Tsutui K, Tsutui K, Nakagawa S (2010) The matrix protein hnRNP U is required for chromosomal localization of Xist RNA. Dev Cell 19(3):469–476

He Y, Vogelstein B, Velculescu VE, Papadopoulos N, Kinzler KW (2008) The antisense transcriptomes of human cells. Science 322(5909):1855–1857

Hendrich BD, Brown CJ, Willard HF (1993) Evolutionary conservation of possible functional domains of the human and murine XIST genes. Hum Mol Genet 2(6):663–672

Hoki Y, Kimura N, Kanbayashi M, Amakawa Y, Ohhata T, Sasaki H, Sado T (2009) A proximal conserved repeat in the Xist gene is essential as a genomic element for X-inactivation in mouse. Development 136(1):139–146

Hong YK, Ontiveros SD, Strauss WM (2000) A revision of the human XIST gene organization and structural comparison with mouse Xist. Mamm Genome 11(3):220–224

Huarte M, Guttman M, Feldser D, Garber M, Koziol MJ, Kenzelmann-Broz D, Khalil AM, Zuk O, Amit I, Rabani M et al (2010) A large intergenic noncoding RNA induced by p53 mediates global gene repression in the p53 response. Cell 142(3):409–419

Iioka H, Loiselle D, Haystead TA, Macara IG (2011) Efficient detection of RNA-protein interactions using tethered RNAs. Nucleic Acids Res 39(8):e53

Janicki SM, Tsukamoto T, Salghetti SE, Tansey WP, Sachidanandam R, Prasanth KV, Ried T, Shav-Tal Y, Bertrand E, Singer RH et al (2004) From silencing to gene expression: real-time analysis in single cells. Cell 116(5):683–698

Jeck WR, Sorrentino JA, Wang K, Slevin MK, Burd CE, Liu J, Marzluff WF, Sharpless NE (2013) Circular RNAs are abundant, conserved, and associated with ALU repeats. RNA 19(2): 141–157

Jeon Y, Lee JT (2011) YY1 tethers Xist RNA to the inactive X nucleation center. Cell 146(1): 119–133

Jones BK, Levorse JM, Tilghman SM (1998) Igf2 imprinting does not require its own DNA methylation or H19 RNA. Genes Dev 12(14):2200–2207

Jonkers I, Barakat TS, Achame EM, Monkhorst K, Kenter A, Rentmeester E, Grosveld F, Grootegoed JA, Gribnau J (2009) RNF12 is an X-Encoded dose-dependent activator of X chromosome inactivation. Cell 139(5):999–1011

Kalantry S, Mills KC, Yee D, Otte AP, Panning B, Magnuson T (2006) The Polycomb group protein Eed protects the inactive X-chromosome from differentiation-induced reactivation. Nat Cell Biol 8(2):195–202

Kaneko S, Li G, Son J, Xu CF, Margueron R, Neubert TA, Reinberg D (2010) Phosphorylation of the PRC2 component Ezh2 is cell cycle-regulated and up-regulates its binding to ncRNA. Genes Dev 24(23):2615–2620

Kanhere A, Viiri K, Araujo CC, Rasaiyaah J, Bouwman RD, Whyte WA, Pereira CF, Brookes E, Walker K, Bell GW et al (2010) Short RNAs are transcribed from repressed polycomb target genes and interact with polycomb repressive complex-2. Mol Cell 38(5):675–688

Kapranov P, Cheng J, Dike S, Nix DA, Duttagupta R, Willingham AT, Stadler PF, Hertel J, Hackermuller J, Hofacker IL et al (2007) RNA maps reveal new RNA classes and a possible function for pervasive transcription. Science 316(5830):1484–1488

Katayama S, Tomaru Y, Kasukawa T, Waki K, Nakanishi M, Nakamura M, Nishida H, Yap CC, Suzuki M, Kawai J et al (2005) Antisense transcription in the mammalian transcriptome. Science 309(5740):1564–1566

Keene JD, Komisarow JM, Friedersdorf MB (2006) RIP-Chip: the isolation and identification of mRNAs, microRNAs and protein components of ribonucleoprotein complexes from cell extracts. Nat Protoc 1(1):302–307

Khalil AM, Guttman M, Huarte M, Garber M, Raj A, Rivea Morales D, Thomas K, Presser A, Bernstein BE, van Oudenaarden A et al (2009) Many human large intergenic noncoding RNAs associate with chromatin-modifying complexes and affect gene expression. Proc Natl Acad Sci U S A 106(28):11667–11672

Kishore S, Jaskiewicz L, Burger L, Hausser J, Khorshid M, Zavolan M (2011) A quantitative analysis of CLIP methods for identifying binding sites of RNA-binding proteins. Nat Methods 8(7):559–564

Kiss T (2001) Small nucleolar RNA-guided post-transcriptional modification of cellular RNAs. EMBO J 20(14):3617–3622

Klattenhoff CA, Scheuermann JC, Surface LE, Bradley RK, Fields PA, Steinhauser ML, Ding H, Butty VL, Torrey L, Haas S et al (2013) Braveheart, a long noncoding RNA required for cardiovascular lineage commitment. Cell 152(3):570–583

Kogo R, Shimamura T, Mimori K, Kawahara K, Imoto S, Sudo T, Tanaka F, Shibata K, Suzuki A, Komune S et al (2011) Long noncoding RNA HOTAIR regulates polycomb-dependent chromatin modification and is associated with poor prognosis in colorectal cancers. Cancer Res 71(20):6320–6326

Konig J, Zarnack K, Rot G, Curk T, Kayikci M, Zupan B, Turner DJ, Luscombe NM, Ule J (2010) iCLIP reveals the function of hnRNP particles in splicing at individual nucleotide resolution. Nat Struct Mol Biol 17(7):909–915

Ku M, Koche RP, Rheinbay E, Mendenhall EM, Endoh M, Mikkelsen TS, Presser A, Nusbaum C, Xie X, Chi AS et al (2008) Genomewide analysis of PRC1 and PRC2 occupancy identifies two classes of bivalent domains. PLoS Genet 4(10):e1000242

Kuzmichev A, Nishioka K, Erdjument-Bromage H, Tempst P, Reinberg D (2002) Histone methyltransferase activity associated with a human multiprotein complex containing the Enhancer of Zeste protein. Genes Dev 16(22):2893–2905

Lai F, Orom UA, Cesaroni M, Beringer M, Taatjes DJ, Blobel GA, Shiekhattar R (2013) Activating RNAs associate with Mediator to enhance chromatin architecture and transcription. Nature 494(7438):497–501

Lebedeva S, Jens M, Theil K, Schwanhausser B, Selbach M, Landthaler M, Rajewsky N (2011) Transcriptome-wide analysis of regulatory interactions of the RNA-binding protein HuR. Mol Cell 43(3):340–352

Lee JT, Davidow LS, Warshawsky D (1999) Tsix, a gene antisense to Xist at the X-inactivation centre. Nat Genet 21(4):400–404

Lee JT, Lu N (1999) Targeted mutagenesis of Tsix leads to nonrandom X inactivation. Cell 99(1):47–57

Leung AK, Young AG, Bhutkar A, Zheng GX, Bosson AD, Nielsen CB, Sharp PA (2011) Genome-wide identification of Ago2 binding sites from mouse embryonic stem cells with and without mature microRNAs. Nat Struct Mol Biol 18(2):237–244

Li Y, Altman S (2002) Partial reconstitution of human RNase P in HeLa cells between its RNA subunit with an affinity tag and the intact protein components. Nucleic Acids Res 30(17):3706–3711

Licatalosi DD, Mele A, Fak JJ, Ule J, Kayikci M, Chi SW, Clark TA, Schweitzer AC, Blume JE, Wang X et al (2008) HITS-CLIP yields genome-wide insights into brain alternative RNA processing. Nature 456(7221):464–469

Lyon MF (1998) X-chromosome inactivation: a repeat hypothesis. Cytogenet Cell Genet 80(1–4): 133–137

Ma Z, Swigut T, Valouev A, Rada-Iglesias A, Wysocka J (2011) Sequence-specific regulator Prdm14 safeguards mouse ESCs from entering extraembryonic endoderm fates. Nat Struct Mol Biol 18(2):120–127

Maenner S, Blaud M, Fouillen L, Savoye A, Marchand V, Dubois A, Sanglier-Cianferani S, Van Dorsselaer A, Clerc P, Avner P et al (2010a) 2-D structure of the A region of Xist RNA and its implication for PRC2 association. PLoS Biol 8(1):e1000276

Maenner S, Blaud M, Fouillen L, Savoye A, Marchand V, Dubois A, Sanglier-Cianferani S, Van Dorsselaer A, Clerc P, Avner P et al (2010b) 2-D structure of the A region of Xist RNA and its implication for PRC2 association. PLoS Biol 8(1):e1000276

Maherali N, Sridharan R, Xie W, Utikal J, Eminli S, Arnold K, Stadtfeld M, Yachechko R, Tchieu J, Jaenisch R et al (2007) Directly reprogrammed fibroblasts show global epigenetic remodeling and widespread tissue contribution. Cell Stem Cell 1(1):55–70

Mao YS, Sunwoo H, Zhang B, Spector DL (2011) Direct visualization of the co-transcriptional assembly of a nuclear body by noncoding RNAs. Nat Cell Biol 13(1):95–101

Martianov I, Ramadass A, Serra Barros A, Chow N, Akoulitchev A (2007) Repression of the human dihydrofolate reductase gene by a non-coding interfering transcript. Nature 445(7128): 666–670

Matouk IJ, DeGroot N, Mezan S, Ayesh S, Abu-lail R, Hochberg A, Galun E (2007) The H19 noncoding RNA is essential for human tumor growth. PLoS One 2(9):e845

Memczak S, Jens M, Elefsinioti A, Torti F, Krueger J, Rybak A, Maier L, Mackowiak SD, Gregersen LH, Munschauer M et al (2013) Circular RNAs are a large class of animal RNAs with regulatory potency. Nature 495(7441):333–338

Min J, Zhang Y, Xu RM (2003) Structural basis for specific binding of Polycomb chromodomain to histone H3 methylated at Lys 27. Genes Dev 17(15):1823–1828

Nagano T, Mitchell JA, Sanz LA, Pauler FM, Ferguson-Smith AC, Feil R, Fraser P (2008) The Air noncoding RNA epigenetically silences transcription by targeting G9a to chromatin. Science 322(5908):1717–1720

Navarro P, Chambers I, Karwacki-Neisius V, Chureau C, Morey C, Rougeulle C, Avner P (2008) Molecular coupling of Xist regulation and pluripotency. Science 321(5896):1693–1695

Navarro P, Oldfield A, Legoupi J, Festuccia N, Dubois A, Attia M, Schoorlemmer J, Rougeulle C, Chambers I, Avner P (2010) Molecular coupling of Tsix regulation and pluripotency. Nature 468(7322):457–460

Navarro P, Pichard S, Ciaudo C, Avner P, Rougeulle C (2005) Tsix transcription across the Xist gene alters chromatin conformation without affecting Xist transcription: implications for X-chromosome inactivation. Genes Dev 19(12):1474–1484

Nesterova TB, Senner CE, Schneider J, Alcayna-Stevens T, Tattermusch A, Hemberger M, Brockdorff N (2011) Pluripotency factor binding and Tsix expression act synergistically to repress Xist in undifferentiated embryonic stem cells. Epigenetics Chromatin 4(1):17

O'Neill MJ (2005) The influence of non-coding RNAs on allele-specific gene expression in mammals. Hum Mol Genet 14(Spec No 1):R113–R120

Ogawa Y, Lee JT (2003) Xite, X-inactivation intergenic transcription elements that regulate the probability of choice. Mol Cell 11(3):731–743

Ogawa Y, Sun BK, Lee JT (2008) Intersection of the RNA interference and X-inactivation pathways. Science 320(5881):1336–1341

Okamoto I, Otte AP, Allis CD, Reinberg D, Heard E (2004) Epigenetic dynamics of imprinted X inactivation during early mouse development. Science 303(5658):644–649

Okazaki Y, Furuno M, Kasukawa T, Adachi J, Bono H, Kondo S, Nikaido I, Osato N, Saito R, Suzuki H et al (2002) Analysis of the mouse transcriptome based on functional annotation of 60,770 full-length cDNAs. Nature 420(6915):563–573

Okutsu T, Kuroiwa Y, Kagitani F, Kai M, Aisaka K, Tsutsumi O, Kaneko Y, Yokomori K, Surani MA, Kohda T et al (2000) Expression and imprinting status of human PEG8/IGF2AS, a paternally expressed antisense transcript from the IGF2 locus, in Wilms' tumors. J Biochem 127(3):475–483

Orom UA, Derrien T, Beringer M, Gumireddy K, Gardini A, Bussotti G, Lai F, Zytnicki M, Notredame C, Huang Q et al (2010) Long noncoding RNAs with enhancer-like function in human cells. Cell 143(1):46–58

Pageau GJ, Hall LL, Ganesan S, Livingston DM, Lawrence JB (2007) The disappearing Barr body in breast and ovarian cancers. Nat Rev Cancer 7(8):628–633

Pandey RR, Mondal T, Mohammad F, Enroth S, Redrup L, Komorowski J, Nagano T, Mancini-Dinardo D, Kanduri C (2008) Kcnq1ot1 antisense noncoding RNA mediates lineage-specific transcriptional silencing through chromatin-level regulation. Mol Cell 32(2):232–246

Penny GD, Kay GF, Sheardown SA, Rastan S, Brockdorff N (1996) Requirement for Xist in X chromosome inactivation. Nature 379(6561):131–137

Plath K, Fang J, Mlynarczyk-Evans SK, Cao R, Worringer KA, Wang H, de la Cruz CC, Otte AP, Panning B, Zhang Y (2003) Role of histone H3 lysine 27 methylation in X inactivation. Science 300(5616):131–135

Popova BC, Tada T, Takagi N, Brockdorff N, Nesterova TB (2006) Attenuated spread of X-inactivation in an X;autosome translocation. Proc Natl Acad Sci U S A 103(20):7706–7711

Preker P, Nielsen J, Kammler S, Lykke-Andersen S, Christensen MS, Mapendano CK, Schierup MH, Jensen TH (2008) RNA exosome depletion reveals transcription upstream of active human promoters. Science 322(5909):1851–1854

Raj A, van den Bogaard P, Rifkin SA, van Oudenaarden A, Tyagi S (2008) Imaging individual mRNA molecules using multiple singly labeled probes. Nat Methods 5(10):877–879

Rastan S (1983) Non-random X-chromosome inactivation in mouse X-autosome translocation embryos—location of the inactivation centre. J Embryol Exp Morphol 78:1–22

Rastan S, Robertson EJ (1985) X-chromosome deletions in embryo-derived (EK) cell lines associated with lack of X-chromosome inactivation. J Embryol Exp Morphol 90:379–388

Riley KJ, Steitz JA (2013) The "Observer Effect" in genome-wide surveys of protein-RNA interactions. Mol Cell 49(4):601–604

Rinn JL, Chang HY (2012) Genome regulation by long noncoding RNAs. Annu Rev Biochem 81:145–166

Rinn JL, Kertesz M, Wang JK, Squazzo SL, Xu X, Brugmann SA, Goodnough LH, Helms JA, Farnham PJ, Segal E et al (2007) Functional demarcation of active and silent chromatin domains in human HOX loci by noncoding RNAs. Cell 129(7):1311–1323

Royce-Tolland ME, Andersen AA, Koyfman HR, Talbot DJ, Wutz A, Tonks ID, Kay GF, Panning B (2010) The A-repeat links ASF/SF2-dependent Xist RNA processing with random choice during X inactivation. Nat Struct Mol Biol 17(8):948–954
Sado T, Hoki Y, Sasaki H (2005) Tsix silences Xist through modification of chromatin structure. Dev Cell 9(1):159–165
Sahoo T, del Gaudio D, German JR, Shinawi M, Peters SU, Person RE, Garnica A, Cheung SW, Beaud AL, et al (2008) Prader-Willi phenotype caused by paternal deficiency for the HBII-85 C/D box small nucleolar RNA cluster. Nat Genet 40(6):719–721
Salmena L, Poliseno L, Tay Y, Kats L, Pandolfi PP (2011) A ceRNA hypothesis: the Rosetta Stone of a hidden RNA language? Cell 146(3):353–358
Salzman J, Gawad C, Wang PL, Lacayo N, Brown PO (2012) Circular RNAs are the predominant transcript isoform from hundreds of human genes in diverse cell types. PLoS One 7(2):e30733
Sanford JR, Wang X, Mort M, Vanduyn N, Cooper DN, Mooney SD, Edenberg HJ, Liu Y (2009) Splicing factor SFRS1 recognizes a functionally diverse landscape of RNA transcripts. Genome Res 19(3):381–394
Sarma K, Levasseur P, Aristarkhov A, Lee JT (2010) Locked nucleic acids (LNAs) reveal sequence requirements and kinetics of Xist RNA localization to the X chromosome. Proc Natl Acad Sci U S A 107(51):22196–22201
Schmitz KM, Mayer C, Postepska A, Grummt I (2010) Interaction of noncoding RNA with the rDNA promoter mediates recruitment of DNMT3b and silencing of rRNA genes. Genes Dev 24(20):2264–2269
Schoeftner S, Sengupta AK, Kubicek S, Mechtler K, Spahn L, Koseki H, Jenuwein T, Wutz A (2006) Recruitment of PRC1 function at the initiation of X inactivation independent of PRC2 and silencing. EMBO J 25(13):3110–3122
Schor IE, Llleres D, Risso GJ, Pawellek A, Ule J, Lamond AI, Kornblihtt AR (2012) Perturbation of chromatin structure globally affects localization and recruitment of splicing factors. PLoS One 7(11):e48084
Shamovsky I, Ivannikov M, Kandel ES, Gershon D, Nudler E (2006) RNA-mediated response to heat shock in mammalian cells. Nature 440(7083):556–560
Shibata S, Lee JT (2003) Characterization and quantitation of differential Tsix transcripts: implications for Tsix function. Hum Mol Genet 12(2):125–136
Shin J, Bossenz M, Chung Y, Ma H, Byron M, Taniguchi-Ishigaki N, Zhu X, Jiao B, Hall LL, Green MR et al (2010) Maternal Rnf12/RLIM is required for imprinted X-chromosome inactivation in mice. Nature 467(7318):977–981
Silva J, Mak W, Zvetkova I, Appanah R, Nesterova TB, Webster Z, Peters AH, Jenuwein T, Otte AP, Brockdorff N (2003) Establishment of histone h3 methylation on the inactive X chromosome requires transient recruitment of Eed-Enx1 polycomb group complexes. Dev Cell 4(4):481–495
Simon JA, Kingston RE (2013) Occupying chromatin: polycomb mechanisms for getting to genomic targets, stopping transcriptional traffic, and staying put. Mol Cell 49(5):808–824
Simon MD, Wang CI, Kharchenko PV, West JA, Chapman BA, Alekseyenko AA, Borowsky ML, Kuroda MI, Kingston RE (2011) The genomic binding sites of a noncoding RNA. Proc Natl Acad Sci U S A 108(51):20497–20502
Srisawat C, Engelke DR (2001) Streptavidin aptamers: affinity tags for the study of RNAs and ribonucleoproteins. RNA 7(4):632–641
Srisawat C, Goldstein IJ, Engelke DR (2001) Sephadex-binding RNA ligands: rapid affinity purification of RNA from complex RNA mixtures. Nucleic Acids Res 29(2):E4
Sun BK, Deaton AM, Lee JT (2006) A transient heterochromatic state in Xist preempts X inactivation choice without RNA stabilization. Mol Cell 21(5):617–628
Sunwoo H, Dinger ME, Wilusz JE, Amaral PP, Mattick JS, Spector DL (2009) MEN epsilon/beta nuclear-retained non-coding RNAs are up-regulated upon muscle differentiation and are essential components of paraspeckles. Genome Res 19(3):347–359
Surralles J, Natarajan AT (1998) Position effect of translocations involving the inactive X chromosome: physical linkage to XIC/XIST does not lead to long-range de novo inactivation in human differentiated cells. Cytogenet Cell Genet 82(1–2):58–66

Tian D, Sun S, Lee JT (2010) The long noncoding RNA, Jpx, is a molecular switch for X chromosome inactivation. Cell 143(3):390–403

Tollervey JR, Curk T, Rogelj B, Briese M, Cereda M, Kayikci M, Konig J, Hortobagyi T, Nishimura AL, Zupunski V et al (2011) Characterizing the RNA targets and position-dependent splicing regulation by TDP-43. Nat Neurosci 14(4):452–458

Tripathi V, Ellis JD, Shen Z, Song DY, Pan Q, Watt AT, Freier SM, Bennett CF, Sharma A, Bubulya PA et al (2010) The nuclear-retained noncoding RNA MALAT1 regulates alternative splicing by modulating SR splicing factor phosphorylation. Mol Cell 39(6):925–938

Tsai MC, Manor O, Wan Y, Mosammaparast N, Wang JK, Lan F, Shi Y, Segal E, Chang HY (2010) Long noncoding RNA as modular scaffold of histone modification complexes. Science 329(5992):689–693

Ule J, Jensen KB, Ruggiu M, Mele A, Ule A, Darnell RB (2003) CLIP identifies Nova-regulated RNA networks in the brain. Science 302(5648):1212–1215

Ulitsky I, Shkumatava A, Jan CH, Sive H, Bartel DP (2011) Conserved function of lincRNAs in vertebrate embryonic development despite rapid sequence evolution. Cell 147(7):1537–1550

Urlaub H, Hartmuth K, Luhrmann R (2002) A two-tracked approach to analyze RNA-protein crosslinking sites in native, nonlabeled small nuclear ribonucleoprotein particles. Methods 26(2):170–181

Vasudevan S, Steitz JA (2007) AU-rich-element-mediated upregulation of translation by FXR1 and Argonaute 2. Cell 128(6):1105–1118

Walker SC, Scott FH, Srisawat C, Engelke DR (2008) RNA affinity tags for the rapid purification and investigation of RNAs and RNA-protein complexes. Methods Mol Biol 488:23–40

Wang H, Wang L, Erdjument-Bromage H, Vidal M, Tempst P, Jones RS, Zhang Y (2004) Role of histone H2A ubiquitination in Polycomb silencing. Nature 431(7010):873–878

Wang KC, Yang YW, Liu B, Sanyal A, Corces-Zimmerman R, Chen Y, Lajoie BR, Protacio A, Flynn RA, Gupta RA et al (2011) A long noncoding RNA maintains active chromatin to coordinate homeotic gene expression. Nature 472(7341):120–124

Wang X, Arai S, Song X, Reichart D, Du K, Pascual G, Tempst P, Rosenfeld MG, Glass CK, Kurokawa R (2008) Induced ncRNAs allosterically modify RNA-binding proteins in cis to inhibit transcription. Nature 454(7200):126–130

Wang Z, Kayikci M, Briese M, Zarnack K, Luscombe NM, Rot G, Zupan B, Curk T, Ule J (2010) iCLIP predicts the dual splicing effects of TIA-RNA interactions. PLoS Biol 8(10):e1000530

Wilusz JE, Freier SM, Spector DL (2008) 3′ end processing of a long nuclear-retained noncoding RNA yields a tRNA-like cytoplasmic RNA. Cell 135(5):919–932

Wilusz JE, JnBaptiste CK, Lu LY, Kuhn CD, Joshua-Tor L, Sharp PA (2012) A triple helix stabilizes the 3′ ends of long noncoding RNAs that lack poly(A) tails. Genes Dev 26(21): 2392–2407

Wilusz JE, Sunwoo H, Spector DL (2009) Long noncoding RNAs: functional surprises from the RNA world. Genes Dev 23:1494–1504

Wutz A, Rasmussen TP, Jaenisch R (2002) Chromosomal silencing and localization are mediated by different domains of Xist RNA. Nat Genet 30(2):167–174

Xu M, Medvedev S, Yang J, Hecht NB (2009) MIWI-independent small RNAs (MSY-RNAs) bind to the RNA-binding protein, MSY2, in male germ cells. Proc Natl Acad Sci U S A 106(30):12371–12376

Yang L, Duff MO, Graveley BR, Carmichael GG, Chen LL (2011a) Genomewide characterization of non-polyadenylated RNAs. Genome Biol 12(2):R16

Yang L, Lin C, Liu W, Zhang J, Ohgi KA, Grinstein JD, Dorrestein PC, Rosenfeld MG (2011b) ncRNA- and Pc2 methylation-dependent gene relocation between nuclear structures mediates gene activation programs. Cell 147(4):773–788

Yao H, Brick K, Evrard Y, Xiao T, Camerini-Otero RD, Felsenfeld G (2010) Mediation of CTCF transcriptional insulation by DEAD-box RNA-binding protein p68 and steroid receptor RNA activator SRA. Genes Dev 24(22):2543–2555

Yap KL, Li S, Munoz-Cabello AM, Raguz S, Zeng L, Mujtaba S, Gil J, Walsh MJ, Zhou MM (2010) Molecular interplay of the noncoding RNA ANRIL and methylated histone H3 lysine 27 by polycomb CBX7 in transcriptional silencing of INK4a. Mol Cell 38(5):662–674

Yeo GW, Coufal NG, Liang TY, Peng GE, Fu XD, Gage FH (2009) An RNA code for the FOX2 splicing regulator revealed by mapping RNA-protein interactions in stem cells. Nat Struct Mol Biol 16(2):130–137

Yildirim E, Kirby JE, Brown DE, Mercier FE, Sadreyev RI, Scadden DT, Lee JT (2013) Xist RNA is a potent suppressor of hematologic cancer in mice. Cell 152(4):727–742

Yin QF, Yang L, Zhang Y, Xiang JF, Wu YW, Carmichael GG, Chen LL (2012) Long noncoding RNAs with snoRNA ends. Mol Cell 48(2):219–230

Yu BD, Hess JL, Horning SE, Brown GA, Korsmeyer SJ (1995) Altered Hox expression and segmental identity in Mll-mutant mice. Nature 378(6556):505–508

Yu W, Gius D, Onyango P, Muldoon-Jacobs K, Karp J, Feinberg AP, Cui H (2008) Epigenetic silencing of tumour suppressor gene p15 by its antisense RNA. Nature 451(7175):202–206

Zarnack K, Konig J, Tajnik M, Martincorena I, Eustermann S, Stevant I, Reyes A, Anders S, Luscombe NM, Ule J (2013) Direct competition between hnRNP C and U2AF65 protects the transcriptome from the exonization of Alu elements. Cell 152(3):453–466

Zhang C, Frias MA, Mele A, Ruggiu M, Eom T, Marney CB, Wang H, Licatalosi DD, Fak JJ, Darnell RB (2010) Integrative modeling defines the Nova splicing-regulatory network and its combinatorial controls. Science 329(5990):439–443

Zhao J, Ohsumi TK, Kung JT, Ogawa Y, Grau DJ, Sarma K, Song JJ, Kingston RE, Borowsky M, Lee JT (2010) Genome-wide identification of polycomb-associated RNAs by RIP-seq. Mol Cell 40(6):939–953

Zhao J, Sun BK, Erwin JA, Song JJ, Lee JT (2008) Polycomb proteins targeted by a short repeat RNA to the mouse X chromosome. Science 322(5902):750–756

Zhou W, Zhu P, Wang J, Pascual G, Ohgi KA, Lozach J, Glass CK, Rosenfeld MG (2008) Histone H2A monoubiquitination represses transcription by inhibiting RNA polymerase II transcriptional elongation. Mol Cell 29(1):69–80

Zhou Z, Licklider LJ, Gygi SP, Reed R (2002) Comprehensive proteomic analysis of the human spliceosome. Nature 419(6903):182–185

Zhu S, Zhang X-O, Yang L (2013) Panning for Long Noncoding RNAs. Biomolecules 3(1):226–241

Zuo T, Wang L, Morrison C, Chang X, Zhang H, Li W, Liu Y, Wang Y, Liu X, Chan MW et al (2007) FOXP3 is an X-linked breast cancer suppressor gene and an important repressor of the HER-2/ErbB2 oncogene. Cell 129(7):1275–1286

Zisoulis DG, Lovci MT, Wilbert ML, Hutt KR, Liang TY, Pasquinelli AE, Yeo GW (2010) Comprehensive discovery of endogenous Argonaute binding sites in Caenorhabditis elegans. Nat Struct Mol Biol 17(2):173–179

Chapter 5
Piwi Proteins and piRNAs Step onto the Systems Biology Stage

Josef P. Clark and Nelson C. Lau

Abstract Animal germ cells are totipotent because they maintain a highly unique and specialized epigenetic state for its genome. To accomplish this, germ cells express a rich repertoire of specialized RNA-binding protein complexes such as the Piwi proteins and Piwi-interacting RNAs (piRNAs): a germ-cell branch of the RNA interference (RNAi) phenomenon which includes microRNA and endogenous small interfering RNA pathways. Piwi proteins and piRNAs are deeply conserved in animal evolution and play essential roles in fertility and regeneration. Molecular mechanisms for how these ribonucleoproteins act upon the transcriptome and genome are only now coming to light with the application of systems-wide approaches in both invertebrates and vertebrates. Systems biology studies on invertebrates have revealed that transcriptional and heritable silencing is a main mechanism driven by Piwi proteins and piRNA complexes. In vertebrates, Piwi-targeting mechanisms and piRNA biogenesis have progressed, while the discovery that the nuclease activity of Piwi protein is essential for vertebrate germ cell development but not completely required in invertebrates highlights the many complexities of this pathway in different animals. This review recounts how recent systems-wide approaches have rapidly accelerated our appreciation for the broad reach of the Piwi pathway on germline genome regulation and what questions facing the field await to be unraveled.

Keywords Piwi pathway • Small regulatory RNA • Germ cell development

1 Introduction: RNA Interference (RNAi) Pathways in Animal Germ Cells

Gene expression control is the sum of both gene activation and gene repression, and in nearly all animal cells, RNAi is a central pathway for cells to execute broad and rapid gene silencing at the transcriptional and posttranscriptional level.

J.P. Clark • N.C. Lau (✉)
Department of Biology and Rosenstiel Biomedical Research Center, Brandeis University, 415 South Street, Waltham, MA 02454, USA
e-mail: nlau@brandeis.edu

G.W. Yeo (ed.), *Systems Biology of RNA Binding Proteins*, Advances in Experimental Medicine and Biology 825, DOI 10.1007/978-1-4939-1221-6_5

Each cell type expresses specific repertoires of genome-encoded small regulatory RNAs that become incorporated into ribonucleoprotein (RNP) complexes. These small RNAs then serve as guides to direct the RNP complexes to search out target transcripts and genomic loci, thereby providing a dynamic closed circuit for gene regulation. In essence, the activation of a small RNA-producing gene leads to the repression of a target gene with base-pairing homology to the small RNA.

In animal cells, the most ubiquitous arm of RNAi is the microRNA (miRNA) pathway. The 20–23-nt-long miRNAs are incorporated into Argonaute (AGO) proteins and have evolved to search for messenger RNA (mRNA) targets using the complementarity of the first 2–9 base pairs in the 5′ end of the miRNA to "seed" an interaction before locking the interaction in through a combination of mismatches and pairings with the rest of the miRNA (Bartel 2009). The AGO-miRNA RNP forms the core of a larger, less defined RNA-induced silencing complex (RISC) that typically seeks the 3′ untranslated regions of target mRNAs and can induce inhibition of mRNA translation as well as mRNA destabilization. Although animal genomes encode several hundreds of different individual miRNA sequences, different cell types can express specific sets of miRNAs because each miRNA derives from a single small hairpin-structured precursor (~60–100 bp) that can sit in the middle of an intron or a longer noncoding transcript made by RNA Polymerase II (Pol II) (Carthew and Sontheimer 2009). Despite being short, some miRNAs have deep conservation through their entire mature miRNA sequence, such as miR-1 and miR-Let-7, which may be attributed to the broad number of mRNA targets regulated by these miRNAs (Ambros 2011).

A second arm of RNAi is the endogenous small interfering RNA (endo-siRNA) pathway which is found in invertebrate somatic cells and only mammalian oocytes, cells which do not express vertebrate innate immunity factors that drive cellular shutdown in the presence of long double-stranded RNA (dsRNA) (Okamura and Lai 2008). Although endo-siRNAs are generally ~21 nt long, they are different from miRNAs because they are thought to derive from longer (>100 bp) dsRNA precursors forming from very long fold-back structures, from two RNAs from different loci interacting in *trans*, or from the direct conversion of an mRNA into dsRNA by an RNA-dependent RNA Polymerase (RdRP). In flies, endo-siRNAs preferentially load into Ago2 as opposed to miRNAs tending to load into Ago1. In nematodes, endo-siRNAs partner with a myriad of AGO homologs; however in mammals, the distinction between miRNA- and endo-siRNA-AGOs is unclear. The target selection mechanisms for endo-siRNAs are presumed to entail mainly complete complementarity towards genes, repetitive elements such as transposable elements (TEs), and viral transcripts (Ghildiyal and Zamore 2009). The physiological role for endo-siRNAs in animal development remains unclear because mutants that specifically disrupt endo-siRNA accumulation in *Drosophila* have subtle phenotypes, whereas in mammals there is only one Dicer enzyme that processes both miRNA and endo-siRNAs, thus complicating the analysis of endo-siRNAs alone. However, endo-siRNAs generated via RdRPs are likely involved in important gene regulatory pathways like nematode dauer formation (Hall et al. 2013), and antiviral responses in flies (Goic et al. 2013).

2 The Piwi Pathway: A Germ Cell Innovation

The third arm of RNAi is the Piwi-interacting RNA (piRNA) pathway, which is distinct from miRNAs and endo-siRNAs because piRNA biogenesis does not depend on Dicer, the key enzyme that matures miRNAs and endo-siRNAs (Ishizu et al. 2012). Although Piwi proteins form a distinct subclade of the AGO protein family, Piwi proteins and piRNAs are mainly enriched in animal germ cells while miRNAs bound by AGO proteins are ubiquitous in somatic cells. Furthermore, piRNAs appear to derive from single-stranded transcript precursors that are noncoding with no annotated features, are transcripts that correspond to the 3′UTR of certain protein-coding genes, or are transcripts that bear an unusually high concentration of transposable element (TE) sequences. Somehow, these transcripts are selected and processed into a diverse array of piRNAs for which the pattern of piRNAs sometimes appears erratic yet nonrandom (Betel et al. 2007), such as some piRNAs that repeatedly accumulate more abundantly, and many preferentially begin with a 5′ uridine or having an adenine at position 10 (see below about the "ping-pong" cycle, and Siomi et al. 2011).

Despite the characteristics mentioned above, our understanding of piRNA biogenesis is still in its infancy. We know that piRNAs are longer in length (25–31 nt) than miRNAs and endo-siRNAs (20–23 nt), the 3′ terminal 2′ oxygen of piRNAs is methylated by the Hen1 enzyme, and there are many putative helicases, putative endonucleases, and tudor-domain-containing proteins genetically required to generate and/or stabilize piRNAs (Ishizu et al. 2012). However, our mechanistic picture of the Piwi pathway is still not fully defined: the only in vitro piRNA biogenesis activity known is a "Trimmer" activity observed in *Bombyx* gonadal cell extracts which can trim a long 5′ phosphorylated transcript bound by SIWI into a mature piRNA (Kawaoka et al. 2012). Although genetic studies that placed artificial sequences like GFP into a piRNA precursor demonstrated artificial piRNA production (Kawaoka et al. 2012; Muerdter et al. 2012), we still cannot predict exactly which piRNAs would be generated most abundantly from these piRNA-generating transgenes.

However, piRNAs are clearly essential for the proper development of germ cells and fertility, because mutations that ablate Piwi protein function or piRNA biogenesis in both invertebrates and vertebrates result in germ cell death, gonadal atrophy, and ultimately sterility. One molecular consequence of losing piRNAs is the volatile expression of TE transcripts severalfold above the negligible level in wild-type animals, and TE mobilization is thought to be damaging germ cell DNA and to cause de novo mutations that result in germ cell apoptosis or failed fertilization. Since a major proportion of metazoan piRNAs are perfectly complementary to the coding sequence of TEs, the role of piRNAs in taming TEs is obvious. However, many animals have large complements of piRNAs that lack homology to TEs, and given the extraordinary diversity of piRNA populations in germ cells compared to miRNAs, the challenge is to determine the full breadth of targets that are regulated by piRNAs.

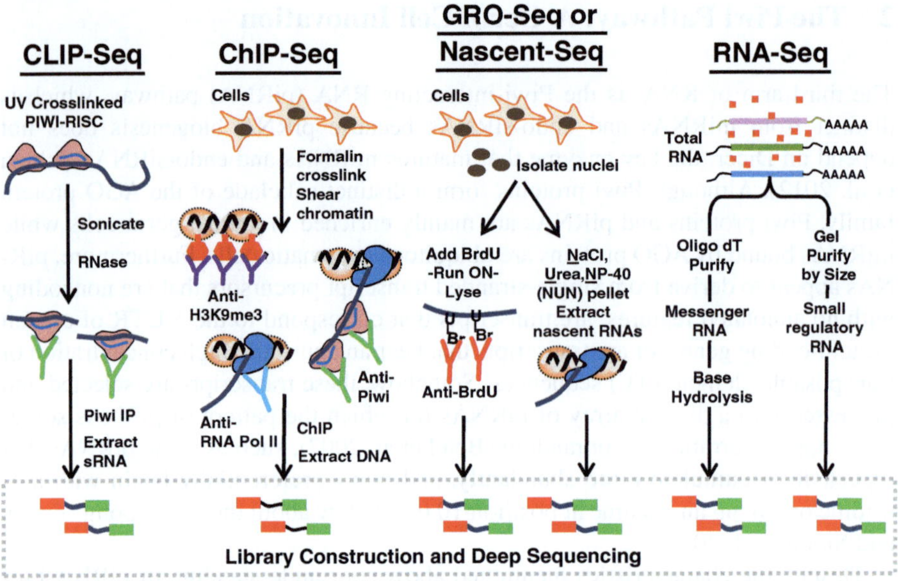

Fig. 5.1 Systems biology approaches applied to study Piwi pathways. Four main biochemical techniques now enable systems-wide analyses of the Piwi-piRNA pathway. Cross-linking immunoprecipitation (CLIP) enriches transcripts associated with Piwi proteins. Chromatin immunoprecipitation (ChIP) enriches genomic DNA regulated by Piwi complexes. Global run on (GRO) and nascent RNA analyses reinforce a transcriptional gene silencing role by Piwi, while messenger RNA and piRNA analyses were facilitated by high-throughput deep sequencing of cDNA libraries

In this chapter, we cover the monumental progress in recent years towards dissecting the functions and mechanisms of Piwi protein complexes and piRNA biogenesis factors, due in large part to the application of systems-wide approaches to this biomedically and biologically important RNA-binding protein pathway. These systems-wide approaches have been built around the extensions of biochemical techniques that recover RNAs and genomic DNAs and characterizing the nucleic acids extremely thoroughly with high-throughput deep sequencing technologies (Fig. 5.1). Supporting these approaches have been heavy dependence on available sequenced and assembled animal genomes and the application of new bioinformatics tools and infrastructure to handle large datasets.

In this review, we first focus on piRNAs from the nematode *Caenorhabditis elegans*, which are quite distinct from other metazoan piRNAs, but appear to influence epigenetic inheritance. Next, we discuss new insight into piRNA biogenesis and targeting mechanisms uncovered in *Drosophila* mutants, ovarian cell line studies, and genome-wide knockdown screens. Finally, we examine the latest findings of vertebrate Piwi protein function and biogenesis that highlight what may be specific distinctions of the Piwi pathway that differ between vertebrates and invertebrates. Because we will be covering mainly the latest progress on how systems biology approaches have propelled the Piwi pathway field, we refer readers to several recent reviews that provide additional detailed and historical coverage of the Piwi field, from early genetics and cell biology to the advent of deep sequencing technology that have provided comprehensive views of piRNA diversity (Lau 2010; Juliano et al. 2011; Siomi et al. 2011; Ishizu et al. 2012).

3 Astounding Argonaute Diversity in *C. elegans*

The nematode *C. elegans* has the distinction of being the first animal where RNAi was described (Fire et al. 1998), where the first two miRNAs were cloned (Lee et al. 1993; Reinhart et al. 2000), and where its small RNAs were first deeply sequenced by next-generation sequencing technology (Ruby et al. 2006). The conservation of Piwi protein function in animal germline development was also first demonstrated in *C. elegans* via RNAi knockdown more than a decade ago (Cox et al. 1998). However, thorough understanding of *C. elegans* Piwi protein mutations has now progressed significantly due to the investigations of the expanded family of other AGO genes in *C. elegans*: 21 worm-specific AGOs (WAGOs and others) accompany the metazoan AGO-related *alg-1 and -2* and the Piwi-related genes *prg-1*, *prg-2*, and *ergo-1*. As such, *C. elegans* has a smorgasbord of different small RNAs: miRNAs, exogenous siRNAs, endogenous siRNAs that come in two flavors (22G-RNAs and 26G-RNAs named by their length and most common first nucleotide), and piRNAs that are also known as 21U-RNAs (Fischer 2010; Ketting 2011). Although miRNAs partition into ALG-1 and ALG-2 complexes, and 21U-RNAs partition into PRG-1 complexes, the line is blurry amongst which assorted worm AGO proteins load which particular 22G- and 26G-RNAs (Lau 2010; Claycomb 2012).

Before the systems biology revolution swept over *C. elegans*, it was assumed at first that the miRNA, piRNA, and endo-siRNA pathways were somewhat distinct pathways with just a few shared enzyme factors. Although Dicer is required for the biogenesis of miRNAs, primary exogenous siRNAs that feed into RDE-1, and the 26G-RNAs, there seems to be a separated generation of secondary endo-siRNAs by different RdRPs. RRF-3 is required for 26G-RNA biogenesis whereas RRF-1 and EGO-1 are required for 22G-RNA biogenesis. Mutations in *prg-1/prg-2* that affect 21U-RNA biogenesis do not affect overall endo-siRNA function (Batista et al. 2008; Das et al. 2008), while mutations that affect 22G-RNA and 26G-RNA accumulations do not impact 21U-RNA production (Han et al. 2009; Pavelec et al. 2009).

In fact, most mutations that affect endo-siRNA formation have not displayed overt phenotypes other than an enhanced RNAi (ERI) response when exogenous dsRNAs are used to target endogenous genes. However, notable exceptions are mutations that perturb 22G-RNA production in the germline (*csr-1*, *ego-1*, *drh-3*, *ekl-1*, and *cde-1*), and *eri-3* and *alg-3/alg-4* mutations that affect sperm-dependent 26G-RNAs (Pavelec et al. 2009; Conine et al. 2010). The *csr-1* and *ego-1* mutant embryos have severely compromised cell division cycles so that they arrest early in embryonic development (Claycomb et al. 2009; She et al. 2009), whereas *eri-3* and *alg3/4* mutants display decreased sperm count at elevated temperatures compared to wild type (Pavelec et al. 2009; Conine et al. 2010). Gonadal development defects were also observed in *prg-1* mutants particularly at elevated temperatures, but at below the standard rearing temperature of 20 °C, these mutants can sustain gametogenesis despite lacking 21U- and 26G-RNAs (Batista et al. 2008; Das et al. 2008). In particular, *ergo-1* is a PIWI-subclade AGO gene whose mutants fail to generate 26G-RNAs yet appear nearly normal in development (Vasale et al. 2010). These mild phenotypes under laboratory conditions seen in mutants lacking most endo-siRNAs and piRNAs populations underscore the mystery of how nematodes can compensate so well when other metazoan germ cells are so dependent on the Piwi pathway.

4 *C. elegans* piRNAs Are Different from Other Metazoan piRNAs

Although the ERGO-1 and PRG-1/PRG-2 proteins have sequence homology to other metazoan Piwi proteins, we believe that *C. elegans* piRNAs are highly distinct from metazoan piRNAs, contrary to the impressions of similarity given from phylogenetic trees. First, the 21U-RNAs bound by PRG-1/PRG-2 are quite sharply restricted to 21 nt long and shorter than the 24–31 nt length of other animal piRNAs. Second, although the 26G-RNAs bound by ERGO-1 are more similar in length to other animal piRNAs, they require Dicer and the RdRP RRF-3 for biogenesis (Vasale et al. 2010), whereas general animal piRNA biogenesis in germ cells is independent of Dicer and RdRP activity (Vagin et al. 2006; Houwing et al. 2007). Third, many other animal piRNAs mature as a complex cluster of overlapping small RNAs derived from a long intergenic transcript (see sections below), but 21U-RNA and 26G-RNA genomic loci are also strikingly different in their configuration.

Rather than a single long transcript that gives rise to thousands of overlapping piRNAs in other animals, 21U-RNAs arise from thousands of individual miniature loci that consist of a small GTTTC-containing motif and other signatures existing upstream of the sequence that is putatively transcribed as a short transcript (Ruby et al. 2006). Transcription of 21U-RNAs may be triggered by worm-specific Forkhead box transcription factors binding at or near this motif to recruit RNA Pol II (Cecere et al. 2012), and then an ~26-nt-long capped transcript is then somehow processed at both the 5′ and 3′ ends to yield a mono-phosphorylated 21U-RNA (Gu et al. 2012). In the genomes of *C. elegans* and the related species *C. briggsae*, the 21U-RNA loci are all clustered together in two main regions of chromosome IV and chromosome I, suggesting an evolutionary requirement of this arrangement perhaps to facilitate epigenetic control (Ruby et al. 2006; Shi et al. 2013). However, the close proximity of each 21U-RNA locus to each other in *C. elegans* genome is not required for 21U-RNA biogenesis, because an autonomous transgene with a single 21U RNA locus can efficiently produce an exogenous 21U-RNA (Billi et al. 2012). This indicates that we do not fully understand the transcription termination control and the genetic requirements for the concentration of 21U-RNA loci into clusters.

Comparative genomics of small RNA populations between the soil nematodes of the *Caenorhabditis* genus (Shi et al. 2013) and other parasitic nematodes such as *Ascaris* further highlight the conundrum of the functional role of nematode piRNAs (21U-RNAs). The total number of 21U-RNAs in *C. elegans* may still be in flux between ~10,000 and ~24,000 individual sequences as different approaches to library construction, greater sequencing depth, and bioinformatics predictions sort out the final tally (Gu et al. 2012; Shi et al. 2013). However, a recent study reported that the 21U-RNA gene number may be doubled in *C. remanei* and *C. brenneri* compared to *C. elegans* and *C. briggsae*. The Shi et al. study suggested that 21U-RNA diversification may be more important for germline development in gonochoristic nematodes (male and female sexes, *C. remanei* and *C. brenneri*) compared to androdioecious nematodes (male and hermaphrodite sexes, *C. elegans* and *C. briggsae*). However, the parasitic nematode *Ascaris suum* is also gonochoristic yet appears to completely lack PRG-1/PRG-2 homologs and 21U-RNAs all together, even though *Ascaris* produce miRNAs and 22G- and 26G-RNAs (Wang et al. 2011). *Ascaris* gonocyte production is extremely prolific, and there is also a chromosome diminution process in somatic cells of the embryos which eliminates DNA sequences that are typically only expressed in the germline. Although chromosome diminution and maintenance of the germline in protozoans are dependent on small RNA pathways, it currently seems that *Ascaris* small RNAs are not involved in chromosome diminution. Notwithstanding that the miRNA pathway is conserved in *Ascaris*, we conjecture that nematode versions of the piRNA and endo-siRNA pathways are much more extensively evolved to suit a soil or a gut environment, such that their RNAi pathways are quite distinct from their counterparts in other metazoans.

5 Worm Piwi Pathways: More Interconnected than We Thought

Despite the differences between *C. elegans* piRNAs and other animal piRNAs, the field has also searched for commonality amongst these piRNAs. Since the Piwi pathway clearly suppresses TEs in fly and mammalian germ cells, TEs were examined in the *prg-1* and *prg-2* null mutants, and only a limited number of TEs such as Tc3 showed a robust up-regulation of its transcript in mutants despite little evidence of TE mobilization (Batista et al. 2008; Das et al. 2008). Another shared feature with general animal piRNAs is that 21U- and a subset of the 26G-RNAs are specifically methylated at the 3′ terminal 2′ hydroxyl by the HENN1 RNA methylase, similar to fly, fish, and mammalian piRNAs and endo-siRNAs being methylated by HEN1 orthologs (Billi et al. 2012; Kamminga et al. 2012; Montgomery et al. 2012). Despite its evolutionary conservation, the functional importance of Hen-1 for general animal piRNAs and endo-siRNAs is not yet clear because in the fly *hen-1* mutant there is only a modest decrease in piRNA and endo-siRNA levels and few obvious developmental defects (Horwich et al. 2007; Saito et al. 2007). However, *C. elegans* 21U-RNA and 26G-RNA levels are clearly perturbed and diminished, respectively, in *henn-1* null mutants (Billi et al. 2012; Kamminga et al. 2012; Montgomery et al. 2012). Although there are indications that secondary endo-siRNAs like 22G-RNAs are not modified with a 2′-O-methyl mark (Montgomery et al. 2012), the gene-silencing function of the 22G-RNAs is affected in *henn-1* mutants due to the genetic connection between 26G-RNAs and 22G-RNAs.

The genetic connection between 22G- and 26G-RNAs has been previously appreciated with the isolation and characterization of several mutants that display consistent loss of 22G-RNA accumulation and function whenever 26G-RNAs were also affected (Conine et al. 2010; Gent et al. 2010; Vasale et al. 2010; Zhang et al. 2011a). As mentioned earlier, the expanded variety of worm AGO proteins mirrors the diversification of 26G- and 22G-RNAs: some small RNAs are germline or soma specific (Yigit et al. 2006; Conine et al. 2010; Vasale et al. 2010), sperm or oocyte specific (Han et al. 2009; Gent et al. 2010), perhaps even specific to developmental stages and environmental responses (Hall et al. 2013). The emerging model is that 26G-RNAs are the primary endo-siRNA trigger that initiates pairing to a target transcript, which then subsequently stimulates 22G-RNA production and amplification of the silencing process through worm-specific RdRPs and AGO proteins (Fig. 5.2).

Although this 26G- to 22G-RNA link is now apparent, earlier Northern blots indicated that 21U-RNA accumulation was not linked to bulk 26G-RNA nor 22G-RNA biogenesis, which predated our understanding that there are two overarching cohorts of 22G-RNAs: the CSR-1-specific 22G-RNAs versus the WAGO-specific 22G-RNAs (Claycomb et al. 2009; Gu et al. 2009). By constructing *C. elegans* strains containing integrated transgene reporters that could be targeted and silenced by a 21U-RNA and PRG-1, three groups performing small RNA deep sequencing discovered that exogenous, transgene-specific 22G-RNAs were generated in a *prg-1-dependent* manner (Ashe et al. 2012; Bagijn et al. 2012; Buckley et al. 2012; Lee et al. 2012; Luteijn et al. 2012; Shirayama et al. 2012). The most abundant sets of these exogenous 22G-RNAs accumulated proximally from the 21U-RNA-binding site. Interestingly, exogenous 22G-RNAs and transgene silencing were maintained even with two mismatches between the 21U-RNA and the binding site (Bagijn et al. 2012). Since few perfectly complementary targets to 21U-RNAs had been detected, the search was broadened to 21U-RNA-binding site with up to three mismatches and a strong correlation could now be seen between some populations of endogenous 22G-RNAs and predicted 21U-RNA-binding sites on the genome (Bagijn et al. 2012; Lee et al. 2012). Two genes detected as being strongly mis-regulated in the *prg-1* mutant were *bath-45* and *F54F2.2b* which lost almost all 22G-RNAs in the *prg-1* mutant. However, other genes dependent on *prg-1* to produce antisense 22G-RNAs were rather modestly up-regulated while many other genes and TEs that have associated 22G-RNAs retain them regardless of *prg-1*. Why did *prg-1* have such a strong influence on a subset of 22G-RNAs against certain genes and not have much effect on other 22G-RNAs, sometimes against the same gene or elements like Tc3?

To resolve this conundrum, two groups tracked derepression of their 21U-RNA fluorescent reporter transgene in various mutant backgrounds, and discovered that

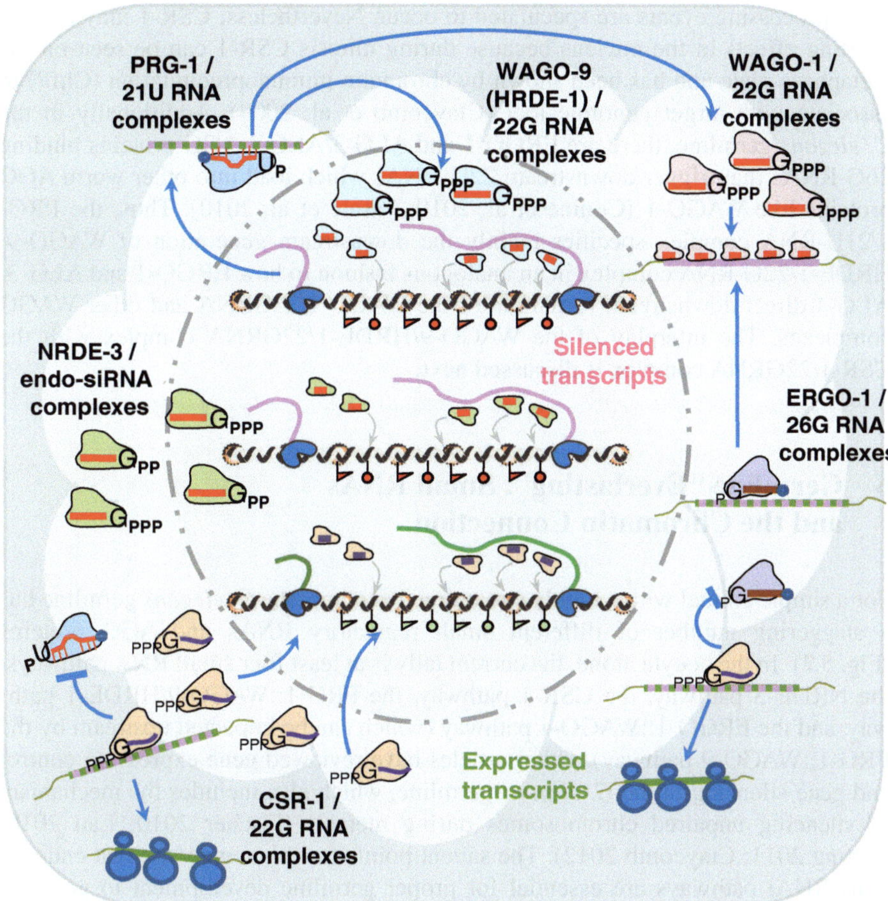

Fig. 5.2 The *C. elegans* Piwi pathway is genetically connected to other endogenous RNAi pathways. This diagram represents a developing germ cell in *C. elegans* with several small RNA pathways operating concurrently, some in the nucleus like the PRG-1 and WAGO-9 complexes and the NRDE-3 and CSR-1 complexes, while ERGO-1 and WAGO-1 are speculated to work in the cytoplasm. These current models depict these small RNA complexes in either silencing or maintaining gene expression in the *C. elegans* germline. Many important biogenesis factors are omitted (i.e., RdRPs, Dicer, helicases) because this diagram has been simplified to focus on the AGO-small RNA complexes

prg-1-mediated silencing in the germline was mainly linked genetically with WAGO-9/HRDE-1, a nuclear-localized worm-specific AGO (WAGO) protein that binds 22G-RNAs generated from upstream factors like *rrf-1*, *mut-7*, *drh-3*, and *rde-2* (Bagijn et al. 2012; Lee et al. 2012). These WAGO-bound 22G-RNAs are distinct from CSR-1-bound 22G-RNAs even though biogenesis factors like *drh-3* may be shared and both CSR-1 and WAGO-9/HRDE-1 are abundantly expressed in the germline. Whereas WAGO-9/HRDE-1 is primarily nuclear, CSR-1 is mostly concentrated in the cytoplasm and perinuclear organelles called P-granules where some

RNA processing events are speculated to occur. Nevertheless, CSR-1 may also be exerting effects in the nucleus because during mitosis CSR-1 can be seen on the metaphase plate and has been shown by chromatin immunoprecipitation (ChIP) to associate with target genomic loci (Claycomb et al. 2009). Additionally in the *C. elegans* germline, there are ERGO-1 and ALG-3/ALG-4 AGO proteins binding 26G-RNAs that trigger downstream 22G-RNAs which load into other worm AGO proteins like WAGO-1 (Conine et al. 2010; Vasale et al. 2010). Thus, the PRG-1/21U-RNA complex specifies mainly the downstream generation of WAGO-9/HRDE-1/22G-RNA complex in an analogous fashion to how ERGO-1 and ALG-3/ALG-4 direct downstream formation of the WAGO-1/22GRNA and other WAGO complexes. The interplay of the WAGO-9/HRDE-1/22GRNA complex with the CSR-1/22GRNA complex is discussed next.

6 Germline "Everlasting": Small RNAs and the Chromatin Connection.

For a simple animal with a highly streamlined genome, the *C. elegans* germline has a staggering number of different small regulatory RNAs and AGO proteins (Fig. 5.2). In the oocyte alone, the current tally is at least four small RNA pathways: the NRDE-3 pathway, the CSR-1 pathway, the PRG-1::WAGO-9/HRDE-1 pathway, and the ERGO-1::WAGO-1 pathway (which can be impacted upstream by the PRG-1::WAGO-9 pathway). Other articles have reviewed gene expression control and gene silencing in the *C. elegans* germline, which also includes the mechanism of silencing unpaired chromosomes during meiosis (Fischer 2010; Lau 2010; Ketting 2011; Claycomb 2012). The salient points for this review are that endogenous RNAi pathways are essential for proper germline development to generate gametes that form the totipotent zygote, and *C. elegans* embryogenesis is extracorporeal, so maternal stores of mRNAs, proteins, and lipids must be deposited and organized in the oocytes.

To highlight one essential maternal factor, the CSR-1 complex is absolutely required for proper embryonic cell division and chromosome segregation events, and molecular studies suggest that 22G-RNAs guide CSR-1 to specific chromatin domains in order to help establish the boundaries needed to segregate the holocentric chromosomes of *C. elegans* (Claycomb et al. 2009; van Wolfswinkel et al. 2009). Thus, CSR-1 represents one important layer of epigenetic control in the germline.

However, the future challenge will be to understand why there are so many additional concurrent small RNA-AGO complexes in the *C. elegans* germ cell that are acting either redundantly or antagonistically. For example, one group has proposed that CSR-1 may serve an anti-silencing role in antagonizing the gene-silencing activities of the WAGO complexes, but the details are obscure for how these two complexes bind similar 22G-RNAs yet partition and regulate different genes ((Shirayama et al. 2012), Fig. 5.2). Furthermore, in addition to the nucleus-directed

PRG-1::WAGO-9/HRDE-1 pathway, there is also the NRDE-3 pathway that is triggered by exogenous siRNAs and also leads to gene silencing at the chromatin level in the nucleus (Guang et al. 2010). Although the *nrde-1*, *nrde-2*, and *nrde-4* genes are required for gene silencing by the NRDE-3/endo-siRNA complex and the WAGO-9/HRDE-1/22G-RNA complex, NRDE-3 is not required for the inherited silencing by WAGO-9/HRDE-1 (Ashe et al. 2012; Buckley et al. 2012). We summarize that there is a cross talk in common genetic factor as well as partitioning of targeting functions and primary triggers in *C. elegans* germline RNAi pathways, and much further study is needed to sort all these pathways out.

A new paradigm proposed in these recent studies is that these multiple germline RNAi processes establish a mechanism for controlling and discriminating nonself versus self transcripts (Ashe et al. 2012; Shirayama et al. 2012). This paradigm argues that when a complex containing PRG-1 and a 21U-RNA fortuitously binds to a foreign transgene, a TE, or a virus (via base mismatches analogous to miRNA recognition), this event licenses WAGO-9/HRDE-1 to mount a seemingly permanent silencing of the nonself locus. This model is compelling for artificial transgenes, but is less able to explain why most *C. elegans* TEs and viruses are not massively mis-regulated in *prg-1* and *wago-9/hrde-1* mutants. Perhaps these nonself elements are redundantly controlled by other WAGO/endo-siRNA complexes like NRDE-3 or WAGO-1.

While *prg-1* and *wago-9/hrde-1* mutant germlines appear as fertile as wild type, perhaps reflecting the modest expression changes in limited numbers of genes (Batista et al. 2008; Das et al. 2008; Bagijn et al. 2012; Lee et al. 2012), a possible explanation to this conundrum is that the *wago-9/hrde-1* mutants lose germline immortality. In other words, although the brood sizes of these mutants are similar to wild type at the F1 progeny (the first descendants of a mutant in-bred cross), the F2 brood size begins to drop and reaches near complete sterility by the F5 progeny (Buckley et al. 2012). This outcome suggests that the germline may be highly sensitive to subtle yet compounding gene expression perturbations from multiple generations.

Perhaps germline immortality requires redundant levels of small RNAs for regulating key protein expression to levels that we cannot yet pinpoint as important for fertility. In fact, *prg-1* mutants only display their brood size deficiencies at elevated temperatures (the phenotype is masked at cooler temperatures), where perhaps germline transcriptional enzyme activity or other gene-regulatory processes may be more unbridled. Therefore, we propose that another natural function of the *prg-1/wago-9/hrde-1* pathway may be to keep the self transcriptome and proteome in the germline in check, perhaps to regulate endogenous soma-expressed genes that may be promiscuously expressed in the totipotent epigenetic state of the germline. This idea may also be related to the fact that the general Piwi/piRNA pathways in all other animals from flies to humans are primarily enriched in rapidly dividing germ cells, which display the widest diversity in its transcriptome (Yeo et al. 2004; Pao et al. 2006; Ravasi et al. 2006).

7 Flies Push Piwi/piRNA Insights into New Heights

Drosophila germline genetic screens pioneered the discovery of the Piwi pathway (Schupbach and Wieschaus 1989; Schupbach and Wieschaus 1991; Wilson et al. 1996; Lin and Spradling 1997) and the *Drosophila* female germline has continued to be the most fruitful system to study this pathway. Basic features of *Drosophila* piRNA biogenesis steps have been introduced here and in other reviews (Juliano et al. 2011; Siomi et al. 2011; Ishizu et al. 2012); therefore this section discusses how recent advances in our recent understanding of piRNA biogenesis and Piwi regulation mechanisms have benefited from systems biology approaches now being applied on *Drosophila* Piwi pathway mutants and *Drosophila* gonadal cell lines.

Starting with the beginning of the pathway, there has been recent progress in understanding the regulation of piRNA precursor transcription from the large intergenic piRNA clusters within pericentromeric heterochromatin. The two prototypical major piRNA clusters in *Drosophila* are the *Flamenco* locus on the X chromosome and the *42AB* piRNA cluster locus on chromosome 2R (Fig. 5.3). In addition

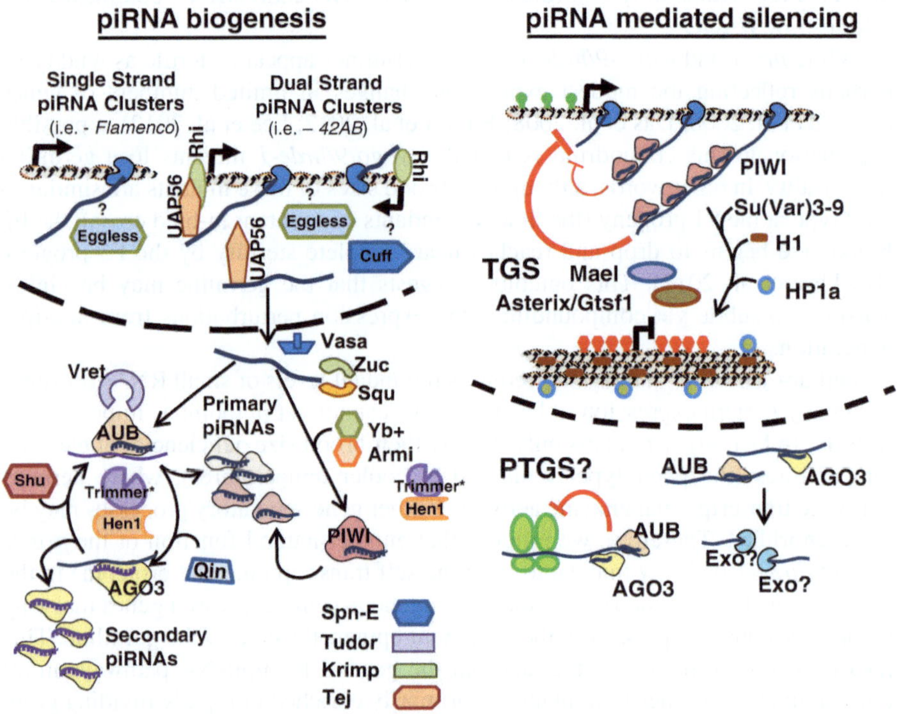

Fig. 5.3 Current models of piRNA biogenesis and piRNA-mediated gene silencing mechanisms in *Drosophila*. (*Left*) Depiction of known roles for factors genetically implicated in *Drosophila* piRNA biogenesis. More factors are known to regulate dual-strand than single-strand piRNA clusters. The piRNA precursor transcripts either are directly processed into primary piRNAs bound by AUB and

to location, these clusters are distinct because *Flamenco* generates transcripts from one genomic strand and is mainly expressed in the somatic follicle cells that surround the egg chamber, whereas *42AB* generates transcripts from both genomic strands and is mainly expressed in the nurse cells of the female germline and these transcripts are then deposited into the oocyte (Juliano et al. 2011; Siomi et al. 2011). Both clusters, however, are highly concentrated in genomic sequences corresponding to TE relics, mutated copies of TEs that have landed in these clusters to become templates for the piRNAs which can then silence the active TE copies elsewhere in the genome.

For piRNAs to be expressed abundantly in gonadal cells there is exquisite regulation of the piRNA cluster loci chromatin. The histone methyltransferase *Eggless* (also referred to as *Egg* and dSETDB1) may broadly impact transcription of both *42AB* and *Flamenco* clusters by specifically tri-methylating histone H3 on lysine 9 (leaving a H3K9me3 mark) (Rangan et al. 2011). One protein that binds the H3K9me3 mark is heterochromatin protein-1 (HP1, also known as *Su(Var)2-05*) which is genetically linked to the Piwi pathway (see below). However, a gonad-specific HP1 homolog called *Rhino* (*Rhi*) is enriched at and is required for transcription of the *42AB* piRNA cluster (Klattenhoff et al. 2009). Associated with *Rhi* at the *42AB* cluster is *Cutoff* (*Cuff*), a germ cell-specific protein homologous to the Rai1 nuclease, and *Cuff* is required for expression of both *42AB* cluster expression and a single-strand piRNA cluster called the *20A* cluster (Pane et al. 2011). Despite appearing as general chromatin-associated factors, *Rhi* or *Cuff* is required only for *42AB* piRNA expression but not *Flamenco* piRNA expression, and questions remain as to what specifies RHI and CUFF to only localize at one piRNA cluster rather than being a general factor like EGG.

After transcription, RNA processing and transport are likely to play a special role in handling piRNA precursor transcripts. Vasa is one well-known germ cell-specific RNA-binding protein with putative helicase activity because it contains a DEAD-box domain, and mutations in this gene abrogate *42AB* piRNA expression (Malone et al. 2009). Recently, a second DEAD-box containing protein, UAP56, was found to be required for piRNA expression, is localized near Rhi foci on nurse cell chromatin, and appears to bind newly transcribed piRNA-precursor transcripts from the *42AB* locus (Zhang et al. 2012). The model from this study is that UAP56 may shuttle a piRNA precursor transcript to the nuclear membrane for a hand off to Vasa for further processing at the nuclear periphery (Zhang et al. 2012).

Fig. 5.3 (continued) PIWI or engage in a subsequent secondary piRNA "ping-pong" amplification by AUB and AGO3. *The identity of "Trimmer" has not been genetically determined. (*Right*) Since AUB and AGO3 are in perinuclear nuage and cytoplasm, they may mediate posttranscriptional gene silencing. The nuclear PIWI binds to nascent TE transcripts, where it can recruit chromatin-modifying factors to induce a repressive chromatin state. Many other factors not shown in this diagram that may affect TE silencing have been identified in knockdown screens, but placement into the model is unclear, as is the role for tudor-domain-containing factors like Spn-E, Tudor, Krimp, and Tej

8 Pulling Back the Layers of piRNA Biogenesis

Some hints to the puzzle of piRNA biogenesis have come from understanding how the front and back termini of piRNAs are formulated. Despite the immense diversity of piRNA sequences, two common features define the termini of piRNAs—a preference for uridine at the 5′ end, and a highly conserved 2′O-methylation modification added to the 3′ end of piRNAs by Hen1 (Horwich et al. 2007; Saito et al. 2007). Recently, a biochemical extract from silkworm cells overexpressing the Piwi protein SIWI was able to recapitulate the activity of piRNA 3′ end formation and SIWI loading in vitro when given an ~50 bp 5′ radioactive phosphate-labeled RNA (Kawaoka et al. 2011). The Tomari group also demonstrated that the 3′ end trimming was coordinated with Hen-1-directed 2′O-methylation; however efforts to purify and identify the putative ribonuclease for "Trimmer" were stymied by the Trimmer activity being restricted to the insoluble fraction (Kawaoka et al. 2011).

Nevertheless, one factor, Zucchini (Zuc), is a tempting candidate for being a Trimmer enzyme because it is absolutely required for piRNA biogenesis and has homology to bacterial endonucleases (Pane et al. 2007). In addition, Zuc's metazoan homologs are mitochondrial phospholipase D (also known as *Pld6* in mouse) (Huang et al. 2011a; Watanabe et al. 2011; Anand and Kai 2012) and are localized to the surface of mitochondrial membranes that might exhibit an insoluble characteristic in extracts. Two recent studies solved the structures of recombinant forms of Zuc and have confirmed that Zuc has endoribonuclease activity in vitro, but the nature of the phospholipase D function remains unclear (Ipsaro et al. 2012; Nishimasu et al. 2012). Structure and biochemical data indicate that Zuc cleaves single-stranded DNA and RNA, and does so without requiring magnesium, a typical ion required by many ribonucleoprotein enzymatic activities including Argonaute's Slicer activity. The catalytic activity of Zuc was not enhanced against a 5′ uridine (5′U), the preferred starting base of many piRNAs; however its RNase activity was only enhanced with the absence of sodium chloride, which is an interesting reflection of perhaps lower ionic strengths in a membraneous, less soluble microenvironment where Zuc may act to process piRNA precursors.

The "ping-pong" model of piRNA biogenesis is one potential mechanism for specifying the 5′ end of many piRNAs. In brief, one Piwi-family protein like aubergine (AUB) prefers to bind 5′U-starting piRNAs derived from piRNA cluster loci and are antisense to the coding strand of TEs, while Argonaute-3 (AGO3) prefers to bind piRNAs that have a preference of adenine at position 10 and are derived from TE-coding strands (Brennecke et al. 2007; Gunawardane et al. 2007). In this model, AUB and its piRNA bind perfectly complementary to a TE transcript and slice the transcript to yield a cleavage product with a free 5′ phosphate. The 5′ end of this cleavage product is then bound by AGO3 and "trimming" of this transcript leads to mature piRNAs. Vice versa in fashion, AGO3 and its piRNA specify the 5′ end and loading the next AUB and piRNA complex (Fig. 5.3). While the "ping-pong" model is not yet proven biochemically, there is genetic support from *ago3* mutations that cause the collapse of AUB piRNAs (Li et al. 2009), and signatures of the ping-pong cycle appear to be conserved in vertebrates as well (Houwing et al. 2007; Aravin et al. 2008; Lau et al. 2009). Recently, a newly identified factor called Qin, which has E3 ligase and TUDOR domains, was shown to modulate the biased loading of specific ping-pong piRNAs into AUB or AGO3. Although *qin* mutants (also known as *kumo* (Anand and Kai 2012)) still produce piRNAs, *qin* mutations perturb normal AGO3 piRNA loading and affect the protein-protein interactions between AUB and AGO3 (Zhang et al. 2011b).

The piRNA "ping-pong" cycle was originally hypothesized to be hierarchical such that primary piRNAs primed the amplification loop for secondary piRNAs (Brennecke et al. 2007; Gunawardane et al. 2007). This connection is supported by mutations in protein factors that abolish primary piRNAs from *Flamenco* which also ablate the secondary piRNAs such as the ones from the *42AB* cluster. The mutations of processing factors that destroy primary and secondary piRNA biogenesis include *armi*, *zuc*, *vert*, *shu*, and *gasz* (Saito et al. 2009; Haase et al. 2010; Zamparini et al. 2011; Olivieri et al. 2012; Preall et al. 2012) (Handler et al. 2011; Handler et al. 2013). In contrast are mutations that only affect *42AB* piRNAs but leave *Flamenco* piRNAs relatively unaffected, such as *spn-E*, *krim*, *vasa*, *rhi*, *UAP56*, *cuff*, *kumo*, and *tejas* (Lim and Kai 2007; Klattenhoff et al. 2009; Malone et al. 2009; Olivieri et al. 2010; Patil and Kai 2010; Anand and Kai 2012; Zhang et al. 2012). *Yb* is an exceptional factor because its loss of function mainly affects only the *Flamenco* piRNAs without much effect on *42AB* piRNAs, analogous to the *Flamenco* promoter mutations that only affect *Flamenco* piRNAs (Malone et al. 2009; Olivieri et al. 2010; Handler et al. 2011). Although this list of factors can be divided based on their roles in either primary or secondary piRNA biogenesis, it is still unclear how 5′ ends of primary piRNAs are defined and why the 3′UTR of certain genic transcripts are selected for primary piRNA biogenesis. Perhaps this gap is attributed to an incompleteness of the list of genes that do impinge upon the Piwi pathway.

Addressing this issue are recent reports by the Hannon and Brennecke labs on genome-wide knockdown screens in flies and cultured follicle cell lines that have now greatly expanded this gene list (Czech et al. 2013; Handler et al. 2013; Muerdter et al. 2013). Many previously identified genetic factors implicated in piRNA biogenesis were recovered and validated in these powerful screens, and gratifyingly several new factors have been identified to impact piRNA biogenesis, such as the *Drosophila* GasZ homolog CG2183, nuclear pore component Nxt1, and the sumolyation E1 ligase Uba2, just to name a few. These screens provided between ~50 and ~90 strongly validated new factors that impact the piRNA pathway in the *Drosophila* female germline, either at the biogenesis or the gene silencing effector level. Many future studies will ensue to place each of these factors in more discreet positions in the pathway.

9 Multi-talented Piwi Proteins Do TGS (and PTGS?)

Insights into how Piwi proteins carry out their effecter role in gene silencing have also dramatically increased recently due to systems approaches in *Drosophila*. Besides the *Drosophila* Piwi proteins themselves (PIWI, AUB, AGO3), a couple of Piwi pathway factors have been defined as effectors of Piwi-mediated gene silencing because their loss of function did not affect piRNA accumulation but did allow for TE transcript up-regulation. Squash (*squ*) and Maelstrom (*mael*) were two of the first known effectors because mutations or knockdowns greatly impacted fertility yet most piRNAs were sustained in these animals or gonadal cells (Pane et al. 2007; Malone et al. 2009; Haase et al. 2010). Squ has some homology to an RNase HII protein, but its biochemical activity is still not well understood, whereas Mael can associate with the microtubule-organizing center and serves a role in specifying the polarity of *Drosophila* oocytes (Sato et al. 2011). A third effecter named Asterix/Gstf1 was recently recovered in the genome-wide RNAi screens as crucial for TE repression by the Piwi pathway despite little change in piRNA production (Donertas et al. 2013; Muerdter et al. 2013; Ohtani et al. 2013). We can expect more effecters to emerge as the scrutiny turns to how the PIWI complex with piRNAs regulates targets for silencing.

With *Drosophila* PIWI being the founding member of this protein sub-clade, earlier studies had pointed to a potential transcriptional gene silencing (TGS) role for PIWI. PIWI was known to be nuclear, genetically and biochemically interacted with HP1, and was localized on polytene chromosomes in proximity to HP1 (Cox et al. 2000; Pal Bhadra et al. 2006; Brower-Toland et al. 2007). In addition, Piwi pathway mutations genetically interacted with Polycomb complex genes (Grimaud et al. 2006) and have been shown to modulate heterochromatin formation on transgenic loci harboring TE sequences as well as telomeres and insulated elements (Haynes et al. 2006; Yin and Lin 2007; Moshkovich and Lei 2010; Sentmanat and Elgin 2012). But the question remained as to how directly PIWI was involved in instigating TGS because these studies examined somatic cell phenotypes that were thought to be established by *piwi* in the germline.

Several recent systems-wide studies interrogating chromatin and nascent RNAs following perturbation of *piwi* function now clearly indicate that PIWI intimately directs TGS in gonadal cells. These studies corroborate that the loss of PIWI results in strong reduction of HP1 deposition, histone H3, and lysine 9 tri-methylation (H3K9me3), and concomitant increases in RNA Pol II occupancy at TE loci (Sienski et al. 2012; Huang et al. 2013; Le Thomas et al. 2013; Rozhkov et al. 2013). These signatures were consistent with the response of increased levels of steady-state and nascent TE transcripts across the *Drosophila* genome (Sienski et al. 2012; Sytnikova et al. 2014). These data suggest that PIWI association at TE chromatin helps recruit H3K9me3 and HP1 marks that prevent the transcription machinery from engaging. However, the answer to whether H3K9me3 and HP1 marks are the cause or the consequence of gene silencing may be complicated because the Brennecke group observed that the loss of the effecter, Mael, which allows dramatic TE up-regulation like the loss of PIWI, had very little reduction in H3K9me3 and HP1 (Sienski et al. 2012). Future studies may clarify other notable exceptions to the link between HP1 and PIWI, such as continued HP1 deposition at certain piRNA-targeted loci in *piwi* mutants (Moshkovich and Lei 2010), and PIWI transgenes that delete putative HP1-binding motifs yet can rescue TE silencing (Wang and Elgin 2011).

Another open question regarding PIWI triggering TGS is whether PIWI binds chromatin directly or through nascent transcripts. Although one study was able to achieve a chromatin immunoprecipitation of PIWI (Huang et al. 2013), another study that performed CLIP-Seq on PIWI instead revealed that TE transcripts are the most heavily associated transcripts bound by PIWI, dominating a smaller fraction of genic transcripts (Sytnikova et al. 2014). Although coding gene expression changes have not been detected in fly ovaries deficient of *piwi* (Le Thomas et al. 2013), several genes did become strongly up-regulated in OSS and OSC cells upon *piwi* knockdown on the account of a de novo TE insertion in close proximity to the gene (Sienski et al. 2012; Sytnikova et al. 2014). By using the NUN-pellet protocol to obtain nascent transcripts, the Lau group was able to detect nascent TE transcripts up-regulated during PIWI knockdown that were independently transcribed from the nearby affected gene, and this TE transcript was complementary to the TE piRNAs. Furthermore, they showed with crafted reporter genes that PIWI and piRNAs must pair with the nascent transcript and not the DNA, and a threshold number of perfectly complementary piRNAs to the target transcript are required in order to initiate the TGS mechanism (Post et al. 2014).

Defining how far PIWI-mediated TGS can spread and what chromatin modulators are required for this spread will be an important next question to address. The reporter gene data point to a nascent transcript association requirement and is in agreement with mass spectrometry analysis of alternative splicing and nuclear RNA processing factors associated with PIWI (Le Thomas et al. 2013), indicating that PIWI/piRNA complex can rapidly interact with a TE nascent transcript and promote a transcriptionally repressed state that can spread at least a few kilobases away. As the scrutiny turns to other chromatin factors, one recent report suggests that depletion of the linker histone H1 can alter H3K9 methylation and in turn unleash TE expression as well as strangely increase the accumulation of piRNA-like species (Lu et al. 2013).

Although the recent attention of TGS mechanisms has been focused on PIWI, other diverse mechanisms of gene regulation may exist for the other Piwi family members. For instance, *Aub* can regulate *Nanos* (also known as *Nos*) mRNA expression by genetically determining its localization to the posterior pole of the *Drosophila* oocyte where it helps direct the formation of the pole plasm in the embryo (Harris and Macdonald 2001; Megosh et al. 2006). The biased transport of *Nanos* mRNA is required for anterior-posterior patterning and primordial germ cell specification during embryogenesis (Lehmann and Nusslein-Volhard 1991; Wang et al. 1994). To achieve this, AUB can bind the 3′UTR of *Nanos* mRNA directly with possible help from the Rump protein (Becalska et al. 2011). However, 4 h after fertilization, *Nanos* mRNA is also degraded; and recently *Aub* was implicated by associating with the general mRNA degradation factors like Smg and the CCR4 de-adenylase complex (Rouget et al. 2010). Furthermore, it was suggested that the Nanos 3′UTR contained imperfect complementary to some TE piRNAs to specifically recruit AUB (Rouget et al. 2010); however this targeting model does not yet resolve how *Nanos* gets discriminately selected over other transcripts.

Interestingly, *Nanos* mRNA posterior pole localization also depends on the function of protein chaperones like Hsp90 (Song et al. 2007), and genetic phenomenon like canalization has been proposed to connect the role of Hsp90 chaperones and the piRNA pathway (Sato and Siomi 2010). The term canalization describes how organisms maintain developmental robustness despite so much variability from environmental pressures and genetic variation. Hsp90 chaperones are thought to be one level of this canalization because they help mutated or misfolded proteins still fold into functional enzymes, thereby "buffering" phenotypes to remain similar across a population of varying genotypes. However, to explain how Hsp90 perturbations allowed genetic variation to magnify the connection to piRNAs was only recently made in a study showing that genetic and chemical disruptions of Hsp90 activity can negatively affect piRNA populations and therefore yield derepression of TEs (Specchia et al. 2010). This was further supported by data showing that Hsp90 may foster posttranslational modifications of PIWI necessary for TE repression (Gangaraju et al. 2011), and a new linking of the Piwi pathway to Shu/FKBP6 as additional chaperones that may have functions related to Hsp90 (Olivieri et al. 2012; Preall et al. 2012; Xiol et al. 2012). These findings reflect how beautifully complex and intertwined these genetic and biochemical pathways are to ensure proper animal germ cell development.

10 The Role piRNAs Play in Hybrid Dysgenesis and Species Evolution

A key evolutionary pressure for the Piwi and piRNA pathway is to ensure fertility within a species, but it seems that the Piwi pathway is also a potent licensing system that opposes the formation of certain hybrids. In the hybrid dysgenesis phenomenon, daughter animals resulting from a mating of two hybrids seem to suffer from poorer health compared to parents, such as sterility despite normal development to adulthood. For intraspecies hybrid dysgenesis of *D. melanogaster* strains (Shpiz and Kalmykova 2009), dysgenesis has a pattern that suggests that a maternally deposited epigenetic factor in the embryo is a key determinant to ensure fertility, while the lack of the maternal factor means that the embryo cannot respond to a harmful paternal factor. A breakthrough in understanding hybrid dysgenesis systems was placing piRNAs as the maternally deposited factor required to respond to a particular TE imparted by the paternal chromosomes (Brennecke et al. 2008; Malone et al. 2009). Compellingly, in a *Drosophila* mother with an entirely intact Piwi pathway and diverse piRNA populations, lacking just the small complement of piRNAs against a novel TE that invades the father's germline can spell trouble in the progeny.

Interestingly, with time dysgenic daughters can begin to regain some fertility, but by what mechanism? By carefully tracking piRNAs and re-sequencing progeny genomes in the P-element system of hybrid dysgenesis, the Theurkauf group showed that P-element TEs were indeed mobilizing in the dysgenic progeny, and likely causing genomic damage in the germ cells that result in sterility (Khurana et al. 2011). But when the mobile TE lands back into a piRNA cluster, new P-element piRNAs begin to emerge in aging dysgenic progeny that also begin to regain some (but not all) fertility. This observation was bolstered by other studies demonstrating an analogous recovery from complete infertility in dysgenic progeny from an I-R element hybrid system (Grentzinger et al. 2012) and in a paramutation-based genetic system for P-elements (de Vanssay et al. 2012). Together, these studies indicate that primary piRNA biogenesis systems are insufficient to handle novel TEs contributed by the paternal genome. Instead, the gain of TEs landing into specific piRNA clusters that yield maternally deposited piRNAs is needed to stimulate some aspect of the "ping-pong cycle" for sustaining piRNA function through embryogenesis and into larval gonad development.

Although the study of the Piwi pathway in dysgenic strains of *D. melanogaster* has been extremely insightful, these strains are nonnatural products of careful laboratory rearing practices, whereas in the wild the selection pressures are stronger and more rapid on the Piwi pathway to maintain fertility. One evolutionary study has found that the piRNA compositions in *D. melanogaster* populations certainly provide evidence for a positive selection signal (Lu and Clark 2010); however there also seems to be a tolerance for *Drosophila* to allow TEs more leeway to expand in the genome, perhaps due to the canalization roles that Piwi proteins and piRNAs play. The role of piRNA incompatibility with general species barriers has also been proposed (Shpiz and Kalmykova 2009), but the first study to examine piRNAs in an actual interspecies hybrid was in a special cross of a *D. melanogaster Hmr* mutant with the closest relative *D. simulans*, in which hybrid progeny development can proceed to adulthood (Kelleher et al. 2012). Although these interspecies hybrid progeny eventually share similar phenotypic deficiencies with intraspecies dysgenic hybrids, such as distorted populations of maternally deposited piRNAs, sterility, and loss of repression for a broad number of TEs, the Barbash group could not detect a clear correlation between TEs derepressed highly in the hybrid and decreases or divergences in sequences between parental piRNAs and hybrid progeny piRNAs. Furthermore, by showing that a transgene bearing the *D. simulans Aub* was unable to rescue the *D. melanogaster aub* mutant, the authors suggest that there may be underlying deficiencies or incompatibilities between the Piwi pathway protein components of different species rather than just differences in piRNAs may account for dysgenesis in interspecies hybrids (Kelleher et al. 2012). Another evolutionary study of Piwi pathway gene codon bias may support this notion amongst *Drosophilids* (Castillo et al. 2011). These fly experiments are initial forays of evolutionary studies applied to the Piwi pathway. However, a greater diversity of model systems like vertebrates must be brought to bear to see how pervasive these genetic and evolutionary phenomena extend into animal evolution.

11 Regulating the Vertebrate piRNA Pathway

Similar to flies, mammalian genomes encode at least three clear Piwi homologs, and nomenclature for vertebrate Piwi homologs is unfortunately complicated. The first letter of the animal denotes the Piwi homologs' origin; hence the mouse encodes Miwi, Miwi2, and Mili (Miwi-like), while humans encode Hiwi, Hiwi2, and Hili. A Miwi knockout mouse was one of the first mutants demonstrating the conserved function of Piwi pathway in vertebrates for germ cell development (Deng and Lin 2002). However, as the Piwi pathway in mammals and other model vertebrates comes into better focus (Fig. 5.4), it is apparent that there are important differences between vertebrates and flies in the function and requirement of Piwi pathway components. For example, although knockout mutations in any of the loci of Miwi, Mili, or Miwi2 render male mice completely sterile because of spermatogenic arrest and atrophy of the testes, all female homozygous mutants are superficially as fertile as heterozygous littermates. This contrasts with flies where severe Piwi pathway mutations impair fertility in males and females (Cox et al. 2000), and contrast with zebrafish where Ziwi and Zili null mutants in both genders cannot form rudimentary gonads and become sterile masculinized fish (Houwing et al. 2007; Houwing et al. 2008). Although Mili and some piRNAs are expressed in mouse oocytes, the proportion of piRNAs is comparably lower with respect to abundant endo-siRNAs in mouse eggs (Tam et al. 2008; Watanabe et al. 2008), and mammalian oogenesis is typically restricted to very few mitotic divisions after birth while fish and fly ovaries continuously engage in mitotic cell divisions of oogonia. We speculate that the Piwi pathway may be most essential in germ cells that must engage in continuous rounds of mitosis during gametogenesis, but formally testing this hypothesis is challenging.

Mouse Piwi protein expression is temporally regulated during spermatogenesis, such that Miwi2 is expressed earliest and only temporarily in the embryonic gonad, whereas Miwi is expressed latest after 10 days postpartum (dpp) when the first synchronized wave of spermatogenesis approaches the pachytene stage of meiosis. Mili's expression is most pervasive, starting at 12.5 days postcoitus in male primordial germ cells and extending through adulthood. The temporal regulation of the Piwi proteins themselves coincides with three rough categories of piRNA expression: pre-natal piRNAs, pre-pachytene piRNAs, and post-pachytene piRNAs, which differ on the level of how many TE sequences and how strong a ping-pong cycle signature is present. In prenatal piRNAs, a ping-pong cycle signature exists between Miwi2 and Mili piRNAs which tend to be more enriched in TE sequences such as LINE1, the most prevalent TE class in mammalian genomes. After birth but before 12 dpp, when only Mili dominates, these pre-pachytene piRNAs have some enrichment of TE sequences suggesting that Mili can perform the ping-pong cycle with itself as well as a strong enrichment of 3′UTR genic piRNAs (Robine et al. 2009). Finally, post-pachytene piRNAs are incorporated into both Miwi and Mili, are depleted in TE sequences, almost devoid of ping-pong cycle signatures (Beyret et al. 2012), and derive from over a hundred intergenic and genic clusters that presumably express the piRNA precursor as single-stranded RNA (Robine et al. 2009).

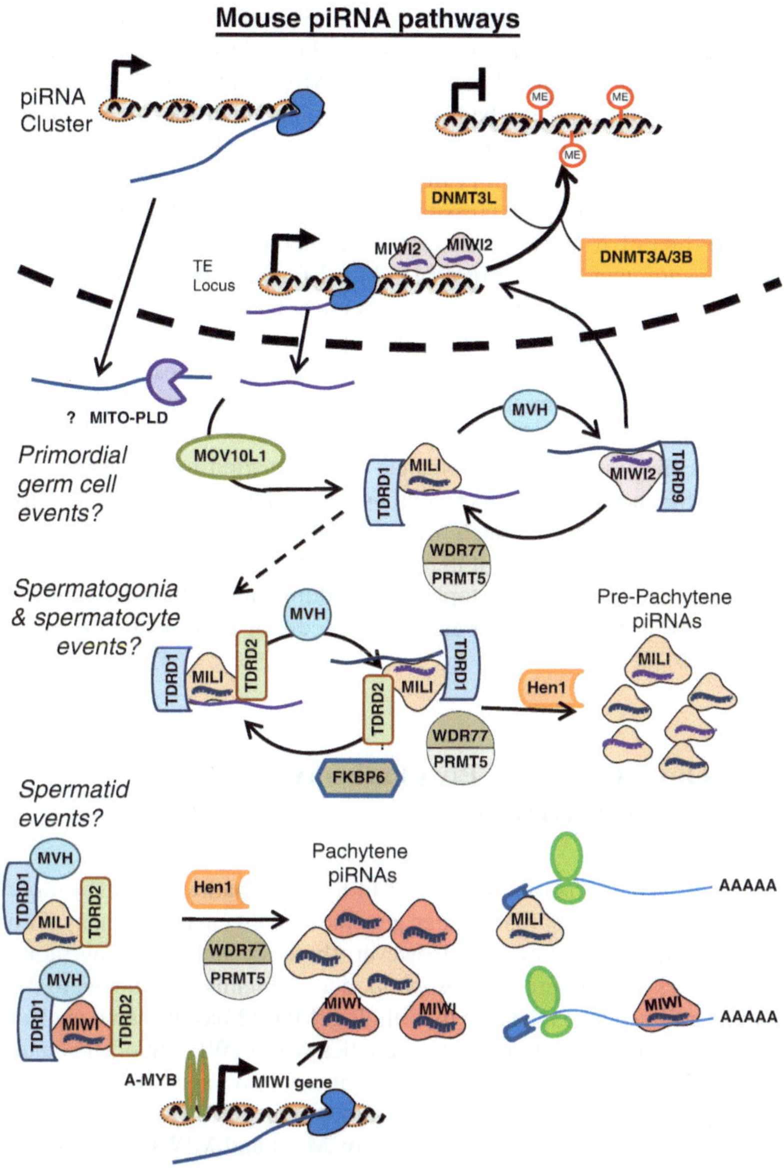

Fig. 5.4 The mouse piRNA pathway is tightly regulated throughout spermatogenic development. During early spermatogenesis, primary pre-pachytene piRNAs are produced from TE and genic transcripts. These MILI-bound, sense-oriented piRNAs can then direct "ping-pong" cycle production of MIWI2-bound or MILI-bound antisense piRNAs in primordial germ cells and spermatogonia, respectively. MIWI2-piRNA complexes are then able to translocate to the nucleus to initiate transcriptional gene silencing by promoting de novo DNA methylation. After meiosis and during spermatid maturation, non-repetitive pachytene piRNAs begin to dominate and load into both MILI and MIWI with the assistance of tudor-domain-containing proteins. The post-pachytene MIWI and MILI complexes are hypothesized to regulate target transcripts posttranscriptionally

Although reasons are still unknown for the temporal expression of mammalian Piwi pathway genes, an important transcription factor has recently been discovered to play a role in this temporal expression of Miwi and post-pachytene piRNAs. A mouse mutagenesis screen had previously pinpointed the transcription factor A-MYB being required for spermatogenesis as well as breast tissue development (Toscani et al. 1997). Recently, A-MYB was implicated in promoting pachytene piRNA and Miwi transcript expression in spermatocytes and spermatids (Li et al. 2013). ChIP-Seq of A-MYB found it at the promoters of Miwi and other piRNA biogenesis factor genes like MitoPLD, Mov10L, and Tdrd6, the promoters of pachytene piRNA clusters, and its own promoter. Therefore, A-MYB may engage in a feed-forward circuit that allows it to ramp up transcription of both piRNA precursors and piRNA biogenesis machinery. However, the presence of A-MYB in breast tissue, which lacks mouse Piwi protein and piRNA expression, also suggests that A-MYB itself is not sufficient to drive the expression of the entire Piwi pathway. Perhaps the Piwi pathway demands a combination of transcription factors, such as the USF and NF-Y factors recently suggested to also affect Miwi expression (Hou et al. 2012). Finally, although Miwi and piRNAs are most highly expressed in round spermatids, they are absent in mature sperm, and this turnover may either be part of the standard process of cytoplasmic shedding during the final stages of spermatogenesis, or it may be part of an active process of turnover by the proteasome (Zhao et al. 2013).

12 Not Quite the Same: Piwi Pathway Nuances in Vertebrates

With so many genetic factors implicated in the Piwi pathways in *Drosophila*, we might expect that the larger mammalian genomes would encode as many if not more Piwi pathway factors as possible. However, there are only as many mouse Piwi protein effectors as there are in *Drosophila*, but it remains a challenge to designate which mouse Piwi gene is the true fly ortholog. MIWI2 has the same nuclear localization and chromatin regulation characteristics with PIWI, while MILI and MIWI are cytoplasmic like AUB and AGO3. Furthermore, PIWI and MIWI mainly engage in primary piRNA biogenesis and not the ping-pong cycle, while MILI and MIWI2 can engage in the cycle in the mouse. Finally, MILI and MIWI2 have been genetically connected to directing de novo DNA methylation of TE elements during embryogenesis (see below), whereas in flies there is only minor evidence that DNA methylation has an influence on *Drosophila* development and no evidence yet for a link to PIWI, AUB, or AGO3.

Amongst the mammalian Piwi pathway factors that can be clearly assigned to orthologs in *Drosophila*, there appears to be a difference in the degree that mutations in mouse genes affect piRNA biogenesis and gametogenesis compared to the fly genes. Many mutations in the *Drosophila* Piwi pathways are extremely severe, causing complete sterility and major piRNA loss. However, other than the mouse

Piwi genes themselves, the only other gene mutation to cause complete loss of mouse piRNAs in the testes is the putative helicase Mov10-Like (Mov10L), which is distinct from the Mov10 paralog that is required for miRNA maturation into the RISC (Meister et al. 2005; Frost et al. 2010). Mov10L is the mammalian ortholog of *Drosophila armi*, and the *Mov10L* mutant male is infertile with shrunken testes that are devoid of functional sperm due to changes in spermatogonia and spermatocyte organelle morphology and an increased buildup of mitochondria (Zheng et al. 2010).

In contrast are several mouse Piwi factors where the mutations still cause spermatogenic arrest and male infertility, but their effect on global piRNA production during the initial waves of spermatogenesis is surprisingly mild. For example, whereas *Drosophila* vasa mutant ovaries are deficient of most piRNAs (Malone et al. 2009), mutations in Mouse Vasa Homolog (MVH) seem to only cause loss of Miwi2 piRNAs in the fetal testes, while Mili piRNAs can still accumulate (Kuramochi-Miyagawa et al. 2010). Mouse Maelstrom (ortholog of *Drosophila mael*) resides most frequently in the piP (piRNA Processing) bodies, associates with the Tdrd9 and Tdrd1 factors, and influences Miwi2 piRNAs but not Mili piRNAs (Soper et al. 2008; Aravin et al. 2009). Since the suppression of mouse TEs is affected in mouse *mael* mutants without loss of piRNAs, the effector role of silencing rather than piRNA biogenesis may be conserved between mouse and fly *mael* orthologs. Although mutations and knockdowns of fly Zuc severely compromise *Drosophila* piRNAs, knockout mutants in the mouse ortholog MitoPLD seem to affect the production of some piRNAs, but not others, particularly SINE element piRNAs bound by Mili (Huang et al. 2011a; Watanabe et al. 2011). Mouse GasZ knockout mutants (Ma et al. 2009) were created before knockdowns of fly GasZ (CG2183) were studied (Czech et al. 2013; Handler et al. 2013), and whereas TEs are derepressed in both fly and mouse GasZ mutants, piRNA biogenesis is much more severely impacted in the fly GasZ mutant. Knockdowns and mutations of fly Shutdown (Shu) strongly impair piRNA biogenesis (Olivieri et al. 2012; Preall et al. 2012), but knockouts of the mouse ortholog, FKBP6, only cause some loss of Miwi2 piRNAs but effects are quite modest against Mili piRNAs (Xiol et al. 2012). Finally, knockdown of the glycerol-3-phosphate acyltransferase 2 (GPAT2), a mitochondrial outer membrane protein, in a germinal stem cell mouse line was shown to affect piRNA loading into Mili (Shiromoto et al. 2013) and its fly homolog CG5508 also recovered in one of the RNAi screens for being required in TE silencing (Czech et al. 2013) but role of CG5508 in piRNA biogenesis is not yet known.

A large group of genes linked to the Piwi pathway and also conserved throughout animal evolution are the tudor-domain-containing proteins which are specifically expressed in germ cells. The tudor domain was named after the founding member, the *Drosophila Tudor* (*Tud*) gene whose closest mammalian homolog is TuDoR-Domain 6 (TDRD6). Tud and TDRD6 are the largest Tudor proteins with 11 and 8 repeated tudor domain in the fly and mouse genes, respectively. Tud mutants display the grandchildless phenotype where the daughters develop infertility, similar to hybrid dysgenesis and to the PRMT5 arginine methylase mutants that deposit symmetrical dimethyl arginines onto Aub and AGO3 to facilitate its interaction with Tud (Kirino et al. 2009; Nishida et al. 2009). Some amount of TE derepression and piRNA decrease is observed in these mutants, but these phenotypes are mild compared to more severe phenotypes of Piwi protein mutations themselves or other biogenesis factors. Like Tud in flies, Tdrd6 is also required for spermatogenesis in mouse and associates with Miwi and Mili (Vasileva et al. 2009). The tudor domain is hypothesized to foster protein-protein docking interactions, so the presence of just multiple tudor domains in Tud and Tdrd6 suggests that their main function is to act as scaffolds.

Connected to the Piwi pathway are three other tudor-domain-containing genes with clear orthologs between flies and mice: *papi*, whose mouse homolog is Tdrd2 (also called Tdrkh); *tejas* (tej), whose mouse homolog is Tdrd5; and *spindle-E* (Spn-E), whose mouse homolog is Tdrd9. These genes differ from Tud because they only contain a single tudor domain, but Papi has transmembrane domains, Tej has a LOTUS domain, and Spn-E has helicase and RNA-binding domains (Handler et al. 2011). Papi associates with Ago3 and its mutant may display higher TE expression (Liu et al. 2011), but RNAi knockdown of Papi has not replicated this phenotype (Czech et al. 2013; Handler et al. 2013; Muerdter et al. 2013), and Papi's influence on piRNA production is not clear. Tdrd2 interacts strongly with Miwi (Chen et al. 2009), and a recent knockout mouse of Tdrd2 shows LINE1 derepression, spermatogenic arrest before pachytene is reached, and a very interesting lengthening of Mili-bound piRNAs to ~6–8 nt longer (Saxe et al. 2013). Although Tejas and Tdrd5 are essential for female and male germline development, respectively, only Tejas has been shown to be important for piRNA biogenesis and TE silencing (Patil and Kai 2010; Yabuta et al. 2011). Classic female sterile fly mutations of Spn-E suffer severe loss of piRNAs (Vagin et al. 2006; Malone et al. 2009), but while TDRD9 knockout mice are male sterile, they have overall similar levels of piRNAs in prepachytene stage testis to wild type, with just minor differences in a few specific piRNA sequences (Chuma et al. 2006; Shoji et al. 2009). Although TDRD9 associates with Miwi2 and its localization in spermatogonia is influenced by Mili (Shoji et al. 2009), it is enigmatic why in contrast to Spn-E that mouse Tdrd9 is not critical for piRNA biogenesis.

For other tudor-domain-containing genes in fly and mouse, the homologs have diverged such that either the obvious ortholog is not apparent or designating the ortholog bioinformatically may not be obvious. For example, Tdrd1 has been proposed to the closest homolog to *Drosophila* Qin in one study (Siomi et al. 2010) or to CG9925 in another study which suggested that Qin was more related to Tdrd4 (also known as RNF17) (Pan et al. 2005; Handler et al. 2011). Tdrd1 is essential for male germ cell development and TE silencing via promoting efficient piRNA biogenesis in Mili, but whether Tdrd1 has a role in the ping-pong cycle like Qin remains open because there are still significant piRNAs present in the Tdrd1 –/– mutant (Reuter et al. 2009; Vagin et al. 2009; Wang et al. 2009). A purification of zebrafish Tdrd1 identified longer transcripts which may be putative piRNA precursors (Huang et al. 2011b), perhaps analogous to the putative piRNA precursors found associated in an ARMI purification from *Drosophila* OSC cells (Saito et al. 2010). Although the Tdrd4 knockout mouse has not been reported yet, knockdown of CG9925 did not elicit TE derepression (Handler et al. 2011); thus it remains open which is the clear *Drosophila* ortholog of Tdrd1. Then there is the example of *Drosophila* Krimper, which lacks any mammalian ortholog (Siomi et al. 2010; Handler et al. 2011), but clearly has a strong effect on the Piwi pathway in flies when mutated or knocked down (Lim and Kai 2007; Malone et al. 2009; Handler et al. 2011).

13 Mammalian Piwi Proteins Need "Slicing" Activity to Do Their Jobs

Despite the progress in dissecting the Piwi pathway in flies and mice, there are still open questions regarding the silencing mechanisms of the Piwi proteins. Like certain AGO proteins, the Piwi-group proteins retain the key amino acid residues required for endonucleolytic slicing activity that occurs upon a target transcript substrate paired against a guide small RNA in the protein. The slicing activity is central to the ping-pong cycle model (Brennecke et al. 2007; Gunawardane et al. 2007; Li et al. 2009); however it has not been reported whether AUB and AGO3 transgenes mutated in catalytic residues can rescue the *aub* and *ago3* mutants. Although PIWI has slicing activity in vitro (Saito et al. 2006), two studies using a Piwi transgene mutated in the catalytic residues indicate that slicing activity is dispensable for TE silencing in *Drosophila* OSC cells and in the female germline (Saito et al. 2010; Darricarrere et al. 2013). With the revelation that *Drosophila* Piwi mediates TGS, the dispensability of the slicing activity is surprising because Ago1 from fission yeast requires slicing activity in order to manifest its TGS on heterochromatic loci (Irvine et al. 2006).

Two mouse Piwi proteins, Mili and Miwi2, have also been implicated to promote TGS upon TEs in the early development phases of fetal and prepubertal sperm. In Mili and Miwi2 knockout mice, there is a substantial loss of DNA methylation upon chromatin containing LINE1 sequences (Aravin et al. 2007; Carmell et al. 2007; Aravin et al. 2008; Kuramochi-Miyagawa et al. 2008), mirroring the phenotype of *dnmt3l* mutants, the gene responsible for de novo establishment of this silencing mark on TEs during mammalian embryogenesis, after parental epigenetic imprints are erased (Lane et al. 2003). The genetic function of Mili and Miwi2 in specifying DNA methylation of TEs is compelling, and may even extend to imprinted loci like *Rasgrf1* (Watanabe et al. 2011). However, the biochemical mechanism and questions of how directly do Mili and Miwi2 direct DNA methylation remain obscure because *dnmt3l* mutants have no effect on piRNA biogenesis (Aravin et al. 2008), and proteomic analysis of Mili and Miwi2 complexes has not identified associated DNA methyltransferases (Vagin et al. 2009). In addition, not all TEs are equally repressed at the TGS level, such as the IAP TE which is not affected in Miwi2 mutants (Kuramochi-Miyagawa et al. 2008).

In contrast to Miwi2 and Mili being genetically linked to TGS, Miwi has mainly been thought to regulate target transcripts posttranscriptionally, although demonstrating that this mechanism has not been fully addressed. It is clear that Miwi is mainly cytoplasmic and concentrated in the perinuclear organelle called the chromatoid body, a dense structure believed to be a transit hub for mRNAs (Kotaja and Sassone-Corsi 2007). Miwi can also bind and perhaps stabilize some mRNAs (Deng and Lin 2002) as well as pachytene piRNA precursor transcripts (Vourekas et al. 2012), but whether Miwi has sequence-binding specificity or how mechanistically Miwi can pair with certain transcripts is unclear (Robine et al. 2009). A translational regulation role of Miwi and Mili has mainly been portrayed from density gradient fractionations of adult testis extracts that show Miwi and Mili co-fractionating with polyribosomes (Grivna et al. 2006; Unhavaithaya et al. 2009). It has been challenging to formally test the translational regulatory activity of Miwi and Mili upon synthetic mRNAs and transgenes in testis extracts competent for in vitro translation. However the recent report of a mouse gonadal cell culture line (Germline Stem Cells, or GSCs) (Shiromoto et al. 2013) with endogenous Mili and piRNAs might possibly open a door to more rigorous biochemistry if these cells can be cultured in bulk.

Mammalian Piwi proteins were first demonstrated to exhibit Slicer activity in vitro (Lau et al. 2006), and this activity was thought to sustain a ping-pong cycle signature between Miwi2 and Mili (Aravin et al. 2008). However, only mammalian Ago2 retains slicer activity to form the RISC capable for target RNA cleavage, while Ago1, -3, and -4 are incapable of slicing activity (Liu et al. 2004) but can regulate targets through binding and recruiting translation-repressing factors. Although Ago2's catalytic residues are critical for mouse hematopoiesis because Ago2 slicing is required to process the essential miR-451 miRNA (Cheloufi et al. 2010; Cifuentes et al. 2010), a mutated Ago2 transgene lacking slicing activity can rescue hematopoiesis in Ago2 −/− mice, suggesting that the Ago2 slicing activity is highly specialized (O'Carroll et al. 2007).

To address the importance of the slicing activity in mouse Piwi proteins, point mutations that disrupted the catalytic residues of Miwi (Reuter et al. 2011), Mili, and Miwi2 (De Fazio et al. 2011) were generated and revealed some surprising and perplexing genetic effects. Slicing-inactive Mili mutant males ($Mili^{DAH}$) were infertile and displayed up-regulated LINE1 TE expression and decreased DNA methylation, but overall piRNA production within Mili was comparable to wild type with mainly a mild reduction of piRNAs complementary to LINE1 sequences (De Fazio et al. 2011). Interestingly, it was Miwi2 piRNA loading that was perturbed in $Mili^{DAH}$ mutants, suggesting that Mili slicing was required for specifying a ping-pong cycle mechanism to load piRNAs into Miwi2. Perplexingly, inactivation of slicing residues in Miwi2 failed to reveal a reciprocal defect in piRNA production neither in Mili nor in Miwi2 itself, with TE silencing also intact (De Fazio et al. 2011). Whereas catalytic residue mutations in Mili and Miwi2 are genetically recessive, the catalytic residue mutation in Miwi ($Miwi^{ADH}$) turned out to be a dominant negative in causing male sterility in the heterozygous state (Reuter et al. 2011). If one speculates that Miwi proteins form dimers as an explanation of the dominant negative phenotype, then it is mysterious that pachytene piRNA accumulation and general testis transcriptome profiles were largely unaffected in the $Miwi^{ADH}$ mutant. Indeed, this study displayed LINE1 TE up-regulation in $Miwi^{ADH}$ mutants and some evidence that Miwi may use certain piRNAs to slice LINE1 TE transcripts posttranscriptionally as a means to suppress TE mobilization (Reuter et al. 2011). Recently, conditional Mili knockout and conditional slicing Mili^{DAH} mutants were shown to still exhibit mouse male sterility despite functional Mili capable of establishing proper DNA methylation of TE loci (Di Giacomo et al. 2013), thus leading to the model that LINE1 TEs reanimate during pachytene stages of meiosis but require both Mili and Miwi protein to prevent germ cell damage. Future work is required to tease the mechanistic details for primary piRNA biogenesis in mouse spermatocytes, which apparently can proceed even when slicing activity is compromised. These genetic studies have raised more open questions and new conundrums to the Piwi pathway that await future novel approaches and additional model organisms and germ cell systems to fully dissect this pathway.

14 Conclusion: Piwi "on the Brain" and Going Beyond the Germline?

The Piwi pathway genes and piRNAs are most influential and abundantly expressed in animal gonads, but the detection and genetic influence of this pathway in other somatic animal tissues continue to be debated. It is possible that Piwi-mediated chromatin modifications determined in the germ cells and early embryo are later perpetuated in a Piwi-independent fashion via a "cellular memory" to somatic tissues like eye pigmentation or salivary gland polytene chromosomes (Pal-Bhadra et al. 2002; Brower-Toland et al. 2007; Moshkovich and Lei 2010). However, caution should also be heeded in determining whether bona fide piRNAs are indeed expressed in somatic cells, because several abundant and stable structural RNA fragments have been misclassified as potential piRNAs in somatic cells without the confirmation that they are indeed loaded into a Piwi protein complex (Janic et al. 2010; Lee et al. 2011; Yan et al. 2011).

Nevertheless, there is a convergence of data implicating that some invertebrate animal neurons may harbor their own repository of piRNAs and Piwi machinery. The sea slug *Aplysia* is a classic model system for neurophysiology because of its exceptionally large, well-characterized ganglion comprising many cell types including neurons (Bailey et al. 2004). It was recently shown that *Aplysia* ganglions contain putative piRNAs (Rajasethupathy et al. 2012), but characterization of the *Aplysia* piRNAs, whether they are derived from large intergenic clusters or target TEs, is hampered by the rough draft stage of the *Aplysia* genome (Moroz et al. 2009; Rajasethupathy et al. 2012). Injection of mimics and inhibitors against a single putative *Aplysia* piRNA was somehow able to instill neurophysiological responses and DNA methylation changes in these ganglions (Rajasethupathy et al. 2012). A focused scrutiny of specific neurons in the fly brain has also suggested that Piwi proteins may be expressed and playing a role in controlling TE mobilization in the fly brain (Perrat et al. 2013). As a possible mechanism for fly neuroplasticity, this study hypothesized that certain neurons lacking AUB and AGO3 would allow TE activity to yield heterogeneity within the *Drosophila* brain that could be useful for learning and memory (Perrat et al. 2013). Although biochemical analysis of specific fly neurons is technically challenging because of the small amount of tissue, genetic and chemical ablation of miRNAs and endo-siRNAs has enabled the detection of putative piRNAs in fly heads (Ghildiyal et al. 2008). Finally, there is a recent report suggesting that piRNAs are expressed in human induced pluripotent stem cells (Marchetto et al. 2013). So as technology in this arena advances, including more efficient and cost-effective systems biology approaches, we expect to uncover new avenues where the Piwi pathway and piRNAs may influence animal behavior and other processes beyond the animal germline.

The greatest progress in applying systems approaches to the Piwi pathway has occurred in flies and mice because of the ease of genetically manipulating flies and the large knowledgebase for mice as a model mammal of humans. As such, both organisms have the strongest foundation of genomics resources and largest community of researchers. But understanding the Piwi pathway fully will demand going beyond the fly and mouse and making use of other vertebrate and invertebrate models. For example, to ask if vertebrates are susceptible to hybrid dysgenesis like in flies where a mother lacking just one class of piRNAs will produce sterile daughters, mice may not be an optimal system because mammalian oogenesis does not appear to depend on the Piwi pathway. There is some controversy whether mammalian sperm may transmit piRNAs as epigenetic factors (Krawetz et al. 2011), because there is also Miwi turnover during sperm maturation (Sytnikova and Lau 2013; Zhao et al. 2013). However, the frog *Xenopus tropicalis* has well-characterized piRNA clusters configured in an analogous fashion to mammalian clusters (Armisen et al. 2009; Lau et al. 2009), as well as highly abundant levels of Xiwi in the oocyte that appears to be concentrated asymmetrically in the germplasm (Lau et al. 2009), specialized cytoplasm that can specify the formation of primordial germ cells (Houston and King 2000). We propose that *X. tropicalis* will be a useful vertebrate model to compare and contrast the Piwi pathway with respect to mammals and flies.

In addition to gametogenesis, the Piwi pathway may also be connected to regeneration of lost body parts in certain animals. Planarians and cnidarians contain a large and dispersed population of special stem cells called neoblasts that express Piwi proteins and piRNAs (Reddien et al. 2007; Palakodeti et al. 2008). These neoblasts endow these animals with extraordinary regenerative capabilities such as restoring complete body plans from nearly any piece of severed tissue (Petersen and Reddien 2009). In fact, the major small RNA populations in these animals are not miRNAs but rather piRNAs (Grimson et al. 2008; Palakodeti et al. 2008). Amongst vertebrates, the axolotl salamander is prominently one of a few vertebrates capable of regenerating a severed limb during adulthood, and recently it was suggested that axolotl Piwi transcripts might be triggered for expression in the regenerating blastema (Zhu et al. 2012). Perhaps somatic cells in the blastema are reanimating the Piwi pathway as a form of induced dedifferentiation similar to induced pluripotency, and indeed mutations of a tumor-suppressor gene in flies may possibly be one route to achieve this dedifferentiation state (Janic et al. 2010). We speculate that the majority of other vertebrates where somatic cells lack the expression of the Piwi pathway may explain our limited regenerative potential.

Finally, we discuss lastly (but not the least) that the Piwi pathway is not the sole purview of multicellular organisms: the single-celled protozoans have co-opted the Piwi pathway for reproductive and growth functions. *Tetrahymena* encodes at least eight genes that are more similar to metazoan Piwi genes than Ago genes, and the *Tetrahymen*a Piwi genes (TWI) have very essential roles in growth (Couvillion et al. 2010) and nuclear RNA metabolism (Couvillion et al. 2010). Notably, *Tetrahymena* express scnRNAs that are just one form of protozoan piRNAs, which are required for meiosis, conjugation, and a complex process of DNA elimination from the macronucleus (Mochizuki and Gorovsky 2004a; Schoeberl et al. 2012). The *Tetrahymena* scnRNAs appear to direct chromatin modification events that lead to the removal of large swaths of DNA to yield a slimmed macronucleus perhaps optimized for genic transcription (Mochizuki et al. 2002; Mochizuki and Gorovsky 2004b). But in opposite fashion, another protozoan, *Oxytricha*, uses its piRNAs to instead specify the retention of DNA elements and prevent genes from being lost when its genome rearranges into thousands of tiny chromosomes (Fang et al. 2012; Swart et al. 2013). These quirky protozoan Piwi pathways are a testament to the rich diversity of natural biological processes, and should motivate us to continue casting a wide net of systems biology approaches across multiple model organisms.

Acknowledgments We are grateful to members of the Lau lab, Dianne Schwarz and Julie Claycomb, for discussions and critical reading of this manuscript. This work was supported by the Searle Scholars foundation to NCL and an NIH genetics training grant to JPC.

References

Ambros V (2011) MicroRNAs and developmental timing. Curr Opin Genet Dev 21:511–517

Anand A, Kai T (2012) The tudor domain protein kumo is required to assemble the nuage and to generate germline piRNAs in Drosophila. EMBO J 31:870–882

Aravin AA, Sachidanandam R, Bourc'his D, Schaefer C, Pezic D, Toth KF, Bestor T, Hannon GJ (2008) A piRNA pathway primed by individual transposons is linked to de novo DNA methylation in mice. Mol Cell 31:785–799

Aravin AA, Sachidanandam R, Girard A, Fejes-Toth K, Hannon GJ (2007) Developmentally regulated piRNA clusters implicate MILI in transposon control. Science 316:744–747

Aravin AA, van der Heijden GW, Castaneda J, Vagin VV, Hannon GJ, Bortvin A (2009) Cytoplasmic compartmentalization of the fetal piRNA pathway in mice. PLoS Genet 5:e1000764

Armisen J, Gilchrist MJ, Wilczynska A, Standart N, Miska EA (2009) Abundant and dynamically expressed miRNAs, piRNAs, and other small RNAs in the vertebrate Xenopus tropicalis. Genome Res 19:1766–1775

Ashe A, Sapetschnig A, Weick EM, Mitchell J, Bagijn MP, Cording AC, Doebley AL, Goldstein LD, Lehrbach NJ, Le Pen J et al (2012) piRNAs can trigger a multigenerational epigenetic memory in the germline of C. elegans. Cell 150:88–99

Bagijn MP, Goldstein LD, Sapetschnig A, Weick EM, Bouasker S, Lehrbach NJ, Simard MJ, Miska EA (2012) Function, targets, and evolution of Caenorhabditis elegans piRNAs. Science 337:574–578

Bailey CH, Kandel ER, Si K (2004) The persistence of long-term memory: a molecular approach to self-sustaining changes in learning-induced synaptic growth. Neuron 44:49–57

Bartel DP (2009) MicroRNAs: target recognition and regulatory functions. Cell 136:215–233

Batista PJ, Ruby JG, Claycomb JM, Chiang R, Fahlgren N, Kasschau KD, Chaves DA, Gu W, Vasale JJ, Duan S et al (2008) PRG-1 and 21U-RNAs interact to form the piRNA complex required for fertility in C. elegans. Mol Cell 31:67–78

Becalska AN, Kim YR, Belletier NG, Lerit DA, Sinsimer KS, Gavis ER (2011) Aubergine is a component of a nanos mRNA localization complex. Dev Biol 349:46–52

Betel D, Sheridan R, Marks DS, Sander C (2007) Computational analysis of mouse piRNA sequence and biogenesis. PLoS Comput Biol 3:e222

Beyret E, Liu N, Lin H (2012) piRNA biogenesis during adult spermatogenesis in mice is independent of the ping-pong mechanism. Cell Res 22:1429–1439

Billi AC, Alessi AF, Khivansara V, Han T, Freeberg M, Mitani S, Kim JK (2012) The Caenorhabditis elegans HEN1 ortholog, HENN-1, methylates and stabilizes select subclasses of germline small RNAs. PLoS Genet 8:e1002617

Brennecke J, Aravin AA, Stark A, Dus M, Kellis M, Sachidanandam R, Hannon GJ (2007) Discrete small RNA-generating loci as master regulators of transposon activity in Drosophila. Cell 128:1089–1103

Brennecke J, Malone CD, Aravin AA, Sachidanandam R, Stark A, Hannon GJ (2008) An epigenetic role for maternally inherited piRNAs in transposon silencing. Science 322:1387–1392

Brower-Toland B, Findley SD, Jiang L, Liu L, Yin H, Dus M, Zhou P, Elgin SC, Lin H (2007) Drosophila PIWI associates with chromatin and interacts directly with HP1a. Genes Dev 21:2300–2311

Buckley BA, Burkhart KB, Gu SG, Spracklin G, Kershner A, Fritz H, Kimble J, Fire A, Kennedy S (2012) A nuclear Argonaute promotes multigenerational epigenetic inheritance and germline immortality. Nature 489:447–451

Carmell MA, Girard A, van de Kant HJ, Bourc'his D, Bestor TH, de Rooij DG, Hannon GJ (2007) MIWI2 is essential for spermatogenesis and repression of transposons in the mouse male germline. Dev Cell 12(4):503–514

Carthew RW, Sontheimer EJ (2009) Origins and mechanisms of miRNAs and siRNAs. Cell 136:642–655

Castillo DM, Mell JC, Box KS, Blumenstiel JP (2011) Molecular evolution under increasing transposable element burden in Drosophila: a speed limit on the evolutionary arms race. BMC Evol Biol 11:258

Cecere G, Zheng GX, Mansisidor AR, Klymko KE, Grishok A (2012) Promoters recognized by forkhead proteins exist for individual 21U-RNAs. Mol Cell 47:734–745

Cheloufi S, Dos Santos CO, Chong MM, Hannon GJ (2010) A dicer-independent miRNA biogenesis pathway that requires Ago catalysis. Nature 465:584–589

Chen C, Jin J, James DA, Adams-Cioaba MA, Park JG, Guo Y, Tenaglia E, Xu C, Gish G, Min J et al (2009) Mouse Piwi interactome identifies binding mechanism of Tdrkh Tudor domain to arginine methylated Miwi. Proc Natl Acad Sci U S A 106:20336–20341

Chuma S, Hosokawa M, Kitamura K, Kasai S, Fujioka M, Hiyoshi M, Takamune K, Noce T, Nakatsuji N (2006) Tdrd1/Mtr-1, a tudor-related gene, is essential for male germ-cell differentiation and nuage/germinal granule formation in mice. Proc Natl Acad Sci U S A 103:15894–15899

Cifuentes D, Xue H, Taylor DW, Patnode H, Mishima Y, Cheloufi S, Ma E, Mane S, Hannon GJ, Lawson ND et al (2010) A novel miRNA processing pathway independent of Dicer requires Argonaute2 catalytic activity. Science 328:1694–1698

Claycomb JM (2012) Caenorhabditis elegans small RNA pathways make their mark on chromatin. DNA Cell Biol 31(Suppl 1):S17–33

Claycomb JM, Batista PJ, Pang KM, Gu W, Vasale JJ, van Wolfswinkel JC, Chaves DA, Shirayama M, Mitani S, Ketting RF et al (2009) The Argonaute CSR-1 and its 22G-RNA cofactors are required for holocentric chromosome segregation. Cell 139:123–134

Conine CC, Batista PJ, Gu W, Claycomb JM, Chaves DA, Shirayama M, Mello CC (2010) Argonautes ALG-3 and ALG-4 are required for spermatogenesis-specific 26G-RNAs and thermotolerant sperm in Caenorhabditis elegans. Proc Natl Acad Sci U S A 107:3588–3593

Couvillion MT, Sachidanandam R, Collins K (2010) A growth-essential Tetrahymena Piwi protein carries tRNA fragment cargo. Genes Dev 24:2742–2747

Cox DN, Chao A, Baker J, Chang L, Qiao D, Lin H (1998) A novel class of evolutionarily conserved genes defined by piwi are essential for stem cell self-renewal. Genes Dev 12:3715–3727

Cox DN, Chao A, Lin H (2000) piwi encodes a nucleoplasmic factor whose activity modulates the number and division rate of germline stem cells. Development 127:503–514

Czech B, Preall JB, McGinn J, Hannon GJ (2013) A transcriptome-wide RNAi screen in the Drosophila ovary reveals factors of the germline piRNA pathway. Mol Cell 50(5):749–761

Darricarrere N, Liu N, Watanabe T, Lin H (2013) Function of Piwi, a nuclear Piwi/Argonaute protein, is independent of its slicer activity. Proc Natl Acad Sci U S A 110:1297–1302

Das PP, Bagijn MP, Goldstein LD, Woolford JR, Lehrbach NJ, Sapetschnig A, Buhecha HR, Gilchrist MJ, Howe KL, Stark R et al (2008) Piwi and piRNAs act upstream of an endogenous siRNA pathway to suppress Tc3 transposon mobility in the Caenorhabditis elegans germline. Mol Cell 31:79–90

De Fazio S, Bartonicek N, Di Giacomo M, Abreu-Goodger C, Sankar A, Funaya C, Antony C, Moreira PN, Enright AJ, O'Carroll D (2011) The endonuclease activity of Mili fuels piRNA amplification that silences LINE1 elements. Nature 480:259–263

de Vanssay A, Bouge AL, Boivin A, Hermant C, Teysset L, Delmarre V, Antoniewski C, Ronsseray S (2012) Paramutation in Drosophila linked to emergence of a piRNA-producing locus. Nature 490:112–115

Deng W, Lin H (2002) miwi, a murine homolog of piwi, encodes a cytoplasmic protein essential for spermatogenesis. Dev Cell 2:819–830

Di Giacomo M, Comazzetto S, Saini H, De Fazio S, Carrieri C, Morgan M, Vasiliauskaite L, Benes V, Enright AJ, O'Carroll D (2013) Multiple epigenetic mechanisms and the piRNA pathway enforce LINE1 silencing during adult spermatogenesis. Mol Cell 50:601–608

Donertas D, Sienski G, Brennecke J (2013) Drosophila Gtsf1 is an essential component of the Piwi-mediated transcriptional silencing complex. Genes Dev 27:1693–1705

Fang W, Wang X, Bracht JR, Nowacki M, Landweber LF (2012) Piwi-interacting RNAs protect DNA against loss during Oxytricha genome rearrangement. Cell 151:1243–1255

Fire A, Xu S, Montgomery MK, Kostas SA, Driver SE, Mello CC (1998) Potent and specific genetic interference by double-stranded RNA in Caenorhabditis elegans. Nature 391:806–811

Fischer SE (2010) Small RNA-mediated gene silencing pathways in C. elegans. Int J Biochem Cell Biol 42:1306–1315

Frost RJ, Hamra FK, Richardson JA, Qi X, Bassel-Duby R, Olson EN (2010) MOV10L1 is necessary for protection of spermatocytes against retrotransposons by Piwi-interacting RNAs. Proc Natl Acad Sci U S A 107:11847–11852

Gangaraju VK, Yin H, Weiner MM, Wang J, Huang XA, Lin H (2011) Drosophila Piwi functions in Hsp90-mediated suppression of phenotypic variation. Nat Genet 43:153–158

Gent JI, Lamm AT, Pavelec DM, Maniar JM, Parameswaran P, Tao L, Kennedy S, Fire AZ (2010) Distinct phases of siRNA synthesis in an endogenous RNAi pathway in C. elegans soma. Mol Cell 37:679–689

Ghildiyal M, Seitz H, Horwich MD, Li C, Du T, Lee S, Xu J, Kittler EL, Zapp ML, Weng Z et al (2008) Endogenous siRNAs derived from transposons and mRNAs in Drosophila somatic cells. Science 320:1077–1081

Ghildiyal M, Zamore PD (2009) Small silencing RNAs: an expanding universe. Nat Rev Genet 10:94–108

Goic B, Vodovar N, Mondotte JA, Monot C, Frangeul L, Blanc H, Gausson V, Vera-Otarola J, Cristofari G, Saleh MC (2013) RNA-mediated interference and reverse transcription control the persistence of RNA viruses in the insect model Drosophila. Nat Immunol 14:396–403

Grentzinger T, Armenise C, Brun C, Mugat B, Serrano V, Pelisson A, Chambeyron S (2012) piRNA-mediated transgenerational inheritance of an acquired trait. Genome Res 22:1877–1888

Grimaud C, Bantignies F, Pal-Bhadra M, Ghana P, Bhadra U, Cavalli G (2006) RNAi components are required for nuclear clustering of Polycomb group response elements. Cell 124:957–971

Grimson A, Srivastava M, Fahey B, Woodcroft BJ, Chiang HR, King N, Degnan BM, Rokhsar DS, Bartel DP (2008) Early origins and evolution of microRNAs and Piwi-interacting RNAs in animals. Nature 455:1193–1197

Grivna ST, Pyhtila B, Lin H (2006) MIWI associates with translational machinery and PIWI-interacting RNAs (piRNAs) in regulating spermatogenesis. Proc Natl Acad Sci U S A 103:13415–13420

Gu W, Lee HC, Chaves D, Youngman EM, Pazour GJ, Conte D Jr, Mello CC (2012) CapSeq and CIP-TAP identify Pol II start sites and reveal capped small RNAs as C. elegans piRNA precursors. Cell 151:1488–1500

Gu W, Shirayama M, Conte D Jr, Vasale J, Batista PJ, Claycomb JM, Moresco JJ, Youngman EM, Keys J, Stoltz MJ et al (2009) Distinct argonaute-mediated 22G-RNA pathways direct genome surveillance in the C. elegans germline. Mol Cell 36:231–244

Guang S, Bochner AF, Burkhart KB, Burton N, Pavelec DM, Kennedy S (2010) Small regulatory RNAs inhibit RNA polymerase II during the elongation phase of transcription. Nature 465:1097–1101

Gunawardane LS, Saito K, Nishida KM, Miyoshi K, Kawamura Y, Nagami T, Siomi H, Siomi MC (2007) A slicer-mediated mechanism for repeat-associated siRNA 5′ end formation in Drosophila. Science 315:1587–1590

Haase AD, Fenoglio S, Muerdter F, Guzzardo PM, Czech B, Pappin DJ, Chen C, Gordon A, Hannon GJ (2010) Probing the initiation and effector phases of the somatic piRNA pathway in Drosophila. Genes Dev 24:2499–2504

Hall SE, Chirn GW, Lau NC, Sengupta P (2013) RNAi pathways contribute to developmental history-dependent phenotypic plasticity in C. elegans. RNA 19(3):306–319

Han T, Manoharan AP, Harkins TT, Bouffard P, Fitzpatrick C, Chu DS, Thierry-Mieg D, Thierry-Mieg J, Kim JK (2009) 26G endo-siRNAs regulate spermatogenic and zygotic gene expression in Caenorhabditis elegans. Proc Natl Acad Sci U S A 106:18674–18679

Handler D, Meixner K, Pizka M, Lauss K, Schmied C, Gruber FS, Brennecke J (2013) The genetic makeup of the Drosophila piRNA pathway. Mol Cell 50(5):762–777

Handler D, Olivieri D, Novatchkova M, Gruber FS, Meixner K, Mechtler K, Stark A, Sachidanandam R, Brennecke J (2011) A systematic analysis of Drosophila TUDOR domain-containing proteins identifies Vreteno and the Tdrd12 family as essential primary piRNA pathway factors. EMBO J 30:3977–3993

Harris AN, Macdonald PM (2001) Aubergine encodes a Drosophila polar granule component required for pole cell formation and related to eIF2C. Development 128:2823–2832

Haynes KA, Caudy AA, Collins L, Elgin SC (2006) Element 1360 and RNAi components contribute to HP1-dependent silencing of a pericentric reporter. Curr Biol 16:2222–2227

Horwich MD, Li C, Matranga C, Vagin V, Farley G, Wang P, Zamore PD (2007) The Drosophila RNA methyltransferase, DmHen1, modifies germline piRNAs and single-stranded siRNAs in RISC. Curr Biol 17:1265–1272

Hou Y, Yuan J, Zhou X, Fu X, Cheng H, Zhou R (2012) DNA demethylation and USF regulate the meiosis-specific expression of the mouse Miwi. PLoS Genet 8:e1002716

Houston DW, King ML (2000) Germ plasm and molecular determinants of germ cell fate. Curr Top Dev Biol 50:155–181

Houwing S, Berezikov E, Ketting RF (2008) Zili is required for germ cell differentiation and meiosis in zebrafish. EMBO J 27:2702–2711

Houwing S, Kamminga LM, Berezikov E, Cronembold D, Girard A, van den Elst H, Filippov DV, Blaser H, Raz E, Moens CB et al (2007) A role for Piwi and piRNAs in germ cell maintenance and transposon silencing in Zebrafish. Cell 129:69–82

Huang H, Gao Q, Peng X, Choi SY, Sarma K, Ren H, Morris AJ, Frohman MA (2011a) piRNA-associated germline nuage formation and spermatogenesis require MitoPLD profusogenic mitochondrial-surface lipid signaling. Dev Cell 20:376–387

Huang HY, Houwing S, Kaaij LJ, Meppelink A, Redl S, Gauci S, Vos H, Draper BW, Moens CB, Burgering BM et al (2011b) Tdrd1 acts as a molecular scaffold for Piwi proteins and piRNA targets in zebrafish. EMBO J 30:3298–3308

Huang XA, Yin H, Sweeney S, Raha D, Snyder M, Lin H (2013) A major epigenetic programming mechanism guided by piRNAs. Dev Cell 24:502–516

Ipsaro JJ, Haase AD, Knott SR, Joshua-Tor L, Hannon GJ (2012) The structural biochemistry of Zucchini implicates it as a nuclease in piRNA biogenesis. Nature 491:279–283

Irvine DV, Zaratiegui M, Tolia NH, Goto DB, Chitwood DH, Vaughn MW, Joshua-Tor L, Martienssen RA (2006) Argonaute slicing is required for heterochromatic silencing and spreading. Science 313:1134–1137

Ishizu H, Siomi H, Siomi MC (2012) Biology of PIWI-interacting RNAs: new insights into biogenesis and function inside and outside of germlines. Genes Dev 26:2361–2373

Janic A, Mendizabal L, Llamazares S, Rossell D, Gonzalez C (2010) Ectopic expression of germline genes drives malignant brain tumor growth in Drosophila. Science 330:1824–1827

Juliano C, Wang J, Lin H (2011) Uniting germline and stem cells: the function of Piwi proteins and the piRNA pathway in diverse organisms. Annu Rev Genet 45:447–469

Kamminga LM, van Wolfswinkel JC, Luteijn MJ, Kaaij LJ, Bagijn MP, Sapetschnig A, Miska EA, Berezikov E, Ketting RF (2012) Differential impact of the HEN1 homolog HENN-1 on 21U and 26G RNAs in the germline of Caenorhabditis elegans. PLoS Genet 8:e1002702

Kawaoka S, Izumi N, Katsuma S, Tomari Y (2011) 3′ end formation of PIWI-interacting RNAs in vitro. Mol Cell 43:1015–1022

Kawaoka S, Mitsutake H, Kiuchi T, Kobayashi M, Yoshikawa M, Suzuki Y, Sugano S, Shimada T, Kobayashi J, Tomari Y et al (2012) A role for transcription from a piRNA cluster in de novo piRNA production. RNA 18:265–273

Kelleher ES, Edelman NB, Barbash DA (2012) Drosophila interspecific hybrids phenocopy piRNA-pathway mutants. PLoS Biol 10:e1001428

Ketting RF (2011) The many faces of RNAi. Dev Cell 20:148–161

Khurana JS, Wang J, Xu J, Koppetsch BS, Thomson TC, Nowosielska A, Li C, Zamore PD, Weng Z, Theurkauf WE (2011) Adaptation to P element transposon invasion in Drosophila melanogaster. Cell 147:1551–1563

Kirino Y, Kim N, de Planell-Saguer M, Khandros E, Chiorean S, Klein PS, Rigoutsos I, Jongens TA, Mourelatos Z (2009) Arginine methylation of Piwi proteins catalysed by dPRMT5 is required for Ago3 and Aub stability. Nat Cell Biol 11:652–658

Klattenhoff C, Xi H, Li C, Lee S, Xu J, Khurana JS, Zhang F, Schultz N, Koppetsch BS, Nowosielska A et al (2009) The Drosophila HP1 homolog Rhino is required for transposon silencing and piRNA production by dual-strand clusters. Cell 138:1137–1149

Kotaja N, Sassone-Corsi P (2007) The chromatoid body: a germ-cell-specific RNA-processing centre. Nat Rev Mol Cell Biol 8:85–90

Krawetz SA, Kruger A, Lalancette C, Tagett R, Anton E, Draghici S, Diamond MP (2011) A survey of small RNAs in human sperm. Hum Reprod 26:3401–3412

Kuramochi-Miyagawa S, Watanabe T, Gotoh K, Takamatsu K, Chuma S, Kojima-Kita K, Shiromoto Y, Asada N, Toyoda A, Fujiyama A et al (2010) MVH in piRNA processing and gene silencing of retrotransposons. Genes Dev 24:887–892

Kuramochi-Miyagawa S, Watanabe T, Gotoh K, Totoki Y, Toyoda A, Ikawa M, Asada N, Kojima K, Yamaguchi Y, Ijiri TW et al (2008) DNA methylation of retrotransposon genes is regulated by Piwi family members MILI and MIWI2 in murine fetal testes. Genes Dev 22:908–917

Lane N, Dean W, Erhardt S, Hajkova P, Surani A, Walter J, Reik W (2003) Resistance of IAPs to methylation reprogramming may provide a mechanism for epigenetic inheritance in the mouse. Genesis 35:88–93

Lau NC (2010) Small RNAs in the animal gonad: guarding genomes and guiding development. Int J Biochem Cell Biol 42:1334–1347

Lau NC, Ohsumi T, Borowsky M, Kingston RE, Blower MD (2009) Systematic and single cell analysis of Xenopus Piwi-interacting RNAs and Xiwi. EMBO J 28:2945–2958

Lau NC, Seto AG, Kim J, Kuramochi-Miyagawa S, Nakano T, Bartel DP, Kingston RE (2006) Characterization of the piRNA complex from rat testes. Science 313:363–367

Le Thomas A, Rogers AK, Webster A, Marinov GK, Liao SE, Perkins EM, Hur JK, Aravin AA, Toth KF (2013) Piwi induces piRNA-guided transcriptional silencing and establishment of a repressive chromatin state. Genes Dev 27:390–399

Lee EJ, Banerjee S, Zhou H, Jammalamadaka A, Arcila M, Manjunath BS, Kosik KS (2011) Identification of piRNAs in the central nervous system. RNA 17:1090–1099

Lee HC, Gu W, Shirayama M, Youngman E, Conte D Jr, Mello CC (2012) C. elegans piRNAs mediate the genome-wide surveillance of germline transcripts. Cell 150:78–87

Lee RC, Feinbaum RL, Ambros V (1993) The C. elegans heterochronic gene lin-4 encodes small RNAs with antisense complementarity to lin-14. Cell 75:843–854

Lehmann R, Nusslein-Volhard C (1991) The maternal gene nanos has a central role in posterior pattern formation of the Drosophila embryo. Development 112:679–691

Li C, Vagin VV, Lee S, Xu J, Ma S, Xi H, Seitz H, Horwich MD, Syrzycka M, Honda BM et al (2009) Collapse of germline piRNAs in the absence of Argonaute3 reveals somatic piRNAs in flies. Cell 137:509–521

Li XZ, Roy CK, Dong X, Bolcun-Filas E, Wang J, Han BW, Xu J, Moore MJ, Schimenti JC, Weng Z et al (2013) An ancient transcription factor initiates the burst of piRNA production during early meiosis in mouse testes. Mol Cell 50:67–81

Lim AK, Kai T (2007) Unique germ-line organelle, nuage, functions to repress selfish genetic elements in Drosophila melanogaster. Proc Natl Acad Sci U S A 104:6714–6719

Lin H, Spradling AC (1997) A novel group of pumilio mutations affects the asymmetric division of germline stem cells in the Drosophila ovary. Development 124:2463–2476

Liu J, Carmell MA, Rivas FV, Marsden CG, Thomson JM, Song JJ, Hammond SM, Joshua-Tor L, Hannon GJ (2004) Argonaute2 is the catalytic engine of mammalian RNAi. Science 305: 1437–1441

Liu L, Qi H, Wang J, Lin H (2011) PAPI, a novel TUDOR-domain protein, complexes with AGO3, ME31B and TRAL in the nuage to silence transposition. Development 138:1863–1873

Lu J, Clark AG (2010) Population dynamics of PIWI-interacting RNAs (piRNAs) and their targets in Drosophila. Genome Res 20:212–227

Lu X, Wontakal SN, Kavi H, Kim BJ, Guzzardo PM, Emelyanov AV, Xu N, Hannon GJ, Zavadil J, Fyodorov DV et al (2013) Drosophila H1 regulates the genetic activity of heterochromatin by recruitment of Su(var)3-9. Science 340:78–81

Luteijn MJ, van Bergeijk P, Kaaij LJ, Almeida MV, Roovers EF, Berezikov E, Ketting RF (2012) Extremely stable Piwi-induced gene silencing in Caenorhabditis elegans. EMBO J 31: 3422–3430

Ma L, Buchold GM, Greenbaum MP, Roy A, Burns KH, Zhu H, Han DY, Harris RA, Coarfa C, Gunaratne PH et al (2009) GASZ is essential for male meiosis and suppression of retrotransposon expression in the male germline. PLoS Genet 5:e1000635

Malone CD, Brennecke J, Dus M, Stark A, McCombie WR, Sachidanandam R, Hannon GJ (2009) Specialized piRNA pathways act in germline and somatic tissues of the Drosophila ovary. Cell 137:522–535

Marchetto MC, Narvaiza I, Denli AM, Benner C, Lazzarini TA, Nathanson JL, Paquola AC, Desai KN, Herai RH, Weitzman MD et al (2013) Differential L1 regulation in pluripotent stem cells of humans and apes. Nature 503(7477):525–529

Megosh HB, Cox DN, Campbell C, Lin H (2006) The role of PIWI and the miRNA machinery in Drosophila germline determination. Curr Biol 16:1884–1894

Meister G, Landthaler M, Peters L, Chen PY, Urlaub H, Luhrmann R, Tuschl T (2005) Identification of novel argonaute-associated proteins. Curr Biol 15:2149–2155

Mochizuki K, Fine NA, Fujisawa T, Gorovsky MA (2002) Analysis of a piwi-related gene implicates small RNAs in genome rearrangement in tetrahymena. Cell 110:689–699

Mochizuki K, Gorovsky MA (2004a) Conjugation-specific small RNAs in Tetrahymena have predicted properties of scan (scn) RNAs involved in genome rearrangement. Genes Dev 18:2068–2073

Mochizuki K, Gorovsky MA (2004b) RNA polymerase II localizes in Tetrahymena thermophila meiotic micronuclei when micronuclear transcription associated with genome rearrangement occurs. Eukaryot Cell 3:1233–1240

Montgomery TA, Rim YS, Zhang C, Dowen RH, Phillips CM, Fischer SE, Ruvkun G (2012) PIWI associated siRNAs and piRNAs specifically require the Caenorhabditis elegans HEN1 ortholog henn-1. PLoS Genet 8:e1002616

Moroz LL, Ju J, Russo JJ, Puthanveetti S, Kohn A, Medina M, Walsh PJ, Birren B, Lander ES, Kandel ER (2009) Aplysia Genome Project. p. 11x coverage sequencing of *Aplysia californica* genome. Broad Institute
Moshkovich N, Lei EP (2010) HP1 recruitment in the absence of argonaute proteins in Drosophila. PLoS Genet 6:e1000880
Muerdter F, Guzzardo PM, Gillis J, Luo Y, Yu Y, Chen C, Fekete R, Hannon GJ (2013) A genome-wide RNAi screen draws a genetic framework for transposon control and primary piRNA biogenesis in Drosophila. Mol Cell 50(5):736–748
Muerdter F, Olovnikov I, Molaro A, Rozhkov NV, Czech B, Gordon A, Hannon GJ, Aravin AA (2012) Production of artificial piRNAs in flies and mice. RNA 18:42–52
Nishida KM, Okada TN, Kawamura T, Mituyama T, Kawamura Y, Inagaki S, Huang H, Chen D, Kodama T, Siomi H et al (2009) Functional involvement of Tudor and dPRMT5 in the piRNA processing pathway in Drosophila germlines. EMBO J 28:3820–3831
Nishimasu H, Ishizu H, Saito K, Fukuhara S, Kamatani MK, Bonnefond L, Matsumoto N, Nishizawa T, Nakanaga K, Aoki J et al (2012) Structure and function of Zucchini endoribonuclease in piRNA biogenesis. Nature 491:284–287
O'Carroll D, Mecklenbrauker I, Das PP, Santana A, Koenig U, Enright AJ, Miska EA, Tarakhovsky A (2007) A Slicer-independent role for Argonaute 2 in hematopoiesis and the microRNA pathway. Genes Dev 21:1999–2004
Ohtani H, Iwasaki YW, Shibuya A, Siomi H, Siomi MC, Saito K (2013) DmGTSF1 is necessary for Piwi-piRISC-mediated transcriptional transposon silencing in the Drosophila ovary. Genes Dev 27:1656–1661
Okamura K, Lai EC (2008) Endogenous small interfering RNAs in animals. Nat Rev Mol Cell Biol 9:673–678
Olivieri D, Senti KA, Subramanian S, Sachidanandam R, Brennecke J (2012) The cochaperone shutdown defines a group of biogenesis factors essential for all piRNA populations in Drosophila. Mol Cell 47:954–969
Olivieri D, Sykora MM, Sachidanandam R, Mechtler K, Brennecke J (2010) An in vivo RNAi assay identifies major genetic and cellular requirements for primary piRNA biogenesis in Drosophila. EMBO J 29:3301–3317
Pal-Bhadra M, Bhadra U, Birchler JA (2002) RNAi related mechanisms affect both transcriptional and posttranscriptional transgene silencing in Drosophila. Mol Cell 9:315–327
Pal Bhadra M, Bhadra U, Birchler JA (2006) Misregulation of sex-lethal and disruption of male-specific lethal complex localization in Drosophila species hybrids. Genetics 174:1151–1159
Palakodeti D, Smielewska M, Lu YC, Yeo GW, Graveley BR (2008) The PIWI proteins SMEDWI-2 and SMEDWI-3 are required for stem cell function and piRNA expression in planarians. RNA 14:1174–1186
Pan J, Goodheart M, Chuma S, Nakatsuji N, Page DC, Wang PJ (2005) RNF17, a component of the mammalian germ cell nuage, is essential for spermiogenesis. Development 132:4029–4039
Pane A, Jiang P, Zhao DY, Singh M, Schupbach T (2011) The Cutoff protein regulates piRNA cluster expression and piRNA production in the Drosophila germline. EMBO J 30:4601–4615
Pane A, Wehr K, Schupbach T (2007) zucchini and squash encode two putative nucleases required for rasiRNA production in the Drosophila germline. Dev Cell 12:851–862
Pao SY, Lin WL, Hwang MJ (2006) In silico identification and comparative analysis of differentially expressed genes in human and mouse tissues. BMC Genomics 7:86
Patil VS, Kai T (2010) Repression of retroelements in Drosophila germline via piRNA pathway by the Tudor domain protein Tejas. Curr Biol 20(8):724–730
Pavelec DM, Lachowiec J, Duchaine TF, Smith HE, Kennedy S (2009) Requirement for the ERI/DICER complex in endogenous RNA interference and sperm development in Caenorhabditis elegans. Genetics 183:1283–1295
Perrat PN, DasGupta S, Wang J, Theurkauf W, Weng Z, Rosbash M, Waddell S (2013) Transposition-driven genomic heterogeneity in the Drosophila brain. Science 340:91–95
Petersen CP, Reddien PW (2009) Wnt signaling and the polarity of the primary body axis. Cell 139:1056–1068

Post C, Clark J, Sytnikova Y, Chirn GW, Lau NC (2014) The capacity of target silencing by the Drosophila Piwi protein and piRNAs. Submitted

Preall JB, Czech B, Guzzardo PM, Muerdter F, Hannon GJ (2012) Shutdown is a component of the Drosophila piRNA biogenesis machinery. RNA 18:1446–1457

Rajasethupathy P, Antonov I, Sheridan R, Frey S, Sander C, Tuschl T, Kandel ER (2012) A role for neuronal piRNAs in the epigenetic control of memory-related synaptic plasticity. Cell 149: 693–707

Rangan P, Malone CD, Navarro C, Newbold SP, Hayes PS, Sachidanandam R, Hannon GJ, Lehmann R (2011) piRNA production requires heterochromatin formation in Drosophila. Curr Biol 21:1373–1379

Ravasi T, Suzuki H, Pang KC, Katayama S, Furuno M, Okunishi R, Fukuda S, Ru K, Frith MC, Gongora MM et al (2006) Experimental validation of the regulated expression of large numbers of non-coding RNAs from the mouse genome. Genome Res 16:11–19

Reddien PW, Bermange AL, Kicza AM, Sanchez AA (2007) BMP signaling regulates the dorsal planarian midline and is needed for asymmetric regeneration. Development 134:4043–4051

Reinhart BJ, Slack FJ, Basson M, Pasquinelli AE, Bettinger JC, Rougvie AE, Horvitz HR, Ruvkun G (2000) The 21-nucleotide let-7 RNA regulates developmental timing in Caenorhabditis elegans. Nature 403:901–906

Reuter M, Berninger P, Chuma S, Shah H, Hosokawa M, Funaya C, Antony C, Sachidanandam R, Pillai RS (2011) Miwi catalysis is required for piRNA amplification-independent LINE1 transposon silencing. Nature 480:264–267

Reuter M, Chuma S, Tanaka T, Franz T, Stark A, Pillai RS (2009) Loss of the Mili-interacting Tudor domain-containing protein-1 activates transposons and alters the Mili-associated small RNA profile. Nat Struct Mol Biol 16:639–646

Robine N, Lau NC, Balla S, Jin Z, Okamura K, Kuramochi-Miyagawa S, Blower MD, Lai EC (2009) A broadly conserved pathway generates 3′UTR-directed primary piRNAs. Curr Biol 19:2066–2076

Rouget C, Papin C, Boureux A, Meunier AC, Franco B, Robine N, Lai EC, Pelisson A, Simonelig M (2010) Maternal mRNA deadenylation and decay by the piRNA pathway in the early Drosophila embryo. Nature 467:1128–1132

Rozhkov NV, Hammell M, Hannon GJ (2013) Multiple roles for Piwi in silencing Drosophila transposons. Genes Dev 27:400–412

Ruby JG, Jan C, Player C, Axtell MJ, Lee W, Nusbaum C, Ge H, Bartel DP (2006) Large-scale sequencing reveals 21U-RNAs and additional microRNAs and endogenous siRNAs in C. elegans. Cell 127:1193–1207

Saito K, Inagaki S, Mituyama T, Kawamura Y, Ono Y, Sakota E, Kotani H, Asai K, Siomi H, Siomi MC (2009) A regulatory circuit for piwi by the large Maf gene traffic jam in Drosophila. Nature 461:1296–1299

Saito K, Ishizu H, Komai M, Kotani H, Kawamura Y, Nishida KM, Siomi H, Siomi MC (2010) Roles for the Yb body components Armitage and Yb in primary piRNA biogenesis in Drosophila. Genes Dev 24:2493–2498

Saito K, Nishida KM, Mori T, Kawamura Y, Miyoshi K, Nagami T, Siomi H, Siomi MC (2006) Specific association of Piwi with rasiRNAs derived from retrotransposon and heterochromatic regions in the Drosophila genome. Genes Dev 20:2214–2222

Saito K, Sakaguchi Y, Suzuki T, Siomi H, Siomi MC (2007) Pimet, the Drosophila homolog of HEN1, mediates 2′-O-methylation of Piwi- interacting RNAs at their 3′ ends. Genes Dev 21:1603–1608

Sato K, Nishida KM, Shibuya A, Siomi MC, Siomi H (2011) Maelstrom coordinates microtubule organization during Drosophila oogenesis through interaction with components of the MTOC. Genes Dev 25:2361–2373

Sato K, Siomi H (2010) Is canalization more than just a beautiful idea? Genome Biol 11:109

Saxe JP, Chen M, Zhao H, Lin H (2013) Tdrkh is essential for spermatogenesis and participates in primary piRNA biogenesis in the germline. EMBO J 332(13):1869–1885

Schoeberl UE, Kurth HM, Noto T, Mochizuki K (2012) Biased transcription and selective degradation of small RNAs shape the pattern of DNA elimination in Tetrahymena. Genes Dev 26:1729–1742

Schupbach T, Wieschaus E (1989) Female sterile mutations on the second chromosome of Drosophila melanogaster. I Maternal effect mutations. Genetics 121:101–117

Schupbach T, Wieschaus E (1991) Female sterile mutations on the second chromosome of Drosophila melanogaster. II. Mutations blocking oogenesis or altering egg morphology. Genetics 129:1119–1136

Sentmanat MF, Elgin SC (2012) Ectopic assembly of heterochromatin in Drosophila melanogaster triggered by transposable elements. Proc Natl Acad Sci U S A 109:14104–14109

She X, Xu X, Fedotov A, Kelly WG, Maine EM (2009) Regulation of heterochromatin assembly on unpaired chromosomes during caenorhabditis elegans meiosis by components of a small RNA-mediated pathway. PLoS Genet 5:e1000624

Shi Z, Montgomery TA, Qi Y, Ruvkun G (2013) High-throughput sequencing reveals extraordinary fluidity of miRNA, piRNA, and siRNA pathways in nematodes. Genome Res 23:497–508

Shirayama M, Seth M, Lee HC, Gu W, Ishidate T, Conte D Jr, Mello CC (2012) piRNAs initiate an epigenetic memory of nonself RNA in the C. elegans germline. Cell 150:65–77

Shiromoto Y, Kuramochi-Miyagawa S, Daiba A, Chuma S, Katanaya A, Katsumata A, Nishimura K, Ohtaka M, Nakanishi M, Nakamura T et al (2013) GPAT2, a mitochondrial outer membrane protein, in piRNA biogenesis in germline stem cells. RNA 19:803–810

Shoji M, Tanaka T, Hosokawa M, Reuter M, Stark A, Kato Y, Kondoh G, Okawa K, Chujo T, Suzuki T et al (2009) The TDRD9-MIWI2 complex is essential for piRNA-mediated retrotransposon silencing in the mouse male germline. Dev Cell 17:775–787

Shpiz S, Kalmykova A (2009) Epigenetic transmission of piRNAs through the female germline. Genome Biol 10:208

Sienski G, Donertas D, Brennecke J (2012) Transcriptional silencing of transposons by Piwi and maelstrom and its impact on chromatin state and gene expression. Cell 151:964–980

Siomi MC, Mannen T, Siomi H (2010) How does the royal family of Tudor rule the PIWI-interacting RNA pathway? Genes Dev 24:636–646

Siomi MC, Sato K, Pezic D, Aravin AA (2011) PIWI-interacting small RNAs: the vanguard of genome defence. Nat Rev Mol Cell Biol 12:246–258

Song Y, Fee L, Lee TH, Wharton RP (2007) The molecular chaperone Hsp90 is required for mRNA localization in Drosophila melanogaster embryos. Genetics 176:2213–2222

Soper SF, van der Heijden GW, Hardiman TC, Goodheart M, Martin SL, de Boer P, Bortvin A (2008) Mouse maelstrom, a component of nuage, is essential for spermatogenesis and transposon repression in meiosis. Dev Cell 15:285–297

Specchia V, Piacentini L, Tritto P, Fanti L, D'Alessandro R, Palumbo G, Pimpinelli S, Bozzetti MP (2010) Hsp90 prevents phenotypic variation by suppressing the mutagenic activity of transposons. Nature 463:662–665

Swart EC, Bracht JR, Magrini V, Minx P, Chen X, Zhou Y, Khurana JS, Goldman AD, Nowacki M, Schotanus K et al (2013) The Oxytricha trifallax macronuclear genome: a complex eukaryotic genome with 16,000 tiny chromosomes. PLoS Biol 11:e1001473

Sytnikova Y, Lau NC (2013) Does the APC/C Mark MIWI and piRNAs for a final farewell? Dev Cell 24:119–120

Sytnikova Y, Rahmam R, Chirn GW, Post C, Clark J, Lau NC (2014) Transposable element dynamics and PIWI regulation impacts lncRNA and gene expression diversity in Drosophila ovarian cell cultures. Submitted

Tam OH, Aravin AA, Stein P, Girard A, Murchison EP, Cheloufi S, Hodges E, Anger M, Sachidanandam R, Schultz RM et al (2008) Pseudogene-derived small interfering RNAs regulate gene expression in mouse oocytes. Nature 453:534–538

Toscani A, Mettus RV, Coupland R, Simpkins H, Litvin J, Orth J, Hatton KS, Reddy EP (1997) Arrest of spermatogenesis and defective breast development in mice lacking A-myb. Nature 386:713–717

Unhavaithaya Y, Hao Y, Beyret E, Yin H, Kuramochi-Miyagawa S, Nakano T, Lin H (2009) MILI, a PIWI-interacting RNA-binding protein, is required for germ line stem cell self-renewal and appears to positively regulate translation. J Biol Chem 284:6507–6519

Vagin VV, Sigova A, Li C, Seitz H, Gvozdev V, Zamore PD (2006) A distinct small RNA pathway silences selfish genetic elements in the germline. Science 313:320–324

Vagin VV, Wohlschlegel J, Qu J, Jonsson Z, Huang X, Chuma S, Girard A, Sachidanandam R, Hannon GJ, Aravin AA (2009) Proteomic analysis of murine Piwi proteins reveals a role for arginine methylation in specifying interaction with Tudor family members. Genes Dev 23: 1749–1762

van Wolfswinkel JC, Claycomb JM, Batista PJ, Mello CC, Berezikov E, Ketting RF (2009) CDE-1 affects chromosome segregation through uridylation of CSR-1-bound siRNAs. Cell 139: 135–148

Vasale JJ, Gu W, Thivierge C, Batista PJ, Claycomb JM, Youngman EM, Duchaine TF, Mello CC, Conte D Jr (2010) Sequential rounds of RNA-dependent RNA transcription drive endogenous small-RNA biogenesis in the ERGO-1/Argonaute pathway. Proc Natl Acad Sci U S A 107:3582–3587

Vasileva A, Tiedau D, Firooznia A, Muller-Reichert T, Jessberger R (2009) Tdrd6 is required for spermiogenesis, chromatoid body architecture, and regulation of miRNA expression. Curr Biol 19:630–639

Vourekas A, Zheng Q, Alexiou P, Maragkakis M, Kirino Y, Gregory BD, Mourelatos Z (2012) Mili and Miwi target RNA repertoire reveals piRNA biogenesis and function of Miwi in spermiogenesis. Nat Struct Mol Biol 19:773–781

Wang C, Dickinson LK, Lehmann R (1994) Genetics of nanos localization in Drosophila. Dev Dyn 199:103–115

Wang J, Czech B, Crunk A, Wallace A, Mitreva M, Hannon GJ, Davis RE (2011) Deep small RNA sequencing from the nematode Ascaris reveals conservation, functional diversification, and novel developmental profiles. Genome Res 21:1462–1477

Wang J, Saxe JP, Tanaka T, Chuma S, Lin H (2009) Mili interacts with tudor domain-containing protein 1 in regulating spermatogenesis. Curr Biol 19:640–644

Wang SH, Elgin SC (2011) Drosophila Piwi functions downstream of piRNA production mediating a chromatin-based transposon silencing mechanism in female germ line. Proc Natl Acad Sci U S A 108:21164–21169

Watanabe T, Chuma S, Yamamoto Y, Kuramochi-Miyagawa S, Totoki Y, Toyoda A, Hoki Y, Fujiyama A, Shibata T, Sado T et al (2011) MITOPLD is a mitochondrial protein essential for nuage formation and piRNA biogenesis in the mouse germline. Dev Cell 20:364–375

Watanabe T, Totoki Y, Toyoda A, Kaneda M, Kuramochi-Miyagawa S, Obata Y, Chiba H, Kohara Y, Kono T, Nakano T et al (2008) Endogenous siRNAs from naturally formed dsRNAs regulate transcripts in mouse oocytes. Nature 453:539–543

Wilson JE, Connell JE, Macdonald PM (1996) aubergine enhances oskar translation in the Drosophila ovary. Development 122:1631–1639

Xiol J, Cora E, Koglgruber R, Chuma S, Subramanian S, Hosokawa M, Reuter M, Yang Z, Berninger P, Palencia A et al (2012) A role for Fkbp6 and the chaperone machinery in piRNA amplification and transposon silencing. Mol Cell 47:970–979

Yabuta Y, Ohta H, Abe T, Kurimoto K, Chuma S, Saitou M (2011) TDRD5 is required for retrotransposon silencing, chromatoid body assembly, and spermiogenesis in mice. J Cell Biol 192:781–795

Yan Z, Hu HY, Jiang X, Maierhofer V, Neb E, He L, Hu Y, Hu H, Li N, Chen W et al (2011) Widespread expression of piRNA-like molecules in somatic tissues. Nucleic Acids Res 39:6596–6607

Yeo G, Holste D, Kreiman G, Burge CB (2004) Variation in alternative splicing across human tissues. Genome Biol 5:R74

Yigit E, Batista PJ, Bei Y, Pang KM, Chen CC, Tolia NH, Joshua-Tor L, Mitani S, Simard MJ, Mello CC (2006) Analysis of the C. elegans Argonaute family reveals that distinct Argonautes act sequentially during RNAi. Cell 127:747–757

Yin H, Lin H (2007) An epigenetic activation role of Piwi and a Piwi-associated piRNA in Drosophila melanogaster. Nature 450:304–308

Zamparini AL, Davis MY, Malone CD, Vieira E, Zavadil J, Sachidanandam R, Hannon GJ, Lehmann R (2011) Vreteno, a gonad-specific protein, is essential for germline development and primary piRNA biogenesis in Drosophila. Development 138:4039–4050

Zhang C, Montgomery TA, Gabel HW, Fischer SE, Phillips CM, Fahlgren N, Sullivan CM, Carrington JC, Ruvkun G (2011a) mut-16 and other mutator class genes modulate 22G and 26G siRNA pathways in Caenorhabditis elegans. Proc Natl Acad Sci U S A 108:1201–1208

Zhang F, Wang J, Xu J, Zhang Z, Koppetsch BS, Schultz N, Vreven T, Meignin C, Davis I, Zamore PD et al (2012) UAP56 couples piRNA clusters to the perinuclear transposon silencing machinery. Cell 151:871–884

Zhang Z, Xu J, Koppetsch BS, Wang J, Tipping C, Ma S, Weng Z, Theurkauf WE, Zamore PD (2011b) Heterotypic piRNA Ping-Pong requires qin, a protein with both E3 ligase and Tudor domains. Mol Cell 44:572–584

Zhao S, Gou LT, Zhang M, Zu LD, Hua MM, Hua Y, Shi HJ, Li Y, Li J, Li D et al (2013) piRNA-triggered MIWI ubiquitination and removal by APC/C in late spermatogenesis. Dev Cell 24:13–25

Zheng K, Xiol J, Reuter M, Eckardt S, Leu NA, McLaughlin KJ, Stark A, Sachidanandam R, Pillai RS, Wang PJ (2010) Mouse MOV10L1 associates with Piwi proteins and is an essential component of the Piwi-interacting RNA (piRNA) pathway. Proc Natl Acad Sci U S A 107:11841–11846

Zhu W, Pao GM, Satoh A, Cummings G, Monaghan JR, Harkins TT, Bryant SV, Randal Voss S, Gardiner DM, Hunter T (2012) Activation of germline-specific genes is required for limb regeneration in the Mexican axolotl. Dev Biol 370:42–51

Chapter 6
Manipulation of RNA Using Engineered Proteins with Customized Specificity

Rajarshi Choudhury and Zefeng Wang

Abstract A large number of RNA-binding proteins play critical roles in controlling eukaryotic gene expression at multiple RNA-processing steps. Many of these proteins have modular configuration, containing a RNA binding domain to recognize their target and functional module to affect RNA metabolism. This simple configuration motivated the design of artificial factors that specifically manipulate RNA. While significant progress has been made since 1990s to engineer DNA binding proteins with designed specificity, design of analogous RNA binding factors was not practical until recently. With the increasing complexity of biological pathways involving RNA regulation, engineering RNA binding factors with customized specificity and function has become an emerging field of research. Such factors can serve as novel method to manipulate RNA metabolism and thus are very useful in basic biological and medical research. Here we discuss the current advances in engineering RNA binding proteins, with emphasis on the design principles and their potential applications as new therapeutic reagents and basic biological tools.

Keywords Modular activity • RNA binding protein • Artificial RNA binding factors • Modular RNA binding code • Engineered splicing factor • Artificial site-specific RNA endonuclease • PUF scaffold • Pentatricopeptide repeat

1 Introduction

The central dogma of molecular biology proposed that genetic information is transferred from DNAs to RNAs and then to proteins (Crick 1970). Although some variations of this liner flow of information have been discovered, RNAs are universally required in all living cells as intermediates for transferring genetic information. In eukaryotes, the newly synthesized RNAs undergo a number of processing steps to

R. Choudhury • Z. Wang (✉)
Department of Pharmacology, University of North Carolina, 4113 Genetic Medicine, 120 Mason Farm Rd, Chapel Hill, NC 27599, USA
e-mail: zefeng@med.unc.edu

G.W. Yeo (ed.), *Systems Biology of RNA Binding Proteins*, Advances in Experimental Medicine and Biology 825, DOI 10.1007/978-1-4939-1221-6_6

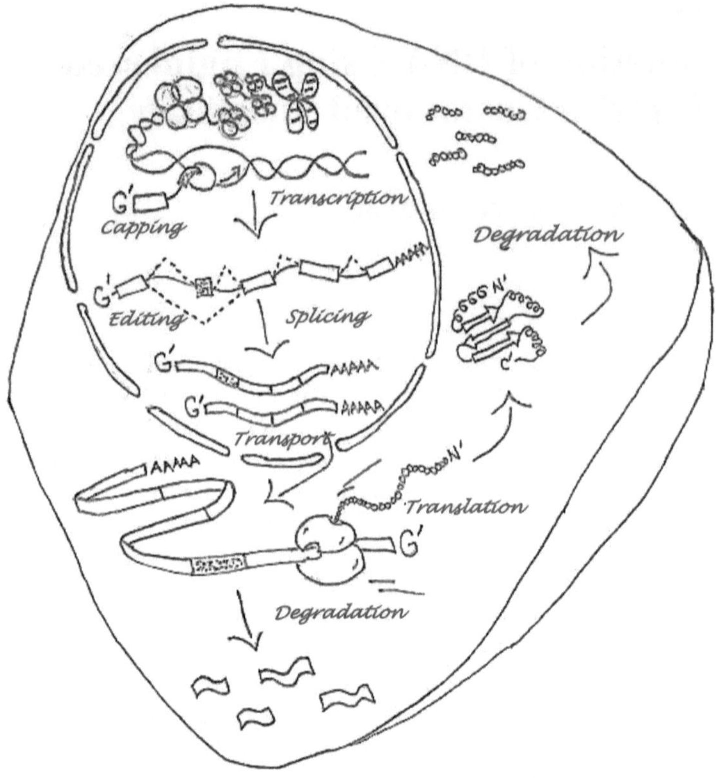

Fig. 6.1 Schematic diagram of RNA processing pathways during eukaryotic gene expression. The sketch is drawn by Mr. Russell Maxwell

generate mature mRNAs that are translated into proteins. These RNA processing steps, including capping, splicing, polyadenylation, editing, translocation, and degradation, are exquisitely regulated to ensure correct expression of genes (Fig. 6.1). In addition to serving as a messenger that conveys genetic information from DNA to protein, there are many types of noncoding RNAs that have diverse functions by themselves, including tRNA, snRNA, snoRNA, microRNA, long noncoding RNA, and so on. Therefore, RNA is often referred as the central component of the central dogma, and manipulation of RNA using synthetic biology approach presents a unique opportunity to control cellular function.

RNA-binding proteins (RBPs) play essential roles in every aspect of RNA metabolism. Most RBPs that specifically recognize RNAs have modular configuration: they usually contain various RNA binding domains (RBDs) that directly interacts with RNA and different functional domains that modulate RNA metabolism (Auweter et al. 2006a). Common domains that specifically bind single stranded RNA (ssRNA) include RNA recognition motifs (RRMs), pentatricopeptide repeats (PPRs), K homology (KH domains), zinc-finger, Pumilio/FBF (PUFs), and cold-shock domains (CSDs). In the past two decades, significant progress has been made in determining the relationships between RBP structures and functions, providing a systematic knowledge that serves as the foundation for the design of artificial RBPs with customized specificity and activity. Such artificial RBPs can be generated with a simple modular design that consists of a target recognition module which specifically binds RNA, a functional module that catalyzes RNA metabolism and a linker region to assemble the two modules in correct conformation (Mackay et al. 2011).

This modular design principle was first utilized in engineering artificial proteins that specifically bind DNA. Successful selection of zinc fingers with new DNA-binding specificities (Rebar and Pabo 1994) made it possible to engineer transcription factors with novel sequence specificity (Pomerantz et al. 1995). Subsequently a large number of artificial transcription factors and DNA nuclease were developed using zinc fingers to specifically recognize its cognate DNA substrate and a functional domain to modulate DNAs, which opened a new door for specific manipulation of gene expression or genome editing (Jamieson et al. 2003; Perez-Pinera et al. 2012; Urnov et al. 2010; Blancafort et al. 2004; Klug 2010). A limitation of these factors is that the DNA binding specificity of zinc finger is not fully reprogrammable, and certain screen schemes have to be applied to select for zinc fingers that specifically recognize a given sequence (Greisman and Pabo 1997; Sepp and Choo 2005). Recently a new generation of DNA binding factors has been engineered using the programmable DNA binding domain in TAL (transcription activator-like) effectors (Zhang et al. 2011), and the TALE nucleases were widely used in genome editing to facilitate in vivo DNA recombination (Kim et al. 2013; Miller et al. 2011).

The development of analogous factors that modulate RNA metabolism has long been proposed and recently become a reality due to the determination of a modular RNA-binding code in Pumilio/FBF (PUF) (Cheong and Hall 2006). Compared to DNA, RNA is involved in more diverse range of biological processes, thus the engineered RBPs will have a broad range of activities when combining with various functional domains. In addition, manipulation of RNA metabolism by artificial factors is usually transient and reversible; therefore, this technology may be more useful as potential human therapeutics as it does not incur permanent changes in the genome. In this chapter, we discuss current advances in engineering RBPs, with emphasis on the design principles and their potential applications. We cover several RBDs that have potential to be used in protein engineering, and the readers are encouraged to refer to a more detailed review on other RBDs (Auweter et al. 2006a; Lunde et al. 2007; Chen and Varani 2013).

2 Design Principle

Leonardo da Vinci once stated that "simplicity is the ultimate sophistication", which captured the exact principle in protein engineering. The artificial RNA-binding factor can be simply constructed by combining an RBD, a functional domain, and a short peptide linker between two domains. The key of such design is to use a programmable RBD that specifically recognize given targets with limited off-target effect. The ideal functional domain (aka effector domain) needs to be small and have well-defined biological function *in trans* when tethered to RNA. Additional parameters, including the relative size of the two domains, relative orientations of the two domains, and the length and conformation of inter-domain linker will also need to be considered. Thus, a successful design of artificial factors depends on intelligent combinations of RBD with customized specificity with a variety of effector domains in the right conformation.

3 RNA Binding Domains

3.1 RNA Recognition Motif (RRM)

The RRMs are one of the most abundant structural motifs in human proteome and are present in over 50 % of RBPs (Maris et al. 2005). A typical RRM domain contains 80–90 amino acids that fold into a β1α1β2β3α2β4 topology where the four antiparallel beta-sheets and the two additional alpha helices create ample surface that interacts with RNA (Daubner et al. 2013). The most conserved region of the RRM consists of two short sites (6-8 aa) in β1 and β3 (named RNP-2 and RNP-1 respectively) that are crucial for RNA interaction (Birney et al. 1993; Caceres and Krainer 1993; Sickmier et al. 2006) (Fig. 6.2a). However, recent structures of various RRMs bound by their cognate RNA show that RRMs may interact with RNA through diverse mechanisms. For example, hnRNP I (polypyrimidine tract binding protein or PTB) has four RRMs with similar specificities. The β3 of each RRM (RNP-1 site) contributes only weakly to RNA binding, whereas the side chains in β2 are responsible for binding to RNA bases through hydrophobic interactions (Oberstrass et al. 2005) (Fig. 6.2b). In the RRM of Fox-1, the loop region connecting β1 and α1 is also crucial for target recognition in that the Phe 126 of the loop is stacked between two RNA bases to provide additional interaction (Auweter et al. 2006b) (Fig. 6.2c). In some cases like hnRNP F, interactions between the RNA target and the RRM were found mainly in the loop region rather than in the β-sheet of the RRM (Dominguez et al. 2010) (Fig. 6.2d).

Single RRM usually recognize a short RNA element of 2–8 nt (typically 4 nt), and multiple RRMs are often found in some RBPs. The tandem RRMs in same RBP can either bind to similar RNA sequences and function cooperatively (Oberstrass et al. 2005; Dominguez et al. 2010; Sickmier et al. 2006), or have very different RNA binding activities/specificities (Burd et al. 1991), or only one/some of the RRMs are functional while the others do not exhibit RNA binding (Safaee et al. 2012). Therefore, in RBPs with multiple RRMs, the general rules for how each RRM contributes to specificity are largely unclear. In addition, the RNA-binding specificity of a RRM cannot be predicted based on the amino acid sequence and thus cannot be reprogrammed, restricting its application as an RNA recognition

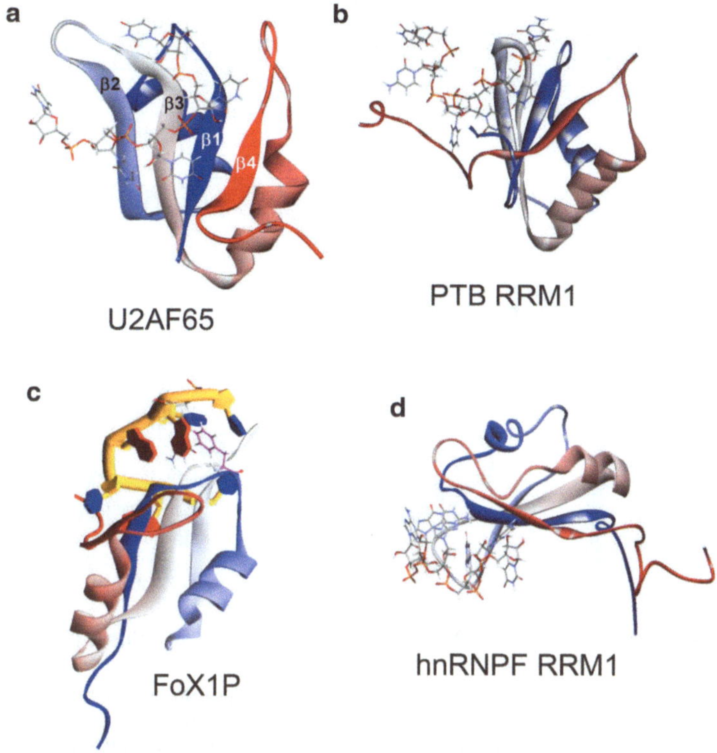

Fig. 6.2 Structure of RRMs in complex with cognate target RNAs. The proteins are shown as *ribbons*, colored with *blue* to *red* from N to C terminus. The RNAs are shown as *stick* model. (**a**) The canonical recognition mode of the RRM domain in U2AF65 with its RNA target (PDB ID: 2YH1). The four β-sheets are labeled. (**b**) Structure of the first RRM of PTB in complex with RNA target (PDB ID: 2AD9). (**c**) Structure of the first RRM of Fox-1 in complex with cognate RNA, the interaction between Phe 126 (shown in *purple*) in loop region with RNA base is highlighted (PDB ID: 2ERR). (**d**) Interaction of third qRRM domain in hnRNP F with cognate RNA (PDB ID: 2KG1)

module in designer RBPs. However, since there are a large number of natural RRM domains that recognize a broad range of sequences, it is still possible, at least in theory, to select a combination of RRMs and link them in tandem to impart specificity to certain sequences.

3.2 *The K Homology Domain (KH Domain)*

The KH domains are another highly abundant structure that recognizes both ssRNA and ssDNA (Valverde et al. 2008). These domains are involved in the regulation of a wide variety of processes including splicing, transcription, and translation. The domains are generally ~70 aa long and includes a signature of I/L/V/

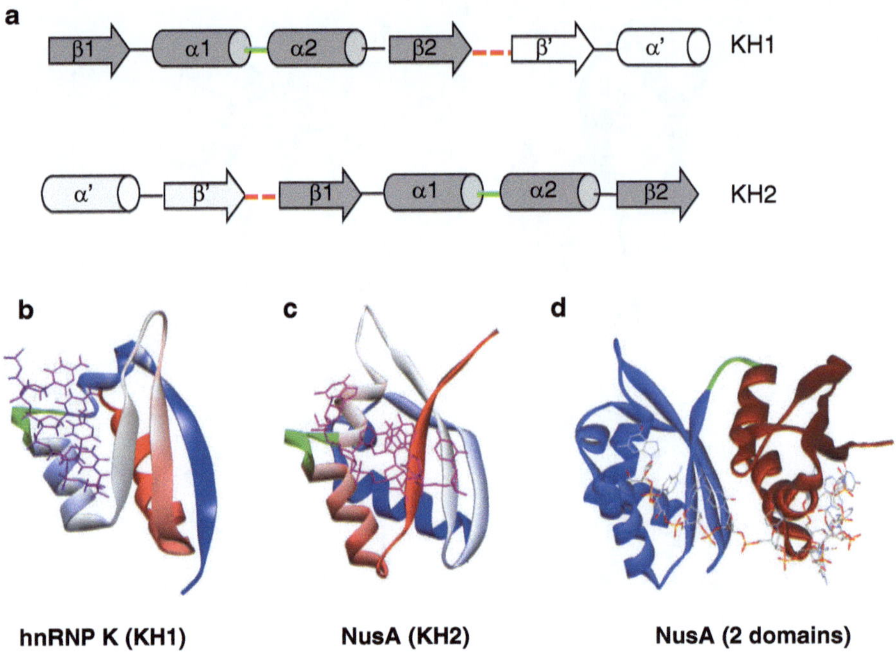

Fig. 6.3 Interaction of KH domain and RNA target. (**a**) Topology of domain arrangements in KH1 and KH2. The GXXG loop is shown in *green*, and the variable loop is indicated by *dotted line*. (**b**) Structure of KH1 type KH domain of hnRNP K in complex with RNA (PDB ID: 1J5K). (**c**) Structure of KH of NUSA which is type 2 KH domain in complex with RNA (PDB ID: 2ATW). (**d**) Structure of tandem KH domain of NUSA in complex with RNA (PDB ID: 2ASB)

IGXXGXXI/L/V. Structurally the KH domains are classified as two different types called type I and type II (Grishin 2001). The type I domains are mainly found in higher eukaryotes, while the type II domains are predominantly found in prokaryotes. The type I KH domains have a three stranded antiparallel β-sheet packed against three α-helices in a topology of β1α1α2β2β′α′, in which the first four alpha helix and beta sheets (β1α1α2β2) are known as minimal KH motif. In the type II KH domain, the minimal KH domain is present at the C-terminal with a topology of α′β′β1α1α2β2 (Fig. 6.3a).

A typical KH domain binds 4-nt RNA with the first 3 nt spreading on the surface of the domain (Fig. 6.3b, c). For both classes of KH domains, the four nucleotides are recognized in a cleft that is formed by the two consecutive α-helices (α1 and α2), the GXXG loop between α1 and α2, the subsequent β-strand (β2) and a variable loop between α2 and β2. The base of N1 is often stacked onto a peptide bond between the conserved glycine and the following residue in the GXXG loop, while the rest of the bases interact with a hydrophobic surface in the binding cleft. Unlike other types of ssRNA binding motifs, the KH domain does not contain aromatic amino acid in their RNA binding pocket. Therefore, the stacking between RNA base and the aromatic side chain of protein is lacking in the RNA–KH domain interaction (Valverde et al. 2008).

KH domains are usually found in multiple copies in a protein. In some cases, these tandem KH domains are linked via a long linker and thus bind ssDNA or RNA independently (Braddock et al. 2002). Alternatively, the tandem KH domains can be linked with as short linker (six residues in NusA) that causes two KH domain closely packed together to form an extended and continuous surface for binding of 11-nt RNA (Beuth et al. 2005) (Fig. 6.3d). In both situations, the coupling of two RNA-binding domains will effectively increase the specificity and affinity of RNA–protein interaction, although the two distinct binding modes have different effect on the length of RNA bound by the protein.

3.3 *Pumilio/FBF (PUF) Domain*

The founding members of the PUF proteins are pumilio in fly and fem-3 mRNA binding factor (FBF) in worm, which share a conserved RNA-binding domain with eight ~36 amino acid sequence repeats (Macdonald 1992; Zamore et al. 1997; Barker et al. 1992; Zhang et al. 1997). The native PUF proteins were found to regulate stability or translation by recruiting effector protein complexes to their target RNAs (Van Etten et al. 2012; Blewett and Goldstrohm 2012; Weidmann and Goldstrohm 2012; Goldstrohm et al. 2006, 2007). In contrast to hundreds of proteins with RRMs or KH domains that were found in an organism, PUF proteins comprise a relatively small family of RNA-binding proteins with few members in any organism. Most PUF proteins are predicted to have eight PUF repeats that are highly conserved. However, the RNA target sequences of these proteins are considerably more diverse than would be expected from the strong conservation. The core target sequences begin with a UGUR sequence, but may contain from 8 to 10 bases with differing three sequences.

The crystal structure of human Pumilio1 in complex with an RNA ligand has revealed a unique RNA-protein binding mode, in which each of the eight PUF repeats contain three α-helices and the amino acid side chains at specific positions of the second α-helix determine the RNA binding specificity (Fig. 6.4a) (Wang et al. 2001, 2002). The second helix in each PUF repeat includes a 5-residue sequence, designated here as 12xx5, where the side chain at the second position stacks with the recognized base and the side chains at the first and fifth positions interact specifically with edge of the RNA base (Wang et al. 2002; Lu and Hall 2011). There is a remarkably simple code for base recognition: glutamate and serine at the first and fifth positions recognize guanine; glutamine and cysteine/serine recognize adenine; and glutamine and asparagine recognize uracil (Fig. 6.4b) (Cheong and

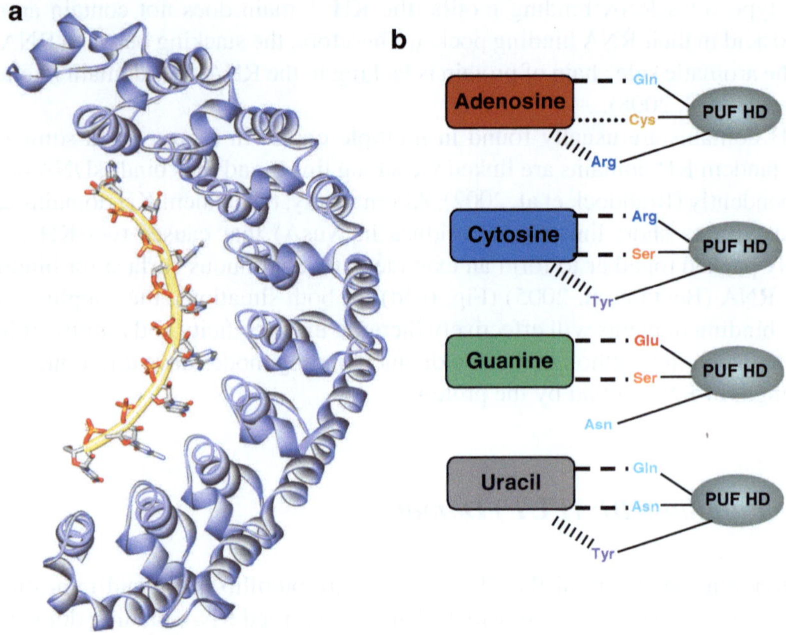

Fig. 6.4 Unique RNA recognition mode of PUF domain. (**a**) Structure of human Pum1 in complex with cognate RNA (PDB ID: 3Q0P). The protein has eight repeats and each repeat consists of three alpha helices. (**b**) Modular RNA binding code of PUF for all RNA bases. Two amino acid from each repeat contacts Watson–Crick edge of the RNA base while there is minimal side chain to backbone interaction. Most interactions are hydrogen bonded (indicated with "*dashed lines*") while van der Waal's interaction (indicated with "*dotted lines*") as well as stacking interactions (indicated with "*vertical lines*") also play a big role in base selection

Hall 2006; Wang et al. 2002). The Pumilio1 binds RNA with high affinity due to the stacking interactions between protein side chains and RNA bases (Cheong and Hall 2006). With this RNA recognition code, in principle, the RNA-binding specificity of human Pumilio1 can be manipulated by site-directed mutagenesis to recognize various RNA targets containing guanine, adenine, and uracil.

Although no recognition code for cytosine was found in nature, two groups succeeded in identifying combinations of amino acid side chains that specify cytosine recognition by a PUF repeat, expanding the PUF recognition code and permitting the creation of PUFs that recognize any 8-base RNA sequence (Filipovska et al. 2011; Dong et al. 2011). Both groups used yeast three-hybrid screening of a PUF domain library with random sequences at the first and fifth positions of the RNA interaction motif in repeat 6 to select proteins that bound a cytosine at the corresponding third base of the RNA. Dong et al. found that a combination of serine at the first position and arginine at the fifth position specify cytosine (Dong et al. 2011). This binding code was further validated by a crystal structure that showed hydrogen bonding between the side chain of arginine and the O2 and N3 positions of the cytosine. The serine residue positions the arginine, but does not contact the RNA. Using a similar strategy, Filipovska et al. selected five PUF variants that specify cytosine binding (Filipovska et al. 2011). Each variant pairs an arginine at the fifth position with a small side chain at the first position. The "C code" could be transferred to different PUF repeats, and tyrosine, histidine, or arginine side chains at the second position of the RNA interaction sequence are capable of forming stacking interactions with cytosine (Dong et al. 2011).

The completion of the PUF-RNA binding code for all bases makes it possible to design PUF domains that bind any 8-base RNA target sequence, increasing the potential application of engineered PUF proteins. Currently PUF is the only RNA binding module that has been successfully used in engineering RBPs with designed specificity, and additional discussion on this topic can be found in a detailed review of engineered PUF factors (Wang et al. 2013).

3.4 Pentapeptide Repeat (PPR)

Pentapeptide repeats proteins were first discovered in *Arabidopsis thaliana* and characterized by having degenerate 35 aa motif (Small and Peeters 2000). Although closely related to the tetratricopeptide repeats that mediate protein–protein interactions, the PPR protein seems to mediate RNA–protein interactions. These proteins are mainly found in the mitochondria and chloroplasts, where they specifically interact with RNAs and are involved in various posttranscriptional processes such as splicing, editing, processing, and translation (Delannoy et al. 2007; Saha et al. 2007; Rackham and Filipovska 2012). The PPR proteins usually contain 2–26 tandem repeat motifs, each containing 35 amino acids that fold into a helix-turn-helix structure (Howard et al. 2012; Ringel et al. 2011). The crystal structures for PPR proteins suggested that this domain might bind RNA in a similar fashion to the TALE-DNA interaction, as the inner face of the domain is positively charged to provide an interaction surface to nucleic acids (Kobayashi et al. 2012). It was proposed through molecular modeling that the hinge region between the two repeats directly interact with RNA through the amino acid side chain (Delannoy et al. 2007). Using the conservation information of PPR proteins, a putative combinatorial amino acid code for RNA recognition has been recently proposed (Takenaka et al. 2013; Barkan et al. 2012) (Fig. 6.5c). In one of the study, the putative code was further validated by modifying the PPR protein sequence to bind a new RNA, suggesting that PPR protein may indeed bind RNA in a modular fashion with the combination of two amino acids in each repeat (Barkan et al. 2012). The structure of PPR–RNA complex will help to fully decipher the RNA-binding code of PPR.

Compared to PUF domains that typically have eight repeats, the number of 35-aa repeats in PPR proteins have a wide range from 2 to 30, suggesting a more extended RNA-binding surface that may bind longer RNA. However, it is not clear if every consecutive repeat in PPR can interact with RNA base, which together with the lack of definitive RNA-binding code has limited the application of PPR in engineered RBP.

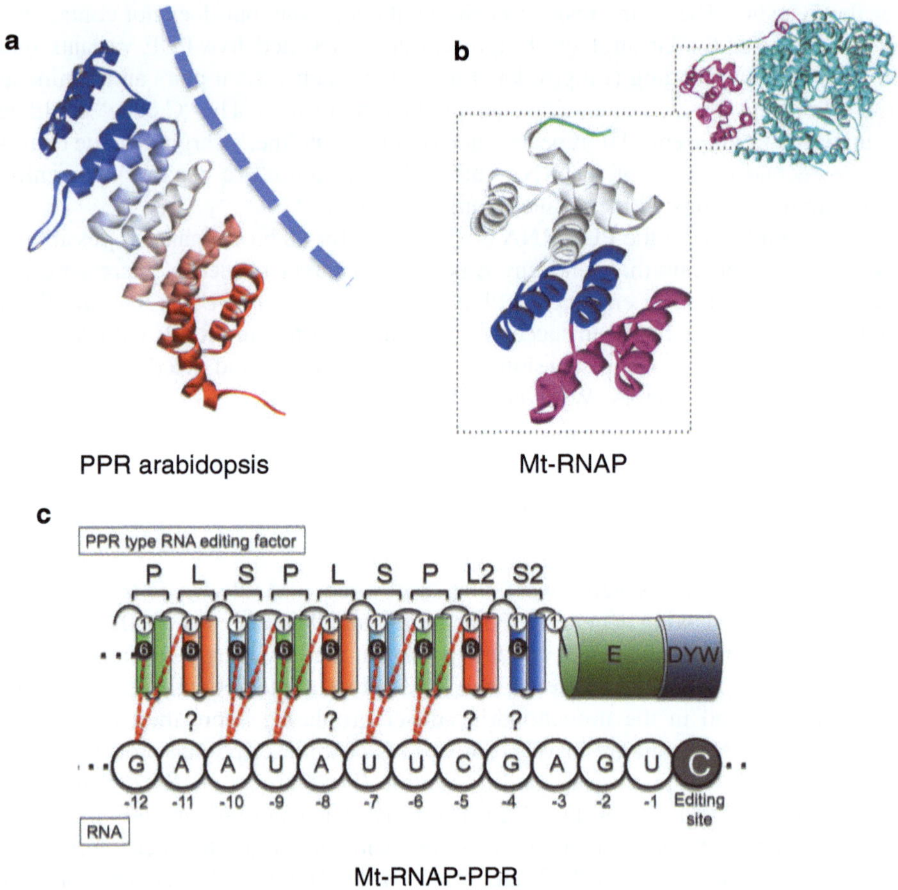

Pentapeptide repeats of arabidopsis and human mitochondrial RNA polymerase are shown. Co-crystallized RNA substrates are not available for this molecule

Fig. 6.5 (**a**) Structure of PPR domain of the mitochondrial ribonuclease P in Arabidopsis. The predicted surface for RNA binding is indicated with a *dashed line* (PDB ID: 4G23). (**b**) Structure of human mitochondrial RNA polymerase with the PPR domain enlarged PDB ID: 3SPA). (**c**) A putative model of modular interaction between PPR and cognate RNA. This figure is adopted from Takenaka et al. (2013)

4 Assembly of Artificial RBPs: Design Considerations

4.1 Selection of RBDs with Customized Specificity

Nature has additional RBDs that recognize RNA through specific structures or sequences, including cold-shock domains, Zn fingers, Zn knuckles, Sm or Sm-like domains, and double-stranded RNA binding domains. For the purpose of engineering RBPs with designer specificity, the four domains discussed earlier provided most promises. The RRM and KH domains recognize short RNA motifs (2–8 nt for RRM and 4-nt for KH), and their binding specificity cannot be fully reprogrammed, limiting its application in RBP engineering. However, it is conceivable that several RRMs or KH domains can be joined tandemly to recognize a longer RNA. Both RRMs and KH domain are very abundant in human genome, providing a large library of domains with different specificities. It may be possible to use certain screen approach or in vitro evolution methods to screen a large combinatorial library for RRMs or KH domains that recognize specific RNA targets, as similar screens were successfully applied in selection of zinc fingers that bind to customized DNAs (Blancafort et al. 2004; Blancafort and Beltran 2008).

The RNA binding specificity and/or affinity can also be affected by the relative orientation of tandem RRMs or KH domains with respect to the inter-domain linker, which presents a major challenge in design and construction of RNA-binding scaffold. For example, yeast Hrp1 protein has two tandem RRMs connected by a short linker in close proximity, the two RRMs behave as independent rigid bodies in the absence of RNA (Perez-Canadillas 2006). However, the binding of RNA target lead to a conformational rearrangements in the linker region, which facilitate the formation of a large RNA binding surface by two RRMs (Perez-Canadillas 2006). Such cooperative binding is often found when tandem RRMs connected by a short inter-domain linker, making it difficult to predict the behavior of multiple domains from the RNA-binding specificity of individual domains (Shamoo et al. 1995). In order to use tandem RRMs or KH domains as RNA binding module in engineered proteins, this change of RNA-binding property induced by connecting multiple domains has to be addressed with correct design of inter-domain linkers.

Due to the unique configuration of modular repeats, PUF domains provide an RNA-scaffold whose specificity can be individually modified for each position of RNA target. Therefore, it is an ideal RNA binding module in engineering RBPs with customized specificity and various functions. The main disadvantage of PUF is that the native protein contains only eight repeats that recognize 8-nt target; however, it is possible to change the length of PUF recognition sites (discussed later). Similar to the PUF domain, the PPR domain can recognize RNA in a modular fashion and several researchers have begun to make progress in solving its modular binding code (Barkan et al. 2012). Once the binding code is fully validated, the PPR may be used similarly to PUF as a modular RNA binding scaffold in engineered factors. Since the native PPR proteins contain 2–30 copies of the repeats, it is likely that an engineered PPR domain can be constructed with different number of repeats to recognize the RNA targets of different length (Takenaka et al. 2013). However, The PUF and PPR domain required 35 or 36 amino acids to recognize a single RNA base, there might be a size limit for engineered PPR or PUF with increasing number of repeats to recognize long RNA targets.

4.2 Inter-domain Linkers and Domain Arrangements

The inter-domain linker is an important consideration when engineering RBPs with multiple domains. A linker that is too flexible will lead to effector domain action independent of RNA binding or impart protein–protein interaction, while on the other hand a rigid inter-domain linker might affect structural and functional integrity by blocking independent domain movements (Bhaskara et al. 2012). However, designing the linker is a major challenge due to lack of prior knowledge of domain flexibility imparted by these inter domain structures. The inter-domain linkers of many RBPs can affect the orientation of tandem RBDs, and thus may determine how the target RNAs are recognized by individual RBD vs. multiple tandem RBDs. This was exemplified by the cooperative binding of RNA by proteins with multiple RRMs or KH domains (Beuth et al. 2005; Perez-Canadillas 2006). Typically an inter-domain linker has a length of 3–25 amino acid, and different length distributions were found by independent groups with different protein database (Bhaskara et al. 2012; George and Heringa 2002), suggesting the linker length may vary in different types of proteins. A simple solution applied in the past is to use a long flexible Gly-rich peptide, which is considered as an inert linker as it is unlikely to interfere with structures of other domain. A more sophistic design will be to test different natural linkers in engineered RBPs to determine how they can affect tandem RBDs in heterogeneous contexts, which may achieve the cooperative RNA-binding found in natural RBPs.

Another consideration is to design inter-domain liners that connect the RNA-binding module with the functional module. Traditionally little attention was paid in this issue because most tethering experiments use a longer flexible linker to connect the protein of interest with a small RNA binding domain (like MS2) that binds to a well-defined RNA structure. The engineered RBPs that modulate RNA processing (such as splicing, translation) may be able to use the linkers previously tested by tether experiments; however, a more careful design will be needed for artificial RNA enzymes that are more sensitive to the conformational orientation of individual domains. For example, we have used linker or linker(s) with mixed helical propensity in design of artificial RNA endonucleases (Choudhury et al. 2012). We found the selection of different linkers can affect the catalytic activity of this enzyme: a linker that is longer than necessary to connect two domains may be too flexible and thus could act as an energetic or structural nuisance to interfere activity (Argos 1990), whereas a short linker may generate a structural barrier that prevents simultaneous contact of the two domains with RNA. Based on a database of known linkers (George and Heringa 2002), we designed linkers with various lengths and structural propensities and inserted them between PUF and a nonspecific RNA endonuclease domain (PIN) (Choudhury et al. 2012). We found that the tripeptide linker had very low activity, whereas a considerably higher activity was observed with linker lengths of 7 aa or 12 aa. However, nonspecific cleavage became apparent with 12 aa linker that approach ~20 Å in length, probably due to excess flexibility that allowed the PIN endonuclease domain to recognize and cleave any RNA

(Choudhury et al. 2012). We found linkers with α-helical or helix-coil-helix structures produced the most active ASREs, whereas linkers with 3_{10} helix structures reduced the activity. These experiments presented a particular case for linker design, and we also assumed that the inter-domain linker in fusion protein would adopt similar structure to that of their native contexts. It will be interesting to see if such design principle is still valid for other engineered RBPs.

5 Applications of Engineered RBPs

The unique RNA-binding mode of the PUF domain makes it possible to engineer artificial RBPs that modulate various processes in RNA metabolism. Here we will discuss the engineered RBPs using PUF as an example of RNA binding scaffold, and the same modular design principle can be applied by fusing these functional domains to other RNA binding modules. Compared with other RBDs, the PUF domain has the advantage of having a fully determined modular binding code. Varying the choice of functional modules, an assortment of PUF factors has been engineered successfully with multiple applications since 2007, and more factors with novel activities are expected to be generated in the future.

5.1 Engineered RBPs as RNA Probes

The artificial PUF proteins were first used as a fluorescent marker to visualize RNAs in live cells (Ozawa et al. 2007). Traditional approach to visualize RNAs in live cells is to use a green fluorescent protein (GFP) fused to bacteriophage coat protein MS2 that recognizes reporter RNAs containing multiple copies (e.g., 24 copies) of MS2 hairpin sequences (Lionnet et al. 2011; Park et al. 2012) (reviewed in Querido and Chartrand (2008)). However, the GFP-MS2 fusion protein requires artificial RNA targets with MS2 hairpins. To visualize endogenous RNA, Ozawa et al. designed a split GFP system that comprises two fusion proteins, each containing a PUF domain and either an N-terminal or C-terminal fragment of GFP (Fig. 6.6a). When the PUF domains recognize the adjacent 8-nt sequences in the same RNA, the two fusion proteins are brought together so that the split GFP fragments assemble and fluoresce. The specificity of this system is increased by requiring binding to two 8-nt sites separated by a small spacer. This design has been used to visualize mitochondrial RNA in cultured mammalian cells (Ozawa et al. 2007), cytosolic β-actin mRNA (Yamada et al. 2011), and a modified version using intact GFP fused with two PUFs was also used to detect cytosolic β-actin mRNA (Yoshimura et al. 2012). A similar approach was also developed by Tilsner et al., who combined PUF domains with split mCitrine and used it to reveal plant viral RNA localization (Tilsner et al. 2009). Another variation of this method used PUF domains fused with split luciferase to detect ssRNA in vitro (Furman et al. 2010); however, this approach has not yet been tested in live cells.

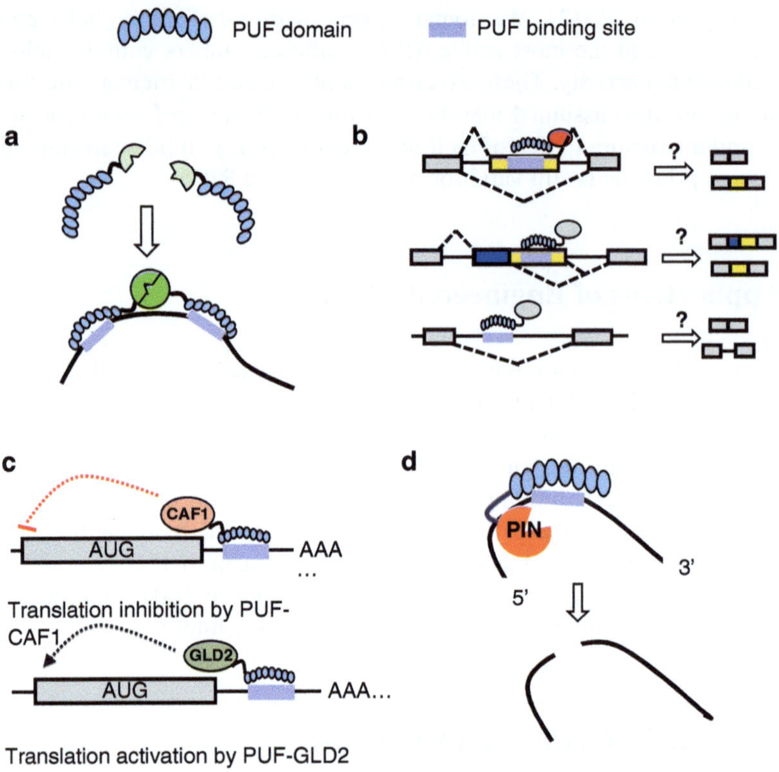

Fig. 6.6 Engineered PUF factors. (**a**) Fluorescent probe for in vivo RNA labeling. Combination of split GFP (or other fluorescent protein) with PUF scaffold generated an RNA probe to visualize RNA in live cells. (**b**) Engineered splicing factors. Combination of a PUF scaffold with a splicing regulatory domain (e.g., RS domain or a Gly-rich domain) generated splicing factors that control various types of alternative splicing events. (**c**) Modulation of translation. Fusion of GLD2 or CAF1 with PUF domain produced novel factors that can activate or inhibit mRNA translation. (**d**) Artificial site-specific RNA endonuclease. Combination of PUF domain with a nonspecific RNA endonuclease (PIN domain) can produce a new class of enzymes that specifically recognize and cleave RNA

The use of split fluorescent protein reduces the noise level of RNA detection, because the fluorescent protein is assembled by the RNA target. However, because of the limited efficiency in the co-folding of two GFP fragments and only one GFP per RNA, this system produces lower fluorescent signal compared to the GFP-MS2. On the other hand, the PUF-split GFP can detect a single endogenous RNA target, whereas the GFP-MS2 requires a reporter RNA with multiple MS2 target sites to produce a higher signal over the diffuse background noise of unbound fluorescent protein. For some special cases, it should be possible to use the GFP-PUF fusion protein directly. Like the GFP-MS2 fusion protein, the GFP-PUF will produce a weak fluorescence background. However, if the target contains multiple copies of the recognition site, the GFP-PUF may form bright foci on the target RNA, like the

GFP-MS2 protein forms a bright spot on reporter RNAs with consecutive MS2 hairpins (Park et al. 2012). Several trinucleotide expansions have been found to cause neurodegenerative diseases (such as CAG repeats in Huntington disease and CUG repeats in myotonic dystrophy, reviewed in Galka-Marciniak et al. (2012)). These pathogenic RNA repeats usually contain hundreds of copies of the trinucleotide, thus a GFP-PUF recognizing these repeated sequences may provide a new way to visualize the RNA and study the dynamics of such pathogenic RNA in live cells.

5.2 *Engineered RBPs as Splicing Factors*

Most human genes undergo alternative splicing to produce multiple isoforms with distinct activities. This process is tightly regulated, and the misregulation of splicing is a common cause of human diseases (Singh and Cooper 2012). The specific manipulation of alternative splicing will provide a useful way to fine-tune gene function. Many known splicing factors have modular activities in which they recognize a short RNA element in the target with an RNA-binding module (e.g., RRM or KH domain) and alter splicing with a functional module. The best-known functional domains for splicing modulation are the RS domains and Gly-rich domains that can promote or inhibit exon inclusion when recruited to an alternative exon (Wang and Burge 2008).

By fusing a PUF domain with Arg/Ser-rich (RS) domains of SRSF1 or the glycine-rich (Gly) domain of hnRNP A1, Wang et al. developed engineered splicing factors (ESFs) (Fig. 6.6b) (Wang et al. 2009). The resulting ESFs can function either as splicing activators (PUF-RS) or inhibitors (PUF-Gly), and can specifically control different types of alternative splicing. Specifically, PUF-RS promotes exon inclusion when binding to an alternative exon, but inhibits splicing when binding downstream of an alternative exon. PUF-Gly, on the other hand, inhibits exon inclusion when binding to the target RNA either inside or downstream of the alternative exon (Wang et al. 2009, 2013b). When binding to the region between two alternative 5′ or 3′ splice sites, PUF-RS promotes use of the proximal splice site to favor the longer splicing isoform, whereas PUF-Gly inhibits proximal site usage to induce the shorter isoform (Wang et al. 2009). Using this general approach, a PUF-Gly ESF was generated to shift splicing of the Bcl-x pre-mRNA from the antiapoptotic long isoform (Bcl-xL) to the pro-apoptotic short isoform (Bcl-xS). The altered splice isoform distribution was sufficient to sensitize several cancer cell lines to multiple anticancer drugs (Wang et al. 2009). With the identification of the C code, both PUF-RS and PUF-Gly ESFs were engineered to shift splicing of VEGF-A pre-mRNA from the angiogenic to antiangiogenic isoform (Dong et al. 2011).

In addition to the RS domains or Gly-rich domains, other domains in splicing regulatory factors may have distinct activities in regulating splicing. We found that the C-terminal domain of several SR proteins can activate or inhibit splicing when binding to different pre-mRNA regions (Wang et al. 2012a), the alanine-rich motif of RBM4 can inhibit splicing, and the proline-rich motif of DAZAP1 can enhance splicing (Wang et al. 2012b). These new functional domains can be used to derive additional types of artificial splicing factors to fine-tune alterative splicing (Fig. 6.6b).

5.3 Engineered RBPs to Control Translation

Since many RNA processing pathways are controlled by protein factors with separable RNA recognition and functional domains, a common method to study the function of RNA regulatory proteins is to tether the protein or a fragment to a specific RNA target and analyze the consequences. Using this engineering principle, artificial regulatory factors can be derived by combining a PUF domain with different effector domains. An elegant use of the modular configuration for RNA regulation was developed by Cooke et al. to alter translation (Cooke et al. 2011). They attached a PUF domain to either a translational activator, GLD2, or a translational repressor, CAF1 (Fig. 6.6c). The resulting PUF-GLD2 engineered proteins specifically recognized their RNA target and activated translation and induced polyadenylation in *Xenopus* oocytes, whereas the PUF-CAF1 fusion protein repressed translation and directed deadenylation. The coordinate regulation of translation and/or stability of mRNA by PUF proteins and miRNAs (Kedde et al. 2010; Leibovich et al. 2010; Miles et al. 2012; Nolde et al. 2007) suggests that engineered PUF proteins may be designed to use as antagonizers or enhancers of miRNA regulation.

5.4 Engineered RNA Endonucleases

Specific cleavage of RNAs is critical for in vitro manipulation of RNA and for in vivo gene silencing. However, a simple enzyme that cleaves RNA in a sequence-specific manner has not been found in nature despite extensive investigations. Most RNA endonucleases either have limited sequence specificity (e.g., RNase A or RNase T1 (Marshall et al. 2008; Yoshida 2001; Beintema and Kleineidam 1998)) or recognize their targets by specific structures (e.g., tRNA splicing endonuclease (Calvin and Li 2008)) or through guide RNA that pairs with target (e.g., Argonaute proteins (Parker 2010)). By fusing a PUF domain with a nonspecific endoribonuclease domain (PIN domain of Smg6p), Choudhury et al. engineered artificial site-specific RNA endonucleases (ASREs) that specifically recognize RNA substrates and efficiently cleave near the binding sites both in vitro and in cultured cells (Fig. 6.6d) (Choudhury et al. 2012). The digested products have 5′-phosphate and 3′-hydroxyl groups, making it possible to religate. Two ASREs were designed to silence specifically the expression of a bacterial gene or human mitochondrial mRNAs that contain one or two binding sites of designer PUFs. Since PUF domains recognize their targets through an 8-nt sequence, comparable length to the seed match of siRNA, engineered ASREs may serve as an RNA silencing tool complementary to RNAi, which will be effective in organisms or cellular compartments where RNAi machinery is not present.

6 Future Applications of PUF-Based Designer Factors

Mammalian gene expression contains multiple RNA processing stages including RNA splicing, polyadenylation, editing, translocation, translation, and degradation, which are under tight control by a variety of RNA-binding proteins. The reprogrammable RNA-binding scaffold of the PUF domain makes it possible, and tempting, to create artificial factors that can specifically modulate target RNAs at each processing step. A benefit of manipulating gene expression at the RNA level rather than the DNA level is that the effect is often nonpermanent and reversible. The factors can be engineered by fusing designer PUF domains with a protein factor (or its functional domain) with known regulatory activity on RNA metabolism. In theory, any protein that has demonstrated activity in a tethering experiment can be fused with PUF domains to produce novel artificial factors that target endogenous RNA targets. Minimal functional domains can be determined by fusing fragments of known regulatory factors to PUF domains and testing activity. Alternatively, new functional modules can be identified by constructing a cDNA library of PUF domain fusions and screening for RNA regulation. A previous review has proposed several ideas for potential applications of engineered RNA-binding proteins (Mackay et al. 2011), some of which have been achieved successfully using the PUF scaffold (Fig. 6.6). We propose other potential applications using the engineered PUF factors in humans, and similar applications can be adopted to other model organisms.

6.1 Engineering RNA Editing Enzymes

RNA editing is commonly found in many eukaryotes to generate specific sequence substitutions and changes in gene expression levels, resulting in increase of RNA and protein diversity. Adenosine-to-inosine (A-to-I) editing represents the most important class of RNA editing in human and affects function of many genes, especially in the central nervous system (recently reviewed in Nishikura (2010)). The native substrate for A-to-I editing is a double-stranded RNA (dsRNA) region, recognized by the adenosine deaminase acting on RNA (ADAR). Canonical ADAR proteins have a conserved catalytic deaminase domain and one or more dsRNA-binding domains (dsRBDs); however, there is no report on whether these domains function in a modular fashion (Nishikura 2010). Thus one can replace the dsRBDs with a PUF domain and test whether the resulting fusion factors have deaminase activities against single-stranded RNA (Fig. 6.7a). With optimization, the engineered enzyme may specifically recognize an RNA target by sequence and deaminate a nearby adenosine.

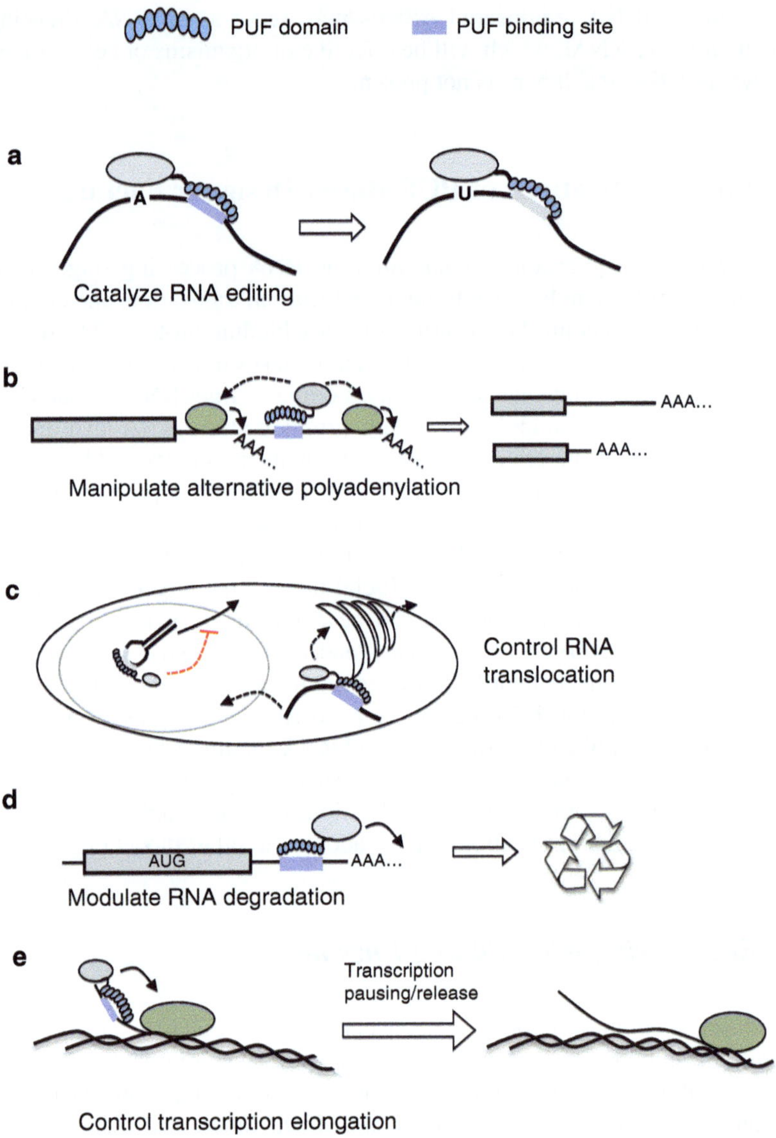

Fig. 6.7 Future engineered factors containing PUF scaffold. (**a**) Novel enzyme for single-stranded RNA editing. (**b**) Engineered factors to manipulate alternative polyadenylation. By combining a PUF domain with regulatory proteins/domains for alternative polyadenylation sites, artificial factors may be constructed to inhibit or activate certain polyadenylation sites and thus change the 3′ UTR of mRNA. (**c**) Engineered RNA transporters. New proteins can be engineered by combining a PUF domain with protein translocation signals to transport RNA into different cellular compartments. (**d**) Engineered factors to control RNA degradation. The PUF domain can be linked with additional domains to specifically recruit RNA stabilization or destabilization proteins, thus controlling RNA half-life. (**e**) Engineered factors to control transcription elongation. By specifically recognizing the nascent RNA transcript, new PUF factors can be engineered to induce transcription pausing or release paused RNA polymerases

6.2 Engineered Factors to Control Alternative Polyadenylation

Many human genes contain multiple polyadenylation sites. The use of different polyadenylation sites creates mRNA isoforms with different 3′ UTRs that regulate the translation efficiency and stability of mRNA. Like alternative splicing, control of alternative polyadenylation can have multifaceted effects on biological processes, and shortening of the 3′ UTR through proximal polyadenylation site usage is closely regulated in cell proliferation and differentiation (Ji and Tian 2009; Sandberg et al. 2008; Thomsen et al. 2010). Thus reprogramming the 3′ UTR region by modulating polyadenylation site selection may provide another way to manipulate gene expression. Several recent reports suggested that alternative polyadenylation is regulated by *trans*-acting factors that recognize *cis*-elements near polyadenylation sites (Hall-Pogar et al. 2007; Newnham et al. 2010; Peterson et al. 2006). Some of these factors can control both splicing and polyadenylation (e.g., U1A, PTB) (Hall-Pogar et al. 2007; Peterson et al. 2006). Therefore, we expect engineered PUF fusion proteins to be similarly successful in modulating the use of alternative polyadenylation sites (Fig. 6.7b).

6.3 Engineered Factors to Control RNA Localization

The transport of RNA to correct intracellular locations is an essential step for the expression of both coding and noncoding messages, as the function of RNA is highly localized. While there are many known RNA elements that govern their transport and final localization, other recognition elements are unknown and the detailed mechanisms are unclear. On the other hand, protein localization signals are better understood and well-characterized signal peptides or fragments have been established that are responsible for protein localization in nuclear or cytoplasmic compartments or for protein secretion. Therefore, by combining these established protein localization signals with PUF domains, one can devise engineered factors to specifically transport RNA (Fig. 6.7c). For example, a PUF domain fused with a nuclear localization sequence (NLS) may be created to block nuclear-to-cytoplasmic RNA transport. Such engineered proteins can be designed to recognize the loop-region of a specific pre-miRNA, thus retaining the pre-miRNA in the nucleus and preventing its processing to mature form in the cytoplasm, thereby inhibiting the miRNA function.

6.4 Engineered PUF Factors to Affect RNA Stability

Turnover of mRNA is continuous in the cytoplasm and often regulated by protein factors that recognize the 3′ UTR region. A well-studied case is directed by AU-rich elements in the 3′ UTR of short-lived mRNA, which destabilize the messages (reviewed in von Roretz et al. (2011)). Several factors positively or negatively regulate this process, and PUF fusion proteins with fragments of these factors may function as engineered factors to modulate RNA stability (Fig. 6.7d).

6.5 Engineered PUF Transcriptional Regulators

RNA transcription can be controlled at the initiation-to-elongation switch of RNA polymerase II through a transcriptional pausing step after the initiation and synthesis of a small RNA fragment (reviewed recently in Adelman and Lis (2012)). Such pausing is released to continue transcription, which is often controlled by transcription elongation factors. Using PUF engineered factors that recognize the nascent transcripts, it may be possible to design factors that specifically control transcription at the initiation-to-elongation switch by promoting or releasing the paused RNA polymerase II (Fig. 6.7e). For example, the release of paused RNA polymerase II can be mediated by CDK9 and Cyclin T1, which form a tight complex to promote transcription elongation (Larochelle et al. 2012). It may be possible to recruit CDK9 through a PUF-CDK9 fusion protein to the nascent transcripts, which in turn releases paused polymerase II and activates transcription elongation.

7 Additional Challenges for Engineered FUF Factors

7.1 Expanding Binding Specificity of PUF

Most naturally occurring PUF proteins have eight RNA-binding repeats that recognize 8 bases in target mRNA. Some PUF domains can accommodate one or two extra bases; however, the specificity of the extra bases cannot be reprogrammed. The sequence specificity of engineered PUF factors is largely determined by an 8-nt PUF binding site, which is comparable to the specificity of miRNA that recognize targets by 7-nt seed match. However, like miRNA, designer PUF factors may have off-target effects, since on average any given 8-nt sequence will occur by chance in sequences with 4^8 bases or 64 kb in length.

Nevertheless it is possible to increase the specificity of engineered PUF factors. One obvious approach is to use two PUF domains in combination, which in theory provides a recognition site of 16 nts. This design was used first for the visualization of RNA in vivo with two fusion proteins of PUF domains and split GFP fragments (Ozawa et al. 2007) and later was used by creating a fusion protein containing two PUF domains separated by an intact GFP protein (Yoshimura et al. 2012). Specificity for a 16-nt sequence should ensure the recognition of a unique target in the human transcriptome. However, engineered factors with tandem PUF domains may recognize an 8-nt "half-site" with sufficient affinity for functional activity. An alternative approach is to construct an expanded PUF domain by inserting more PUF repeats into a native PUF domain. For example, a PUF domain with 16 repeats was demonstrated to bind cognate RNA target in a yeast three-hybrid assay (Filipovska et al. 2011), and it should be possible to construct PUF domains with different numbers of repeats using similar design. We have generated expanded PUF domains with 9, 10, or 12 PUF repeats and found that these novel PUF scaffolds can indeed recognize cognate targets (Zhang and Wang, unpublished results).

While most applications will benefit from a PUF domain with increased specificity, in certain cases the engineered PUF factors may need to recognize multiple diverse targets with a degenerate binding code (i.e., a decreased specificity). This requires generating PUF domains that recognize shorter target sequences. Two strategies may solve this problem. One is to identify amino acid side chain combinations in PUF repeats that can tolerate multiple or all bases to produce a binding code for "N." This degenerate code can be used in combination with the specific base recognition code to make new PUF domains that specifically recognize sequences shorter than 8-nt while maintaining binding affinity. We suggest it may be possible to screen for a degenerate code of PUF-RNA interaction. Another approach is to construct a PUF scaffold with fewer than 8 PUF repeats. The minimum number of repeats in a stable PUF scaffold is unknown, but we speculate that a minimal number may be required for correct folding. Using a PUF scaffold with fewer repeats also decreases the size of the engineered protein, which may be beneficial for in vivo applications and simplify the design.

7.2 Minimizing Off-Target Effects

The control of off-target effects is an important concern for any method that specifically targets endogenous gene expression, especially when the engineered PUF factors are used in vivo. In addition to using a PUF scaffold with higher specificity (e.g., a scaffold containing additional repeats), careful experimental design should minimize off-target effects.

The first opportunity to minimize off-target effects is in the design stage by identifying a target sequence with fewer off-target hits in unrelated RNAs. Since the sequence of the human genome is not random, some sequence patterns are more common than others. When selecting PUF recognition sites in RNA targets, it will be helpful to search the transcriptome and choose relatively rare sites. While off-target effects are impossible to eliminate, such a search will estimate the frequency of possible off-target sites. Another concern in selecting target sequences is evaluating how RNA structure may affect PUF binding. While the PUF scaffold can recognize an RNA target sequence in a double-stranded region (Filipovska et al. 2011; Kedde et al. 2010) (Wang, unpublished result), the occurrence of extensive hairpin structures in the binding site greatly decreases binding affinity (Wang, unpublished result). Therefore, we should select target sites in less structured regions, whereas possible off-target sites in more structured regions may not pose a problem.

Another method to reduce off-target effects is to control the expression of the engineered PUF factors by expressing them at the same time and place as their targets. For example, by designing an ASRE with mitochondrial targeting signals to silence mitochondrial RNA, we produced a protein with low off-target effects against nuclear or cytoplasmic RNA, as the ASRE is undetectable outside mitochondria (Choudhury et al. 2012). A wide array of gene expression tools have been developed over the years to fine-tune expression of exogenous proteins (such as cell type specific vectors). They should be helpful tools to fine-tune the expression pattern of the engineered PUF factors.

7.3 In Vivo Delivery of Engineered PUF Factors

Since engineered PUF factors are typical proteins, they should be deliverable to live animals using established gene therapy vectors. For application in cultured cells or animal models, the lentiviral system is a good choice for ease of use and robust expression in most cell types. For therapeutic applications in humans, adeno-associated virus (AAV) is a favorable choice in gene therapy, because it lacks known pathogenicity, can infect non-dividing cells, and has many serotypes that allow specific gene delivery into different tissues. The AAV can stably integrate into the host cell genome at a specific site (designated AAVS1) in human chromosome 19 (Kotin et al. 1990; Surosky et al. 1997); however, the non-integrating AAV vectors are used for all gene therapy applications. The drawback of AAV is its small packing capacity, but the current engineered PUF factors are small enough to be packed by AAV.

Several nonviral methods may be used to deliver DNAs or proteins into animals, and some of these methods are under testing for use in humans. These methods include different nanoparticles and cell penetration peptides, and more in-depth information on in vivo delivery methods can be found in related reviews (Li et al. 2012; Elzoghby et al. 2012; Koren and Torchilin 2012). These methods can be applied, at least in principle, for in vivo delivery of the engineered PUF factors. Since the application of engineered PUF factors is still in early stages, we expect the technology for protein delivery to advance ahead of the in vivo application of these factors in live animals or humans.

8 Final Comments

Nature has provided simple design rule for multi-domain proteins and a large variety of domains that either bind RNA or affect different processes of RNA biology. Thus far engineered factors combining a PUF scaffold with a functional module have proven to be versatile tools for manipulating and understanding RNA metabolism. The advances in structural biology and biochemistry allowed design for general RNA-binding scaffold to recognize any RNA sequence. We believe that the lesson learned from engineered PUF factors can be extended to other types of RBDs, and thus expanding the potential to apply this technology to regulate RNA. Starting with the natural design, continued adherence to the rule of simplicity may be prudent in moving this exciting field forward.

References

Adelman K, Lis JT (2012) Promoter-proximal pausing of RNA polymerase II: emerging roles in metazoans. Nat Rev Genet 13(10):720–731

Argos P (1990) An investigation of oligopeptides linking domains in protein tertiary structures and possible candidates for general gene fusion. J Mol Biol 211(4):943–958

Auweter SD, Oberstrass FC, Allain FH (2006a) Sequence-specific binding of single-stranded RNA: is there a code for recognition? Nucleic Acids Res 34(17):4943–4959

Auweter SD, Fasan R, Reymond L, Underwood JG, Black DL, Pitsch S et al (2006b) Molecular basis of RNA recognition by the human alternative splicing factor Fox-1. EMBO J 25(1):163–173, PMCID: 1356361

Barkan A, Rojas M, Fujii S, Yap A, Chong YS, Bond CS et al (2012) A combinatorial amino acid code for RNA recognition by pentatricopeptide repeat proteins. PLoS Genet 8(8):e1002910, PMCID: 3420917

Barker DD, Wang C, Moore J, Dickinson LK, Lehmann R (1992) Pumilio is essential for function but not for distribution of the Drosophila abdominal determinant Nanos. Genes Dev 6(12A): 2312–2326

Beintema JJ, Kleineidam RG (1998) The ribonuclease A superfamily: general discussion. Cell Mol Life Sci 54(8):825–832

Beuth B, Pennell S, Arnvig KB, Martin SR, Taylor IA (2005) Structure of a Mycobacterium tuberculosis NusA-RNA complex. EMBO J 24(20):3576–3587, PMCID: 1276712

Bhaskara RM, de Brevern AG, Srinivasan N (2012) Understanding the role of domain-domain linkers in the spatial orientation of domains in multi-domain proteins. J Biomol Struct Dyn 31(12):1467–1480

Birney E, Kumar S, Krainer AR (1993) Analysis of the RNA-recognition motif and RS and RGG domains: conservation in metazoan pre-mRNA splicing factors. Nucleic Acids Res 21(25): 5803–5816

Blancafort P, Beltran AS (2008) Rational design, selection and specificity of artificial transcription factors (ATFs): the influence of chromatin in target gene regulation. Comb Chem High Throughput Screen 11(2):146–158

Blancafort P, Segal DJ, Barbas CF III (2004) Designing transcription factor architectures for drug discovery. Mol Pharmacol 66(6):1361–1371

Blewett NH, Goldstrohm AC (2012) A eukaryotic translation initiation factor 4E-binding protein promotes mRNA decapping and is required for PUF repression. Mol Cell Biol 32(20): 4181–4194, PMCID: 3457345

Braddock DT, Louis JM, Baber JL, Levens D, Clore GM (2002) Structure and dynamics of KH domains from FBP bound to single-stranded DNA. Nature 415(6875):1051–1056

Burd CG, Matunis EL, Dreyfuss G (1991) The multiple RNA-binding domains of the mRNA poly(A)-binding protein have different RNA-binding activities. Mol Cell Biol 11(7): 3419–3424, PMCID: 361068

Caceres JF, Krainer AR (1993) Functional analysis of pre-mRNA splicing factor SF2/ASF structural domains. EMBO J 12(12):4715–4726

Calvin K, Li H (2008) RNA-splicing endonuclease structure and function. Cell Mol Life Sci 65(7–8):1176–1185

Chen Y, Varani G (2013) Engineering RNA-binding proteins for biology. FEBS J 280(16):3734–3754

Cheong CG, Hall TMT (2006) Engineering RNA sequence specificity of Pumilio repeats. Proc Natl Acad Sci U S A 103(37):13635–13639

Choudhury R, Tsai YS, Dominguez D, Wang Y, Wang Z (2012) Engineering RNA endonucleases with customized sequence specificities. Nat Commun 3:1147

Cooke A, Prigge A, Opperman L, Wickens M (2011) Targeted translational regulation using the PUF protein family scaffold. Proc Natl Acad Sci U S A 108(38):15870–15875

Crick F (1970) Central dogma of molecular biology. Nature 227(5258):561–563

Daubner GM, Clery A, Allain FH (2013) RRM-RNA recognition: NMR or crystallography…and new findings. Curr Opin Struct Biol 23(1):100–108

Delannoy E, Stanley WA, Bond CS, Small ID (2007) Pentatricopeptide repeat (PPR) proteins as sequence-specificity factors in post-transcriptional processes in organelles. Biochem Soc Trans 35(Pt 6):1643–1647

Dominguez C, Fisette JF, Chabot B, Allain FH (2010) Structural basis of G-tract recognition and encaging by hnRNP F quasi-RRMs. Nat Struct Mol Biol 17(7):853–861

Dong S, Wang Y, Cassidy-Amstutz C, Lu G, Bigler R, Jezyk MR et al (2011) Specific and modular binding code for cytosine recognition in Pumilio/FBF (PUF) RNA-binding domains. J Biol Chem 286(30):26732–26742

Elzoghby AO, Samy WM, Elgindy NA (2012) Protein-based nanocarriers as promising drug and gene delivery systems. J Control Release 161(1):38–49

Filipovska A, Razif MF, Nygard KK, Rackham O (2011) A universal code for RNA recognition by PUF proteins. Nat Chem Biol 7(7):425–427

Furman JL, Badran AH, Ajulo O, Porter JR, Stains CI, Segal DJ et al (2010) Toward a general approach for RNA-templated hierarchical assembly of split-proteins. J Am Chem Soc 132(33):11692–11701

Galka-Marciniak P, Urbanek MO, Krzyzosiak WJ (2012) Triplet repeats in transcripts: structural insights into RNA toxicity. Biol Chem 393(11):1299–1315

George RA, Heringa J (2002) An analysis of protein domain linkers: their classification and role in protein folding. Protein Eng 15(11):871–879

Goldstrohm AC, Hook BA, Seay DJ, Wickens M (2006) PUF proteins bind Pop2p to regulate messenger RNAs. Nat Struct Mol Biol 13(6):533–539

Goldstrohm AC, Seay DJ, Hook BA, Wickens M (2007) PUF protein-mediated deadenylation is catalyzed by Ccr4p. J Biol Chem 282(1):109–114

Greisman HA, Pabo CO (1997) A general strategy for selecting high-affinity zinc finger proteins for diverse DNA target sites. Science 275(5300):657–661

Grishin NV (2001) KH domain: one motif, two folds. Nucleic Acids Res 29(3):638–643, PMCID: 30387

Hall-Pogar T, Liang S, Hague LK, Lutz CS (2007) Specific trans-acting proteins interact with auxiliary RNA polyadenylation elements in the COX-2 3′-UTR. RNA 13(7):1103–1115

Howard MJ, Lim WH, Fierke CA, Koutmos M (2012) Mitochondrial ribonuclease P structure provides insight into the evolution of catalytic strategies for precursor-tRNA 5′ processing. Proc Natl Acad Sci U S A 109(40):16149–16154, PMCID: 3479547

Jamieson AC, Miller JC, Pabo CO (2003) Drug discovery with engineered zinc-finger proteins. Nat Rev Drug Discov 2(5):361–368

Ji Z, Tian B (2009) Reprogramming of 3' untranslated regions of mRNAs by alternative polyadenylation in generation of pluripotent stem cells from different cell types. PLoS One 4(12):e8419

Kedde M, van Kouwenhove M, Zwart W, Oude Vrielink JA, Elkon R, Agami R (2010) A Pumilio-induced RNA structure switch in p27-3′ UTR controls miR-221 and miR-222 accessibility. Nat Cell Biol 12(10):1014–1020

Kim Y, Kweon J, Kim A, Chon JK, Yoo JY, Kim HJ et al (2013) A library of TAL effector nucleases spanning the human genome. Nat Biotechnol 31(3):251–258

Klug A (2010) The discovery of zinc fingers and their applications in gene regulation and genome manipulation. Annu Rev Biochem 79:213–231

Kobayashi K, Kawabata M, Hisano K, Kazama T, Matsuoka K, Sugita M et al (2012) Identification and characterization of the RNA binding surface of the pentatricopeptide repeat protein. Nucleic Acids Res 40(6):2712–2723, PMCID: 3315335

Koren E, Torchilin VP (2012) Cell-penetrating peptides: breaking through to the other side. Trends Mol Med 18(7):385–393

Kotin RM, Siniscalco M, Samulski RJ, Zhu XD, Hunter L, Laughlin CA et al (1990) Site-specific integration by adeno-associated virus. Proc Natl Acad Sci U S A 87(6):2211–2215, PMCID: 53656

Larochelle S, Amat R, Glover-Cutter K, Sanso M, Zhang C, Allen JJ et al (2012) Cyclin-dependent kinase control of the initiation-to-elongation switch of RNA polymerase II. Nat Struct Mol Biol 19(11):1108–1115

Leibovich L, Mandel-Gutfreund Y, Yakhini Z (2010) A structural-based statistical approach suggests a cooperative activity of PUM1 and miR-410 in human 3′-untranslated regions. Silence 1(1):17

Li Y, Wang J, Satterle A, Wu Q, Liu F (2012) Gene transfer to skeletal muscle by site-specific delivery of electroporation and ultrasound. Biochem Biophys Res Commun 424(2):203–207

Lionnet T, Czaplinski K, Darzacq X, Shav-Tal Y, Wells AL, Chao JA et al (2011) A transgenic mouse for in vivo detection of endogenous labeled mRNA. Nat Methods 8(2):165–170

Lu G, Hall TMT (2011) Alternate modes of cognate RNA recognition by human PUMILIO proteins. Structure 19(3):361–367

Lunde BM, Moore C, Varani G (2007) RNA-binding proteins: modular design for efficient function. Nat Rev Mol Cell Biol 8(6):479–490

Macdonald PM (1992) The Drosophila pumilio gene: an unusually long transcription unit and an unusual protein. Development 114(1):221–232

Mackay JP, Font J, Segal DJ (2011) The prospects for designer single-stranded RNA-binding proteins. Nat Struct Mol Biol 18(3):256–261

Maris C, Dominguez C, Allain FH (2005) The RNA recognition motif, a plastic RNA-binding platform to regulate post-transcriptional gene expression. FEBS J 272(9):2118–2131

Marshall GR, Feng JA, Kuster DJ (2008) Back to the future: ribonuclease A. Biopolymers 90(3):259–277

Miles WO, Tschop K, Herr A, Ji JY, Dyson NJ (2012) Pumilio facilitates miRNA regulation of the E2F3 oncogene. Genes Dev 26(4):356–368

Miller JC, Tan S, Qiao G, Barlow KA, Wang J, Xia DF et al (2011) A TALE nuclease architecture for efficient genome editing. Nat Biotechnol 29(2):143–148

Newnham CM, Hall-Pogar T, Liang S, Wu J, Tian B, Hu J et al (2010) Alternative polyadenylation of MeCP2: influence of cis-acting elements and trans-acting factors. RNA Biol 7(3):361–372

Nishikura K (2010) Functions and regulation of RNA editing by ADAR deaminases. Annu Rev Biochem 79:321–349

Nolde MJ, Saka N, Reinert KL, Slack FJ (2007) The Caenorhabditis elegans pumilio homolog, puf-9, is required for the 3′UTR-mediated repression of the let-7 microRNA target gene, hbl-1. Dev Biol 305(2):551–563

Oberstrass FC, Auweter SD, Erat M, Hargous Y, Henning A, Wenter P et al (2005) Structure of PTB bound to RNA: specific binding and implications for splicing regulation. Science 309(5743):2054–2057

Ozawa T, Natori Y, Sato M, Umezawa Y (2007) Imaging dynamics of endogenous mitochondrial RNA in single living cells. Nat Methods 4(5):413–419

Park HY, Trcek T, Wells AL, Chao JA, Singer RH (2012) An unbiased analysis method to quantify mRNA localization reveals its correlation with cell motility. Cell Rep 1(2):179–184

Parker JS (2010) How to slice: snapshots of Argonaute in action. Silence 1(1):3, PMCID: 2835997

Perez-Canadillas JM (2006) Grabbing the message: structural basis of mRNA 3′UTR recognition by Hrp1. EMBO J 25(13):3167–3178, PMCID: 1500993

Perez-Pinera P, Ousterout DG, Gersbach CA (2012) Advances in targeted genome editing. Curr Opin Chem Biol 16(3–4):268–277, PMCID: 3424393

Peterson ML, Bingham GL, Cowan C (2006) Multiple features contribute to the use of the immunoglobulin M secretion-specific poly(A) signal but are not required for developmental regulation. Mol Cell Biol 26(18):6762–6771

Pomerantz JL, Sharp PA, Pabo CO (1995) Structure-based design of transcription factors. Science 267(5194):93–96

Querido E, Chartrand P (2008) Using fluorescent proteins to study mRNA trafficking in living cells. Methods Cell Biol 85:273–292

Rackham O, Filipovska A (2012) The role of mammalian PPR domain proteins in the regulation of mitochondrial gene expression. Biochim Biophys Acta 1819(9–10):1008–1016

Rebar EJ, Pabo CO (1994) Zinc finger phage: affinity selection of fingers with new DNA-binding specificities. Science 263(5147):671–673

Ringel R, Sologub M, Morozov YI, Litonin D, Cramer P, Temiakov D (2011) Structure of human mitochondrial RNA polymerase. Nature 478(7368):269–273

Safaee N, Kozlov G, Noronha AM, Xie J, Wilds CJ, Gehring K (2012) Interdomain allostery promotes assembly of the poly(A) mRNA complex with PABP and eIF4G. Mol Cell 48(3): 375–386

Saha D, Prasad AM, Srinivasan R (2007) Pentatricopeptide repeat proteins and their emerging roles in plants. Plant Physiol Biochem 45(8):521–534

Sandberg R, Neilson JR, Sarma A, Sharp PA, Burge CB (2008) Proliferating cells express mRNAs with shortened 3′ untranslated regions and fewer microRNA target sites. Science 320(5883): 1643–1647

Sepp A, Choo Y (2005) Cell-free selection of zinc finger DNA-binding proteins using in vitro compartmentalization. J Mol Biol 354(2):212–219

Shamoo Y, Abdul-Manan N, Williams KR (1995) Multiple RNA binding domains (RBDs) just don't add up. Nucleic Acids Res 23(5):725–728, PMCID: 306750

Sickmier EA, Frato KE, Shen H, Paranawithana SR, Green MR, Kielkopf CL (2006) Structural basis for polypyrimidine tract recognition by the essential pre-mRNA splicing factor U2AF65. Mol Cell 23(1):49–59, PMCID: 2043114

Singh RK, Cooper TA (2012) Pre-mRNA splicing in disease and therapeutics. Trends Mol Med 18(8):472–482

Small ID, Peeters N (2000) The PPR motif - a TPR-related motif prevalent in plant organellar proteins. Trends Biochem Sci 25(2):46–47

Surosky RT, Urabe M, Godwin SG, McQuiston SA, Kurtzman GJ, Ozawa K et al (1997) Adeno-associated virus Rep proteins target DNA sequences to a unique locus in the human genome. J Virol 71(10):7951–7959, PMCID: 192153

Takenaka M, Zehrmann A, Brennicke A, Graichen K (2013) Improved computational target site prediction for pentatricopeptide repeat RNA editing factors. PLoS One 8(6):e65343, PMCID: 3675099

Thomsen S, Azzam G, Kaschula R, Williams LS, Alonso CR (2010) Developmental RNA processing of 3'UTRs in Hox mRNAs as a context-dependent mechanism modulating visibility to microRNAs. Development 137(17):2951–2960

Tilsner J, Linnik O, Christensen NM, Bell K, Roberts IM, Lacomme C et al (2009) Live-cell imaging of viral RNA genomes using a Pumilio-based reporter. Plant J 57(4):758–770

Urnov FD, Rebar EJ, Holmes MC, Zhang HS, Gregory PD (2010) Genome editing with engineered zinc finger nucleases. Nat Rev Genet 11(9):636–646

Valverde R, Edwards L, Regan L (2008) Structure and function of KH domains. FEBS J 275(11): 2712–2726

Van Etten J, Schagat TL, Hrit J, Weidmann CA, Brumbaugh J, Coon JJ et al (2012) Human Pumilio proteins recruit multiple deadenylases to efficiently repress messenger RNAs. J Biol Chem 287(43):36370–36383, PMCID: 3476303

von Roretz C, Di Marco S, Mazroui R, Gallouzi IE (2011) Turnover of AU-rich-containing mRNAs during stress: a matter of survival. Wiley Interdiscip Rev RNA 2(3):336–347

Wang Z, Burge CB (2008) Splicing regulation: from a parts list of regulatory elements to an integrated splicing code. RNA 14(5):802–813, PMCID: 2327353

Wang X, Zamore PD, Hall TMT (2001) Crystal structure of a Pumilio homology domain. Mol Cell 7(4):855–865
Wang X, McLachlan J, Zamore PD, Hall TMT (2002) Modular recognition of RNA by a human pumilio-homology domain. Cell 110(4):501–512
Wang Y, Cheong CG, Hall TMT, Wang Z (2009) Engineering splicing factors with designed specificities. Nat Methods 6(11):825–830
Wang Y, Xiao X, Zhang J, Choudhury R, Robertson A, Li K et al (2012a) A complex network of factors with overlapping affinities represses splicing through intronic elements. Nat Struct Mol Biol 20(1):36–45, PMCID: 3537874
Wang Y, Ma M, Xiao X, Wang Z (2012b) Intronic splicing enhancers, cognate splicing factors and context-dependent regulation rules. Nat Struct Mol Biol 19(10):1044–1052
Wang Y, Wang Z, Tanaka Hall TM (2013) Engineered proteins with Pumilio/fem-3 mRNA binding factor scaffold to manipulate RNA metabolism. FEBS J 280(16):3755–3767
Weidmann CA, Goldstrohm AC (2012) Drosophila Pumilio protein contains multiple autonomous repression domains that regulate mRNAs independently of Nanos and brain tumor. Mol Cell Biol 32(2):527–540, PMCID: 3255780
Yamada T, Yoshimura H, Inaguma A, Ozawa T (2011) Visualization of nonengineered single mRNAs in living cells using genetically encoded fluorescent probes. Anal Chem 83(14): 5708–5714
Yoshida H (2001) The ribonuclease T1 family. Methods Enzymol 341:28–41
Yoshimura H, Inaguma A, Yamada T, Ozawa T (2012) Fluorescent probes for imaging endogenous beta-actin mRNA in living cells using fluorescent protein-tagged pumilio. ACS Chem Biol 7(6):999–1005
Zamore PD, Williamson JR, Lehmann R (1997) The Pumilio protein binds RNA through a conserved domain that defines a new class of RNA-binding proteins. RNA 3(12):1421–1433, PMCID: 1369583
Zhang B, Gallegos M, Puoti A, Durkin E, Fields S, Kimble J et al (1997) A conserved RNA-binding protein that regulates sexual fates in the C elegans hermaphrodite germ line. Nature 390(6659):477–484
Zhang F, Cong L, Lodato S, Kosuri S, Church GM, Arlotta P (2011) Efficient construction of sequence-specific TAL effectors for modulating mammalian transcription. Nat Biotechnol 29(2):149–153, PMCID: 3084533

Chapter 7
Genetic Variation and RNA Binding Proteins: Tools and Techniques to Detect Functional Polymorphisms

Rachel Soemedi, Hugo Vega, Judson M. Belmont, Sohini Ramachandran, and William G. Fairbrother

Abstract At its most fundamental level the goal of genetics is to connect genotype to phenotype. This question is asked at a basic level evaluating the role of genes and pathways in genetic model organism. Increasingly, this question is being asked in the clinic. Genomes of individuals and populations are being sequenced and compared. The challenge often comes at the stage of analysis. The variant positions are analyzed with the hope of understanding human disease. However after a genome or exome has been sequenced, the researcher is often deluged with hundreds of potentially relevant variations. Traditionally, amino-acid changing mutations were considered the tractable class of disease-causing mutations; however, mutations that disrupt noncoding elements are the subject of growing interest. These noncoding changes are a major avenue of disease (e.g., one in three hereditary disease alleles are predicted to affect splicing). Here, we review some current practices of medical

R. Soemedi
Center for Computational Molecular Biology, Brown University, Providence, RI, USA

Molecular and Cellular Biology and Biochemistry, Brown University, Providence, RI, USA

H. Vega
Molecular and Cellular Biology and Biochemistry, Brown University, Providence, RI, USA

NIH Undiagnosed Diseases Program, NIH Office of Rare Diseases Research and NHGRI, Bethesda, MD, USA

J.M. Belmont
Molecular and Cellular Biology and Biochemistry, Brown University, Providence, RI, USA

S. Ramachandran
Center for Computational Molecular Biology, Brown University, Providence, RI, USA

Ecology and Evolutionary Biology, Brown University, Providence, RI, USA

W.G. Fairbrother (✉)
Center for Computational Molecular Biology, Brown University, Providence, RI, USA

Molecular and Cellular Biology and Biochemistry, Brown University, Providence, RI, USA

Center for Biomedical Engineering, Brown University, Providence, RI, USA
e-mail: William_fairbrother@brown.edu

G.W. Yeo (ed.), *Systems Biology of RNA Binding Proteins*, Advances in Experimental Medicine and Biology 825, DOI 10.1007/978-1-4939-1221-6_7

genetics, the basic theory behind biochemical binding and functional assays, and then explore technical advances in how variations that alter RNA protein recognition events are detected and studied. These advances are advances in scale—high-throughput implementations of traditional biochemical assays that are feasible to perform in any molecular biology laboratory. This chapter utilizes a case study approach to illustrate some methods for analyzing polymorphisms. The first characterizes a functional intronic SNP that deletes a high affinity PTB site using traditional low-throughput biochemical and functional assays. From here we demonstrate the utility of high-throughput splicing and spliceosome assembly assays for screening large sets of SNPs and disease alleles for allelic differences in gene expression. Finally we perform three pilot drug screens with small molecules (G418, tetracycline, and valproic acid) that illustrate how compounds that rescue specific instances of differential pre-mRNA processing can be discovered.

Keywords Pre-mRNA splicing • Human genetic variation • Single nucleotide polymorphism (SNP) • RNA binding protein • Polypyrimidine track binding protein (PTB)

1 Introduction

1.1 Brief History of Medical Genetics

Until recently, hereditary diseases were analyzed by classical genetic methods. These approaches were remarkably similar to the original, unrecognized work of Gregor Mendel that was published in an obscure journal in 1865 where it lay dormant for 35 years before being rediscovered independently by multiple labs in 1900. An early medical application of the science of transmission of genetic traits was provided by the characterization of ABO blood groups by Landsteiner (Jorde et al. 1996). The modes of inheritance of diseases such as phenylketonuria, Huntington's disease, cystic fibrosis, and sickle-cell anemia were all understood before the recognition of DNA as the hereditary molecule. The degree to which our recognition of the chemical basis of hereditary lagged behind our understanding of genetic mechanisms of transmission is difficult to appreciate today. It was not until 1955 that the structure of DNA was reported; in 1956 the correct number of human chromosomes identified. The genetic code was deduced in the 1960s and in the 1970s the complex intron–exon architecture of genes was reported. The completion of the human genome sequencing project in 2003 marked the completion of an era of raw discovery and the development of a new genomic-scale in science that leverages this information (Lander et al. 2001). One type of experiment that leverages the existence of a reference genome is deeper sampling within the human population to study variation and reconstruct the events of recent human evolution (Sabeti et al. 2002; Reich et al. 2002; Sachidanandam et al. 2001).

1.2 Population Genetics: The Landscape of Variation in Humans

Following the publication of the human reference genome sequence, the promise of the genomic era has been to make personalized medicine—where diagnoses and therapies for patients are tailored towards their genome—a reality (Manolio et al. 2013). Mutations underlying Mendelian diseases have been identified based on patients presenting with a phenotype and the patients' pedigrees being examined to identify candidate loci for gene mapping experiments. Complex diseases (diseases influenced by multiple mutations of varying effect size such as heart disease and diabetes) also segregate in families. One hypothesis, The Common Disease/ Common Variant hypothesis predicts that genetic risk for common diseases will be due to disease-predisposing alleles with relatively high frequencies (e.g., (Lander 1996)). This hypothesis underlies the genome-wide association (GWA) study framework, where disease state (i.e., 0 implies healthy control, 1 implies patient for a disease of interest) is regressed onto the number of copies of a SNP allele and principal components representing genome-wide ancestry (Price et al. 2006) and SNPs explaining disease status are identified after correcting for the multiple tests carried out.

Over 1,600 GWA studies have been conducted since 2005 (see www.genome.gov/gwastudies), but, disappointingly, most variants identified in GWA studies confer small increments in disease risk (Manolio et al. 2009; Dickson et al. 2010). A further problem is that 96 % of individuals genotyped for GWA studies were of European descent (Bustamante et al. 2011). Since very few studies have been replicated in individuals of multiple ancestries (Casto and Feldman 2011), it is unknown whether risk alleles identified in those studies are relevant for the human population as a whole. As population geneticists turn from GWA studies with common variants (frequency >5 %) to studying the association of rare variants with complex diseases, a biased focus on individuals of European descent will limit the impact of association studies and their applicability to all humans. Rare variants tend to be population-specific, and a rare variant in one population is likely to have an even smaller frequency in another population (Gravel et al. 2011).

Much of what we know and will learn about genomic variation comes from the industrial scale sequencing of individual genomes, which are then compared to the reference sequence. This flood of data has ushered in renaissance for population genetics, association studies and created a new field of personalized medicine.

1.3 Personalized Genomics: The Variation Landscape in a Typical Human Being

High-throughput sequencing is performed by a rapidly changing suite of technologies. At the time of this writing, individual genomes are characterized by either whole genome sequencing (WGS) or by exome sequencing—a capture approach that utilizes a library of biotinylated capture probes designed to enrich exon reads(Gonzaga-Jauregui et al. 2012). Data summarized from the sequencing of individual genomes and the large scale pilot analysis from the 1,000 genome project illustrate several general features of genetic variation in humans(Abecasis et al. 2010, 2012). Although the degree of variation with respect to the reference genome varies with loci and ancestry, a single individual will have, in total, about three million positions of variation with respect to the reference sequence. While most of these variants are selectively neutral, some fractions of this total represent functional polymorphisms. As predicted by models of purifying selection, exonic variations and variations predicted to disrupt gene function are rare relative to neutral variants. However, an average human carries 50–100 variants classified by the Human Gene Mutation Database as causing a hereditary disease. Each individual carries 340–400 alleles that would be expected to result in a loss of function (e.g., nonsense mutations, splice site disruptions, frameshift mutations). On average, an individual would be a compound heterozygote for about five of these loss-of-function mutations. In addition, there are 10,000–11,000 non-synonymous and 10,000–12,000 synonymous variations that could alter protein or gene expression and cause disease or affect fitness. Frequently, genome wide association studies can be useful in identifying regions that contain causal disease alleles (GWAs). However because of the extended haplotype structure observed in the human population, a GWAs study does not necessarily identify the causal allele but rather, narrows the search to the GWAs SNPs and the several dozen neighboring SNPs in strong linkage disequilibrium with the GWAs SNP. The need to sort through a set of variants and prioritize them by likelihood of being functional is a very important problem in medical genetics.

Perhaps the most straightforward situation is the variant that can be reliably predicted to alter or eliminates the coding potential of an expressed gene. An individual is typically heterozygous for 67–100 SNPs that create nonsense mutations, 192–280 small frameshifting indels and several large deletions that encompass 33–49 genes (Abecasis et al. 2010, 2012). A great many more non-synonymous variations alter the coding region with a less certain effect on gene function. For synonymous variants and other transcript changes that occur outside the CDS, a biological effect could be mediated by an alteration in gene expression (e.g., a change in transcript level, stability, translatability or isoform usage). For these types of molecular phenotypes, it is likely that the variants effect is mediated by an allelic difference in the binding of a *trans*-acting factor. While about three quarters of the population variation in human transcript level is estimated to be explained by transcriptional differences, noncoding variants that function through an RNA-based mechanism like splicing have an additional potential to not only change the transcript level but to create new isoforms that are more deleterious than modest changes in transcript level. While splice altering variants make a minority of polymorphisms, they comprise more than 90 % of the noncoding hereditary disease-causing alleles reported in the Human Gene Mutation Database (HGMD) (Stenson et al. 2003). Typically the identification of such variants comes after a series of filters and annotations set in place to process deep sequencing data. Such an analysis pipeline is often limited to a set of candidate genes involved in biological processes relevant to the disease. Although one in three mutations are thought to have some effect on splicing, in most workflows only the most obvious class of RNA functional variations, the splice site mutations, are considered to be potential causal variants(Lim et al. 2011).

1.4 Specific Considerations for RNA Variations

In the next section, we discuss an overview of in silico screens for noncoding functional RNA variants that can be used to narrow a list of observed variants to a set of candidate causal SNPs. To validate the effect of variation on RNA–protein interactions, protein binding and function assays need to be applied to both allelic version of the region of interest. High-throughput implementations of these biochemical assays will be discussed alongside recent computational approaches of characterizing RNA-level variation. Both the computational and biochemical methods are largely adaptations of earlier techniques to characterize protein–DNA binding methods (Das and Dai 2007; Siddharthan 2007). Despite the obvious similarities between DNA and RNA, the different properties and biological roles of RNA create several very important distinctions.

1.4.1 Functional Noncoding Variants in DNA Disrupt Protein Binding but for RNA, Not Always

Variations that have some sort of noncoding biological impact that is mediated through DNA almost always involve the creation or disruption of DNA protein recognition events. For RNA, this is not necessary the case. There are noncoding RNAs where variations in RNA could directly affect a catalytic function. A variation could also alter the structure of an RNA, changing the access of a RBP to its binding site. This indirect mechanism could potentially alter binding some distance away from the position of variation. High-throughput binding assays that record the relative affinity of PTB on thousands of oligos demonstrate a preference for PTB binding oligos with the high affinity sites in open conformations (Reid et al. 2009). For RBP ligands, the additional parameter of structure along with sequence likely reduces the effectiveness of traditional motif finding and binding target prediction. Below we address this issue by presenting a computational framework for including structural and sequence information into binding models. There is some indication that siRNAs and miRNAs accessibility to their targets are also modulated by secondary structure (Long et al. 2007). There are online tools that predict secondary structure, explore its effect on access and the potential for sequence variations to alter secondary structure (see Table 7.1).

Table 7.1 Online tools to predict splicing elements

Tools	Description	URL
Program to evaluate canonical splice sites		
Spliceman (Lim and Fairbrother 2012)	Spliceman is used to predict the likelihood that distant mutations around annotated splice sites were to disrupt splicing based on the concept that splicing elements have signature positional distributions around constitutively spliced exons. Mutations that exhibit higher distances disrupted splicing, and those with smaller distances have no effect on splicing.	http://fairbrother.biomed.brown.edu/spliceman/index.cgi
MaxEntScan (Yeo and Burge 2004)	MaxEntScan is used to score exon–intron junctions by simultaneously accounts for non-adjacent and adjacent dependencies between positions. It is based on the "Maximum Entropy Principle" and generalizes probabilistic models of sequence motifs such as position weight matrix and inhomogeneous Markov models.	http://genes.mit.edu/burgelab/maxent/Xmaxentscan_scoreseq.html
ASSEDA (Goren et al. 2006)	Automated Splice Site And Exon Definition Analyses is used to predict the effects of mutations that cause aberrant splicing in human diseases. The program evaluates changes in splice site strength based on information theory-based models of donor and acceptor splice sites.	http://splice.uwo.ca
NetGene2 (Hebsgaard et al. 1996; Brunak et al. 1991)	The NetGene2 server provides neural network splice site predictions in human, *C. elegans*, and *A. thaliana*.	http://www.cbs.dtu.dk/services/NetGene2/
HSF (Desmet et al. 2009)	The Human Splicing Finder is an online bioinformatics tool to predict the effects of mutations on splicing as well as to identify splicing motifs in any human sequence.	http://www.umd.be/HSF/

(continued)

Table 7.1 (continued)

Tools	Description	URL
Cryp-Skip (Divina et al. 2009; Kralovicova and Vorechovsky 2007; Buratti et al. 2007; Vorechovsky 2006)	CRYP-SKIP uses multiple logistic regression to predict the two aberrant transcripts from the primary sequence. It takes sequence of the mutated exon together with flanking intronic sequences and provides the overall probability of cryptic splice-site activation as opposed to exon skipping.	http://cryp-skip.img.cas.cz
NNSplice (Reese et al. 1997)	Splice Site Prediction by Neural Network for drosophila and human.	http://www.fruitfly.org/seq_tools/splice.html
Program to evaluate splicing regulatory elements		
RESCUE-ESE (Fairbrother et al. 2004)	RESCUE-ESE is a tool to identify exonic sequence enhancer (ESE) activity. A given hexanucleotide is predicted to be an ESE if it has a significantly higher occurrence in the exons than in introns as well as in exons with weak splice sites as compared to exons with strong splice sites.	http://genes.mit.edu/burgelab/rescue-ese/
ESEFinder (Smith et al. 2006; Cartegni et al. 2003)	ESEfinder is a Web-based resource that facilitates rapid identification of exon sequences to identify putative ESEs responsive to the human SR proteins SF2/ASF, SC35, SRp40, and SRp55, and to predict whether exonic mutations disrupt such elements.	http://rulai.cshl.edu/cgi-bin/tools/ESE3/esefinder.cgi?process=home
PESX (Zhang and Chasin 2004; Zhang et al. 2005b)	PESX (Putative Exonic Splicing Enhancers/Silencers) is a tool to predict exonic splicing signals.	http://cubweb.biology.columbia.edu/pesx/
ExonScan	ExonScan is a Web tool that takes primary transcript sequence as input. It scores splice sites using MaxEntScan (Yeo and Burge 2004), predict ESE ESS using previously established datasets (Wang et al. 2004; Fairbrother et al. 2004).	http://genes.mit.edu/exonscan/

(continued)

Table 7.1 (continued)

Tools	Description	URL
ESRsearch	ESRsearch is used to identify putative ESRs (exonic splicing regulatory elements) in the input sequence by using four previously published datasets (Wang et al. 2004; Fairbrother et al. 2004; Goren et al. 2006; Zhang and Chasin 2004).	http://esrsearch.tau.ac.il
Programs to evaluate RNA secondary structure		
Mfold (Reese et al. 1997)	Mfold is a Web application to predict the secondary structure of single stranded nucleic acids.	http://mfold.rna.albany.edu/?q=mfold/RNA-Folding-Form
Pfold (Knudsen and Hein 1999, 2003)	Pfold takes an alignment of RNA sequences as input and predicts a common structure for all the input sequences.	http://daimi.au.dk/~compbio/pfold/
Sfold (Ding et al. 2004)	Sfold is a Web tool that predicts RNA secondary structure. It offers multiple modules including Sirna for siRNAs prediction, Soligo for antisense oligos, Sribo for ribozymes, Srna for comprehensive features for statistical representation of sampled structures.	http://sfold.wadsworth.org/cgi-bin/index.pl
SNPfold (Halvorsen et al. 2010)	SNPfold is a Web application that computes the potential effect of SNPs on RNA structure.	http://ribosnitch.bio.unc.edu/snpfold/SNPfold.html

1.4.2 RNA Binding Proteins Bind with Low Specificity

RNA binding proteins are generally perceived as binding with lower specificity than DNA binding proteins. Double stranded DNA presents a rigid, well-defined surface for protein contact through which distinct binding specificities can arise. Many DNA binding proteins bind palindromic sequence as dimers, an arrangement that extends the region of contact between the nucleic acid and the protein. Typically RNA is at least partially structured and can adopt multiple conformations. Biochemically, the conformational flexibility of single stranded RNA creates a degenerate set of surfaces for protein–nucleic acid interactions. This lack of a single stable conformation in the target RNA combined with the flexibility of off-target RNAs presumably narrows the difference between the binding energies of specific versus nonspecific interactions and likely accounts for the lower specificity of RNA binding proteins.

1.4.3 DNA Is Present in Fixed Stoichiometry, RNA Is Ephemeral

Finally RNAs are a transient unstable species that are expressed to different copy number levels in different cells. RNAs are in constant turnover and so RNA protein complexes have no analogy for the long term chromatin states observed on DNA. In addition to its stability, all sites in DNA are represented in a fixed stoichiometry. Chip-seq signal can serve as a reasonable proxy measurement for occupancy at a particular locus. For RNA, this is not the case—mRNA sequence is often hundreds of times more abundant than pre-mRNA. Weak interactions in the exon may return more signal than stronger binding to a less abundant intronic site (Reid et al. 2009). This difference also seems to have implications for how a nucleic acid binding protein is distributed amongst the ensemble of potential sites in a cell. The landscape of DNA binding sites is static in the sense that the DNA binding surfaces available to proteins in a cell is more or less constant. The cellular pool of bound protein will partition across the defined set of all possible genomic sites according to (a) the affinity, (b) accessibility of the DNA. In contrast, the cellular ligand reservoir for RBPs is not defined but can vary dramatically across tissues, timepoints and conditions with the differential expression RNA transcripts. Taking this scenario to an extreme, it is possible for a "sponge" RNAs to titrate RNA binding proteins away from other sites without changing the overall level of protein.

1.5 The Signals and Mechanisms of Pre-mRNA Splicing: The Largest Class of RNA-Level Functional Variants

Pre-mRNA splicing is an essential feature of eukaryotic gene regulation, whereby introns are excised and exons are joined together to make meaningful transcripts that can be translated into functional proteins. The majority of human transcriptome is alternatively spliced in highly regulated cell-specific or developmental stage-specific manner (Pan et al. 2008; Wang et al. 2008). The core splicing signals comprise of consensus sequences that span the intron–exon junction at −14 to +1 position (termed 3′ splice site; 3′ss), the exon–intron junction at −3 to +6 position (termed 5′ splice site; 5′ss) and the branch point sequence. However, these core signals are extremely diverse. With the exception of the AG and GU dinucleotides that precede and follow the exon, respectively, the highest frequency of occurrence of a particular base in any given position is in the range of 35–80 %.

Therefore, pseudo splice sites occur in greater abundance than the actual splice sites (Sun and Chasin 2000; Roca et al. 2012). This problem is thought to be mostly overcome by the auxiliary splicing signals in the form of a complex interlay of enhancing and silencing elements that are recognized mainly by two protein families: SR proteins and hnRNPs, which exert antagonistic effects; the balance of which determines the fate of splicing. These secondary signals are embedded in the exons (i.e., exonic splicing enhancer or silencer; ESE or ESS) as well as in the introns (i.e., intronic splicing enhancer and silencer; ISE and ISS) (Fairbrother and Chasin 2000; Fairbrother et al. 2002; Wang et al. 2004). They mostly occur within 50 nucleotides of the splice site boundaries, but long distance elements also exist. For example, a 4-base deletion in intron 20 of the *ATM* gene that is situated 2 and 0.6 kb from the flanking splice sites causes ataxia telangiectasia and led to the discovery of an unknown splicing element that when mutated activates a nearby pseudo exon (Zhang et al. 2005a; Pagani et al. 2002). These auxiliary splicing signals are known to be composite, context-dependent and position-dependent (Pagani et al. 2003; Ule et al. 2006; Kanopka et al. 1996). There is also compelling evidence suggesting that the architecture of the genes further determines how all these core and auxiliary splicing elements play their roles in the assembly of the splicing machinery. When a given exon is considerably shorter than its flanking introns, as in the case of most human genes, the splicing machinery is thought to assemble across the exon, and splicing is called to be "exon defined." The opposite happens in the case of large exons that are flanked by short introns, in which case they are called to be "intron defined" (Berget 1995; Robberson et al. 1990; Chen and Chasin 1994; Fox-Walsh et al. 2005).

The splicing reaction of most human introns is catalyzed by the spliceosome, arguably the most complex and dynamic macromolecular machine. The majority of the human spliceosome consists of five small nuclear RNAs (U1, U2, U4, U5, and U6) and over 300 different proteins (Jurica and Moore 2003). A secondary splicing apparatus has also been identified in human. This minor form of the human spliceosome contains U11, U12, U4atac, U5 and U6atac snRNAs (Steitz et al. 2008). Biochemical studies of the major human and yeast spliceosomes assembled in vitro revealed an ordered and stepwise formation of five distinct subcomplexes. The H (heterogenous) complex contains different hnRNP proteins that can bind on any largely single-stranded RNAs. The E (early) complex is formed when U1 snRNA binds to the consensus 5′ splice site sequence in an ATP-independent manner. This is followed by the formation of an ATP-dependent A complex with the base-pairing of U2 snRNA with the branchpoint sequence. Then the tri-snRNP (U5, U4, and U6) is recruited and form the B complex, which further catalyzes multiple structural rearrangements needed to form the activated B* spliceosome, where the two chemical steps of splicing are to occur. First, the branch point sequence carries out a nucleophilic attack on the 5′ splice site and generate a free 5′ exon and a lariat intermediate; this forms the C1 complex. In the second step, the free 3′ hydroxyl group of the 5′ exon attacks the phosphodiester bond at the 3′ splice site in the C2 complex and generates ligated exons and an intronic lariat species.

These spliceosomal complexes are recycled after each round of splicing (Hoskins and Moore 2012; Jurica and Moore 2002).

An increasing number of studies have indicated an unexpectedly high fraction of diseases being caused by genetic variations that alter splicing (Cartegni et al. 2002; Wang and Cooper 2007). Aberrant splicing is the most common cause of some Mendelian diseases, such as neurofibromatosis type 1 (*NF1*) and ataxia telangiectasia (*ATM*) (Ars et al. 2000; Teraoka et al. 1999). Genetic variations that act as disease modifiers or as disease susceptibility factors have also been identified; most of them do so by altering splicing efficiency of specific exons (Kralovicova et al. 2004; Steiner et al. 2004; Nielsen et al. 2007). Variants that disrupt GU or AG dinucleotides of the consensus splice signals constitute ~15 % of disease-causing point mutations that have been reported at HGMD (Krawczak et al. 1992). Causative variants that disrupt the branchpoint sequence likewise have been reported in a number of congenital disorders (Burrows et al. 1998; Crotti et al. 2009; Maslen et al. 1997; Webb et al. 1996), although large-scale studies of such variants are not yet visible due to the limited number of branch sites that have been characterized in humans (Taggart et al. 2012). But the greatest fraction of splice-altering variants can be expected to result from disruption of auxiliary splicing codes, which represent a much larger mutational target (more base coverage). A recent survey examined ESE and ESS gains/losses that result from 27,681 causative exonic mutations that have been reported at HGMD and 8,601 exonic SNPs with minor allele frequency (MAF) > 0.18. Significant enrichment for ESE loss and ESS gain was found in disease causing variants as compared to common SNPs that are assumed to be neutral (Sterne-Weiler et al. 2011). The authors concluded that 25 % of the reported missense and nonsense mutations known to cause Mendelian inherited disorders are functional splice-altering variants that exert their effects by obliterating or weakening existing ESEs or by creating de novo ESSs. Furthermore, as mentioned in Sect. 1.3, while synonymous mutations are routinely classified as non-disease-causing in clinical genetics practices, some of them may also be functional splicing variants. Comprehensive analysis of exon 9 and 12 of *CFTR*, the causative gene for cystic fibrosis, revealed that a quarter of the synonymous substitutions resulted in splicing defects (Pagani et al. 2005). Mutations that disrupt highly regulated developmental stage-specific alternative splicing likewise can cause human hereditary disorders. CUG repeat expansion in the 3′ UTR of *DMPK* or CCTG repeat expansion in intron 1 of *ZNF9* cause myotonic dystrophy by generating "toxic RNAs" that disrupt the activities of the splicing factors that regulate the expression of postnatal splicing patterns of a subset of developmentally regulated genes (Ranum and Cooper 2006; Cho and Tapscott 2007). In addition, recent data showed that approximately 80 % of pre-mRNA splicing occur co-transcriptionally (Girard et al. 2012). Therefore it is not surprising that nucleosome positioning and histone modifications, factors that determine the rate of RNA polymerase II elongation, were found to significantly influence the efficiency of splice site recognition (Schwartz and Ast 2010). These represent another class of mutational target that may cause aberrant splicing that can lead to human disease.

2 Computational Methods for Studying Variation that Function at the RNA Level

There are a variety of mechanisms through which variants could alter gene function at the RNA level. The general approach most often applied to considering the effect of a variant involves annotating wild type and variant alleles with models of splicing elements. Biochemical dissection of the sequence determinants of splicing has revealed a 5′ss (recognized by U1snRNP), a branchpoint (recognized first by SF1, then U2snRNP), a polypyrimidine site (recognized by U2AF65) and the 3′ss AG (recognized by U2AF35 and then u6snRNA) (Hastings and Krainer 2001; Auweter et al. 2006; Matlin and Moore 2007). The best defined signals are the splice sites themselves—each intron contains a 5′ss and 3′ss motif. Initially identified as a GT in the first two and an AG in the last two intronic positions, the splice sites are now defined by more sophisticated probabilistic binding models. Hidden Markov models are used to capture context effects that greatly improve the performance of these classifiers. There are often physical explanation to these context effects. For example if the penultimate base in the region of complementarity between the 5′ss and u1snRNP is disrupted the last position of complementarity is unimportant because an isolated Watson–Crick base pair is not stable (Carmel et al. 2004). Several Web implementations of these algorithms allow for batch submission of sequences for annotation (Table 7.1). Most programs output a log odd score where a value of zero indicates an input sequence that is as equally likely to derive from the set of real sites as from the background, a positive score indicates input that is more likely to derive from the set of real sites. Scores can be evaluated by comparison to the distribution of scores observed from real sites or as the probability of occurring in the background (i.e., *P* values).

We have suggested that predictions of secondary structure for pre-mRNA can be used to improve models of RNA–protein binding by incorporating structure and motif combinations into a predictive model. One form that such a predictive model could take is to extend the model used by pattern search programs such as Patser (Hertz and Stormo 1999) with an additional scoring component related to the degree of single strandedness of the target site. For example,

$$\text{Score}_i = \sum_{w=0}^{l-1} \log\left(\frac{p_{n(i+w)}}{p_{B(i+w)}}\right) + \log\left(\frac{p_{u(N(i))}}{p_{o(N(i))}}\right);$$

where l is the length of the site, $p_{n(i+w)}, p_{B(i+w)}$ are the probabilities of the nucleotide from a position weight matrix and background model, respectively, at relative position w in the site starting at i. $N(i)$ is the number of unpaired bases in the site starting at i and $p_{u(N(i))}$ represents the probability of $N(i)$unpaired bases. $P_o(N(i))$ represents a similar probability for the overall collection of motif sites. For example, $P_u(N(i))$ can be obtained from the counts of unpaired bases in the motif sites in experimentally verified binding substrates.

In addition, to splice sites there is a host of auxiliary elements that modulate splice site usage. There are a variety of enhancer elements required for high fidelity splicing that represent a mutational target. Estimates for the fraction of exonic mutations that alter splicing range from 22 to 25 % (Lim et al. 2011; Sterne-Weiler et al. 2011). Unlike transcriptional control elements, RNA elements tend to function in a highly position specific manner. This behavior can be seen in the major class of splicing activator proteins, the SR proteins. The SR proteins increase splicing by binding purine-rich exonic splicing enhancers (Manley and Tacke 1996; Sciabica and Hertel 2006; Lin and Fu 2007). Interestingly, these factors repress splicing when bound at intronic loci presumably through a decoy process that recruits other factors like U2AF or U1snRNP into nonproductive complexes (Kanopka et al. 1996; McNally and McNally 1996; Lareau et al. 2007). Another class of *trans-acting* factors, the hnRNP proteins bind pre-mRNA at intronic positions. While some of these factors can enhance splicing from the intron, they have been implicated in repressing splicing when bound at exonic locations (Damgaard et al. 2002; Martinez-Contreras et al. 2006; Wang et al. 2006). Binding specificities have been determined for some individual SR and hnRNP proteins and there are some online tools developed to score input sequences for their agreement to these sequences (Table 7.1 (Cartegni et al. 2003)). However, by themselves, these tools developed to characterize the ligand of a particular *trans*-acting factor are limited by their inability to evaluate all variants for any kind of splicing defect.

There are several computational tools that enable a more complete annotation of splicing substrates. The first such tool, RESCUE-ESE scores sequence for exonic splicing enhancers(Fairbrother et al. 2002). This algorithm assumes exonic splicing enhancers will be more enriched in exons with sub-optimal splice sites than exons with strong splice sites. Hexamers that were simultaneously enriched in (a) exons relative to introns and in (b) exons adjacent to weak sites relative to exons adjacent to strong sites. This method produced k-mer motifs that were predictive of known splicing mutants in the *hprt* gene. The 238 hexamers identified as ESEs were later shown to be under purifying selection especially near exon boundaries in human transcripts (Fairbrother et al. 2004). Additional approaches utilized k-mers enriched in real exons relative to pseudoexons (intronic regions adjacent to strong matches to splice sites) to identify putative exonic splicing enhancers (PESE) and also negatively acting putative exonic splicing silencers (pESS) (Zhang and Chasin 2004). High-throughput minigene experiments similar to a functional SELEX approach were used to derive k-mer motifs that functioned negatively from an exonic context (Wang et al. 2004). It has been observed that, due to the proliferation of studies that define splicing control elements, three quarters of all exonic sequences are predicted to be an enhancer by at least one of these methods (Chasin 2007). This underscores the need for unified approaches that attempts to integrate different motif sets and the diverse modes that sequence determinants can function in splicing. One such tool, exon scan integrates several of these exonic k-mer sets and one type of intronic enhancer element along with splice sites models to direct an in silico spliceosome (Table 7.1). The output of this program is a prediction of whether the spliceosome would recognize an internal segment in the input string as an exon—for positive predictions the boundaries are returned.

Another comprehensive approach, Spliceman, is a specific tool to assay the difference between the affect two allelic variants can have on splicing. Spliceman exploits the observation that many elements that function in splicing are highly position dependent—ESE relocated to introns are inhibitory and vice versa. As splicing motifs function at particular positions relative to splice sites, it may be useful to organize k-mers according to their positional distribution (i.e., their distribution around splice sites in the genome). The Spliceman algorithm assumes that k-mers with a similar distribution will play a similar role in splicing. Spliceman considers a variation and compares the global positional distribution of the k-mers in the reference sequence to the k-mer in the variant sequence. In vivo testing of mutations demonstrated that this approach was predictive of exonic splicing mutations (Lim et al. 2011). In a small scale validation of Spliceman's predictions of exonic splicing mutants, four out of six human disease alleles originally classified as "missense" in the literature exhibited splicing defects when tested in vivo. Spliceman returns two types of output (Lim and Fairbrother 2012). The raw output of Spliceman is L1 distance score which is an indicator of the difference in positional distribution between the wild type and mutant allele (i.e., the area between two normalized distribution plots). In addition, Spliceman outputs relative measure of allelic differences (percentile ranks of the input variation relative to all possible single nucleotide changes in sequences). The latter allows the researcher to prioritize a dataset of variations for their likelihood of disrupting splicing. It is often necessary to confirm these predictions with a biochemical and functional assays some of which are described below.

3 Experimental Methods for Studying Variation that Function at the RNA Level

3.1 Low-Throughput RNA Binding Assay

There are several different methods for analyzing protein nucleic acid interactions that can be effectively used to assay allele specific binding: (a) electrophoretic mobility shift assay, (b) filter binding assay, and (c) various co-immunoprecipitation and label transfer approaches (Chodosh 2001). While in vivo assays of endogenous transcripts are more likely to offer more physiological models of variant function, they occur in complex genetic backgrounds which sometimes obscure which variant is causing the phenotype. In vitro binding assays are a useful reductionist tool to control for these effects and detect any intrinsic difference in affinity between RNA alleles. Many of the variants of interest affect fitness and are often quite rare in the human population and so another benefit of an in vitro binding assay is the convenience of being able to engineer the variant into a synthetic oligonucleotide rather than trying to obtain a biopsied sample with the rare genotype. Another common situation where the availability of an appropriate sample is limiting for in vivo assays are variants in genes that are only expressed in human tissues that cannot be readily sampled. An example of such a situation is the in/del, rs3215227 that encompasses a perfect match to a PTB binding site in intron 13 of the neuronally expressed RGS9 gene.

3.1.1 Case Study: rs3215227 Deletes an Intronic PTB Binding Site in the RGS9 Gene

This case study considers a polymorphism that encompasses a perfect five nucleotide deletion of a PTB binding site, TTTCT, in intron 13 of the RGS9 gene (Reid et al. 2009). Alignment to other primate genomes (e.g., chimpanzee, rhesus macaque) indicates that the polymorphism arose through a deletion that occurred in the human lineage sometime prior to the human–chimpanzee divergence (5 Mya). While most mutations that negatively affect fitness are rapidly eliminated from the population other variants are adaptive or are maintained through complex modes of selection. To explore a potential biological function for rs3215227, we assayed RNA transcribed from both alleles for differences in binding to PTB in nuclear extract. The binding affinity can be measured through a variety of technologies. A gel shift (also called gel retardation or electrophoretic mobility shift assay (EMSA)) is an option for measuring protein RNA affinity. Because of the intrinsic flexibility of RNA this leads to a loss of resolution during electrophoresis and can be a problematic choice for probes longer than 50 nucleotides. Filter binding assays would involve the incubation of PTB with radiolabelled RNA probe. Passing the mixture through a protein binding nitrocellulose filter retains only the RNA that is in complex with PTB. While this is a rapid and convenient assay, it must be performed with isolated recombinant proteins and does not allow for the combinatorial interactions that occur in extract. The binding affinity of PTB to these allelic RNA substrates was determined with a standard UV cross-linking reaction. Briefly this assay captures RNA-protein binding events through a label transfer strategy. RNA is transcribed with incorporating P^{32} labeled and incubated in HeLa nuclear extract to allow RNA protein complexes to form. UV exposure triggers covalent attachment between the internally labeled RNA and proteins in the complex. After cross-linking, RNAses are used to "trim" the covalently attached RNA to a few nucleotides effectively serving as a radiolabel for interacting proteins. The mobility of these labeled factors can be measured by PAGE and their identity can be confirmed by immuno-precipitation. Both the RNA alleles bound a 58 kDa protein—the wild type sequence more strongly than the deletion variant (Fig. 7.1a). As the cross-linking efficiency and degree of label transfer can vary from substrate to substrate a quantitative estimate of alleles can only be achieved by using a single well-characterized ligand of PTB for the binding assay and then testing the ability of each (unlabeled) RNA to compete with this binding. We implemented this strategy with a previously published PTB ligand, S11, in the presence of a 5–50 molar excess of unlabeled wild type and deletion allele competitor (Fig. 7.1b).

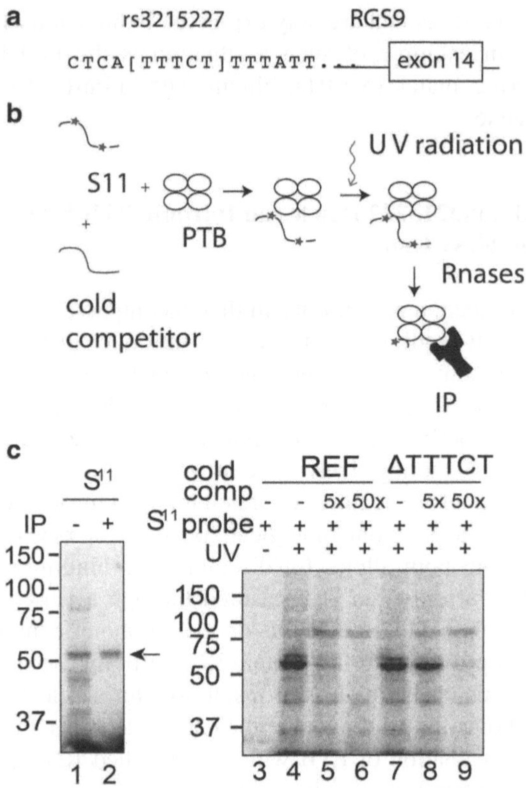

Fig. 7.1 (**a**) The reference genome (reference allele) at rs3215527 contains a match to the binding site of the splicing repressor, PTB, while the ΔTTTCT allele represents a complete deletion of the predicted site. (**b**) Generalized scheme for UV cross-linking in vitro binding assay between a protein extract and RNA oligos ^{32}P radiolabelled at internal positions (*red stars*). (**c**) In vitro PTB binding assay detects allelic bias in PTB binding: The well-characterized PTB ligand, "S11" was radiolabelled, incubated in HeLa extract and UV cross-linked. A prominent 58KD band can be immunoprecipitated with antibodies against PTB (lanes *1* and *2*). This binding reaction serves as a standardized PTB cross-linking assay in lanes *3–9*. The relative strength of REF or the ΔTTTCT allele's ability to bind PTB is inferred from its ability to compete with the S11 ligand in the standardized PTB cross-linking assay

Cross-linking followed by immunoprecipitation confirmed that PTB is dominant binding protein for S11 in complex extract (note the prominent 58 kDa is precipitated by anti-PTB) Repeating this binding assay with progressively higher concentrations of unlabeled wild type and deletion variant of RGS9 demonstrates that a tenfold higher concentration of the deletion variant is required to achieve the same amount of competition as the wild type allele (Fig. 7.1b, compare lane 5 with 9). This data suggests that the wild type sequence binds PTB with a tenfold higher affinity than the deletion variant.

3.2 Low-Throughput Minigene Splicing Assay

While it is important to understand the mechanism of allelic difference in binding events, a researcher's initial goal is usually to determine if a particular variant is functional. Often this line of experimentation starts with an observed difference in processing of the endogenous transcript. The goal is then to test a SNP in isolation of other nearby variants or possible genetic background effects to assign causality to a single variant. A minigene splicing reporter construct tests a portion of an endogenous gene in a chimeric context. There are many types of reporters designed for different purposes (Cooper 2005). Most minigene splicing reporters contain a backbone that contains all signals necessary for transcription and proper 3′ end formation and an internal cloning site. Some reporters are specially designed to test a small region for a particular type of effect on splicing. For example, pSXN tests 32mer in an exon for an enhancing effect. The test exon has a default skipping behavior and so the degree to which the test sequence rescue's exon inclusion becomes the measure of enhancer activity(Coulter et al. 1997). In general it is better to test larger insert region to better recapitulate the endogenous transcript—an entire exon with flanking introns or an entire intron with flanking exons. A common design results in a reporter with a first exon, a last exon and a "test exon" that contains the exon in question along with an appropriate amount of endogenous flanking intron sequence. An appropriate insert window would include all the *cis*-elements required to recapitulate the endogenous splicing behavior. While the size of this window will vary from loci to loci, it is typically sufficient to include 200 nucleotides on either side of the exon. If insert size is limiting, an asymmetric window may be appropriate as the obligatory splicing signals (branchpoints, polypyrimidine tract, and 3′ss) are broadly distributed upstream of the exon than downstream. Some estimate of the upstream boundary of spliceosome substrate recognition can be learned from the location of the branchpoint. About 85 % of all branchpoints are within 35 nucleotides of the 3′ss and less than 1 % are further than 100 nucleotides away. There are exceptions to this compact arrangement—alternatively spliced exons have on average more distal branchpoints with some extreme examples located more than 200 nucleotides away from their 3′ss (Gooding et al. 2006). Sequence conservation suggests the presence of functional elements further into the intron for alternatively spliced exons relative to constitutively spliced elements (Yeo et al. 2005). In general, a larger window asymmetric around the exon is more conservative than a smaller window. After cloning a reference and variant version of the minigene, the constructs are usually transfected in triplicates into a tissue culture cell line. In general manufacturer's transfection protocols are optimized for maximum transfection efficiency. As very little signal is required for most splicing assays, transfection protocols should be optimized to minimize cellular stress. Analysis is typically performed by RT-PCR with primers in flanking exon regions design to give an inclusion product between 300 and 700 nucleotides in length. As the most common alternate processing pathway is exon skipping, an alteration in splicing usually presents as two spliced product that are differ by length of the exon. The convention for reporting the ratio of these two products is "percent spliced in" (p.s.i.) often assigned the Greek letter psi (Ψ). While this technique is often called semiquantitative PCR, we have found by performing mixing experiments with total RNA from transfections of constructs that splice exclusively to inclusion or skipping that the measured ratio of isoforms can be highly quantitative (Fairbrother et al. 2002). As with any reporter experiment performed in transient transfection experiments in tissue culture, there could be numerous confounding features such as promoter effects, tissue specific splicing factors, chromatin effects and intron size effects that prevent the minigene from replicating the endogenous behavior. However despite this potential for false negatives, the minigene affords the researcher the chance to make focused comparisons between two alleles and understand whether there is a potential for allele specific splicing.

3.2.1 Case Study: The Deletion Variant of rs3215227 Exhibits Reduced Sensitivity to PTB Suppression in a RGS9 Minigene

In Sect. 3.1.1 we analyzed a five-nucleotide deletion variant, rs3215227, that encompassed a perfect copy of a high affinity PTB site and demonstrated the deletion allele had a tenfold lower affinity to PTB than the reference allele. In the second part of this case study we use a minigene approach to test the hypothesis that the exon downstream of rs3215227 (exon 14, RGS9 gene) is less sensitive to PTB mediated repression than the reference allele. To see whether rs3215227's allelic difference in affinity translates into a difference in splicing, we constructed two splicing reporter constructs and transfected them into HeLa tissue culture cells. This example replicates both alleles of intron 13 in a CMV promoter driven backbone (Fig. 7.2). Together the insert and reporter produce a chimeric RGS9 minigene composed of four exons with sequence from intron 13 and portions of exon 13 and 14 of the RGS9 gene between reporter exons 2 and 3. These constructs were introduced into HeLa cells by transient transfection and total RNA isolated 48 h post transfection was analyzed by RT-PCR. Exon inclusion corresponded to 451 nt product whereas the skipping of exon immediately downstream of the polymorphism would result in a ~400 nt band. Our initial comparison between these two constructs resulted in near complete exon inclusion suggesting that there is no effect of this intronic variant in HeLa cells. However, it possible that the tissues in which RGS9 is expressed have a higher level of PTB expression or a splicing environment that is more sensitized to PTB repression. To increase the assays sensitivity to PTB, we repeated the experiment under elevated levels of PTB by co-transfecting PTB overexpression constructs with the reporters (Fig. 7.2b, lanes 1 and 2). Co-transfection of PTB cloned into an expression vector led to a clear allelic difference—two additional PCR products corresponding to distinct alternatively spliced mRNA products

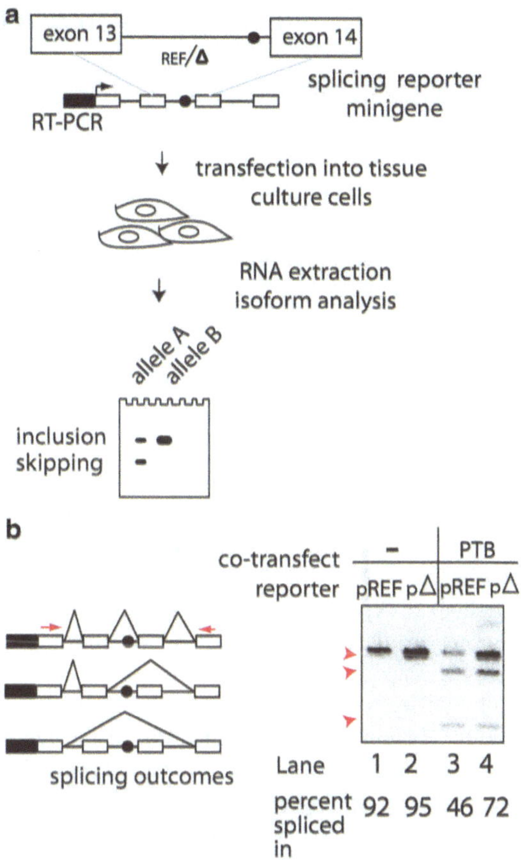

Fig. 7.2 In vivo minigene experiment reveals allelic difference in sensitivity to PTB repression. (**a**) Two minigene reporters are designed by cloning both allelic versions of intron 13 of RGS9 gene into PZW4. (**b**) Both of the resulting four-exon minigenes are transfected into HeLa cells where their isoform usage is analyzed by RT-PCR amplified from total RNA. (**c**) Correct splicing is measure from RT-PCR following transient transfection experiments. *Arrows* label inclusion and exon skipping products that result from splicing repression. Co-transfection of PTB overexpression construct induces exon skipping in REF and to a lesser degree in pΔ

(Fig. 7.2b lane 3 and 4) were seen, mostly in the processing of the reference allele. A ~400 bp band was detected consistent with a splicing product in which the adjacent exon downstream of the RGS9 intronic sequence was skipped. This would result if experimental exons 1 and 2 were spliced correctly but exon 2 was spliced to exon 4, skipping exon 3. A third product deleted exon 2 and 3. While both wild type and deletion polymorphism splice exclusively to product a while transfected, the co-transfection of PTB results in allelic biased repression of splicing. This repressive effect of PTB is almost twice as strong on the reference sequence which contains an extra PTB binding site than on the deletion variant (Fig. 7.2b). This result suggests that the reference allele is disproportionately bound by PTB in vivo and can be subject to a greater level of repression by PTB than the deletion allele. However, whether this situation is actually occurring in any relevant cell type or condition in vivo is a question for future study.

3.3 High-Throughput RNA–Protein Binding Assays

While the above methods are informative about single variations, they are limited to case by case analysis. However, when these mechanistic characterizations are applied downstream of genomic technologies (like sequencing), the number of candidate variants often exceeds the capacity of low-throughput methods. In the following three sections we will review high-throughput in vitro binding assays, occupancy assays and splicing/spliceosome assembly assays. In the final segment of this section we cover a case study of a high-throughput implementation of the in vivo minigene assay that was used to screen 236 disease alleles.

In each of the technologies covered below a pool of synthetic oligonucleotides are purchased from a vender. They are synthesized on a solid phase in an ordered array. Initially, pools were generated from heat treating and scouring custom oligonucleotide arrays (Ferraris et al. 2011a, b; Yajima et al. 2012; Tantin et al. 2008; Chang et al. 2009; Reid et al. 2009; Watkins et al. 2009). More recently commercially synthesized mixed pools of custom designed oligonucleotides have become available for capture probes or for use as overlapping templates in large scale gene synthesis. Oligos can be ordered up to limits of 200-mers in length and 50,000 in batch size. Typically oligos are designed with universal primers flanking (variable) regions of interest so the pool can be amplified in a single PCR reaction. In certain cases library application can be problematic and some practitioners recommend high fidelity polymerases (Dr Wilson Agilent, personal communication) (Reid et al. 2009). Through PCR amplification of the DNA library, T7 promoter is added via tailed primers. The resulting amplified product are used as templates for RNA synthesis. These RNA products can then be used as the input for binding studies. In the following section we describe a co-immunoprecipitation approach illustrated with a case study using PTB (Fig. 7.3b). One downside of this approach is that this technique requires a candidate factor for immunoprecipitation. Most human genetics applications will be concerned with identifying all functional variants, not just the subset that act through the differential binding of a specific RBP. To address this concern, we also describe an occupancy assay that identifies ligands that form stable RNPs without regard for the identity of the proteins in the complex (Fig. 7.3c) (Ferraris et al. 2011a). Both of these assays have some similarity to SELEX (Djordjevic 2007; Gopinath 2007; Stoltenburg et al. 2007). Large pools of diverse sequence are synthesized and allowed to compete for limiting protein. Here the similarity ends. While SELEX tries to infer specificity by starting with random sequence and trying to detect binding motifs in the selected fraction, the analysis described below test a less complex mixture of real sequences and ranks each according to their enrichment in the bound fraction. For the analysis of variants, the pool consists of thousands of reference and variant alleles synthesized with some flanking sequence to present both alleles in their natural sequence context for the competitive binding assay. The bound fraction is physically separated from the unbound fraction either by magnetic antibody/streptavidin Dynabeads (Fig. 7.3b) or by adherence to nitrocellulose filter in a dot blot assay (Fig. 7.3c). Once the bound

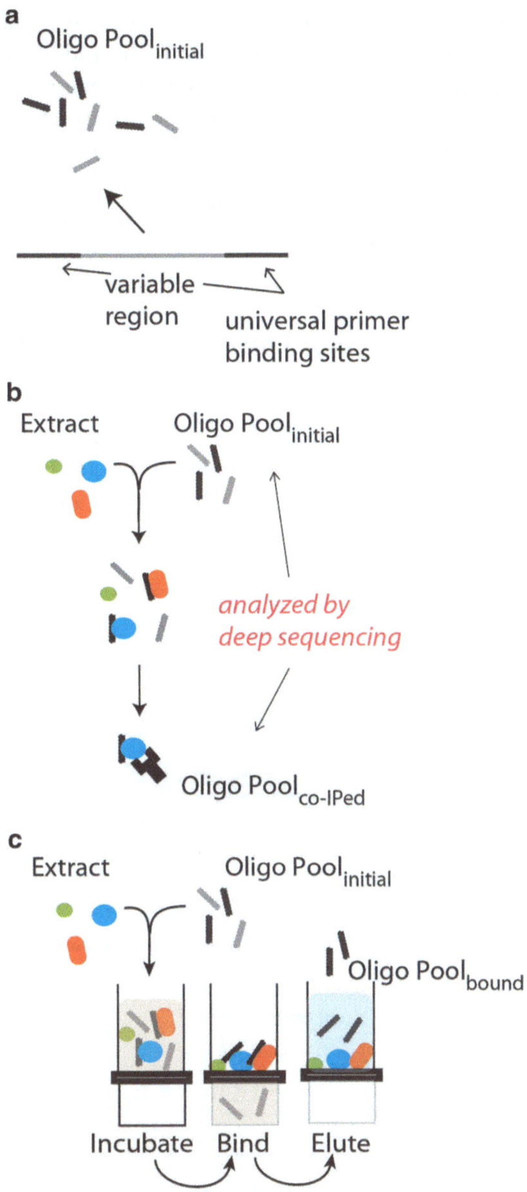

Fig. 7.3 Overview of high-throughput binding assays. (**a**) Oligonucleotide libraries are chemically synthesized and delivered as a mixed pool. Oligos correspond to short windows of genomic sequences transcribed to RNA and screened for binding (*black line* = ligand; *grey* = non-ligand). Each oligo is flanked by regions complementary to universal primers allowing PCR amplification of library. (**b**) Oligo pool is incubated in nuclear extract. RNPs are immunoprecipitated and analyzed by sequencing or microarray. Relative enrichment of oligos in the bound fraction is compared to relative enrichment in the starting pool. (**c**) Oligo library is incubated in extract as above. RNP are retained on nitrocellulose filter, eluted and analyzed as described above

fraction is reverse-transcribed it can be either analyzed or re-enriched by repeating the protocol. The analysis can be performed either by array (as in the case study below) or, for a SNP application that requires genotyping at single nucleotide resolution, by deep sequencing (case study 3.5). The library preparation for deep sequencing can be facilitated by PCR mediated introduction of Illumina sequencing adapters with special tailed primers. While this is fairly straightforward library prep, care must be taken to avoid excessive amplification prior to deep sequencing and to maintain familiarity with the rapidly changing sequencing protocol. For example, it is currently necessary to ensure that the first 4 nucleotides adjacent to the Illumina sequencing primer are balanced in sequence. For protocols that require the sequencing of PCR products this random 4-mer should be added as part of the primer as failure to start a read with a near random 4-mer in Illumina's current approach to establishing clusters can dramatically reduce the read yield of a sequencing run.

3.3.1 Case Study: Large Scale Determination PTB Binding Affinities on Pre-mRNA

We present a co-immunoprecipitation study (Fig. 7.3a) using oligonucleotides designed by tiling through about 4,000 alternatively spliced exon/intron regions ("alternate spliced," Fig. 7.4a) and an oligonucleotide library that tiles through

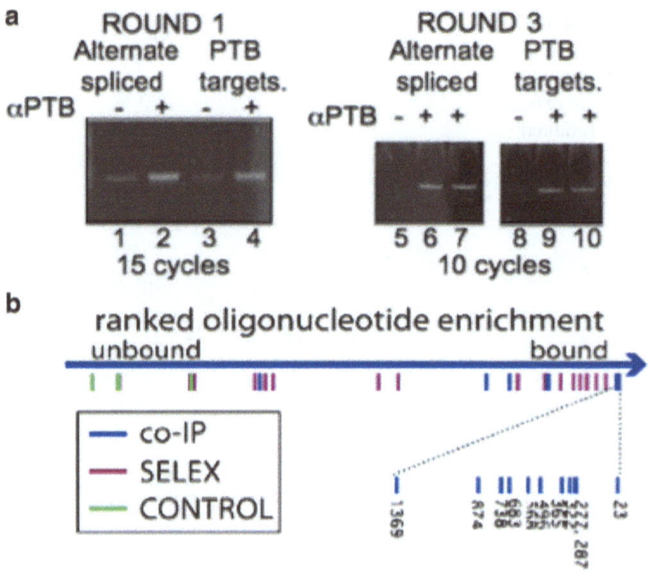

Fig. 7.4 Enrichment of the RNA oligonucleotide pool after multiple rounds of PTB binding selection. An oligonucleotide pool that tiles through ~300 intron–exon–intron alternatively spliced exons was enriched for PTB binding activity by multiple rounds of co-immunoprecipitation (lanes *2*, *6* and *7*). RT-PCR signal was dependent on antibody in IP (lanes *1* versus *2*) and required fewer cycles after three rounds of enrichment. Similar enrichment was performed on a oligonucleotide pool that tiles through pre-mRNAs identified as in vivo targets of PTB (lanes *4*, *9* and *10*). (**b**) Oligonucleotides were ranked by microarray determined binding enrichment. Enrichment rankings for oligonucleotides cloned from after three rounds of co-IP (i.e., lane *9*) were marked as *blue lines* and compared to CONTROL oligos (*green lines*) cloned from the starting pool and positive control oligos identified by earlier SELEX studies (*magenta*)

in vivo PTB targets identified by RNAi depletion studies (Boutz et al. 2007). We have also introduced PTB SELEX winners as "spike-in" positive controls and we then enrich over three rounds of selection (Singh et al. 1995). The technique of iterative selection could be introduced to increase the sensitivity of a rank difference measurement of two oligonucleotides at the expense of an accurate measure of their relative affinities. After three rounds of co-IP the signal amplifies in an entirely antibody dependant fashion, requiring five fewer cycles than after only one round of co-IP (Fig. 7.4a lanes 4 vs. 9). This demonstrates a rapid increase in the average affinity of the pool through the course of this iterative binding enrichment protocol.

Additionally, we compared different measures of oligonucleotide enrichment of the oligonucleotide pool. Enrichment of each oligonucleotide in the starting pool was compared to its enrichment after three rounds of selection by two color microarray. The lookup table generated from these array measurements was used to calculate the enrichment of randomly cloned oligonucleotides after three rounds of IP. Eleven of these sixteen randomly cloned oligonucleotides scored above the top 0.5 % percentile of the entire oligonucleotide pool. Spiking known PTB ligands (SELEX) into the starting oligonucleotide pool as internal controls allows the data to be related to prior reports in the literature. Interestingly, these top 0.5 % of the oligonucleotides that are enriched in the PTB bound fraction scored well above all of the SELEX winner oligonucleotides that were generated after 11 rounds of selection (Perez et al. 1997). While PTB has typically been regarded as an RNA binding protein of low specificity, this result demonstrates the existence of a significant population of ultra high affinity PTB ligands in natural pre-mRNA sequence.

3.4 High-Throughput In Vitro Functional Assays

While the classic minigene approach to measuring the effect of variation on function are extremely useful, it does not have the capacity to process genomic scale data. To address this, we developed a high-throughput splicing and spliceosome assembly assay to investigate the impact of genetic variation in RNA splicing phenotype. This assay utilizes RNA substrates that incorporate a library containing thousands of potential and known disease causing alleles that can be genotyped with next generation sequencing technology

We designed a 200-mer oligo library that contains about 6,000 disease-causing sequence variation that have been reported in HGMD as well as polymorphic SNPs that may contribute to the susceptibility of disease traits. These library sequences are flanked by universal primer pair sequence for PCR manipulation purposes. Each of these entries in the library represents an allelic pair—for every mutant allele there is a reference sequence. This paired design enables the measurement of wild-type–mutant allelic ratio in the starting pool and later at various stages of splicing cycle. The observation of an allelic skew at some stage of splicing suggests that a variant is affecting splicing.

3.4.1 High-Throughput In Vitro Splicing Assay

In vitro splicing assay has been introduced since in the early 1980s (Furneaux et al. 1985; Ruskin et al. 1984; Lin et al. 1985) and it has been proven invaluable in allowing biochemical characterization of the splicing components that cannot be achieved otherwise and result in our current understanding of the mechanism of pre-mRNA splicing. This assay is carried out in a whole cell extract or more commonly, in nuclear extract that has been prepared to ensure competence for RNA splicing (Dignam et al. 1983; Krainer et al. 1984; Folco et al. 2012). The preparation of splicing-viable nuclear extract consists of harvesting the cells, lysing the cells while leaving the nuclei intact, followed by nuclei isolation, extraction of salt-soluble splicing complexes and dialysis. The pre-mRNA substrate used for this assay is normally obtained from in vitro transcription assay that utilizes prokaryotic phage promoter (T7, T3, or SP6) with linearized plasmid or PCR product as DNA template. This ensures sufficient quantity of pre-mRNA substrates that can be used for in vitro splicing assay. Uniform RNA labeling is usually achieved by incorporating [α-^{32}P]NTP into the in vitro transcription reaction. Alternatively, RNA can be labeled post-transcription. 5′ end RNA labeling can be done by first removing the 5′ phosphate using calf intestinal phosphatase, followed by ^{32}P transfer from [γ-^{32}P] ATP to the 5′ end of the RNA molecule by T4 polynucleotide kinase. 3′ end RNA labeling can also be done by ligating [^{32}P]pCp to the 3′ end of the RNA with T4 RNA ligase. Labeled RNA substrate can then be incubated in nuclear extract to allow formation of the intermediates and products of splicing and RNAs are resolved in denaturing PAGE for visualization. Figure 7.5a shows in vitro splicing assay of a previously described Adenovirus major late derivative substrate (pHMS81; Ad81) (Gozani et al. 1994) in HeLa nuclear extract. The products of first step (lariat intermediates and free 5′ exon) and second step of splicing (ligated exons and lariat product) can be visualized after 30 min incubation at 30 °C in HeLa nuclear extract (Fig. 7.5a).

We have generated library-containing pre-mRNA substrates suitable for splicing in HeLa nuclear extract. The DNA template was constructed by adding the first exon and part of the intron of Ad81 (Gozani et al. 1994) to the 5′ end of the library by PCR. T7 sequence flanking the forward primer was used to amplify the full-length library-containing substrate template (see Fig. 7.5b) for subsequent in vitro transcription using commercial T7 RNA Polymerase and buffer supplemented with [m7G(5′)ppp(5′)G] cap analog and [α-^{32}P]UTP. The transcribed RNA was purified in denaturing PAGE, phenol–chloroform-extracted and precipitated in ethanol. The purified library-containing pre-mRNA substrate was incubated in HeLa nuclear extract to allow splicing reaction to occur. After 30 and 60 min incubation at 30 °C, lariat intermediates and the free 5′ exon (the intermediate species resulting from the first step of splicing) can be visualized in denaturing PAGE (see Fig. 7.5c). The different splicing intermediates can be further purified and analyzed by deep sequencing. This allows us to gain insights into how sequence variation affects the efficiency of specific steps of splicing. Downstream applications of this assay may also include identification of pathogenic variants that can be rescued by small molecules that can

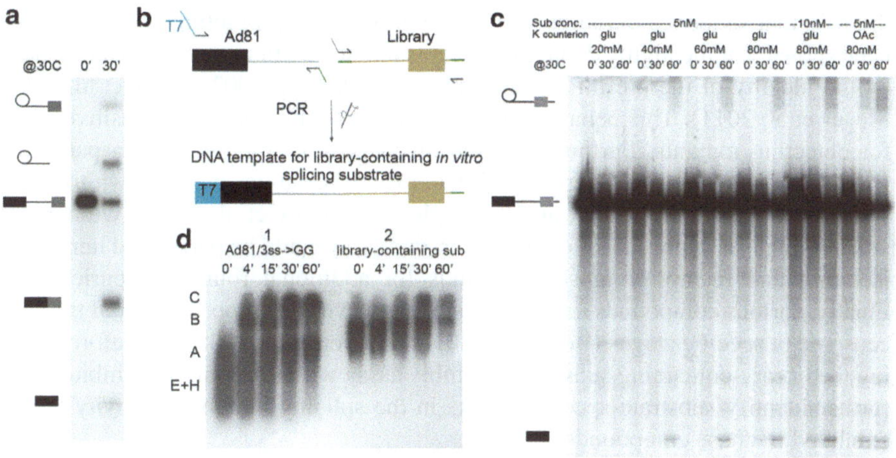

Fig. 7.5 High-throughput in vitro splicing and assembly assays. (**a**) In vitro splicing reaction of Ad81 substrate. After 30 min incubation with HeLa nuclear extract, lariat intermediate and free 5′ exon (intermediates from the first step of splicing) as well as ligated exons and lariat product (from the second step of splicing) can be visualized on 15 % denaturing PAGE. (**b**) DNA template for library-containing in vitro splicing substrate was constructed by PCR. The first exon and part of the intron of Ad81 was amplified using T7 sequence flanking forward primer. The *green* flanking sequence in the library indicates the common region that serves as universal primer sequence for PCR amplification. Equimolar of Ad81 and library amplification products were joined by PCR to generate DNA template for in vitro transcription to generate library-containing pre-mRNA substrate. (**c**) In vitro splicing reactions of library-containing RNA substrate are shown at various time points with varying substrate concentrations, potassium concentrations and two forms of potassium counterions (glutamate and acetate). Heterogeneous species of lariat intermediates are formed and can be subsequently gel purified and analyzed with RT-PCR followed by deep sequencing. (**d**) In vitro spliceosome assembly assay of Ad81 with its 3′ splice site mutated from AG to GG (1) and library containing substrate (2) at 0, 4, 15, 30, and 60 min time points

then serve as candidates for drug development. The high-throughput in vitro splicing assay can be complemented with the spliceosome assembly assay as described in the next section.

3.4.2 High-Throughput In Vitro Spliceosome Assembly Assay

The different spliceosomal complexes can be captured by performing in vitro splicing reaction (as described in Sect. 3.4.1) with increment time points, followed by treatment with heparin and visualization in native gel. In Fig. 7.5d, we show that our library containing RNA substrate is capable of forming spliceosomal complexes in vitro. The library-containing RNA substrate species that are associated with the different subcomplexes can be recovered using a variety of methods including glycerol gradient fractionation, column fractionation, as well as gel purification. By purifying these spliceosomal subcomplexes and identifying RNA species that are associated with them with deep sequencing, we can obtain mechanistic insights in the role that sequence variation exerts in inhibiting or promoting splicing. It has been shown that the highly dynamic spliceosomal interactions are highly influenced by the identity of the pre-mRNA substrates (Clark et al. 2002; Pleiss et al. 2007; Meyer et al. 2011). This behavior suggests that assembly may be a sensitive assay for detecting mutants. Furthermore, the mechanism of inhibition of a particular variant can sometimes be learned—mutated RNA substrates were shown to cause spliceosome assembly to halt at B* complex (Bessonov et al. 2010). We demonstrate a single assembly reaction; however, this assay could be performed iteratively isolating and reamplifying shifted bands for additional rounds of enrichment. Finally, various compounds that inhibit splicing have been reported to stall the spliceosome at specific stages (Berg et al. 2012; O'Brien et al. 2008). Therefore, treating the library-containing substrate assembly assay with these known inhibitors can further identify substrate-specific effects in the splicing inhibitory activity that is exhibited by these compounds.

3.5 High-Throughput In Vivo Functional Assays

An excellent complement to the in vitro splicing and assembly assays is a high-throughput implementation of the in vivo minigene reporter assay described in Sect. 3.2. This assay is similar to the in vitro splicing assay in that it provides a test of the function consequence of variation on gene processing but has the advantage of more closely modeling the endogenous splicing reactions. In vitro splicing is uncoupled from other processes, so a comparison between the performances of two alleles can be regarded as a pure measure of allele specific splicing activity. The high-throughput in vivo splicing assay, on the other hand, compares steady state exon inclusion levels of alleles in mRNA which is mostly driven by differences in splicing but could be influenced by other process like transcription or stability. For the purpose of screening variants, this more comprehensive screen may help capture changes in expression that would be missed by other methods.

There are a few technical issues that have been discovered through earlier SELEX experiments. One issue is the difficulty of maintaining library representation during propagation and plasmid amplification in *E. coli*. It has been observed that minigene libraries that contain short inserts tend to undergo a dramatic loss of representation for certain sequences during library construction (Chen and Chasin 1994; Ke et al. 2011). Whether this phenomena occurs in all vectors is poorly understood. An effective workaround utilizes PCR ligation and overlap techniques to make library without transforming *E. coli* (Ke et al. 2011). Linear PCR libraries can be transfected into tissue culture cells, expressed and spliced. There are a variety of construction strategies that add flank to one side of the insert and then the other (Fig. 7.6). After construction of the linear minigene library it is necessary to record the initial representation of each allele of each loci in the starting library and then compare this representation to the spliced library. A skew in the allele ratio observed in the mRNA relative to the unspliced RNA (i.e., the input library) can be interpreted as an allelic

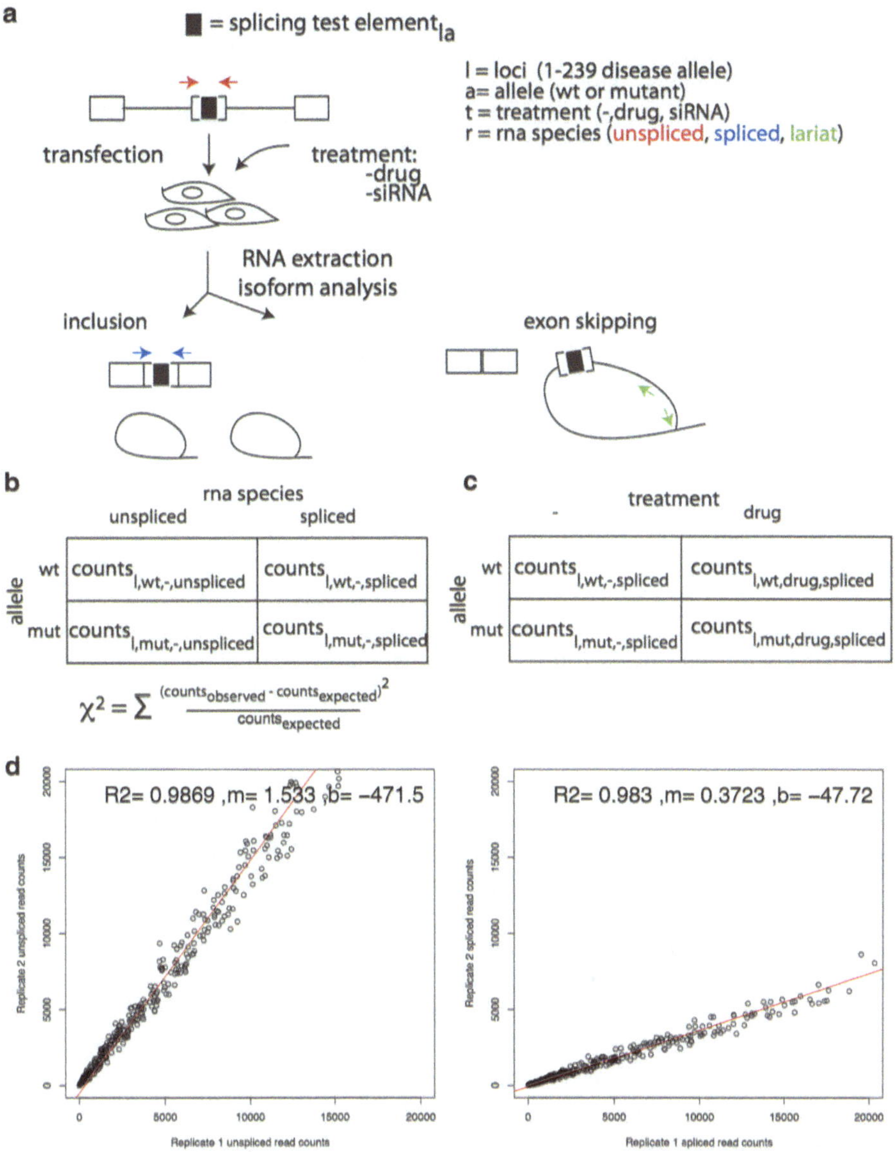

Fig. 7.6 Workflow for high-throughput in vivo splicing assay. This scheme cover a pilot experiment that screened 236 wild type/mutant 30-mer windows for allelic differences in their function as a splicing element. (**a**) The pilot library contains wild type and mutant alleles ("a" in notation) of 236 regions or loci known to be mutated in hereditary disease ("l" in notation) for a total of 478 different sequences. The library is transfected into tissue culture cells possibly in combination with drug or siRNA treatment. Total RNA extracted and individual species (unspliced, spliced, lariat) are analyzed by RT-PCR with the described arrangement of primers. Gel purified RT-PCR products are then analyzed by deep sequencing. (**b**) Allelic ratios determined from read counts are compared in unspliced and spliced species with the chi squared statistic. Allelic skew is suggestive of a variant that alters splicing. (**c**) Comparing splicing outcomes in the presence of drugs identifies compounds that influence splicing, potentially restoring allelic balance to transcripts disrupted by the variants identified above. (**d**) Comparison across replicates of counts for unspliced and spliced RNA species

difference in gene expression. Once such a system is established, it can become a screening platform for compounds that affect splicing. A traditional screening platform optimizes a biological assay to screen thousands of different compounds against one disease. The high-throughput minigene splicing assay screens thousands of diseases against one compound. Candidate compounds can be selected based on properties like their ability to bind RNA, previous reports of altering splicing or affecting kinases/phosphatases that can modify RNA splicing factors.

3.5.1 Case study: Pilot Project to Analyze 236 Disease Alleles

About 15 % of point mutations that cause hereditary disease localize to the canonical splice sites (Krawczak et al. 2000). Studies by our group and others suggest as many as 25 % of missense mutations also have a splicing defect (Lim et al. 2011). To screen a hereditary disease alleles for changes in splicing, we constructed a library in the minigene reporter pZW4. Conceptually this assay is identical to the approach described in Sect. 3.2. This backbone splices a test exon via a predominantly an inclusion phenotype making it most sensitive to detecting splicing silencers (Fairbrother and Chasin 2000). Therefore not all types of allelic differences will be returned by this assay. To be detected in this background, either the wild type or variant allele must reduce the inclusion of the constitutively spliced test exon.

The insert library consisted of 750 allelic pairs that were synthesized, amplified and cloned into the XhoI and ApaI sites in the test exon of pZW4. A successful cloning event places the 31 nucleotides window (i.e., missense mutation with 15 nucleotides of upstream and downstream flank) in the middle of the test exon. The test exon with insert is 98 nucleotides in length with the variant located at 46th nucleotide position downstream from the 3′ss. The pilot library was constructed by ligation and transformation into *E. coli*. Deep sequencing of the pooled library confirmed 236 successful wild type–mutant pairs, a value far lower than predicted (consistent with an *E. coli* propagation bias seen in other libraries). However, despite the nonuniform representation of the initial library separate sequencing experiments returned highly correlated biological replicates of the unspliced and spliced libraries (Fig. 7.6b)

The library was then transfected into tissue culture cells and assayed by RT-PCR. Three different primer sets can record allelic representation in starting library (i.e., Fig. 7.6a, unspliced) and the lariat and mRNA species that represent the various outcomes of splicing. Deep sequencing of these PCR products provides an estimate of allelic ratio for each of the 236 loci as they pass through different stages of the splicing reaction. The identification of a skew in allele ratio indicates the variant is altering the activity of a *cis*-element that is involved in splicing. The significance of a skew in allelic ratio is evaluated by a statistical test such as the chi-square test. A preliminary analysis of 236 missense alleles revealed that 13.5 % significantly altered splicing in the direction detectable by this reporter (Table 7.2). In most (19) of these 32 cases, it was the mutant allele that appeared to create some sort of negative sequence. In the diseases caused by these variations, it is reasonable to ask whether these mutations can be reversed by therapeutic treatment.

Table 7.2 Summary of high-throughput minigene screen of 236 disease alleles

Outcome	Number of pairs
Total Allelic pairs in library	236
Differentially spliced[a]	32 (13.6 %)
Mutations that increase splicing efficiency[b]	13 (5.5 %)
Mutations that decrease splicing efficiency[c]	19 (8.1 %)
Allelic difference induced by G418 treatment[d]	85 (36.0 %)
Allelic difference exacerbated by G418 treatment[e]	0
Allelic difference reduced by G418 treatment[f]	3 (1.3 %)
Allelic difference induced by tetracycline treatment	0
Allelic difference exacerbated by tetracycline treatment	0
Allelic difference reduced by tetracycline treatment	0
Allelic difference induced by valproic acid treatment	0
Allelic difference exacerbated by valproic acid treatment	0
Allelic difference reduced by valproic acid treatment	0

[a]Allelic difference in splicing ratios detected by χ^2 test at $q<0.05$
[b]Differentially spliced and enrichment index < 0
[c]Differentially spliced and enrichment index > 0
[d]No differential splicing in untreated samples; allele-dependent differential recovery of treated and untreated spliced products detected by χ^2 test at $q<0.05$
[e]Allelic differences detected in untreated samples; drug treatment increased difference between mutant and untreated wild-type splicing ratios
[f]Allelic differences detected in untreated samples; drug treatment decreases difference between mutant and untreated wild-type splicing ratios

To test the response of this set of 236 allele pairs to small molecules, we repeated the transfection in the presence of C50 doses of G418, tetracycline, and valproic acid. Each of these three treatments had demonstrated some ability to bind RNA or alter splicing. The RNA was extracted from each treated cell culture and sequenced. The allelic ratio in the mRNA of the treated versus untreated was recorded for each of three compounds. The preliminary result from this type of screening returned G418 as an active compound in this assay and tetracycline and valproic acid as having no significant effects. Sublethal concentration of the translational inhibitor G418 affected over one third of all allelic pairs in some way. Three of these alterations occurred in a direction that reduced the allelic expression bias observed in the initial screen.

4 Discussion

This chapter covered methods of studying variations that affect molecular events at the RNA level. We review computational and experimental methods for detecting changes in RNA protein interaction. We also cover several high-throughput functional assays: spliceosome formation assay, in vitro and in vivo splicing.

This suite of protocols could be used to screen all disease alleles and common variants for sequence changes that may alter splicing. In many cases the approaches are complimentary and could be combined for cross validation or further characterization. Secondary libraries that contain only allele specific events could be used to screen compounds for drugs that reverse splicing defects. In vitro splicing assays could confirm a drug acts directly on the RNA.

In addition, these technologies could be used to report molecular phenotypes of a variant. We presented a series of characterizations of a deletion of a PTB binding site in the RGS9 transcript. This variation appears to be associated with a significant but modest phenotype at the organismal level. The RGS9 gene results in significant body mass difference in rodents. The SNP in the RGS9 in humans has been reported to be associated with slight differences in body mass index (Waugh et al. 2011). Interestingly rs3215227 is a largely private to populations of Asian ancestry. The deletion occurs in Caucasian and African American populations with a low frequency (<1 %) but was much more common in samples drawn from multiple Asian populations (69 %). How to fit these types of observations into a broader population genetic framework is an important challenge. Due to the recent expansion of modern humans out of Africa and the reduction in genetic diversity through bottlenecks as all continents were peopled (Ramachandran et al. 2005), we might expect disease incidence to vary across ethnicities. Currently no framework exists to explain variable disease incidence across ethnic backgrounds although some GWA studies have been replicated in multiple populations. Waters et al. (2010) conducted association studies within various ethnic groups (European Americans, African Americans, Latinos, Japanese Americans, and Native Hawaiians) for type II diabetes. Of the 19 SNPs associated with type II diabetes in European Americans, 13 were associated with disease state in other ethnic groups but not at genome-wide significance. Furthermore, 5 of the 19 variants were not associated with diabetes in non-Europeans. The limitations of these approaches underscore the importance of molecular characterization of causal alleles for a better understanding of disease in all human populations. We describe high-throughput implementations of mature technologies that enable a system wide study of RBP ligand interaction. These approaches bring biochemical validation to a genomic scale and open novel avenues for learning disease mechanism and drug discovery for this understudied class of variants.

5 Protocol Appendix

Section 3.2 described low-throughput binding and functional assays that are fairly mature technologies with excellent protocols described elsewhere. Here we present a brief protocol for the high-throughput binding and occupancy assay described in Sect. 3.3; the high-throughput in vitro assembly and splicing assays described in Sect. 3.4; and the high-throughput minigene assay described in Sect. 3.5 and the case study 3.5.1.

5.1 Library Design and Oligonucleotide Synthesis

A complex pool of 60-mer oligonucleotides was synthesized to tile through about 4,000 alternatively spliced exons. This pool was synthesized as a custom oligonucleotide microarray in the 244K format (Agilent). Each oligonucleotide was designed as a tiled genomic 30-mer flanked by the common sequences that were used as the universal primer pair. Oligonucleotides were liberated from the slide by boiling at 99 °C for 1 h. They were recovered by gently pipetting 500 µl dH_2O to the slide followed by thorough scouring and resuspending with a sterile 25-gauge hypodermic needle or the array was sandwiched and placed in a hybridization chamber at 99 °C for 1 h. This is followed by sonication at 50 % amplitude for three 5 s pulses in a Sonic Dismembrator Model 500 (Fisher). Oligonucleotide pools were then amplified with low cycle PCR (1 min at 94 °C, 20 s at 55 °C, 1 min at 72 °C at the first round; 10, 20, 10 s at each respective temperature for subsequent rounds; and final elongation of 5 min at 72 °C). While this is the most economical way to obtain oligonucleotide libraries, there are commercial venders (e.g., Agilent and MYcroarray) who will synthesize sets of longer oligonucleotides as mixed pools. These oligonucleotide pools were amplified by PCR primer pair that includes flanking T7 promoter sequence in the forward primer, and subsequently used as template for in vitro RNA synthesis using Ambion MEGAshortscript™ kit.

5.2 PTB Binding Analysis (Sect. 3.3)

A 1:1 ratio of Protein A and Protein G Dynabeads (Invitrogen) mixture was incubated with 200 ng of RNA, 5 µl of mab BB7 hybridoma supernatant (ATCC# CRL-2501), and 120 µg of HeLa nuclear extract at 4 °C for 1 h. RNA was recovered by boiling in 1 % SDS. The subsequent enrichment was quantified by two-color microarray analysis using standard protocols or alternatively, they may be analyzed by deep sequencing.

5.3 Dot Blots (Sect. 3.3)

Samples were prepared in 30 µl solution containing 0.6× Buffer D, 50 ng/ml Poly dI·dC, 1 mM DTT, 1 µg/µl BSA, varying amounts of sonicated herring sperm (Invitrogen), 50 ng of probes, and 5 µg of ES whole cell extract. Nitrocellulose was soaked for 5 min in extraction buffer. Dot blots were blocked with 100 µl of 2 mg/ml sonicated herring sperm for 15 min. Samples were incubated for 30 min at room temperature and after addition of 100 µl with 0.6× extraction buffer, the samples were loaded and blotted. Nitrocellulose was washed with cold 1× PBS at room temperature for 15 min. Dot blots were subsequently imaged using Phosphorimaging screens and the Typhoon 9410 (GE). Dots were cut out and placed in 1.5 ml eppendorf tubes. Then 400 μl 7 M urea and 200 μl phenol were added to each tube and incubated for 30 min at room temperature. Then 200 μl chloroform was added and samples were spun down and the aqueous layer was removed. The phenol–chloroform extraction was repeated one more time before samples were used for subsequent PCR and next rounds of selection.

5.4 Protocol for In Vitro Splicing Assay (Sect. 3.4)

In vitro splicing assay of library containing substrate was performed by assembling reaction on ice containing 10 nM labeled RNA substrate, 40 % of HeLa nuclear extract, 2 mM magnesium acetate, 20–80 mM potassium glutamate (or 80 mM potassium acetate when indicated), 1 mM ATP, 5 mM creatine phosphate, and nuclease-free H_2O. Note that different nuclear extract preparations may require different potassium concentration for optimal splicing. Reaction was incubated at 30 °C and placed on ice after each indicated time points (0, 30, and 60 min). Splicing reaction was stopped by addition of 5–10× reaction volume of splicing dilution buffer containing 100 mM Tris (pH 7.5), 10 mM EDTA, 1 % SDS, 150 mM sodium chloride and 300 mM sodium acetate. This is followed by phenol–chloroform extraction and ethanol precipitation. RNA was subsequently separated by denaturing 15 % PAGE (29:1 acrylamide–bis-acrylamide).

5.5 Protocol for In Vitro Spliceosome Assembly Assay (Sect. 3.4)

In vitro splicing reaction of library-containing pre-mRNA substrate was performed as indicated above, except after reactions were quenched on ice after 0, 4, 15, 30, and 60 min time points, heparin was added to a final concentration of 0.5 mg/ml, and followed by additional incubation at 30 °C for 5 min. Spliceosomal complexes can then be separated in 2.1 % Ultrapure™ (Invitrogen) low melting point agarose (Fig. 7.5d) or 4 % native PAGE (80:1 acrylamide–bis-acrylamide) in Tris-glycine buffer. Note that different time points maybe required for different pool of pre-mRNA substrates to visualize the different spliceosomal complexes. As discussed in Sect. 3.4, each of the subcomplexes maybe isolated using various purification techniques and RNA associated with each of the subcomplexes can be analyzed by deep sequencing.

5.6 Protocol for Minigene Experiment (Sect. 3.5)

Oligonucleotide pool of wild type/mutant 30-mers was cloned into the pZW4 minigene construct at the XhoI and ApaI sites situated at +28 from the 3′ss and -39 from the 5′ss of exon 2 as shown in Fig. 7.6a. DNA from the pool of resulting colonies is transiently transfected into Hek293 cells, which may include drug or siRNA treatment, and the total RNA is extracted after 48 h. Alternatively, the ligation reaction can also be transfected directly into cultured cells to avoid sequence bias in bacterial propagation. The unspliced, spliced and lariat species can be analyzed with RT-PCR with the appropriate primers (see Fig. 7.6a).

A barcode system was used for multiplexing different experimental conditions. We first excluded reads that did not retain intact barcodes. All the remaining reads were classified in four categories: wild type spliced, wild type unspliced, mutant spliced, and mutant unspliced, as shown in the contingency table at Fig. 7.6b. Allelic pairs with low read counts (<5) in more than one of the four categories were further excluded. The remaining allelic pairs were subjected to χ^2 testing, and we accepted tests with a FDR < 5 % after multiple testing corrections as evidence for differential splicing. Additionally, we calculated an enrichment index (EI) for each allelic pair, defined as the ratio of unspliced to spliced read counts of the wild type allele over the unspliced to spliced ratio of the mutant.

To screen for allele-specific splicing effects under various drug treatments, we compared allelic ratios of spliced products from untreated and drug-treated cells (Fig. 7.6c). For each pair that exhibited an allelic effect on splicing and an additional drug-dependent effect, we determined if the treatment increased or reduced the skew in mutant splicing (see Table 7.2). This is done by recalculating the EI and replacing the mutant non-treated spliced counts with the mutant drug-treated spliced counts.

References

Abecasis GR, Altshuler D, Auton A, Brooks LD, Durbin RM, Gibbs RA, Hurles ME, McVean GA (2010) A map of human genome variation from population-scale sequencing. Nature 467(7319):1061–1073. doi:10.1038/nature09534

Abecasis GR, Auton A, Brooks LD, DePristo MA, Durbin RM, Handsaker RE, Kang HM, Marth GT, McVean GA (2012) An integrated map of genetic variation from 1,092 human genomes. Nature 491(7422):56–65. doi:10.1038/nature11632

Ars E, Serra E, Garcia J, Kruyer H, Gaona A, Lazaro C, Estivill X (2000) Mutations affecting mRNA splicing are the most common molecular defects in patients with neurofibromatosis type 1. Hum Mol Genet 9(2):237–247

Auweter SD, Oberstrass FC, Allain FHT (2006) Sequence-specific binding of single-stranded RNA: is there a code for recognition? Nucleic Acids Res 34(17):4943–4959. doi:10.1093/nar/gkl620

Berg MG, Wan L, Younis I, Diem MD, Soo M, Wang C, Dreyfuss G (2012) A quantitative high-throughput in vitro splicing assay identifies inhibitors of spliceosome catalysis. Mol Cell Biol 32(7):1271–1283. doi:10.1128/MCB.05788-11

Berget SM (1995) Exon recognition in vertebrate splicing. J Biol Chem 270(6):2411–2414

Bessonov S, Anokhina M, Krasauskas A, Golas MM, Sander B, Will CL, Urlaub H, Stark H, Luhrmann R (2010) Characterization of purified human Bact spliceosomal complexes reveals compositional and morphological changes during spliceosome activation and first step catalysis. RNA 16(12):2384–2403. doi:10.1261/rna.2456210

Boutz PL, Stoilov P, Li Q, Lin CH, Chawla G, Ostrow K, Shiue L, Ares M Jr, Black DL (2007) A post-transcriptional regulatory switch in polypyrimidine tract-binding proteins reprograms alternative splicing in developing neurons. Genes Dev 21(13):1636–1652

Brunak S, Engelbrecht J, Knudsen S (1991) Prediction of human mRNA donor and acceptor sites from the DNA sequence. J Mol Biol 220(1):49–65

Buratti E, Chivers M, Kralovicova J, Romano M, Baralle M, Krainer AR, Vorechovsky I (2007) Aberrant 5′ splice sites in human disease genes: mutation pattern, nucleotide structure and comparison of computational tools that predict their utilization. Nucleic Acids Res 35(13):4250–4263. doi:10.1093/nar/gkm402

Burrows NP, Nicholls AC, Richards AJ, Luccarini C, Harrison JB, Yates JR, Pope FM (1998) A point mutation in an intronic branch site results in aberrant splicing of COL5A1 and in Ehlers-Danlos syndrome type II in two British families. Am J Hum Genet 63(2):390–398. doi:10.1086/301948

Bustamante CD, Burchard EG, De la Vega FM (2011) Genomics for the world. Nature 475(7355):163–165. doi:10.1038/475163a

Carmel I, Tal S, Vig I, Ast G (2004) Comparative analysis detects dependencies among the 5′ splice-site positions. RNA 10(5):828–840

Cartegni L, Chew SL, Krainer AR (2002) Listening to silence and understanding nonsense: exonic mutations that affect splicing. Nat Rev Genet 3(4):285–298. doi:10.1038/nrg775

Cartegni L, Wang J, Zhu Z, Zhang MQ, Krainer AR (2003) ESEfinder: A web resource to identify exonic splicing enhancers. Nucleic Acids Res 31(13):3568–3571

Casto AM, Feldman MW (2011) Genome-wide association study SNPs in the human genome diversity project populations: does selection affect unlinked SNPs with shared trait associations? PLoS Genet 7(1):e1001266. doi:10.1371/journal.pgen.1001266

Chang B, Levin J, Thompson WA, Fairbrother WG (2009) High-throughput binding analysis determines the binding specificity of ASF/SF2 on alternatively spliced human pre-mRNAs. Comb Chem High Throughput Screen 13(3):242–252

Chasin LA (2007) Searching for splicing motifs. Adv Exp Med Biol 623:85–106

Chen IT, Chasin LA (1994) Large exon size does not limit splicing in vivo. Mol Cell Biol 14(3):2140–2146

Cho DH, Tapscott SJ (2007) Myotonic dystrophy: emerging mechanisms for DM1 and DM2. Biochim Biophys Acta 1772(2):195–204. doi:10.1016/j.bbadis.2006.05.013

Chodosh LA (2001) UV crosslinking of proteins to nucleic acids. Curr Protoc Mol Biol Chapter 12: Unit 12 15

Clark TA, Sugnet CW, Ares M Jr (2002) Genomewide analysis of mRNA processing in yeast using splicing-specific microarrays. Science 296(5569):907–910. doi:10.1126/science.1069415

Cooper TA (2005) Use of minigene systems to dissect alternative splicing elements. Methods 37(4):331–340

Coulter LR, Landree MA, Cooper TA (1997) Identification of a new class of exonic splicing enhancers by in vivo selection. Mol Cell Biol 17(4):2143–2150

Crotti L, Lewandowska MA, Schwartz PJ, Insolia R, Pedrazzini M, Bussani E, Dagradi F, George AL Jr, Pagani F (2009) A KCNH2 branch point mutation causing aberrant splicing contributes to an explanation of genotype-negative long QT syndrome. Heart Rhythm 6(2):212–218. doi:10.1016/j.hrthm.2008.10.044

Damgaard CK, Tange TO, Kjems J (2002) hnRNP A1 controls HIV-1 mRNA splicing through cooperative binding to intron and exon splicing silencers in the context of a conserved secondary structure. RNA 8(11):1401–1415

Das MK, Dai HK (2007) A survey of DNA motif finding algorithms. BMC Bioinformatics 8(Suppl 7):S21, doi:1471-2105-8-S7-S21 [pii] 10.1186/1471-2105-8-S7-S21

Desmet FO, Hamroun D, Lalande M, Collod-Beroud G, Claustres M, Beroud C (2009) Human splicing finder: an online bioinformatics tool to predict splicing signals. Nucleic Acids Res 37(9):e67. doi:10.1093/nar/gkp215

Dickson SP, Wang K, Krantz I, Hakonarson H, Goldstein DB (2010) Rare variants create synthetic genome-wide associations. PLoS Biol 8(1):e1000294. doi:10.1371/journal.pbio.1000294

Dignam JD, Lebovitz RM, Roeder RG (1983) Accurate transcription initiation by RNA polymerase II in a soluble extract from isolated mammalian nuclei. Nucleic Acids Res 11(5):1475–1489

Ding Y, Chan CY, Lawrence CE (2004) Sfold web server for statistical folding and rational design of nucleic acids. Nucleic Acids Res 32(Web Server issue):W135–W141, doi:10.1093/nar/gkh449

Divina P, Kvitkovicova A, Buratti E, Vorechovsky I (2009) Ab initio prediction of mutation-induced cryptic splice-site activation and exon skipping. Eur J Hum Genet 17(6):759–765. doi:10.1038/ejhg.2008.257

Djordjevic M (2007) SELEX experiments: new prospects, applications and data analysis in inferring regulatory pathways. Biomol Eng 24(2):179–189

Fairbrother WG, Chasin LA (2000) Human genomic sequences that inhibit splicing. Mol Cell Biol 20(18):6816–6825

Fairbrother WG, Yeh RF, Sharp PA, Burge CB (2002) Predictive identification of exonic splicing enhancers in human genes. Science 297(5583):1007–1013. doi:10.1126/science.1073774

Fairbrother WG, Yeo GW, Yeh R, Goldstein P, Mawson M, Sharp PA, Burge CB (2004) RESCUE-ESE identifies candidate exonic splicing enhancers in vertebrate exons. Nucleic Acids Res 32(Web Server issue):W187–W190, doi:10.1093/nar/gkh393

Ferraris L, Stewart AP, Gemberling MP, Reid DC, Lapadula MJ, Thompson WA, Fairbrother WG (2011a) High-throughput mapping of protein occupancy identifies functional elements without the restriction of a candidate factor approach. Nucleic Acids Res 39(6):e33. doi:10.1093/nar/gkq1213

Ferraris L, Stewart AP, Kang J, DeSimone AM, Gemberling M, Tantin D, Fairbrother WG (2011b) Combinatorial binding of transcription factors in the pluripotency control regions of the genome. Genome Res 21(7):1055–1064. doi:10.1101/gr.115824.110

Folco EG, Lei H, Hsu JL, Reed R (2012) Small-scale nuclear extracts for functional assays of gene-expression machineries. J Vis Exp. (64). doi:10.3791/4140

Fox-Walsh KL, Dou Y, Lam BJ, Hung SP, Baldi PF, Hertel KJ (2005) The architecture of pre-mRNAs affects mechanisms of splice-site pairing. Proc Natl Acad Sci U S A 102(45):16176–16181. doi:10.1073/pnas.0508489102

Furneaux HM, Perkins KK, Freyer GA, Arenas J, Hurwitz J (1985) Isolation and characterization of two fractions from HeLa cells required for mRNA splicing in vitro. Proc Natl Acad Sci U S A 82(13):4351–4355

Girard C, Will CL, Peng J, Makarov EM, Kastner B, Lemm I, Urlaub H, Hartmuth K, Luhrmann R (2012) Post-transcriptional spliceosomes are retained in nuclear speckles until splicing completion. Nat Commun 3:994. doi:10.1038/ncomms1998

Gonzaga-Jauregui C, Lupski JR, Gibbs RA (2012) Human genome sequencing in health and disease. Annu Rev Med 63:35–61. doi:10.1146/annurev-med-051010-162644

Gooding C, Clark F, Wollerton MC, Grellscheid SN, Groom H, Smith CW (2006) A class of human exons with predicted distant branch points revealed by analysis of AG dinucleotide exclusion zones. Genome Biol 7(1):R1

Gopinath SC (2007) Methods developed for SELEX. Anal Bioanal Chem 387(1):171–182

Goren A, Ram O, Amit M, Keren H, Lev-Maor G, Vig I, Pupko T, Ast G (2006) Comparative analysis identifies exonic splicing regulatory sequences–The complex definition of enhancers and silencers. Mol Cell 22(6):769–781. doi:10.1016/j.molcel.2006.05.008

Gozani O, Patton JG, Reed R (1994) A novel set of spliceosome-associated proteins and the essential splicing factor PSF bind stably to pre-mRNA prior to catalytic step II of the splicing reaction. EMBO J 13(14):3356–3367

Gravel S, Henn BM, Gutenkunst RN, Indap AR, Marth GT, Clark AG, Yu F, Gibbs RA, Bustamante CD (2011) Demographic history and rare allele sharing among human populations. Proc Natl Acad Sci U S A 108(29):11983–11988. doi:10.1073/pnas.1019276108

Halvorsen M, Martin JS, Broadaway S, Laederach A (2010) Disease-associated mutations that alter the RNA structural ensemble. PLoS Genet 6(8):e1001074. doi:10.1371/journal.pgen.1001074

Hastings ML, Krainer AR (2001) Pre-mRNA splicing in the new millennium. Curr Opin Cell Biol 13(3):302–309

Hebsgaard SM, Korning PG, Tolstrup N, Engelbrecht J, Rouze P, Brunak S (1996) Splice site prediction in Arabidopsis thaliana pre-mRNA by combining local and global sequence information. Nucleic Acids Res 24(17):3439–3452

Hertz GZ, Stormo GD (1999) Identifying DNA and protein patterns with statistically significant alignments of multiple sequences. Bioinformatics 15(7–8):563–577

Hoskins AA, Moore MJ (2012) The spliceosome: a flexible, reversible macromolecular machine. Trends Biochem Sci 37(5):179–188. doi:10.1016/j.tibs.2012.02.009

Jorde LB, Carey JC, White RL (1996) Medical genetics. Mosby Inc., Orlando

Jurica MS, Moore MJ (2002) Capturing splicing complexes to study structure and mechanism. Methods 28(3):336–345

Jurica MS, Moore MJ (2003) Pre-mRNA splicing: awash in a sea of proteins. Mol Cell 12(1):5–14

Kanopka A, Muhlemann O, Akusjarvi G (1996) Inhibition by SR proteins of splicing of a regulated adenovirus pre-mRNA. Nature 381(6582):535–538. doi:10.1038/381535a0

Ke S, Shang S, Kalachikov SM, Morozova I, Yu L, Russo JJ, Ju J, Chasin LA (2011) Quantitative evaluation of all hexamers as exonic splicing elements. Genome Res 21(8):1360–1374. doi:10.1101/gr.119628.110

Knudsen B, Hein J (1999) RNA secondary structure prediction using stochastic context-free grammars and evolutionary history. Bioinformatics 15(6):446–454

Knudsen B, Hein J (2003) Pfold: RNA secondary structure prediction using stochastic context-free grammars. Nucleic Acids Res 31(13):3423–3428

Krainer AR, Maniatis T, Ruskin B, Green MR (1984) Normal and mutant human beta-globin pre-mRNAs are faithfully and efficiently spliced in vitro. Cell 36(4):993–1005

Kralovicova J, Vorechovsky I (2007) Global control of aberrant splice-site activation by auxiliary splicing sequences: evidence for a gradient in exon and intron definition. Nucleic Acids Res 35(19):6399–6413. doi:10.1093/nar/gkm680

Kralovicova J, Houngninou-Molango S, Kramer A, Vorechovsky I (2004) Branch site haplotypes that control alternative splicing. Hum Mol Genet 13(24):3189–3202. doi:10.1093/hmg/ddh334

Krawczak M, Reiss J, Cooper DN (1992) The mutational spectrum of single base-pair substitutions in mRNA splice junctions of human genes: causes and consequences. Hum Genet 90(1–2):41–54

Krawczak M, Ball EV, Fenton I, Stenson PD, Abeysinghe S, Thomas N, Cooper DN (2000) Human gene mutation database-a biomedical information and research resource. Hum Mutat 15(1):45–51

Lander ES (1996) The new genomics: global views of biology. Science 274(5287):536–539

Lander ES, Linton LM, Birren B, Nusbaum C, Zody MC, Baldwin J, Devon K, Dewar K, Doyle M, FitzHugh W, Funke R, Gage D, Harris K, Heaford A, Howland J, Kann L, Lehoczky J, LeVine R, McEwan P, McKernan K, Meldrim J, Mesirov JP, Miranda C, Morris W, Naylor J, Raymond C, Rosetti M, Santos R, Sheridan A, Sougnez C, Stange-Thomann N, Stojanovic N, Subramanian A, Wyman D, Rogers J, Sulston J, Ainscough R, Beck S, Bentley D, Burton J, Clee C, Carter N, Coulson A, Deadman R, Deloukas P, Dunham A, Dunham I, Durbin R, French L, Grafham D, Gregory S, Hubbard T, Humphray S, Hunt A, Jones M, Lloyd C, McMurray A, Matthews L, Mercer S, Milne S, Mullikin JC, Mungall A, Plumb R, Ross M, Shownkeen R, Sims S, Waterston RH, Wilson RK, Hillier LW, McPherson JD, Marra MA, Mardis ER, Fulton LA, Chinwalla AT, Pepin KH, Gish WR, Chissoe SL, Wendl MC, Delehaunty KD, Miner TL, Delehaunty A, Kramer JB, Cook LL, Fulton RS, Johnson DL,

Minx PJ, Clifton SW, Hawkins T, Branscomb E, Predki P, Richardson P, Wenning S, Slezak T, Doggett N, Cheng JF, Olsen A, Lucas S, Elkin C, Uberbacher E, Frazier M, Gibbs RA, Muzny DM, Scherer SE, Bouck JB, Sodergren EJ, Worley KC, Rives CM, Gorrell JH, Metzker ML, Naylor SL, Kucherlapati RS, Nelson DL, Weinstock GM, Sakaki Y, Fujiyama A, Hattori M, Yada T, Toyoda A, Itoh T, Kawagoe C, Watanabe H, Totoki Y, Taylor T, Weissenbach J, Heilig R, Saurin W, Artiguenave F, Brottier P, Bruls T, Pelletier E, Robert C, Wincker P, Smith DR, Doucette-Stamm L, Rubenfield M, Weinstock K, Lee HM, Dubois J, Rosenthal A, Platzer M, Nyakatura G, Taudien S, Rump A, Yang H, Yu J, Wang J, Huang G, Gu J, Hood L, Rowen L, Madan A, Qin S, Davis RW, Federspiel NA, Abola AP, Proctor MJ, Myers RM, Schmutz J, Dickson M, Grimwood J, Cox DR, Olson MV, Kaul R, Shimizu N, Kawasaki K, Minoshima S, Evans GA, Athanasiou M, Schultz R, Roe BA, Chen F, Pan H, Ramser J, Lehrach H, Reinhardt R, McCombie WR, de la Bastide M, Dedhia N, Blocker H, Hornischer K, Nordsiek G, Agarwala R, Aravind L, Bailey JA, Bateman A, Batzoglou S, Birney E, Bork P, Brown DG, Burge CB, Cerutti L, Chen HC, Church D, Clamp M, Copley RR, Doerks T, Eddy SR, Eichler EE, Furey TS, Galagan J, Gilbert JG, Harmon C, Hayashizaki Y, Haussler D, Hermjakob H, Hokamp K, Jang W, Johnson LS, Jones TA, Kasif S, Kaspryzk A, Kennedy S, Kent WJ, Kitts P, Koonin EV, Korf I, Kulp D, Lancet D, Lowe TM, McLysaght A, Mikkelsen T, Moran JV, Mulder N, Pollara VJ, Ponting CP, Schuler G, Schultz J, Slater G, Smit AF, Stupka E, Szustakowski J, Thierry-Mieg D, Thierry-Mieg J, Wagner L, Wallis J, Wheeler R, Williams A, Wolf YI, Wolfe KH, Yang SP, Yeh RF, Collins F, Guyer MS, Peterson J, Felsenfeld A, Wetterstrand KA, Patrinos A, Morgan MJ, de Jong P, Catanese JJ, Osoegawa K, Shizuya H, Choi S, Chen YJ (2001) Initial sequencing and analysis of the human genome. Nature 409(6822): 860–921

Lareau LF, Inada M, Green RE, Wengrod JC, Brenner SE (2007) Unproductive splicing of SR genes associated with highly conserved and ultraconserved DNA elements. Nature 446(7138):926–929, doi:nature05676 [pii] 10.1038/nature05676

Lim KH, Fairbrother WG (2012) Spliceman—a computational web server that predicts sequence variations in pre-mRNA splicing. Bioinformatics 28(7):1031–1032. doi:10.1093/bioinformatics/bts074

Lim KH, Ferraris L, Filloux ME, Raphael BJ, Fairbrother WG (2011) Using positional distribution to identify splicing elements and predict pre-mRNA processing defects in human genes. Proc Natl Acad Sci U S A 108(27):11093–11098. doi:10.1073/pnas.1101135108

Lin S, Fu XD (2007) SR proteins and related factors in alternative splicing. Adv Exp Med Biol 623:107–122

Lin RJ, Newman AJ, Cheng SC, Abelson J (1985) Yeast mRNA splicing in vitro. J Biol Chem 260(27):14780–14792

Long D, Lee R, Williams P, Chan CY, Ambros V, Ding Y (2007) Potent effect of target structure on microRNA function. Nat Struct Mol Biol 14(4):287–294

Manley JL, Tacke R (1996) SR proteins and splicing control. Genes Dev 10(13):1569–1579

Manolio TA, Collins FS, Cox NJ, Goldstein DB, Hindorff LA, Hunter DJ, McCarthy MI, Ramos EM, Cardon LR, Chakravarti A, Cho JH, Guttmacher AE, Kong A, Kruglyak L, Mardis E, Rotimi CN, Slatkin M, Valle D, Whittemore AS, Boehnke M, Clark AG, Eichler EE, Gibson G, Haines JL, Mackay TF, McCarroll SA, Visscher PM (2009) Finding the missing heritability of complex diseases. Nature 461(7265):747–753. doi:10.1038/nature08494

Manolio TA, Chisholm RL, Ozenberger B, Roden DM, Williams MS, Wilson R, Bick D, Bottinger EP, Brilliant MH, Eng C, Frazer KA, Korf B, Ledbetter DH, Lupski JR, Marsh C, Mrazek D, Murray MF, O'Donnell PH, Rader DJ, Relling MV, Shuldiner AR, Valle D, Weinshilboum R, Green ED, Ginsburg GS (2013) Implementing genomic medicine in the clinic: the future is here. Genet Med 15(4):258–267. doi:10.1038/gim.2012.157

Martinez-Contreras R, Fisette JF, Nasim FU, Madden R, Cordeau M, Chabot B (2006) Intronic binding sites for hnRNP A/B and hnRNP F/H proteins stimulate pre-mRNA splicing. PLoS Biol 4(2):e21

Maslen C, Babcock D, Raghunath M, Steinmann B (1997) A rare branch-point mutation is associated with missplicing of fibrillin-2 in a large family with congenital contractural arachnodactyly. Am J Hum Genet 60(6):1389–1398. doi:10.1086/515472

Matlin AJ, Moore MJ (2007) Spliceosome assembly and composition. Adv Exp Med Biol 623:14–35

McNally LM, McNally MT (1996) SR protein splicing factors interact with the Rous sarcoma virus negative regulator of splicing element. J Virol 70(2):1163–1172

Meyer M, Plass M, Perez-Valle J, Eyras E, Vilardell J (2011) Deciphering 3´ss selection in the yeast genome reveals an RNA thermosensor that mediates alternative splicing. Mol Cell 43(6):1033–1039. doi:10.1016/j.molcel.2011.07.030

Nielsen KB, Sorensen S, Cartegni L, Corydon TJ, Doktor TK, Schroeder LD, Reinert LS, Elpeleg O, Krainer AR, Gregersen N, Kjems J, Andresen BS (2007) Seemingly neutral polymorphic variants may confer immunity to splicing-inactivating mutations: a synonymous SNP in exon 5 of MCAD protects from deleterious mutations in a flanking exonic splicing enhancer. Am J Hum Genet 80(3):416–432. doi:10.1086/511992

O'Brien K, Matlin AJ, Lowell AM, Moore MJ (2008) The biflavonoid isoginkgetin is a general inhibitor of Pre-mRNA splicing. J Biol Chem 283(48):33147–33154. doi:10.1074/jbc.M805556200

Pagani F, Buratti E, Stuani C, Bendix R, Dork T, Baralle FE (2002) A new type of mutation causes a splicing defect in ATM. Nat Genet 30(4):426–429. doi:10.1038/ng858

Pagani F, Stuani C, Tzetis M, Kanavakis E, Efthymiadou A, Doudounakis S, Casals T, Baralle FE (2003) New type of disease causing mutations: the example of the composite exonic regulatory elements of splicing in CFTR exon 12. Hum Mol Genet 12(10):1111–1120

Pagani F, Raponi M, Baralle FE (2005) Synonymous mutations in CFTR exon 12 affect splicing and are not neutral in evolution. Proc Natl Acad Sci U S A 102(18):6368–6372. doi:10.1073/pnas.0502288102

Pan Q, Shai O, Lee LJ, Frey BJ, Blencowe BJ (2008) Deep surveying of alternative splicing complexity in the human transcriptome by high-throughput sequencing. Nat Genet 40(12):1413–1415. doi:10.1038/ng.259

Perez I, Lin CH, McAfee JG, Patton JG (1997) Mutation of PTB binding sites causes misregulation of alternative 3´ splice site selection in vivo. RNA 3(7):764–778

Pleiss JA, Whitworth GB, Bergkessel M, Guthrie C (2007) Transcript specificity in yeast pre-mRNA splicing revealed by mutations in core spliceosomal components. PLoS Biol 5(4):e90. doi:10.1371/journal.pbio.0050090

Price AL, Patterson NJ, Plenge RM, Weinblatt ME, Shadick NA, Reich D (2006) Principal components analysis corrects for stratification in genome-wide association studies. Nat Genet 38(8):904–909. doi:10.1038/ng1847

Ramachandran S, Deshpande O, Roseman CC, Rosenberg NA, Feldman MW, Cavalli-Sforza LL (2005) Support from the relationship of genetic and geographic distance in human populations for a serial founder effect originating in Africa. Proc Natl Acad Sci U S A 102(44):15942–15947. doi:10.1073/pnas.0507611102

Ranum LP, Cooper TA (2006) RNA-mediated neuromuscular disorders. Annu Rev Neurosci 29:259–277. doi:10.1146/annurev.neuro.29.051605.113014

Reese MG, Eeckman FH, Kulp D, Haussler D (1997) Improved splice site detection in Genie. J Comput Biol 4(3):311–323

Reich DE, Schaffner SF, Daly MJ, McVean G, Mullikin JC, Higgins JM, Richter DJ, Lander ES, Altshuler D (2002) Human genome sequence variation and the influence of gene history, mutation and recombination. Nat Genet 32(1):135–142

Reid DC, Chang BL, Gunderson SI, Alpert L, Thompson WA, Fairbrother WG (2009) Next-generation SELEX identifies sequence and structural determinants of splicing factor binding in human pre-mRNA sequence. RNA 15(12):2385–2397. doi:10.1261/rna.1821809

Robberson BL, Cote GJ, Berget SM (1990) Exon definition may facilitate splice site selection in RNAs with multiple exons. Mol Cell Biol 10(1):84–94

Roca X, Akerman M, Gaus H, Berdeja A, Bennett CF, Krainer AR (2012) Widespread recognition of 5′ splice sites by noncanonical base-pairing to U1 snRNA involving bulged nucleotides. Genes Dev 26(10):1098–1109. doi:10.1101/gad.190173.112

Ruskin B, Krainer AR, Maniatis T, Green MR (1984) Excision of an intact intron as a novel lariat structure during pre-mRNA splicing in vitro. Cell 38(1):317–331

Sabeti PC, Reich DE, Higgins JM, Levine HZ, Richter DJ, Schaffner SF, Gabriel SB, Platko JV, Patterson NJ, McDonald GJ, Ackerman HC, Campbell SJ, Altshuler D, Cooper R, Kwiatkowski D, Ward R, Lander ES (2002) Detecting recent positive selection in the human genome from haplotype structure. Nature 419(6909):832–837

Sachidanandam R, Weissman D, Schmidt SC, Kakol JM, Stein LD, Marth G, Sherry S, Mullikin JC, Mortimore BJ, Willey DL, Hunt SE, Cole CG, Coggill PC, Rice CM, Ning Z, Rogers J, Bentley DR, Kwok PY, Mardis ER, Yeh RT, Schultz B, Cook L, Davenport R, Dante M, Fulton L, Hillier L, Waterston RH, McPherson JD, Gilman B, Schaffner S, Van Etten WJ, Reich D, Higgins J, Daly MJ, Blumenstiel B, Baldwin J, Stange-Thomann N, Zody MC, Linton L, Lander ES, Altshuler D (2001) A map of human genome sequence variation containing 1.42 million single nucleotide polymorphisms. Nature 409(6822):928–933

Schwartz S, Ast G (2010) Chromatin density and splicing destiny: on the cross-talk between chromatin structure and splicing. EMBO J 29(10):1629–1636. doi:10.1038/emboj.2010.71

Sciabica KS, Hertel KJ (2006) The splicing regulators Tra and Tra2 are unusually potent activators of pre-mRNA splicing. Nucleic Acids Res 34(22):6612–6620

Siddharthan R (2007) Parsing regulatory DNA: general tasks, techniques, and the PhyloGibbs approach. J Biosci 32(5):863–870

Singh R, Valcarcel J, Green MR (1995) Distinct binding specificities and functions of higher eukaryotic polypyrimidine tract-binding proteins. Science 268(5214):1173–1176

Smith PJ, Zhang C, Wang J, Chew SL, Zhang MQ, Krainer AR (2006) An increased specificity score matrix for the prediction of SF2/ASF-specific exonic splicing enhancers. Hum Mol Genet 15(16):2490–2508. doi:10.1093/hmg/ddl171

Steiner B, Truninger K, Sanz J, Schaller A, Gallati S (2004) The role of common single-nucleotide polymorphisms on exon 9 and exon 12 skipping in nonmutated CFTR alleles. Hum Mutat 24(2):120–129. doi:10.1002/humu.20064

Steitz JA, Dreyfuss G, Krainer AR, Lamond AI, Matera AG, Padgett RA (2008) Where in the cell is the minor spliceosome? Proc Natl Acad Sci U S A 105(25):8485–8486. doi:10.1073/pnas.0804024105

Stenson PD, Ball EV, Mort M, Phillips AD, Shiel JA, Thomas NS, Abeysinghe S, Krawczak M, Cooper DN (2003) Human Gene Mutation Database (HGMD): 2003 update. Hum Mutat 21(6):577–581

Sterne-Weiler T, Howard J, Mort M, Cooper DN, Sanford JR (2011) Loss of exon identity is a common mechanism of human inherited disease. Genome Res 21(10):1563–1571. doi:10.1101/gr.118638.110

Stoltenburg R, Reinemann C, Strehlitz B (2007) SELEX–a (r)evolutionary method to generate high-affinity nucleic acid ligands. Biomol Eng 24(4):381–403, doi:S1389-0344(07)00066-4 [pii] 10.1016/j.bioeng.2007.06.001

Sun H, Chasin LA (2000) Multiple splicing defects in an intronic false exon. Mol Cell Biol 20(17):6414–6425

Taggart AJ, DeSimone AM, Shih JS, Filloux ME, Fairbrother WG (2012) Large-scale mapping of branchpoints in human pre-mRNA transcripts in vivo. Nat Struct Mol Biol 19(7):719–721. doi:10.1038/nsmb.2327

Tantin D, Gemberling M, Callister C, Fairbrother W (2008) High-throughput biochemical analysis of in vivo location data reveals novel distinct classes of POU5F1(Oct4)/DNA complexes. Genome Res 18(4):631–639

Teraoka SN, Telatar M, Becker-Catania S, Liang T, Onengut S, Tolun A, Chessa L, Sanal O, Bernatowska E, Gatti RA, Concannon P (1999) Splicing defects in the ataxia-telangiectasia gene, ATM: underlying mutations and consequences. Am J Hum Genet 64(6):1617–1631. doi:10.1086/302418

Ule J, Stefani G, Mele A, Ruggiu M, Wang X, Taneri B, Gaasterland T, Blencowe BJ, Darnell RB (2006) An RNA map predicting Nova-dependent splicing regulation. Nature 444(7119):580–586. doi:10.1038/nature05304

Vorechovsky I (2006) Aberrant 3′ splice sites in human disease genes: mutation pattern, nucleotide structure and comparison of computational tools that predict their utilization. Nucleic Acids Res 34(16):4630–4641. doi:10.1093/nar/gkl535

Wang GS, Cooper TA (2007) Splicing in disease: disruption of the splicing code and the decoding machinery. Nat Rev Genet 8(10):749–761. doi:10.1038/nrg2164

Wang Z, Rolish ME, Yeo G, Tung V, Mawson M, Burge CB (2004) Systematic identification and analysis of exonic splicing silencers. Cell 119(6):831–845. doi:10.1016/j.cell.2004.11.010

Wang Z, Xiao X, Van Nostrand E, Burge CB (2006) General and specific functions of exonic splicing silencers in splicing control. Mol Cell 23(1):61–70. doi:10.1016/j.molcel.2006.05.018

Wang ET, Sandberg R, Luo S, Khrebtukova I, Zhang L, Mayr C, Kingsmore SF, Schroth GP, Burge CB (2008) Alternative isoform regulation in human tissue transcriptomes. Nature 456(7221):470–476. doi:10.1038/nature07509

Waters KM, Stram DO, Hassanein MT, Le Marchand L, Wilkens LR, Maskarinec G, Monroe KR, Kolonel LN, Altshuler D, Henderson BE, Haiman CA (2010) Consistent association of type 2 diabetes risk variants found in Europeans in diverse racial and ethnic groups. PLoS Genet 6(8). doi:10.1371/journal.pgen.1001078

Watkins KH, Stewart A, Fairbrother W (2009) A rapid high-throughput method for mapping ribonucleoproteins (RNPs) on human pre-mRNA. J Vis Exp. (34)

Waugh JL, Celver J, Sharma M, Dufresne RL, Terzi D, Risch SC, Fairbrother WG, Neve RL, Kane JP, Malloy MJ, Pullinger CR, Gu HF, Tsatsanis C, Hamilton SP, Gold SJ, Zachariou V, Kovoor A (2011) Association between regulator of G protein signaling 9-2 and body weight. PLoS One 6(11):e27984. doi:10.1371/journal.pone.0027984

Webb JC, Patel DD, Shoulders CC, Knight BL, Soutar AK (1996) Genetic variation at a splicing branch point in intron 9 of the low density lipoprotein (LDL)-receptor gene: a rare mutation that disrupts mRNA splicing in a patient with familial hypercholesterolaemia and a common polymorphism. Hum Mol Genet 5(9):1325–1331

Yajima M, Fairbrother WG, Wessel GM (2012) ISWI contributes to ArsI insulator function in development of the sea urchin. Development 139(19):3613–3622. doi:10.1242/dev.081828

Yeo G, Burge CB (2004) Maximum entropy modeling of short sequence motifs with applications to RNA splicing signals. J Comput Biol 11(2–3):377–394. doi:10.1089/1066527041410418

Yeo GW, Van Nostrand E, Holste D, Poggio T, Burge CB (2005) Identification and analysis of alternative splicing events conserved in human and mouse. Proc Natl Acad Sci U S A 102(8):2850–2855, doi:0409742102 [pii] 10.1073/pnas.0409742102

Zhang XH, Chasin LA (2004) Computational definition of sequence motifs governing constitutive exon splicing. Genes Dev 18(11):1241–1250. doi:10.1101/gad.1195304

Zhang XH, Leslie CS, Chasin LA (2005a) Dichotomous splicing signals in exon flanks. Genome Res 15(6):768–779. doi:10.1101/gr.3217705

Zhang XH, Leslie CS, Chasin LA (2005b) Computational searches for splicing signals. Methods 37(4):292–305. doi:10.1016/j.ymeth.2005.07.011

Chapter 8
Genome-Wide Activities of RNA Binding Proteins That Regulate Cellular Changes in the Epithelial to Mesenchymal Transition (EMT)

Thomas W. Bebee, Benjamin W. Cieply, and Russ P. Carstens

Abstract The epithelial to mesenchymal transition (EMT) and reverse mesenchymal to epithelial transition (MET) are developmentally conserved processes that are essential for patterning of developing embryos and organs. The EMT/MET are further utilized in wound healing, but they can also be hijacked by cancer cells to promote tumor progression and metastasis. The molecular pathways governing these processes have historically focused on the transcriptional regulation and networks that control them. Indeed, global profiling of transcriptional changes has provided a wealth of information into how these networks are regulated, the downstream targets, and functional consequence of alterations to the global transcriptome. However, recent evidence has revealed that the posttranscriptional landscape of the cell is also dramatically altered during the EMT/MET and contributes to changes in cell behavior and phenotypes. While studies of this aspect of EMT biology are still in their infancy, recent progress has been achieved by the identification of several RNA binding proteins (RBPs) that regulate splicing, polyadenylation, mRNA stability, and translational control during EMT. This chapter focuses on the global impact of RBPs that regulate mRNA maturation as well as outlines the functional impact of several key posttranscriptional changes during the EMT. The growing evidence of RBP involvement in the cellular transformation during EMT underscores that a coordinated regulation of both transcriptional and posttranscriptional changes is essential for EMT. Furthermore, new discoveries into these events will paint a more detailed picture of the transcriptome during the EMT/MET and provide novel molecular targets for treatment of human diseases.

Keywords RNA binding proteins (RBP) • Epithelial to mesenchymal transition (EMT) • Mesenchymal to epithelial transition (MET) • Alternative splicing • Polyadenylation • Translational control • mRNA stability

---

T.W. Bebee • B.W. Cieply • R.P. Carstens (✉)
Department of Medicine, University of Pennsylvania, Perelman School of Medicine, Philadelphia, PA, USA
e-mail: russcars@upenn.edu

G.W. Yeo (ed.), *Systems Biology of RNA Binding Proteins*, Advances in Experimental Medicine and Biology 825, DOI 10.1007/978-1-4939-1221-6_8

Abbreviations

RBP	RNA binding proteins
EMT	Epithelial to mesenchymal transition
MET	Mesenchymal to epithelial transition
EMP	Epithelial-mesenchymal plasticity
AS	Alternative splicing
APA	Alternative polyadenylation
miRNA	microRNA
AS-NMD	Alternative splicing-mediated nonsense-mediated decay

1 Introduction

The epithelial to mesenchymal transition (EMT) and the reverse process of MET are fundamental to invertebrate and vertebrate development, driving morphogenesis of a complex body plan and formation of specialized organs (Shook and Keller 2003; Acloque et al. 2009; Thiery et al. 2009; Nieto 2011; Lim and Thiery 2012). During EMT, epithelial cells lose cell–cell junctions comprising tight junctions, adherens junctions, and desmosomes that establish apical basal polarity. The resulting mesenchymal cells exhibit front-rear polarity, migrate along the extracellular matrix, and invade through basement membranes to reach more distal sites. During development cells that undergo EMT often re-epithelialize after migration through MET to form epithelial cell barriers. EMT/MET transitions have also been implicated in human diseases, including cancer, where both have been implicated in metastasis through tumor cell invasion, anoikis and apoptosis resistance, formation of cancer stem cells (CSCs), and colonization in distant organs (Yang and Weinberg 2008; Gunasinghe et al. 2012; Scheel and Weinberg 2012; van Denderen and Thompson 2013). There has therefore been intense interest in defining global molecular and cellular mechanisms that underlie epithelial-mesenchymal plasticity (EMP) and its regulation. Transcriptional gene regulatory networks that control EMT are now well described and are characterized by global decreases in the expression of epithelial cell markers and activation of mesenchymal markers at the total transcript and protein level. These coordinated changes in gene expression contribute to a loss of cell-cell adhesion and apical-basal polarity and the acquisition of migratory and invasive cell behaviors (Nieto and Cano 2012). For example, transcriptional inactivation of genes encoding components of epithelial tight junctions and adherens junctions, such as ZO-1 and E-cadherin, leads to dissolution of these subcellular structures. At the same time, upregulation of mesenchymal genes such as fibronectin and matrix metalloproteases (MMPs) leads to changes in cell/matrix adhesion and invasion across basement membranes. The activation of a number of signaling pathways in EMT, such as TGF-β and Ras/Map kinase, induces the expression of mesenchymal transcription factors such as the Snail, Zeb, and Twist family members that repress the expression of epithelial genes and promote mesenchymal genes (Peinado et al. 2007). More recent work has also characterized epithelial specific transcription factors, such as the grainyhead-like genes, that promote MET (Cieply et al. 2012; De Craene and Berx 2013). While these transcription factors and the networks they direct are essential for the regulation of EMP, more recent studies have begun to define the role of posttranscriptional gene regulation in EMT. These posttranscriptional changes affect gene expression at numerous steps in the life cycle of a gene transcript such as splicing, polyadenylation, stability, translation, and localization. These layers of regulation are primarily directed by RNA binding proteins (RBPs), with noncoding RNAs contributing to additional functions. Analogous to the roles of transcription factors in coordinating gene expression in different cell contexts or cellular responses, there are now numerous examples in which RBPs have been shown to regulate the expression of mRNAs that encode functionally related proteins in biologically coherent "regulons." (Keene 2007). It has also become increasingly apparent that RBPs are multifunctional proteins where the same RBP can regulate posttranscriptional gene expression at distinct steps in both the nucleus and the cytoplasm (Wilkinson and Shyu 2001; Sawicka et al. 2008; Wang et al. 2012). Thus, at a systems level, the identification of RBPs and the posttranscriptional networks they control in EMT will lead to a more integrated understanding of the individual gene products as well as complexes and pathways that affect the distinct morphologies and functions of epithelial and mesenchymal cell subtypes. In this review we summarize our still emerging understanding of the RBPs and posttranscriptional networks that regulate and impact EMP during development as well as in pathophysiologic conditions. While numerous RBPs have been implicated in EMT through regulation of specific transcripts, we primarily focus on the limited examples in which there have been studies that have expanded their focus to define coordinated networks of regulation. We also begin to explore models for how transcriptional and posttranscriptional networks, though regulating distinct gene transcripts, nonetheless converge to coordinately regulate pathways and cell behaviors that are involved in EMT.

2 The Important Functions of the EMT and Reverse Process of MET in Development, Physiology, and Regenerative Medicine

2.1 *EMT/MET and Development*

During invertebrate and vertebrate development the EMT and MET events are conserved processes that are fundamental for the establishment of the body plan and organogenesis (reviewed in Baum et al. 2008; Acloque et al. 2009; Thiery et al. 2009; Lim and Thiery 2012). In mammals, there are several sequential rounds of EMT and MET that give rise to the structure and organization of the tissues and organs.

The best described primary EMTs occur during gastrulation and neural crest delamination. During gastrulation in mammals, epithelial cells of the epiblast that migrate through the primitive streak undergo EMT to form the mesoderm and definitive endoderm layers and thus account for the establishment of the three primary germ layers (Tam and Behringer 1997; Shook and Keller 2003; Arnold and Robertson 2009; Lim and Thiery 2012). Roles for several signaling pathways, including those of the Wnt, TGF-β, and FGF families in the EMT, have been well characterized during gastrulation as well as roles of the Snail transcription factors to downregulate expression of E-cadherin at the primitive streak. During the EMT associated with neural crest delamination cells at the junction of neural and non-neural ectoderm migrate to different regions in the embryo where they give rise to the cranial bones and cartilage, most peripheral nerves, and melanocytes (Kerosuo and Bronner-Fraser 2012). Many of the same signaling pathways and changes in transcription associated with gastrulation have also been shown to have common functions in the EMT associated with neural crest. Subsequently, secondary EMT as well as MET events account for further development giving rise to the complex organization of numerous tissues and organs (Thiery et al. 2009). While these secondary and tertiary EMTs and METs have been detailed elsewhere, we note that perhaps one of the best described examples of a developmental MET occurs during renal organogenesis where cells of the metanephric mesenchyme give rise to the renal tubular epithelial cells that are essential for kidney function (Dressler 2006). Given the numerous steps at which EMP impacts mammalian development, it is now also evident that a more comprehensive understanding of the gene expression programs that underlie these transitions is needed to achieve the promises of regenerative medicine. Numerous studies have demonstrated how somatic cells can be manipulated either into induced pluripotent stem cells (iPSCs) that can be differentiated down specific cell lineages or directly transdifferentiated into desired cell types for therapeutic purposes in human diseases (Lengner 2010; Stadtfeld and Hochedlinger 2010; Vierbuchen and Wernig 2012). Indeed, many of the protocols that are being used to differentiate cells into clinically desired cell types make use of signaling molecules such as Wnt, TGF-β, and FGF family members that were originally characterized in systems of developmental EMT or MET (Murry and Keller 2008). While the original studies used reprogramming transcription factors such as Oct4, Klf4, Sox2, and c-Myc (OKSM) to induce pluripotency, changes in posttranscriptional regulation via microRNAs have now also been shown to generate iPSCs as well as to direct differentiation into desired cell types (Bao et al. 2013). Recent studies have further demonstrated that a mesenchymal to epithelial transition (MET) is a critical initiating event during the reprogramming of iPSCs from mouse embryonic fibroblasts (MEFs) (Li et al. 2010; Samavarchi-Tehrani et al. 2010). Taken together, it is apparent that studies of developmental EMT and MET transitions, including transcriptional and posttranscriptional networks of regulation, will provide fundamentally relevant findings that will guide development of future therapies.

2.2 EMT in Wound Healing

Changes in cellular plasticity associated with EMT have also been described during wound healing in the skin. Although there are complex cellular changes that occur during this process, an initial inflammatory response activates growth factors and signaling pathways that induce keratinocytes at the wound margins to undergo EMT, migrate, and re-epithelialize to close the wound. This process involves cellular changes associated with EMT and has been best described as a partial and reversible EMT wherein keratinocytes at the leading edge undergo migration while also retaining some of the cell–cell junctions needed to maintain epidermal integrity. Many of these cellular responses are shared with EMTs associated with developmental and pathophysiologic EMTs, but with some differences (Leopold et al. 2012). Increased expression of the transcription factor Slug (also known as Snai2) in keratinocytes has been shown to play a central role in cutaneous wound healing and mice that lack Slug display defective wound healing (Savagner et al. 2005; Hudson et al. 2009). Induction of Slug and partial EMT has further been shown to occur at least in part through an EGFR/Erk5-mediated pathway (Arnoux et al. 2008).

3 Pathophysiological Consequences of EMT/MET

3.1 EMT/MET in Cancer

Most human cancers are carcinomas, which originate from epithelial cells in tissues such as the mammary gland, intestine, lung, and prostate. The movements of epithelial cells typically occur while maintaining cell-cell adhesion and are restricted to lateral movements along the basement membrane without penetrating into the extracellular matrix (ECM). These properties of epithelial cells within a tumor thus present a fundamental barrier to cancer invasion and metastasis. While collective epithelial cell invasion can occur locally the EMT induces phenotypic changes in carcinoma cells that enhance their cell-autonomous invasive and metastatic potential (Weinberg 2008; Yang and Weinberg 2008). It also induces resistance to multiple lines of cancer therapies contributing to posttreatment disease recurrence (Singh and Settleman 2010). Because of these malignant properties that are associated with the EMT, it has become an intensely investigated topic in the cancer field.

The tightly associated cell–cell junctions at the lateral membrane of epithelial cells play a significant role in preventing invasion of carcinoma. This is supported by multiple lines of evidence showing that antagonism of E-cadherin, a required component of epithelial junctions, is sufficient to induce invasion (Vleminckx et al. 1991; Kinsella et al. 1993). By downregulating E-cadherin as well as other critical components of epithelial cell–cell junctions, the EMT can enhance the invasive capacity of carcinoma (Peinado et al. 2004; Brabletz et al. 2005; Thiery et al. 2009). The EMT program also up-regulates matrix metalloproteinases, which are secreted enzymes that dissolve the basement membrane (Miyoshi et al. 2004, 2005; Jorda et al. 2005), and it reorganizes the actin cytoskeleton to promote single-cell motility (Thiery et al. 2009). Carcinoma cells that undergo EMT thereby acquire the ability to disassociate from adjacent epithelial cells, breach the basement membrane, and invade nearby tissues and access the bloodstream. However, when epithelial cells detach from their appropriate cell/matrix interactions, they are programmed to undergo apoptosis, a context of apoptosis known as anoikis (Frisch and Francis 1994). EMT solves this fundamental barrier to tumor progression as well by turning off the anoikis sensitivity machinery of the cell (Onder et al. 2008; Klymkowsky and Savagner 2009; Drasin et al. 2011; Tiwari et al. 2012; Frisch et al. 2013). The precise mechanism of this remains to be elucidated, but has been proposed to involve diverse and complex cell signaling coupled to gene expression changes (Frisch et al. 2013; Hogstrand et al. 2013; Paoli et al. 2013). Therefore, EMT enhances the ability of carcinoma cells to survive anchorage independently in the lymphatic and/or circulatory vessels as well as in other suboptimal microenvironments of distant sites, allowing them to potentially seed a metastatic tumor (Yang et al. 2004; Rennebeck et al. 2005; Eccles and Welch 2007; Onder et al. 2008; Geiger and Peeper 2009).

Metastatic colonization is believed to often involve a reversal of EMT or an MET, as supported by the evidence that metastatic tumors are typically epithelial in nature like the primary tumor. In fact, MET is believed to potentially facilitate colonization and maintenance of EMT can inhibit metastasis in some contexts (Kang and Massague 2004; Yang et al. 2004; Ocana et al. 2012; van Denderen and Thompson 2013). For example, the sequential roles of EMT and MET in the metastatic cascade were elegantly displayed using a mouse model with an inducible tet-ON Twist transgene; it was shown that a transient and localized Twist1 expression at the primary tumor site led to increased metastasis of squamous cell carcinoma via EMT-mediated enhanced local invasion. On the other hand constitutive Twist expression failed to promote metastasis because an epithelial phenotype, which Twists inhibited, was required for metastatic colonization (Tsai et al. 2012).

EMT can promote additional malignant properties of cancer cells that are not implicitly related to metastasis. The enhanced cell survival associated with EMT can induce resistance to multiple lines of cancer treatment including radiation, cytotoxic chemotherapy, and even some targeted therapies, for example, EGFR inhibitors (Thomson et al. 2005; Woodward et al. 2007; Barr et al. 2008; Gupta et al. 2009). Because of this, while the bulk of a given tumor may be epithelial and initially sensitive to these therapies, a subset of cells that had undergone EMT could survive, leading to disease recurrence.

Related to the enhanced survival and tumor recurrence phenotypes, EMT has also been proposed to produce "cancer stem cells (CSCs)" also known as tumor-initiating cells (TICs), defined as individual tumor cells that have the capacity to seed a nascent tumor. This concept was first described in the context of breast cancer when it was found that only a minute fraction of the cells in a given tumor had the ability to seed a new tumor when injected into immune-compromised mice; these TIC/CSCs could be enriched for by FACS sorting heterogeneous tumor populations for the $CD44^{HIGH}/CD24^{LOW}$ cell surface antigens (Al-Hajj et al. 2003). Understanding the origin and defining characteristics of CSC/TICs are of great interest to the field of cancer as they are potentially the source of metastasis and post-therapy disease recurrence. A major stride was made to this end when it was discovered that the EMT switched mammary epithelial cells from $CD44^{LOW}/CD24^{HIGH}$ to $CD44^{HIGH}/CD24^{LOW}$ phenotype that was associated with enhanced tumorigenicity (Mani et al. 2008; Morel et al. 2008). These and other observations suggested the possibility that EMT can promote tumorigenicity, but also potentially provide tangible characteristics that describe an otherwise poorly defined population of tumor cells. While the relationship between EMT and CSC/TIC phenotypes has been extensively investigated (Drasin et al. 2011; May et al. 2011; Scheel and Weinberg 2012; Tiwari et al. 2012; Hollier et al. 2013) there has been some evidence to refute this model (Celia-Terrassa et al. 2012).

Given the established role of EMT in enhancing invasion, metastasis, drug resistance, and tumor recurrence, it will be important to identify mesenchymal specific "druggable" targets that could potentially be used to prevent or curb tumor progression. Drugs that target the malignant characteristics of mesenchymal tumor cells rather than reversing the EMT would be ideal, in theory, to achieve these treatment goals. But these targets remain elusive and require further characterization of the molecular mechanisms that define epithelial and mesenchymal cells. Given the metastatic colonization role of MET, EMT-reversing/MET-inducing treatment strategies would seem to be most logically used in conjunction with other cytotoxic and/or targeted agents with the goal being to sensitize the refractory mesenchymal tumor cells to these treatments. Indeed, establishing the molecular targets and contexts for EMT-specific cancer therapies will be a subject of continued investigations.

3.2 EMT and Fibrosis

EMT has also been implicated in fibrotic diseases of the kidney, liver, and lung (Iwano et al. 2002; Kalluri and Neilson 2003; Kalluri and Weinberg 2009; Zeisberg and Neilson 2009). Based on several disease models it was proposed that organ fibrosis can occur when epithelial cells undergo EMT, cross the basement membrane into the interstitium, and generate myofibroblasts that secrete collagens and other matrix proteins that cause fibrosis and organ failure. However, while both in vivo and in vitro studies provided evidence that epithelial cells could undergo EMT to produce myofibroblasts, several recent studies have questioned whether EMT plays a significant role in organ fibrosis in vivo and there is now debate surrounding the contention that EMT induces tissue fibrosis (Iwano et al. 2002; Lin et al. 2008; Duffield 2010; Humphreys et al. 2010; Zeisberg and Duffield 2010; Kriz et al. 2011). More recently it was shown that while EMT provides one source of myofibroblasts that induce fibrosis in a model of kidney fibrosis, they constituted a smaller fraction than suggested in previous studies (Lebleu et al. 2013). Thus, while more investigation is needed to resolve this controversy, it may be that the relative contribution of EMT in causing tissue fibrosis may differ in different organs or depending upon the mechanisms of injury that induce fibrosis in different cell or tissue contexts.

4 Transcriptional Regulation of EMT/MET

Genome-wide profiles of total mRNA transcript levels using traditional microarrays have identified numerous genes that are differentially expressed in epithelial vs. mesenchymal cells, many of which demonstrate complete on or off switches in expression during EMT (LaGamba et al. 2005; Blick et al. 2008; Taube et al. 2010; Cieply et al. 2012). Therefore the gene expression programs and specific genes that are regulated at the whole transcript level during EMT have received the majority of the attention in the field. Many genes that are transcriptionally inactivated during EMT regulate epithelial specific characteristics such as cell-cell adhesion and apico-basolateral polarity, including E-cadherin, desmosomal components (DSG3, DSP), tight junction components (CLDN4/7), and the apical polarity regulator CRB. Mesenchymal genes that are turned on during EMT contribute to single-cell motility and invasion such as Vimentin and MMPs 2 and 9. The biological significance and molecular mechanisms of these epithelial and mesenchymal genes with respect to EMT have been reviewed elsewhere as have the transcription factors that mediate these switches in expression (Kalluri and Weinberg 2009; Thiery et al. 2009; De Craene and Berx 2013). However, there is clear evidence that changes in gene expression that are coordinately regulated at the transcriptional level control the changes in cell morphology and function that are hallmarks of EMT (Fig. 8.1).

The most well-characterized EMT inducing transcription factors include ZEB1/2, SNAI1/2, TWIST1/2, GSC, and FOXC2 (Thiery et al. 2009). While these factors have all been shown to induce an EMT when ectopically expressed, in many cases they also activate the endogenous expression of one or more of the other EMT-inducing factors (Guaita et al. 2002; Taube et al. 2010; Dave et al. 2011). This suggests that they comprise a network or a cascade of coordinated transcriptional regulation during EMT. A common EMT-associated transcriptional network is supported further by the data showing that individual ectopic expression of TWIST1, SNAI11, GSC, or FOXC2 (as well as addition of TGF-β or the use of RNA interference to downregulate E-cadherin) induced an EMT gene signature (Taube et al. 2010). This expression signature included the on/off switches mentioned above, as well as many other genes whose relevance to EMT has yet to be discovered. The precise hierarchy, minimal requirements, and/or interdependence between EMT-inducing transcription factors remain to be elucidated.

TGF-β/TGFBRI/RII signaling is a well-established signaling pathway that converges on and functions together with transcription factors to induce EMT, often via canonical Smads 3 and 4 (Vincent et al. 2009; Fuxe and Karlsson 2012). TGF-β can also affect transcription and EMT via Smad-independent signaling mechanisms. For example, TGFBRI/RII-mediated phosphorylation of PAR6 leads to the

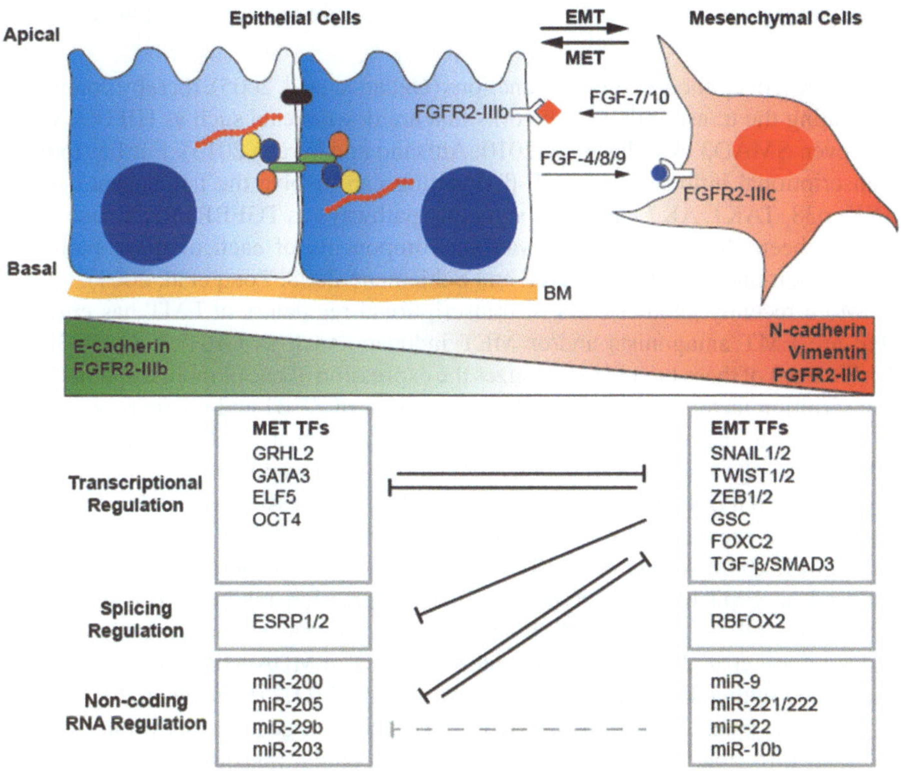

Fig. 8.1 Transcriptional and posttranscriptional networks regulate the EMT. Epithelial cells are associated with cell–cell junctions and polarize cells with an apical-basal orientation, providing a barrier to transport of molecules and functional separation of luminal and basal compartments. Epithelial cells express E-cadherin (depicted in *green*), a transmembrane component of adherens junctions, along with other components of the adherens junction p120-catenin, β-catenin, and α-catenin interacts with the actin cytoskeleton. Conversely, mesenchymal cells express N-cadherin and Vimentin and lack adherens junctions. The EMT process requires the dissolution of the adherens junction associated with loss of E-cadherin and expression of mesenchymal markers (N-cadherin and Vimentin), a transition from the cuboidal morphology of epithelial cells to the more spindle-like mesenchymal cell, and the ability to migrate and proliferate independent of basement membrane (BM) attachment (while the MET requires the inverse). Alternative splicing of the FGF receptor *FGFR2* is a hallmark of epithelial and mesenchymal distinction, wherein epithelial cells express FGFR2-IIIb that responds to mesenchymally secreted FGFs, and mesenchymal cells express FGFR2-IIIc which responds to epithelially secreted FGFs. This provides for epithelial/mesenchymal communication and coordinated interactions. Regulation of the EMT/MET processes has been described at the transcriptional, splicing, and noncoding RNA level. Transcriptional regulation comprises the best understood network of molecular players including the EMT-inducing transcription factors (EMT TFs) and MET-inducing transcription factors (MET TFs). All examples of EMT are associated with the loss of E-cadherin and expression of one or more EMT TFs. Newly identified global splicing regulatory networks have been profiled during the EMT and are primarily associated with the epithelial restricted epithelial splicing regulatory proteins 1 and 2 (ESRP1/2) and the more ubiquitously expressed but mesenchymally enriched RBFOX2 regulatory splicing factor. Noncoding RNAs (ncRNAs) comprising microRNAs account for regulatory feedback loops that potentiate epithelial or mesenchymal fates as well as regulation of E-cadherin and transcription factors associated with the EMT/MET. Each of these regulatory mechanisms represents a component of the global regulatory network of the EMT/MET at the transcriptional and posttranscriptional level, with associated cross talk and regulatory feedback loops as schematized by *arrows* (examples of translational regulation and mRNA stability are outlined in Fig. 8.2)

breakdown of epithelial cell–cell junctions (Ozdamar et al. 2005), thereby potentially regulating the transcriptional networks that are downstream such as HIPPO, WNT, and even SMAD3 (Varelas et al. 2010; Attisano and Wrana 2013). Further indirect transcriptional influences of TGF-β signaling can involve the target genes of the JNK, p38, TAK1, AKT, and ERK signaling pathways as TGFBRI/RII kinase activity has been shown to regulate various components of each of these pathways (Moustakas and Heldin 2005; Neil and Schiemann 2008; Zouq et al. 2009).

More recently another class of transcriptional regulators of EMT has emerged that are EMT antagonists and/or MET inducers, such as GRHL2, GATA3, and ELF5. Each of these factors antagonizes the expression of one or more EMT-inducing transcription factors. GRHL2 is a direct repressor of ZEB1 and microarray analysis of the profile induced by ectopic GRHL2 indicated that it inhibited/reversed the EMT gene signature. GRHL2 downregulation is a component of the EMT gene signature and it is a direct target of ZEB1-mediated repression, indicating reciprocal antagonistic regulation by these transcription factors (Taube et al. 2010; Cieply et al. 2012; 2013; Werner et al. 2013). Knockout of Elf5 in mouse mammary glands induced an EMT-like phenotype and ELF5 expression inhibited TGF-β induced EMT as well as induced MET in vitro by directly suppressing Snail2/Slug (Chakrabarti et al. 2012). GATA3 induced an MET in MDA-MB-231??? which was associated with a downregulation of Snail1/2 and ZEB1/2 (Chou et al. 2013).

Once the balance of epithelial versus mesenchymal gene transcription is tipped in either direction, it is easy to understand how cellular transitions ensue based on the known regulatory loops detailed above. However, what remains unclear is how different cellular milieus determine whether induction of a specific signaling pathway or transcription factor is sufficient to induce EMT or MET. Most epithelial cell lines that express classical epithelial cell markers such as E-cadherin are not "EMT permissive" in the sense that they are resistant to EMT even in the presence of well-established ectopic EMT-inducing transcription factors and extracellular stimuli Argast et al. (2010; De Craene and Berx 2013). Indeed there remain only a handful of cell lines commonly used as model systems of EMT, reflecting the fact that most cell lines cannot be induced to undergo EMT in response to a single stimulus. This would suggest that while there are well-established models for transcriptional regulation of EMT, they do not paint the entire picture of its regulation. Such observations thus raise the question as to what other cellular processes, including post-transcriptional regulation through alternative pre-mRNA splicing, microRNA regulation, or regulation of mRNA stability or translation, contribute to the coordinated networks that define epithelial and mesenchymal cell phenotypes. While examples of epigenetic changes, including DNA methylation and histone modifications, that occur in EMT have been described, there is still limited information about genome-wide epigenetic changes during EMT (reviewed in Wang and Shang 2013). It is therefore increasingly apparent that these layers of transcriptional regulation likely function together with posttranscriptional regulation to achieve the changes in cell phenotype and behavior that accompany EMT or MET.

5 Regulation of EMT/MET by Noncoding RNAs

In the past few years a number of microRNAs have been established as regulators of EMT and cancer metastasis. Because recent comprehensive reviews have described such microRNAs, here we focus on several microRNAs that are most directly tied to EMT and the regulation of EMT transcription factors and markers (Zhang and Ma 2012; Lamouille et al. 2013). Mature microRNAs are 22-nucleotide RNAs that contain a short (6–8) base "seed" sequence that is complementary to their target RNAs. In most cases, binding of microRNAs to the 3′ UTR of mRNAs leads to translation inhibition or degradation by directing the RNA binding proteins of the RISC complex (Krol et al. 2010). MicroRNA regulation represents an additional level of gene expression regulation that contributes to the maintenance of both epithelial and mesenchymal cell phenotypes. Several of the EMT-inducing transcription factors as well as other mesenchymal genes, which are regulated at the transcriptional level, are also targets of epithelial specific microRNAs. Also, microRNAs that target epithelial-specific transcripts to promote mesenchymal properties are becoming evident (Fig. 8.1).

The epithelial-specific microRNA-200 family, consisting of miR-200a, miR-200b, miR-200c, miR-141, and miR-429, have been the best characterized microRNAs with a central role in EMT (Burk et al. 2008, Gregory et al. 2008a, Korpal et al. 2008; Park et al. 2008). These key regulators were identified by several groups who showed that their expression is lost during the EMT. Furthermore, ectopic expression of miR-200s in mesenchymal cells is sufficient to induce an MET and mir-200 antagonism can induce EMT. Key target transcripts that are directly downregulated by the miR-200s are ZEB1 and -2. Because the ZEB transcription factors repress epithelial-specific genes such as E-cadherin expression, the miR-200s thus indirectly induce epithelial-specific gene expression, including the epithelial-splicing factors ESRP1 and ESRP2 (Horiguchi et al. 2012). Interestingly, the miR-200 clusters are themselves targets for direct transcriptional repression mediated by ZEB1/2, which establishes a reciprocal antagonistic loop that regulates EMT and MET. Therefore it is firmly established that the miR-200s are one of the core regulators of epithelial cell identity and maintenance. Their role in EMT and cancer phenotypes has been reviewed extensively elsewhere (Gregory et al. 2008; Hill et al. 2013). A genome-wide determination of miR-200 target transcripts has thus far not been carried out, but will undoubtedly provide novel insights into posttranscriptional programs that maintain epithelial identity and that are subverted in EMT (Fig. 8.2).

Additional microRNAs that target mesenchymal transcription factors and other genes have also emerged as key modulators of EMT/MET. It is well established that the Gata3 transcription factor is a driver of luminal cell differentiation during mammary gland development (Kouros-Mehr et al. 2006; Chou et al. 2010). It has also been shown to induce MET in the widely used mesenchymal MDA-MB-231 breast cancer cell model (Yan et al. 2010). Very recently it was revealed that much of the effect of Gata3 in enforcing epithelial properties could be attributed to its

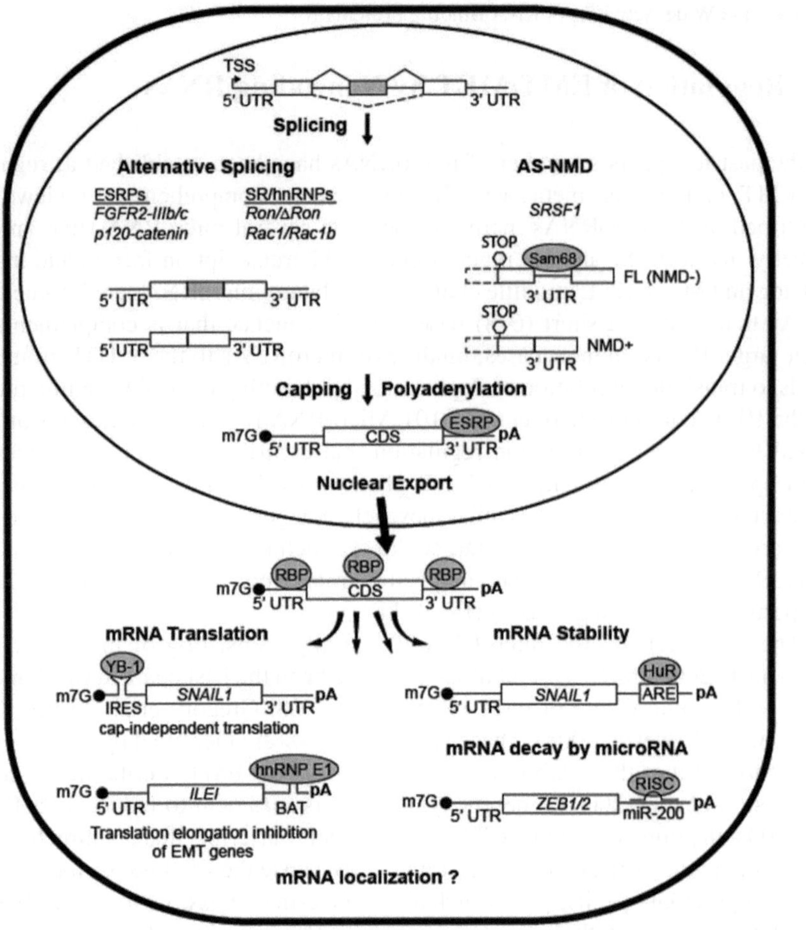

Fig. 8.2 RNA binding proteins regulate the posttranscriptional landscape of EMT mRNAs. pre-mRNAs comprise noncoding (5′ and 3′ UTRs and introns, depicted as *lines*) and coding sequences (exons are depicted as *boxes*). Processing of pre-mRNA to mature mRNA requires splicing, 5′ capping, and polyadenylation. These processes are mediated by RBPs and may include alternative inclusion of exons (alternative splicing) or alternative polyadenylation site usage. Alternative splicing in EMT is regulated by splicing factors such as the ESRPs, RBFOX2, and SR and hnRNP proteins, which target the generation of EMT-associated transcripts such as *FGFR2*, *p120-catenin*, *Ron*, and *Rac1*. Alternative splicing-associated nonsense-mediated decay (AS-NMD) can also alter the level of SR proteins, for example through Sam68 regulation of alternative splicing in the 3′ UTR of SRSF1. Following nuclear export, the AS-NMD isoform is degraded in the cytosol. The ESRPs have also been implicated in changes in alternative splicing-associated polyadenylation site usage. Following nuclear export of mature mRNAs, cytosolic regulation of mRNAs is also mediated by RBPs, directing mRNA translation, stability, and decay, and potentially localization of mRNAs to distant sites for translation. The RBPs YB-1 and hnRNP E1 regulate global networks of transcripts, directing EMT-induced translation. YB-1 promotes the translation of the EMT transcription factor *SNAIL1*, and hnRNP E1 represses translation elongation of *ILEI* prior to TGF-β-induced EMT. HuR stabilizes *SNAIL1* mRNA by binding to an ARE element in the 3′ UTR and the miR-200 family members target the EMT transcription factors ZEB1/2. The coordinated regulation of EMT-associated transcripts by a posttranscriptional network of RBPs mediates both maintenance of epithelial and mesenchymal cellular states and initiation and potentiation of the EMT/MET processes

transcriptional activation of miR-29b, which targeted key mesenchymal genes including MMP9 and TGF-β2 and -3 (Chou et al. 2013); miR-29b was also shown to inhibit EMT and invasion in prostate cancer cells (Ru et al. 2012).

miR-30a, -182, and -203 are regulators of the Snail family of transcription factors. miR-30a induces a substantial downregulation of Snail1, which promotes epithelial properties; and its expression is inversely correlated with Snail's in a panel of tumor cell lines (Kumarswamy et al. 2012). miR-182 and -203 have been shown to induce a dramatic MET in multiple prostate cancer cell lines including a global alteration of gene expression patterns (Qu et al. 2013). miR-203 was also shown to induce an MET and inhibit the invasive and experimental metastatic potential of MDA-MB-231 (Ding et al. 2013) and it exhibits an epithelial specific expression pattern that mirrors that of the miR-200s (Park et al. 2008). The METs induced by miR-203 and -182 were dependent on their direct antagonism of Snail2 expression, as ectopic Snail2 could reverse/inhibit the MET induced by miR-203 and -182 (Qu et al. 2013). In tune with the recurrent theme of reciprocal cross-regulation during EMT, Snail2 as well as ZEB1 can directly repress miR-203 expression (Wellner et al. 2009; Ding et al. 2013) These collective studies illustrate a complex network of EMT regulation that comprises miR-200s, and miR-203 to regulate ZEB1 and Snail2 through cross talk between transcriptional and posttranscriptional layers of gene regulation.

The intricate networks of gene regulation that are involved in EMT and MET also involve microRNAs that are expressed in mesenchymal cells and antagonize epithelial gene expression. miR-9 targets and downregulates E-cadherin leading to enhanced mesenchymal properties including invasion and metastasis (Ma et al. 2010). miR-10b is highly expressed in the mesenchymal mammary tumor line MDA-MB-231 relative to the epithelial MCF-7 and the EMT-inducing transcription factor Twist activated its expression leading to enhanced invasion (Ma et al. 2007). miR-22 was found to be specifically expressed in mesenchymal cells and it indirectly antagonized miR-200 expression by repressing TET family-mediated hydroxymethylation of the miR-200 family promoters, leading to EMT and tumor progression (Park et al. 2008; Song et al. 2013). miR-221/222 were shown to be highly expressed in a panel of mesenchymal tumor cell lines relative to epithelial lines and they induced an EMT when ectopically expressed. Mechanistically, they antagonized the expression of TRPS1, which is a repressor of ZEB2 (Stinson et al. 2011), thus leading to ZEB2 transcriptional activation.

The examples of miR-22 and miR-221/222, both of which induce EMT by antagonizing repressors of ZEB1/2, vastly expand the scope and complexity of transcriptional coupled to posttranscriptional gene regulatory networks that influence epithelial and mesenchymal phenotypes. For example, repression of miR-200s by miR-22 can theoretically lead to the upregulation of ZEB1/2 in turn feeding into the mir-203/Snail regulation discussed above, thereby activating Snail2 expression to continue the EMT cascade. Of interest for future investigations will be the identification of factors that regulate these mesenchymal microRNAs as well as directly testing the interdependence between the various components of the microRNA-EMT regulatory network. MicroRNA-mediated regulation of gene expression can also be fine-tuned by alternative splicing and polyadenylation to produce transcripts with different 3′ UTRs or 3′ UTRs of differing length that can profoundly alter microRNA target sites (Shen et al. 2008; Di Giammartino et al. 2011; Shi 2012; Tian and Manley 2013). It will therefore be of future interest to investigate the intersection of networks of regulation at several different posttranscriptional steps to regulate EMT.

While our understanding of miRNAs in the regulation of the transcriptional network controlling the EMT/MET processes has become clearer, it is also likely that long noncoding RNAs (lncRNAs) will be shown to play a prominent role. However, at the present time there is very limited understanding of lncRNAs in these cellular transitions, and no epithelial or mesenchymal cell type-specific examples have been described.

6 Regulators of Alternative Splicing in EMT/MET

6.1 General Mechanisms of Splicing Regulation

While several recent reviews have detailed the mechanisms of alternative splicing regulation by both ubiquitous and tissue-specific RBP splicing regulators, a brief introduction is provided here (Chen and Manley 2009; Nilsen and Graveley 2010). The most common type of alternative splicing consists of an alternative cassette exon that can either be included in the final mRNA transcript or skipped. Splicing is carried out by the spliceosome, a dynamic macromolecular machine consisting of the small nuclear ribonucleoproteins (snRNPs) and a large set of additional proteins. Components of the spliceosome are recruited to the consensus splicing sequence elements at the upstream intron/exon boundary (the 3′ splice site and upstream branch point sequence) and the downstream exon/intron boundary (the 5′ splice site). In addition to the 3′ and 5′ splice sites, additional "auxiliary" cis-acting determinants of splicing are also present both within an exon and the flanking introns. These sequence elements can either act to promote splicing of the exon, in which case they are known as intronic or exonic splicing enhancers (ISEs or ESEs), or they can promote skipping of the exon (intronic or exonic splicing silencers, ISSs or ESSs). A general model of regulation of alternative cassette exon splicing is that RBPs bind to generally degenerate sequence motifs within splicing enhancers and silencers and either promote or inhibit recruitment of the spliceosome to the splice sites. While the generally abundant hnRNP and SR protein families of RBPs have been the best characterized splicing regulators to date, additional examples of both ubiquitously expressed and cell- or tissue-specific splicing regulatory RBPs continue to emerge (Chen and Manley 2009). Whether a given alternative exon is spliced or skipped is determined by the combinatorial functions of RBPs that either promote or inhibit splicing (Hertel 2008). Alternative splicing can have major functional consequences since the different splice variants can encode protein isoforms with very distinct and even opposing functions (see below) (Kelemen et al. 2013).

From a systems viewpoint it is important to emphasize that the global patterns of alternative splicing that differ at different developmental stages, in different cell types, and those regulated by cell- or tissue-specific splicing have been shown to be functionally coordinated (Ule et al. 2005; Kalsotra et al. 2008; Kalsotra and Cooper 2011). Thus, splicing regulatory networks (SRNs) regulated by different *trans-factors* operate to shape biologically coherent programs of alternative splicing in which the different splice isoforms cooperate to promote appropriate cell type-specific behaviors (Braunschweig et al. 2013).

6.2 *Identification of Alternative Splicing in EMT*

The sentinel alternative splicing event that was first linked to EMT was a splicing switch in fibroblast growth factor receptor 2 (FGFR2) in rat bladder carcinoma cells (Savagner et al. 1994). Splicing of two mutually exclusive alternative cassette exons, known as exons IIIb and IIIc, in the extracellular ligand-binding domain of FGFR2 produces receptor isoforms known as FGFR2-IIIb and FGFR2-IIIc in epithelial and mesenchymal cells, respectively. This splicing choice is exquisitely cell type specific and the resulting receptors have distinct ligand-binding preferences (Fig. 8.1) (Zhang et al. 2006). These differences are of profound functional importance during mammalian development and this remains one of the best examples where the functional consequences of alternative splicing have been well characterized molecularly and in vivo (De Moerlooze et al. 2000; Revest et al. 2001). Years later it was shown that there was also a switch in splicing of tandem cassette exons in p120-catenin (also known as delta catenin, CTNND1) in Snail-induced EMT in MDCK cells (Ohkubo and Ozawa 2004) (Fig. 8.2). While later studies also identified epithelial-specific splice variants in CD44 and Mena (also known as ENAH), it has only been in the last few years that global splicing regulatory networks that are altered during EMT have begun to emerge (Cooper et al. 1992; Di Modugno et al. 2007). While we and others have recently published reviews on some of these more recent studies, here we focus on several splicing regulators implicated in splicing switches associated with EMT for which global networks have begun to be characterized (Biamonti et al. 2012; Warzecha and Carstens 2012).

6.3 *Epithelial Splicing Regulatory Proteins 1 and 2 (ESRP1 and ESRP2)*

The ESRPs are paralogous RBPs that were first identified in a cell-based screen for regulators of *FGFR2* splicing (Warzecha et al. 2009a). Both ESRP1 and ESRP2 (previously known as RBM35A and RBM35B) are expressed specifically in epithelial cells and a combined knockdown of both transcripts in an epithelial cell line induced a nearly complete switch in *FGFR2* splicing from the epithelial to the mesenchymal isoform. The ESRPs were also shown to be required for predominant expression of the epithelial isoforms of *p120-catenin*, *CD44*, and *ENAH* (Warzecha et al. 2009; Brown et al. 2011). A direct connection between the ESRPs and EMT was demonstrated by the observation that both ESRP1 and ESRP2 are nearly completely transcriptionally inactivated in Twist-induced EMT in a human mammary epithelial cell line (Warzecha et al. 2009). Inactivation of ESRP expression has now been shown to be a general phenomenon in EMT, whether induced by Twist, TGF-β, Snail, ZEB1, ZEB2, or knockdown of E-cadherin (Onder et al. 2008; Taube et al. 2010; Shapiro et al. 2011; Thomson et al. 2011; Horiguchi et al. 2012). In the case of ZEB1 and Snail, recent studies have furthermore shown that they directly repress transcription of the ESRPs (Horiguchi et al. 2012; Reinke et al. 2012). Collectively, these studies strongly implicated the ESRPs as central players in a broader set of splicing switches that were suspected to occur during EMT. Therefore, further studies used a combination of splicing-sensitive microarrays and high-throughput sequencing (RNA-Seq) to define ESRP-regulated transcripts, using either ESRP1 and ESRP2 knockdown in epithelial cells or ectopic expression of ESRP1 in mesenchymal cells (Warzecha et al. 2009, 2010; Dittmar et al. 2012). Together these studies identified hundreds of alternative splicing events that are regulated by the ESRPs and in all cases tested, the expected change in splicing of confirmed ESRP targets was also observed in Twist-induced EMT. Additional evidence that the ESRPs are key players in EMT was also provided by a study that directly profiled splicing changes in Twist-induced EMT using RNA-Seq (Shapiro et al. 2011). In addition to further verifying that the ESRPs were the RBPs with the most striking changes in EMT, the splicing switches they identified strongly overlapped with those that were shown to be ESRP targets. Consistent with the concept that RBPs coordinately regulate biologically relevant target transcripts, the ESRP targets were further shown to be enriched for genes that function in processes associated with the EMT such as cell-cell adhesion, cell motility, regulation of the cytoskeleton, and regulators and effectors of RhoGTPases (Warzecha et al. 2010; Dittmar et al. 2012). Similar processes were also noted among the set of EMT-associated splicing switches (Shapiro et al. 2011). In addition, there were several intriguing examples of ESRP-regulated transcripts in which the literature has provided evidence for distinct isoform-specific functions that are central to EMT. For example, the epithelial isoform of p120-catenin was shown to promote cell-cell adhesion through a direct interaction with E-cadherin, whereas the mesenchymal isoform had the opposite effect of promoting cell motility and invasion (Yanagisawa et al. 2008). However, for the vast majority of transcripts with EMT-associated splicing switches the functional consequences of these changes and their relation to EMT remain to be determined (Warzecha and Carstens 2012). In addition, while the ESRPs are clearly important players in EMT-associated splicing switches, there is also evidence for combinatorial roles of other RBPs in the regulation of many of the splicing changes in EMT.

6.4 RBFOX2

RBFOX2 (also known as Fox-2, RBM9) has been implicated in the regulation of EMT-associated splicing targets in several studies. RBFOX2 was shown to play a role in the regulation of *FGFR2* splicing, where it, like the ESRPs, promoted the epithelial splice form. The consensus binding site for RBFOX2, (U)GCAUG, was shown to be enriched in the introns flanking a set of ESRP-regulated cassette exons, suggesting that they coordinate the regulation of an overlapping set of transcripts (Warzecha et al. 2010). A subsequent study using splicing-sensitive junction arrays to profile alternative splicing across a series of breast cancer cell lines with distinct epithelial vs. mesenchymal gene expression programs showed that many exons differentially spliced in epithelial and mesenchymal cells are regulated by RBFOX2 (Lapuk et al. 2010). Many of the same targets were also shown in our group's work to be ESRP-regulated exons (Warzecha et al. 2010; Dittmar et al. 2012). While there was a modest increase in RBFOX2 levels in Twist-induced EMT, several exons shown to switch splicing were likewise shown to have altered splicing in response to RBFOX2 depletion. In addition, the knockdown of RBFOX2 in cells that had undergone EMT induced a partial reversion towards an epithelial phenotype (Shapiro et al. 2011). Another study also showed modest increases in RBFOX2 in two mouse EMT models and similarly suggested that RBFOX2-regulated splicing generally promotes mesenchymal cell type behaviors (Braeutigam et al. 2013). Another study used a high-throughput PCR platform to profile splicing differences between colonic epithelial and mesenchymal cells, many of which were also shown to switch splicing in response to RBFOX2 knockdown (Venables et al. 2013). These data together with a generally higher level of RBFOX2 in epithelial vs. mesenchymal cells led these investigators to suggest that RBFOX2 is a driver of mesenchymal-specific splicing. We also showed that there was significant overlap between the ESRP-regulated SRN and an RBFOX2 SRN as defined by several groups (Zhang et al. 2008; Venables et al. 2009; Yeo et al. 2009; Warzecha et al. 2010; Dittmar et al. 2012). While there were many cases in which the ESRPs and RBFOX2 promoted opposite changes in splicing, there were also many events in which they promote the same change in splicing, or where RBFOX2 promotes the epithelial splicing pattern. Taken together, it is clear that RBFOX2 is a major component of the combinatorial regulation of splicing during EMT, but also that the integration of ESRP and RBFOX2 networks is complex and not always directly antagonistic.

6.5 SRSF1 (Previously Known as ASF/SF2 or SF2/ASF)

The SR protein SRSF1 was first implicated in EMT as a regulator of splicing of exon 11 in the Ron (also known as MST1R) proto-oncogene which is a receptor tyrosine kinase for macrophage-stimulating protein. Ron is involved in cell motility and actin dynamics associated with invasive growth and cell scattering. The alternatively spliced isoform ΔRon lacks exon 11, does not undergo proteolytic cleavage, and functions as a constitutively active kinase in the absence of its ligand (Collesi, Santoro et al. 1996). In turn, ΔRon is able to induce an invasive phenotype in transfected cells and is associated with metastatic gastric tumors (Collesi et al. 1996; Zhou et al. 2003). SRSF1 enhances the skipping of Ron exon 11 through a splicing enhancer in exon 12, and thereby was shown to induce an EMT in gastric tumor cells (Ghigna et al. 2005). SRSF1 is also associated with an EMT-associated splicing switch in Rac1, a member of the Rho family of small GTPases that regulates actin cytoskeleton dynamics. The inclusion of exon 3b, a 57-nt exon encoding 19 amino acids, generates a constitutively active Rac1b isoform, which is associated with cellular transformation and is upregulated in colon and breast cancer (Jordan et al. 1999; Schnelzer et al. 2000; Singh et al. 2004). *Rac1b* alternative splicing is regulated by the antagonistic effects of SRSF1, which promotes exon 3b splicing and SRSF3 (SRp20) which promotes skipping. SRSF1 expression is stimulated by PI3K signaling and SRSF3 by Wnt/β-catenin signaling, and inhibition of these pathways resulted in the associated changes in *Rac1b* splicing (Goncalves et al. 2009). Splicing of exon 3b to generate the *Rac1b* isoform was observed in matrix metalloproteinase-3 (MMP-3)-induced EMT and was shown to cause the accumulation of reactive oxygen species and induce the expression of Snail and Vimentin (Radisky et al. 2005). In this system Rac1b was further shown to promote a reversible migratory phenotype necessary for response to the local tumor microenvironment, and this provides a mechanism for tumor progression and invasion through EMT. SRSF1 was further shown to be upregulated in human cancers and to function as a proto-oncogene through regulation of splicing events in several genes involved in transformation and apoptosis (Karni et al. 2007). Interestingly, SRSF1 was shown to promote splicing of exon 12A in *Bin1*, and isoforms that contain this exon do not bind to and inhibit the function of myc. In contrast, isoforms that skip exon 12A function as a tumor suppressor through inhibition of myc activity and the ESRPs induce exon 12A skipping (Ge et al. 1999; Warzecha et al. 2010). These observations thus suggest a broader role for SRSF1 to promote splicing changes that promote EMT through antagonism of ESRP-mediated regulation and it will be of interest to determine the overlap between their targets on a genome-wide level.

6.6 Muscleblind Proteins (MBNL1, MBNL2, MBNL3)

The muscleblind proteins are splicing factors that have been shown to play a role in muscle and heart development and are implicated in splicing defects associated with myotonic dystrophy (DM) (Kalsotra et al. 2008; Du et al. 2010; Charizanis et al. 2012; Suenaga et al. 2012; Wang et al. 2012). Genome-wide networks of muscleblind-regulated alternative splicing have been identified, mainly for MBNL1 and MBNL2. Knockdown of MBNL1 in developing chicken heart increased TGF-β-mediated EMT in endocardial cushions suggesting that it is a negative regulator of EMT. A downregulation of MBNL2 was noted in Twist-induced EMT whereas knockdown of MBNL1 in a mesenchymal cell line suggested that it cooperates with RBFOX2 to induce mesenchymal splicing (Shapiro et al. 2011; Venables et al. 2013). Further studies are needed to determine the broader role of muscleblind-regulated targets and their overlap with splicing changes that occur in EMT.

6.7 Polypyrimidine Tract-Binding Protein (PTB, PTBP1)

PTB is one of the very best characterized splicing regulators and work from several groups has led to a determination of genome-wide PTB targets (Boutz et al. 2007; Xue et al. 2009; Llorian et al. 2010; Hall et al. 2013). An ~2.5-fold reduction in PTB was observed in Twist-mediated EMT and there was enrichment for PTB-binding sites that were determined using UV cross-linking immunoprecipitation coupled with high-throughput sequencing (CLIP-Seq) (Xue et al. 2009; Shapiro et al. 2011). While a subset of exons that underwent changes in splicing during EMT were also affected by PTBP1 (together with PTBP2) knockdown, the direction of splicing changes resulting from PTB depletion was both positively and negatively correlated with EMT (Venables et al. 2013). Therefore, although PTB also combinatorially impacts splicing changes during EMT, its ability to promote or inhibit EMT-associated splicing switches is context dependent.

6.8 Additional Splicing Factors Implicated in EMT

The regulation of EMT-associated alternative splicing is also mediated by both cooperative and antagonistic functions of additional SR and hnRNP proteins. Changes in the level and activation of splicing factors are associated with cancer and can alter global splicing changes that mediate the EMT and pro-tumorigenic protein isoforms. The levels of several SR proteins, including SRSF1, have been shown to be regulated by the mRNA surveillance process of nonsense-mediated decay (NMD) (reviewed in Lareau et al. 2007; Isken and Maquat 2008). While SRSF1 promotes the skipping of Ron exon 11 to generate ΔRon, the levels of SRSF1 are also regulated by another splicing factor during the EMT, Sam68 (also known as KHDRBS1). Sam68 regulates this process by promoting the inclusion of a 3′ UTR alternative intron in *SRSF1* leading to escape from NMD. However when Sam68 is inactive the intron is spliced and SRSF1 levels are reduced (Valacca et al. 2010). ΔRon is also promoted by hnRNP A2/B1 and is an oncogenic driver in glioblastoma (Golan-Gerstl et al. 2011). Another member of the hnRNP family, hnRNP H, also promotes the skipping of Ron exon 11 (ΔRon) in gliomas, in this case through binding of an exonic splicing silencer in exon 11 (Lefave et al. 2011). hnRNP A1 promotes Ron exon 11 inclusion by inhibiting the binding of SRSF1 and thus promoting an MET (Bonomi et al. 2013). hnRNP A1 also regulates splicing of Rac1b via binding to an exonic splicing silencer in exon 3b, and MMP-3-induced EMT also reduces binding of hnRNP A1 to exon 3b, thereby promoting *Rac1b* expression. The splicing of Δ*Ron* and *Rac1b* together outlines a scenario wherein signaling pathways involved in the EMT can mediate changes in SR and hnRNP splicing factor levels or activity and direct the splicing of EMT-regulated exons. The regulation of Δ*Ron* splicing exemplifies the use of several layers of posttranscriptional regulation, namely AS and AS-NMD, by splicing factors as key components in the EMT/MET conversion. Moreover, the combinatorial impact of SR and hnRNPs on the production of Δ*Ron* and *Rac1b* reveals the complex regulation of only a handful of splicing isoforms in EMT, and since SR and hnRNP regulatory splicing factors regulate many splicing events, there is likely a global regulation of alternative splicing during EMT via coordinated SR and hnRNP interaction.

6.9 Potential Roles of RBPs That Regulate Alternative Splicing Coupled with Alternative Polyadenylation in EMT

An area of investigation that remains unexplored in EMT involves the roles of RBPs in the regulation of alternative polyadenylation (polyA) site selection (APA) (Di Giammartino et al. 2011). In many cases alternative polyA sites occur in alternative 3′ terminal exons in which case both the C-terminal coding sequence as well as 3′ UTRs can be regulated in different cellular contexts. In other cases, alternative polyA sites within the same 3′ UTR can yield shorter or longer 3′ UTRs. Several RBPs that have been shown to regulate alternative splicing, including the ESRPs and hnRNP H among others, also regulate alternative polyA site selection (Fig. 8.2) (Katz et al. 2010; Dittmar et al. 2012). One example of such regulation that is likely to impact EMT involves the FERM protein EPB41L5, which is required for the EMT in development (Lee et al. 2007; Hirano et al. 2008). The Esrps regulate EPB41L5 splicing and the mesenchymal isoform of this protein contains an extended C-terminus that includes a paxillin-binding domain not present in the shorter epithelial isoform. While not yet investigated, these two different protein isoforms most likely have significant differences in function that impact EMT. Hence, future work to investigate RBPs and mechanisms governing regulated APA and the functional consequences of these events for EMT is needed.

7 RBPs Involved in the Regulation of EMT by Translational Control and RNA Stability

In addition to splicing and polyadenylation, the processes of mRNA stability and translational control provide additional layers of regulation by RBPs that can alter the posttranscriptional landscape during EMT. Changes in gene expression regulated at these cytoplasmic steps can dynamically activate early cellular responses to extracellular stimuli and cellular signaling. The mechanism for translational regulation by RBPs can include direct translational activation or inactivation. The regulation of protein levels in the cytoplasm can also be influenced by changes in mRNA stability and mRNA localization (Fig. 8.2). The role of RBPs in these processes during EMT has only recently begun to be investigated, but they have the potential to further coordinate global changes to the posttranscriptional regulatory network of the EMT. Although we will describe the limited examples of RBPs for which EMT-relevant posttranscriptional networks have been investigated, there are also other RBPs that have been shown to regulate gene expression of EMT-relevant transcripts at these steps. It is therefore likely that future studies will reveal additional networks that impact EMT via stability and translation regulation.

7.1 Y-Box-Binding Protein-1 (YB-1)

The process of EMT in cancer is often associated with reduced proliferation and migration of mesenchymal cells; accordingly an RBP that regulates proliferation may also mediate an EMT. The Y-box-binding protein-1 (YB-1; also known as YBX1) is a DNA/RNA binding protein member of a family of proteins with an evolutionarily conserved cold-shock domain (Evdokimova et al. 2012). YB-1 has been implicated in cancer progression via the nuclear function of YB-1 as a transcriptional activator of proliferative genes, through binding to the Y-box promoter element (Kohno et al. 2003). YB-1 has also been shown to regulate alternative splicing via direct interactions with pre-mRNAs (Chansky et al. 2001; Stickeler et al. 2001; Allemand et al. 2007; Dutertre et al. 2010; Wei et al. 2012). Conversely, cytosolic YB-1 was observed to function as a tumor suppressor and was found in complex with translationally inactive mRNA transcripts near the 5′ mRNA cap structure (Evdokimova et al. 2006; Bader and Vogt 2008). To address the function of cytoplasmic YB-1 in the EMT, the Sorenson group utilized the immortalized human mammary epithelial cell line, MCF10A, and oncogenic H-Ras-transformed premalignant subclone, MCF10AT. Ectopic expression of YB-1 in the MCF10AT cells induced an EMT with characteristic changes in morphology from cuboidal to spindle-like cells, downregulation of E-cadherin, and activation of mesenchymal markers. Furthermore, orthotopic injections of MCF10AT-YB-1 cells into the mammary fat pad of mice promoted tumor invasion and metastasis, and this observation was further extended to human breast cancer, wherein cells expressing YB-1 and lacking E-cadherin exhibited an invasive phenotype (Evdokimova et al. 2009).

The conversion of MCF10AT cells by YB-1 was also associated with a reduction in proliferative capacity, and failure to reenter the cell cycle following starvation. Conversely, knockdown of YB-1 in MCF10AT-YB-1 cells was able to reverse the cellular morphology, and these cells were able to reenter the cell cycle efficiently. The antiproliferative role of YB-1 in promoting EMT was associated with cytosolic expression of YB-1 and implicated YB-1 in translational regulation of EMT-inducing genes. To identify transcripts that were not altered transcriptionally, but rather at the translational level by expression of YB-1 in the MCF10AT cells, total RNA levels were compared to RNAs isolated from translationally active polysomes and inactive monosome cellular fractions. This approach verified the reduction in mRNAs of epithelial genes including E-cadherin, and upregulation of mesenchymal transcripts in YB-1-expressing cells. The cell cycle regulators cyclin B1, D1, and D3 were among those transcripts without changes in mRNA levels, but that were reduced in the polysome fraction of YB-1-expressing cells, suggesting that reduced proliferation and failure to reenter the cell cycle were due to translational silencing by YB-1. However, another set of transcripts were enriched in the polysome fraction of YB-1-expressing cells and included the EMT inducers *Snail1* and *Zeb2/Sip1*, as well as other EMT factors. Of note, the EMT transcription factor Twist was increased at both the transcriptional and translational levels in the YB-1-expressing cells. Using the *Snail1* 5′ UTR as a model, the authors demonstrated a potential mechanism in which YB-1 induced cap-independent translation of Snail1 via a cellular internal ribosome entry site (IRES) (Evdokimova et al. 2009). These investigators proposed a model by which YB-1 coordinates IRES-mediated cap-independent translation of *Snail1* and a broader set of other transcripts to promote EMT via translational regulation. This report thus suggested that YB-1-mediated translational activation regulates a global network of EMT factors and cell cycle regulators, to coordinate regulation of translation during EMT.

7.2 hnRNP E1 Regulates a TGF-β-Induced EMT Translational Network

TGF-β is a primary driver of EMT both during development and in cancer (Thiery et al. 2009; Nieto 2011; Heldin et al. 2012). Another example of translational regulation involved in EMT was provided in models of TGF-β-induced EMT. Two EMT-associated genes, interleukin-like EMT inducer (ILEI) (Waerner et al. 2006) and Disabled-2 (Dab2) (Prunier and Howe 2005), were shown to be upregulated at the protein, but not mRNA, level after TGF-β treatment to induce EMT (Chaudhury et al. 2010). The Howe group utilized murine mammary epithelial NMuMG cells which can be induced to undergo a rapid EMT following TGF-β treatment. In unstimulated cells, both *Dab2* and *ILEI* mRNAs were abundant but were not associated with actively translating polysomes. Upon TGF-β treatment they rapidly redistributed to polysomes. RNA structural analysis of the 3′ UTR of *Dab2* and *ILEI* identified a structurally conserved stem–loop with an asymmetrical bulge. In vitro translation assays using luciferase reporters confirmed that the structural sequence was responsive to and necessary for TGF-β-induced translation. These elements were termed the TGF-β-activated translational element (BAT) and were utilized for in vitro binding assays to identify interacting proteins (Chaudhury et al. 2010; Hussey et al. 2011). hnRNP E1 (also known as PCBP1 or alpha-CP), a member of the family of heterogeneous ribonuclear proteins that functions in translation regulation (Ostareck et al. 1997), was one of two proteins that disassociated from the BAT element upon TGF-β treatment (Chaudhury et al. 2010). Also present in the BAT mRNP complex was the eukaryotic elongation factor-1 A1 (eEF1A1), a translation elongation factor. The authors outline a model of translational regulation wherein binding of hnRNP E1 and eEF1A1 is stabilized by the BAT element and prevents translation. TGF-β treatment in turn induces Akt2 phosphorylation of hnRNP E1 and release from the BAT, allowing eEF1A1 to promote translation elongation (Hussey et al. 2011). To identify a broader set of hnRNP E1-regulated mRNA targets these investigators used microarrays to identify mRNAs targeted to translationally active polysome fractions in an hnRNPE1-dependent manner upon TGF-β treatment. These datasets led to a determination of a group of 36 transcripts translationally upregulated in TGF-β-induced EMT via an hnRNPE1-dependent mechanism through binding to similar RNA "BAT" structural elements. As in the case of ILEI, total RNA levels of the newly identified BAT genes did not change but the transcripts were polysome associated and protein expression was increased by TGF-β treatment (Hussey et al. 2012). The network of genes regulated by hnRNP E1 under TGF-β stimulation represents another systems-level analysis of global regulation of translation. It is noteworthy that hnRNPE1 (PCBP1) also functions to regulate splicing, mRNA stability, and polyadenylation, with genome-wide targets of this RBP recently identified in different cell contexts (Ji et al. 2007, 2011, 2013). It is thus possible that the gene regulatory networks controlled by this protein include a subset of targets that impact EMT at more than one posttranscriptional regulatory step.

7.3 Additional Examples of Regulation by RBPs in the Cytoplasm That Impact EMT

The identification of truly global regulatory networks that impact EMT through the regulation of translation and mRNA stability has only just begun to emerge. However, there are several other examples of stability or translation regulatory RBPs that target key transcripts involved in EMT, although not at a genome-wide level in EMT model systems. HuR (also known as ELAVL1) is a ubiquitous member of the embryonic lethal abnormal vision (ELAV) family of RBPs, which primarily functions in regulating AU-rich element (ARE)-mediated mRNA stability (reviewed in Brennan and Steitz 2001; Srikantan and Gorospe 2012). In response to a variety of cellular stimuli HuR can relocalize to the cytoplasm, bind AREs of mRNAs, and regulate mRNA stability and translation. Upregulation of the EMT transcription factor Snail1 was observed in a hydrogen peroxide-induced model of EMT (Radisky et al. 2005). Based on the identification of AREs in the 3′ UTR of *Snail1*, it was shown that HuR promotes mRNA stability of *Snail1* and thus contributes to EMT. Knockdown of HuR resulted in a reduced *Snail1* mRNA half-life, increased E-cadherin, and was mediated through ARE elements residing in the 3′ UTR. Knockdown of HuR also inhibited hydrogen peroxide-induced EMT via reduced *Snail* mRNA levels (Dong et al. 2007). Genome-wide targets of HuR have been identified in numerous studies and it has also been shown to regulate splicing (Lebedeva et al. 2011; Mukherjee et al. 2011). However, it remains to be seen whether other components of HuR-regulated networks are centrally involved in EMT based upon changes in HuR regulatory activities.

Tristetraprolin (TTP) (also known as ZFP36) is a member of the TTP/TIS11 family of RBPs and also regulates mRNA stability by binding to AREs of a wide variety of transcripts (Ciais et al. 2013). In a TGF-β-induced model of EMT in Eph4 murine mammary cells, miR-29a was the most upregulated miRNA and TTP was identified as a target of miR-29a. Overexpression of miR-29a in parental epithelial cells induced an EMT, which was also observed by knockdown of TTP. Moreover, reintroduction of TTP in TTP knockdown cells reestablished E-cadherin levels and localization, reminiscent of an MET (Gebeshuber et al. 2009), suggesting that TTP can regulate EMT/MET conversion in part through regulation by miR-29a. This stands as an example of cross talk between posttranscriptional regulation of gene expression by miRNAs and implicating TTP-regulated mRNA stability in EMT. As TTP also regulates a wide range of mRNA transcripts, it is possible that TTP may also influence mRNA stability globally during EMT. While both HuR and TTP may impact EMT, however, it bears mention that global changes in the expression or activity of these regulators and their targets in EMT have not been shown.

The examples outlined in this section represent the early identification of RBP-regulated networks involving translation and mRNA stability during the EMT process. Evaluation of these processes in global networks provides an additional layer of gene regulation to the EMT in development and disease, similar to the previously documented transcriptional and newly discovered splicing networks. As many of the RBPs outlined here are associated with a large number of target transcripts, there is a high likelihood for additional coordinated global modifications of the translational landscape yet to be uncovered. Moreover, many of these RBPs have known nuclear and cytoplasmic functions and thus the potential for regulation at multiple levels of mRNA maturation is possible. hnRNP A2/B1, which was described in the splicing section as a regulator of ΔRon, was also shown to mediate an EMT in lung cancer cells. While alternative splicing was not evaluated, alterations in transcript levels were observed and may implicate other mechanisms of mRNA metabolism such as stability in hnRNP A2/B1-mediated EMT (Tauler et al. 2010). In addition to the examples cited, there have also been several other key genes involved in EMT including E-cadherin that have been shown to be regulated posttranscriptionally at the level of stability and/or translation (reviewed in Aparicio et al. 2013). However, the RBPs involved and the mechanism of this regulation have not been well defined.

8 Conclusions

While there is a wealth of literature regarding global programs of transcriptional regulation in EMT, the role of posttranscriptional regulation in this process has only recently begun to emerge. The integration of gene regulatory networks that impact EMT at each step in the life cycle of mRNA transcripts is needed to fully understand the pathways and mechanisms that underlie EMT/MET transitions in development and disease. For example, while the genes that are regulated at the transcript level are distinct from those that are regulated at the level of alternative splicing in EMT, both gene sets are enriched in functional categories that are known to be central to epithelial-mesenchymal plasticity. The recent identification of epithelial-specific microRNAs and splicing regulators has led to clear examples as to how distinct epithelial vs. mesenchymal gene expression programs are controlled posttranscriptionally. The roles of RBPs that affect posttranscriptional regulation at all stages are now beginning to come into view. However, the global networks of regulation controlled by RBPs and how their interrelationships bring about the changes associated with these cellular transitions are largely unknown. We also note that while several of the RBPs described here are clearly regulated during EMT (transcriptionally or posttranscriptionally), some of them are not known to undergo dynamic changes during EMT. However, due to the combinatorial functions of RBPs in posttranscriptional control, the contribution of these proteins cannot be ignored. It will therefore also be a challenge to decipher the roles of RBPs with central functions in EMT from those that have more auxiliary functions in the regulation of shared transcripts. There are also additional steps of regulation, including mRNA localization and localized translation, that have not been linked to EMT/MET. However, examples of localized mRNAs encoding proteins involved in epithelial cell polarity and cell migration have been described, such as the tight junction protein ZO-1 (TJP1) (Lecuyer et al. 2007, 2009; Medioni et al. 2012; Nagaoka et al. 2012). Networks of RBPs that direct mRNAs to be locally translated at sites that maintain polarity or induce cell migration are therefore also likely to play a role in EMT. Finally, while this review describes specific examples of RBPs and their posttranscriptional regulation of specific processes, it is becoming apparent that most RBPs are multifunctional proteins. Thus, many RBPs that had originally been studied in specific regulatory processes, such as splicing or mRNA stability, have been shown to impact gene expression at multiple steps (e.g., HuR, PTB, MBNL1/2). Therefore a broader systems view of these regulatory networks will also have to account for this added level of complexity. While there currently are limited numbers of RBP-regulated networks described in EMT/MET transitions, the few examples described in this review illustrate how the definition of broader posttranscriptional networks regulated by specific RBPs will lead to a more integrated understanding of EMT, as well as novel approaches in targeting EMT in disease.

Acknowledgements We thank members of the Carstens Lab for thoughtful suggestions and input. We also thank Peter Stoilov and Rik Thompson for helpful suggestions during the preparation of the manuscript. Work in the Carstens Lab is supported by NIH grants R01-GM088809 and R21-HG006892.

References

Acloque H, Adams MS, Fishwick K, Bronner-Fraser M, Nieto MA (2009) Epithelial-mesenchymal transitions: the importance of changing cell state in development and disease. J Clin Invest 119(6):1438–1449

Al-Hajj M, Wicha MS, Benito-Hernandez A, Morrison SJ, Clarke MF (2003) Prospective identification of tumorigenic breast cancer cells. Proc Natl Acad Sci U S A 100(7):3983–3988

Allemand E, Hastings ML, Murray MV, Myers MP, Krainer AR (2007) Alternative splicing regulation by interaction of phosphatase PP2Cgamma with nucleic acid-binding protein YB-1. Nat Struct Mol Biol 14(7):630–638

Aparicio LA, Abella V, Valladares M, Figueroa A (2013) Posttranscriptional regulation by RNA-binding proteins during epithelial-to-mesenchymal transition. Cell Mol Life Sci 70(23): 4463–4477

Argast GM, Mercado P, Mulford IJ, O'Connor M, Keane DM, Shaaban S, Epstein DM, Pachter JA, Kan JL (2011) Cooperative signaling between oncostatin M, hepatocyte growth factor and transforming growth factor-beta enhances epithelial to mesenchymal transition in lung and pancreatic tumor models. Cells Tissues Organs 193(1–2):114–132

Arnold SJ, Robertson EJ (2009) Making a commitment: cell lineage allocation and axis patterning in the early mouse embryo. Nat Rev Mol Cell Biol 10(2):91–103

Arnoux V, Nassour M, L'Helgoualc'h A, Hipskind RA, Savagner P (2008) Erk5 controls Slug expression and keratinocyte activation during wound healing. Mol Biol Cell 19(11):4738–4749

Attisano L, Wrana JL (2013) Signal integration in TGF-beta, WNT, and Hippo pathways. F1000Prime Rep 5:17

Bader AG, Vogt PK (2008) Phosphorylation by Akt disables the anti-oncogenic activity of YB-1. Oncogene 27(8):1179–1182

Bao X, Zhu X, Liao B, Benda C, Zhuang Q, Pei D, Qin B, Esteban MA (2013) MicroRNAs in somatic cell reprogramming. Curr Opin Cell Biol 25(2):208–214

Barr S, Thomson S, Buck E, Russo S, Petti F, Sujka-Kwok I, Eyzaguirre A, Rosenfeld-Franklin M, Gibson NW, Miglarese M, Epstein D, Iwata KK, Haley JD (2008) Bypassing cellular EGF receptor dependence through epithelial-to-mesenchymal-like transitions. Clin Exp Metastasis 25(6):685–693

Baum B, Settleman J, Quinlan MP (2008) Transitions between epithelial and mesenchymal states in development and disease. Semin Cell Dev Biol 19(3):294–308

Biamonti G, Bonomi S, Gallo S, Ghigna C (2012) Making alternative splicing decisions during epithelial-to-mesenchymal transition (EMT). Cell Mol Life Sci 69(15):2515–2526

Blick T, Widodo E, Hugo H, Waltham M, Lenburg ME, Neve RM, Thompson EW (2008) Epithelial mesenchymal transition traits in human breast cancer cell lines. Clin Exp Metastasis 25(6): 629–642

Bonomi S, Di Matteo A, Buratti E, Cabianca DS, Baralle FE, Ghigna C, Biamonti G (2013) HnRNP A1 controls a splicing regulatory circuit promoting mesenchymal-to-epithelial transition. Nucleic Acids Res 41(18):8665–8679

Boutz PL, Chawla G, Stoilov P, Black DL (2007) MicroRNAs regulate the expression of the alternative splicing factor nPTB during muscle development. Genes Dev 21(1):71–84

Brabletz T, Hlubek F, Spaderna S, Schmalhofer O, Hiendlmeyer E, Jung A, Kirchner T (2005) Invasion and metastasis in colorectal cancer: epithelial-mesenchymal transition, mesenchymal-epithelial transition, stem cells and beta-catenin. Cells Tissues Organs 179(1–2):56–65

Braeutigam C, Rago L, Rolke A, Waldmeier L, Christofori G, Winter J (2013) The RNA-binding protein Rbfox2: an essential regulator of EMT-driven alternative splicing and a mediator of cellular invasion. Oncogene 33(9):1082–1092

Braunschweig U, Gueroussov S, Plocik AM, Graveley BR, Blencowe BJ (2013) Dynamic Integration of Splicing within Gene Regulatory Pathways. Cell 152(6):1252–1269

Brennan CM, Steitz JA (2001) HuR and mRNA stability. Cell Mol Life Sci 58(2):266–277

Brown RL, Reinke LM, Damerow MS, Perez D, Chodosh LA, Yang J, Cheng C (2011) CD44 splice isoform switching in human and mouse epithelium is essential for epithelial-mesenchymal transition and breast cancer progression. J Clin Invest 121(3):1064–1074

Burk U, Schubert J, Wellner U, Schmalhofer O, Vincan E, Spaderna S, Brabletz T (2008) A reciprocal repression between ZEB1 and members of the miR-200 family promotes EMT and invasion in cancer cells. EMBO Rep 9(6):582–589

Celia-Terrassa T, Meca-Cortes O, Mateo F, de Paz AM, Rubio N, Arnal-Estape A, Ell BJ, Bermudo R, Diaz A, Guerra-Rebollo M, Lozano JJ, Estaras C, Ulloa C, Alvarez-Simon D, Mila J, Vilella R, Paciucci R, Martinez-Balbas M, de Herreros AG, Gomis RR, Kang Y, Blanco J, Fernandez PL, Thomson TM (2012) Epithelial-mesenchymal transition can suppress major attributes of human epithelial tumor-initiating cells. J Clin Invest 122(5):1849–1868

Chakrabarti R, Hwang J, Andres Blanco M, Wei Y, Lukacisin M, Romano RA, Smalley K, Liu S, Yang Q, Ibrahim T, Mercatali L, Amadori D, Haffty BG, Sinha S, Kang Y (2012) Elf5 inhibits the epithelial-mesenchymal transition in mammary gland development and breast cancer metastasis by transcriptionally repressing Snail2. Nat Cell Biol 14(11):1212–1222

Chansky HA, Hu M, Hickstein DD, Yang L (2001) Oncogenic TLS/ERG and EWS/Fli-1 fusion proteins inhibit RNA splicing mediated by YB-1 protein. Cancer Res 61(9):3586–3590

Charizanis K, Lee KY, Batra R, Goodwin M, Zhang C, Yuan Y, Shiue L, Cline M, Scotti MM, Xia G, Kumar A, Ashizawa T, Clark HB, Kimura T, Takahashi MP, Fujimura H, Jinnai K, Yoshikawa H, Gomes-Pereira M, Gourdon G, Sakai N, Nishino S, Foster TC, Ares M Jr, Darnell RB, Swanson MS (2012) Muscleblind-like 2-mediated alternative splicing in the developing brain and dysregulation in myotonic dystrophy. Neuron 75(3):437–450

Chaudhury A, Hussey GS, Ray PS, Jin G, Fox PL, Howe PH (2010) TGF-beta-mediated phosphorylation of hnRNP E1 induces EMT via transcript-selective translational induction of Dab2 and ILEI. Nat Cell Biol 12(3):286–293

Chen M, Manley JL (2009) Mechanisms of alternative splicing regulation: insights from molecular and genomics approaches. Nat Rev Mol Cell Biol 10(11):741–754

Chou J, Lin JH, Brenot A, Kim JW, Provot S, Werb Z (2013) GATA3 suppresses metastasis and modulates the tumour microenvironment by regulating microRNA-29b expression. Nat Cell Biol 15(2):201–213

Chou J, Provot S, Werb Z (2010) GATA3 in development and cancer differentiation: cells GATA have it! J Cell Physiol 222(1):42–49

Ciais D, Cherradi N, Feige JJ (2013) Multiple functions of tristetraprolin/TIS11 RNA-binding proteins in the regulation of mRNA biogenesis and degradation. Cell Mol Life Sci 70(12): 2031–2044

Cieply B, Farris J, Denvir J, Ford H, Frisch SM (2013) A reciprocal feedback loop between Grainyhead-like-2 and ZEB1 controls EMT and tumor suppression. Cancer Res 73(20): 6299–6309

Cieply B, Riley PT, Pifer PM, Widmeyer J, Addison JB, Ivanov AV, Denvir J, Frisch SM (2012) Suppression of the epithelial-mesenchymal transition by Grainy head-like-2. Cancer Res 72(9): 2440–2453

Collesi C, Santoro MM, Gaudino G, Comoglio PM (1996) A splicing variant of the RON transcript induces constitutive tyrosine kinase activity and an invasive phenotype. Mol Cell Biol 16(10):5518–5526

Cooper DL, Dougherty G, Harn HJ, Jackson S, Baptist EW, Byers J, Datta A, Phillips G, Isola NR (1992) The complex CD44 transcriptional unit; alternative splicing of three internal exons generates the epithelial form of CD44. Biochem Biophys Res Commun 182(2):569–578

Dave N, Guaita-Esteruelas S, Gutarra S, Frias A, Beltran M, Peiro S, de Herreros AG (2011) Functional cooperation between Snail1 and twist in the regulation of ZEB1 expression during epithelial to mesenchymal transition. J Biol Chem 286(14):12024–12032

De Craene B, Berx G (2013) Regulatory networks defining EMT during cancer initiation and progression. Nat Rev Cancer 13(2):97–110

De Moerlooze L, Spencer-Dene B, Revest J, Hajihosseini M, Rosewell I, Dickson C (2000) An important role for the IIIb isoform of fibroblast growth factor receptor 2 (FGFR2) in

mesenchymal-epithelial signalling during mouse organogenesis. Development 127(3): 483–492
Di Giammartino DC, Nishida K, Manley JL (2011) Mechanisms and consequences of alternative polyadenylation. Mol Cell 43(6):853–866
Di Modugno F, DeMonte L, Balsamo M, Bronzi G, Nicotra MR, Alessio M, Jager E, Condeelis JS, Santoni A, Natali PG, Nistico P (2007) Molecular cloning of hMena (ENAH) and its splice variant hMena+11a: epidermal growth factor increases their expression and stimulates hMena+11a phosphorylation in breast cancer cell lines. Cancer Res 67(6):2657–2665
Ding X, Park SI, McCauley LK, Wang CY (2013) Signaling between transforming growth factor beta (TGF-beta) and transcription factor SNAI2 represses expression of microRNA miR-203 to promote epithelial-mesenchymal transition and tumor metastasis. J Biol Chem 288(15): 10241–10253
Dittmar KA, Jiang P, Park JW, Amirikian K, Wan J, Shen S, Xing Y, Carstens RP (2012) Genome-wide determination of a broad ESRP-regulated posttranscriptional network by high-throughput sequencing. Mol Cell Biol 32(8):1468–1482
Dong R, Lu JG, Wang Q, He XL, Chu YK, Ma QJ (2007) Stabilization of Snail by HuR in the process of hydrogen peroxide induced cell migration. Biochem Biophys Res Commun 356(1): 318–321
Drasin DJ, Robin TP, Ford HL (2011) Breast cancer epithelial-to-mesenchymal transition: examining the functional consequences of plasticity. Breast Cancer Res 13(6):226
Dressler GR (2006) The cellular basis of kidney development. Annu Rev Cell Dev Biol 22: 509–529
Du H, Cline MS, Osborne RJ, Tuttle DL, Clark TA, Donohue JP, Hall MP, Shiue L, Swanson MS, Thornton CA, Ares M Jr (2010) Aberrant alternative splicing and extracellular matrix gene expression in mouse models of myotonic dystrophy. Nat Struct Mol Biol 17(2):187–193
Duffield JS (2010) Epithelial to mesenchymal transition in injury of solid organs: fact or artifact? Gastroenterology 139(4):1081–1083, 1083 e1081–1085
Dutertre M, Sanchez G, De Cian MC, Barbier J, Dardenne E, Gratadou L, Dujardin G, Le Jossic-Corcos C, Corcos L, Auboeuf D (2010) Cotranscriptional exon skipping in the genotoxic stress response. Nat Struct Mol Biol 17(11):1358–1366
Eccles SA, Welch DR (2007) Metastasis: recent discoveries and novel treatment strategies. Lancet 369(9574):1742–1757
Evdokimova V, Ruzanov P, Anglesio MS, Sorokin AV, Ovchinnikov LP, Buckley J, Triche TJ, Sonenberg N, Sorensen PH (2006) Akt-mediated YB-1 phosphorylation activates translation of silent mRNA species. Mol Cell Biol 26(1):277–292
Evdokimova V, Tognon C, Ng T, Ruzanov P, Melnyk N, Fink D, Sorokin A, Ovchinnikov LP, Davicioni E, Triche TJ, Sorensen PH (2009) Translational activation of snail1 and other developmentally regulated transcription factors by YB-1 promotes an epithelial-mesenchymal transition. Cancer Cell 15(5):402–415
Evdokimova V, Tognon CE, Sorensen PH (2012) On translational regulation and EMT. Semin Cancer Biol 22(5–6):437–445
Frisch SM, Francis H (1994) Disruption of epithelial cell-matrix interactions induces apoptosis. J Cell Biol 124(4):619–626
Frisch SM, Schaller M, Cieply B (2013) Mechanisms that link the oncogenic epithelial-mesenchymal transition to suppression of anoikis. J Cell Sci 126(Pt 1):21–29
Fuxe J, Karlsson MC (2012) TGF-beta-induced epithelial-mesenchymal transition: a link between cancer and inflammation. Semin Cancer Biol 22(5–6):455–461
Ge K, DuHadaway J, Du W, Herlyn M, Rodeck U, Prendergast GC (1999) Mechanism for elimination of a tumor suppressor: aberrant splicing of a brain-specific exon causes loss of function of Bin1 in melanoma. Proc Natl Acad Sci U S A 96(17):9689–9694
Gebeshuber CA, Zatloukal K, Martinez J (2009) miR-29a suppresses tristetraprolin, which is a regulator of epithelial polarity and metastasis. EMBO Rep 10(4):400–405
Geiger TR, Peeper DS (2009) Metastasis mechanisms. Biochim Biophys Acta 1796(2):293–308

Ghigna C, Giordano S, Shen H, Benvenuto F, Castiglioni F, Comoglio PM, Green MR, Riva S, Biamonti G (2005) Cell motility is controlled by SF2/ASF through alternative splicing of the Ron protooncogene. Mol Cell 20(6):881–890

Golan-Gerstl R, Cohen M, Shilo A, Suh SS, Bakacs A, Coppola L, Karni R (2011) Splicing factor hnRNP A2/B1 regulates tumor suppressor gene splicing and is an oncogenic driver in glioblastoma. Cancer Res 71(13):4464–4472

Goncalves V, Matos P, Jordan P (2009) Antagonistic SR proteins regulate alternative splicing of tumor-related Rac1b downstream of the PI3-kinase and Wnt pathways. Hum Mol Genet 18(19):3696–3707

Gregory PA, Bert AG, Paterson EL, Barry SC, Tsykin A, Farshid G, Vadas MA, Khew-Goodall Y, Goodall GJ (2008a) The miR-200 family and miR-205 regulate epithelial to mesenchymal transition by targeting ZEB1 and SIP1. Nat Cell Biol 10(5):593–601

Gregory PA, Bracken CP, Bert AG, Goodall GJ (2008b) MicroRNAs as regulators of epithelial-mesenchymal transition. Cell Cycle 7(20):3112–3118

Guaita S, Puig I, Franci C, Garrido M, Dominguez D, Batlle E, Sancho E, Dedhar S, De Herreros AG, Baulida J (2002) Snail induction of epithelial to mesenchymal transition in tumor cells is accompanied by MUC1 repression and ZEB1 expression. J Biol Chem 277(42):39209–39216

Gunasinghe NP, Wells A, Thompson EW, Hugo HJ (2012) Mesenchymal-epithelial transition (MET) as a mechanism for metastatic colonisation in breast cancer. Cancer Metastasis Rev 31(3–4):469–478

Gupta PB, Onder TT, Jiang G, Tao K, Kuperwasser C, Weinberg RA, Lander ES (2009) Identification of selective inhibitors of cancer stem cells by high-throughput screening. Cell 138(4):645–659

Hall MP, Nagel RJ, Fagg WS, Shiue L, Cline MS, Perriman RJ, Donohue JP, Ares M Jr (2013) Quaking and PTB control overlapping splicing regulatory networks during muscle cell differentiation. RNA 19(5):627–638

Heldin CH, Vanlandewijck M, Moustakas A (2012) Regulation of EMT by TGFbeta in cancer. FEBS Lett 586(14):1959–1970

Hertel KJ (2008) Combinatorial control of exon recognition. J Biol Chem 283(3):1211–1215

Hill L, Browne G, Tulchinsky E (2013) ZEB/miR-200 feedback loop: at the crossroads of signal transduction in cancer. Int J Cancer 132(4):745–754

Hirano M, Hashimoto S, Yonemura S, Sabe H, Aizawa S (2008) EPB41L5 functions to post-transcriptionally regulate cadherin and integrin during epithelial-mesenchymal transition. J Cell Biol 182(6):1217–1230

Hogstrand C, Kille P, Ackland ML, Hiscox S, Taylor KM (2013) A mechanism for epithelial-mesenchymal transition and anoikis resistance in breast cancer triggered by zinc channel ZIP6 and signal transducer and activator of transcription 3 (STAT3). Biochem J 455(2):229–237

Hollier BG, Tinnirello AA, Werden SJ, Evans KW, Taube JH, Sarkar TR, Sphyris N, Shariati M, Kumar SV, Battula VL, Herschkowitz JI, Guerra R, Chang JT, Miura N, Rosen JM, Mani SA (2013) FOXC2 expression links epithelial-mesenchymal transition and stem cell properties in breast cancer. Cancer Res 73(6):1981–1992

Horiguchi K, Sakamoto K, Koinuma D, Semba K, Inoue A, Inoue S, Fujii H, Yamaguchi A, Miyazawa K, Miyazono K, Saitoh M (2012) TGF-beta drives epithelial-mesenchymal transition through deltaEF1-mediated downregulation of ESRP. Oncogene 31(26):3190–3201

Hudson LG, Newkirk KM, Chandler HL, Choi C, Fossey SL, Parent AE, Kusewitt DF (2009) Cutaneous wound reepithelialization is compromised in mice lacking functional Slug (Snai2). J Dermatol Sci 56(1):19–26

Humphreys BD, Lin SL, Kobayashi A, Hudson TE, Nowlin BT, Bonventre JV, Valerius MT, McMahon AP, Duffield JS (2010) Fate tracing reveals the pericyte and not epithelial origin of myofibroblasts in kidney fibrosis. Am J Pathol 176(1):85–97

Hussey GS, Chaudhury A, Dawson AE, Lindner DJ, Knudsen CR, Wilce MC, Merrick WC, Howe PH (2011) Identification of an mRNP complex regulating tumorigenesis at the translational elongation step. Mol Cell 41(4):419–431

Hussey GS, Link LA, Brown AS, Howley BV, Chaudhury A, Howe PH (2012) Establishment of a TGFbeta-induced post-transcriptional EMT gene signature. PLoS One 7(12):e52624

Isken O, Maquat LE (2008) The multiple lives of NMD factors: balancing roles in gene and genome regulation. Nat Rev Genet 9(9):699–712
Iwano M, Plieth D, Danoff TM, Xue C, Okada H, Neilson EG (2002) Evidence that fibroblasts derive from epithelium during tissue fibrosis. J Clin Invest 110(3):341–350
Ji X, Kong J, Carstens RP, Liebhaber SA (2007) The 3' untranslated region complex involved in stabilization of human alpha-globin mRNA assembles in the nucleus and serves an independent role as a splice enhancer. Mol Cell Biol 27(9):3290–3302
Ji X, Kong J, Liebhaber SA (2011) An RNA-protein complex links enhanced nuclear 3′ processing with cytoplasmic mRNA stabilization. EMBO J 30(13):2622–2633
Ji X, Wan J, Vishnu M, Xing Y, Liebhaber SA (2013) alphaCP Poly(C) binding proteins act as global regulators of alternative polyadenylation. Mol Cell Biol 33(13):2560–2573
Jorda M, Olmeda D, Vinyals A, Valero E, Cubillo E, Llorens A, Cano A, Fabra A (2005) Upregulation of MMP-9 in MDCK epithelial cell line in response to expression of the Snail transcription factor. J Cell Sci 118(Pt 15):3371–3385
Jordan P, Brazao R, Boavida MG, Gespach C, Chastre E (1999) Cloning of a novel human Rac1b splice variant with increased expression in colorectal tumors. Oncogene 18(48):6835–6839
Kalluri R, Neilson EG (2003) Epithelial-mesenchymal transition and its implications for fibrosis. J Clin Invest 112(12):1776–1784
Kalluri R, Weinberg RA (2009) The basics of epithelial-mesenchymal transition. J Clin Invest 119(6):1420–1428
Kalsotra A, Cooper TA (2011) Functional consequences of developmentally regulated alternative splicing. Nat Rev Genet 12(10):715–729
Kalsotra A, Xiao X, Ward AJ, Castle JC, Johnson JM, Burge CB, Cooper TA (2008) A postnatal switch of CELF and MBNL proteins reprograms alternative splicing in the developing heart. Proc Natl Acad Sci U S A 105(51):20333–20338
Kang Y, Massague J (2004) Epithelial-mesenchymal transitions: twist in development and metastasis. Cell 118(3):277–279
Karni R, de Stanchina E, Lowe SW, Sinha R, Mu D, Krainer AR (2007) The gene encoding the splicing factor SF2/ASF is a proto-oncogene. Nat Struct Mol Biol 14(3):185–193
Katz Y, Wang ET, Airoldi EM, Burge CB (2010) Analysis and design of RNA sequencing experiments for identifying isoform regulation. Methods, Nat
Keene JD (2007) RNA regulons: coordination of post-transcriptional events. Nat Rev Genet 8(7):533–543
Kelemen O, Convertini P, Zhang Z, Wen Y, Shen M, Falaleeva M, Stamm S (2013) Function of alternative splicing. Gene 514(1):1–30
Kerosuo L, Bronner-Fraser M (2012) What is bad in cancer is good in the embryo: importance of EMT in neural crest development. Semin Cell Dev Biol 23(3):320–332
Kinsella AR, Green B, Lepts GC, Hill CL, Bowie G, Taylor BA (1993) The role of the cell-cell adhesion molecule E-cadherin in large bowel tumour cell invasion and metastasis. Br J Cancer 67(5):904–909
Klymkowsky MW, Savagner P (2009) Epithelial-mesenchymal transition: a cancer researcher's conceptual friend and foe. Am J Pathol 174(5):1588–1593
Kohno K, Izumi H, Uchiumi T, Ashizuka M, Kuwano M (2003) The pleiotropic functions of the Y-box-binding protein, YB-1. Bioessays 25(7):691–698
Korpal M, Lee ES, Hu G, Kang Y (2008) The miR-200 family inhibits epithelial-mesenchymal transition and cancer cell migration by direct targeting of E-cadherin transcriptional repressors ZEB1 and ZEB2. J Biol Chem 283(22):14910–14914
Kouros-Mehr H, Slorach EM, Sternlicht MD, Werb Z (2006) GATA-3 maintains the differentiation of the luminal cell fate in the mammary gland. Cell 127(5):1041–1055
Kriz W, Kaissling B, Le Hir M (2011) Epithelial-mesenchymal transition (EMT) in kidney fibrosis: fact or fantasy? J Clin Invest 121(2):468–474
Krol J, Loedige I, Filipowicz W (2010) The widespread regulation of microRNA biogenesis, function and decay. Nat Rev Genet 11(9):597–610

Kumarswamy R, Mudduluru G, Ceppi P, Muppala S, Kozlowski M, Niklinski J, Papotti M, Allgayer H (2012) MicroRNA-30a inhibits epithelial-to-mesenchymal transition by targeting Snai1 and is downregulated in non-small cell lung cancer. Int J Cancer 130(9):2044–2053

LaGamba D, Nawshad A, Hay ED (2005) Microarray analysis of gene expression during epithelial-mesenchymal transformation. Dev Dyn 234(1):132–142

Lamouille S, Subramanyam D, Blelloch R, Derynck R (2013) Regulation of epithelial-mesenchymal and mesenchymal-epithelial transitions by microRNAs. Curr Opin Cell Biol 25(2):200–207

Lapuk A, Marr H, Jakkula L, Pedro H, Bhattacharya S, Purdom E, Hu Z, Simpson K, Pachter L, Durinck S, Wang N, Parvin B, Fontenay G, Speed T, Garbe J, Stampfer M, Bayandorian H, Dorton S, Clark TA, Schweitzer A, Wyrobek A, Feiler H, Spellman P, Conboy J, Gray JW (2010) Exon-level microarray analyses identify alternative splicing programs in breast cancer. Mol Cancer Res 8(7):961–974

Lareau LF, Inada M, Green RE, Wengrod JC, Brenner SE (2007) Unproductive splicing of SR genes associated with highly conserved and ultraconserved DNA elements. Nature 446(7138):926–929

Lebedeva S, Jens M, Theil K, Schwanhausser B, Selbach M, Landthaler M, Rajewsky N (2011) Transcriptome-wide analysis of regulatory interactions of the RNA-binding protein HuR. Mol Cell 43(3):340–352

Lebleu VS, Taduri G, O'Connell J, Teng Y, Cooke VG, Woda C, Sugimoto H, Kalluri R (2013) Origin and function of myofibroblasts in kidney fibrosis. Nat Med 19(8):1047–1053

Lecuyer E, Yoshida H, Krause HM (2009) Global implications of mRNA localization pathways in cellular organization. Curr Opin Cell Biol 21(3):409–415

Lecuyer E, Yoshida H, Parthasarathy N, Alm C, Babak T, Cerovina T, Hughes TR, Tomancak P, Krause HM (2007) Global analysis of mRNA localization reveals a prominent role in organizing cellular architecture and function. Cell 131(1):174–187

Lee JD, Silva-Gagliardi NF, Tepass U, McGlade CJ, Anderson KV (2007) The FERM protein Epb4.1l5 is required for organization of the neural plate and for the epithelial-mesenchymal transition at the primitive streak of the mouse embryo. Development 134(11):2007–2016

Lefave CV, Squatrito M, Vorlova S, Rocco GL, Brennan CW, Holland EC, Pan YX, Cartegni L (2011) Splicing factor hnRNPH drives an oncogenic splicing switch in gliomas. EMBO J 30(19):4084–4097

Lengner CJ (2010) iPS cell technology in regenerative medicine. Ann N Y Acad Sci 1192:38–44

Leopold PL, Vincent J, Wang H (2012) A comparison of epithelial-to-mesenchymal transition and re-epithelialization. Semin Cancer Biol 22(5–6):471–483

Li R, Liang J, Ni S, Zhou T, Qing X, Li H, He W, Chen J, Li F, Zhuang Q, Qin B, Xu J, Li W, Yang J, Gan Y, Qin D, Feng S, Song H, Yang D, Zhang B, Zeng L, Lai L, Esteban MA, Pei D (2010) A mesenchymal-to-epithelial transition initiates and is required for the nuclear reprogramming of mouse fibroblasts. Cell Stem Cell 7(1):51–63

Lim J, Thiery JP (2012) Epithelial-mesenchymal transitions: insights from development. Development 139(19):3471–3486

Lin SL, Kisseleva T, Brenner DA, Duffield JS (2008) Pericytes and perivascular fibroblasts are the primary source of collagen-producing cells in obstructive fibrosis of the kidney. Am J Pathol 173(6):1617–1627

Llorian M, Schwartz S, Clark TA, Hollander D, Tan LY, Spellman R, Gordon A, Schweitzer AC, de la Grange P, Ast G, Smith CW (2010) Position-dependent alternative splicing activity revealed by global profiling of alternative splicing events regulated by PTB. Nat Struct Mol Biol 17(9):1114–1123

Ma L, Teruya-Feldstein J, Weinberg RA (2007) Tumour invasion and metastasis initiated by microRNA-10b in breast cancer. Nature 449(7163):682–688

Ma L, Young J, Prabhala H, Pan E, Mestdagh P, Muth D, Teruya-Feldstein J, Reinhardt F, Onder TT, Valastyan S, Westermann F, Speleman F, Vandesompele J, Weinberg RA (2010) miR-9, a MYC/MYCN-activated microRNA, regulates E-cadherin and cancer metastasis. Nat Cell Biol 12(3):247–256

Mani SA, Guo W, Liao MJ, Eaton EN, Ayyanan A, Zhou AY, Brooks M, Reinhard F, Zhang CC, Shipitsin M, Campbell LL, Polyak K, Brisken C, Yang J, Weinberg RA (2008) The epithelial-mesenchymal transition generates cells with properties of stem cells. Cell 133(4):704–715

May CD, Sphyris N, Evans KW, Werden SJ, Guo W, Mani SA (2011) Epithelial-mesenchymal transition and cancer stem cells: a dangerously dynamic duo in breast cancer progression. Breast Cancer Res 13(1):202

Medioni C, Mowry K, Besse F (2012) Principles and roles of mRNA localization in animal development. Development 139(18):3263–3276

Miyoshi A, Kitajima Y, Kido S, Shimonishi T, Matsuyama S, Kitahara K, Miyazaki K (2005) Snail accelerates cancer invasion by upregulating MMP expression and is associated with poor prognosis of hepatocellular carcinoma. Br J Cancer 92(2):252–258

Miyoshi A, Kitajima Y, Sumi K, Sato K, Hagiwara A, Koga Y, Miyazaki K (2004) Snail and SIP1 increase cancer invasion by upregulating MMP family in hepatocellular carcinoma cells. Br J Cancer 90(6):1265–1273

Morel AP, Lievre M, Thomas C, Hinkal G, Ansieau S, Puisieux A (2008) Generation of breast cancer stem cells through epithelial-mesenchymal transition. PLoS One 3(8):e2888

Moustakas A, Heldin CH (2005) Non-Smad TGF-beta signals. J Cell Sci 118(Pt 16):3573–3584

Mukherjee N, Corcoran DL, Nusbaum JD, Reid DW, Georgiev S, Hafner M, Ascano M Jr, Tuschl T, Ohler U, Keene JD (2011) Integrative Regulatory Mapping Indicates that the RNA-Binding Protein HuR Couples Pre-mRNA Processing and mRNA Stability. Mol Cell 43(3):327–339

Murry CE, Keller G (2008) Differentiation of embryonic stem cells to clinically relevant populations: lessons from embryonic development. Cell 132(4):661–680

Nagaoka K, Udagawa T, Richter JD (2012) CPEB-mediated ZO-1 mRNA localization is required for epithelial tight-junction assembly and cell polarity. Nat Commun 3:675

Neil JR, Schiemann WP (2008) Altered TAB1:I kappaB kinase interaction promotes transforming growth factor beta-mediated nuclear factor-kappaB activation during breast cancer progression. Cancer Res 68(5):1462–1470

Nieto MA (2011) The ins and outs of the epithelial to mesenchymal transition in health and disease. Annu Rev Cell Dev Biol 27:347–376

Nieto MA, Cano A (2012) The epithelial-mesenchymal transition under control: global programs to regulate epithelial plasticity. Semin Cancer Biol 22(5–6):361–368

Nilsen TW, Graveley BR (2010) Expansion of the eukaryotic proteome by alternative splicing. Nature 463(7280):457–463

Ocana OH, Corcoles R, Fabra A, Moreno-Bueno G, Acloque H, Vega S, Barrallo-Gimeno A, Cano A, Nieto MA (2012) Metastatic colonization requires the repression of the epithelial-mesenchymal transition inducer Prrx1. Cancer Cell 22(6):709–724

Ohkubo T, Ozawa M (2004) The transcription factor Snail downregulates the tight junction components independently of E-cadherin downregulation. J Cell Sci 117(Pt 9):1675–1685

Onder TT, Gupta PB, Mani SA, Yang J, Lander ES, Weinberg RA (2008) Loss of E-cadherin promotes metastasis via multiple downstream transcriptional pathways. Cancer Res 68(10): 3645–3654

Ostareck DH, Ostareck-Lederer A, Wilm M, Thiele BJ, Mann M, Hentze MW (1997) mRNA silencing in erythroid differentiation: hnRNP K and hnRNP E1 regulate 15-lipoxygenase translation from the 3′ end. Cell 89(4):597–606

Ozdamar B, Bose R, Barrios-Rodiles M, Wang HR, Zhang Y, Wrana JL (2005) Regulation of the polarity protein Par6 by TGFbeta receptors controls epithelial cell plasticity. Science 307(5715):1603–1609

Paoli P, Giannoni E, Chiarugi P (2013) Anoikis molecular pathways and its role in cancer progression. Biochim Biophys Acta 1833(12):3481–3498

Park SM, Gaur AB, Lengyel E, Peter ME (2008) The miR-200 family determines the epithelial phenotype of cancer cells by targeting the E-cadherin repressors ZEB1 and ZEB2. Genes Dev 22(7):894–907

Peinado H, Olmeda D, Cano A (2007) Snail, Zeb and bHLH factors in tumour progression: an alliance against the epithelial phenotype? Nat Rev Cancer 7(6):415–428

Peinado H, Portillo F, Cano A (2004) Transcriptional regulation of cadherins during development and carcinogenesis. Int J Dev Biol 48(5–6):365–375

Prunier C, Howe PH (2005) Disabled-2 (Dab2) is required for transforming growth factor beta-induced epithelial to mesenchymal transition (EMT). J Biol Chem 280(17):17540–17548

Qu Y, Li WC, Hellem MR, Rostad K, Popa M, McCormack E, Oyan AM, Kalland KH, Ke XS (2013) MiR-182 and miR-203 induce mesenchymal to epithelial transition and self-sufficiency of growth signals via repressing SNAI2 in prostate cells. Int J Cancer 133(3):544–555

Radisky DC, Levy DD, Littlepage LE, Liu H, Nelson CM, Fata JE, Leake D, Godden EL, Albertson DG, Nieto MA, Werb Z, Bissell MJ (2005) Rac1b and reactive oxygen species mediate MMP-3-induced EMT and genomic instability. Nature 436(7047):123–127

Reinke LM, Xu Y, Cheng C (2012) Snail represses the splicing regulator epithelial splicing regulatory protein 1 to promote epithelial-mesenchymal transition. J Biol Chem 287(43):36435–36442

Rennebeck G, Martelli M, Kyprianou N (2005) Anoikis and survival connections in the tumor microenvironment: is there a role in prostate cancer metastasis? Cancer Res 65(24): 11230–11235

Revest JM, Spencer-Dene B, Kerr K, De Moerlooze L, Rosewell I, Dickson C (2001) Fibroblast growth factor receptor 2-IIIb acts upstream of Shh and Fgf4 and is required for limb bud maintenance but not for the induction of Fgf8, Fgf10, Msx1, or Bmp4. Dev Biol 231(1):47–62

Ru P, Steele R, Newhall P, Phillips NJ, Toth K, Ray RB (2012) miRNA-29b suppresses prostate cancer metastasis by regulating epithelial-mesenchymal transition signaling. Mol Cancer Ther 11(5):1166–1173

Samavarchi-Tehrani P, Golipour A, David L, Sung HK, Beyer TA, Datti A, Woltjen K, Nagy A, Wrana JL (2010) Functional genomics reveals a BMP-driven mesenchymal-to-epithelial transition in the initiation of somatic cell reprogramming. Cell Stem Cell 7(1):64–77

Savagner P, Kusewitt DF, Carver EA, Magnino F, Choi C, Gridley T, Hudson LG (2005) Developmental transcription factor slug is required for effective re-epithelialization by adult keratinocytes. J Cell Physiol 202(3):858–866

Savagner P, Valles AM, Jouanneau J, Yamada KM, Thiery JP (1994) Alternative splicing in fibroblast growth factor receptor 2 is associated with induced epithelial-mesenchymal transition in rat bladder carcinoma cells. Mol Biol Cell 5(8):851–862

Sawicka K, Bushell M, Spriggs KA, Willis AE (2008) Polypyrimidine-tract-binding protein: a multifunctional RNA-binding protein. Biochem Soc Trans 36(Pt 4):641–647

Scheel C, Weinberg RA (2012) Cancer stem cells and epithelial-mesenchymal transition: concepts and molecular links. Semin Cancer Biol 22(5–6):396–403

Schnelzer A, Prechtel D, Knaus U, Dehne K, Gerhard M, Graeff H, Harbeck N, Schmitt M, Lengyel E (2000) Rac1 in human breast cancer: overexpression, mutation analysis, and characterization of a new isoform, Rac1b. Oncogene 19(26):3013–3020

Shapiro IM, Cheng AW, Flytzanis NC, Balsamo M, Condeelis JS, Oktay MH, Burge CB, Gertler FB (2011) An EMT-driven alternative splicing program occurs in human breast cancer and modulates cellular phenotype. PLoS Genet 7(8):e1002218

Shen Y, Ji G, Haas BJ, Wu X, Zheng J, Reese GJ, Li QQ (2008) Genome level analysis of rice mRNA 3′-end processing signals and alternative polyadenylation. Nucleic Acids Res 36(9):3150–3161

Shi Y (2012) Alternative polyadenylation: new insights from global analyses. RNA 18(12): 2105–2117

Shook D, Keller R (2003) Mechanisms, mechanics and function of epithelial-mesenchymal transitions in early development. Mech Dev 120(11):1351–1383

Singh A, Karnoub AE, Palmby TR, Lengyel E, Sondek J, Der CJ (2004) Rac1b, a tumor associated, constitutively active Rac1 splice variant, promotes cellular transformation. Oncogene 23(58):9369–9380

Singh A, Settleman J (2010) EMT, cancer stem cells and drug resistance: an emerging axis of evil in the war on cancer. Oncogene 29(34):4741–4751

Song SJ, Poliseno L, Song MS, Ala U, Webster K, Ng C, Beringer G, Brikbak NJ, Yuan X, Cantley LC, Richardson AL, Pandolfi PP (2013) MicroRNA-antagonism regulates breast cancer stemness and metastasis via TET-family-dependent chromatin remodeling. Cell 154(2):311–324

Srikantan S, Gorospe M (2012) HuR function in disease. Front Biosci (Landmark Ed) 17:189–205

Stadtfeld M, Hochedlinger K (2010) Induced pluripotency: history, mechanisms, and applications. Genes Dev 24(20):2239–2263

Stickeler E, Fraser SD, Honig A, Chen AL, Berget SM, Cooper TA (2001) The RNA binding protein YB-1 binds A/C-rich exon enhancers and stimulates splicing of the CD44 alternative exon v4. EMBO J 20(14):3821–3830

Stinson S, Lackner MR, Adai AT, Yu N, Kim HJ, O'Brien C, Spoerke J, Jhunjhunwala S, Boyd Z, Januario T, Newman RJ, Yue P, Bourgon R, Modrusan Z, Stern HM, Warming S, De Sauvage FJ, Amler L, Yeh RF, Dornan D (2011) miR-221/222 targeting of trichorhinophalangeal 1 (TRPS1) promotes epithelial-to-mesenchymal transition in breast cancer. Sci Signal 4(186):pt5

Suenaga K, Lee KY, Nakamori M, Tatsumi Y, Takahashi MP, Fujimura H, Jinnai K, Yoshikawa H, Du H, Ares M Jr, Swanson MS, Kimura T (2012) Muscleblind-like 1 knockout mice reveal novel splicing defects in the myotonic dystrophy brain. PLoS One 7(3):e33218

Tam PP, Behringer RR (1997) Mouse gastrulation: the formation of a mammalian body plan. Mech Dev 68(1–2):3–25

Taube JH, Herschkowitz JI, Komurov K, Zhou AY, Gupta S, Yang J, Hartwell K, Onder TT, Gupta PB, Evans KW, Hollier BG, Ram PT, Lander ES, Rosen JM, Weinberg RA, Mani SA (2010) Core epithelial-to-mesenchymal transition interactome gene-expression signature is associated with claudin-low and metaplastic breast cancer subtypes. Proc Natl Acad Sci U S A 107(35):15449–15454

Tauler J, Zudaire E, Liu H, Shih J, Mulshine JL (2010) hnRNP A2/B1 modulates epithelial-mesenchymal transition in lung cancer cell lines. Cancer Res 70(18):7137–7147

Thiery JP, Acloque H, Huang RY, Nieto MA (2009) Epithelial-mesenchymal transitions in development and disease. Cell 139(5):871–890

Thomson S, Buck E, Petti F, Griffin G, Brown E, Ramnarine N, Iwata KK, Gibson N, Haley JD (2005) Epithelial to mesenchymal transition is a determinant of sensitivity of non-small-cell lung carcinoma cell lines and xenografts to epidermal growth factor receptor inhibition. Cancer Res 65(20):9455–9462

Thomson S, Petti F, Sujka-Kwok I, Mercado P, Bean J, Monaghan M, Seymour SL, Argast GM, Epstein DM, Haley JD (2011) A systems view of epithelial-mesenchymal transition signaling states. Clin Exp Metastasis 28(2):137–155

Tian B, Manley JL (2013) Alternative cleavage and polyadenylation: the long and short of it. Trends Biochem Sci 38(6):312–320

Tiwari N, Gheldof A, Tatari M, Christofori G (2012) EMT as the ultimate survival mechanism of cancer cells. Semin Cancer Biol 22(3):194–207

Tsai JH, Donaher JL, Murphy DA, Chau S, Yang J (2012) Spatiotemporal regulation of epithelial-mesenchymal transition is essential for squamous cell carcinoma metastasis. Cancer Cell 22(6):725–736

Ule J, Ule A, Spencer J, Williams A, Hu JS, Cline M, Wang H, Clark T, Fraser C, Ruggiu M, Zeeberg BR, Kane D, Weinstein JN, Blume J, Darnell RB (2005) Nova regulates brain-specific splicing to shape the synapse. Nat Genet 37(8):844–852

Valacca C, Bonomi S, Buratti E, Pedrotti S, Baralle FE, Sette C, Ghigna C, Biamonti G (2010) Sam68 regulates EMT through alternative splicing-activated nonsense-mediated mRNA decay of the SF2/ASF proto-oncogene. J Cell Biol 191(1):87–99

van Denderen BJ, Thompson EW (2013) Cancer: the to and fro of tumour spread. Nature 493(7433):487–488

Varelas X, Samavarchi-Tehrani P, Narimatsu M, Weiss A, Cockburn K, Larsen BG, Rossant J, Wrana JL (2010) The Crumbs complex couples cell density sensing to Hippo-dependent control of the TGF-beta-SMAD pathway. Dev Cell 19(6):831–844

Venables JP, Brosseau JP, Gadea G, Klinck R, Prinos P, Beaulieu JF, Lapointe E, Durand M, Thibault P, Tremblay K, Rousset F, Tazi J, Abou Elela S, Chabot B (2013) RBFOX2 is an important regulator of mesenchymal tissue-specific splicing in both normal and cancer tissues. Mol Cell Biol 33(2):396–405

Venables JP, Klinck R, Koh C, Gervais-Bird J, Bramard A, Inkel L, Durand M, Couture S, Froehlich U, Lapointe E, Lucier JF, Thibault P, Rancourt C, Tremblay K, Prinos P, Chabot B, Elela SA (2009) Cancer-associated regulation of alternative splicing. Nat Struct Mol Biol 16(6): 670–676

Vierbuchen T, Wernig M (2012) Molecular roadblocks for cellular reprogramming. Mol Cell 47(6):827–838

Vincent T, Neve EP, Johnson JR, Kukalev A, Rojo F, Albanell J, Pietras K, Virtanen I, Philipson L, Leopold PL, Crystal RG, de Herreros AG, Moustakas A, Pettersson RF, Fuxe J (2009) A SNAIL1-SMAD3/4 transcriptional repressor complex promotes TGF-beta mediated epithelial-mesenchymal transition. Nat Cell Biol 11(8):943–950

Vleminckx K, Vakaet L Jr, Mareel M, Fiers W, van Roy F (1991) Genetic manipulation of E-cadherin expression by epithelial tumor cells reveals an invasion suppressor role. Cell 66(1):107–119

Waerner T, Alacakaptan M, Tamir I, Oberauer R, Gal A, Brabletz T, Schreiber M, Jechlinger M, Beug H (2006) ILEI: a cytokine essential for EMT, tumor formation, and late events in metastasis in epithelial cells. Cancer Cell 10(3):227–239

Wang ET, Cody NA, Jog S, Biancolella M, Wang TT, Treacy DJ, Luo S, Schroth GP, Housman DE, Reddy S, Lecuyer E, Burge CB (2012) Transcriptome-wide regulation of pre-mRNA splicing and mRNA localization by muscleblind proteins. Cell 150(4):710–724

Wang Y, Shang Y (2013) Epigenetic control of epithelial-to-mesenchymal transition and cancer metastasis. Exp Cell Res 319(2):160–169

Warzecha CC, Carstens RP (2012) Complex changes in alternative pre-mRNA splicing play a central role in the epithelial-to-mesenchymal transition (EMT). Semin Cancer Biol 22(5–6): 417–427

Warzecha CC, Jiang P, Amirikian K, Dittmar KA, Lu H, Shen S, Guo W, Xing Y, Carstens RP (2010) An ESRP-regulated splicing programme is abrogated during the epithelial-mesenchymal transition. EMBO J 29(19):3286–3300

Warzecha CC, Sato TK, Nabet B, Hogenesch JB, Carstens RP (2009a) ESRP1 and ESRP2 are epithelial cell-type-specific regulators of FGFR2 splicing. Mol Cell 33(5):591–601

Warzecha CC, Shen S, Xing Y, Carstens RP (2009b) The epithelial splicing factors ESRP1 and ESRP2 positively and negatively regulate diverse types of alternative splicing events. RNA Biol 6(5):546–562

Wei WJ, Mu SR, Heiner M, Fu X, Cao LJ, Gong XF, Bindereif A, Hui J (2012) YB-1 binds to CAUC motifs and stimulates exon inclusion by enhancing the recruitment of U2AF to weak polypyrimidine tracts. Nucleic Acids Res 40(17):8622–8636

Weinberg RA (2008) Mechanisms of malignant progression. Carcinogenesis 29(6):1092–1095

Wellner U, Schubert J, Burk UC, Schmalhofer O, Zhu F, Sonntag A, Waldvogel B, Vannier C, Darling D, Zur Hausen A, Brunton VG, Morton J, Sansom O, Schuler J, Stemmler MP, Herzberger C, Hopt U, Keck T, Brabletz S, Brabletz T (2009) The EMT-activator ZEB1 promotes tumorigenicity by repressing stemness-inhibiting microRNAs. Nat Cell Biol 11(12): 1487–1495

Werner S, Frey S, Riethdorf S, Schulze C, Alawi M, Kling L, Vafaizadeh V, Sauter G, Terracciano L, Schumacher U, Pantel K, Assmann V (2013) Dual roles of the transcription factor grainyhead-like 2 (GRHL2) in breast cancer. J Biol Chem 288(32):22993–23008

Wilkinson MF, Shyu AB (2001) Multifunctional regulatory proteins that control gene expression in both the nucleus and the cytoplasm. Bioessays 23(9):775–787

Woodward WA, Chen MS, Behbod F, Alfaro MP, Buchholz TA, Rosen JM (2007) WNT/beta-catenin mediates radiation resistance of mouse mammary progenitor cells. Proc Natl Acad Sci U S A 104(2):618–623

Xue Y, Zhou Y, Wu T, Zhu T, Ji X, Kwon YS, Zhang C, Yeo G, Black DL, Sun H, Fu XD, Zhang Y (2009) Genome-wide analysis of PTB-RNA interactions reveals a strategy used by the general splicing repressor to modulate exon inclusion or skipping. Mol Cell 36(6):996–1006

Yan W, Cao QJ, Arenas RB, Bentley B, Shao R (2010) GATA3 inhibits breast cancer metastasis through the reversal of epithelial-mesenchymal transition. J Biol Chem 285(18):14042–14051

Yanagisawa M, Huveldt D, Kreinest P, Lohse CM, Cheville JC, Parker AS, Copland JA, Anastasiadis PZ (2008) A p120 catenin isoform switch affects Rho activity, induces tumor cell invasion, and predicts metastatic disease. J Biol Chem 283(26):18344–18354

Yang J, Mani SA, Donaher JL, Ramaswamy S, Itzykson RA, Come C, Savagner P, Gitelman I, Richardson A, Weinberg RA (2004) Twist, a master regulator of morphogenesis, plays an essential role in tumor metastasis. Cell 117(7):927–939

Yang J, Weinberg RA (2008) Epithelial-mesenchymal transition: at the crossroads of development and tumor metastasis. Dev Cell 14(6):818–829

Yeo GW, Coufal NG, Liang TY, Peng GE, Fu XD, Gage FH (2009) An RNA code for the FOX2 splicing regulator revealed by mapping RNA-protein interactions in stem cells. Nat Struct Mol Biol 16(2):130–137

Zeisberg M, Duffield JS (2010) Resolved: EMT produces fibroblasts in the kidney. J Am Soc Nephrol 21(8):1247–1253

Zeisberg M, Neilson EG (2009) Biomarkers for epithelial-mesenchymal transitions. J Clin Invest 119(6):1429–1437

Zhang C, Zhang Z, Castle J, Sun S, Johnson J, Krainer AR, Zhang MQ (2008) Defining the regulatory network of the tissue-specific splicing factors Fox-1 and Fox-2. Genes Dev 22(18): 2550–2563

Zhang J, Ma L (2012) MicroRNA control of epithelial-mesenchymal transition and metastasis. Cancer Metastasis Rev 31(3–4):653–662

Zhang X, Ibrahimi OA, Olsen SK, Umemori H, Mohammadi M, Ornitz DM (2006) Receptor specificity of the fibroblast growth factor family. The complete mammalian FGF family. J Biol Chem 281(23):15694–15700

Zhou YQ, He C, Chen YQ, Wang D, Wang MH (2003) Altered expression of the RON receptor tyrosine kinase in primary human colorectal adenocarcinomas: generation of different splicing RON variants and their oncogenic potential. Oncogene 22(2):186–197

Zouq NK, Keeble JA, Lindsay J, Valentijn AJ, Zhang L, Mills D, Turner CE, Streuli CH, Gilmore AP (2009) FAK engages multiple pathways to maintain survival of fibroblasts and epithelia: differential roles for paxillin and p130Cas. J Cell Sci 122(Pt 3):357–367

Chapter 9
Antisense Oligonucleotide-Based Therapies for Diseases Caused by pre-mRNA Processing Defects

Frank Rigo, Punit P. Seth, and C. Frank Bennett

Abstract Before a messenger RNA (mRNA) is translated into a protein in the cytoplasm, its pre-mRNA precursor is extensively processed through capping, splicing and polyadenylation in the nucleus. Defects in the processing of pre-mRNAs due to mutations in RNA sequences often cause disease. Traditional small molecules or protein-based therapeutics are not well suited for correcting processing defects by targeting RNA. However, antisense oligonucleotides (ASOs) designed to bind RNA by Watson-Crick base pairing can target most RNA transcripts and have emerged as the ideal therapeutic agents for diseases that are caused by pre-mRNA processing defects. Here we review the diverse ASO-based mechanisms that can be exploited to modulate the expression of RNA. We also discuss how advancements in medicinal chemistry and a deeper understanding of the pharmacokinetic and toxicological properties of ASOs have enabled their use as therapeutic agents. We end by describing how ASOs have been used successfully to treat various pre-mRNA processing diseases in animal models.

Keywords Antisense oligonucleotide • Pre-mRNA processing • Splicing • RNA binding protein • RNase H • siRNA • Myotonic dystrophy • Spinal muscular atrophy • Hutchinson-Gilford progeria syndrome • Usher syndrome • Fukuyama congenital muscular dystrophy

1 Introduction

The production of messenger RNAs (mRNAs) in eukaryotes is a complex, multistep process that begins with the initiation of transcription in the nucleus and culminates in the export of the mature message to the cytoplasm. During this journey, precursor

F. Rigo (✉) • P.P. Seth • C.F. Bennett (✉)
Isis Pharmaceuticals, 2855 Gazelle Court, Carlsbad, CA, USA
e-mail: FRigo@isisph.com; PSeth@isisph.com; FBennett@isisph.com

G.W. Yeo (ed.), *Systems Biology of RNA Binding Proteins*, Advances in Experimental Medicine and Biology 825, DOI 10.1007/978-1-4939-1221-6_9

mRNAs (pre-mRNAs) are processed through capping, splicing, polyadenylation, and sometimes editing. These modifications are necessary for generating an mRNA that can be translated into protein. The molecular machineries responsible for these modifications are intimately interconnected both with each other and with the transcriptional apparatus to maximize the efficiency of each step and for the purpose of regulation (Braunschweig et al. 2013; Kornblihtt et al. 2004; Maniatis and Reed 2002; Perales and Bentley 2009; Proudfoot 2011). The importance of pre-mRNA processing to cellular function is underscored by the fact that a significant fraction of the genome, hundreds of proteins, is devoted to this process (Jurica and Moore 2003; Shatkin and Manley 2000).

Arguably, the most important step in pre-mRNA processing is splicing. During this process, sequences that code for amino acids (exons), are ligated together and the sequences that connected them (introns) are discarded. The final product, the mature mRNA, is then translated into a protein (Black 2003). Remarkably, it is often the case that adjacent exons are not ligated together. Instead, an exon can be spliced to another exon further downstream. This alternative splicing allows shuffling of protein-coding domains and occurs in >90 % of multi-exon genes to dramatically increase the proteomic output of the genome (Nilsen and Graveley 2010). Furthermore, alternative splicing is highly regulated in order to achieve cell- and tissue-specific expression of protein isoforms during and after development (Calarco et al. 2011; Kalsotra and Cooper 2011).

The importance of pre-mRNA processing for proper biological function is even more apparent from the many mutations that cause diseases as a result of processing defects (Anthony and Gallo 2010; Calandra et al. 2011; Castello et al. 2013; Cooper et al. 2009; Danckwardt et al. 2008; Havens et al. 2013; Mordes et al. 2006; Tazi et al. 2009; Wang and Cooper 2007). The most frequent mutations can either create or disrupt sequence motifs (cis-acting) that control splicing, or impair the function of splicing factors (trans-acting). The lower estimate for the fraction of disease-causing, proximal cis-acting mutations is often cited as 15 % (Krawczak et al. 1992). However, a more recent study concluded that cis-acting mutations account for up to 33 % of all disease-causing mutations (Lim et al. 2011). Trans-acting mutations in splicing factors have already been shown to cause disease and this class of mutations is expected to grow (Castello et al. 2012; Kim et al. 2013; Mordes et al. 2006; Padgett 2012; Polymenidou et al. 2012). In addition to causing disease directly, mutations that affect splicing can also act as disease modifiers or affect disease susceptibility (Wang and Cooper 2007). Considering its important role in pre-mRNA processing, it is not surprising that mutations that affect polyadenylation also result in disease (Danckwardt et al. 2008).

Historically, biological science has centered almost entirely, and for good reason, on elucidating the function of proteins. Therefore, it is not surprising that since its inception the modern pharmaceutical industry has focused almost exclusively on proteins as drug targets. The technology platforms that have delivered a large number of drug products are small molecules and, more recently, the protein-based therapeutics (Overington et al. 2006). However, since RNA was recognized in eukaryotic cells by Jean Brachet over 70 years ago (Brachet 1941), a wealth of data supports its central role in regulating cellular function, beyond simply serving as a template for producing proteins. Furthermore, RNA has emerged as an important therapeutic target (Bennett and Swayze 2010; Cooper et al. 2009).

With the exception of naturally occurring antibiotics and analogs thereof, which target highly structured regions in bacterial ribosomal RNA (present at low micromolar concentration in the cytoplasm), identification of other classes of safe and efficacious RNA-targeting small-molecule drugs has remained elusive (Guan and Disney 2012). The majority of RNA in the cell is not an ideal target for small-molecule drugs as it does not possess well-defined, long-lived structured pockets for binding small-molecule ligands. Moreover, RNA itself is transiently expressed in the cell at low copy numbers. However, almost any RNA in the cell can be readily targeted using antisense oligonucleotides (ASOs).

For the purpose of this review we define an antisense oligonucleotide (ASO) as a synthetic molecule composed of a stretch of nucleotides or nucleotide analogs (2–50) that can bind complementary nucleic acids with high specificity through Watson-Crick base pairing. Upon binding, an ASO can modulate the intermediary metabolism of RNA by harnessing a multitude of antisense mechanisms. Below we describe the array of ASO-based mechanisms that have been developed, and we elaborate on several examples where ASOs have been used successfully in animal models of pre-mRNA processing diseases.

2 Factors That May Influence Antisense Oligonucleotide Activity

For an ASO to be active, it must bind to its target RNA with sufficient affinity (Monia et al. 1992). This affinity stems from the hydrogen bonds formed between the bases of the ASO and the bases of the RNA, in addition to hydrophobic contacts between adjacent base pairs. Guanine and cytosine can form three hydrogen bonds versus adenine and thymine or uracil, which form two. Therefore, the affinity of an ASO for its target is proportional to the length of the ASO and its base composition (Freier et al. 1992). When the ASO is too long this relationship can break down, in part, because of the higher propensity to form intra- and intermolecular base pairs rather than base pairs with the target (Matveeva et al. 2003). To maximize its affinity, the ASO also needs to be fully complementary to its target since mismatches can have a large destabilizing effect (Freier et al. 1992). Importantly, the affinity of the ASO for its target can be increased by the incorporation of certain chemical modifications (see Sect. 4). Even though affinity is required for activity, it is certainly not sufficient since there are many additional factors that can affect activity.

Affinity is not only determined by the ASO sequence itself, but can also be affected by the structure of the target RNA (Chastain and Tinoco 1993; Ecker 1993). RNA can fold into a variety of secondary structures by intramolecular base pairing of the primary sequence or it can form tertiary structures by interactions between secondary structures or between secondary structures and single-stranded regions (Batey et al. 1999). These structures can decrease the binding affinity, binding rate (Lima et al. 1992), and the activity of an ASO (Vickers et al. 2000).

In contrast to the well-documented role of posttranslational chemical modifications of amino acids for modulating protein function, the extent to which nucleotides are chemically modified remains largely unknown. Nevertheless, studies have already started to document the prevalence of modifications such as adenosine-to-inosine conversion (RNA editing), and adenosine or cytosine methylation for the entire transcriptome (Meyer et al. 2012; Peng et al. 2012; Squires et al. 2012). It will be interesting to learn how these modifications affect the activity of ASOs.

In addition to the propensity to form structured RNA, RNA is also bound by a large number of RNA-binding proteins (RBPs) (Baltz et al. 2012; Castello et al. 2012). The exact constellation of RBPs that bind to RNAs at any given time is not known. However, the transcriptome-wide binding landscape of many RBPs, in addition to their sequence preferences, has already been determined (Kapeli and Yeo 2012; Licatalosi and Darnell 2010; Ray et al. 2013). There are a few examples where ASOs have been shown to prevent proteins from binding to transcripts (Rigo et al. 2012a; Wheeler et al. 2009). However, virtually nothing is known about the competition between RBPs and ASOs for binding to their targets and to what extent RBPs inhibit ASO activity. It is also important to note that RBPs may even enhance ASO activity (Rigo et al. 2012a).

One could expect that the expression level of a target RNA is an important factor for ASO activity. However, at least for ASOs that use the RNase H mechanism this has, so far, not proven to be the case (Miraglia et al. 2000). Very scarce RNAs like enhancer RNAs (eRNAs) (Lam et al. 2013) or very abundant ones like the U1 small nuclear RNA (*U1 snRNA*) (Liang et al. 2011) or metastasis-associated lung adenocarcinoma transcript 1 (*MALAT1*) (Gutschner et al. 2013) can be targeted with ASOs effectively in cultured cells and animals. However, the localization of RNA can determine how well it is targeted by an ASO. For example, if an ASO is designed to utilize a mechanism that operates exclusively in one cellular location (e.g., splice-switching ASOs acting on pre-mRNA in the nucleus), target RNAs found at a different cellular location will be spared (e.g., mRNAs in the cytoplasm).

The number of mechanisms that can be employed by ASOs continues to grow and usually more than one option is available for achieving a desired outcome (Bennett and Swayze 2010; Crooke et al. 2007; Kole et al. 2012). However, not every mechanism will provide the same level of activity and it is the responsibility of the experimenter to assess the strengths and weaknesses of each.

Regardless of the mechanism that is selected, the experimenter also has to select the sites on a transcript where ASOs should be targeted. Computational tools can be used to increase the probability of identifying active ASOs versus random selection (Aartsma-Rus 2012; Matveeva et al. 2007; Sipes and Freier 2008). However, these tools do not adequately predict the activity of ASOs. Therefore, screening ASOs for activity in cell culture is highly encouraged. This is also very important when one desires to evolve the design of an ASO so as to maximize its activity for a particular mechanism.

3 Antisense Oligonucleotide-Based Mechanisms

ASOs, as defined above, encompass a wide variety of designs and mechanisms of action. However, this definition does not include oligonucleotides that form triple-helix structures with DNA, double-stranded DNAs that bind transcription factors, oligonucleotides that bind proteins, or polynucleotides that undergo trans-splicing. ASO-based mechanisms can be broadly classified as (1) occupancy-mediated RNA cleavage, and (2) occupancy-only mediated. In many instances ASOs that operate through these mechanisms are found naturally, in living cells, but have been optimized for therapeutic purposes. In other instances, the ASOs are purely the result of rational design, yet often founded on biological concepts. Since more than one ASO-based mechanism can theoretically be used to correct a given pre-mRNA processing defect, we elaborate on a wide gamut of options.

3.1 Occupancy-Mediated RNA Cleavage Mechanisms

3.1.1 RNase H ssASOs

The potential of ASOs as therapeutic agents was first appreciated in 1978 with the demonstration of antisense activity in human cells using a DNA single-stranded ASO (ssASO) (Zamecnik and Stephenson 1978). During the early development of antisense therapeutics, a proposed ASO mechanism of action was the RNase H-mediated degradation of the target RNA (Helene and Toulme 1990). However, it was not until 2004 that Crooke and colleagues unequivocally demonstrated that DNA-like ssASOs reduce the expression of RNAs in human cells by harnessing RNase H1 (Wu et al. 2004). Cellular functions of the RNase H enzymes include removing Okazaki fragments during DNA replication and resolving R-loops that form during transcription (Cerritelli and Crouch 2009). Human RNase H1 binds to the duplex formed between a DNA containing ssASO and a target RNA through its RNA-binding domain located at its N-terminus. In order to cleave the RNA in the duplex, the RNase H1 catalytic domain needs at least 5 consecutive DNA/RNA base pairs, and cleavage usually occurs within 7–10 nucleotides from the 5′-end of the RNA. After cleavage, the exposed phosphate on the 5′-end and the hydroxyl on the 3′-end are recognized, and the RNA is subsequently degraded by cellular nucleases by virtue of the fact that RNA is significantly depleted (Lima et al. 2007). However, more work will reveal the identity of the exact nucleases that are involved in this process. At some point after RNase H1 cleaves the RNA, the ssASO is released and is, presumably, available to re-engage another transcript.

RNase H1 resides primarily in the nucleus and mitochondria (Cerritelli and Crouch 2009; Lima et al. 2007). In the nucleus it can destroy pre-mRNAs that are maturing (Vickers et al. 2003) and noncoding RNAs (ncRNAs) (Liang et al. 2011; Prasanth et al. 2005; Tripathi et al. 2010). However, since RNase H1 is also present in the cytoplasm (Lima et al. 2007; Wu et al. 2004), mRNAs that escape the nucleus intact may be degraded in the cytoplasm. The RNase H mechanism is arguably the most robust ASO-based mechanism available and has amassed a wealth of positive data in cells, animals, and humans (Bennett 2007; Bennett and Swayze 2010; Lee et al. 2013; Lightfoot and Hall 2012).

3.1.2 miRNA, siRNA, and ss-siRNA

microRNAs (miRNAs) are naturally occurring ASOs, transcribed as long primary transcripts that are subsequently processed in the nucleus by Drosha to yield 60–70-nucleotide-long stem-loop-structured precursor miRNAs (pre-miRNAs). These are exported to the cytoplasm where they encounter Dicer, which further processes them into 21–23-nucleotide-long double-stranded RNAs (dsRNAs). Both strands are separated and one strand, the mature miRNA, is bound to an Argonaute protein (AGO). This miRNA/AGO complex assembles with other cofactors to form the RNA-induced silencing complex (RISC), which is guided to a target mRNA by hybridizing with the miRNA (Bartel 2004; Carthew and Sontheimer 2009). Most human miRNAs do not base pair with full complementarity to their target sequences in the 3′-UTRs of transcripts and therefore do not promote their cleavage (Bartel 2004, 2009; Carthew and Sontheimer 2009) (see Sect. 3.2.3). A few human miRNAs do have extensive pairing complementarity and promote cleavage of their targets like siRNAs.

Synthetic 21–23-nucleotide-long dsRNAs called siRNAs (siRNAs) that do not undergo endogenous Dicer processing can be administered to cells to reduce the expression of mRNAs, a mechanism coined RNA interference (RNAi). The perfectly complementary guide strand of the siRNA loads into an AGO and after the assembly of RISC, cleavage of the target mRNA ensues (Ghildiyal and Zamore 2009; Meister and Tuschl 2004). The use of synthetic siRNAs was first reported in human cells by the Tuschl lab in 2001 (Elbashir et al. 2001). siRNAs are another example of an ASO-based mechanism that triggers RNA degradation through an endogenous enzyme. The enzyme responsible for cleaving the mRNA in human cells is Argonaute 2 (AGO2), which uses an RNase H-like enzymatic mechanism to cleave between nucleotides 10 and 11 from the 5′-end of the guide strand (Patel et al. 2006). The cleaved mRNA is degraded by cytoplasmic exonucleases (Orban and Izaurralde 2005) and RISC, with its loaded guide strand, is presumably free to engage another target (Hutvagner and Zamore 2002).

siRNAs have proven to be very effective tools for research in cultured cells and animals (Bumcrot et al. 2006). Early clinical trials using nanoparticulate formulations of siRNAs are showing promising results (Castanotto and Rossi 2009; Vaishnaw et al. 2010). For various reasons, the double-stranded nature of siRNAs is an added complexity that makes their development as therapeutic agent more challenging (Crooke et al. 2007). To overcome this limitation, single-stranded siRNAs (ss-siRNAs) have been designed that work through the RNAi pathway in cells and animals (Lima et al. 2012; Yu et al. 2012).

Although it is well documented that the RNAi pathway is operant in the cytoplasm, it is also possible to use siRNAs to inhibit gene expression in the nucleus through a mechanism called transcriptional gene silencing (TGS). siRNAs

(Janowski et al. 2005; Morris et al. 2004) and miRNAs (Kim et al. 2008; Younger and Corey 2011), referred to as antigene RNAs (agRNAs), targeted to promoter sequences have been found to inhibit gene expression. It appears that the inhibition requires promoter-associated RNA (paRNA) (Han et al. 2007; Schwartz et al. 2008) and recruitment of either AGO1 or AGO2 (Chu et al. 2010; Kim et al. 2006; Napoli et al. 2009). It is still not clear if TGS requires enzymatic cleavage of the target RNA. In fact, the recruitment of AGO is thought to trigger the accumulation of repressive chromatin marks, but more work is required to have a better understanding of this silencing mechanism (Gagnon and Corey 2012; Turner and Morris 2010).

3.1.3 Ribozyme, DNAzyme, and Chemical Warhead ASOs

Transcripts can also be degraded by ASOs that do not depend on endogenous enzymes for their activity. Ribozymes, first discovered by Cech and Altman, are RNA molecules that form an enzymatically active site and catalyze reactions independent of proteins (Cech 1988). Many ribozymes are self-cleaving but others can act on RNA substrates in trans. These trans-ribozymes are ~30 nucleotides in length and contain a central catalytic core of ~15 nucleotides that is flanked by sequences that are complementary to a target RNA (Usman and Blatt 2000; Vitiello et al. 2000). DNAzymes are DNA molecules engineered to have a similar architecture as ribozymes and, like them, can support catalytic cleavage of RNA (Breaker and Joyce 1994). Chemical warhead ASOs are ASOs with a metallonuclease appended to cleave the targeted RNA (Baker et al. 1999; Morrow and Iranzo 2004; Riguet et al. 2004).

3.2 *Occupancy-Only Mediated Mechanisms*

3.2.1 Splicing Modulation

Splice-switching oligonucleotides (SSOs) are ssASOs that are typically 15–25 nucleotides long and because of their chemical design they do not recruit RNA-cleaving enzymes such as RNase H and AGO2. Therefore, SSOs act purely as steric blockers. The first demonstration of the use of an SSO to correct a splicing mutation was by the Kole lab in 1993 using an in vitro splicing system (Dominski and Kole 1993). Subsequently, SSOs were shown to work in cells (Dunckley et al. 1998; Sierakowska et al. 1996), animals (Mann et al. 2001; Sazani et al. 2002), and humans (Kinali et al. 2009; van Deutekom et al. 2007). The precise mechanism by which SSOs modulate splicing is dependent on the constellation of core splicing factors, enhancers, and silencers that control the splicing of a particular exon. SSOs have been demonstrated to modulate splicing in the nucleus, on chromatin (Rigo et al. 2012a), by blocking splice sites and sequences that bind enhancer and silencers located both in exons and introns (Bauman et al. 2009; Havens et al. 2013; Spitali and Aartsma-Rus 2012). Interestingly, an SSO was used to modulate splicing, not by acting as a steric blocker, but by favoring the formation of a secondary structure important for splicing (Peacey et al. 2012). In addition to using SSOs to correct splicing, these ASOs can also be used to inactivate proteins by skipping specific exons (Siwkowski et al. 2004; Wancewicz et al. 2010). Surprisingly, siRNAs can also modulate splicing, either through an AGO2-dependent, cleavage-independent mechanism, presumably by acting as steric blockers (Liu et al. 2011), or by mediating changes in histone modifications that are thought to reduce the polymerase elongation rate (Allo et al. 2009).

SSOs, called bifunctional SSOs, have also been used to correct splicing (Sazani et al. 2007). Bifunctional SSOs are typically 40–50 nucleotides long and are composed of an antisense sequence that hybridizes to the pre-mRNA and an appended sequence motif that can recruit either a splicing enhancer protein (coined TOES for targeted oligonucleotide enhancer of splicing) (Skordis et al. 2003) or a splicing silencer protein (coined TOSS for targeted oligonucleotide silencer of splicing) (Villemaire et al. 2003), depending on the sequence that is used. Bifunctional SSOs have also been designed with an appended peptide motif derived from a splicing enhancer protein (coined ESSENCE for exon-specific splicing enhancement by small chimeric effectors) (Cartegni and Krainer 2003; Wilusz et al. 2005). Surprisingly, SSOs can also be engineered to recruit proteins to the pre-mRNA without using appendages. This is accomplished by simply incorporating specific chemical modification into the SSO (Rigo et al. 2012a).

One outcome of modulating pre-mRNA splicing is the production of transcripts that undergo nonsense-mediated decay (NMD). NMD is an endogenous RNA surveillance mechanism that recognizes transcripts that contain premature translation termination codons (PTCs) and targets them for degradation. This mechanism is in place to ensure that transcripts with PTCs do not express truncated proteins. The inclusion or the exclusion of an exon with a length that is not a multiple of 3 (a codon) will generate an alternate reading frame and will introduce a PTC in an exon further downstream. Transcripts with PTCs are eliminated by the NMD machinery in the cytoplasm (Schoenberg and Maquat 2012; Schweingruber et al. 2013). SSOs can be used to modulate the splicing of a particular exon to produce a transcript that contains a PTC and is a substrate for NMD, a mechanism named forced splicing-dependent NMD (FSD-NMD) (Lefave et al. 2011; Zammarchi et al. 2011). Even though SSOs have been used to introduce PTCs and reduce the steady-state levels of mRNAs, it has not formally been proven that their degradation is actually through the NMD pathway.

3.2.2 Polyadenylation Modulation and U1 Interference

The level of transcripts can also be reduced by modulating polyadenylation. A steric blocker ssASO targeted to a polyadenylation signal in the 3′-UTR of a pre-mRNA can block its polyadenylation and reduce its expression. Interestingly, blocking polyadenylation of a pre-mRNA at one site can activate an alternative polyadenylation signal that is upstream to increase the expression of RNA and protein.

Utilization of the upstream polyadenylation signal results in the removal of the downstream 3′-UTR sequence, which is known to harbor RNA-destabilizing elements (Vickers et al. 2001). In addition, siRNAs targeted to certain polyadenylation sites can also inhibit polyadenylation or promote deadenylation independent of AGO proteins (Vickers and Crooke 2012).

Bifunctional ASOs called U1 adaptors can also inhibit polyadenylation through a mechanism dubbed U1 interference. U1 adaptors are 25-nucleotide-long ssASOs that contain a 15 nucleotide antisense sequence on the 5′-end that is complementary to the 3′-UTR of a pre-mRNA, and a 10-nucleotide antisense sequence on the 3′-end that is complementary to the 5′-end of the U1 snRNA (Goraczniak et al. 2009). This mechanism is based on the ability of the U1 snRNP to block polyadenylation of some pre-mRNAs by binding to a 5′-splice-site-like sequence in the 3′-UTR and inhibiting the activity of poly(A) polymerase (PAP) to prevent polyadenylation and promote pre-mRNA degradation (Gunderson et al. 1998).

3.2.3 miRNA Mimics

Most endogenous miRNAs bind to their target mRNAs with partial complementarity, through a seed region that extends 6–8 nucleotides from the 5′-end of the guide strand. These miRNAs repress gene expression, not by promoting mRNA cleavage directly, but by recruiting factors such as decapping and deadenylation enzymes that trigger RNA degradation or by inhibiting translation (Huntzinger and Izaurralde 2011). miRNAs have been shown to be an important class of regulatory RNAs that play important roles in biological processes (Bartel 2009). Synthetic siRNAs can be designed to operate through a miRNA mechanism by introducing mismatches in the seed region of the guide strand (Doench et al. 2003). Synthetic siRNAs and ss-siRNAs, designed to mimic miRNAs, targeted to repetitive sequences in transcripts (e.g., CAG repeats in huntingtin transcripts) have been shown to repress their translation (Fiszer et al. 2011; Hu et al. 2010; Yu et al. 2012).

3.2.4 Anti-miRNA ASOs

miRNA-mediated repression can be relieved by using a steric blocker ssASO that hybridizes to a miRNA and prevents it from interacting with its target mRNAs. This has been demonstrated in cell culture (Esau et al. 2004; Meister et al. 2004), in animals (Elmen et al. 2008; Esau et al. 2006; Krutzfeldt et al. 2005), and recently in humans (Janssen et al. 2013). These anti-miRNA ASOs (AMOs) probably act by blocking the function of miRNAs without affecting their steady-state levels (Davis et al. 2009). The function of a miRNA can also be inhibited by using a steric blocker ssASO (miR-mask) to occlude the miRNA-binding site in the 3′-UTR of an mRNA (Choi et al. 2007; Xiao et al. 2007).

3.2.5 Translation Inhibition

Several non-miRNA mechanisms can also be used to inhibit translation. These can be divided into: (1) blocking the scanning of the 40S subunit on the mRNA; (2) blocking assembly of the 40S and 60S ribosomal subunits; and (3) blocking ribosome translocation down the mRNA after translation initiation. The 5′-terminal nucleotide of a steric blocker ssASO, which binds adjacent to the mRNA 5′-cap, can hybridize with the inverted guanosine of the 5′-cap to inhibit the binding of the eukaryotic translation initiation factor 4H (EIF4H) (Baker et al. 1992). Steric blocker ssASOs that bind adjacent to the 5′-cap can also interfere with translation initiation by inhibiting the binding of the 40S ribosome subunit (Baker et al. 1997). These ssASOs can also inhibit translation by binding to the AUG start codon or to sites in the open reading frame (Arora et al. 2000; Brown-Driver et al. 1999; Hanecak et al. 1996; Hu et al. 2009; Iversen 2007).

3.2.6 Gene Expression Upregulation Through Targeting NATs and paRNAs

The vast majority of the noncoding portion of the genome can be transcribed into short and long ncRNAs (Djebali et al. 2012) that do not generate protein products (Guttman et al. 2013). A large number of ncRNAs are detected at steady state probably because they have been captured before being targeted for degradation and may have no function (Belostotsky 2009). However, it is becoming increasingly clear that many ncRNAs (other than tRNA, snRNA, snoRNA, miRNA, rRNA, etc.) participate in a meaningful way to regulate gene expression and may contribute to disease pathogenesis (Mattick 2009; Qureshi et al. 2010; Rinn and Chang 2012; Ulitsky and Bartel 2013). Many ncRNAs called natural antisense transcripts (NATs) overlap with sense transcripts produced from protein-coding genes (He et al. 2008; Katayama et al. 2005; Yelin et al. 2003). Furthermore, ncRNAs can be targeted by ASOs to manipulate the expression of genes (Wahlestedt 2013). NATs can be targeted with siRNAs, or even with steric blocker ssASOs called antagoNATs, to increase the expression of the overlapping sense transcripts (Faghihi et al. 2010; Modarresi et al. 2012). It is not clear to what extent, if any, cleavage of NATs by AGO contributes to the upregulation. Several mechanisms could account for this observation (Faghihi and Wahlestedt 2009), but ASO-induced removal of repressive histone modifications is the preferred model (Modarresi et al. 2012).

Regions around promoters also generate an abundance of sense and antisense transcripts called paRNAs (Preker et al. 2009; Seila et al. 2009). agRNAs such as siRNAs (Janowski et al. 2007; Li et al. 2006) and miRNAs (Place et al. 2008) targeted to promoter sequences are also capable of activating gene expression. The mechanism of upregulation still needs to be investigated further, but it was shown to be dependent on the presence of paRNAs (Morris et al. 2008; Schwartz et al. 2008) and on the recruitment of AGO2 (Chu et al. 2010; Li et al. 2006). Perhaps the recruitment of AGO2 to promoter regions creates an epigenetic landscape that favors gene activation.

3.2.7 Blocking PRC2 Binding

One function of ncRNAs is to serve as a tether for recruiting chromatin-modifying enzymes to nearby genes to modulate their expression (Guttman and Rinn 2012). The recruitment of polycomb repressive complex 2 (PRC2) through ncRNA to inactivate the X-chromosome has been well studied (Lee 2012). In addition, PRC2 has been found to be associated with many loci (Zhao et al. 2010). Steric blocker ssASOs can be used to inhibit the interaction of the PRC2 complex with RNA (Sarma et al. 2010) for the purpose of upregulating the expression of a gene (McSwiggen 2013).

3.2.8 Disruption of Secondary Structure

RNA forms a variety of secondary and tertiary structures that play important roles in gene expression (McManus and Graveley 2011; Wan et al. 2011). Steric blocker ssASOs can be used to invade RNA structures to disrupt them and interfere with their natural function. As an example, ssASOs that bind the HIV trans-activation response element (TAR) disrupt its stem-loop structure, preventing the recruitment of trans-activator of transcription (Tat) and the expression of HIV transcripts (Ivanova et al. 2007; Vickers et al. 1991).

3.2.9 RNA Localization Modulation

mRNAs can be mislocalized in disease as exemplified by the nuclear retention of dystrophia myotonica protein kinase (*DMPK*) mRNAs that contain an expanded number of CUG repeats (see Sect. 7.1). *DMPK* mRNAs with CUG repeats are retained in nuclear foci by the binding of a splicing factor called muscleblind-like 1 (MBNL1). *DMPK* mRNAs can be liberated from nuclear foci into the cytoplasm by using a steric blocker ssASO that binds to the CUG repeats so as to compete with MBNL1 (Wheeler et al. 2009). Some mRNAs also contain localization sequences in their 3′-untranslated regions (3′-UTRs) called zip codes that direct them to specific sites within a cell (Martin and Ephrussi 2009). The localization of mRNAs can be modulated by using steric blocker ssASOs that inhibit the function of zip codes (Kislauskis et al. 1994).

3.2.10 Modulating RNA Editing

The use of ASOs to modulate chemical modifications of RNA nucleotides is still in its infancy since more needs to be learned about where, when, and how these modifications are placed. Nevertheless, steric blocker ssASOs can be used to inhibit the conversion of adenosine to inosine by adenosine deaminases acting on RNA (ADARs) (Mizrahi et al. 2013). It may be even possible to promote editing of specific nucleotides with ASOs (Woolf et al. 1995).

For both occupancy-mediated RNA cleavage and occupancy-only-mediated ASO-based mechanisms it is important to realize that any given mechanism can be used to modulate a target RNA in a way that was not anticipated a priori. As described above, an ASO can modulate a target RNA through a variety of mechanisms. For example, an siRNA can cleavage a target RNA and trigger its degradation, but it can also modulate splicing or even upregulate gene expression. Therefore, it is the responsibility of the experimenter to prove that an ASO is actually operating through the mechanism that is desired.

4 Chemistry of Antisense Oligonucleotides

Natural nucleic acids, DNA and RNA, are extremely labile in blood and undergo rapid nuclease-mediated degradation upon administration to animals. They also have poor intrinsic pharmacokinetic properties due to their weak binding to plasma proteins, which allows them to be rapidly filtered by the glomerulus and excreted in the urine after systemic administration. Moreover, given that much of the RNA in a cell forms secondary and tertiary structures, and that RNA/RNA duplexes are more stable than RNA/DNA duplexes, shorter DNA-based ASOs do not possess sufficient affinity to invade these structures and form productive drug-target interactions. As a result, ASOs are chemically modified to improve their "drug-like" properties. This topic has been extensively reviewed previously (Prakash 2011; Rettig and Behlke 2012; Swayze and Bhat 2007).

From the perspective of a medicinal chemist, the use of chemical modifications to improve the drug-like properties of ASOs has to accomplish at least four distinct objectives, without compromising the specificity imposed by Watson-Crick base pairing: (1) improve stability to protect against nuclease-mediated digestion; (2) impart favorable pharmacokinetic properties; (3) improve affinity for complementary RNA; and (4) maintain or confer the ability to modulate the target to produce a functional response.

4.1 Designing Chemically Modified Antisense Oligonucleotides

The precise positioning of chemical modifications within an ASO fundamentally determines its mechanism of action. For example, ssASOs that contain >5 continuous DNA nucleotides can activate the RNase H pathway while siRNAs or chemically modified ss-siRNAs can harness the RISC pathway. Chemically modified SSOs can modulate splicing if directed to the appropriate site on a pre-mRNA. SSOs are modified so that they do not activate the RNA-degrading pathways, which would lead to the unwanted reduction of all splice variants. Similarly, AMOs that antagonize miRNAs are also designed to avoid activating RNA-degrading mechanisms. In the following section, we briefly discuss select chemical modifications and designs (Fig. 9.1) that are most commonly used to enhance the drug-like properties of ASOs.

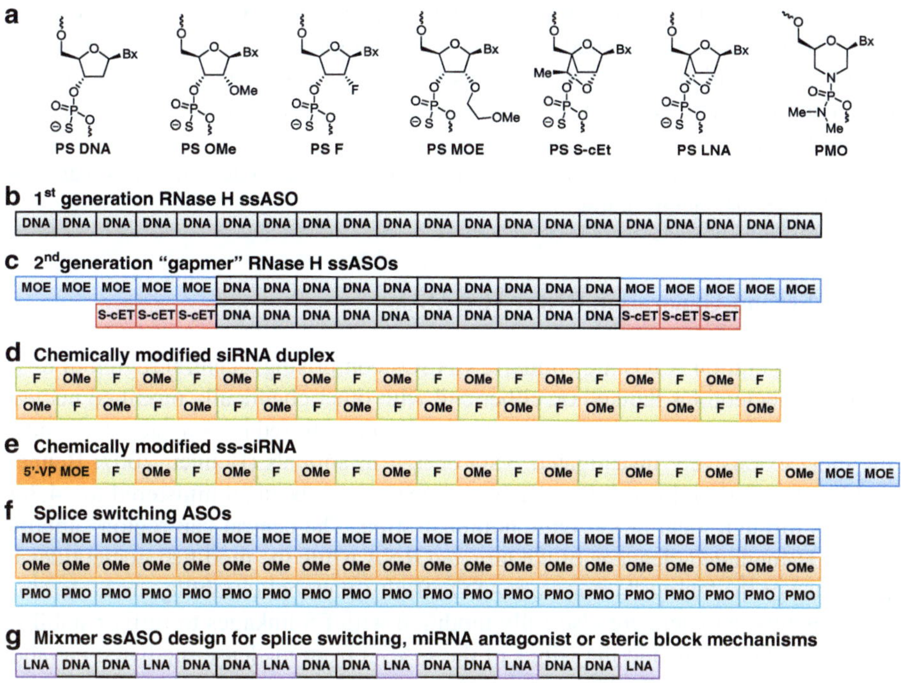

Fig. 9.1 Illustrations of commonly used ASO (**a**) modifications, and (**b–g**) designs

4.1.1 First-Generation RNase H ssASOs

First-generation ASOs based on the RNase H mechanism are ~20-mer ssASOs composed entirely of DNA (Fig. 9.1b), and uniformly modified with the phosphorothioate (PS) linkage, in which one of the non-bridging oxygen atoms of the natural phosphodiester linkage is replaced with sulfur (Eckstein 2000). The PS linkage presents several features that makes it attractive for ASO therapeutics (Geary 2009). It improves the stability of the ssASO to protect it against nuclease-mediated degradation, it does not significantly impair the ability of the ssASO to form duplexes with RNA unlike other backbone modifications, and perhaps most importantly it supports RNase H-mediated cleavage (Bachelin et al. 1998).

The PS modification increases nonspecific binding of ssASOs to various proteins (Eckstein 2000). Improved binding of PS-modified ssASOs to plasma proteins increases their apparent molecular weight, preventing the ssASOs from being excreted by the kidney, thereby allowing distribution to peripheral tissues (Geary 2009). The PS backbone also promotes ssASO binding to cell surface proteins. This allows the ssASO to be internalized into the cell without the aid of any specialized delivery vehicles (Koller et al. 2011; Stein et al. 2010). However, the incorporation of the PS modification can be a double-edged sword. The promiscuous protein-binding properties of PS-modified ssASOs can result in undesirable effects such as nonspecific activation of the immune system (Henry et al. 2007a). The PS modification also reduces ssASO-binding affinity for complementary RNA, −0.5 °C per modification, which is a liability for DNA ssASOs (Eckstein 2000). Furthermore, the metabolic stability of PS DNA is not optimal to support infrequent (weekly) dosing in the clinic (Bennett 2007). These limitations of first-generation PS DNA ssASOs led to the introduction of second-generation ssASO designs, which are discussed in greater detail below.

4.1.2 Second-Generation RNase H ssASOs

Second-generation ASOs are chimeric ssASOs comprising a central region of 8–10 PS DNA nucleotides flanked on either end with other chemically modified nucleotides (Fig. 9.1c) (Monia et al. 1993). This design is referred to as a "gapmer" and is the most extensively investigated ASO design in animals (Bennett 2007; Bennett and Swayze 2010). In addition, gapmer ssASOs have been administered to ~4,300 humans through local or systemic administration. The central gap region supports RNase H-mediated cleavage of the targeted RNA while the modified nucleotides in the flanks or the "wings" improve ssASO affinity, metabolic stability, and tolerability. Typically, gapmers are also fully modified with PS linkages to further stabilize the central DNA gap and the wings of the ssASO from nuclease-mediated metabolism (Bennett 2007).

A diverse collection of modified nucleotides are tolerated in the wings of RNase H gapmer ssASOs. The earliest gapmer designs used 2′-*O*-methyl (2′-OMe) (Shibahara et al. 1987) or 2′-fluoro (2′-F) RNA (Monia et al. 1993) modified nucleotides in the wings. However, these designs did not provide sufficient metabolic stability and were supplanted with 2′-*O*-methoxyethyl RNA (2′-MOE) gapmers (McKay et al. 1999). The 2′-MOE group is bulkier in comparison to a 2′-OMe or 2′-F group and improves ASO metabolic stability. In addition, the 2′-MOE group enhances ssASO affinity for targeted RNA and also traps water molecules in proximity to the PS backbone (Teplova et al. 1999). The improved hydration reduces nonspecific interactions of the ssASO with proteins, which translates into an improved overall safety profile in animal and in humans. Kynamro®, a systemically administered 2′-MOE gapmer (5-10-5 design) targeted to apolipoprotein B (*APOB*) transcripts, was recently approved by the US Food and Drug Administration (USFDA) for the treatment of familial hypercholesterolemia (Lee et al. 2013).

The potency of gapmer ssASOs can be further enhanced by using 2′,4′-bridged nucleic acids (BNA, also commonly referred to as locked nucleic acid or LNA) (Fig. 9.1c) (Vester and Wengel 2004), which provide a further enhancement in RNA-binding affinity. However, LNA gapmers increase the risk for causing hepatotoxicity (Swayze et al. 2007), which can be mitigated, to a large extent, by using other bicyclic nucleotides such as constrained MOE (cMOE) and constrained ethyl (cEt) (Seth et al. 2009), carbocyclic LNA (Seth et al. 2010), and α-L-LNA (Seth et al. 2012a). In addition, the 2′-modified nucleotides or bicyclic nucleotides can also be replaced with tricyclo DNA (Murray et al. 2012) and fluorinated hexitol (Egli et al. 2011) or cyclohexenyl (Seth et al. 2012b) nucleic acids. The use of affinity-enhancing nucleotides in the flanks allows for the use of shorter oligonucleotides (Seth et al. 2009; Straarup et al. 2010). This may also contribute to the enhanced activity of these ssASOs, perhaps because the reduced charge and lower molecular weight enable more efficient escape of the ssASO from endosomal compartments. A *S*-cEt gapmer (16-mer, 3-10-3 design) (Burel et al. 2013) targeting signal transducer and activator of transcription 3 (*STAT3*) has recently shown promising activity in a phase 1 trial for treating B-cell lymphomas (Hong et al. 2013).

4.1.3 siRNA and ss-siRNA

As discussed above, unmodified dsRNA is unstable in blood and has poor drug-like properties. As a result, chemical modifications are used to stabilize dsRNA and nanoparticulate formulations or conjugates are used to improve their pharmacokinetic properties (Rettig and Behlke 2012). Cationic lipids are efficient at delivering dsRNA to liver tissue in rodents and in humans (Jayaraman et al. 2012). However, formulated dsRNA has pro-inflammatory effects that often necessitate the use of high doses of corticosteroids, nonsteroidal anti-inflammatory agents, and anti-histamines. Moreover, the metabolic fate of the cationic lipid has yet to be determined conclusively. As a result, alternative delivery strategies are being explored such as coinjection with N-acetylgalactosamine-conjugated melittin-like peptide (NAG-MLP), a peptide found in bee venom, which facilitates release of the siRNA from endosomes (Wooddell et al. 2013). Alternatively, the sense strand of the siRNA can be conjugated to a small molecule such as N-acetylgalactosamine (GalNAc) that enhances its uptake into liver hepatocytes (Rajeev et al. 2013). While cationic lipids can protect and deliver unmodified or minimally modified nucleic acids into the cytosol, successful siRNA delivery using other strategies like conjugation to GalNAc clusters is critically dependent upon using chemical modifications to stabilize the siRNA in plasma, and within endosomal and lysosomal compartments prior to transit into the cytosol.

A number of chemical modifications have been investigated for siRNA-based gene silencing approaches (Fig. 9.1d). One of the most successful modifications to date is 2′-fluoro RNA (2′-F) (Manoharan et al. 2011). The 2′-F group is a polar hydrophobic substituent that polarizes the anti-periplanar nucleobase and increases affinity for complementary nucleic acids (Pallan et al. 2011). 2′-F is also well tolerated within the AGO2-binding pocket as opposed to bulkier substituents such as 2′-MOE (Manoharan et al. 2011). However, 2′-F is incorporated into genomic DNA and RNA raising concerns about potential long-term toxicity, which will require additional studies to resolve (Peng and Damha 2008).

2′-OMe is a naturally occurring RNA modification that is also compatible with the RISC mechanism when positioned appropriately. For example, introducing 2′-OMe at position 2 in the antisense strand reduces off-target effects of siRNAs, albeit at the expense of potency (Jackson et al. 2006). The most successful use of 2′-OMe in siRNAs is in conjunction with 2′-F (Allerson et al. 2005). This siRNA design is currently in human trials for the treatment of transthyretin amyloidosis (Rajeev et al. 2013).

For siRNAs, the 5′-phosphate of a guide RNA is important for the activity of RISC since it enhances the binding affinity for AGO2. However, ss-siRNAs with natural 5′-phosphates are not active in animals because the 5′-phosphates are removed by enzymes in vivo. In a tour de force, Lima and colleagues designed ss-siRNAs with a metabolically stable 5′-(*E*)-vinylphosphate (5′-VP) that could efficiently bind AGO2 and potently cleave target mRNAs in cells and animals (Fig. 9.1e) (Lima et al. 2012).

4.1.4 Splice-Switching Oligonucleotides

Commonly used designs for SSOs include uniformly PS-modified 2′-OMe, 2′-MOE, and morpholino ssASOs (Fig. 9.1f). For 2′-OMe and 2′-MOE SSOs, the PS modification is required to further enhance the nuclease stability and pharmacokinetic properties of SSOs. The absence of a continuous stretch of DNA nucleotides ensures that the SSO does not activate RNase H-mediated degradation of the targeted transcript. A uniformly modified PS 2′-OMe SSO, drisapersen, that promotes exon skipping to restore the open reading of dystrophin transcript is currently being investigated in human trials for the treatment of Duchene muscular dystrophy (DMD) (Goemans et al. 2011).

A uniformly modified PS 2′-MOE SSO, which promotes inclusion of exon 7 in the *SMN2* gene, has shown promise in early-stage human trials for the treatment of SMA (Rigo et al. 2012b). Interestingly, the profile of this SSO can be changed from exon inclusion to exon skipping by replacing 2′-MOE nucleotides with 2′-F (Rigo et al. 2012a). In this case, the 2′-F SSO binds to the pre-mRNA and the resulting heteroduplex recruits the ILF2/3 heterodimer to cause the omission of *SMN2* exon 7 from the mature transcript.

Phosphorodiamidate-linked morpholinos (PMOs) represent another chemical class of SSOs that are commonly used for splicing modulation. In PMOs, the furanose ring in natural nucleic acids is replaced with a morpholine ring and the negatively charged phosphodiester backbone is replaced with a neutral phosphorodiamidate backbone. PMOs show significantly reduced protein-binding properties as compared to PS backbone-modified ssASOs. This translates into improved tolerability in animal experiments (Moulton and Moulton 2010). However, poor protein binding causes the drug to be rapidly eliminated in the urine and high doses of the oligonucleotide (50 mg/kg in human trials) are required to elicit a pharmacological response in animals (Yokota et al. 2009). A PMO SSO, eteplirsen, is currently being tested in the clinic for the treatment of DMD (Cirak et al. 2011).

Peptide nucleic acids (PNAs) where the sugar-phosphate backbone of natural nucleic acids has been completely replaced with peptide linkages have been extensively investigated for splicing modulation. Unfortunately, PNA SSOs show surprisingly poor activity in animal models, which may be attributed to poor tissue distribution, poor solubility, and their hydrophobic nature (Yin et al. 2010). The solubility of PNAs can be improved by conjugation with cationic or anionic peptides, but this does not improve their pharmacological profile in animals (Wancewicz et al. 2010).

4.1.5 Anti-miRNA ASOs

A common AMO design is a full PS-modified LNA/DNA mixmer (Fig. 9.1g). Miravirsen, an LNA/DNA 15-mer ASO targeting miRNA-122, is being investigated in early-stage human trials for the treatment of HCV (Janssen et al. 2013). The LNA/DNA mixmer design has been successfully employed to antagonize miRNAs in tissues other than liver and AMOs as short as 8-mers have shown pharmacological efficacy in animal experiments (Obad et al. 2011).

2′-F/2′-MOE mixmer is another AMO design that has been employed to antagonize miRNAs in animals (Davis et al. 2009). Interestingly, the 2′-F/2′-MOE AMO was significantly more active than the sequence-matched full 2′-MOE AMO in animal models. This improved activity may be due to the ability of 2′-F-modified oligomers to interact differentially with cell-surface and with intracellular proteins (Rigo et al. 2012a). This ability for fluorinated chemical modifications to improve ASO activity in animals has also been demonstrated with 2′-F-modified siRNA (Manoharan et al. 2011), and with fluorinated hexitol (Egli et al. 2011) and cyclohexenyl (Seth et al. 2012b) nucleic acid-modified gapmer ssASOs.

4.1.6 Translational Inhibition

The Corey lab has recently investigated chemically modified ssASOs for allele-selective inhibition of mutant huntingtin (HTT) protein synthesis for the treatment of Huntington's disease (HD) (Hu et al. 2009). The ssASOs target the CAG repeat region, which is expanded on the mutant allele in comparison to the wild-type allele. The allele selectivity may result from cooperative binding of multiple ssASOs to the repeat expansion, which inhibits ribosomal translocation through the repeats in the mutant transcript. A number of chemical-modified ssASOs with 2′-modified nucleotides, BNA and PNA, were investigated for their ability to induce allele-selective inhibition in cell culture. However, activity in animal models was not disclosed (Gagnon et al. 2010).

The ASO chemistries and designs, discussed above, used for splicing modulation, antagonizing miRNAs, and inhibiting translation operate through the occupancy-only-mediated mechanism, as steric blockers. Theoretically, these types of ASOs can be used for any other mechanism that involves blocking an interaction (e.g., polyadenylation modulation, blocking PRC2 binding). However, oftentimes, the specific chemical modifications and designs that will be most ideal are difficult to predict a priori, so the experimenter has to determine this empirically.

5 Pharmacokinetics of Antisense Oligonucleotides

To be effective, an ASO must engage its target in the nucleus or the cytoplasm. To achieve this, the ASO must transit intact from the site of delivery to the cell surface, cross the plasma membrane into the cytoplasm, and then reach its target RNA. This is a significant challenge for an ASO with a typical molecular weight of ~7,000 Da that contains numerous solvent-exposed negative charges. A substantial amount of information about ASO pharmacokinetics is available for RNase H ssASOs containing PS backbone modifications, which has been reviewed previously (Geary et al. 2007; Levin et al. 2007).

ssASOs are typically delivered to cells in culture by using cationic lipids or electroporation (Bennett et al. 1993; Koller et al. 2011). However, several primary cells such as hepatocytes, macrophages, and neurons and a limited number of transformed cells retain the ability to internalize ssASOs, without the need of transfection reagents, to productively engage their target RNA (Carroll et al. 2011; Koller et al. 2011; Stein et al. 2010). This productive ssASO uptake pathway is still poorly defined. However, it has been shown to consist of a vesicular transport pathway that is clathrin and caveolin independent, but dependent on the adaptor-related protein complex 2, mu 1 subunit (AP2M1). Unfortunately, the majority of ssASOs taken up by cells are retained in endosomal and lysosomal compartments where it has no access to the target RNA (Koller et al. 2011). Vesicular transport can be modulated to enhance the activity of ssASOs in cells. Inhibition of retrograde trafficking from endosomes to the trans-Golgi network enhances ssASO activity (Ming et al. 2013) as does inhibition of the endosomal sorting complex required for transport (ESCRT) machinery (Koller et al., manuscript submitted). During natural productive uptake, a small fraction of the ssASO escapes the endocytotic pathway, at some ill-defined point, into the cytoplasm where it can access RNA. It is still not clear if trafficking of the ssASO to the nucleus is by diffusion or if it is facilitated by binding to proteins that shuttle to the nucleus (Lorenz et al. 2000). More work is still needed to better define the pathways by which ssASOs are taken up into cells and end in the productive compartment.

Double-stranded ASOs such as unconjugated siRNAs that are administered to cells without a transfection reagent are not efficiently taken up in a productive manner (Koller et al. 2011). Therefore, siRNAs need to be delivered to cells after being conjugated or encapsulated in lipids or lipid nanoparticles. siRNAs delivered in this manner are taken up by cells through clathrin-mediated endocytosis and macropinocytosis (Gilleron et al. 2013; Sahay et al. 2013), and appear to escape the vesicular pathway in an early-late hybrid endocytotic compartment that needs further definition (Gilleron et al. 2013).

PS-modified ssASOs bind plasma proteins with high affinity. Therefore, following peripheral administration (subcutaneous or intravenous), these ssASOs transiently circulate bound to plasma proteins, which limits their rapid filtration and excretion into the urine (Crooke et al. 1996; Watanabe et al. 2006). Unconjugated siRNAs (Braasch et al. 2004) and uncharged singe-stranded PNA and PMO ASOs (Iversen 2007; McMahon et al. 2002) exhibit low plasma protein binding and are rapidly excreted in the urine. Thus, it is important for the ssASO to maintain adequate binding to plasma proteins in order to facilitate delivery to tissues. However, the type of plasma proteins to which the ssASO binds is also important. Binding to high-affinity plasma proteins results in greater delivery of ssASO to the liver but, paradoxically, does not result in better activity presumably because the ssASO ends up in a nonproductive compartment (Geary et al. 2009).

After peripheral administration, >90 % of the administered ssASOs distribute from circulation into tissues within 2 h. ssASOs accumulate in multiple tissues with the highest concentration found in kidney, liver, and spleen (Levin et al. 2007). However, tissues that accumulate less ssASO such as adipose tissue (Weismann et al. 2011) and muscle (Hua et al. 2008; Wheeler et al. 2012) also demonstrate good antisense effects following peripheral administration. Interestingly, even though ssASOs accumulate in tissues relatively quickly, it takes 24–72 h to achieve maximum efficacy (Yu et al. 2001). The duration of pharmacologic effects depends on the target tissue. In liver it is 2–4 weeks (Bennett 2007; Levin et al. 2007), but in muscle it is much longer, with a duration that is greater than 6 months (Wheeler et al. 2012).

ssASOs that are administered peripherally do not exhibit pharmacologic effects in the brain or the spinal cord because they cannot traverse an intact brain–blood barrier (BBB) (Geary 2009). Therefore, ssASOs need to be administered directly to the cerebrospinal fluid (CSF) via an injection or an infusion pump (Devos and Miller 2013). Central administration of ssASOs results in broad distribution throughout the spinal cord and brain of animals with accumulation in neuronal and non-neuronal cells. Consistent with the ssASO distribution, robust antisense activity is also observed in the spinal cord and brain (Butler et al. 2005; Kordasiewicz et al. 2012; Koval et al. 2013; Passini et al. 2011; Smith et al. 2006). For RNase H ssASOs, the half-life of antisense effects is typically several months (Kordasiewicz et al. 2012), and over 6 months for an SSO (Hua et al. 2010).

6 Toxicology of Antisense Oligonucleotides

ASOs, as do virtually all drugs, exhibit dose-dependent toxicities. Owing to the vast preclinical and clinical experience amassed over the last 20 years, the toxicology of PS ssASOs, mostly of the RNase H type, is the best understood and has been reviewed extensively (Henry et al. 2007b; Kwoh 2007). Less is known about the toxicological properties of other types of ASOs such as siRNAs, but these are being defined (Castanotto and Rossi 2009; Jackson and Linsley 2010).

Toxicities in tissues where ASOs accumulate can be classified as either hybridization dependent or hybridization independent. Hybridization-based toxicities can be further divided into on-target or off-target. Hybridization-based on-target toxicity can result if the modulation of the target transcript itself produces a side effect. This can be avoided or minimized by wisely selecting the target based on its known biological role and by assessing the liabilities of modulating the target in rodents and nonhuman primates. Hybridization-based off-target toxicity can occur if the ASO modulates a target/s, through an unintended hybridization-based mechanism. Since the genomes of animals used for preclinical research and humans are known, one can use computational tools to assess the propensity of a given ASO to hybridize with off-targets. ASOs that are predicted to fully hybridize with many off-targets can be immediately excluded from further investigation or the modulation of predicted off-targets can be determined experimentally. It is even possible to characterize the hybridization-based off-target effects in an unbiased way by using microarrays or next-generation sequencing technologies. However, ultimately it is the careful toxicological evaluation of an ASO in preclinical species that will satisfy potential off-target concerns.

The best characterized toxicities of ASOs are those that are hybridization independent. These are the so-called aptameric toxicities that occur because of the inappropriate interaction of ASOs with proteins (Henry et al. 2007b). It is not surprising then that these toxicities depend on the particular design/chemical composition of an ASO. Examples include interactions of ASOs with Toll-like receptors (Krieg 2008; Marques and Williams 2005) and other receptors (Burel et al. 2012; Kleinman et al. 2008; Senn et al. 2005) that result in the activation of the innate immune system (Henry et al. 2007b), interactions with complement factor H that results in complement activation (Henry et al. 1997), and interactions with coagulation factors that reduce clotting time (Sheehan and Phan 2001). Since these toxicities have been well characterized they can be easily avoided by careful screening in cells or animals, or managed in the clinic.

7 Pharmacologic Effects of Antisense Oligonucleotides

ASO-based mechanisms have been utilized in numerous instances to correct defects in pre-mRNA processing in cell culture models of disease. We refer the reader to reviews that cover these success stories (Bauman et al. 2009; Havens et al. 2013; Spitali and Aartsma-Rus 2012). Here we concentrate on the ASO-based mechanisms, both occupancy-only-mediated and occupancy-mediated RNA cleavage, which have been tested in mouse models of pre-mRNA processing diseases. Even though ASOs that act through an occupancy-only mechanism are being tested in the clinic for the treatment of DMD, these are not covered here since DMD is not caused primarily by a pre-mRNA processing defect. We refer the reader to reviews that have thoroughly covered the progress made in developing a therapy for DMD (Koo and Wood 2013; Muntoni and Wood 2011).

7.1 Myotonic Dystrophy Type I

Myotonic dystrophy type 1 (DM1) is typically an adult-onset neuromuscular disease with clinical features including myotonia, skeletal muscle weakness and wasting, cardiac arrhythmias, and neuropsychiatric dysfunction. In addition, there is a multisystem dysfunction that ranges from gastrointestinal symptoms to insulin resistance and ocular cataracts. The most common cause of death is respiratory failure or sudden cardiac arrest (Harper et al. 2001). DM1 belongs to the class of triplet repeat diseases and is inherited in an autosomal dominant manner (Gatchel and Zoghbi 2005). The mutation that causes DM1 is a multi-copy insertion of the trinucleotide sequence CTG in the 3′-UTR *DMPK* (Fig. 9.2a) (Brook et al. 1992; Mahadevan et al. 1992).

In DM1 patients the number of CTG insertions in the 3′-UTR of *DMPK* ranges from 50 to greater than 1,500 repeats, compared to the normal range of 5 to 38 repeats (Ranum and Cooper 2006). mRNAs that are produced from a *DMPK* allele with the CUG expansion cause disease through a toxic RNA gain-of-function mechanism (Mankodi et al. 2000; Orengo et al. 2008; Seznec et al. 2001). The CUG repeat-containing mRNAs accumulate in nuclear foci (Davis et al. 1997; Taneja et al. 1995). These foci sequester the splicing factor MBNL1, which is titrated out of the nucleoplasm (Fig. 9.2a) (Fardaei et al. 2001; Mankodi et al. 2001; Miller et al. 2000). In addition, the level of a second splicing factor called CUG-binding protein 1 (CUGBP1) is inappropriately elevated, likely because it is stabilized by protein kinase C-mediated phosphorylation, which is triggered by the CUG repeat-containing RNA (Kuyumcu-Martinez et al. 2007). Changes in the levels of MBNL1 and CUGBP are thought to drive, in large part, the widespread misregulation of alternative splicing in genes whose deficiency is linked to the clinical features of DM1 (Du et al. 2010; Echeverria and Cooper 2012; Jiang et al. 2004; Kanadia et al. 2006; Lin et al. 2006; Orengo et al. 2008; Osborne et al. 2009; Ward et al. 2010). This defines DM1 as a disease that occurs due to RNA-processing defects.

Since the preponderance of the evidence indicated that DM1 is due to a toxic gain-of-function mechanism, several approaches were developed to either neutralize or reduce the expression of the CUG repeat-containing *DMPK* mRNA (Gao and Cooper 2013). siRNAs (Krol et al. 2007), and several antisense-producing, vector-based approaches such as shRNAs (Langlois et al. 2005), miRNAs derived from introns (Seow et al. 2012), and U7 snRNAs engineered to contain an antisense sequence targeting the CUG repeat expansion (Francois et al. 2011) have shown positive results in cell culture. Below, we describe the ASO-based approaches that have been evaluated in animals.

Initial reports used ASOs designed to act through an occupancy-only-mediated mechanism (Fig. 9.2a) (Mulders et al. 2009; Wheeler et al. 2009). In the first study, a PMO steric blocker ssASO targeted to the CUG repeat expansion prevented the binding or displaced MBNL1 from the transcript in an in vitro binding assay. In addition, when this ssASO was administered to DM1 mice (HSA^{LR}) by intramuscular (IM) injection, it resulted in a drastic reduction of nuclear foci, release of MBNL1 from the foci, and normalization of the splicing misregulation in the injected muscle fiber. Restoration of proper chloride channel, voltage-sensitive 1 (*CLCN1*) splicing markedly reduced the myotonia (Wheeler et al. 2009). In a second study, a 2′-OMe steric blocker ssASO also targeted to the CUG repeat expansion reduced the presence of nuclear foci and reversed splicing defects in a cell culture and a mouse model of DM1 (HSA^{LR}) after IM injection (Mulders et al. 2009). In both the Wheeler and Mulders reports it was surprising to observe a significant reduction of CUG-containing transcripts after administration of ssASOs since these were not designed to act through an RNase H mechanism. The reason for this is still not clear but may be explained, in part, by the ssASO-mediated release of transcripts from an environment, nuclear foci, that may not be accessible to enzymes that degrade RNA.

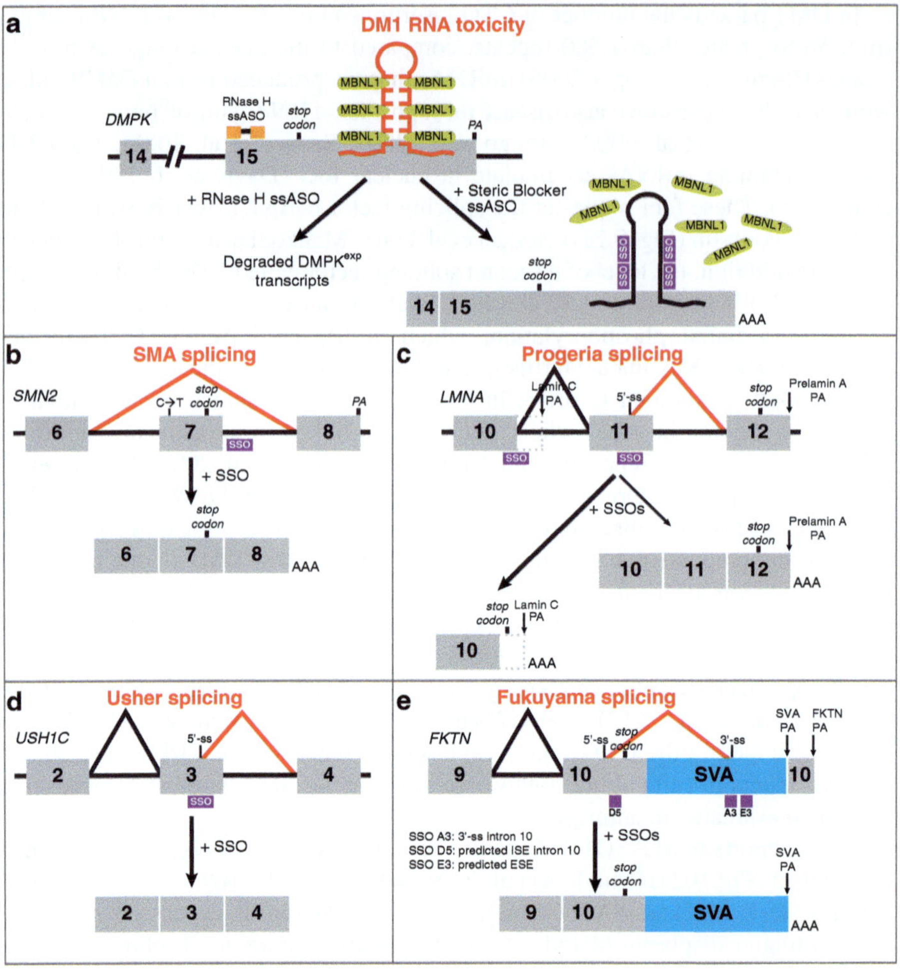

Fig. 9.2 Schematic representations of aberrant pre-mRNA processing in (**a**) myotonic dystrophy type 1, (**b**) spinal muscular atrophy, (**c**) Hutchinson-Gilford progeria syndrome, (**d**) Usher syndrome type 1, and (**e**) Fukuyama congenital muscular dystrophy. Each panel also depicts the ASO-mediated correction that has been achieved in animal models. *Numbered boxes*, *horizontal lines*, and *inverted Vs* (*black*, normal; *red*, aberrant) represent exons, introns, and splicing, respectively. 5′-ss, 5′-splice site; PA, polyadenylation site

Since an siRNA targeted to the CUG repeat expansion was shown to lower the expression of *DMPK* mRNAs in cell culture (Krol et al. 2007), this strategy was also tested in DM1 mice. HSA^{LR} mice were treated with a CUG repeat-targeted siRNA by IM injection and this resulted in a reduction of repeat-containing transcripts, a reduction in the number of intranuclear RNA foci with release of MBNL1, and correction of aberrant splicing (Sobczak et al. 2013).

RNase H ssASOs have also been used to target the CUG repeat expansion directly. In a cell culture model of DM1, an RNase H ssASO targeted to the repeat expansion strongly reduced the levels of RNA containing greater than 40 CUG repeats, but not transcripts containing 12 CUG repeats. ssASO administration to a DM1 mouse model (EpA960/HSA-Cre) by IM injection reduced the expression of CUG-containing RNA and accumulation of nuclear foci. Importantly the expression of other transcripts containing CUG repeats, that were examined, was not affected. However, CUG-targeted ssASOs resulted in muscle damage, perhaps raising concerns about targeting the repeat directly with RNase H ssASOs (Lee et al. 2012). A separate study reported that RNase H ssASOs targeted to the CUG repeat expansion lowered expanded transcripts with the added property of ameliorating the somatic instability observed in the muscle of a DM1 mouse model (DM300-328-XXL) (Nakamori et al. 2011).

The inherent limitation of IM delivery is that the therapeutic agent does not have access to the numerous muscle groups afflicted in DM1. To overcome this limitation a single-stranded peptide-conjugated PMO (PPMO) targeted to the CUG repeat expansion was administered peripherally by intravenous delivery to HSA^{LR} mice. This resulted in a reduction of nuclear foci containing CUG repeat RNA and MBNL1 in muscle. As expected, treatment with the PPMO resulted in the correction of splicing misregulation and no signs of myotonia in the hindlimb muscles were observed (Leger et al. 2013). Even though mice did not show any signs of toxicity upon PPMO exposure (Leger et al. 2013), a different PPMO did show toxicity in the kidneys of nonhuman primates (NHPs) (Kole et al. 2012). Therefore, a PPMO strategy will require further optimization and evaluation before it can be used in DM1 patients.

Targeting sequences outside of the CUG repeat expansion with RNase H ssASOs (Fig. 9.2a) has shown very encouraging result in HSA^{LR} mice. Peripheral administration of an RNase H ssASO resulted in 80 % reduction of actin, alpha 1, skeletal muscle (*ACTA1*) transcripts containing the CUG expansion in hindlimb muscles. This was accompanied by a reduction in nuclear foci and normalization of MBNL1-dependent splicing events. These molecular improvements translated into the elimination of myotonia in the hindlimb muscles and an improvement in muscle pathology. Interestingly, 1 year after the discontinuation of treatment, reduction of *ACTA1* mRNA was still greater than 50 %, indicating that the activity of RNase H ssASOs in muscle is long-lasting (Wheeler et al. 2012). In these studies, RNase H ssASOs were targeted to the *ACTA1* mRNA produced from the transgene for the purpose of achieving proof of concept. RNase H ssASOs were also shown to significantly reduce human *DMPK* transcripts containing a CUG expansion by greater than 50 % in the hindlimb muscles of transgenic mice (Wheeler et al. 2012).

In addition to the ASO-based approaches that are aimed at reducing the expression of transcripts containing a CUG repeat expansion, it is also possible to correct aberrant splicing directly. The myotonia observed in HSA^{LR} mice is due to a splicing defect in the *CLCN1* pre-mRNA that results in the inclusion of exon 7a and the introduction of a PTC in the *CLCN1* mRNA (Mankodi et al. 2002). A PMO SSO targeting the intron 6/exon 7a junction administered to HSA^{LR} mice by an IM injection suppressed the inclusion of exon 7a in the *CLCN1* pre-mRNA. Correction of *CLCN1* splicing was followed by a marked improvement of myotonia in the muscle (Wheeler et al. 2007).

7.2 Spinal Muscular Atrophy

Spinal muscular atrophy (SMA) is a neuromuscular disorder, which in its most severe form is the leading cause of infant mortality (Lunn and Wang 2008). Due to the degeneration of motor neurons in the anterior horn of the spinal cord, patients experience severe muscle weakness and atrophy of the voluntary muscles of the limbs and trunk, resulting in paralysis and respiratory failure (Crawford and Pardo 1996). SMA is inherited in an autosomal recessive manner due to deletions or, more rarely, mutations in the survival motor neuron 1 (*SMN1*) gene (Lefebvre et al. 1995) that result in the functional loss of SMN protein (Coovert et al. 1997; Lefebvre et al. 1997). Even though it is not known how the loss of SMN protein results in SMA, several scenarios have been suggested. The first stems from the well-documented role of the SMN protein in directing the assembly of small nuclear ribonucleoproteins (snRNPs), which are essential components of the splicing machinery (Yong et al. 2004). It is thought that loss of SMN protein results in reduced levels of snRNPs (Wan et al. 2005), which can no longer fulfill their role in the splicing of genes that are important for neuronal function (Baumer et al. 2009; Gabanella et al. 2007). The second scenario, which is not mutually exclusive with the first, posits that the SMN protein facilitates the transport of mRNAs along axons and without this function neurons perish (Burghes and Beattie 2009; Fallini et al. 2012). How the loss of SMN protein causes SMA is still not clear, but it is widely accepted that elevating the levels of SMN protein in patients should provide a therapeutic benefit (Burghes and Beattie 2009).

Humans have a paralog of the *SMN1* gene, *SMN2*, which in the absence of *SMN1* is the primary modifier of SMA. The severity of SMA is correlated with the number of *SMN2* copies present. Patients who have a greater number of copies have less severe forms of the disease (Feldkotter et al. 2002; Prior et al. 2004). A key difference between the *SMN1* and *SMN2* genes is a C-to-T transition in exon 7 of *SMN2* that results in an increase in the production of mRNAs that exclude exon 7 (Fig. 9.2b) (Lorson et al. 1999; Monani et al. 1999). These mRNAs are translated into an unstable protein that is truncated at the C-terminus and is rapidly degraded (Cho and Dreyfuss 2010; Lorson and Androphy 2000). The *SMN2* gene produces a small amount of full-length mRNA that is translated into a functional, full-length SMN protein. An increase in the production of SMN either by having more *SMN2* copies or mutations that enhance the inclusion of exon 7, results in a milder form of SMA in rodents (Monani et al. 2000) and humans (Feldkotter et al. 2002; Prior et al. 2004, 2009; Vezain et al. 2010).

SMA can be considered an RNA-processing disease since the depletion of SMN protein has been shown to cause splicing misregulation of other transcripts and because a splicing defect in *SMN2* is unmasked by the inactivation of *SMN1*. Since the discovery of the mutation in *SMN2* exon 7 that results in increased exon 7 skipping, numerous ASO-based approaches have been developed to correct *SMN2* splicing and increase the levels of SMN protein. These include bifunctional SSOs (Cartegni and Krainer 2003; Skordis et al. 2003), SSOs (Hua et al. 2007, 2008; Lim and Hertel 2001; Miyajima et al. 2002; Singh et al. 2006) targeted to various sites on the *SMN2* pre-mRNA, and various antisense-producing, vector-based strategies (Coady and Lorson 2010; Geib and Hertel 2009; Meyer et al. 2009). Below we describe the ASO-based mechanisms that have proven to be effective in animal models of SMA.

A bifunctional SSO was one of the first ASO-based mechanisms tested in an animal model of SMA. This bifunctional SSO consisted of a 2′-OMe-modified antisense sequence targeted to the 5′-end of *SMN2* exon 8 with an RNA appendage that contained a sequence motif for the splicing repressor heterogeneous nuclear ribonucleoprotein A1 (hnRNPA1). Administration of the bifunctional SSO by ICV bolus injection to severe SMA mice, SMAΔ7 [$Smn^{-/-}$; $SMN2^{+/+}$; $SMN\Delta7^{+/+}$ (Le et al. 2005)], at postnatal day 1 (PND1) resulted in a small qualitative increase in SMN protein in the brain (Dickson et al. 2008). Subsequently, the same group tested a 2′-OMe bifunctional SSO targeting an intronic splicing silencer in intron 6 with a Tra2β1 enhancer sequence attached. This bifunctional SSO was administered by ICV bolus injection to SMAΔ7 mice at PND1 and a twofold increase in SMN protein in the brain and spinal cord was observed. When the bifunctional SSO was administered to another severe mouse model of SMA [$Smn^{-/-}$; $SMN2^{+/+}$ (Monani et al. 2000)] by ICV bolus injection at PND1 and 3 the mean survival was increased by 1 day from the normal life-span of 6 days (Baughan et al. 2009). 2′-OMe bifunctional SSOs with the sequence motif for either splicing factor, arginine/serine-rich 10 (SFRS10) or 1 (SFRS1), were targeted to an intronic splicing silencer (ISS-N1) located slightly downstream of *SMN2* exon 7 (Osman et al. 2012; Singh et al. 2006). Each was administered by ICV bolus injection to the SMAΔ7 mice at PND1, 3, and 5. This treatment resulted in a 2–3.5-fold increase in SMN protein in the brain and spinal cord. This increase was accompanied by a significant improvement in the righting reflex and an increase in survival from the normal median of 11 days to ~20 days (Osman et al. 2012). Even though bifunctional SSOs have demonstrated some efficacy in animal models, more work needs to be done before such an approach is ready for the clinic. The toxicity of bifunctional SSOs is not well characterized and the manufacturing of molecules of this size (40–50 nucleotides) at large scale will be challenging.

SSO-mediated inclusion of *SMN2* exon 7 has emerged as the most promising strategy for treating SMA in mouse models of disease (Fig. 9.2b). The target sequence for all the SSOs tested is ISS-N1 (Singh et al. 2006), which was shown to contribute to the repression of *SMN2* exon 7 by binding hnRNPA1/A2 (Hua et al. 2008; Rigo et al. 2012a). A 20-mer, 2′-OMe SSO delivered to the SMAΔ7 mice by ICV bolus injection on PND1, 3, 5, 7, and 10 for a total of 10 μg was reported to increase the levels of SMN in the brain and lumbar spinal cord by 2–3.5-fold. However, the level of *SMN2* transcripts that included exon 7 was only increased by ~1.4-fold. This was accompanied by an improvement in the righting reflex but no improvement in survival was observed (Williams et al. 2009). Subsequently it was found that there was no correction of *SMN2* splicing in the brain or the spinal cord when the 20-mer 2′-OMe SSO was administered to adult mild SMA Taiwan mice with four copies of the *SMN2* transgene [$Smn^{-/-}$; $SMN2^{+/+}$ (Hsieh-Li et al. 2000)] by continuous infusion using an osmotic pump (Hua et al. 2010). This was also the case when the 2′-OMe SSO was administered by an ICV bolus injection to adult mild SMA mice (Rigo et al., manuscript submitted).

In contrast to the poor performance of the 2′-OMe SSO in the CNS of mild SMA Taiwan mice, administration of an 18-mer 2′-MOE SSO (ASO-10-27, also known as ISIS-SMN_{Rx}) resulted in a greater than 90 % *SMN2* exon 7 inclusion and a ~4 fold increase in SMN protein in the spinal cord. The SSO was shown to accumulate in spinal cord motor neurons where it also produced SMN protein. Surprisingly, the effects of the SSO in the CNS were durable with *SMN2* splicing correction still at maximal levels 6 months after the end of the infusion (Hua et al. 2010).

When ASO-10-27 was administered to the SMAΔ7 mice by a 4 μg ICV bolus injection at PND0 there was a 4–5.5-fold increase in *SMN2* exon 7 inclusion and an increase in SMN protein in the spinal cord to 55–90 % of the level in WT mice. Interestingly, *SMN2* splicing correction peaked 16 days after the ICV bolus injection and then returned to baseline at day 30, possibly because the cellular depletion of ASO-10-27 was faster in the growing neonate. ASO-mediated increase of SMN protein in the CNS of SMA mice partially prevented motor neuron loss in the spinal cord, almost restored myofiber size and the neuromuscular junction (NMJ) architecture of the intercostal and quadriceps muscles, and improved the motor performance of the mice. ASO-10-27 treatment increased the median survival of SMA mice from 16 to 26 days (Passini et al. 2011). ASO-10-27 was also tested in the severe SMA Taiwan mice [$Smn^{-/-}$; $SMN2^{+/0}$ (Gogliotti et al. 2010; Hsieh-Li et al. 2000)] by a 20 μg ICV bolus injection at PND1. This treatment increased *SMN2* exon 7 inclusion to 84 % and SMN protein by ~4.5-fold in the spinal cord, and increased the median survival from 10 to 16 days (Hua et al. 2011).

Central administration of ASO-10-27 resulted in almost complete correction of *SMN2* splicing in the spinal cord (Hua et al. 2011; Passini et al. 2011), yet the prolongation in survival was significantly less than that observed after systemic *SMN* gene therapy (Dominguez et al. 2011; Foust et al. 2010; Valori et al. 2010). Various reports had suggested that a number of pathological changes in peripheral tissues were contributing to the early demise of severe SMA mice, in addition to their motor neuron pathology (Bevan et al. 2010; Cifuentes-Diaz et al. 2001; Heier et al. 2010; Shababi et al. 2010; Vitte et al. 2004; Voigt et al. 2010). Therefore, Hua and colleagues asked if peripheral administration of ASO-10-27 could enhance the function and survival of severe SMA Taiwan mice. Remarkably, two 50 mg/kg subcutaneous bolus injections administered between PND0 and 3 increased the median survival from 10 to 108 days. As expected, adding central administration (20 μg ICV bolus injection at PND1) to this dosing paradigm further increased the median survival to 173 days. SMA mice that were dosed subcutaneously with ASO-10-27 demonstrated marked improvements in motor neuron counts in the spinal cord, myofiber size of the rectus femoris muscle, NMJ architecture, muscle strength, and motor function, demonstrating that systemic treatment of neonatal mice did improve motor neuron health. In addition, significant improvements in heart weight and thickness of the intraventricular septum and the left ventricular wall were also noted. Interestingly, SMA mice were found to have a depletion of circulating levels of insulin-like growth factor 1 (IGF-1), attributed to a hepatic deficiency in the production of IGF-binding protein acid-labile subunit (IGFALS), which was reversed by peripheral treatment with the ASO (Hua et al. 2011).

The strongest *SMN2* splicing correction after the systemic administration of ASO-10-27, compared to untreated animals, was in the liver, skeletal muscle, and heart (Hua et al. 2011). Splicing correction in these tissues is, very likely, important for the increased longevity of the severe SMA mice. However, *SMN2* spicing correction, albeit to a lesser extent, was also noted in the brain and spinal cord. This may be due to incomplete closure of the brain–blood barrier (BBB) in neonatal mice at the time of dosing. Therefore, *SMN2* splicing correction in CNS tissue after peripheral administration probably contributes to the extended survival of the severe SMA mice.

Since the surprising results of Hua and colleagues, the idea that mouse models of severe SMA have a multitude of peripheral defects has gained traction (Hamilton and Gillingwater 2013; Schreml et al. 2012; Shababi et al. 2013). It remains to be determined if peripheral administration would provide a therapeutic benefit to patients. Although the cardiac changes and peripheral vascular perfusion abnormalities reported in mice have occasionally been reported in severely affected SMA patients (Araujo Ade et al. 2009; El-Matary et al. 2004), they do not appear to be a common feature of the disease. A more careful evaluation of defects in peripheral organ systems in patients, perhaps in the most severe SMA cases, is required to determine if a peripheral SSO therapy should be considered.

Isis Pharmaceuticals has advanced ASO-10-27 (ISIS-SMN_{Rx}) into clinical development. Preclinical studies in NHPs demonstrated broad distribution of ASO-10-27 throughout the spinal cord tissue with accumulation in motor neurons following intrathecal (IT) infusion or bolus injection (Passini et al. 2011) (Rigo et al., manuscript submitted). NHPs do not have the *SMN2* gene (Rochette et al. 2001) and splicing correction cannot be assessed. Experiments in *SMN2* transgenic mice indicated that an ASO-10-27 spinal cord tissue concentrations of 1–2 μg per gram of tissue results in a 50 % correction of *SMN2* splicing (Hua et al. 2010; Passini et al. 2011), and thus provide a target tissue concentration for clinical trials. When ASO-10-27 was administered to NHPs by IT bolus injection, these tissue concentrations were exceeded even at the lowest dose tested (1 mg) (Rigo et al., manuscript submitted). A single-dose, phase 1 study of ISIS SMN_{Rx} in patients with SMA type 2 and 3 demonstrated that ISIS-SMN_{Rx} is well tolerated after administration by IT bolus injection (Chirboga et al. 2013).

PMO SSOs targeted to ISS-N1 have also yielded very encouraging results in severe SMA mice (Mitrpant et al. 2013; Porensky et al. 2011; Zhou et al. 2013). A single 54 μg ICV bolus injection of a 20-mer PMO SSO (PMO-20) targeted to the ISS-N1 administered at PND0 increased *SMN2* exon 7 inclusion and SMN protein by ~3.5-fold and ~2-fold, respectively, in the spinal cord. The median survival of treated SMAΔ7 mice increased from 15 days to 104 days, and showed a marked improvement in motor function. In contrast to ASO-10-27, the PMO-20 was able to partially correct *SMN2* splicing at 65 days after the single injection at PND0, which may account, in part, for its improved survival benefit (Porensky et al. 2011). More recently a 25-mer PMO SSO (PMO-25), also targeting ISS-N1, was tested in the severe SMA Taiwan mice. Administration of 20 μg of the PMO-25 SSO by ICV bolus on PND0 resulted in an increase in the levels of *SMN2* transcripts containing exon 7 and SMN protein in CNS tissue. At the dose of 20 μg the median survival increased from 10 to 43 days, and a maximal median survival of 86 days was attained with a dose of 40 μg (Zhou et al. 2013). As was observed for the PMO-20 SSO (Porensky et al. 2011), the PMO-25 SSO-mediated *SMN2* splicing correction was also long-lasting, returning to baseline levels ~40 days after the injection (Zhou et al. 2013). In the latest report, it was determined that the PMO-25 and PMO-20 SSOs extend the survival of SMA mice to a similar extent when administered centrally (Mitrpant et al. 2013).

Although the PMO SSOs have not been compared directly to ASO-10-27 in the severe SMA mouse models after central administration, it appears that they do confer a greater increase in survival. We compared the potencies of ASO-10-27 and the PMO-20 SSO in adult mild SMA Taiwan mice and found that ASO-10-27 is more potent for *SMN2* splicing correction compared to the PMO-20 SSO ASO (Rigo et al., manuscript submitted). Several possibilities can account for the greater activity of the PMO SSOs in the severe SMA mouse models. These include the longer duration of *SMN2* splicing correction in a growing neonatal mouse, a higher exposure to peripheral tissues after CNS administration, or enhanced activity in important cell subpopulations in the CNS. Additional studies are necessary to explain the difference in activity between the PMO SSOs and ASO-10-27 in the severe SMA mice.

7.3 Hutchinson-Gilford Progeria Syndrome

Hutchinson-Gilford progeria syndrome (HGPS) is a disease of accelerated aging and is characterized by multiple anomalies, including growth impairment, hair loss, osteoporosis, and depleted subcutaneous fat deposits. Cardiovascular complications are frequently the cause of death before the age of 15 (Merideth et al. 2008). HGPS is an autosomal dominant disease that is usually caused by a mutation in exon 11 of the lamin A (*LMNA*) gene. The mutation is a C-to-T spontaneous substitution that does not change the amino acid of the corresponding codon, but results in the activation of a 5′-splice site in exon 11 (Fig. 9.2c). Utilization of the activated splice site results in the production of a prelamin A protein with an in-frame deletion of 50 amino acids near its C-terminus, a protein commonly referred to as progerin (De Sandre-Giovannoli et al. 2003; Eriksson et al. 2003).

Normally, the prelamin A protein is processed posttranslationally by several enzymes. In the final step, mature lamin A is released from its tether to the inner nuclear membrane by the zinc metallopeptidase STE24 (ZMPSTE24). The deletion of 50 amino acids caused by the activation of the 5′-splice site in exon 11 deletes the region normally targeted by ZMPSTE24, causing progerin to remain farnesylated and membrane bound (Worman et al. 2009). The continued binding of progerin to the nuclear membrane disrupts the nuclear lamina in a dominant negative fashion.

Precisely how the disruption of the nuclear membrane results in HGPS is not known but there are several possibilities including altered transcription, epigenetic changes, genomic instability, and mitotic defects (Dechat et al. 2008).

ssASOs have been used to inhibit the production of progerin or prelamin A in cell culture. A PMO SSO targeted to the HGPS-activated splice site reduced its utilization and the level of progerin (Scaffidi and Misteli 2005). An RNase H ssASO reduced prelamin A without affecting the level of lamin C in Zmpste24$^{-/-}$ fibroblasts and reduced the number of cells with misshapen nuclei (Fong et al. 2006). SSOs have also shown promising results in a new mouse model of HGPS. This mouse model has a knock-in of the HGPS mutation (*Lmn*$^{G609G/G609G}$) and has a phenotype with many similarities to the human condition (Osorio et al. 2011). HGPS mice were treated with a cocktail of two vivo-morpholinos (vPMOs), which are PMOs that are conjugated to an octa-guanidine dendrimer for enhanced delivery into tissues (Moulton and Jiang 2009). The cocktail consisted of a vPMO targeted to the HGPS mutation in exon 11 to reduce progerin mRNA (Scaffidi and Misteli 2005) and a vPMO targeted to the exon 10/intron 10 junction to reduce progerin and prelamin A mRNAs (Fig. 9.2c) (Osorio et al. 2011). The mechanism for the reduction of progerin and prelamin A mRNAs is not entirely clear but may be largely due to the preferential production of lamin C mRNA. Lamin C mRNA is produced from the *LMNA* gene by the use of a polyadenylation signal in intron 10 and this precludes the production of prelamin A mRNA (Fig. 9.2c) (Lin and Worman 1993). Although not demonstrated, the vPMO probably triggers polyadenylation in intron 10 by blocking the recruitment of the U1 snRNP to the intron 10, 5′-splice site (Kaida et al. 2010; Vorlova et al. 2011). Peripheral administration of the vPMOs to adult HGPS mice demonstrated significant reductions in progerin mRNA and protein in the liver, kidney, and heart, but not the muscle. This was accompanied by an increased body weight, reduced lordokyphosis, more subcutaneous fat, increased serum glucose, and extended median survival from 111 to 155 days (Osorio et al. 2011).

7.4 Usher Syndrome

Usher syndrome type 1 (USH1) is characterized by profound hearing impairment and vestibular dysfunction at birth and progressive blindness starting in early adolescence. USH1 is an autosomal recessive disease that is genetically heterogeneous (Petit 2001). Mutations in the Usher syndrome 1C (*USH1C*) gene, which encodes for a protein called harmonin, account for 6–8 % of USH1 cases (Ouyang et al. 2005; Verpy et al. 2000). A mutation that has recently been targeted therapeutically in an animal model is the *USH1C216G>A* mutation (Bitner-Glindzicz et al. 2000). This mutation creates a new 5′-splice site in exon 3 that is used preferentially over the authentic 5′-splice site further downstream. This aberrant splicing deletes the 3′-end of exon 3 resulting in a translation frameshift and a truncated harmonin protein (Fig. 9.2d) (Lentz et al. 2005). Harmonin is an important scaffold protein that plays a role in the proper development and function of stereocilia and photoreceptors, and without functional harmonin these cells perish (Kremer et al. 2006).

A mouse model of USH1 that faithfully recapitulates the hearing and vestibular dysfunction observed in patients has been generated (Lentz et al. 2010). The USH1 mice have a knock-in of the human *USH1C216G>A* mutation (Lentz et al. 2007). A 2′-MOE SSO targeted to the aberrant 5′-splice site in exon 3 (Fig. 9.2d) was administered to neonatal USH1 mice at PND3 by intraperitoneal (IP) bolus injection. This resulted in a low mount (2-5 %) of correct exon 3 spliced mRNA in the cochlea. However, correctly spliced mRNA was still present ~180 days after the injection. This finding mirrors the long duration of activity observed in the CNS and muscle discussed previously, and is perhaps due to decreased ASO metabolic activity or due to the post-mitotic state of the hair cells in the cochleae. Surprisingly, the level of harmonin in the cochlea was similar to the level in the cochlea of mice with one wild-type allele, indicating that harmonin was restored to ~50 % of the level of wild-type mice. The robust increase in harmonin, despite the low level of correctly spliced mRNA, can be explained if harmonin has a long half-life in the cochlea. The harmonin produced by SSO treatment was properly expressed, as it was localized to the tips of outer hair cell stereocilia bundles. In addition, the number of outer hair cells and hair cell morphology in the apical-mid regions of the cochlea were rescued (Lentz et al. 2013).

Since treatment with the SSO rescued hair cells, Lentz and colleagues determined if the hearing impairment of the USH1 mice was ameliorated by measuring their startle response and auditory evoked brain response (ABR). SSO-treated USH1 had a normal startle response to loud sounds and had ABR thresholds typical of mice with normal hearing. The mice could hear low- and mid- but not high-frequency sounds, which is in agreement with the observed rescue of outer hair cells in the apical-mid regions of the cochlea. Consistent with the long duration of splicing correction, treated mice could hear at 6 months of age. Furthermore, the treated USH1 mice did not show the characteristic circling behavior that is indicative of a vestibular dysfunction (Lentz et al. 2013).

In humans, as opposed to mice, the development of the ear and hearing is completed in utero (Hall 2000). Therefore, more work is required in animals to determine if in utero delivery of the SSO can correct splicing in the cochlea, and if the approach is safe. The analysis of vision in the USH1C mice was not performed in this study but it will be exciting to ascertain if vision can also be corrected by either peripheral or direct administration of the SSO to the eye.

7.5 Fukuyama Congenital Muscular Dystrophy

Fukuyama congenital muscular dystrophy (FCMD) is an infant-onset neuromuscular disease with clinical feature including generalized muscle weakness and wasting, cardiac abnormalities, joint contractures, visual impairment, and motor and neuropsychiatric dysfunction. The disease results in severe disability and premature death in childhood or adolescence due to aspiration pneumonia and heart complications (Fukuyama et al. 1981). The disease is inherited in an autosomal recessive manner and results from insertion of a SINE-VNTR-Alu (SVA) retrotransposon into the 3′-UTR of the fukutin (*FKTN*) gene (Toda and Kobayashi 1999). The retrotransposon carries a 3′-splice site that is used in conjunction with a 5′-splice site in the last exon, exon 10, which is otherwise inactive. The new splicing event results in the elimination of the FKTN C-terminal 38 amino acids and is replaced by the amino acids encoded by the retrotransposon sequence (Fig. 9.1e) (Taniguchi-Ikeda et al. 2011). FKTN is involved in the glycosylation of the α-dystroglycan (α-DG) protein (Michele et al. 2002). It is thought that the mutated FKTN is mislocalized to the endoplasmic reticulum so that it cannot participate in glycosylation (Taniguchi-Ikeda et al. 2011). Hypoglycosylated α-DG dystroglycan loses its ability to bind to extracellular matrix proteins such as laminin, agrin, and neurexin (Barresi and Campbell 2006).

SSOs have been tested in a mouse model of FCMD, which has one copy of the *Fktn* gene with a humanized exon 10 and the SVA insertion. First, a cocktail of 2′-OMe SSOs targeting the 3′-splice site of intron 10, and predicted intronic and exonic splicing enhancers in intron 10 and exon 11, respectively (Fig. 9.1e), transfected into patient-derived cells restored normal *FKTN* mRNA to greater than 50 % of the level in normal cells. In addition, FKTN protein, α-DG glycosylation, and its binding to laminin were significantly restored. IM injection or peripheral administration of a cocktail of the SSOs made as vPMOs to FCMD mice also increased normal *Fktn* mRNA in skeletal muscle, albeit to a lesser extent as observed in cultured cells. It was also encouraging to see increased FKTN protein, α-DG glycosylation, and binding to laminin (Taniguchi-Ikeda et al. 2011).

8 Conclusions and Outlook

The processing of pre-mRNA is a fundamental step in the expression of genes. Accordingly, it is highly regulated and this contributes in a significant way to added biological complexity. Therefore, it is not surprising that defects in pre-mRNA processing are often the culprits of human disease. Since the first therapeutic implementation of an ASO in 1978 by Zamecnik and Stevenson (1978), it took over 30 years for this technology to transform into a platform technology. This occurred as a result of gradual improvements in ASO medicinal chemistry and screening methods, in addition to acquiring a deeper understanding of pharmacokinetics, toxicology, and pharmacologic properties of ASO drugs in animals and humans. Due to the fact that ASOs target RNA, these have emerged as the ideal therapeutic agents for diseases that are caused by pre-mRNA processing defects. Clearly, ASOs have been used effectively to correct pre-mRNA processing defects and treat diseases in animal models, and in the near future we will learn if this is also true for human beings that have SMA.

Even though much progress has been made in the ASO field, there are still many questions that remain unanswered. We have a better appreciation of the factors that limit the activity of ASOs, such as their affinity or the structure of the target RNA, but these cannot be used a priori to predict where on a transcript an ASO should be targeted to for maximal activity. Hundreds to thousands of ASOs are routinely screened for activity in cell culture in order to find the ones that are most active. In this genomic revolution, where massive amounts of data from RNA structure to RNA/protein interactions and chromatin modifications are being generated for the entire transcriptome it may be possible to build more sophisticated computational tools to better predict ASO activity. In addition, there is also a large gap in our understanding of how ASOs distribute to tissues, are taken up by cells, and find their targets in a crowded environment. A more in-depth understanding of these steps may provide clues for how to further optimize ASOs to achieve greater activity and safety through medicinal chemistry, conjugation, or formulations.

Due to the central role of pre-mRNA processing in gene expression, it is likely that new disease mutations that cause processing defects will be discovered. Perhaps this has already occurred with the finding that a hexanucleotide (GGGGCC) repeat expansion in the first intron of chromosome 9 open reading frame 72 (*C9orf72*) is the most prevalent mutation in sporadic and familial amyotrophic lateral sclerosis (ALS) and frontotemporal dementia (FTD) (DeJesus-Hernandez et al. 2011; Renton et al. 2011). It is still not clear how this mutation causes ALS and FTD. However, neuronal cells from deceased patients showed accumulation of RNAs with the hexanucleotide repeat expansion in nuclear foci. It is tempting to speculate that, as is the case for DM1, perhaps RBPs are sequestered in the foci and this results in pre-mRNA processing defects. If this is indeed the case, then an ASO-based intervention may be a good therapeutic strategy to treat C9-ALS/FTD.

Even though there are a number of pre-mRNA processing diseases for which there is a sufficient number of patients to conduct robust clinical trials, this is not the case for many others. In fact, even within a given disease there are patients with unique mutations that cannot be treated with the therapy that is available for the most prevalent mutation. In the current framework of drug discovery and development it is very difficult to use ASOs for personalized medicine. Conceivably, future changes in the regulatory requirements will make it more feasible to develop tailored ASO-based therapies for patients who are in dire need.

Acknowledgements We apologize to those authors whose work could not be cited due to space limitations. We thank Tracy Reigle for the illustrations.

References

Aartsma-Rus A (2012) Overview on AON design. Methods Mol Biol 867:117–129

Allerson CR, Sioufi N, Jarres R, Prakash TP, Naik N, Berdeja A, Wanders L, Griffey RH, Swayze EE, Bhat B (2005) Fully 2′-modified oligonucleotide duplexes with improved in vitro potency and stability compared to unmodified small interfering RNA. J Med Chem 48:901–904

Allo M, Buggiano V, Fededa JP, Petrillo E, Schor I, de la Mata M, Agirre E, Plass M, Eyras E, Elela SA et al (2009) Control of alternative splicing through siRNA-mediated transcriptional gene silencing. Nat Struct Mol Biol 16:717–724

Anthony K, Gallo JM (2010) Aberrant RNA processing events in neurological disorders. Brain Res 1338:67–77

Araujo Ade Q, Araujo M, Swoboda KJ (2009) Vascular perfusion abnormalities in infants with spinal muscular atrophy. J Pediatr 155:292–294

Arora V, Knapp DC, Smith BL, Statdfield ML, Stein DA, Reddy MT, Weller DD, Iversen PL (2000) c-Myc antisense limits rat liver regeneration and indicates role for c-Myc in regulating cytochrome P-450 3A activity. J Pharmacol Exp Ther 292:921–928

Bachelin M, Hessler G, Kurz G, Hacia JG, Dervan PB, Kessler H (1998) Structure of a stereoregular phosphorothioate DNA/RNA duplex. Nat Struct Biol 5:271–276

Baker BF, Lot SS, Condon TP, Cheng-Flournoy S, Lesnik EA, Sasmor HM, Bennett CF (1997) 2′-O-(2-Methoxy)ethyl-modified anti-intercellular adhesion molecule 1 (ICAM-1) oligonucleotides selectively increase the ICAM-1 mRNA level and inhibit formation of the ICAM-1 translation initiation complex in human umbilical vein endothelial cells. J Biol Chem 272:11994–12000

Baker BF, Lot SS, Kringel J, Cheng-Flournoy S, Villiet P, Sasmor HM, Siwkowski AM, Chappell LL, Morrow JR (1999) Oligonucleotide-europium complex conjugate designed to cleave the 5′ cap structure of the ICAM-1 transcript potentiates antisense activity in cells. Nucleic Acids Res 27:1547–1551

Baker BF, Miraglia L, Hagedorn CH (1992) Modulation of eucaryotic initiation factor-4E binding to 5′-capped oligoribonucleotides by modified anti-sense oligonucleotides. J Biol Chem 267:11495–11499

Baltz AG, Munschauer M, Schwanhausser B, Vasile A, Murakawa Y, Schueler M, Youngs N, Penfold-Brown D, Drew K, Milek M et al (2012) The mRNA-bound proteome and its global occupancy profile on protein-coding transcripts. Mol Cell 46:674–690

Barresi R, Campbell KP (2006) Dystroglycan: from biosynthesis to pathogenesis of human disease. J Cell Sci 119:199–207

Bartel DP (2004) MicroRNAs: genomics, biogenesis, mechanism, and function. Cell 116: 281–297

Bartel DP (2009) MicroRNAs: target recognition and regulatory functions. Cell 136:215–233

Batey RT, Rambo RP, Doudna JA (1999) Tertiary Motifs in RNA Structure and Folding. Angewandte Chemie 38:2326–2343

Baughan TD, Dickson A, Osman EY, Lorson CL (2009) Delivery of bifunctional RNAs that target an intronic repressor and increase SMN levels in an animal model of spinal muscular atrophy. Hum Mol Genet 18(9):1600–1611

Bauman J, Jearawiriyapaisarn N, Kole R (2009) Therapeutic potential of splice-switching oligonucleotides. Oligonucleotides 19(1):1–13

Baumer D, Lee S, Nicholson G, Davies JL, Parkinson NJ, Murray LM, Gillingwater TH, Ansorge O, Davies KE, Talbot K (2009) Alternative splicing events are a late feature of pathology in a mouse model of spinal muscular atrophy. PLoS Genet 5:e1000773

Belostotsky D (2009) Exosome complex and pervasive transcription in eukaryotic genomes. Curr Opin Cell Biol 21:352–358

Bennett CF (2007) Pharmacological properties of 2′-O-methoxyethyl-modified oligonucleotides. In: Crooke ST (ed) Antisense drug technology: principles, strategies, and applications. CRC, Boca Raton, FL, pp 273–303

Bennett CF, Chiang MY, Chan H, Grimm S (1993) Use of cationic lipids to enhance the biological activity of antisense oligonucleotides. J Liposome Res 3:85–102

Bennett CF, Swayze EE (2010) RNA targeting therapeutics: molecular mechanisms of antisense oligonucleotides as a therapeutic platform. Annu Rev Pharmacol Toxicol 50:259–293

Bevan AK, Hutchinson KR, Foust KD, Braun L, McGovern VL, Schmelzer L, Ward JG, Petruska JC, Lucchesi PA, Burghes AH et al (2010) Early heart failure in the SMNDelta7 model of spinal muscular atrophy and correction by postnatal scAAV9-SMN delivery. Hum Mol Genet 19:3895–3905

Bitner-Glindzicz M, Lindley KJ, Rutland P, Blaydon D, Smith VV, Milla PJ, Hussain K, Furth-Lavi J, Cosgrove KE, Shepherd RM et al (2000) A recessive contiguous gene deletion causing

infantile hyperinsulinism, enteropathy and deafness identifies the Usher type 1C gene. Nat Genet 26:56–60
Black DL (2003) Mechanisms of alternative pre-messenger RNA splicing. Annu Rev Biochem 72:291–336
Braasch DA, Paroo Z, Constantinescu A, Ren G, Oz OK, Mason RP, Corey DR (2004) Biodistribution of phosphodiester and phosphorothioate siRNA. Bioorg Med Chem Lett 14:1139–1143
Brachet J (1941) La détection histochimique et le microdosage des acides pentosenucléiques (tissus animaux—développement embryonnaire des amphibiens). Enzymologia 10:87–96
Braunschweig U, Gueroussov S, Plocik AM, Graveley BR, Blencowe BJ (2013) Dynamic integration of splicing within gene regulatory pathways. Cell 152:1252–1269
Breaker RR, Joyce GF (1994) A DNA enzyme that cleaves RNA. Chem Biol 1:223–229
Brook JD, McCurrach ME, Harley HG, Buckler AJ, Church D, Aburatani H, Hunter K, Stanton VP, Thirion JP, Hudson T et al (1992) Molecular basis of myotonic dystrophy: expansion of a trinucleotide (CTG) repeat at the 3′ end of a transcript encoding a protein kinase family member. Cell 68:799–808
Brown-Driver V, Eto T, Lesnik E, Anderson KP, Hanecak RC (1999) Inhibition of translation of hepatitis C virus RNA by 2-modified antisense oligonucleotides. Antisense Nucleic Acid Drug Dev 9:145–154
Bumcrot D, Manoharan M, Koteliansky V, Sah DW (2006) RNAi therapeutics: a potential new class of pharmaceutical drugs. Nat Chem Biol 2:711–719
Burel SA, Han SR, Lee HS, Norris DA, Lee BS, Machemer T, Park SY, Zhou T, He G, Kim Y et al (2013) Preclinical evaluation of the toxicological effects of a novel constrained ethyl modified antisense compound targeting signal transducer and activator of transcription 3 in mice and cynomolgus monkeys. Nucleic Acid Ther 23:213–227
Burel SA, Machemer T, Ragone FL, Kato H, Cauntay P, Greenlee S, Salim A, Gaarde WA, Hung G, Peralta R et al (2012) Unique O-methoxyethyl ribose-DNA chimeric oligonucleotide induces an atypical melanoma differentiation-associated gene 5-dependent induction of type I interferon response. J Pharmacol Exp Ther 342:150–162
Burghes AH, Beattie CE (2009) Spinal muscular atrophy: why do low levels of survival motor neuron protein make motor neurons sick? Nat Rev Neurosci 10:597–609
Butler M, Hayes CS, Chappell A, Murray SF, Yaksh TL, Hua XY (2005) Spinal distribution and metabolism of 2′-O-(2-methoxyethyl)-modified oligonucleotides after intrathecal administration in rats. Neuroscience 131:705–715
Calandra S, Tarugi P, Bertolini S (2011) Altered mRNA splicing in lipoprotein disorders. Curr Opin Lipidol 22:93–99
Calarco JA, Zhen M, Blencowe BJ (2011) Networking in a global world: Establishing functional connections between neural splicing regulators and their target transcripts. RNA 17(5): 775–791
Carroll JB, Warby SC, Southwell AL, Doty CN, Greenlee S, Skotte N, Hung G, Bennett CF, Freier SM, Hayden MR (2011) Potent and selective antisense oligonucleotides targeting single-nucleotide polymorphisms in the Huntington disease gene/allele-specific silencing of mutant huntingtin. Mol Ther: J Am Soc Gene Therapy 19:2178–2185
Cartegni L, Krainer AR (2003) Correction of disease-associated exon skipping by synthetic exon-specific activators. Nat Struct Biol 10:120–125
Carthew RW, Sontheimer EJ (2009) Origins and mechanisms of miRNAs and siRNAs. Cell 136:642–655
Castanotto D, Rossi JJ (2009) The promises and pitfalls of RNA-interference-based therapeutics. Nature 457:426–433
Castello A, Fischer B, Eichelbaum K, Horos R, Beckmann BM, Strein C, Davey NE, Humphreys DT, Preiss T, Steinmetz LM et al (2012) Insights into RNA biology from an atlas of mammalian mRNA-binding proteins. Cell 149:1393–1406
Castello A, Fischer B, Hentze MW, Preiss T (2013) z. Trends Genet 29:318–327
Cech TR (1988) Ribozymes and their medical implications. JAMA 260:3030–3034

Cerritelli SM, Crouch RJ (2009) Ribonuclease H: the enzymes in eukaryotes. FEBS J 276: 1494–1505
Chastain M, Tinoco I Jr (1993) RNA structure as related to antisense drugs. In: Lebleu STCAB (ed) Antisense research and applications. CRC, Boca Raton, FL, pp 55–66
Chirboga C, Swoboda K, Darras B, Iannaccone S, Montes J, Allen H, Parad R, Johnson S, De Vivo D, Norris D et al (2013) Results of an open-label, escalating dose study to assess the safety, tolerability, and dose range finding of a single intrathecal dose of ISIS-SMNRx in patients with spinal muscular atrophy. In 65th American Academy of Neurology Annual Meeting, abstract S36002
Cho S, Dreyfuss G (2010) A degron created by SMN2 exon 7 skipping is a principal contributor to spinal muscular atrophy severity. Genes Dev 24:438–442
Choi WY, Giraldez AJ, Schier AF (2007) Target protectors reveal dampening and balancing of Nodal agonist and antagonist by miR-430. Science 318:271–274
Chu Y, Yue X, Younger ST, Janowski BA, Corey DR (2010) Involvement of argonaute proteins in gene silencing and activation by RNAs complementary to a non-coding transcript at the progesterone receptor promoter. Nucleic Acids Res 38:7736–7748
Cifuentes-Diaz C, Frugier T, Tiziano FD, Lacene E, Roblot N, Joshi V, Moreau MH, Melki J (2001) Deletion of murine SMN exon 7 directed to skeletal muscle leads to severe muscular dystrophy. J Cell Biol 152:1107–1114
Cirak S, Arechavala-Gomeza V, Guglieri M, Feng L, Torelli S, Anthony K, Abbs S, Garralda ME, Bourke J, Wells DJ et al (2011) Exon skipping and dystrophin restoration in patients with Duchenne muscular dystrophy after systemic phosphorodiamidate morpholino oligomer treatment: an open-label, phase 2, dose-escalation study. Lancet 378:595–605
Coady TH, Lorson CL (2010) Trans-splicing-mediated improvement in a severe mouse model of spinal muscular atrophy. J Neurosci 30:126–130
Cooper TA, Wan L, Dreyfuss G (2009) RNA and disease. Cell 136:777–793
Coovert DD, Le TT, McAndrew PE, Strasswimmer J, Crawford TO, Mendell JR, Coulson SE, Androphy EJ, Prior TW, Burghes AH (1997) The survival motor neuron protein in spinal muscular atrophy. Hum Mol Genet 6:1205–1214
Crawford TO, Pardo CA (1996) The neurobiology of childhood spinal muscular atrophy. Neurobiol Dis 3:97–110
Crooke ST, Graham MJ, Zuckerman JE, Brooks D, Conklin BS, Cummins LL, Greig MJ, Guinosso CJ, Kornbrust D, Manoharan M et al (1996) Pharmacokinetic properties of several novel oligonucleotide analogs in mice. J Pharmacol Exp Ther 277:923–937
Crooke ST, Vickers T, Lima W, Wu H (2007) Mechanisms of antisense drug action, an introduction. In: Crooke ST (ed) Antisense drug technology: principles, strategies, and applications. CRC, Boca Raton, FL, pp 3–46
Danckwardt S, Hentze MW, Kulozik AE (2008) 3′ end mRNA processing: molecular mechanisms and implications for health and disease. EMBO J 27:482–498
Davis BM, McCurrach ME, Taneja KL, Singer RH, Housman DE (1997) Expansion of a CUG trinucleotide repeat in the 3′ untranslated region of myotonic dystrophy protein kinase transcripts results in nuclear retention of transcripts. Proc Natl Acad Sci U S A 94:7388–7393
Davis S, Propp S, Freier SM, Jones LE, Serra MJ, Kinberger G, Bhat B, Swayze EE, Bennett CF, Esau C (2009) Potent inhibition of microRNA in vivo without degradation. Nucleic Acids Res 37:70–77
De Sandre-Giovannoli A, Bernard R, Cau P, Navarro C, Amiel J, Boccaccio I, Lyonnet S, Stewart CL, Munnich A, Le Merrer M et al (2003) Lamin a truncation in Hutchinson-Gilford progeria. Science 300:2055
Dechat T, Pfleghaar K, Sengupta K, Shimi T, Shumaker DK, Solimando L, Goldman RD (2008) Nuclear lamins: major factors in the structural organization and function of the nucleus and chromatin. Genes Dev 22:832–853
DeJesus-Hernandez M, Mackenzie IR, Boeve BF, Boxer AL, Baker M, Rutherford NJ, Nicholson AM, Finch NA, Flynn H, Adamson J et al (2011) Expanded GGGGCC hexanucleotide repeat in noncoding region of C9ORF72 causes chromosome 9p-linked FTD and ALS. Neuron 72:245–256

Devos SL, Miller TM (2013) Direct intraventricular delivery of drugs to the rodent central nervous system. J Vis Exp 75:e50326

Dickson A, Osman E, Lorson CL (2008) A negatively acting bifunctional RNA increases survival motor neuron both in vitro and in vivo. Hum Gene Ther 19:1307–1315

Djebali S, Davis CA, Merkel A, Dobin A, Lassmann T, Mortazavi A, Tanzer A, Lagarde J, Lin W, Schlesinger F et al (2012) Landscape of transcription in human cells. Nature 489:101–108

Doench JG, Petersen CP, Sharp PA (2003) siRNAs can function as miRNAs. Genes Dev 17:438–442

Dominguez E, Marais T, Chatauret N, Benkhelifa-Ziyyat S, Duque S, Ravassard P, Carcenac R, Astord S, de Moura AP, Voit T et al (2011) Intravenous scAAV9 delivery of a codon-optimized SMN1 sequence rescues SMA mice. Hum Mol Genet 20:681–693

Dominski Z, Kole R (1993) Restoration of correct splicing in thalassemic pre-mRNA by antisense oligonucleotides. Proc Natl Acad Sci U S A 90:8673–8677

Du H, Cline MS, Osborne RJ, Tuttle DL, Clark TA, Donohue JP, Hall MP, Shiue L, Swanson MS, Thornton CA et al (2010) Aberrant alternative splicing and extracellular matrix gene expression in mouse models of myotonic dystrophy. Nat Struct Mol Biol 17:187–193

Dunckley MG, Manoharan M, Villiet P, Eperon IC, Dickson G (1998) Modification of splicing in the dystrophin gene in cultured Mdx muscle cells by antisense oligoribonucleotides. Hum Mol Genet 7:1083–1090

Echeverria GV, Cooper TA (2012) RNA-binding proteins in microsatellite expansion disorders: mediators of RNA toxicity. Brain Res 1462:100–111

Ecker DJ (1993) Strategies for invasion of RNA secondary structure. In: Lebleu STCAB (ed) Antisense research and applications. CRC, Boca Raton, FL, pp 387–400

Eckstein F (2000) Phosphorothioate oligodeoxynucleotides: what is their origin and what is unique about them? Antisense Nucleic Acid Drug Dev 10:117–121

Egli M, Pallan PS, Allerson CR, Prakash TP, Berdeja A, Yu J, Lee S, Watt A, Gaus H, Bhat B et al (2011) Synthesis, improved antisense activity and structural rationale for the divergent RNA affinities of 3′-fluoro hexitol nucleic acid (FHNA and Ara-FHNA) modified oligonucleotides. J Am Chem Soc 133:16642–16649

El-Matary W, Kotagiri S, Cameron D, Peart I (2004) Spinal muscle atrophy type 1 (Werdnig-Hoffman disease) with complex cardiac malformation. Eur J Pediatr 163:331–332

Elbashir SM, Harborth J, Lendeckel W, Yalcin A, Weber K, Tuschl T (2001) Duplexes of 21-nucleotide RNAs mediate RNA interference in cultured mammalian cells. Nature 411:494–498

Elmen J, Lindow M, Schutz S, Lawrence M, Petri A, Obad S, Lindholm M, Hedtjarn M, Hansen HF, Berger U et al (2008) LNA-mediated microRNA silencing in non-human primates. Nature 452:896–899

Eriksson M, Brown WT, Gordon LB, Glynn MW, Singer J, Scott L, Erdos MR, Robbins CM, Moses TY, Berglund P et al (2003) Recurrent de novo point mutations in lamin A cause Hutchinson-Gilford progeria syndrome. Nature 423:293–298

Esau C, Davis S, Murray SF, Yu XX, Pandey SK, Pear M, Watts L, Booten SL, Graham M, McKay R et al (2006) miR-122 regulation of lipid metabolism revealed by in vivo antisense targeting. Cell Metab 3:87–98

Esau C, Kang X, Peralta E, Hanson E, Marcusson EG, Ravichandran LV, Sun Y, Koo S, Perera RJ, Jain R et al (2004) MicroRNA-143 regulates adipocyte differentiation. J Biol Chem 279:52361–52365

Faghihi MA, Kocerha J, Modarresi F, Engstrom PG, Chalk AM, Brothers SP, Koesema E, St Laurent G, Wahlestedt C (2010) RNAi screen indicates widespread biological function for human natural antisense transcripts. PLoS One 5

Faghihi MA, Wahlestedt C (2009) Regulatory roles of natural antisense transcripts. Nat Rev Mol Cell Biol 10:637–643

Fallini C, Bassell GJ, Rossoll W (2012) Spinal muscular atrophy: the role of SMN in axonal mRNA regulation. Brain Res 1462:81–92

Fardaei M, Larkin K, Brook JD, Hamshere MG (2001) In vivo co-localisation of MBNL protein with DMPK expanded-repeat transcripts. Nucleic Acids Res 29:2766–2771

Feldkotter M, Schwarzer V, Wirth R, Wienker TF, Wirth B (2002) Quantitative analyses of SMN1 and SMN2 based on real-time light Cycler PCR: fast and highly reliable carrier testing and prediction of severity of spinal muscular atrophy. Am J Hum Genet 70:358–368

Fiszer A, Mykowska A, Krzyzosiak WJ (2011) Inhibition of mutant huntingtin expression by RNA duplex targeting expanded CAG repeats. Nucleic Acids Res 39:5578–5585

Fong LG, Ng JK, Lammerding J, Vickers TA, Meta M, Cote N, Gavino B, Qiao X, Chang SY, Young SR et al (2006) Prelamin A and lamin A appear to be dispensable in the nuclear lamina. J Clin Invest 116:743–752

Foust KD, Wang X, McGovern VL, Braun L, Bevan AK, Haidet AM, Le TT, Morales PR, Rich MM, Burghes AH et al (2010) Rescue of the spinal muscular atrophy phenotype in a mouse model by early postnatal delivery of SMN. Nat Biotechnol 28:271–274

Francois V, Klein AF, Beley C, Jollet A, Lemercier C, Garcia L, Furling D (2011) Selective silencing of mutated mRNAs in DM1 by using modified hU7-snRNAs. Nat Struct Mol Biol 18:85–87

Freier SM, Lima WF, Sanghvi YS, Vickers T, Zounes M, Cook PD, Ecker DJ (1992) Thermodynamics of antisense oligonucleotide hybridization. In: Izant RPEAJG (ed) Gene regulation: biology of antisense RNA and DNA. Raven, New York, NY, pp 95–107

Fukuyama Y, Osawa M, Suzuki H (1981) Congenital progressive muscular dystrophy of the Fukuyama type—clinical, genetic and pathological considerations. Brain Dev 3:1–29

Gabanella F, Butchbach ME, Saieva L, Carissimi C, Burghes AH, Pellizzoni L (2007) Ribonucleoprotein assembly defects correlate with spinal muscular atrophy severity and preferentially affect a subset of spliceosomal snRNPs. PLoS One 2:e921

Gagnon KT, Corey DR (2012) Argonaute and the nuclear RNAs: new pathways for RNA-mediated control of gene expression. Nucleic Acid Ther 22:3–16

Gagnon KT, Pendergraff HM, Deleavey GF, Swayze EE, Potier P, Randolph J, Roesch EB, Chattopadhyaya J, Damha MJ, Bennett CF et al (2010) Allele-selective inhibition of mutant huntingtin expression with antisense oligonucleotides targeting the expanded CAG repeat. Biochemistry 49:10166–10178

Gao Z, Cooper TA (2013) Antisense oligonucleotides: rising stars in eliminating RNA toxicity in myotonic dystrophy. Hum Gene Ther 24:499–507

Gatchel JR, Zoghbi HY (2005) Diseases of unstable repeat expansion: mechanisms and common principles. Nat Rev Genet 6:743–755

Geary RS (2009) Antisense oligonucleotide pharmacokinetics and metabolism. Expert Opin Drug Metab Toxicol 5:381–391

Geary RS, Wancewicz E, Matson J, Pearce M, Siwkowski A, Swayze E, Bennett F (2009) Effect of dose and plasma concentration on liver uptake and pharmacologic activity of a 2′-methoxyethyl modified chimeric antisense oligonucleotide targeting PTEN. Biochem Pharmacol 78:284–291

Geary RS, Yu RZ, Siwkowski A, Levin AA (2007) Pharmacokinetic/pharmacodynamic properties of phosphorothioate 2′-O-(2-methoxyethyl)-modified antisense oligonucleotides in animals and man. In: Crooke ST (ed) Antisense drug technology: principles, strategies, and applications. CRC, Boca Raton, FL, pp 305–326

Geib T, Hertel KJ (2009) Restoration of full-length SMN promoted by adenoviral vectors expressing RNA antisense oligonucleotides embedded in U7 snRNAs. PLoS One 4:e8204

Ghildiyal M, Zamore PD (2009) Small silencing RNAs: an expanding universe. Nat Rev Genet 10:94–108

Gilleron J, Querbes W, Zeigerer A, Borodovsky A, Marsico G, Schubert U, Manygoats K, Seifert S, Andree C, Stoter M et al (2013) Image-based analysis of lipid nanoparticle-mediated siRNA delivery, intracellular trafficking and endosomal escape. Nat Biotechnol 31:638–646

Goemans NM, Tulinius M, van den Akker JT, Burm BE, Ekhart PF, Heuvelmans N, Holling T, Janson AA, Platenburg GJ, Sipkens JA et al (2011) Systemic administration of PRO051 in Duchenne's muscular dystrophy. N Engl J Med 364:1513–1522

Gogliotti RG, Hammond SM, Lutz C, Didonato CJ (2010) Molecular and phenotypic reassessment of an infrequently used mouse model for spinal muscular atrophy. Biochem Biophys Res Commun 391:517–522

Goraczniak R, Behlke MA, Gunderson SI (2009) Gene silencing by synthetic U1 adaptors. Nat Biotechnol 27:257–263
Guan L, Disney MD (2012) Recent advances in developing small molecules targeting RNA. ACS Chem Biol 7:73–86
Gunderson SI, Polycarpou-Schwarz M, Mattaj IW (1998) U1 snRNP inhibits pre-mRNA polyadenylation through a direct interaction between U1 70K and poly(A) polymerase. Mol Cell 1:255–264
Gutschner T, Hammerle M, Eissmann M, Hsu J, Kim Y, Hung G, Revenko A, Arun G, Stentrup M, Gross M et al (2013) The noncoding RNA MALAT1 is a critical regulator of the metastasis phenotype of lung cancer cells. Cancer Res 73:1180–1189
Guttman M, Rinn JL (2012) Modular regulatory principles of large non-coding RNAs. Nature 482:339–346
Guttman M, Russell P, Ingolia NT, Weissman JS, Lander ES (2013) Ribosome profiling provides evidence that large noncoding RNAs do not encode proteins. Cell 154:240–251
Hall JW 3rd (2000) Development of the ear and hearing. J Perinatol 20:S12–S20
Hamilton G, Gillingwater TH (2013) Spinal muscular atrophy: going beyond the motor neuron. Trends Mol Med 19:40–50
Han J, Kim D, Morris KV (2007) Promoter-associated RNA is required for RNA-directed transcriptional gene silencing in human cells. Proc Natl Acad Sci U S A 104:12422–12427
Hanecak R, Brown-Driver V, Fox MC, Azad RF, Furusako S, Nozaki C, Ford C, Sasmor H, Anderson KP (1996) Antisense oligonucleotide inhibition of hepatitis C virus gene expression in transformed hepatocytes. J Virol 70:5203–5212
Harper PS, Brook JD, Newman E (2001) Myotonic dystrophy. WB Saunders, London, UK
Havens MA, Duelli DM, Hastings ML (2013) Targeting RNA splicing for disease therapy. Wiley Interdiscip Rev RNA 4:247–266
He Y, Vogelstein B, Velculescu VE, Papadopoulos N, Kinzler KW (2008) The antisense transcriptomes of human cells. Science 322:1855–1857
Heier CR, Satta R, Lutz C, DiDonato CJ (2010) Arrhythmia and cardiac defects are a feature of spinal muscular atrophy model mice. Hum Mol Genet 19:3906–3918
Helene C, Toulme JJ (1990) Specific regulation of gene expression by antisense, sense and antigene nucleic acids. Biochim Biophys Acta 1049:99–125
Henry SP, Giclas PC, Leeds J, Pangburn M, Auletta C, Levin AA, Kornbrust DJ (1997) Activation of the alternative pathway of complement by a phosphorothioate oligonucleotide: potential mechanism of action. J Pharmacol Exp Ther 281:810–816
Henry SP, Kim T-W, Kramer-Strickland K, Zanardi TA, Fey RA, Levin AA (2007) Toxicologic properties of 2′O-methoxyethyl chimeric antisense inhibitors in animals and man. In: Crooke ST (ed) Antisense drug technology: principles, strategies and applications. CRC, Boca Raton, FL, pp 327–364
Hong DS, Younes A, Fayad L, Fowler NH, Hagemeister FB, Mistry R, Nemunaitis JJ, Borad MJ, Bryce AH, Yamashita M, et al (2013) A phase I study of ISIS 481464 (AZD9150), a first-in-human, first-in-class, antisense oligonucleotide inhibitor of STAT3, in patients with advanced cancers. In American Society of Clinical Oncology Annual Meeting, abstract 8523
Hsieh-Li HM, Chang JG, Jong YJ, Wu MH, Wang NM, Tsai CH, Li H (2000) A mouse model for spinal muscular atrophy. Nat Genet 24:66–70
Hu J, Liu J, Corey DR (2010) Allele-selective inhibition of huntingtin expression by switching to an miRNA-like RNAi mechanism. Chem Biol 17:1183–1188
Hu J, Matsui M, Gagnon KT, Schwartz JC, Gabillet S, Arar K, Wu J, Bezprozvanny I, Corey DR (2009) Allele-specific silencing of mutant huntingtin and ataxin-3 genes by targeting expanded CAG repeats in mRNAs. Nat Biotechnol 27:478–484
Hua Y, Sahashi K, Hung G, Rigo F, Passini MA, Bennett CF, Krainer AR (2010) Antisense correction of SMN2 splicing in the CNS rescues necrosis in a type III SMA mouse model. Genes Dev 24:1634–1644
Hua Y, Sahashi K, Rigo F, Hung G, Horev G, Bennett CF, Krainer AR (2011) Peripheral SMN restoration is essential for long-term rescue of a severe spinal muscular atrophy mouse model. Nature 478:123–126

Hua Y, Vickers TA, Baker BF, Bennett CF, Krainer AR (2007) Enhancement of SMN2 exon 7 inclusion by antisense oligonucleotides targeting the exon. PLoS Biol 5:e73

Hua Y, Vickers TA, Okunola HL, Bennett CF, Krainer AR (2008) Antisense masking of an hnRNP A1/A2 intronic splicing silencer corrects SMN2 splicing in transgenic mice. Am J Hum Genet 82:834–848

Huntzinger E, Izaurralde E (2011) Gene silencing by microRNAs: contributions of translational repression and mRNA decay. Nat Rev Genet 12:99–110

Hutvagner G, Zamore PD (2002) A microRNA in a multiple-turnover RNAi enzyme complex. Science 297:2056–2060

Ivanova G, Reigadas S, Ittig D, Arzumanov A, Andreola ML, Leumann C, Toulme JJ, Gait MJ (2007) Tricyclo-DNA containing oligonucleotides as steric block inhibitors of human immunodeficiency virus type 1 tat-dependent trans-activation and HIV-1 infectivity. Oligonucleotides 17:54–65

Iversen PL (2007) Morpholinos. In: Crooke ST (ed) Antisense drug technology: principles, strategies, and applications. CRC, Boca Raton, FL, pp 565–582

Jackson AL, Burchard J, Leake D, Reynolds A, Schelter J, Guo J, Johnson JM, Lim L, Karpilow J, Nichols K et al (2006) Position-specific chemical modification of siRNAs reduces "off-target" transcript silencing. RNA 12:1197–1205

Jackson AL, Linsley PS (2010) Recognizing and avoiding siRNA off-target effects for target identification and therapeutic application. Nat Rev Drug Discov 9:57–67

Janowski BA, Huffman KE, Schwartz JC, Ram R, Hardy D, Shames DS, Minna JD, Corey DR (2005) Inhibiting gene expression at transcription start sites in chromosomal DNA with antigene RNAs. Nat Chem Biol 1:216–222

Janowski BA, Younger ST, Hardy DB, Ram R, Huffman KE, Corey DR (2007) Activating gene expression in mammalian cells with promoter-targeted duplex RNAs. Nat Chem Biol 3:166–173

Janssen HL, Reesink HW, Lawitz EJ, Zeuzem S, Rodriguez-Torres M, Patel K, van der Meer AJ, Patick AK, Chen A, Zhou Y et al (2013) Treatment of HCV infection by targeting microRNA. N Engl J Med 368:1685–1694

Jayaraman M, Ansell SM, Mui BL, Tam YK, Chen J, Du X, Butler D, Eltepu L, Matsuda S, Narayanannair JK et al (2012) Maximizing the potency of siRNA lipid nanoparticles for hepatic gene silencing in vivo. Angewandte Chemie 51:8529–8533

Jiang H, Mankodi A, Swanson MS, Moxley RT, Thornton CA (2004) Myotonic dystrophy type 1 is associated with nuclear foci of mutant RNA, sequestration of muscleblind proteins and deregulated alternative splicing in neurons. Hum Mol Genet 13:3079–3088

Jurica MS, Moore MJ (2003) Pre-mRNA splicing: awash in a sea of proteins. Mol Cell 12:5–14

Kaida D, Berg MG, Younis I, Kasim M, Singh LN, Wan L, Dreyfuss G (2010) U1 snRNP protects pre-mRNAs from premature cleavage and polyadenylation. Nature 468:664–668

Kalsotra A, Cooper TA (2011) Functional consequences of developmentally regulated alternative splicing. Nat Rev Genet 12:715–729

Kanadia RN, Shin J, Yuan Y, Beattie SG, Wheeler TM, Thornton CA, Swanson MS (2006) Reversal of RNA missplicing and myotonia after muscleblind overexpression in a mouse poly(CUG) model for myotonic dystrophy. Proc Natl Acad Sci U S A 103:11748–11753

Kapeli K, Yeo GW (2012) Genome-wide approaches to dissect the roles of RNA binding proteins in translational control: implications for neurological diseases. Front Neurosci 6:144

Katayama S, Tomaru Y, Kasukawa T, Waki K, Nakanishi M, Nakamura M, Nishida H, Yap CC, Suzuki M, Kawai J et al (2005) Antisense transcription in the mammalian transcriptome. Science 309:1564–1566

Kim DH, Saetrom P, Snove O Jr, Rossi JJ (2008) MicroRNA-directed transcriptional gene silencing in mammalian cells. Proc Natl Acad Sci U S A 105:16230–16235

Kim DH, Villeneuve LM, Morris KV, Rossi JJ (2006) Argonaute-1 directs siRNA-mediated transcriptional gene silencing in human cells. Nat Struct Mol Biol 13:793–797

Kim HJ, Kim NC, Wang YD, Scarborough EA, Moore J, Diaz Z, MacLea KS, Freibaum B, Li S, Molliex A et al (2013) Mutations in prion-like domains in hnRNPA2B1 and hnRNPA1 cause multisystem proteinopathy and ALS. Nature 495:467–473

Kinali M, Arechavala-Gomeza V, Feng L, Cirak S, Hunt D, Adkin C, Guglieri M, Ashton E, Abbs S, Nihoyannopoulos P et al (2009) Local restoration of dystrophin expression with the morpholino oligomer AVI-4658 in Duchenne muscular dystrophy: a single-blind, placebo-controlled, dose-escalation, proof-of-concept study. Lancet Neurol 8:918–928

Kislauskis EH, Zhu X, Singer RH (1994) Sequences responsible for intracellular localization of beta-actin messenger RNA also affect cell phenotype. J Cell Biol 127:441–451

Kleinman ME, Yamada K, Takeda A, Chandrasekaran V, Nozaki M, Baffi JZ, Albuquerque RJ, Yamasaki S, Itaya M, Pan Y et al (2008) Sequence- and target-independent angiogenesis suppression by siRNA via TLR3. Nature 452:591–597

Kole R, Krainer AR, Altman S (2012) RNA therapeutics: beyond RNA interference and antisense oligonucleotides. Nat Rev Drug Discov 11:125–140

Koller E, Vincent TM, Chappell A, De S, Manoharan M, Bennett CF (2011) Mechanisms of single-stranded phosphorothioate modified antisense oligonucleotide accumulation in hepatocytes. Nucleic Acids Res 39:4795–4807

Koo T, Wood MJ (2013) Clinical trials using antisense oligonucleotides in Duchenne muscular dystrophy. Hum Gene Ther 24:479–488

Kordasiewicz HB, Stanek LM, Wancewicz EV, Mazur C, McAlonis MM, Pytel KA, Artates JW, Weiss A, Cheng SH, Shihabuddin LS et al (2012) Sustained therapeutic reversal of Huntington's disease by transient repression of huntingtin synthesis. Neuron 74:1031–1044

Kornblihtt AR, de la Mata M, Fededa JP, Munoz MJ, Nogues G (2004) Multiple links between transcription and splicing. RNA 10:1489–1498

Koval ED, Shaner C, Zhang P, du Maine X, Fischer K, Tay J, Chau BN, Wu GF, Miller TM (2013) Method for widespread microRNA-155 inhibition prolongs survival in ALS-model mice. Hum Mol Genet 22(20):4127–4135

Krawczak M, Reiss J, Cooper DN (1992) The mutational spectrum of single base-pair substitutions in mRNA splice junctions of human genes: causes and consequences. Hum Genet 90:41–54

Kremer H, van Wijk E, Marker T, Wolfrum U, Roepman R (2006) Usher syndrome: molecular links of pathogenesis, proteins and pathways. Hum Mol Genet 15(Spec No 2):R262–R270

Krieg AM (2008) Toll-like receptor 9 (TLR9) agonists in the treatment of cancer. Oncogene 27:161–167

Krol J, Fiszer A, Mykowska A, Sobczak K, de Mezer M, Krzyzosiak WJ (2007) Ribonuclease dicer cleaves triplet repeat hairpins into shorter repeats that silence specific targets. Mol Cell 25:575–586

Krutzfeldt J, Rajewsky N, Braich R, Rajeev KG, Tuschl T, Manoharan M, Stoffel M (2005) Silencing of microRNAs in vivo with 'antagomirs'. Nature 438:685–689

Kuyumcu-Martinez NM, Wang GS, Cooper TA (2007) Increased steady-state levels of CUGBP1 in myotonic dystrophy 1 are due to PKC-mediated hyperphosphorylation. Mol Cell 28:68–78

Kwoh TJ (2007) An overview of the clinical safety experience of first- and second-generation antisense oligonucleotides. In: Crooke ST (ed) Antisense drug technology: principles, strategies, and applications. CRC, Boca Raton, FL, pp 365–399

Lam MT, Cho H, Lesch HP, Gosselin D, Heinz S, Tanaka-Oishi Y, Benner C, Kaikkonen MU, Kim AS, Kosaka M et al (2013) Rev-Erbs repress macrophage gene expression by inhibiting enhancer-directed transcription. Nature 498:511–515

Langlois MA, Boniface C, Wang G, Alluin J, Salvaterra PM, Puymirat J, Rossi JJ, Lee NS (2005) Cytoplasmic and nuclear retained DMPK mRNAs are targets for RNA interference in myotonic dystrophy cells. J Biol Chem 280:16949–16954

Le TT, Pham LT, Butchbach ME, Zhang HL, Monani UR, Coovert DD, Gavrilina TO, Xing L, Bassell GJ, Burghes AH (2005) SMNDelta7, the major product of the centromeric survival motor neuron (SMN2) gene, extends survival in mice with spinal muscular atrophy and associates with full-length SMN. Hum Mol Genet 14:845–857

Lee JE, Bennett CF, Cooper TA (2012) RNase H-mediated degradation of toxic RNA in myotonic dystrophy type 1. Proc Natl Acad Sci U S A 109:4221–4226

Lee JT (2012) Epigenetic regulation by long noncoding RNAs. Science 338:1435–1439

Lee RG, Crosby J, Baker BF, Graham MJ, Crooke RM (2013) Antisense technology: an emerging platform for cardiovascular disease therapeutics. J Cardiovasc Transl Res 6(6):969–980

Lefave CV, Squatrito M, Vorlova S, Rocco GL, Brennan CW, Holland EC, Pan YX, Cartegni L (2011) Splicing factor hnRNPH drives an oncogenic splicing switch in gliomas. EMBO J 30(19):4084–4097

Lefebvre S, Burglen L, Reboullet S, Clermont O, Burlet P, Viollet L, Benichou B, Cruaud C, Millasseau P, Zeviani M et al (1995) Identification and characterization of a spinal muscular atrophy-determining gene. Cell 80:155–165

Lefebvre S, Burlet P, Liu Q, Bertrandy S, Clermont O, Munnich A, Dreyfuss G, Melki J (1997) Correlation between severity and SMN protein level in spinal muscular atrophy. Nat Genet 16:265–269

Leger AJ, Mosquea LM, Clayton NP, Wu IH, Weeden T, Nelson CA, Phillips L, Roberts E, Piepenhagen PA, Cheng SH et al (2013) Systemic delivery of a peptide-linked morpholino oligonucleotide neutralizes mutant RNA toxicity in a mouse model of myotonic dystrophy. Nucleic Acid Ther 23:109–117

Lentz J, Pan F, Ng SS, Deininger P, Keats B (2007) Ush1c216A knock-in mouse survives Katrina. Mutat Res 616:139–144

Lentz J, Savas S, Ng SS, Athas G, Deininger P, Keats B (2005) The USH1C 216G–>A splice-site mutation results in a 35-base-pair deletion. Hum Genet 116:225–227

Lentz JJ, Gordon WC, Farris HE, MacDonald GH, Cunningham DE, Robbins CA, Tempel BL, Bazan NG, Rubel EW, Oesterle EC et al (2010) Deafness and retinal degeneration in a novel USH1C knock-in mouse model. Dev Neurobiol 70:253–267

Lentz JJ, Jodelka FM, Hinrich AJ, McCaffrey KE, Farris HE, Spalitta MJ, Bazan NG, Duelli DM, Rigo F, Hastings ML (2013) Rescue of hearing and vestibular function by antisense oligonucleotides in a mouse model of human deafness. Nat Med 19:345–350

Levin AA, Yu RZ, Geary RS (2007) Basic principles of the pharmacokinetics of antisense oligonucleotide drugs. In: Crooke ST (ed) Antisense drug technology: principles, strategies, and applications. CRC, Boca Raton, FL, pp 183–215

Li LC, Okino ST, Zhao H, Pookot D, Place RF, Urakami S, Enokida H, Dahiya R (2006) Small dsRNAs induce transcriptional activation in human cells. Proc Natl Acad Sci U S A 103:17337–17342

Liang XH, Vickers TA, Guo S, Crooke ST (2011) Efficient and specific knockdown of small non-coding RNAs in mammalian cells and in mice. Nucleic Acids Res 39:e13

Licatalosi DD, Darnell RB (2010) RNA processing and its regulation: global insights into biological networks. Nat Rev Genet 11:75–87

Lightfoot HL, Hall J (2012) Target mRNA inhibition by oligonucleotide drugs in man. Nucleic Acids Res 40:10585–10595

Lim KH, Ferraris L, Filloux ME, Raphael BJ, Fairbrother WG (2011) Using positional distribution to identify splicing elements and predict pre-mRNA processing defects in human genes. Proc Natl Acad Sci U S A 108:11093–11098

Lim SR, Hertel KJ (2001) Modulation of survival motor neuron pre-mRNA splicing by inhibition of alternative 3′ splice site pairing. J Biol Chem 276:45476–45483

Lima W, Wu H, Crooke ST (2007) The RNase H mechanism. In: Crooke ST (ed) Antisense drug technology: principles, strategies, and applications. CRC, Boca Raton, FL, pp 47–74

Lima WF, Monia BP, Ecker DJ, Freier SM (1992) Implication of RNA structure on antisense oligonucleotide hybridization kinetics. Biochemistry 31:12055–12061

Lima WF, Prakash TP, Murray HM, Kinberger GA, Li W, Chappell AE, Li CS, Murray SF, Gaus H, Seth PP et al (2012) Single-stranded siRNAs activate RNAi in animals. Cell 150:883–894

Lin F, Worman HJ (1993) Structural organization of the human gene encoding nuclear lamin A and nuclear lamin C. J Biol Chem 268:16321–16326

Lin X, Miller JW, Mankodi A, Kanadia RN, Yuan Y, Moxley RT, Swanson MS, Thornton CA (2006) Failure of MBNL1-dependent post-natal splicing transitions in myotonic dystrophy. Hum Mol Genet 15:2087–2097

Liu J, Hu J, Corey DR (2011) Expanding the action of duplex RNAs into the nucleus: redirecting alternative splicing. Nucleic Acids Res 40(3):1240–1250

Lorenz P, Misteli T, Baker BF, Bennett CF, Spector DL (2000) Nucleocytoplasmic shuttling: a novel in vivo property of antisense phosphorothioate oligodeoxynucleotides. Nucleic Acids Res 28:582–592

Lorson CL, Androphy EJ (2000) An exonic enhancer is required for inclusion of an essential exon in the SMA-determining gene SMN. Hum Mol Genet 9:259–265

Lorson CL, Hahnen E, Androphy EJ, Wirth B (1999) A single nucleotide in the SMN gene regulates splicing and is responsible for spinal muscular atrophy. Proc Natl Acad Sci U S A 96:6307–6311

Lunn MR, Wang CH (2008) Spinal muscular atrophy. Lancet 371:2120–2133

Mahadevan M, Tsilfidis C, Sabourin L, Shutler G, Amemiya C, Jansen G, Neville C, Narang M, Barcelo J, O'Hoy K et al (1992) Myotonic dystrophy mutation: an unstable CTG repeat in the 3′ untranslated region of the gene. Science 255:1253–1255

Maniatis T, Reed R (2002) An extensive network of coupling among gene expression machines. Nature 416:499–506

Mankodi A, Logigian E, Callahan L, McClain C, White R, Henderson D, Krym M, Thornton CA (2000) Myotonic dystrophy in transgenic mice expressing an expanded CUG repeat. Science 289:1769–1773

Mankodi A, Takahashi MP, Jiang H, Beck CL, Bowers WJ, Moxley RT, Cannon SC, Thornton CA (2002) Expanded CUG repeats trigger aberrant splicing of ClC-1 chloride channel pre-mRNA and hyperexcitability of skeletal muscle in myotonic dystrophy. Mol Cell 10:35–44

Mankodi A, Urbinati CR, Yuan QP, Moxley RT, Sansone V, Krym M, Henderson D, Schalling M, Swanson MS, Thornton CA (2001) Muscleblind localizes to nuclear foci of aberrant RNA in myotonic dystrophy types 1 and 2. Hum Mol Genet 10:2165–2170

Mann CJ, Honeyman K, Cheng AJ, Ly T, Lloyd F, Fletcher S, Morgan JE, Partridge TA, Wilton SD (2001) Antisense-induced exon skipping and synthesis of dystrophin in the mdx mouse. Proc Natl Acad Sci U S A 98:42–47

Manoharan M, Akinc A, Pandey RK, Qin J, Hadwiger P, John M, Mills K, Charisse K, Maier MA, Nechev L et al (2011) Unique gene-silencing and structural properties of 2′-fluoro-modified siRNAs. Angewandte Chemie 50:2284–2288

Marques JT, Williams BR (2005) Activation of the mammalian immune system by siRNAs. Nat Biotechnol 23:1399–1405

Martin KC, Ephrussi A (2009) mRNA localization: gene expression in the spatial dimension. Cell 136:719–730

Mattick JS (2009) The genetic signatures of noncoding RNAs. PLoS Genet 5:e1000459

Matveeva O, Nechipurenko Y, Rossi L, Moore B, Saetrom P, Ogurtsov AY, Atkins JF, Shabalina SA (2007) Comparison of approaches for rational siRNA design leading to a new efficient and transparent method. Nucleic Acids Res 35:e63

Matveeva OV, Shabalina SA, Nemtsov VA, Tsodikov AD, Gesteland RF, Atkins JF (2003) Thermodynamic calculations and statistical correlations for oligo-probes design. Nucleic Acids Res 31:4211–4217

McKay RA, Miraglia LJ, Cummins LL, Owens SR, Sasmor H, Dean NM (1999) Characterization of a potent and specific class of antisense oligonucleotide inhibitor of human protein kinase C-alpha expression. J Biol Chem 274:1715–1722

McMahon BM, Mays D, Lipsky J, Stewart JA, Fauq A, Richelson E (2002) Pharmacokinetics and tissue distribution of a peptide nucleic acid after intravenous administration. Antisense Nucleic Acid Drug Dev 12:65–70

McManus CJ, Graveley BR (2011) RNA structure and the mechanisms of alternative splicing. Curr Opin Genet Dev 21:373–379

McSwiggen J (2013) Specific gene activation by disruption of PRC2-lncRNA interactions. In The Eighteenth Annual Meeting of the RNA Society, abstract 130

Meister G, Landthaler M, Dorsett Y, Tuschl T (2004) Sequence-specific inhibition of microRNA- and siRNA-induced RNA silencing. RNA 10:544–550
Meister G, Tuschl T (2004) Mechanisms of gene silencing by double-stranded RNA. Nature 431:343–349
Merideth MA, Gordon LB, Clauss S, Sachdev V, Smith AC, Perry MB, Brewer CC, Zalewski C, Kim HJ, Solomon B et al (2008) Phenotype and course of Hutchinson-Gilford progeria syndrome. N Engl J Med 358:592–604
Meyer K, Marquis J, Trub J, Nlend Nlend R, Verp S, Ruepp MD, Imboden H, Barde I, Trono D, Schumperli D (2009) Rescue of a severe mouse model for spinal muscular atrophy by U7 snRNA-mediated splicing modulation. Hum Mol Genet 18:546–555
Meyer KD, Saletore Y, Zumbo P, Elemento O, Mason CE, Jaffrey SR (2012) Comprehensive analysis of mRNA methylation reveals enrichment in 3′ UTRs and near stop codons. Cell 149:1635–1646
Michele DE, Barresi R, Kanagawa M, Saito F, Cohn RD, Satz JS, Dollar J, Nishino I, Kelley RI, Somer H et al (2002) Post-translational disruption of dystroglycan-ligand interactions in congenital muscular dystrophies. Nature 418:417–422
Miller JW, Urbinati CR, Teng-Umnuay P, Stenberg MG, Byrne BJ, Thornton CA, Swanson MS (2000) Recruitment of human muscleblind proteins to (CUG)(n) expansions associated with myotonic dystrophy. EMBO J 19:4439–4448
Ming X, Carver K, Fisher M, Noel R, Cintrat JC, Gillet D, Barbier J, Cao C, Bauman J, Juliano RL (2013) The small molecule Retro-1 enhances the pharmacological actions of antisense and splice switching oligonucleotides. Nucleic Acids Res 41:3673–3687
Miraglia L, Watt AT, Graham MJ, Crooke ST (2000) Variations in mRNA content have no effect on the potency of antisense oligonucleotides. Antisense Nucleic Acid Drug Dev 10:453–461
Mitrpant C, Porensky P, Zhou H, Price L, Muntoni F, Fletcher S, Wilton SD, Burghes AH (2013) Improved antisense oligonucleotide design to suppress aberrant SMN2 gene transcript processing: towards a treatment for spinal muscular atrophy. PLoS One 8:e62114
Miyajima H, Miyaso H, Okumura M, Kurisu J, Imaizumi K (2002) Identification of a cis-acting element for the regulation of SMN exon 7 splicing. J Biol Chem 277:23271–23277
Mizrahi RA, Schirle NT, Beal PA (2013) Potent and selective inhibition of A-to-I RNA editing with 2′-O-methyl/locked nucleic acid-containing antisense oligoribonucleotides. ACS Chem Biol 8:832–839
Modarresi F, Faghihi MA, Lopez-Toledano MA, Fatemi RP, Magistri M, Brothers SP, van der Brug MP, Wahlestedt C (2012) Inhibition of natural antisense transcripts in vivo results in gene-specific transcriptional upregulation. Nat Biotechnol 30:453–459
Monani UR, Lorson CL, Parsons DW, Prior TW, Androphy EJ, Burghes AH, McPherson JD (1999) A single nucleotide difference that alters splicing patterns distinguishes the SMA gene SMN1 from the copy gene SMN2. Hum Mol Genet 8:1177–1183
Monani UR, Sendtner M, Coovert DD, Parsons DW, Andreassi C, Le TT, Jablonka S, Schrank B, Rossoll W, Prior TW et al (2000) The human centromeric survival motor neuron gene (SMN2) rescues embryonic lethality in Smn(−/−) mice and results in a mouse with spinal muscular atrophy. Hum Mol Genet 9:333–339
Monia BP, Johnston JF, Ecker DJ, Zounes MA, Lima WF, Freier SM (1992) Selective inhibition of mutant Ha-ras mRNA expression by antisense oligonucleotides. J Biol Chem 267:19954–19962
Monia BP, Lesnik EA, Gonzalez C, Lima WF, McGee D, Guinosso CJ, Kawasaki AM, Cook PD, Freier SM (1993) Evaluation of 2'-modified oligonucleotides containing 2′-deoxy gaps as antisense inhibitors of gene expression. J Biol Chem 268:14514–14522
Mordes D, Luo X, Kar A, Kuo D, Xu L, Fushimi K, Yu G, Sternberg P Jr, Wu JY (2006) Pre-mRNA splicing and retinitis pigmentosa. Mol Vis 12:1259–1271
Morris KV, Chan SW, Jacobsen SE, Looney DJ (2004) Small interfering RNA-induced transcriptional gene silencing in human cells. Science 305:1289–1292

Morris KV, Santoso S, Turner AM, Pastori C, Hawkins PG (2008) Bidirectional transcription directs both transcriptional gene activation and suppression in human cells. PLoS Genet 4:e1000258
Morrow JR, Iranzo O (2004) Synthetic metallonucleases for RNA cleavage. Curr Opin Chem Biol 8:192–200
Moulton HM, Moulton JD (2010) Morpholinos and their peptide conjugates: therapeutic promise and challenge for Duchenne muscular dystrophy. Biochim Biophys Acta 1798:2296–2303
Moulton JD, Jiang S (2009) Gene knockdowns in adult animals: PPMOs and vivo-morpholinos. Molecules 14:1304–1323
Mulders SA, van den Broek WJ, Wheeler TM, Croes HJ, van Kuik-Romeijn P, de Kimpe SJ, Furling D, Platenburg GJ, Gourdon G, Thornton CA et al (2009) Triplet-repeat oligonucleotide-mediated reversal of RNA toxicity in myotonic dystrophy. Proc Natl Acad Sci U S A 106:13915–13920
Muntoni F, Wood MJ (2011) Targeting RNA to treat neuromuscular disease. Nat Rev Drug Discov 10:621–637
Murray S, Ittig D, Koller E, Berdeja A, Chappell A, Prakash TP, Norrbom M, Swayze EE, Leumann CJ, Seth PP (2012) TricycloDNA-modified oligo-2′-deoxyribonucleotides reduce scavenger receptor B1 mRNA in hepatic and extra-hepatic tissues–a comparative study of oligonucleotide length, design and chemistry. Nucleic Acids Res 40:6135–6143
Nakamori M, Gourdon G, Thornton CA (2011) Stabilization of expanded (CTG)*(CAG) repeats by antisense oligonucleotides. Mol Ther: J Am Soc Gene Ther 19:2222–2227
Napoli S, Pastori C, Magistri M, Carbone GM, Catapano CV (2009) Promoter-specific transcriptional interference and c-myc gene silencing by siRNAs in human cells. EMBO J 28:1708–1719
Nilsen TW, Graveley BR (2010) Expansion of the eukaryotic proteome by alternative splicing. Nature 463:457–463
Obad S, dos Santos CO, Petri A, Heidenblad M, Broom O, Ruse C, Fu C, Lindow M, Stenvang J, Straarup EM et al (2011) Silencing of microRNA families by seed-targeting tiny LNAs. Nat Genet 43:371–378
Orban TI, Izaurralde E (2005) Decay of mRNAs targeted by RISC requires XRN1, the Ski complex, and the exosome. RNA 11:459–469
Orengo JP, Chambon P, Metzger D, Mosier DR, Snipes GJ, Cooper TA (2008) Expanded CTG repeats within the DMPK 3′ UTR causes severe skeletal muscle wasting in an inducible mouse model for myotonic dystrophy. Proc Natl Acad Sci U S A 105:2646–2651
Osborne RJ, Lin X, Welle S, Sobczak K, O'Rourke JR, Swanson MS, Thornton CA (2009) Transcriptional and post-transcriptional impact of toxic RNA in myotonic dystrophy. Hum Mol Genet 18:1471–1481
Osman EY, Yen PF, Lorson CL (2012) Bifunctional RNAs targeting the intronic splicing silencer N1 increase SMN levels and reduce disease severity in an animal model of spinal muscular atrophy. Mol Ther: J Am Soc Gene Ther 20:119–126
Osorio FG, Navarro CL, Cadinanos J, Lopez-Mejia IC, Quiros PM, Bartoli C, Rivera J, Tazi J, Guzman G, Varela I et al (2011) Splicing-directed therapy in a new mouse model of human accelerated aging. Sci Transl Med 3:106ra107
Ouyang XM, Yan D, Du LL, Hejtmancik JF, Jacobson SG, Nance WE, Li AR, Angeli S, Kaiser M, Newton V et al (2005) Characterization of Usher syndrome type I gene mutations in an Usher syndrome patient population. Hum Genet 116:292–299
Overington JP, Al-Lazikani B, Hopkins AL (2006) How many drug targets are there? Nat Rev Drug Discov 5:993–996
Padgett RA (2012) New connections between splicing and human disease. Trends Genet 28:147–154
Pallan PS, Greene EM, Jicman PA, Pandey RK, Manoharan M, Rozners E, Egli M (2011) Unexpected origins of the enhanced pairing affinity of 2′-fluoro-modified RNA. Nucleic Acids Res 39:3482–3495
Passini MA, Bu J, Richards AM, Kinnecom C, Sardi SP, Stanek LM, Hua Y, Rigo F, Matson J, Hung G et al (2011) Antisense oligonucleotides delivered to the mouse CNS ameliorate symptoms of severe spinal muscular atrophy. Sci Transl Med 3:72ra18

Patel DJ, Ma JB, Yuan YR, Ye K, Pei Y, Kuryavyi V, Malinina L, Meister G, Tuschl T (2006) Structural biology of RNA silencing and its functional implications. Cold Spring Harb Symp Quant Biol 71:81–93

Peacey E, Rodriguez L, Liu Y, Wolfe MS (2012) Targeting a pre-mRNA structure with bipartite antisense molecules modulates tau alternative splicing. Nucleic Acids Res 40:9836–9849

Peng CG, Damha MJ (2008) Probing DNA polymerase activity with stereoisomeric 2′-fluoro-β-D-arabinose (2′F-araNTPs) and 2′-fluoro-β-D-ribose (2′F-rNTPs) nucleoside 5′-triphosphates. Can J Chem 86:881–891

Peng Z, Cheng Y, Tan BC, Kang L, Tian Z, Zhu Y, Zhang W, Liang Y, Hu X, Tan X et al (2012) Comprehensive analysis of RNA-Seq data reveals extensive RNA editing in a human transcriptome. Nat Biotechnol 30:253–260

Perales R, Bentley D (2009) "Cotranscriptionality": the transcription elongation complex as a nexus for nuclear transactions. Mol Cell 36:178–191

Petit C (2001) Usher syndrome: from genetics to pathogenesis. Annu Rev Genomics Hum Genet 2:271–297

Place RF, Li LC, Pookot D, Noonan EJ, Dahiya R (2008) MicroRNA-373 induces expression of genes with complementary promoter sequences. Proc Natl Acad Sci U S A 105:1608–1613

Polymenidou M, Lagier-Tourenne C, Hutt KR, Bennett CF, Cleveland DW, Yeo GW (2012) Misregulated RNA processing in amyotrophic lateral sclerosis. Brain Res 1462:3–15

Porensky PN, Mitrpant C, McGovern VL, Bevan AK, Foust KD, Kaspar BK, Wilton SD, Burghes AH (2011) A single administration of morpholino antisense oligomer rescues spinal muscular atrophy in mouse. Hum Mol Genet 21(7):1625–1638

Prakash TP (2011) An overview of sugar-modified oligonucleotides for antisense therapeutics. Chem Biodivers 8:1616–1641

Prasanth KV, Prasanth SG, Xuan Z, Hearn S, Freier SM, Bennett CF, Zhang MQ, Spector DL (2005) Regulating gene expression through RNA nuclear retention. Cell 123:249–263

Preker P, Nielsen J, Schierup MH, Jensen TH (2009) RNA polymerase plays both sides: vivid and bidirectional transcription around and upstream of active promoters. Cell Cycle 8:1106–1107

Prior TW, Krainer AR, Hua Y, Swoboda KJ, Snyder PC, Bridgeman SJ, Burghes AH, Kissel JT (2009) A positive modifier of spinal muscular atrophy in the SMN2 gene. Am J Hum Genet 85:408–413

Prior TW, Swoboda KJ, Scott HD, Hejmanowski AQ (2004) Homozygous SMN1 deletions in unaffected family members and modification of the phenotype by SMN2. Am J Med Genet A 130A:307–310

Proudfoot NJ (2011) Ending the message: poly(A) signals then and now. Genes Dev 25:1770–1782

Qureshi IA, Mattick JS, Mehler MF (2010) Long non-coding RNAs in nervous system function and disease. Brain Res 1338:20–35

Rajeev KG, Zimmermann T, Manoharan M, Maier M, Kuchimanchi S, Charisse K (2013) Small interfering RNAs targeting transthyretin mRNA for treatment of transthyretin-associated diseases (Alnylam Pharmaceuticals, USA), WO2013075035A1

Ranum LP, Cooper TA (2006) RNA-mediated neuromuscular disorders. Annu Rev Neurosci 29:259–277

Ray D, Kazan H, Cook KB, Weirauch MT, Najafabadi HS, Li X, Gueroussov S, Albu M, Zheng H, Yang A et al (2013) A compendium of RNA-binding motifs for decoding gene regulation. Nature 499:172–177

Renton AE, Majounie E, Waite A, Simon-Sanchez J, Rollinson S, Gibbs JR, Schymick JC, Laaksovirta H, van Swieten JC, Myllykangas L et al (2011) A hexanucleotide repeat expansion in C9ORF72 is the cause of chromosome 9p21-linked ALS-FTD. Neuron 72:257–268

Rettig GR, Behlke MA (2012) Progress toward in vivo use of siRNAs-II. Mol Ther: J Am Soc Gene Ther 20:483–512

Rigo F, Hua Y, Chun SJ, Prakash TP, Krainer AR, Bennett CF (2012a) Synthetic oligonucleotides recruit ILF2/3 to RNA transcripts to modulate splicing. Nat Chem Biol 8:555–561

Rigo F, Hua Y, Krainer AR, Bennett CF (2012b) Antisense-based therapy for the treatment of spinal muscular atrophy. J Cell Biol 199:21–25

Riguet E, Tripathi S, Chaubey B, Desire J, Pandey VN, Decout JL (2004) A peptide nucleic acid-neamine conjugate that targets and cleaves HIV-1 TAR RNA inhibits viral replication. J Med Chem 47:4806–4809
Rinn JL, Chang HY (2012) Genome regulation by long noncoding RNAs. Annu Rev Biochem 81:145–166
Rochette CF, Gilbert N, Simard LR (2001) SMN gene duplication and the emergence of the SMN2 gene occurred in distinct hominids: SMN2 is unique to Homo sapiens. Hum Genet 108:255–266
Sahay G, Querbes W, Alabi C, Eltoukhy A, Sarkar S, Zurenko C, Karagiannis E, Love K, Chen D, Zoncu R et al (2013) Efficiency of siRNA delivery by lipid nanoparticles is limited by endocytic recycling. Nat Biotechnol 31:653–658
Sarma K, Levasseur P, Aristarkhov A, Lee JT (2010) Locked nucleic acids (LNAs) reveal sequence requirements and kinetics of Xist RNA localization to the X chromosome. Proc Natl Acad Sci U S A 107:22196–22201
Sazani P, Gemignani F, Kang SH, Maier MA, Manoharan M, Persmark M, Bortner D, Kole R (2002) Systemically delivered antisense oligomers upregulate gene expression in mouse tissues. Nat Biotechnol 20:1228–1233
Sazani P, Graziewicz MA, Kole R (2007) Splice switching oligonucleotides as potential therapeutics. In: Crooke ST (ed) Antisense drug technology: principles, strategies, and applications. CRC, Boca Raton, FL, pp 89–114
Scaffidi P, Misteli T (2005) Reversal of the cellular phenotype in the premature aging disease Hutchinson-Gilford progeria syndrome. Nat Med 11:440–445
Schoenberg DR, Maquat LE (2012) Regulation of cytoplasmic mRNA decay. Nat Rev Genet 13:246–259
Schreml J, Riessland M, Paterno M, Garbes L, Rossbach K, Ackermann B, Kramer J, Somers E, Parson SH, Heller R et al (2012) Severe SMA mice show organ impairment that cannot be rescued by therapy with the HDACi JNJ-26481585. Eur J Human Genet 21(6):643–652
Schwartz JC, Younger ST, Nguyen NB, Hardy DB, Monia BP, Corey DR, Janowski BA (2008) Antisense transcripts are targets for activating small RNAs. Nat Struct Mol Biol 15:842–848
Schweingruber C, Rufener SC, Zund D, Yamashita A, Muhlemann O (2013) Nonsense-mediated mRNA decay—mechanisms of substrate mRNA recognition and degradation in mammalian cells. Biochim Biophys Acta 1829:612–623
Seila AC, Core LJ, Lis JT, Sharp PA (2009) Divergent transcription: a new feature of active promoters. Cell Cycle 8:2557–2564
Senn JJ, Burel S, Henry SP (2005) Non-CpG-containing antisense 2′-methoxyethyl oligonucleotides activate a proinflammatory response independent of Toll-like receptor 9 or myeloid differentiation factor 88. J Pharmacol Exp Ther 314:972–979
Seow Y, Sibley CR, Wood MJ (2012) Artificial mirtron-mediated gene knockdown: functional DMPK silencing in mammalian cells. RNA 18:1328–1337
Seth PP, Allerson CR, Berdeja A, Siwkowski A, Pallan PS, Gaus H, Prakash TP, Watt AT, Egli M, Swayze EE (2010) An exocyclic methylene group acts as a bioisostere of the 2′-oxygen atom in LNA. J Am Chem Soc 132:14942–14950
Seth PP, Jazayeri A, Yu J, Allerson CR, Bhat B, Swayze EE (2012a) Structure activity relationships of alpha-L-LNA modified phosphorothioate gapmer antisense oligonucleotides in animals. Mol Ther Nucleic Acids 1:e47
Seth PP, Siwkowski A, Allerson CR, Vasquez G, Lee S, Prakash TP, Wancewicz EV, Witchell D, Swayze EE (2009) Short antisense oligonucleotides with novel 2′-4′ conformationally restricted nucleoside analogues show improved potency without increased toxicity in animals. J Med Chem 52:10–13
Seth PP, Yu J, Jazayeri A, Pallan PS, Allerson CR, Ostergaard ME, Liu F, Herdewijn P, Egli M, Swayze EE (2012b) Synthesis and antisense properties of fluoro cyclohexenyl nucleic acid (F-CeNA), a nuclease stable mimic of 2′-fluoro RNA. J Org Chem 77:5074–5085
Seznec H, Agbulut O, Sergeant N, Savouret C, Ghestem A, Tabti N, Willer JC, Ourth L, Duros C, Brisson E et al (2001) Mice transgenic for the human myotonic dystrophy region with expanded CTG repeats display muscular and brain abnormalities. Hum Mol Genet 10:2717–2726

Shababi M, Habibi J, Yang HT, Vale SM, Sewell WA, Lorson CL (2010) Cardiac defects contribute to the pathology of spinal muscular atrophy models. Hum Mol Genet 19:4059–4071
Shababi M, Lorson CL, Rudnik-Schoneborn SS (2013) Spinal muscular atrophy: a motor neuron disorder or a multi-organ disease? J Anat 224(1):15–28
Shatkin AJ, Manley JL (2000) The ends of the affair: capping and polyadenylation. Nat Struct Biol 7:838–842
Sheehan JP, Phan TM (2001) Phosphorothioate oligonucleotides inhibit the intrinsic tenase complex by an allosteric mechanism. Biochemistry 40:4980–4989
Shibahara S, Mukai S, Nishihara T, Inoue H, Ohtsuka E, Morisawa H (1987) Site-directed cleavage of RNA. Nucleic Acids Res 15:4403–4415
Sierakowska H, Sambade MJ, Agrawal S, Kole R (1996) Repair of thalassemic human beta-globin mRNA in mammalian cells by antisense oligonucleotides. Proc Natl Acad Sci U S A 93:12840–12844
Singh NK, Singh NN, Androphy EJ, Singh RN (2006) Splicing of a critical exon of human Survival Motor Neuron is regulated by a unique silencer element located in the last intron. Mol Cell Biol 26:1333–1346
Sipes TB, Freier SM (2008) Prediction of antisense oligonucleotide efficacy using aggregate motifs. J Bioinform Comput Biol 6:919–932
Siwkowski AM, Malik L, Esau CC, Maier MA, Wancewicz EV, Albertshofer K, Monia BP, Bennett CF, Eldrup AB (2004) Identification and functional validation of PNAs that inhibit murine CD40 expression by redirection of splicing. Nucleic Acids Res 32:2695–2706
Skordis LA, Dunckley MG, Yue B, Eperon IC, Muntoni F (2003) Bifunctional antisense oligonucleotides provide a trans-acting splicing enhancer that stimulates SMN2 gene expression in patient fibroblasts. Proc Natl Acad Sci U S A 100:4114–4119
Smith RA, Miller TM, Yamanaka K, Monia BP, Condon TP, Hung G, Lobsiger CS, Ward CM, McAlonis-Downes M, Wei H et al (2006) Antisense oligonucleotide therapy for neurodegenerative disease. J Clin Invest 116:2290–2296
Sobczak K, Wheeler TM, Wang W, Thornton CA (2013) RNA interference targeting CUG repeats in a mouse model of myotonic dystrophy. Mol Ther: J Am Soc Gene Ther 21:380–387
Spitali P, Aartsma-Rus A (2012) Splice modulating therapies for human disease. Cell 148:1085–1088
Squires JE, Patel HR, Nousch M, Sibbritt T, Humphreys DT, Parker BJ, Suter CM, Preiss T (2012) Widespread occurrence of 5-methylcytosine in human coding and non-coding RNA. Nucleic Acids Res 40:5023–5033
Stein CA, Hansen JB, Lai J, Wu S, Voskresenskiy A, Hog A, Worm J, Hedtjarn M, Souleimanian N, Miller P et al (2010) Efficient gene silencing by delivery of locked nucleic acid antisense oligonucleotides, unassisted by transfection reagents. Nucleic Acids Res 38:e3
Straarup EM, Fisker N, Hedtjarn M, Lindholm MW, Rosenbohm C, Aarup V, Hansen HF, Orum H, Hansen JB, Koch T (2010) Short locked nucleic acid antisense oligonucleotides potently reduce apolipoprotein B mRNA and serum cholesterol in mice and non-human primates. Nucleic Acids Res 38:7100–7111
Swayze EE, Bhat B (2007) The medicinal chemistry of oligonucleotides. In: Crooke ST (ed) Antisense drug technology: principles, strategies, and applications. CRC, Boca Raton, FL, pp 143–182
Swayze EE, Siwkowski AM, Wancewicz EV, Migawa MT, Wyrzykiewicz TK, Hung G, Monia BP, Bennett CF (2007) Antisense oligonucleotides containing locked nucleic acid improve potency but cause significant hepatotoxicity in animals. Nucleic Acids Res 35:687–700
Taneja KL, McCurrach M, Schalling M, Housman D, Singer RH (1995) Foci of trinucleotide repeat transcripts in nuclei of myotonic dystrophy cells and tissues. J Cell Biol 128:995–1002
Taniguchi-Ikeda M, Kobayashi K, Kanagawa M, Yu CC, Mori K, Oda T, Kuga A, Kurahashi H, Akman HO, DiMauro S et al (2011) Pathogenic exon-trapping by SVA retrotransposon and rescue in Fukuyama muscular dystrophy. Nature 478:127–131
Tazi J, Bakkour N, Stamm S (2009) Alternative splicing and disease. Biochim Biophys Acta 1792:14–26

Teplova M, Minasov G, Tereshko V, Inamati GB, Cook PD, Manoharan M, Egli M (1999) Crystal structure and improved antisense properties of 2′-O-(2-methoxyethyl)-RNA. Nat Struct Biol 6:535–539
Toda T, Kobayashi K (1999) Fukuyama-type congenital muscular dystrophy: the first human disease to be caused by an ancient retrotransposal integration. J Mol Med 77:816–823
Tripathi V, Ellis JD, Shen Z, Song DY, Pan Q, Watt AT, Freier SM, Bennett CF, Sharma A, Bubulya PA et al (2010) The nuclear-retained noncoding RNA MALAT1 regulates alternative splicing by modulating SR splicing factor phosphorylation. Mol Cell 39:925–938
Turner, A.M., and Morris, K.V. (2010). Controlling transcription with noncoding RNAs in mammalian cells. Biotechniques *48*, ix-xvi.
Ulitsky I, Bartel DP (2013) lincRNAs: genomics, evolution, and mechanisms. Cell 154:26–46
Usman N, Blatt LM (2000) Nuclease-resistant synthetic ribozymes: developing a new class of therapeutics. J Clin Invest 106:1197–1202
Vaishnaw AK, Gollob J, Gamba-Vitalo C, Hutabarat R, Sah D, Meyers R, de Fougerolles T, Maraganore J (2010) A status report on RNAi therapeutics. Silence 1:14
Valori CF, Ning K, Wyles M, Mead RJ, Grierson AJ, Shaw PJ, Azzouz M (2010) Systemic delivery of scAAV9 expressing SMN prolongs survival in a model of spinal muscular atrophy. Sci Transl Med 2:35ra42
van Deutekom JC, Janson AA, Ginjaar IB, Frankhuizen WS, Aartsma-Rus A, Bremmer-Bout M, den Dunnen JT, Koop K, van der Kooi AJ, Goemans NM et al (2007) Local dystrophin restoration with antisense oligonucleotide PRO051. N Engl J Med 357:2677–2686
Verpy E, Leibovici M, Zwaenepoel I, Liu XZ, Gal A, Salem N, Mansour A, Blanchard S, Kobayashi I, Keats BJ et al (2000) A defect in harmonin, a PDZ domain-containing protein expressed in the inner ear sensory hair cells, underlies Usher syndrome type 1C. Nat Genet 26:51–55
Vester B, Wengel J (2004) LNA (locked nucleic acid): high-affinity targeting of complementary RNA and DNA. Biochemistry 43:13233–13241
Vezain M, Saugier-Veber P, Goina E, Touraine R, Manel V, Toutain A, Fehrenbach S, Frebourg T, Pagani F, Tosi M et al (2010) A rare SMN2 variant in a previously unrecognized composite splicing regulatory element induces exon 7 inclusion and reduces the clinical severity of spinal muscular atrophy. Hum Mutat 31:E1110–E1125
Vickers T, Baker BF, Cook PD, Zounes M, Buckheit RW Jr, Germany J, Ecker DJ (1991) Inhibition of HIV-LTR gene expression by oligonucleotides targeted to the TAR element. Nucleic Acids Res 19:3359–3368
Vickers TA, Crooke ST (2012) siRNAs targeted to certain polyadenylation sites promote specific, RISC-independent degradation of messenger RNAs. Nucleic Acids Res 40(13):6223–6234
Vickers TA, Koo S, Bennett CF, Crooke ST, Dean NM, Baker BF (2003) Efficient reduction of target RNAs by small interfering RNA and RNase H-dependent antisense agents. A comparative analysis. J Biol Chem 278:7108–7118
Vickers TA, Wyatt JR, Burckin T, Bennett CF, Freier SM (2001) Fully modified 2′ MOE oligonucleotides redirect polyadenylation. Nucleic Acids Res 29:1293–1299
Vickers TA, Wyatt JR, Freier SM (2000) Effects of RNA secondary structure on cellular antisense activity. Nucleic Acids Res 28:1340–1347
Villemaire J, Dion I, Elela SA, Chabot B (2003) Reprogramming alternative pre-messenger RNA splicing through the use of protein-binding antisense oligonucleotides. J Biol Chem 278: 50031–50039
Vitiello D, Pecchia DB, Burke JM (2000) Intracellular ribozyme-catalyzed trans-cleavage of RNA monitored by fluorescence resonance energy transfer. RNA 6:628–637
Vitte JM, Davoult B, Roblot N, Mayer M, Joshi V, Courageot S, Tronche F, Vadrot J, Moreau MH, Kemeny F et al (2004) Deletion of murine Smn exon 7 directed to liver leads to severe defect of liver development associated with iron overload. Am J Pathol 165:1731–1741
Voigt T, Meyer K, Baum O, Schumperli D (2010) Ultrastructural changes in diaphragm neuromuscular junctions in a severe mouse model for Spinal Muscular Atrophy and their prevention by bifunctional U7 snRNA correcting SMN2 splicing. Neuromuscul Disord 20:744–752

Vorlova S, Rocco G, Lefave CV, Jodelka FM, Hess K, Hastings ML, Henke E, Cartegni L (2011) Induction of antagonistic soluble decoy receptor tyrosine kinases by intronic polya activation. Mol Cell 43:927–939

Wahlestedt C (2013) Targeting long non-coding RNA to therapeutically upregulate gene expression. Nat Rev Drug Discov 12:433–446

Wan L, Battle DJ, Yong J, Gubitz AK, Kolb SJ, Wang J, Dreyfuss G (2005) The survival of motor neurons protein determines the capacity for snRNP assembly: biochemical deficiency in spinal muscular atrophy. Mol Cell Biol 25:5543–5551

Wan Y, Kertesz M, Spitale RC, Segal E, Chang HY (2011) Understanding the transcriptome through RNA structure. Nat Rev Genet 12:641–655

Wancewicz EV, Maier MA, Siwkowski AM, Albertshofer K, Winger TM, Berdeja A, Gaus H, Vickers TA, Bennett CF, Monia BP et al (2010) Peptide nucleic acids conjugated to short basic peptides show improved pharmacokinetics and antisense activity in adipose tissue. J Med Chem 53:3919–3926

Wang GS, Cooper TA (2007) Splicing in disease: disruption of the splicing code and the decoding machinery. Nat Rev Genet 8:749–761

Ward AJ, Rimer M, Killian JM, Dowling JJ, Cooper TA (2010) CUGBP1 overexpression in mouse skeletal muscle reproduces features of myotonic dystrophy type 1. Hum Mol Genet 19: 3614–3622

Watanabe TA, Geary RS, Levin AA (2006) Plasma protein binding of an antisense oligonucleotide targeting human ICAM-1 (ISIS 2302). Oligonucleotides 16:169–180

Weismann D, Erion DM, Ignatova-Todorava I, Nagai Y, Stark R, Hsiao JJ, Flannery C, Birkenfeld AL, May T, Kahn M et al (2011) Knockdown of the gene encoding Drosophila tribbles homologue 3 (Trib3) improves insulin sensitivity through peroxisome proliferator-activated receptor-gamma (PPAR-gamma) activation in a rat model of insulin resistance. Diabetologia 54: 935–944

Wheeler TM, Leger AJ, Pandey SK, MacLeod AR, Nakamori M, Cheng SH, Wentworth BM, Bennett CF, Thornton CA (2012) Targeting nuclear RNA for in vivo correction of myotonic dystrophy. Nature 488:111–115

Wheeler TM, Lueck JD, Swanson MS, Dirksen RT, Thornton CA (2007) Correction of ClC-1 splicing eliminates chloride channelopathy and myotonia in mouse models of myotonic dystrophy. J Clin Invest 117:3952–3957

Wheeler TM, Sobczak K, Lueck JD, Osborne RJ, Lin X, Dirksen RT, Thornton CA (2009) Reversal of RNA dominance by displacement of protein sequestered on triplet repeat RNA. Science 325:336–339

Williams JH, Schray RC, Patterson CA, Ayitey SO, Tallent MK, Lutz GJ (2009) Oligonucleotide-mediated survival of motor neuron protein expression in CNS improves phenotype in a mouse model of spinal muscular atrophy. J Neurosci 29:7633–7638

Wilusz JE, Devanney SC, Caputi M (2005) Chimeric peptide nucleic acid compounds modulate splicing of the bcl-x gene in vitro and in vivo. Nucleic Acids Res 33:6547–6554

Wooddell CI, Rozema DB, Hossbach M, John M, Hamilton HL, Chu Q, Hegge JO, Klein JJ, Wakefield DH, Oropeza CE et al (2013) Hepatocyte-targeted RNAi therapeutics for the treatment of chronic hepatitis B virus infection. Mol Ther: J Am Soc Gene Ther 21:973–985

Woolf TM, Chase JM, Stinchcomb DT (1995) Toward the therapeutic editing of mutated RNA sequences. Proc Natl Acad Sci U S A 92:8298–8302

Worman HJ, Fong LG, Muchir A, Young SG (2009) Laminopathies and the long strange trip from basic cell biology to therapy. J Clin Invest 119:1825–1836

Wu H, Lima WF, Zhang H, Fan A, Sun H, Crooke ST (2004) Determination of the role of the human RNase H1 in the pharmacology of DNA-like antisense drugs. J Biol Chem 279:17181–17189

Xiao J, Yang B, Lin H, Lu Y, Luo X, Wang Z (2007) Novel approaches for gene-specific interference via manipulating actions of microRNAs: examination on the pacemaker channel genes HCN2 and HCN4. J Cell Physiol 212:285–292

Yelin R, Dahary D, Sorek R, Levanon EY, Goldstein O, Shoshan A, Diber A, Biton S, Tamir Y, Khosravi R et al (2003) Widespread occurrence of antisense transcription in the human genome. Nat Biotechnol 21:379–386

Yin H, Betts C, Saleh AF, Ivanova GD, Lee H, Seow Y, Kim D, Gait MJ, Wood MJ (2010) Optimization of peptide nucleic acid antisense oligonucleotides for local and systemic dystrophin splice correction in the mdx mouse. Mol Ther: J Am Soc Gene Ther 18:819–827

Yokota T, Lu QL, Partridge T, Kobayashi M, Nakamura A, Takeda S, Hoffman E (2009) Efficacy of systemic morpholino exon-skipping in Duchenne dystrophy dogs. Ann Neurol 65:667–676

Yong J, Wan L, Dreyfuss G (2004) Why do cells need an assembly machine for RNA-protein complexes? Trends Cell Biol 14:226–232

Younger ST, Corey DR (2011) Transcriptional gene silencing in mammalian cells by miRNA mimics that target gene promoters. Nucleic Acids Res 39(13):5682–5691

Yu D, Pendergraff H, Liu J, Kordasiewicz HB, Cleveland DW, Swayze EE, Lima WF, Crooke ST, Prakash TP, Corey DR (2012) Single-stranded RNAs use RNAi to potently and allele-selectively inhibit mutant huntingtin expression. Cell 150:895–908

Yu RZ, Zhang H, Geary RS, Graham M, Masarjian L, Lemonidis K, Crooke R, Dean NM, Levin AA (2001) Pharmacokinetics and pharmacodynamics of an antisense phosphorothioate oligonucleotide targeting Fas mRNA in mice. J Pharmacol Exp Ther 296:388–395

Zamecnik PC, Stephenson ML (1978) Inhibition of Rous sarcoma virus replication and cell transformation by a specific oligodeoxynucleotide. Proc Natl Acad Sci U S A 75:280–284

Zammarchi F, de Stanchina E, Bournazou E, Supakorndej T, Martires K, Riedel E, Corben AD, Bromberg JF, Cartegni L (2011) Antitumorigenic potential of STAT3 alternative splicing modulation. Proc Natl Acad Sci U S A 108(43):17779–17784

Zhao J, Ohsumi TK, Kung JT, Ogawa Y, Grau DJ, Sarma K, Song JJ, Kingston RE, Borowsky M, Lee JT (2010) Genome-wide identification of polycomb-associated RNAs by RIP-seq. Mol Cell 40:939–953

Zhou H, Janghra N, Mitrpant C, Dickinson RL, Anthony K, Price L, Eperon IC, Wilton SD, Morgan J, Muntoni F (2013) A novel morpholino oligomer targeting ISS-N1 improves rescue of severe spinal muscular atrophy transgenic mice. Hum Gene Ther 24:331–342

Chapter 10
RNA-Binding Protein Misregulation in Microsatellite Expansion Disorders

Marianne Goodwin and Maurice S. Swanson

Abstract RNA-binding proteins (RBPs) play pivotal roles in multiple cellular pathways from transcription to RNA turnover by interacting with RNA sequence and/or structural elements to form distinct RNA–protein complexes. Since these complexes are required for the normal regulation of gene expression, mutations that alter RBP functions may result in a cascade of deleterious events that lead to severe disease. Here, we focus on a group of hereditary disorders, the microsatellite expansion diseases, which alter RBP activities and result in abnormal neurological and neuromuscular phenotypes. While many of these diseases are classified as adult-onset disorders, mounting evidence indicates that disruption of normal RNA–protein interaction networks during embryogenesis modifies developmental pathways, which ultimately leads to disease manifestations later in life. Efforts to understand the molecular basis of these disorders has already uncovered novel pathogenic mechanisms, including RNA toxicity and repeat-associated non-ATG (RAN) translation, and current studies suggest that additional surprising insights into cellular regulatory pathways will emerge in the future.

Keywords Amyotrophic lateral sclerosis/frontotemporal dementia • Microsatellite • Myotonic dystrophy • Neurological disease • Oculopharyngeal muscular dystrophy • RNA processing • RNA toxicity • Spinocerebellar ataxia

1 Introduction

1.1 Benefits and Problems with a Repetitive Genome

Nearly half of the human genome consists of repetitive DNA sequences composed mostly of interspersed and transposon-derived repeats but also tandem repeats (TRs) (Gemayel et al. 2010). Microsatellites, also known as TRs and simple

M. Goodwin • M.S. Swanson (✉)
Department of Molecular Genetics and Microbiology, University of Florida, College of Medicine, Cancer Genetics Research Complex, 2033 Mowry Road, Gainesville, FL 32610-3610, USA
e-mail: mswanson@UFL.EDU

G.W. Yeo (ed.), *Systems Biology of RNA Binding Proteins*, Advances in Experimental Medicine and Biology 825, DOI 10.1007/978-1-4939-1221-6_10

sequence repeats (SSRs), are often defined as repeating units of ≤10 base pairs (bp) while larger repeats are referred to as minisatellites (>10 bp) and macrosatellites or megasatellites (>135 bp). Microsatellites, which account for 3–5 % of mammalian genomes, are highly polymorphic since the DNA replication, repair, and recombination machineries have intrinsic problems handling these unusual repetitive sequences due to their tendency to form imperfect hairpins, quadraplex-like and slipped-stranded structures (Lopez Castel et al. 2010; Mirkin 2007). Many microsatellites are bidirectionally transcribed (Batra et al. 2010; Budworth and McMurray 2013) and repeat length polymorphism is common with mutation rates 10–100,000 fold higher than other genomic regions (Jansen et al. 2012). In the overall population, repeat lengths for a given allele vary moderately and unaffected individuals may harbor alleles with a different number of repeats within the normal range. However, once an allele expands beyond a critical length threshold, instability is greatly amplified and the mutation manifests into a pathological state. Because expansions and contractions occur during cell division and error-prone DNA repair, affected patient tissues are composed of cells containing varying numbers of microsatellite repeats in the disease allele, a process termed somatic mosaicism (Lopez Castel et al. 2010; Mirkin 2007). Furthermore, continuous expansions of mutant repeat lengths often occur during aging, which might explain the progressive nature of many of these neurological diseases. Microsatellite expansions and contractions also occur in the germ line, which affects the repeat length passed on from one generation to the next. Comparison of average repeat lengths among patients within pedigrees shows that successive generations often have progressively larger repeats. Moreover, this increase in repeat length often correlates with an increase in disease severity and earlier age-of-onset of disease symptoms, hence providing a genetic explanation for the observation of intergenerational anticipation (Friedman 2011).

Although the remarkable abundance of repeats in the human genome hints at functionality, early reports classified these sequences as an evolutionary artifact or non-functional "junk" DNA (Doolittle and Sapienza 1980; Ohno 1972; Orgel and Crick 1980). More recent studies indicate that repetitive DNA might serve valuable cellular functions. Microsatellites occur in the protein-coding regions of ~17 % of human genes and the 10–20 % of eukaryotic genes that contain microsatellite repeats are often important for cellular regulatory pathways (Gemayel et al. 2010; Jansen et al. 2012). For example, TRs in budding yeast are primarily found within genes encoding cell-surface and important regulatory proteins, including chromatin modification and transcription factors. However, TRs are not restricted to eukaryotes. Indeed, TRs facilitate antigenic variation in pathogenic prokaryotes as a mechanism to evade host defense systems (Gemayel et al. 2010; Mrazek et al. 2007). Repeat unit variations in promoter regions may lead to changes in gene expression while TR variations in coding regions can result in frameshift mutations and the production of truncated proteins. Thus, simple sequence repeats serve regulatory functions and catalyze adaptations beneficial to pathogen survival. Despite these examples of TR functions, functional roles for microsatellite variability in human genes have not been well documented although TRs are enriched in vertebrate genes that control organ and/or body morphology (Gemayel et al. 2010). Due to their relatively high mutation rate, variability in microsatellite repeat number within a gene might offer

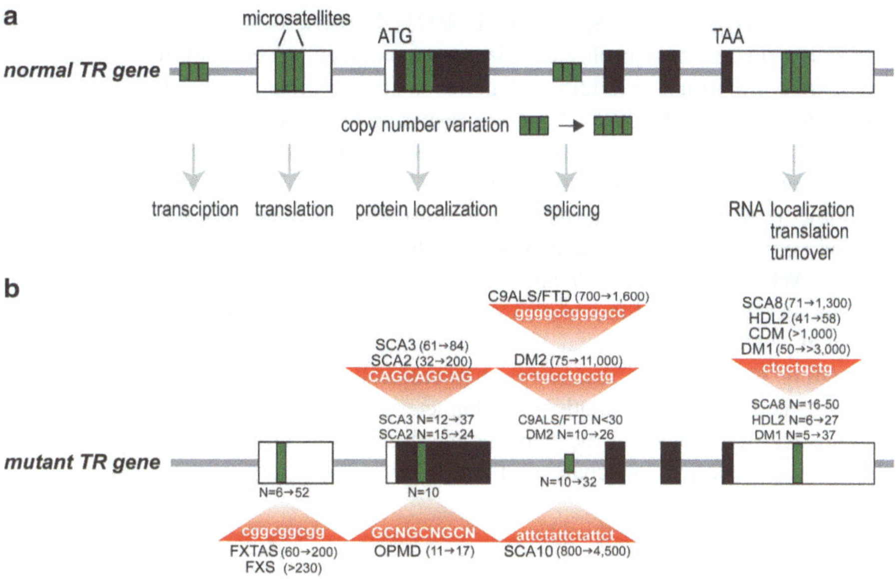

Fig. 10.1 Microsatellites in normal and mutant genes associated with hereditary diseases discussed in this chapter. (**a**) A normal five exon tandem repeat (TR) gene with 5′ and 3′ UTRs (*open boxes*), coding region (*black boxes*), introns (*thick grey line*), and microsatellites (*green boxes*) located in promoter, noncoding (UTRs, intron) and coding regions. Copy number variation, due to errors during DNA replication or repair, may result in a modest variation in TR number (shown here as 3–4 TRs) that might influence specific regulatory pathways depending on the TR location in the allele. (**b**) TR expansions result in the diseases discussed in this chapter. The position of each TR is shown (*single green box*) together with normal (N) allele and mutant (*red triangle*) length ranges

evolutionary and regulatory advantages. For example, variations in GC-rich repeats within the normal range positioned within a promoter could have modest, but advantageous, effects on transcriptional activity while similar repeats in the 5′ untranslated region (5′ UTR) could modulate translation and the number of encoded proteins available during a particular developmental window (Fig. 10.1a). However, most studies have emphasized the detrimental effects of larger microsatellite expansions. How do expanded microsatellites cause disease? Research has revealed several distinct mechanisms, some of which are novel and unique, at least currently, to repeat disease. Relevant to this review, some of these mechanisms involve RNA-binding proteins and altered RNA processing.

1.2 *Microsatellites in Disease*

Over two decades ago, the field of unstable microsatellite disease was initiated by seminal findings on the molecular etiology of fragile X syndrome (FXS), caused by a CGG^{exp} in the *fragile X mental retardation 1* (*FMR1*) gene, and spinal and bulbar muscular atrophy (SBMA), which is due to CAG^{exp} mutations in the *androgen receptor* (*AR*) gene (La Spada et al. 1991; Verkerk et al. 1991) (Fig. 10.1b). Currently, dozens of neurological diseases have been traced back to microsatellite expansion mutations in additional genes with the most recent discovery that a GGGGCC^{exp} in the *C9ORF72* gene is responsible for the most common cause of familial amyotrophic lateral sclerosis (ALS) and frontotemporal dementia (FTD) (C9-linked ALS/FTD) (DeJesus-Hernandez et al. 2011; Renton et al. 2011).

Immediately following the discovery of unstable microsatellites in FXS and SBMA, several perplexing questions emerged regarding how expanded microsatellites cause disease. Why are such unstable and disease-prone repetitive sequences so prevalent in the human genome, and in some cases, conserved during evolution (Buschiazzo and Gemmell 2010; Gemayel et al. 2010). Why are so many microsatellites located in noncoding regions? The discovery of microsatellite expansion disorders has been accompanied by global initiatives to characterize the associated disease pathologies and uncover the molecular mechanisms involved in this group of neurological disorders. Several common mechanistic themes have emerged (Fig. 10.2). Many of the

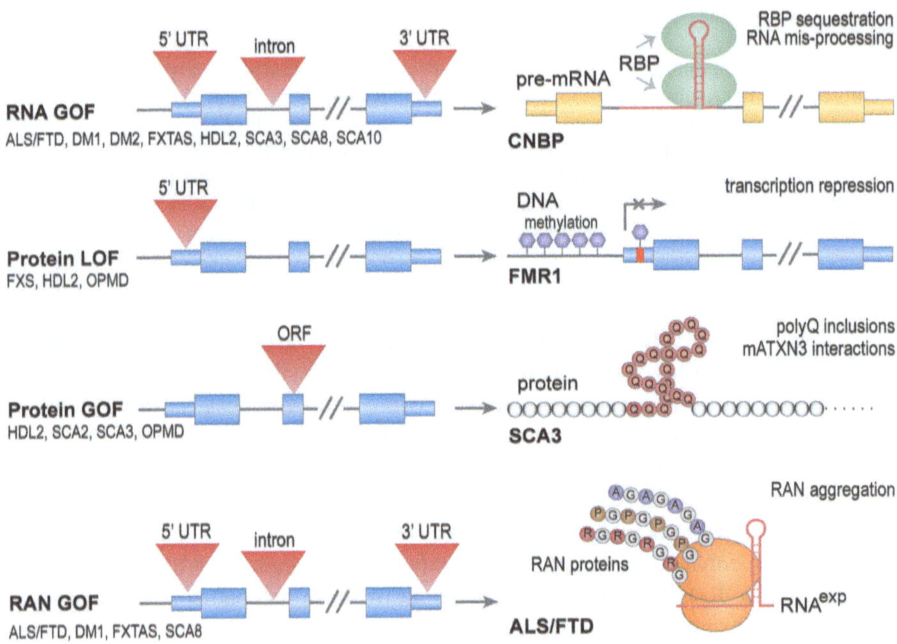

Fig. 10.2 Proposed pathogenic mechanisms for microsatellite repeat diseases. Microsatellite expansions (*red triangles*) in noncoding (UTRs, *small blue boxes*; introns, *black lines*) regions may result in RNA (*yellow boxes*) gain-of-function (RNA GOF), protein (*grey and colored circles*) loss-of-function (protein LOF), or repeat-associated non-ATG (RAN) protein GOF (RAN GOF). In conventional coding regions (open reading frame, ORF), expansions might cause a protein GOF effect such as the CAG expansion in SCA3 which results in a polyGln (polyQ, *brown circles*) expansion that either accumulates in cellular inclusions or alters mutant ATXN3 (mATXN3) interactions. Note that RAN translation of the *C9ORF72* GGGGCC^{exp} repeat in ALS/FTD produces polyGlyAla (GA), polyGlyPro (GP), and polyGlyArg (GR) dipeptide repeat proteins due to the recognition an RNA secondary structure (*red hairpin*) by the ribosome (*orange ellipsoids*)

repeat expansion diseases involve the nervous and musculoskeletal systems, follow a dominant inheritance pattern and show genetic anticipation. Despite widespread expression of certain disease genes, neurological and neuromuscular systems display a particular vulnerability to mutant microsatellite genes. Histological analysis of patient tissue reveals that many of these disorders are characterized by cellular aggregates of homopolymeric proteins, notably polyglutamine (polyQ) inclusions (Orr 2012b; Zoghbi and Orr 2009). In some diseases, patient cells contain nuclear foci harboring RNA expressed from the expanded genes along with specific proteins. Intriguingly, RNA-binding protein misregulation and compromised RNA metabolism have become recurrent disease themes (Echeverria and Cooper 2012; Poulos et al. 2011).

In this chapter, we focus on the molecular mechanisms underlying microsatellite expansion diseases which include: (1) RNA gain-of-function, in which the repeat-containing RNA transcript expressed from the mutant gene sequesters trans-acting factors; (2) protein loss-of-function due to microsatellite expansion mutations that render a protein non-functional or represses expression; (3) protein gain-of-function whereby a repeat expansion in a coding region contributes a deleterious function to the mutant protein; (4) repeat associated non-ATG (RAN) translation that results in homopolymeric or heteropolymeric peptides translated from the repeat region independent of a normal initiation codon. As an additional layer of complexity, bidirectional transcription through repeat DNA enables multiple pathogenic mechanisms from a single locus. Examples of microsatellite expansion diseases that illustrate each of these molecular mechanisms, and their effect on RNA-binding protein function, are the focus of this chapter.

2 RNA Toxicity and Protein Sequestration in Myotonic Dystrophy

2.1 *Tipping the Balance Between Antagonistic RNA Processing Factors*

The discovery that the transcription of some microsatellite expansions results in the synthesis of toxic RNAs originated from studies designed to elucidate the molecular etiology of the most frequent form of muscular dystrophy in adults, myotonic dystrophy (dystrophia myotonica, DM) (Echeverria and Cooper 2012; Poulos et al. 2011; Udd and Krahe 2012). Although DM is classified as a muscular dystrophy, multiple tissues are affected including skeletal muscle (hyperexcitability or myotonia, weakness/wasting), the heart (arrhythmias and conduction block), the visual system (dust-like cataracts), the reproductive system, (testicular atrophy), and the brain (hypersomnia, executive dysfunction, and cerebral atrophy). Interestingly, some disease manifestations, such as brain/muscle atrophy and alopecia (premature balding), resemble the normal aging process (Martin 2005). A distinguishing feature of DM is that microsatellite expansions in two unrelated genes cause disease. DM type 1 (DM1) is associated with a CTG^{exp} in the *dystrophia myotonica protein kinase* (*DMPK*) gene, which encodes a serine–threonine kinase, while DM type 2 is caused by a CCTG^{exp} in *CNBP*, the *cellular nucleic acid binding protein* gene that encodes a factor implicated in both transcription and translation (Brook et al. 1992; Fu et al. 1992; Jansen et al. 1992; Liquori et al. 2001; Mahadevan et al. 1992) (Fig. 10.1b). Unaffected individuals have 5–37 CTG repeats in the *DMPK* gene and 10–26 CCTG repeats in *CNBP*. For DM1, CTG^{exp} lengths expand to 50->3,000 repeats while in DM2 the CCTG^{exp} ranges from 75 to >11,000 repeats. In contrast to DM2, which does not have a congenital form, very large (>1,000 repeats) *DMPK* CTG^{exp} mutations also cause congenital DM1 (CDM) characterized by neonatal hypotonia (floppy baby) and intellectual disability.

Several experimental findings argue that DM is an RNA-mediated, or RNA gain-of-function (GOF), disease (Fig. 10.2). The *DMPK* and *CNBP* expansions are located in noncoding regions and mutant DMPK and CNBP C(C)UG^{exp} RNAs accumulate in RNA foci in the nucleus (Davis et al. 1997; Margolis et al. 2006; Poulos et al. 2011; Ranum and Cooper 2006; Taneja et al. 1995). Attempts to model DM in mouse transgenic models led to the observation that CTG^{exp} mutations cause DM-relevant phenotypes irrespective of gene context and the degree of pathology correlates with transgene expression level (Mankodi et al. 2000; Sicot and Gomes-Pereira 2013). How does C(C)UG^{exp} RNA expression lead to DM? C(C)UG^{exp} RNAs could: (1) promote formation of nuclear RNA foci that impair normal nuclear functions such as RNA processing and nuclear export pathways; (2) fold into a structure that possesses an inherent dominant-negative function; (3) sequester specific proteins leading to loss-of-function of these factors. Interestingly, the current model for DM pathogenesis implicates all of these mechanisms. Transcription of expanded C(C)TG repeats leads to the synthesis of C(C)UG^{exp} RNAs that gain toxic functions, including the recruitment and sequestration of the muscleblind-like (MBNL) proteins and hyperphosphorylation, and increased levels of CUGBP1 and ETR3-like factors (CELF) (Kuyumcu-Martinez et al. 2007; Miller et al. 2000). The MBNL and CELF proteins are antagonistic alternative splicing factors that function during postnatal development by promoting either adult (MBNL) or fetal (CELF) splicing patterns (Charizanis et al. 2012; Du et al. 2010; Ho et al. 2004; Kanadia et al. 2003a, b; Lin et al. 2006; Philips et al. 1998; Timchenko et al. 1996). Thus, DM pathogenesis involves RNA gain-of-function leading to protein loss, and gain, of function and the persistence of, or reversion to, fetal splicing patterns in adult tissues.

Besides MBNL and CELF, other RNA processing factors have been implicated in DM. HnRNP H binds in vitro to DMPK-derived CUG^{exp} RNAs that also contain a splicing branch point distal to the repeat region and siRNA knockdown of this protein rescues nuclear retention of CUG^{exp}-containing RNAs (Kim et al. 2005). HnRNP H protein levels are also elevated in DM1 myoblasts and hnRNP H or CELF1 overexpression leads to the formation of a repressor complex that inhibits splicing of insulin receptor (IR) exon 11 (Paul et al. 2006). HnRNP H exists in a complex with MBNL1 and 9 other proteins (hnRNP H2, H3, F, A2/B1, K, L, DDX5, DDX17, and DHX9) in normal myoblast extracts, but the stoichiometry of these complexes is altered in DM1 extracts (Paul et al. 2011). Staufen1 (STAU1) is a double-stranded (ds)RNA-binding protein that is also misregulated in DM1 skeletal muscle (Ravel-Chapuis et al. 2012). STAU1 levels increase in human DM1, and CUG^{exp} mouse, skeletal muscle, but it is not sequestered in nuclear CUG^{exp} RNA foci. STAU1 promotes nuclear export and translation of CUG^{exp} mRNAs, and its overexpression rescues abnormal splicing of several pre-mRNAs misspliced in DM, so increased levels of STAU1 may be a compensatory response that ameliorates the DM phenotype.

An additional type of RNA-binding protein appears to modulate the interaction of MBNL proteins with pathogenic C(C)UG^{exp} RNAs. MBNL proteins contain tandem zinc finger (ZnF) domains that bind preferentially to YGCY (Y=pyrimidine) RNA elements (Charizanis et al. 2012; Du et al. 2010; Goers et al. 2010; Wang et al. 2012a). Structural analysis of MBNL ZnFs indicates that they target GC steps and induce an antiparallel orientation on these bound elements due to inter-ZnF linker topology (Teplova and Patel 2008). MBNL proteins do not possess a high affinity for uninterrupted Watson–Crick RNA duplexes but instead show preferential binding to imperfect duplexes, particularly with U–U and C–U/U–C mismatches, and steady-state fluorescence quenching analysis confirms that MBNL1 binding alters helical RNA structures (Fu et al. 2012). Since CUG^{exp} RNAs fold into RNA hairpin structures (Krzyzosiak et al. 2012), it is interesting that the RNA helicase p68/DDX5 acts to modulate MBNL1 binding activity (Laurent et al. 2012). The p68/DDX5 helicase colocalizes with nuclear RNA foci and has a stimulatory effect on MBNL1 binding to both CUG^{exp} and MBNL1 splicing target RNA binding sites.

2.2 *Pathogenic RNAs Disrupt Additional Regulatory Pathways Including Pre-miR Processing, mRNA Trafficking, and Translation*

While the impact of C(C)UG^{exp} RNA expression on the regulation of alternative splicing has received considerable attention, these toxic RNAs have also been reported to affect other regulatory pathways. In DM1 skeletal muscle biopsies, microRNA (miR) expression patterns are variable between studies with miR-206 overexpression compared to controls (Gambardella et al. 2010) while another study found that miR-1 and miR-335 were upregulated, miRs 29b,c and miR-33 were downregulated, and miR-1, miR133b and miR-206 were mislocalized (Perbellini et al. 2011). Mis-processing of miR-1 occurs in the DM1 heart and this has been linked to MBNL loss-of-function. MBNL1 normally recognizes a UGC motif in the pre-miR-1 loop to facilitate processing because it competes with LIN28, which blocks Dicer processing via ZCCHC11/TUT4-mediated pre-miR-1 uridylation (Rau et al. 2011).

Several studies have provided evidence that muscle weaknesses and wasting in DM may arise from impaired muscle differentiation due to alterations in CELF1 activity and the resulting effects on translation of specific target mRNAs. As mentioned previously, CELF1 protein levels increase in DM and CELF1 overexpression in transgenic mice inhibits myogenesis and causes MEF2A and p21 overexpression (Timchenko et al. 2004). Several studies have suggested that DM2 is caused by

CNBP haploinsufficiency (Chen et al. 2007; Huichalaf et al. 2009; Raheem et al. 2010) although other groups report that CNBP protein levels are not altered in this disease (Botta et al. 2006; Margolis et al. 2006; Massa et al. 2010). Interestingly, CNBP binds to the 5′ UTRs of terminal oligopyrimidine (TOP) genes encoding a variety of proteins important for translational regulation, including PABPC1, eIF1a, and eIF2, so $CCUG^{exp}$ expression has been proposed to impact the rate of global protein synthesis (Huichalaf et al. 2009; Schneider-Gold and Timchenko 2010).

RNA trafficking is another critical regulatory pathway that is altered by the expansion mutations in DM. Early work provided evidence that MBNL2 regulates the localized expression of integrin α3 to adhesion complexes (Adereth et al. 2005). More recently, transcriptome analysis has indicated that MBNL proteins regulate mRNA localization in vertebrate and invertebrate (*Drosophila*) cells resulting in effects on both translation and protein secretion (Wang et al. 2012a). For future studies, it will be important to link distinct disease manifestations of DM to alterations in the localization of specific RNAs.

2.3 Are RNA Foci Important Features of RNA Toxicity?

Nuclear RNA foci are a hallmark pathological feature of DM as well as other neurological diseases that may be linked to RNA-mediated toxicity (e.g., C9-linked ALS/FTD). However, the exact role of RNA foci in the induction and/or maintenance of DM-associated pathology remains controversial (Junghans 2009; Mahadevan 2012; Wojciechowska and Krzyzosiak 2011). RNA foci might be pathological entities that interfere directly with normal nuclear functions or protective protein–RNA complexes designed to block toxic RNAs from reaching the cytoplasm. For DM1, RNA foci are composed of both MBNL proteins and CUG^{exp} RNAs. Early studies in DM1 cells indicated that RNA foci are insoluble aggregates (Davis et al. 1997; Taneja et al. 1995) and thus formation of MBNL-C(C)UG^{exp} complexes might effectively inhibit MBNL dissociation and enhance disease. An argument against this possibility comes from single-particle tracking, fluorescence recovery after photobleaching (FRAP) and fluorescence loss in photobleaching (FLIP) experiments, which indicate that $(CUG)_{145}$ RNA transcripts expressed in C2C12 myoblasts undergo stochastic aggregation/disaggregation cycles. An early FRAP study provided evidence for rapid exchange between unbound GFP-Mbnl1 and GFP-Mbnl1 bound to aggregates (Ho et al. 2005). However, when GFP-MBNL1 was expressed at levels comparable to endogenous Mbnl1 levels, considerably less freely diffusing GFP-Mbnl1 was observed (Querido et al. 2011). Thus, RNA foci may be dynamic structures but MBNL proteins could be effectively sequestered in non-focal MBNL-C(C)UG^{exp} complexes. In addition, the dynamics of RNA foci formation and MBNL-C(C)UG^{exp} complex stability may be profoundly influenced by the larger repeats in DM1 and DM2 myofibers and neurons compared to other cell types. Irrespective of potential roles in pathogenesis, RNA foci and alternative splicing alternations are important biomarkers of RNA-mediated disease (Cardani et al. 2006, 2009; Nakamori et al. 2013).

3 Dual Disease Mechanisms from a Single Gene: Fragile X and *FMR1*

3.1 FMR1 *Epigenetic Silencing and Misregulated Translation*

Fragile X syndrome (FXS) is the most common form of inherited mental retardation with an incidence of approximately 1 in 5,000 individuals. The symptoms of FXS are variable but include intellectual disability, behavioral abnormalities such as attention deficit hyperactivity disorder (ADHD) and autistic behavior, childhood seizures, connective tissue defects, and macroorchidism (abnormally large testes) (Hernandez et al. 2009; McLennan et al. 2011; Nelson et al. 2013). Unlike many of the other microsatellite diseases, which tend to be inherited in an autosomal dominant pattern, FXS is an X-linked disorder and the mutation creates a fragile site, a locus that is prone to gaps or breaks. FXS arises when the *FMR1* CGG^{exp} repeat, which ranges between 6 and 52 repeats in unaffected individuals, expands beyond 230 repeats (Fig. 10.1b). These mutant repeats create an expanded CpG island in the DNA that becomes aberrantly methylated and coupled with histone deacetylation (Coffee et al. 1999; Sutcliffe et al. 1992) (Fig. 10.2). Thus, the primary molecular basis for this neurological disease is epigenetic silencing and loss of function (LOF) of the encoded protein, FMRP.

FMRP is an RNA-binding protein that interacts with polyribosomes to regulate local protein synthesis (Bhakar et al. 2012; Wang et al. 2012b). FMRP interacts with mRNA via two K homology (KH) domains, KH1 and KH2, as well as an arginine- and glycine-rich (RGG) domain (Siomi et al. 1993). While the RGG domain has been reported to bind to G quartets, FMRP has also been reported to bind to the coding regions of mRNA independent of these G-rich structures (Darnell et al. 2001, 2011). Tight regulation of synaptic protein synthesis in response to neuronal activity is thought to be essential for neuronal processes such as those involved in learning and memory formation. FMRP located in neurons at synapses binds to mRNA targets to fine-tune translation regulation in an activity-dependent manner (Akins et al. 2009; Bassell and Warren 2008). A combination of FMRP binding site analysis in mouse brain using high-throughput sequencing and cross-linking/immunoprecipitation (HITS-CLIP) with a polyribosome-programmed translation system revealed that FMRP binds to transcripts of presynaptic and postsynaptic proteins to stall ribosomes until the appropriate cellular signals are present (Darnell et al. 2011). The current molecular model of FXS suggests dysregulation of local protein synthesis as a consequence of FMRP loss leads to neuronal dysfunction and neurological manifestations of the disease. Consistent with this model, *Fmr1* null mice have increased translation of Fmrp target mRNAs and have similar abnormal neuron morphology and dendritic abnormities as seen in FXS brains (Berman and Willemsen 2009). *Fmr1* null mice also display phenotypes that mirror FXS symptoms (Bhogal and Jongens 2010). Similarly, point mutations in the human *FMR1* gene that disrupt the FMRP RNA binding domain KH2 also result in FXS symptoms (Nelson et al. 2013). Cumulatively, these observations strongly suggest that loss of FMRP underlies the neuronal dysfunction seen in patients and is the primary molecular cause of FXS.

3.2 Fragile X-Associated Tremor/Ataxia Syndrome and RNA Toxicity

In contrast to FXS, *FMR1* transcriptional activity does not decrease in patients with the related disease, fragile X-associated tremor/ataxia syndrome (FXTAS) (Peprah et al. 2010). Whereas FXS results from a full mutation, or $(CGG)_{>230}$, more moderately sized premutation alleles of $(CGG)_{55–200}$ cause FXTAS (Fig. 10.1b). Premutation carriers may suffer from a late-onset neurodegenerative disorder now known as FXTAS, with a typical age of onset in the early 60s, or fragile X-associated primary ovarian insufficiency (FXPOI) with infertility prior to age 40 (Hagerman and Hagerman 2004; Leehey and Hagerman 2012). Clinical features of FXTAS differ from those in FXS and include gait ataxia, progressive action tremor, autonomic dysfunction, and neurodegeneration. The distinct clinical outcomes of FXS and FXTAS result from different pathogenic mechanisms since CGG^{exp} repeats in the premutation range do not repress *FMR1* expression. On the contrary, FXTAS patients have up to eightfold higher FMR1 RNA levels with normal to slightly reduced levels of FMRP (Kenneson et al. 2001; Peprah et al. 2010; Tassone et al. 2000a, b, c). Current evidence points towards $rCGG^{exp}$ RNA toxicity in FXTAS. As in DM, FMR1 repeat-containing RNA accumulates in the nucleus (Tassone et al. 2004a, b). Furthermore, $rCGG^{exp}$ expression is sufficient to cause nuclear inclusions and neurodegeneration irrespective of its context in the *FMR1* gene, as demonstrated by mouse models expressing a CGG^{exp} reporter gene (Hashem et al. 2009). The presence of ribonuclear inclusions led researchers to ask whether FXTAS follows the RNA toxicity paradigm set by DM, in which cellular factors bind to the mutant RNA expansion and are sequestered away from their normal cellular functions (Fig. 10.2, RNA GOF). Indeed, several RNA-binding proteins have been proposed to be sequestered by $rCGG^{exp}$ repeats, including MBNL1, hnRNP A2/B1, hnRNP G, Sam68, and Pur α (Li and Jin 2012; Tassone and Hagerman 2012).

A combination of animal studies and in vitro assays identified hnRNP A2/B1 and Purα as candidate proteins that bind to $rCGG^{exp}$ RNAs and are titrated away from their normal functions (Jin et al. 2007; Muslimov et al. 2011; Sofola et al. 2007). HnRNP A2/B1 is an abundant nuclear RNA-binding protein containing a glycine-rich and two RRM domains and disruption of hnRNP A2/B1 function has been proposed to alter RNA processing in FXTAS. Furthermore, hnRNP A2/B1 tethers another RNA-binding protein to CGG^{exp} RNA, CELF1, previously implicated in DM pathogenesis (Sofola et al. 2007). Overexpression of either hnRNP A2/B1 or CELF1 in a fly model of FXTAS alleviates the neurodegenerative phenotype observed in this model. Furthermore, CGG^{exp} RNA induces mislocalization of hnRNP A2/B1 target RNAs from dendrites to the neuronal cell body, presumably via titration of hnRNP A2/B1 onto the repeats. Delivery of additional hnRNP A2/B1 to neurons restores dendritic localization of target RNA, supporting the proposed role of hnRNP A2/B1 in FXTAS (Muslimov et al. 2011). Purα is an RNA- and DNA-binding protein and has diverse roles in transcription activation, DNA replication and mRNA localization. In the brain, Purα is involved in neuronal cell proliferation and neurodevelopment and plays a role in regulating the dendritic and axonal localization of mRNA targets (Hokkanen et al. 2012; Johnson et al. 2006; Ohashi et al. 2000). Like hnRNP A2/B1, overexpression of Purα suppresses neurodegeneration in the fly model of FXTAS and this protein was identified in nuclear inclusions in FXTAS patient brains, suggesting that sequestration of Pur α may indeed play a role in the disease pathogenesis (Jin et al. 2007).

Another RNA-binding protein found to be present in FXTAS nuclear inclusions is Sam68 (Src-associated substrate during mitosis of 68 kDa) (Sellier et al. 2010). Sam68 is an alternative splicing regulator that has diverse roles in cellular signaling, apoptosis, and neuronal functions. Mice lacking Sam68 display motor coordination defects reminiscent of ataxia (Lukong and Richard 2008; Ramakrishnan and Baltimore 2011; Sellier et al. 2010). Sam68 localizes to nuclear inclusions following transfection of $rCGG^{exp}$ repeats in cell models and, in turn, recruits other RNA-binding proteins, including MBNL1 and hnRNP G (Sellier et al. 2010). Missplicing of pre-mRNA targets of Sam68 has been reported to occur in FXTAS patient brains, and overexpression of a Sam68 mutant lacking CGG binding function in cell culture is sufficient to rescue the splicing abnormalities.

FXTAS-associated $rCGG^{exp}$ nuclear inclusions also contain other proteins, including ubiquitin, heat-shock proteins such as αB-crystallin, lamin A/C, myelin basic protein, and DROSHA/DGCR8 (Greco et al. 2002; Iwahashi et al. 2006; Sellier et al. 2013). DGCR8 is a double-stranded RNA-binding protein and the DROSHA/DGCR8 complex is responsible for the miRNA processing step that converts pri-miRNA into pre-miRNA. DGCR8 binds to $rCGG^{exp}$ RNA and recruits its partner DROSHA into the inclusions. Levels of mature miRNA decrease in FXTAS patient brains, and overexpression of DGCR8 rescues $rCGG^{exp}$-induced neuronal cell death, suggesting that miRNA dysregulation contributes to neuronal dysfunction in FXTAS (Sellier et al. 2013).

Cumulatively, these studies suggest that multiple RNA-binding proteins may be titrated by $rCGG^{exp}$ in FXTAS resulting in widespread dysregulation of RNA processing and neurodegeneration. Larger CGG^{exp} mutations in *FMR1* lead to FMRP loss-of-function and the subsequent loss of translational regulation and synaptic dysfunction seen in FXS. The observation that microsatellite expansions in a non-coding region of a single gene cause several distinct disease outcomes and disrupt multiple molecular pathways has also been observed in DM1/CDM and is likely to be repeated for other unstable microsatellite diseases.

4 Compound Threats: RNA and Protein Toxicity

Microsatellite expansion diseases have been traditionally classified as either or protein- or RNA-mediated. However, these expansions can also pose compound threats highlighted by the recent discoveries of bidirectional transcription through repeat regions and RAN translation (Batra et al. 2010; Zu et al. 2011). Indeed, in vitro studies suggest that CAG•CTG expansion mutations have the potential to produce nine toxic molecules including two pathogenic RNAs (CAG^{exp}, CTG^{exp}) as well as seven homopolymeric proteins produced by conventional and RAN translation (Pearson 2011; Zu et al. 2011). Below, we discuss expansion diseases, including HDL2 and several spinocerebellar ataxias, in which both RNA and protein toxicity have been implicated.

4.1 Huntington Disease-Like 2

Huntington disease-like 2 (HDL2) is a microsatellite expansion disease that is thought to trigger the production of a combination of toxic RNA and protein species (Margolis et al. 2004). HDL2 is a phenocopy of Huntington's disease (HD), both of which are dominantly inherited diseases characterized by motor coordination defects, dementia, cortical and striatal neurodegeneration, and eventual death within decades of diagnosis (Rudnicki et al. 2008). HD is caused by a CAG^{exp} in the coding region of the Huntingtin (*HTT*) gene that results in the expression of a HTT protein with an expanded polyglutamine tract (polyGln or polyQ) (Ha and Fung 2012). HD is a member of the CAG^{exp} neurological disorders characterized by neuronal inclusions of ubiquitinated polyGln. PolyGln is thought to be neurotoxic due to several mechanisms including blockage of the ubiquitin proteasome system (UPS) and disruption of mitochondrial function (Finkbeiner 2011; Orr and Zoghbi 2007).

Although the striking clinical similarities between HD and HDL2 suggest that these diseases have a similar pathogenic mechanism, the mutation responsible for HDL2 was originally described as a CTG^{exp} in the junctophilin-3 (*JPH3*) gene, which is primarily expressed in the brain (Holmes et al. 2001). Whereas unaffected individuals have 6–27 CTG repeats, HDL2 patients have expansions of 41–58 CTG repeats (Fig. 10.1b). The JPH3 protein is involved in formation of a structure that connects the plasma membrane with the endoplasmic reticulum to help control the release of calcium ions during neuronal activity. Loss of JPH3 expression may contribute to HDL2 pathogenesis (Seixas et al. 2012). While the CTG^{exp} mutation is present in alternatively spliced *JPH3* exon 2a, variations in transcription and alternative splicing cause the repeat to be located in different transcript regions. JPH3 transcripts that include exon 2a encode a truncated JPH3 isoform while the CTG^{exp} may result in translation of polyleucine and polyalanine tracts in the truncated protein depending on which exon 2a 3′ splice site is utilized. A third exon 2a splice variant places the CTG^{exp} in the 3′UTR suggesting that toxic CTG^{exp} RNAs could be expressed and CTG^{exp} nuclear RNA foci are detectable in HDL2 neurons (Rudnicki et al. 2007). These foci not only contain CTG^{exp} RNA but also MBNL1, the splicing factor implicated in DM, and several MBNL1-dependent missplicing events seen in DM are also observed in HDL2 neurons.

Despite this data, the RNA-mediated model of HDL2 pathogenesis cannot explain the remarkable commonalities between HDL2 and HD. This discrepancy has now been addressed by the discovery of bidirectional transcription through the repeat that creates an antisense transcript containing a CAG^{exp} encoding polyGln as seen in HD (Wilburn et al. 2011). Early histological analysis of HDL2 patient brains also revealed ubiquitin-positive polyGln nuclear inclusions, reminiscent of those observed in HD, independent of RNA foci (Rudnicki et al. 2008; Walker et al. 2002). Cumulatively, these results indicate that toxic RNA and protein (Fig. 10.2, RNA GOF and protein GOF) may interact synergistically to wreak havoc on neuronal pathways although polyGln toxicity appears to play the predominant role in HDL2.

4.2 *Spinocerebellar Ataxia Types 3, 8 and 10*

A similar interplay of protein and RNA toxicity has been observed in spinocerebellar ataxia types 3, 8, and 10 (SCA3, SCA8, SCA10). The SCAs are a large group of inherited neurological diseases in which neurological dysfunction in the cerebellum and brainstem causes motor coordination defects known as ataxias (Hersheson et al. 2012; Orr 2012a). SCAs can result from mutations in 37 genes (SCA1-37), but intriguingly several of the mutations are microsatellite repeat expansions including coding CAG^{exp} mutations in SCA1, 2, 3, 6, 7, and 17 and noncoding CTG^{exp} in SCA8, CAG^{exp} in SCA12, $ATTCT^{exp}$ in SCA10, TGGAA in SCA31, and $GGCCTG^{exp}$ in SCA36 (Matilla-Duenas et al. 2012; Serrano-Munuera et al. 2013).

4.2.1 SCA3: Toxic ATXN3 Protein and CAG^{exp} RNA

SCA3 (Machado–Joseph disease, MJD) is characterized by late-onset ataxia and neurodegeneration and is the most prevalent SCA worldwide (Orr 2012a). SCA3 is associated with CAG repeats that expand from the normal (12–37) to a pathogenic (61–84) range resulting in an extended polyGln tract in the C-terminus of the ataxin-3 (ATXN3) protein. ATXN3 is a deubiquitinating enzyme (DUB) that is involved in protein homeostasis and transcription and may regulate the expression of genes involved in stress response pathways (Orr 2012a). As with other polyGln diseases, SCA3 neurons contain nuclear inclusions and the extended polyGln tract triggers proteolytic cleavage of mutant ATXN3 protein. These cleavage products are detectable in patient brains and fragment accumulation in transgenic mice is neurotoxic (Goti et al. 2004; Haacke et al. 2006; Paulson et al. 1997; Warrick et al. 1998; Wellington et al. 1998) (Fig. 10.2, protein GOF).

While many studies indicate that SCA3 disease is predominantly caused by the mutant ATXN3 protein (Costa Mdo and Paulson 2012; Orr 2012a), SCA3 $rCAG^{exp}$ RNA has also been implicated in SCA3 pathogenesis. Similar to DM, a toxic RNA-induced protein sequestration hypothesis has been tested in a fly SCA3 model as well as CAG^{exp}-expressing mouse and nematode models (Hsu et al. 2011; Li et al. 2008; Wang et al. 2011). To evaluate the toxicity of CAG^{exp} independent of polyGln, interruption of CAG^{exp} with CAA repeats, which still encode polyGln but do not cause RNA toxicity, mitigates the neurodegeneration observed in the fly SCA3 model (Li et al. 2008). Furthermore, expression of a CAG^{exp} in the 3′UTR of a reporter gene in human cell lines is sufficient to elicit MBNL1-containing nuclear foci and MBNL1-dependent splicing changes previously reported in DM (Mykowska et al. 2011). The RNA-binding protein Orb2, also known as cytoplasmic polyadenylation element (CPE)-binding protein 1 (CPEB1), has been implicated in SCA3 since it is a modifier of neurotoxicity in the SCA3 fly model (Shieh and Bonini 2011). Orb2/CPEB1 contains two RRMs and a zinc finger domain, which it uses to bind the CPE of target mRNAs and regulate translation (Richter 2007). The mRNA targets of Orb2/CPEB1 are enriched for functions related to synaptic plasticity and neuronal growth, suggesting that this protein plays a role in learning and memory. Orb2 overexpression in the SCA3 fly model partially suppresses neurotoxicity hinting that the rCAG^{exp} RNAs might interfere with normal Orb2 functions and lead to the loss of translation regulation in SCA3 neurons (Shieh and Bonini 2011). Similarly, Orb2 colocalizes with CGG^{exp} nuclear foci and is a genetic modifier in a fly FXTAS model (Cziko et al. 2009).

4.2.2 SCA8: Bidirectional Transcription and the Discovery of RAN Translation

SCA8 is characterized by ataxia, slurred speech, and abnormal eye movements (nystagmus) and is associated with a CAG•CTG expansion that undergoes bidirectional transcription to produce a CAG^{exp} from *ATXN8* and a CUG^{exp} RNA from the *ataxin-8 opposite strand* (*AXN8OS*) gene (Ikeda et al. 2008; Moseley et al. 2006). Unaffected individuals have 16–50 CAG•CTG repeats, whereas patients have expansions of 71–1,300 repeats (Fig. 10.1b). Originally, SCA8 pathogenesis was thought to arise primarily from RNA toxicity due to CUG^{exp} expression from the noncoding *AXN8OS* gene (Koob et al. 1999). As predicted, CUG^{exp} RNA foci colocalize with MBNL1 in SCA8 neurons and MBNL1 overexpression reverses SCA8-induced splicing errors (Chen et al. 2009; Daughters et al. 2009). However, *ATXN8* also encodes a short ATXN8 protein that consists primarily of polyGln and SCA8 neurons contain polyGln and ubiquitin-positive inclusions in the nucleus similar to those observed in HD and other polyGln diseases (Moseley et al. 2006) (Fig. 10.3).

Discovery of a third pathogenic mechanism in SCA8 has profound implications not only on the microsatellite expansion and neurological disease fields but also on our understanding of translational regulatory mechanisms. While investigating the role of polyGln in SCA8, ATXN8 expression constructs were generated to remove the ATG initiation codon upstream of the CAG^{exp} encoding polyGln (Cleary and Ranum 2013; Zu et al. 2011). Surprisingly, removal of the start codon fails to ablate translation and polyGln is still expressed. More surprisingly, the translation of the CAG^{exp} repeat occurs in all three open reading frames (ORFs) leading to co-expression of polyGln, polySer, and polyAla in the absence of frameshifting. Immunological evidence as well as direct protein sequencing supports the existence of this repeat-associated non-ATG, or RAN, translation. All three RAN proteins translated from CAG^{exp} RNAs (polyGln, polyAla, polySer) accumulate in

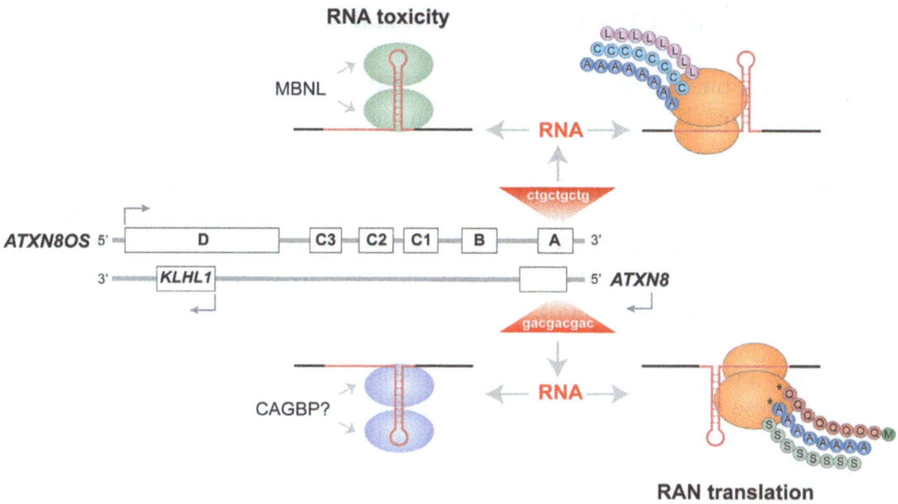

Fig. 10.3 Bidirectional transcription across microsatellite expansions may generate both pathogenic RNAs and proteins. Transcription of a CTG^{exp} mutation in the *ATXN8OS* gene (composed of noncoding exons A–D) produces toxic RNA hairpins (*red*) that either sequester the MBNL proteins or produce polyLeu (L, *pink circles*), polyCys (C, *turquoise*) or polyAla (A, *blue*) by RAN translation. Transcription of the CAG^{exp} on the opposite strand (*ATXN8*) results in RAN-generated polyGln (Q, *orange*), polyAla (A, *blue*) or polySer (S, *green*) homopolymer polypeptides or a toxic RNA that might sequester an unknown rCAG^{exp} binding protein (CAGBP). While all six RAN proteins are observed in transfected cells, only polyGln (*asterisk*), which is initiated with a conventional methionine codon (*green*) in ATXN8, and polyAla (*asterisk*) have been detected in vivo

transfected cells and polyGln and polyAla proteins have been shown to accumulate in vivo in mouse and human SCA8 cerebellar Purkinje cells (Cleary and Ranum 2013; Zu et al. 2011) (Figs. 10.2 and 10.3).

Currently, RAN translation has been implicated in four microsatellite expansion diseases including ALS/FTD, DM1, FXTAS as well as SCA8 (Ash et al. 2013; Mori et al. 2013b; Todd et al. 2013; Zu et al. 2011). For FXTAS, considerable evidence supports the hypothesis that this disease is RNA-mediated, but patient neurons also contain large and ubiquitin-positive nuclear inclusions, a hallmark of protein-mediated disease. This paradox led to the recent discovery of RAN translation in FXTAS with in vitro and in vivo evidence for the accumulation of polyglycine (polyGly) containing aggregates (Todd et al. 2013). Additional studies show that polyAla can also be expressed across CGG^{exp} constructs in transfected cells. Suppression of polyGly expression partially rescues CGG^{exp}-induced toxicity and loss of cell viability in FXTAS models and polyGly accumulates in ubiquitinated intranuclear inclusions in patient brains. Thus, some microsatellite expansion diseases, previously labeled as either RNA mediated or protein mediated, involve multiple pathogenic mechanisms and toxic agents (Ling et al. 2013). Clearly, reexamination of patient brain sections from other microsatellite diseases for RAN protein aggregates should now be a priority.

4.2.3 SCA10: AUUCU^{exp} RNA, hnRNP K and Apoptosis

Another RNA-binding protein, hnRNP K (HNRNPK), is implicated in spinocerebellar ataxia type 10 (SCA10). SCA10 is an autosomal dominant disorder characterized by ataxia, seizures, mild peripheral nerve and cognitive impairment. Disease results from an ATTCT repeat expansion in intron 9 of the *ATXN10* gene (Teive et al. 2011). This pentanucleotide repeat is relatively short in the normal population (10–32) while SCA10 patients have very large expansions of 800–4,500 repeats (Fig. 10.1b). Although the ATTCT^{exp} does not alter *ATXN10* expression levels or splicing (Wakamiya et al. 2006), the AUUCU^{exp} RNA accumulates in nuclear and cytoplasmic foci in SCA10 mouse models and patient fibroblasts (White et al. 2010, 2012). The presence of cytoplasmic foci is a somewhat surprising observation although these structures may be released from the nuclear compartment during mitosis. In vitro, hnRNP K binds to AUUCU^{exp} RNA and colocalizes with AUUCU^{exp}-containing foci (White et al. 2010). HnRNP K contains three KH domains that have a high affinity for C-rich clusters and SCA10 fibroblasts show aberrant splicing of the hnRNP K target β-tropomyosin, suggesting that loss of hnRNP K function contributes to missplicing in SCA10. HnRNPK also acts as a docking protein for multiple factors to modulate chromatin remodeling, transcription, and translation to facilitate cross-talk between multiple gene expression tiers (Bomsztyk et al. 2004). Another study suggests that loss of hnRNP K function triggers apoptosis in AUUCU^{exp}-expressing cells (White et al. 2010). Sequestration of hnRNP K by AUUCU^{exp} RNA blocks its binding to PKCδ, which could allow PKCδ to promote caspase 3-mediated apoptosis. In accordance with this model, overexpression of hnRNP K suppresses the AUUCU^{exp}-induced apoptosis pathway and partially rescues cell viability. Nevertheless, additional cell and animal models must be developed to substantiate the hnRNP K LOF model for SCA10.

5 An Intrinsic Curse: Microsatellite Expansions in RNA-Binding Proteins

5.1 SCA2 and ATXN2

While some expansion mutations indirectly affect RNA-binding protein function via sequestration by repeat expansion RNAs, other diseases stem from expansion mutations in genes encoding RNA-binding proteins. One of the most common types of ataxia, spinocerebellar ataxia type 2 (SCA2), is caused by a CAG^{exp} in *ATNX2*, the gene encoding the RNA-binding protein ataxin-2 (Orr 2012a; Rub et al. 2013). SCA2 patients suffer from cerebellar ataxia in addition to decreased reflexes and polyneuropathy, which affects peripheral nerves throughout the body. Cell death has been observed in both SCA2 motor neurons and Purkinje cells. The genetic basis of SCA2 is an expansion from 15–24 to 32–200 CAG repeats in exon 1 of *ATXN2*, which results in a polyGln expansion in the ATXN2 N-terminus (Fig. 10.1b).

ATXN2 has been implicated in various cellular processes including Golgi-mediated transport, calcium regulation, formation of stress granules and P-bodies, and regulation of RNA post-transcriptional modification and translation (Magana et al. 2013). Several lines of evidence suggest that ATXN2 participates directly and indirectly in RNA metabolism. The ATXN2 protein contains an Sm-like (Lsm) domain, also found in factors that function in pre-RNA processing and mRNA decay (Albrecht et al. 2004), as well as a PAM2 domain that mediates ATXN2 binding to PABPC1, the major cytoplasmic poly(A)-binding protein involved in regulating poly(A) tail length, mRNA stability, and translation (Ralser et al. 2005). ATXN2 and PABPC1 colocalize in cells, assemble onto polyribosomes and are recruited to stress granules, which suggests that ATXN2 and PABPC1 cooperate to sequester mRNAs in stress granules to downregulate their translation during cell stress (Magana et al. 2013; Satterfield and Pallanck 2006).

Another RNA-binding protein, originally identified as a novel protein that interacts with ATXN2, is ataxin-2-binding protein (A2BP1, now known as RBFOX1) (Shibata et al. 2000). RBFOX1 is predominantly expressed in the brain and skeletal muscle and belongs to the RBFOX family of alternative splicing factors (Fogel et al. 2012; Gehman et al. 2011; Underwood et al. 2005). Murine Rbfox1 binds directly to pre-mRNA to activate or repress exon inclusion in a position-dependent manner (Sun et al. 2012). *Rbfox1* null mice have widespread splicing abnormalities in the brain, as well as neuronal excitability and seizures. A combination of splicing microarrays and binding site analysis using individual-nucleotide resolution cross-linking immunoprecipitation (iCLIP) reveals that Rbfox1 regulates the alternative splicing of gene transcripts involved in membrane excitation and synaptic transmission (Gehman et al. 2011). Similarly, knockdown of RBFOX1 in human neuronal cultures results in splicing abnormalities in genes involved in neuronal development (Fogel et al. 2012). The binding of RBFOX1 to ATXN2 suggests that the two proteins interact to regulate alternative splicing in the brain and that RBFOX1 may play a role in SCA2.

Several molecular mechanisms have been proposed for SCA2 pathogenesis, including formation of polyGln aggregates and deleterious gain-of-function of mutant ATXN2. Mutant ATXN2, but not the wild-type protein, has been observed to interact directly with the calcium channel $InsP_3R1$ in neuronal cultures, which may lead to altered calcium signaling and excitotoxic cell death (Liu et al. 2009). Treatment of the cells with an inhibitor of another calcium channel, the ryanodine receptor, reduces cell death highlighting the calcium regulatory pathway as a potential therapeutic target for SCA2. The expanded polyGln tract in mutant ATXN2 also triggers ATXN2 protein misfolding and polyGln aggregation in ubiquinated intranuclear and cytoplasmic inclusions in SCA2 neurons (Huynh et al. 2000; Koyano et al. 1999). In addition, mutant ATXN2 undergoes proteolytic cleavage and the C-terminal fragment might have an altered function compared to the holoprotein while the N-terminal fragment, containing the polyGln expansion, accumulates in potentially toxic aggregates (Huynh et al. 2000). Interestingly, intermediate-sized *ATXN2* expansions are associated with an increased risk for developing amyotrophic lateral sclerosis (ALS) and parkinsonism (Elden et al. 2010; Simon-Sanchez et al. 2005). Supporting a link between ATXN2 and ALS, ATXN2 accumulates in discrete foci, possibly stress granules, in degenerating motor neurons of ALS patients (Li et al. 2013). Furthermore, ATXN2 interacts in an RNA-dependent manner with TDP-43, which plays a key role in ALS pathology, and modifies TDP-43 toxicity in several ALS model systems (Elden et al. 2010). These results underscore the role of ATXN2 in RNA processing and turnover and provide genetic and mechanistic links between ATXN2 and ALS.

5.2 *Oculopharyngeal Muscular Dystrophy and PABPN1*

Another neuromuscular disorder that arises from a microsatellite expansion in the coding region of an RNA-binding protein is oculopharyngeal muscular dystrophy (OPMD) (Brais 2009; Messaed and Rouleau 2009). OPMD is a late-onset muscular dystrophy characterized by muscle weakness, ptosis (eyelid drooping), and dysphagia (difficulty in swallowing that may lead to aspiration pneumonia). OPMD is caused by a GCN^{exp} in *PABPN1*, which encodes the major nuclear polyadenylate-binding protein that plays a vital role in pre-mRNA 3′ end processing and poly(A) tail formation (Banerjee et al. 2013).

Initially, OPMD was associated with GCG^{exp} expansions (Brais et al. 1998), but additional mutations, including point mutations, have also been identified that result in a stretch of GCN codons encoding an expanded polyalanine (polyAla) tract in the N-terminus of PABPN1 (Nakamoto et al. 2002; Robinson et al. 2006). Whereas unaffected individuals typically have 10 GCN repeats, patients have 11–17 repeats, resulting in one to seven additional alanine residues (Fig. 10.1b). Similar to other neurological disorders, a central mystery is why a mutation in an essential and ubiquitously expressed protein causes a late-onset disease that primarily affects specific tissues, in this case facial, tongue, and extremity muscles. Histological analysis of patient skeletal muscle biopsies reveals hallmarks of muscular dystrophies, such as changes in muscle fiber size variability, centralized myonuclei, and an overall loss of myofibers with excessive fibrous and fatty tissue (Tome et al. 1997). Moreover, OPMD muscle fibers have mutant PABPN1 intranuclear inclusions that appear as zones of tubular filaments (Tome et al. 1997; Tome and Fardeau 1980).

Controversy exists over the toxicity of these inclusions and whether they induce toxicity or serve a protective role in confining the mutant protein. Several lines of evidence support the latter hypothesis. In an OPMD cell model, the most toxic form of mutant PABPN1 is soluble and not in insoluble inclusions while inclusion disruption increases cell death (Messaed et al. 2007). Using PABPN1 constructs with variable polyAla length, mutant PABPN1 is most toxic when the expansions are longer although these expanded proteins do not form inclusions (Klein et al. 2008). In a similar vein, an OPMD fly model recapitulates the muscle phenotype observed in the disease, despite the fact that it lacks inclusions (Chartier et al. 2006). PABPN1 forms oligomeric structures prior to inclusion formation so soluble and oligomeric forms of mutant PABPN1 may contribute to the cellular toxicity (Raz et al. 2011). In addition to inclusion-independent models of PABPN1 toxicity, several inclusion-dependent models have been proposed. PABPN1 expansion mutations may cause a dominant-negative effect leading to inclusions that, in turn, sequester other cellular factors and disrupts several cellular pathways. While PABPN1 is an integral constituent of the inclusions, other co-localizing components include transcription factors, polyadenylated RNA, RNA-binding proteins (including CELF1), and components of the UPS (Brais 2009; Corbeil-Girard et al. 2005). As noted previously, UPS dysfunction has been implicated in other neurodegenerative diseases (Ciechanover and Brundin 2003; Dennissen et al. 2012) and addition of a proteasome inhibitor in an OPMD cell model exacerbates inclusion formation in a dose-dependent manner while overexpression of heat shock proteins decreases inclusions (Abu-Baker et al. 2003). Furthermore, transcriptome analysis indicates that genes encoding UPS components are misregulated in OPMD (Anvar et al. 2011). The specificity of the UPS involvement in OPMD is highlighted by the observation that the UPS is consistently misregulated in OPMD while other protein degradation pathways are not. These results have led to a disease model in which an age-related decrease in UPS function causes accumulation of misfolded PABPN1 that aggregates, forms inclusions that, in turn, recruit UPS thus further blocking the UPS in OPMD cells (Raz et al. 2013). In addition to blocking the UPS, recruitment of other cellular factors to aggregates may disrupt additional cellular pathways. To support the toxicity of aggregate formation, oral treatment of anti-aggregation drugs attenuates pathology in an OPMD mouse model (Davies et al. 2005, 2006).

Inclusion-dependent models of OPMD propose that a reduction of soluble PABPN1 due to aggregation leads to PABPN1 loss-of-function. PABPN1 consists of an N-terminal region with the polyAla stretch, a central domain with a conserved RNA recognition motif (RRM) and a C-terminal domain that may also be involved in fibril formation (Winter et al. 2012). PABPN1 was originally characterized as a nuclear factor that increases processivity and efficiency of poly(A) addition by poly(A) polymerase. PABPN1 coats the poly(A) tail, possibly to maintain poly(A) tail length that impacts downstream events including nuclear RNA export, mRNA translation and stability (Banerjee et al. 2013). PABPN1 also appears to shuttle from nucleus to cytoplasm, potentially aiding in nuclear export and translation, and then cooperates with, or is replaced by, its cytoplasmic partner PABPC1 (Lemay et al. 2010). Knockdown of PABPN1 in primary mouse myoblasts prepared from extraocular, pharyngeal, and limb muscles causes shortening of poly(A) tails and accumulation of nuclear poly(A) RNA, supporting a role for PABPN1 in nuclear export (Apponi et al. 2010). Furthermore, loss of PABPN1 in myoblasts results in decreased proliferation and differentiation hinting that myogenesis might be compromised in OPMD. Recently, PABPN1 was found to regulate alternative cleavage and polyadenylation (APA) (de Klerk et al. 2012; Jenal et al. 2012). PABPN1 depletion induces selection of proximal cleavage sites and widespread 3′ UTR shortening. Analysis of OPMD patient cells, as well as an OPMD mouse model, also show 3′ UTR shortening suggesting that mutant PABPN1 and PABPN1 depletion are equivalent phenomena that result in APA dysregulation. These results support a model in which PABPN1 loss-of-function in OPMD leads to widespread molecular abnormalities, including shortening of 3′UTRs, decrease in poly(A) tail length, and blockage of poly(A)-mRNA nuclear export, which results in defects in myogenesis and muscle function. In addition to this mRNA-centric view, PABPN1 also promotes the turnover of a class of long noncoding (lnc)RNAs (Beaulieu et al. 2012). Further investigation into the normal functions of PABPN1 are required to clarify how disruption of these processes results in the distinct pathological phenotypes associated this neuromuscular disease.

6 The Horizon in Microsatellite Expansion Disorders

6.1 C9ORF72 Expansions in ALS/FTD

The microsatellite expansion field has received increased attention with the discovery that a $G_4C_2^{exp}$ mutation in the *C9ORF72* gene is the most common known cause of familial amyotrophic lateral sclerosis (ALS) and frontotemporal dementia (FTD) (DeJesus-Hernandez et al. 2011; Renton et al. 2011). Prior to this discovery, there were indications that these two diseases represent different manifestations of a spectrum disorder with common genetics, pathology, and clinical features. Patients diagnosed with ALS, also known as Lou Gehrig's disease, suffer from muscle wasting and paralysis resulting from denervation caused by the death of upper and lower motor neurons (Van Damme and Robberecht 2013). In contrast, FTD is a condition characterized by loss of neurons in the frontal and temporal lobes of the brain, which results in behavioral changes, dementia, and speech abnormalities or progressive non-fluent aphasia. However, FTD patients may suffer from ALS-like symptoms, such as motor deficits caused by motor neuron dysfunction (Lomen-Hoerth et al. 2002). Conversely, up to 50 % of ALS patients also develop FTD-like symptoms, including dementia and cognitive changes, related to impairment of frontotemporal functions (Giordana et al. 2011).

A positive family history of disease or related symptoms occurs in approximately 10 % of ALS, and 25–50 % of FTD, patients (Graff-Radford and Woodruff 2007; Gros-Louis et al. 2006; Rohrer et al. 2009). These cases are referred to as familial ALS and FTD, while the remainder of cases are sporadic. FTD has been linked to mutations in several genes including microtubule-associated protein tau (*MAPT*), progranulin (*GRN*), valosin-containing protein (*VCP*), charged multivesicular body protein 2B (*CHMP2B*), and *C9ORF72* (Rademakers et al. 2012). Similarly, ALS-linked mutations include superoxide dismutase 1 (*SOD1*), ataxin 2 (*ATXN2*), TATA-binding protein associated factor 15 (*TAF15*), Ewing's sarcoma breakpoint region 1 (*EWSR1*), angiogenin (*ANG*), senataxin (*SETX*), fused in sarcoma (*FUS*), TDP-43 (*TARDBP*), and *C9ORF72* (Robberecht and Philips 2013). Many of the genes have been linked to both ALS and FTD but are more commonly associated with one of these diseases (Al-Chalabi et al. 2012). While the overlapping genetics and clinical features support the view of ALS/FTD is a spectrum disorder, even more striking are the commonalities between FTD and ALS molecular pathology. Most FTD and ALS cases have characteristic neuronal cytoplasmic inclusions, which contain certain

RNA-binding proteins. For example, mutations in FUS and TDP-43 are less common in FTD than ALS, yet both ALS and FTD patient neurons often contain cytoplasmic aggregates of FUS or TDP-43 even without a corresponding gene mutation (Kwiatkowski et al. 2009; Neumann et al. 2006, 2009; Vance et al. 2009). These findings suggest that common molecular pathways may be disrupted by a number of different ALS/FTD-linked genetic variants. In addition to potential effects on RNA processing, RNA editing may be disrupted in ALS due to loss of ADAR2 responsible for A-to-I RNA editing (Aizawa et al. 2010; Kawahara et al. 2004).

The *C9ORF72* expansion mutation is surprisingly common among the genetic variants linked to ALS/FTD, and can cause both sporadic and familial forms of ALS and FTD, strongly supporting the disease spectrum hypothesis. Although the distributions of normal and disease-associated G_4C_2 repeat lengths are still being actively investigated, thus far it appears that unaffected individuals have <30 repeats, generally with 2–5 repeats, whereas ALS/FTD patients have considerably longer expansions of 700–1,600 repeats (DeJesus-Hernandez et al. 2011; Dobson-Stone et al. 2012; Gijselinck et al. 2012; Majounie et al. 2012; Robberecht and Philips 2013) (Fig. 10.1b). The expansion is located in the first intron of the *C9ORF72* gene, but this region is also the promoter of an alternative isoform (Fig. 10.4). The gene encodes three mRNA, and two protein, isoforms. One possibility is that the expansion alters activity of the promoter resulting in decreased gene expression levels. In support of this hypothesis, some studies have found that ALS patients with the $G_4C_2^{exp}$ mutation have reduced levels of C9ORF72 mRNA (DeJesus-Hernandez et al. 2011; Gijselinck et al. 2012). This observation suggests that the repeat could interfere with expression of the C9ORF72 protein and cause haploinsufficiency, however no clear reduction in protein levels has been reported in these ALS/FTD patients.

While the function of the C9ORF72 protein remains obscure, homology searches reveal that it is structurally related to DENN (differentially expressed in normal and neoplasia) domain proteins (Levine et al. 2013). DENN domain proteins are GDP/GTP exchange factors (GEFs) for Rab-GTPases. Therefore, based upon homology, C9ORF72 might be involved in membrane trafficking related to Rab-GTPase switches. C9ORF72 mRNA is expressed in a wide variety of cell types and the C9ORF72 protein is primarily cytoplasmic in neurons (DeJesus-Hernandez et al. 2011). Several immunohistochemical studies have revealed that the $G_4C_2^{exp}$ causes no apparent changes in cellular distribution of the C9ORF72 protein and no abnormal C9ORF72 aggregates, suggesting that altered function of this protein may not be the primary molecular mechanism in ALS/FTD (Rademakers et al. 2012). Although patients have no C9ORF72 protein aggregates, many studies have revealed that patients with *C9ORF72* mutations have the neuropathological hallmark of TDP-43 positive inclusions in the brain and spinal cord, as seen in other ALS/FTD patients. In addition to this characteristic TDP-43 pathology, these patients have a unique feature of neuronal inclusions that are negative for TDP-43, yet positive for UPS-associated proteins such as ubiquitin, ubiquilins, and sequestosome-1 (SQSTM1), also known as ubiquitin-binding protein p62 (p62) (Al-Sarraj et al. 2011). The presence of UPS suggests aberrant accumulation of unidentified molecules marked for degradation, possibly unique to patients with *C9ORF72* mutations.

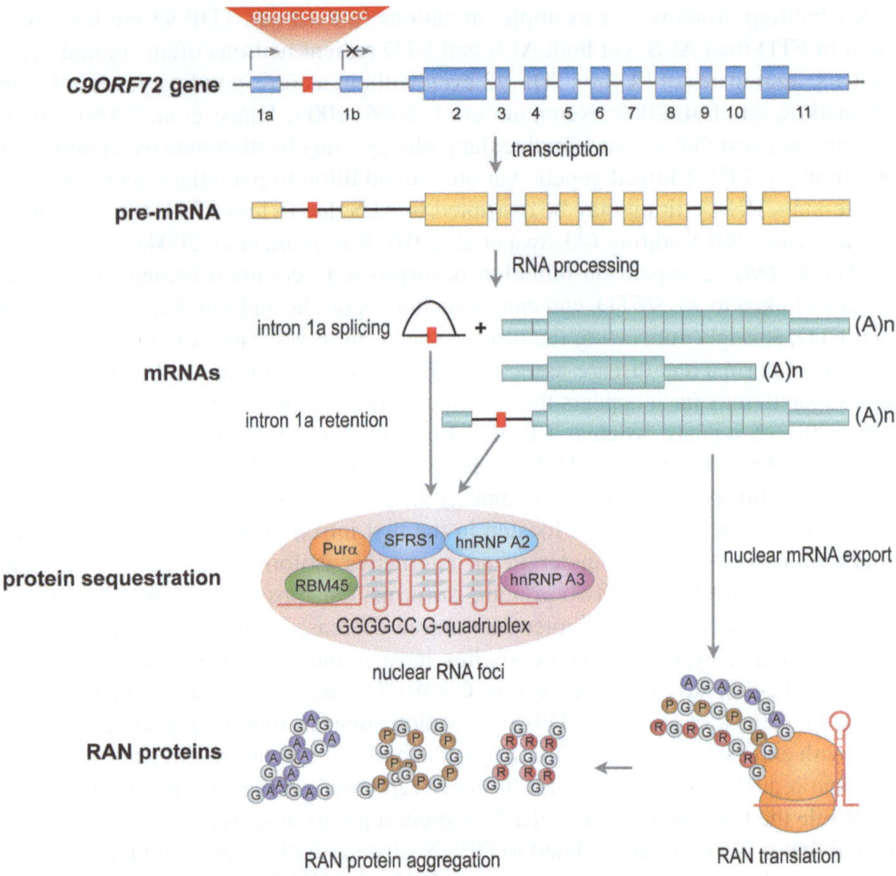

Fig. 10.4 Toxic RNAs and peptides in *C9ORF72* ALS/FTD. The GGGCC^{exp} mutation (*red box*) in *C9ORF72* intron 1a blocks transcription initiation (*arrow*) from exon 1b but promotes initiation from exon 1a. RNA processing generates at least two mRNAs and a released intron 1a lariat (or intron 1a might be retained during alternative splicing) that may fold into a G-quadruplex and accumulate in nuclear foci (*pink ellipsoid*) together with multiple nuclear RNA-binding proteins (RBM45, Purα, SFRS1, hnRNP A2, hnRNP A3). Further repeat expansion may lead to failure of this retention mechanism due to titration of rGGGCC^{exp}-binding factors leading to nuclear export of intron 1a-containing mRNA, RAN translation, and RAN protein aggregation

Several recent reports have shown that RAN translation may be an important pathogenic feature of *C9ORF72* ALS/FTD. The $G_4C_2^{exp}$ was initially predicted to form hairpins but may also fold into G-quadruplex structures (Ash et al. 2013; Fratta et al. 2012; Reddy et al. 2013) (Fig. 10.4). While G-quadruplexes in 5′UTRs often suppress canonical cap-dependent translation, these structures have also been reported to aid in noncanonical IRES-mediated translation initiation (Morris et al. 2010). To address the question of whether RAN translation initiates from $rG_4C_2^{exp}$ RNA, antibodies against the predicted RAN translation products were used to assess the presence of these products in *C9ORF72* patient tissues (Ash et al. 2013; Mori et al. 2013b). As hypothesized, RAN translation products are detectable from all three ORFs resulting in the dipeptide-repeat (DPR) proteins poly(Gly-Ala), poly(Gly-Pro), and poly(Gly-Arg). These DPR proteins form insoluble nuclear and cytoplasmic inclusions in *C9ORF72* patient neurons and are not detectable in healthy controls, ALS/FTD patients negative for the $G_4C_2^{exp}$ mutations or in patients with other neurodegenerative diseases. These inclusions are also distinct from TDP-43 inclusions but colocalize with p62-positive inclusions, suggesting UPS involvement in the turnover of RAN translation products (Ash et al. 2013; Mori et al. 2013b). Whether these protein products are neurotoxic and contribute significantly to ALS/FTD pathology remains a critical question.

Another molecular mechanism potentially involved in *C9ORF72* disease pathogenesis is $rG_4C_2^{exp}$ toxicity, since these mutant RNAs accumulate in nuclear foci in ALS/FTD tissues and patient-derived iPS cells (Almeida et al. 2013; DeJesus-Hernandez et al. 2011) (Fig. 10.4). While several reports have indicated that the total levels of C9ORF72 mRNA are reduced approximately twofold in patient lymphoblasts (DeJesus-Hernandez et al. 2011), frontal lobe (Gijselinck et al. 2012), and cerebellum (Mori et al. 2013b), levels of both sense and antisense intron 1a-containing RNAs increase seven- to eightfold (Mori et al. 2013b). The latter result suggests that *C9ORF72* splicing is impaired and/or unspliced, or partially spliced, pre-mRNAs accumulate in RNA foci in *C9ORF72* ALS/FTD, similar to the accumulation of *CNBP* intron 1 $CCUG^{exp}$ RNAs in DM2 cells (Margolis et al. 2006). Several candidates for factors sequestered by $rG_4C_2^{exp}$ RNAs have been proposed including hnRNP A2/B1 (DeJesus-Hernandez et al. 2011), RBM45 (Collins et al. 2012), hnRNP A3 (Mori et al. 2013a), SRSF1 (ASF/SF2) (Reddy et al. 2013), and Purα (Xu et al. 2013). The next critical step for validation of these putative sequestered factors is to demonstrate that they bind to $rG_4C_2^{exp}$ RNAs in patient's cells and affected tissues. Moreover, appropriate loss-of-function mammalian models must be developed that recapitulate distinct ALS/FTD phenotypes.

6.2 *Conclusion and Perspective*

While the functions of microsatellites within the normal repeat range remain obscure, several common themes have emerged from studies on microsatellite expansions and disease. First, expansions occur frequently, but not exclusively, in GC-rich microsatellites due to their inherent tendency to form imperfect hairpins, quadraplex-like and slipped-stranded structures setting the stage for error-prone DNA replication, recombination and repair. Expansion mechanisms for repeat expansions that are not GC-rich, such as the SCA10 $ATTCT^{exp}$ and the GAA^{exp} in Friedreich's ataxia (FRDA), a recessively inherited neurological disorder caused by frataxin loss-of-function, are less studied. However, abnormal DNA structures may also be involved in these diseases with potential triplexes in FRDA and replication-associated template switching in SCA10 (Cherng et al. 2011; Lopez Castel et al. 2010; Mirkin 2007).

Second, microsatellite expansions preferentially cause neurological and neuromuscular diseases, even when the affected gene is expressed ubiquitously, suggesting that different cell types possess varying sensitivities to these mutations. For example, why does a small increase in a polyalanine stretch in the N-terminal region of the PABPN1 protein cause a late-onset disease characterized by eyelid drooping, swallowing difficulty, and proximal limb weakness? Third, studies on expansion disorders have revealed novel disease mechanisms, including RNA toxicity and RAN translation. RAN translation, or the production of unusual reiterated (homopolymeric and heteropolymeric) proteins from repetitive RNA templates, is reminiscent of the cell-free experiments that employed synthetic RNA polynucleotides to decipher the genetic code (Nirenberg 2004). However, RAN translation produces homopolymeric proteins in all three frames from trinucleotide repeat expansions without frameshifting (Zu et al. 2011). Fourth, bidirectional transcription of repeat expansions generates multiple potentially toxic RNAs as well as conventional mutant, and RAN, proteins. Delineating the relative toxicities of each of these pathogenic entities will be a difficult, although an essential, step towards the development of effective new therapies for these diseases.

Many questions and experimental challenges remain in the unstable microsatellite field. How many additional familial and "sporadic" diseases are caused by tandem repeat expansion mutations? Why are so many of these disorders characterized by a late-onset clinical profile and why are muscle and nervous systems particularly vulnerable? Why are alterations in RNA binding protein functions and aggregation states so prominent in this group of diseases? Are RNA foci and protein inclusions pathogenic, protective, or innocent bystanders? Technological and computational advances in RNA and protein analyses, and the development of more informative cell and animal disease models, should provide additional experimental surprises and mechanistic insights into how microsatellite expansions perturb normal cellular pathways.

Acknowledgements Our studies on microsatellite expansion disease mechanisms are supported by the NIH (AR046799, NS058901), the Keck Foundation, and the Muscular Dystrophy Association. The authors thank T. Ashizawa, J. Cleary, and L. Ranum for comments on the manuscript.

References

Abu-Baker A, Messaed C, Laganiere J, Gaspar C, Brais B, Rouleau GA (2003) Involvement of the ubiquitin-proteasome pathway and molecular chaperones in oculopharyngeal muscular dystrophy. Hum Mol Genet 12:2609–2623

Adereth Y, Dammai V, Kose N, Li R, Hsu T (2005) RNA-dependent integrin alpha3 protein localization regulated by the Muscleblind-like protein MLP1. Nat Cell Biol 7:1240–1247

Aizawa H, Sawada J, Hideyama T, Yamashita T, Katayama T, Hasebe N, Kimura T, Yahara O, Kwak S (2010) TDP-43 pathology in sporadic ALS occurs in motor neurons lacking the RNA editing enzyme ADAR2. Acta Neuropathol 120:75–84

Akins MR, Berk-Rauch HE, Fallon JR (2009) Presynaptic translation: stepping out of the postsynaptic shadow. Front Neural Circuits 3:17

Al-Chalabi A, Jones A, Troakes C, King A, Al-Sarraj S, van den Berg LH (2012) The genetics and neuropathology of amyotrophic lateral sclerosis. Acta Neuropathol 124:339–352

Al-Sarraj S, King A, Troakes C, Smith B, Maekawa S, Bodi I, Rogelj B, Al-Chalabi A, Hortobagyi T, Shaw CE (2011) p62 positive, TDP-43 negative, neuronal cytoplasmic and intranuclear inclusions in the cerebellum and hippocampus define the pathology of C9orf72-linked FTLD and MND/ALS. Acta Neuropathol 122:691–702

Albrecht M, Golatta M, Wullner U, Lengauer T (2004) Structural and functional analysis of ataxin-2 and ataxin-3. Eur J Biochem 271:3155–3170

Almeida S, Gascon E, Tran H, Chou HJ, Gendron TF, Degroot S, Tapper AR, Sellier C, Charlet-Berguerand N, Karydas A, Seeley WW, Boxer AL, Petrucelli L, Miller BL, Gao FB (2013) Modeling key pathological features of frontotemporal dementia with C9ORF72 repeat expansion in iPSC-derived human neurons. Acta Neuropathol 126(3):385–99

Anvar SY, Hoen PA, Venema A, van der Sluijs B, van Engelen B, Snoeck M, Vissing J, Trollet C, Dickson G, Chartier A, Simonelig M, van Ommen GJ, van der Maarel SM, Raz V (2011) Deregulation of the ubiquitin-proteasome system is the predominant molecular pathology in OPMD animal models and patients. Skelet Muscle 1:15

Apponi LH, Leung SW, Williams KR, Valentini SR, Corbett AH, Pavlath GK (2010) Loss of nuclear poly(A)-binding protein 1 causes defects in myogenesis and mRNA biogenesis. Hum Mol Genet 19:1058–1065

Ash PE, Bieniek KF, Gendron TF, Caulfield T, Lin WL, Dejesus-Hernandez M, van Blitterswijk MM, Jansen-West K, Paul JW 3rd, Rademakers R, Boylan KB, Dickson DW, Petrucelli L (2013) Unconventional translation of C9ORF72 GGGGCC expansion generates insoluble polypeptides specific to c9FTD/ALS. Neuron 77:639–646

Banerjee A, Apponi LH, Pavlath GK, Corbett AH (2013) PABPN1: molecular function and muscle disease. FEBS J 280:4230–4250

Bassell GJ, Warren ST (2008) Fragile X syndrome: loss of local mRNA regulation alters synaptic development and function. Neuron 60:201–214

Batra R, Charizanis K, Swanson MS (2010) Partners in crime: bidirectional transcription in unstable microsatellite disease. Hum Mol Genet 19:R77–R82

Beaulieu YB, Kleinman CL, Landry-Voyer AM, Majewski J, Bachand F (2012) Polyadenylation-dependent control of long noncoding RNA expression by the poly(A)-binding protein nuclear 1. PLoS Genet 8:e1003078

Berman RF, Willemsen R (2009) Mouse models of fragile X-associated tremor ataxia. J Investig Med 57:837–841

Bhakar AL, Dolen G, Bear MF (2012) The pathophysiology of fragile X (and what it teaches us about synapses). Annu Rev Neurosci 35:417–443

Bhogal B, Jongens TA (2010) Fragile X syndrome and model organisms: identifying potential routes of therapeutic intervention. Dis Model Mech 3:693–700

Bomsztyk K, Denisenko O, Ostrowski J (2004) hnRNP K: one protein multiple processes. Bioessays 26:629–638

Botta A, Caldarola S, Vallo L, Bonifazi E, Fruci D, Gullotta F, Massa R, Novelli G, Loreni F (2006) Effect of the [CCTG]n repeat expansion on ZNF9 expression in myotonic dystrophy type II (DM2). Biochim Biophys Acta 1762:329–334

Brais B (2009) Oculopharyngeal muscular dystrophy: a polyalanine myopathy. Curr Neurol Neurosci Rep 9:76–82

Brais B, Bouchard JP, Xie YG, Rochefort DL, Chretien N, Tome FM, Lafreniere RG, Rommens JM, Uyama E, Nohira O, Blumen S, Korczyn AD, Heutink P, Mathieu J, Duranceau A, Codere F, Fardeau M, Rouleau GA (1998) Short GCG expansions in the PABP2 gene cause oculopharyngeal muscular dystrophy. Nat Genet 18:164–167

Brook JD, McCurrach ME, Harley HG, Buckler AJ, Church D, Aburatani H, Hunter K, Stanton VP, Thirion JP, Hudson T et al (1992) Molecular basis of myotonic dystrophy: expansion of a trinucleotide (CTG) repeat at the 3′ end of a transcript encoding a protein kinase family member. Cell 68:799–808

Budworth H, McMurray CT (2013) Bidirectional transcription of trinucleotide repeats: roles for excision repair. DNA Repair (Amst) 12:672–684

Buschiazzo E, Gemmell NJ (2010) Conservation of human microsatellites across 450 million years of evolution. Genome Biol Evol 2:153–165

Cardani R, Mancinelli E, Giagnacovo M, Sansone V, Meola G (2009) Ribonuclear inclusions as biomarker of myotonic dystrophy type 2, even in improperly frozen or defrozen skeletal muscle biopsies. Eur J Histochem 53:107–111

Cardani R, Mancinelli E, Rotondo G, Sansone V, Meola G (2006) Muscleblind-like protein 1 nuclear sequestration is a molecular pathology marker of DM1 and DM2. Eur J Histochem 50:177–182

Charizanis K, Lee KY, Batra R, Goodwin M, Zhang C, Yuan Y, Shiue L, Cline M, Scotti MM, Xia G, Kumar A, Ashizawa T, Clark HB, Kimura T, Takahashi MP, Fujimura H, Jinnai K, Yoshikawa H, Gomes-Pereira M, Gourdon G, Sakai N, Nishino S, Foster TC, Ares M Jr, Darnell RB, Swanson MS (2012) Muscleblind-like 2-mediated alternative splicing in the developing brain and dysregulation in myotonic dystrophy. Neuron 75:437–450

Chartier A, Benoit B, Simonelig M (2006) A Drosophila model of oculopharyngeal muscular dystrophy reveals intrinsic toxicity of PABPN1. EMBO J 25:2253–2262

Chen IC, Lin HY, Lee GC, Kao SH, Chen CM, Wu YR, Hsieh-Li HM, Su MT, Lee-Chen GJ (2009) Spinocerebellar ataxia type 8 larger triplet expansion alters histone modification and induces RNA foci. BMC Mol Biol 10:9

Chen W, Wang Y, Abe Y, Cheney L, Udd B, Li YP (2007) Haploinsuffciency for Znf9 in Znf9+/− mice is associated with multiorgan abnormalities resembling myotonic dystrophy. J Mol Biol 368:8–17

Cherng N, Shishkin AA, Schlager LI, Tuck RH, Sloan L, Matera R, Sarkar PS, Ashizawa T, Freudenreich CH, Mirkin SM (2011) Expansions, contractions, and fragility of the spinocerebellar ataxia type 10 pentanucleotide repeat in yeast. Proc Natl Acad Sci U S A 108:2843–2848

Ciechanover A, Brundin P (2003) The ubiquitin proteasome system in neurodegenerative diseases: sometimes the chicken, sometimes the egg. Neuron 40:427–446

Cleary JD, Ranum LP (2013) Repeat-associated non-ATG (RAN) translation in neurological disease. Hum Mol Genet 22(R1):R45–R51

Coffee B, Zhang F, Warren ST, Reines D (1999) Acetylated histones are associated with FMR1 in normal but not fragile X-syndrome cells. Nat Genet 22:98–101

Collins M, Riascos D, Kovalik T, An J, Krupa K, Krupa K, Hood BL, Conrads TP, Renton AE, Traynor BJ, Bowser R (2012) The RNA-binding motif 45 (RBM45) protein accumulates in inclusion bodies in amyotrophic lateral sclerosis (ALS) and frontotemporal lobar degeneration with TDP-43 inclusions (FTLD-TDP) patients. Acta Neuropathol 124:717–732

Corbeil-Girard LP, Klein AF, Sasseville AM, Lavoie H, Dicaire MJ, Saint-Denis A, Page M, Duranceau A, Codere F, Bouchard JP, Karpati G, Rouleau GA, Massie B, Langelier Y, Brais B (2005) PABPN1 overexpression leads to upregulation of genes encoding nuclear proteins that are sequestered in oculopharyngeal muscular dystrophy nuclear inclusions. Neurobiol Dis 18:551–567

Costa Mdo C, Paulson HL (2012) Toward understanding Machado-Joseph disease. Prog Neurobiol 97:239–257

Cziko AM, McCann CT, Howlett IC, Barbee SA, Duncan RP, Luedemann R, Zarnescu D, Zinsmaier KE, Parker RR, Ramaswami M (2009) Genetic modifiers of dFMR1 encode RNA granule components in Drosophila. Genetics 182:1051–1060

Darnell JC, Jensen KB, Jin P, Brown V, Warren ST, Darnell RB (2001) Fragile X mental retardation protein targets G quartet mRNAs important for neuronal function. Cell 107:489–499

Darnell JC, Van Driesche SJ, Zhang C, Hung KY, Mele A, Fraser CE, Stone EF, Chen C, Fak JJ, Chi SW, Licatalosi DD, Richter JD, Darnell RB (2011) FMRP stalls ribosomal translocation on mRNAs linked to synaptic function and autism. Cell 146:247–261

Daughters RS, Tuttle DL, Gao W, Ikeda Y, Moseley ML, Ebner TJ, Swanson MS, Ranum LP (2009) RNA gain-of-function in spinocerebellar ataxia type 8. PLoS Genet 5:e1000600

Davies JE, Sarkar S, Rubinsztein DC (2006) Trehalose reduces aggregate formation and delays pathology in a transgenic mouse model of oculopharyngeal muscular dystrophy. Hum Mol Genet 15:23–31

Davies JE, Wang L, Garcia-Oroz L, Cook LJ, Vacher C, O'Donovan DG, Rubinsztein DC (2005) Doxycycline attenuates and delays toxicity of the oculopharyngeal muscular dystrophy mutation in transgenic mice. Nat Med 11:672–677

Davis BM, McCurrach ME, Taneja KL, Singer RH, Housman DE (1997) Expansion of a CUG trinucleotide repeat in the 3′ untranslated region of myotonic dystrophy protein kinase transcripts results in nuclear retention of transcripts. Proc Natl Acad Sci U S A 94:7388–7393

de Klerk E, Venema A, Anvar SY, Goeman JJ, Hu O, Trollet C, Dickson G, den Dunnen JT, van der Maarel SM, Raz V, t Hoen PA (2012) Poly(A) binding protein nuclear 1 levels affect alternative polyadenylation. Nucleic Acids Res 40:9089–9101

DeJesus-Hernandez M, Mackenzie IR, Boeve BF, Boxer AL, Baker M, Rutherford NJ, Nicholson AM, Finch NA, Flynn H, Adamson J, Kouri N, Wojtas A, Sengdy P, Hsiung GY, Karydas A, Seeley WW, Josephs KA, Coppola G, Geschwind DH, Wszolek ZK, Feldman H, Knopman DS, Petersen RC, Miller BL, Dickson DW, Boylan KB, Graff-Radford NR, Rademakers R (2011) Expanded GGGGCC hexanucleotide repeat in noncoding region of C9ORF72 causes chromosome 9p-linked FTD and ALS. Neuron 72:245–256

Dennissen FJ, Kholod N, van Leeuwen FW (2012) The ubiquitin proteasome system in neurodegenerative diseases: culprit, accomplice or victim? Prog Neurobiol 96:190–207

Dobson-Stone C, Hallupp M, Bartley L, Shepherd CE, Halliday GM, Schofield PR, Hodges JR, Kwok JB (2012) C9ORF72 repeat expansion in clinical and neuropathologic frontotemporal dementia cohorts. Neurology 79:995–1001

Doolittle WF, Sapienza C (1980) Selfish genes, the phenotype paradigm and genome evolution. Nature 284:601–603

Du H, Cline MS, Osborne RJ, Tuttle DL, Clark TA, Donohue JP, Hall MP, Shiue L, Swanson MS, Thornton CA, Ares M Jr (2010) Aberrant alternative splicing and extracellular matrix gene expression in mouse models of myotonic dystrophy. Nat Struct Mol Biol 17:187–193

Echeverria GV, Cooper TA (2012) RNA-binding proteins in microsatellite expansion disorders: mediators of RNA toxicity. Brain Res 1462:100–111

Elden AC, Kim HJ, Hart MP, Chen-Plotkin AS, Johnson BS, Fang X, Armakola M, Geser F, Greene R, Lu MM, Padmanabhan A, Clay-Falcone D, McCluskey L, Elman L, Juhr D, Gruber PJ, Rub U, Auburger G, Trojanowski JQ, Lee VM, Van Deerlin VM, Bonini NM, Gitler AD (2010) Ataxin-2 intermediate-length polyglutamine expansions are associated with increased risk for ALS. Nature 466:1069–1075

Finkbeiner S (2011) Huntington's disease. Cold Spring Harb Perspect Biol 3

Fogel BL, Wexler E, Wahnich A, Friedrich T, Vijayendran C, Gao F, Parikshak N, Konopka G, Geschwind DH (2012) RBFOX1 regulates both splicing and transcriptional networks in human neuronal development. Hum Mol Genet 21:4171–4186

Fratta P, Mizielinska S, Nicoll AJ, Zloh M, Fisher EM, Parkinson G, Isaacs AM (2012) C9orf72 hexanucleotide repeat associated with amyotrophic lateral sclerosis and frontotemporal dementia forms RNA G-quadruplexes. Sci Rep 2:1016

Friedman JE (2011) Anticipation in hereditary disease: the history of a biomedical concept. Hum Genet 130:705–714

Fu Y, Ramisetty SR, Hussain N, Baranger AM (2012) MBNL1-RNA recognition: contributions of MBNL1 sequence and RNA conformation. ChemBioChem 13:112–119

Fu YH, Pizzuti A, Fenwick RG Jr, King J, Rajnarayan S, Dunne PW, Dubel J, Nasser GA, Ashizawa T, de Jong P et al (1992) An unstable triplet repeat in a gene related to myotonic muscular dystrophy. Science 255:1256–1258

Gambardella S, Rinaldi F, Lepore SM, Viola A, Loro E, Angelini C, Vergani L, Novelli G, Botta A (2010) Overexpression of microRNA-206 in the skeletal muscle from myotonic dystrophy type 1 patients. J Transl Med 8:48

Gehman LT, Stoilov P, Maguire J, Damianov A, Lin CH, Shiue L, Ares M Jr, Mody I, Black DL (2011) The splicing regulator Rbfox1 (A2BP1) controls neuronal excitation in the mammalian brain. Nat Genet 43:706–711

Gemayel R, Vinces MD, Legendre M, Verstrepen KJ (2010) Variable tandem repeats accelerate evolution of coding and regulatory sequences. Annu Rev Genet 44:445–477

Gijselinck I, Van Langenhove T, van der Zee J, Sleegers K, Philtjens S, Kleinberger G, Janssens J, Bettens K, Van Cauwenberghe C, Pereson S, Engelborghs S, Sieben A, De Jonghe P, Vandenberghe R, Santens P, De Bleecker J, Maes G, Baumer V, Dillen L, Joris G, Cuijt I, Corsmit E, Elinck E, Van Dongen J, Vermeulen S, Van den Broeck M, Vaerenberg C, Mattheijssens M, Peeters K, Robberecht W, Cras P, Martin JJ, De Deyn PP, Cruts M, Van Broeckhoven C (2012) A C9orf72 promoter repeat expansion in a Flanders-Belgian cohort with disorders of the frontotemporal lobar degeneration-amyotrophic lateral sclerosis spectrum: a gene identification study. Lancet Neurol 11:54–65

Giordana MT, Ferrero P, Grifoni S, Pellerino A, Naldi A, Montuschi A (2011) Dementia and cognitive impairment in amyotrophic lateral sclerosis: a review. Neurol Sci 32:9–16

Goers ES, Purcell J, Voelker RB, Gates DP, Berglund JA (2010) MBNL1 binds GC motifs embedded in pyrimidines to regulate alternative splicing. Nucleic Acids Res 38:2467–2484

Goti D, Katzen SM, Mez J, Kurtis N, Kiluk J, Ben-Haiem L, Jenkins NA, Copeland NG, Kakizuka A, Sharp AH, Ross CA, Mouton PR, Colomer V (2004) A mutant ataxin-3 putative-cleavage fragment in brains of Machado-Joseph disease patients and transgenic mice is cytotoxic above a critical concentration. J Neurosci 24:10266–10279

Graff-Radford NR, Woodruff BK (2007) Frontotemporal dementia. Semin Neurol 27:48–57

Greco CM, Hagerman RJ, Tassone F, Chudley AE, Del Bigio MR, Jacquemont S, Leehey M, Hagerman PJ (2002) Neuronal intranuclear inclusions in a new cerebellar tremor/ataxia syndrome among fragile X carriers. Brain 125:1760–1771

Gros-Louis F, Gaspar C, Rouleau GA (2006) Genetics of familial and sporadic amyotrophic lateral sclerosis. Biochim Biophys Acta 1762:956–972

Ha AD, Fung VS (2012) Huntington's disease. Curr Opin Neurol 25:491–498

Haacke A, Broadley SA, Boteva R, Tzvetkov N, Hartl FU, Breuer P (2006) Proteolytic cleavage of polyglutamine-expanded ataxin-3 is critical for aggregation and sequestration of non-expanded ataxin-3. Hum Mol Genet 15:555–568

Hagerman PJ, Hagerman RJ (2004) Fragile X-associated tremor/ataxia syndrome (FXTAS). Ment Retard Dev Disabil Res Rev 10:25–30

Hashem V, Galloway JN, Mori M, Willemsen R, Oostra BA, Paylor R, Nelson DL (2009) Ectopic expression of CGG containing mRNA is neurotoxic in mammals. Hum Mol Genet 18:2443–2451

Hernandez RN, Feinberg RL, Vaurio R, Passanante NM, Thompson RE, Kaufmann WE (2009) Autism spectrum disorder in fragile X syndrome: a longitudinal evaluation. Am J Med Genet A 149A:1125–1137

Hersheson J, Haworth A, Houlden H (2012) The inherited ataxias: genetic heterogeneity, mutation databases, and future directions in research and clinical diagnostics. Hum Mutat 33:1324–1332

Ho TH, Charlet BN, Poulos MG, Singh G, Swanson MS, Cooper TA (2004) Muscleblind proteins regulate alternative splicing. EMBO J 23:3103–3112

Ho TH, Savkur RS, Poulos MG, Mancini MA, Swanson MS, Cooper TA (2005) Colocalization of muscleblind with RNA foci is separable from mis-regulation of alternative splicing in myotonic dystrophy. J Cell Sci 118:2923–2933

Hokkanen S, Feldmann HM, Ding H, Jung CK, Bojarski L, Renner-Muller I, Schuller U, Kretzschmar H, Wolf E, Herms J (2012) Lack of Pur-alpha alters postnatal brain development and causes megalencephaly. Hum Mol Genet 21:473–484

Holmes SE, O'Hearn E, Rosenblatt A, Callahan C, Hwang HS, Ingersoll-Ashworth RG, Fleisher A, Stevanin G, Brice A, Potter NT, Ross CA, Margolis RL (2001) A repeat expansion in the gene encoding junctophilin-3 is associated with Huntington disease-like 2. Nat Genet 29:377–378

Hsu RJ, Hsiao KM, Lin MJ, Li CY, Wang LC, Chen LK, Pan H (2011) Long tract of untranslated CAG repeats is deleterious in transgenic mice. PLoS One 6:e16417

Huichalaf C, Schoser B, Schneider-Gold C, Jin B, Sarkar P, Timchenko L (2009) Reduction of the rate of protein translation in patients with myotonic dystrophy 2. J Neurosci 29:9042–9049

Huynh DP, Figueroa K, Hoang N, Pulst SM (2000) Nuclear localization or inclusion body formation of ataxin-2 are not necessary for SCA2 pathogenesis in mouse or human. Nat Genet 26:44–50

Ikeda Y, Daughters RS, Ranum LP (2008) Bidirectional expression of the SCA8 expansion mutation: one mutation, two genes. Cerebellum 7:150–158
Iwahashi CK, Yasui DH, An HJ, Greco CM, Tassone F, Nannen K, Babineau B, Lebrilla CB, Hagerman RJ, Hagerman PJ (2006) Protein composition of the intranuclear inclusions of FXTAS. Brain 129:256–271
Jansen A, Gemayel R, Verstrepen KJ (2012) Unstable microsatellite repeats facilitate rapid evolution of coding and regulatory sequences. Genome Dyn 7:108–125
Jansen G, Mahadevan M, Amemiya C, Wormskamp N, Segers B, Hendriks W, O'Hoy K, Baird S, Sabourin L, Lennon G et al (1992) Characterization of the myotonic dystrophy region predicts multiple protein isoform-encoding mRNAs. Nat Genet 1:261–266
Jenal M, Elkon R, Loayza-Puch F, van Haaften G, Kuhn U, Menzies FM, Oude Vrielink JA, Bos AJ, Drost J, Rooijers K, Rubinsztein DC, Agami R (2012) The poly(A)-binding protein nuclear 1 suppresses alternative cleavage and polyadenylation sites. Cell 149:538–553
Jin P, Duan R, Qurashi A, Qin Y, Tian D, Rosser TC, Liu H, Feng Y, Warren ST (2007) Pur alpha binds to rCGG repeats and modulates repeat-mediated neurodegeneration in a Drosophila model of fragile X tremor/ataxia syndrome. Neuron 55:556–564
Johnson EM, Kinoshita Y, Weinreb DB, Wortman MJ, Simon R, Khalili K, Winckler B, Gordon J (2006) Role of Pur alpha in targeting mRNA to sites of translation in hippocampal neuronal dendrites. J Neurosci Res 83:929–943
Junghans RP (2009) Dystrophia myotonia: why focus on foci? Eur J Hum Genet 17:543–553
Kanadia RN, Johnstone KA, Mankodi A, Lungu C, Thornton CA, Esson D, Timmers AM, Hauswirth WW, Swanson MS (2003a) A muscleblind knockout model for myotonic dystrophy. Science 302:1978–1980
Kanadia RN, Urbinati CR, Crusselle VJ, Luo D, Lee YJ, Harrison JK, Oh SP, Swanson MS (2003b) Developmental expression of mouse muscleblind genes Mbnl1, Mbnl2 and Mbnl3. Gene Expr Patterns 3:459–462
Kawahara Y, Ito K, Sun H, Aizawa H, Kanazawa I, Kwak S (2004) Glutamate receptors: RNA editing and death of motor neurons. Nature 427:801
Kenneson A, Zhang F, Hagedorn CH, Warren ST (2001) Reduced FMRP and increased FMR1 transcription is proportionally associated with CGG repeat number in intermediate-length and premutation carriers. Hum Mol Genet 10:1449–1454
Kim DH, Langlois MA, Lee KB, Riggs AD, Puymirat J, Rossi JJ (2005) HnRNP H inhibits nuclear export of mRNA containing expanded CUG repeats and a distal branch point sequence. Nucleic Acids Res 33:3866–3874
Klein AF, Ebihara M, Alexander C, Dicaire MJ, Sasseville AM, Langelier Y, Rouleau GA, Brais B (2008) PABPN1 polyalanine tract deletion and long expansions modify its aggregation pattern and expression. Exp Cell Res 314:1652–1666
Koob MD, Moseley ML, Schut LJ, Benzow KA, Bird TD, Day JW, Ranum LP (1999) An untranslated CTG expansion causes a novel form of spinocerebellar ataxia (SCA8). Nat Genet 21:379–384
Koyano S, Uchihara T, Fujigasaki H, Nakamura A, Yagishita S, Iwabuchi K (1999) Neuronal intranuclear inclusions in spinocerebellar ataxia type 2: triple-labeling immunofluorescent study. Neurosci Lett 273:117–120
Krzyzosiak WJ, Sobczak K, Wojciechowska M, Fiszer A, Mykowska A, Kozlowski P (2012) Triplet repeat RNA structure and its role as pathogenic agent and therapeutic target. Nucleic Acids Res 40:11–26
Kuyumcu-Martinez NM, Wang GS, Cooper TA (2007) Increased steady-state levels of CUGBP1 in myotonic dystrophy 1 are due to PKC-mediated hyperphosphorylation. Mol Cell 28:68–78
Kwiatkowski TJ Jr, Bosco DA, Leclerc AL, Tamrazian E, Vanderburg CR, Russ C, Davis A, Gilchrist J, Kasarskis EJ, Munsat T, Valdmanis P, Rouleau GA, Hosler BA, Cortelli P, de Jong PJ, Yoshinaga Y, Haines JL, Pericak-Vance MA, Yan J, Ticozzi N, Siddique T, McKenna-Yasek D, Sapp PC, Horvitz HR, Landers JE, Brown RH Jr (2009) Mutations in the FUS/TLS gene on chromosome 16 cause familial amyotrophic lateral sclerosis. Science 323:1205–1208
La Spada AR, Wilson EM, Lubahn DB, Harding AE, Fischbeck KH (1991) Androgen receptor gene mutations in X-linked spinal and bulbar muscular atrophy. Nature 352:77–79

Laurent FX, Sureau A, Klein AF, Trouslard F, Gasnier E, Furling D, Marie J (2012) New function for the RNA helicase p68/DDX5 as a modifier of MBNL1 activity on expanded CUG repeats. Nucleic Acids Res 40:3159–3171

Leehey MA, Hagerman PJ (2012) Fragile X-associated tremor/ataxia syndrome. Handb Clin Neurol 103:373–386

Lemay JF, Lemieux C, St-Andre O, Bachand F (2010) Crossing the borders: poly(A)-binding proteins working on both sides of the fence. RNA Biol 7:291–295

Levine TP, Daniels RD, Gatta AT, Wong LH, Hayes MJ (2013) The product of C9orf72, a gene strongly implicated in neurodegeneration, is structurally related to DENN Rab-GEFs. Bioinformatics 29:499–503

Li LB, Yu Z, Teng X, Bonini NM (2008) RNA toxicity is a component of ataxin-3 degeneration in Drosophila. Nature 453:1107–1111

Li Y, Jin P (2012) RNA-mediated neurodegeneration in fragile X-associated tremor/ataxia syndrome. Brain Res 1462:112–117

Li YR, King OD, Shorter J, Gitler AD (2013) Stress granules as crucibles of ALS pathogenesis. J Cell Biol 201:361–372

Lin X, Miller JW, Mankodi A, Kanadia RN, Yuan Y, Moxley RT, Swanson MS, Thornton CA (2006) Failure of MBNL1-dependent post-natal splicing transitions in myotonic dystrophy. Hum Mol Genet 15:2087–2097

Ling SC, Polymenidou M, Cleveland DW (2013) Converging Mechanisms in ALS and FTD: Disrupted RNA and Protein Homeostasis. Neuron 79:416–438

Liquori CL, Ricker K, Moseley ML, Jacobsen JF, Kress W, Naylor SL, Day JW, Ranum LP (2001) Myotonic dystrophy type 2 caused by a CCTG expansion in intron 1 of ZNF9. Science 293:864–867

Liu J, Tang TS, Tu H, Nelson O, Herndon E, Huynh DP, Pulst SM, Bezprozvanny I (2009) Deranged calcium signaling and neurodegeneration in spinocerebellar ataxia type 2. J Neurosci 29:9148–9162

Lomen-Hoerth C, Anderson T, Miller B (2002) The overlap of amyotrophic lateral sclerosis and frontotemporal dementia. Neurology 59:1077–1079

Lopez Castel A, Cleary JD, Pearson CE (2010) Repeat instability as the basis for human diseases and as a potential target for therapy. Nat Rev Mol Cell Biol 11:165–170

Lukong KE, Richard S (2008) Motor coordination defects in mice deficient for the Sam68 RNA-binding protein. Behav Brain Res 189:357–363

Magana JJ, Velazquez-Perez L, Cisneros B (2013) Spinocerebellar ataxia type 2: clinical presentation, molecular mechanisms, and therapeutic perspectives. Mol Neurobiol 47:90–104

Mahadevan M, Tsilfidis C, Sabourin L, Shutler G, Amemiya C, Jansen G, Neville C, Narang M, Barcelo J, O'Hoy K et al (1992) Myotonic dystrophy mutation: an unstable CTG repeat in the 3′ untranslated region of the gene. Science 255:1253–1255

Mahadevan MS (2012) Myotonic dystrophy: is a narrow focus obscuring the rest of the field? Curr Opin Neurol 25:609–613

Majounie E, Renton AE, Mok K, Dopper EG, Waite A, Rollinson S, Chio A, Restagno G, Nicolaou N, Simon-Sanchez J, van Swieten JC, Abramzon Y, Johnson JO, Sendtner M, Pamphlett R, Orrell RW, Mead S, Sidle KC, Houlden H, Rohrer JD, Morrison KE, Pall H, Talbot K, Ansorge O, Chromosome ALSFTDC, French research network on FFA, Consortium I, Hernandez DG, Arepalli S, Sabatelli M, Mora G, Corbo M, Giannini F, Calvo A, Englund E, Borghero G, Floris GL, Remes AM, Laaksovirta H, McCluskey L, Trojanowski JQ, Van Deerlin VM, Schellenberg GD, Nalls MA, Drory VE, Lu CS, Yeh TH, Ishiura H, Takahashi Y, Tsuji S, Le Ber I, Brice A, Drepper C, Williams N, Kirby J, Shaw P, Hardy J, Tienari PJ, Heutink P, Morris HR, Pickering-Brown S, Traynor BJ (2012) Frequency of the C9orf72 hexanucleotide repeat expansion in patients with amyotrophic lateral sclerosis and frontotemporal dementia: a cross-sectional study. Lancet Neurol 11:323–330

Mankodi A, Logigian E, Callahan L, McClain C, White R, Henderson D, Krym M, Thornton CA (2000) Myotonic dystrophy in transgenic mice expressing an expanded CUG repeat. Science 289:1769–1773

Margolis JM, Schoser BG, Moseley ML, Day JW, Ranum LP (2006) DM2 intronic expansions: evidence for CCUG accumulation without flanking sequence or effects on ZNF9 mRNA processing or protein expression. Hum Mol Genet 15:1808–1815

Margolis RL, Holmes SE, Rosenblatt A, Gourley L, O'Hearn E, Ross CA, Seltzer WK, Walker RH, Ashizawa T, Rasmussen A, Hayden M, Almqvist EW, Harris J, Fahn S, MacDonald ME, Mysore J, Shimohata T, Tsuji S, Potter N, Nakaso K, Adachi Y, Nakashima K, Bird T, Krause A, Greenstein P (2004) Huntington's disease-like 2 (HDL2) in North America and Japan. Ann Neurol 56:670–674

Martin GM (2005) Genetic modulation of senescent phenotypes in Homo sapiens. Cell 120:523–532

Massa R, Panico MB, Caldarola S, Fusco FR, Sabatelli P, Terracciano C, Botta A, Novelli G, Bernardi G, Loreni F (2010) The myotonic dystrophy type 2 (DM2) gene product zinc finger protein 9 (ZNF9) is associated with sarcomeres and normally localized in DM2 patients' muscles. Neuropathol Appl Neurobiol 36:275–284

Matilla-Duenas A, Corral-Juan M, Volpini V, Sanchez I (2012) The spinocerebellar ataxias: clinical aspects and molecular genetics. Adv Exp Med Biol 724:351–374

McLennan Y, Polussa J, Tassone F, Hagerman R (2011) Fragile x syndrome. Curr Genomics 12:216–224

Messaed C, Dion PA, Abu-Baker A, Rochefort D, Laganiere J, Brais B, Rouleau GA (2007) Soluble expanded PABPN1 promotes cell death in oculopharyngeal muscular dystrophy. Neurobiol Dis 26:546–557

Messaed C, Rouleau GA (2009) Molecular mechanisms underlying polyalanine diseases. Neurobiol Dis 34:397–405

Miller JW, Urbinati CR, Teng-Umnuay P, Stenberg MG, Byrne BJ, Thornton CA, Swanson MS (2000) Recruitment of human muscleblind proteins to (CUG)(n) expansions associated with myotonic dystrophy. EMBO J 19:4439–4448

Mirkin SM (2007) Expandable DNA repeats and human disease. Nature 447:932–940

Mori K, Lammich S, Mackenzie IR, Forne I, Zilow S, Kretzschmar H, Edbauer D, Janssens J, Kleinberger G, Cruts M, Herms J, Neumann M, Van Broeckhoven C, Arzberger T, Haass C (2013a) hnRNP A3 binds to GGGGCC repeats and is a constituent of p62-positive/TDP43-negative inclusions in the hippocampus of patients with C9orf72 mutations. Acta Neuropathol 125:413–423

Mori K, Weng SM, Arzberger T, May S, Rentzsch K, Kremmer E, Schmid B, Kretzschmar HA, Cruts M, Van Broeckhoven C, Haass C, Edbauer D (2013b) The C9orf72 GGGGCC repeat is translated into aggregating dipeptide-repeat proteins in FTLD/ALS. Science 339:1335–1338

Morris MJ, Negishi Y, Pazsint C, Schonhoft JD, Basu S (2010) An RNA G-quadruplex is essential for cap-independent translation initiation in human VEGF IRES. J Am Chem Soc 132:17831–17839

Moseley ML, Zu T, Ikeda Y, Gao W, Mosemiller AK, Daughters RS, Chen G, Weatherspoon MR, Clark HB, Ebner TJ, Day JW, Ranum LP (2006) Bidirectional expression of CUG and CAG expansion transcripts and intranuclear polyglutamine inclusions in spinocerebellar ataxia type 8. Nat Genet 38:758–769

Mrazek J, Guo X, Shah A (2007) Simple sequence repeats in prokaryotic genomes. Proc Natl Acad Sci U S A 104:8472–8477

Muslimov IA, Patel MV, Rose A, Tiedge H (2011) Spatial code recognition in neuronal RNA targeting: role of RNA-hnRNP A2 interactions. J Cell Biol 194:441–457

Mykowska A, Sobczak K, Wojciechowska M, Kozlowski P, Krzyzosiak WJ (2011) CAG repeats mimic CUG repeats in the misregulation of alternative splicing. Nucleic Acids Res 39:8938–8951

Nakamori M, Sobczak K, Puwanant A, Welle S, Eichinger K, Pandya S, Dekdebrun J, Heatwole CR, McDermott MP, Chen T, Cline M, Tawil R, Osborne RJ, Wheeler TM, Swanson M, Moxley RT 3rd, Thornton CA (2013) Splicing biomarkers of disease severity in myotonic dystrophy. Ann Neurol 59(3):474–477

Nakamoto M, Nakano S, Kawashima S, Ihara M, Nishimura Y, Shinde A, Kakizuka A (2002) Unequal crossing-over in unique PABP2 mutations in Japanese patients: a possible cause of oculopharyngeal muscular dystrophy. Arch Neurol 59:474–477

Nelson DL, Orr HT, Warren ST (2013) The unstable repeats—three evolving faces of neurological disease. Neuron 77:825–843

Neumann M, Rademakers R, Roeber S, Baker M, Kretzschmar HA, Mackenzie IR (2009) A new subtype of frontotemporal lobar degeneration with FUS pathology. Brain 132:2922–2931

Neumann M, Sampathu DM, Kwong LK, Truax AC, Micsenyi MC, Chou TT, Bruce J, Schuck T, Grossman M, Clark CM, McCluskey LF, Miller BL, Masliah E, Mackenzie IR, Feldman H, Feiden W, Kretzschmar HA, Trojanowski JQ, Lee VM (2006) Ubiquitinated TDP-43 in frontotemporal lobar degeneration and amyotrophic lateral sclerosis. Science 314:130–133

Nirenberg M (2004) Historical review: deciphering the genetic code—a personal account. Trends Biochem Sci 29:46–54

Ohashi S, Kobayashi S, Omori A, Ohara S, Omae A, Muramatsu T, Li Y, Anzai K (2000) The single-stranded DNA- and RNA-binding proteins pur alpha and pur beta link BC1 RNA to microtubules through binding to the dendrite-targeting RNA motifs. J Neurochem 75:1781–1790

Ohno S (1972) So much "junk" DNA in our genome. Brookhaven Symp Biol 23:366–370

Orgel LE, Crick FH (1980) Selfish DNA: the ultimate parasite. Nature 284:604–607

Orr HT (2012a) Cell biology of spinocerebellar ataxia. J Cell Biol 197:167–177

Orr HT (2012b) Polyglutamine neurodegeneration: expanded glutamines enhance native functions. Curr Opin Genet Dev 22:251–255

Orr HT, Zoghbi HY (2007) Trinucleotide repeat disorders. Annu Rev Neurosci 30:575–621

Paul S, Dansithong W, Jog SP, Holt I, Mittal S, Brook JD, Morris GE, Comai L, Reddy S (2011) Expanded CUG repeats dysregulate RNA splicing by altering the stoichiometry of the muscleblind 1 complex. J Biol Chem 286:38427–38438

Paul S, Dansithong W, Kim D, Rossi J, Webster NJ, Comai L, Reddy S (2006) Interaction of muscleblind, CUG-BP1 and hnRNP H proteins in DM1-associated aberrant IR splicing. EMBO J 25:4271–4283

Paulson HL, Perez MK, Trottier Y, Trojanowski JQ, Subramony SH, Das SS, Vig P, Mandel JL, Fischbeck KH, Pittman RN (1997) Intranuclear inclusions of expanded polyglutamine protein in spinocerebellar ataxia type 3. Neuron 19:333–344

Pearson CE (2011) Repeat associated non-ATG translation initiation: one DNA, two transcripts, seven reading frames, potentially nine toxic entities! PLoS Genet 7:e1002018

Peprah E, He W, Allen E, Oliver T, Boyne A, Sherman SL (2010) Examination of FMR1 transcript and protein levels among 74 premutation carriers. J Hum Genet 55:66–68

Perbellini R, Greco S, Sarra-Ferraris G, Cardani R, Capogrossi MC, Meola G, Martelli F (2011) Dysregulation and cellular mislocalization of specific miRNAs in myotonic dystrophy type 1. Neuromuscul Disord 21:81–88

Philips AV, Timchenko LT, Cooper TA (1998) Disruption of splicing regulated by a CUG-binding protein in myotonic dystrophy. Science 280:737–741

Poulos MG, Batra R, Charizanis K, Swanson MS (2011) Developments in RNA splicing and disease. Cold Spring Harb Perspect Biol 3:a000778

Querido E, Gallardo F, Beaudoin M, Menard C, Chartrand P (2011) Stochastic and reversible aggregation of mRNA with expanded CUG-triplet repeats. J Cell Sci 124:1703–1714

Rademakers R, Neumann M, Mackenzie IR (2012) Advances in understanding the molecular basis of frontotemporal dementia. Nat Rev Neurol 8:423–434

Raheem O, Olufemi SE, Bachinski LL, Vihola A, Sirito M, Holmlund-Hampf J, Haapasalo H, Li YP, Udd B, Krahe R (2010) Mutant (CCTG)n expansion causes abnormal expression of zinc finger protein 9 (ZNF9) in myotonic dystrophy type 2. Am J Pathol 177:3025–3036

Ralser M, Albrecht M, Nonhoff U, Lengauer T, Lehrach H, Krobitsch S (2005) An integrative approach to gain insights into the cellular function of human ataxin-2. J Mol Biol 346:203–214

Ramakrishnan P, Baltimore D (2011) Sam68 is required for both NF-kappaB activation and apoptosis signaling by the TNF receptor. Mol Cell 43:167–179

Ranum LP, Cooper TA (2006) RNA-mediated neuromuscular disorders. Annu Rev Neurosci 29:259–277

Rau F, Freyermuth F, Fugier C, Villemin JP, Fischer MC, Jost B, Dembele D, Gourdon G, Nicole A, Duboc D, Wahbi K, Day JW, Fujimura H, Takahashi MP, Auboeuf D, Dreumont N, Furling D,

Charlet-Berguerand N (2011) Misregulation of miR-1 processing is associated with heart defects in myotonic dystrophy. Nat Struct Mol Biol 18:840–845

Ravel-Chapuis A, Belanger G, Yadava RS, Mahadevan MS, DesGroseillers L, Cote J, Jasmin BJ (2012) The RNA-binding protein Staufen1 is increased in DM1 skeletal muscle and promotes alternative pre-mRNA splicing. J Cell Biol 196:699–712

Raz V, Abraham T, van Zwet EW, Dirks RW, Tanke HJ, van der Maarel SM (2011) Reversible aggregation of PABPN1 pre-inclusion structures. Nucleus 2:208–218

Raz V, Butler-Browne G, van Engelen B, Brais B (2013) 191st ENMC international workshop: recent advances in oculopharyngeal muscular dystrophy research: from bench to bedside 8–10 June 2012, Naarden, The Netherlands. Neuromuscul Disord 23:516–523

Reddy K, Zamiri B, Stanley SY, Macgregor RB Jr, Pearson CE (2013) The disease-associated r(GGGGCC)n repeat from the C9orf72 gene forms tract length-dependent uni- and multimolecular RNA G-quadruplex structures. J Biol Chem 288:9860–9866

Renton AE, Majounie E, Waite A, Simon-Sanchez J, Rollinson S, Gibbs JR, Schymick JC, Laaksovirta H, van Swieten JC, Myllykangas L, Kalimo H, Paetau A, Abramzon Y, Remes AM, Kaganovich A, Scholz SW, Duckworth J, Ding J, Harmer DW, Hernandez DG, Johnson JO, Mok K, Ryten M, Trabzuni D, Guerreiro RJ, Orrell RW, Neal J, Murray A, Pearson J, Jansen IE, Sondervan D, Seelaar H, Blake D, Young K, Halliwell N, Callister JB, Toulson G, Richardson A, Gerhard A, Snowden J, Mann D, Neary D, Nalls MA, Peuralinna T, Jansson L, Isoviita VM, Kaivorinne AL, Holtta-Vuori M, Ikonen E, Sulkava R, Benatar M, Wuu J, Chio A, Restagno G, Borghero G, Sabatelli M, Consortium I, Heckerman D, Rogaeva E, Zinman L, Rothstein JD, Sendtner M, Drepper C, Eichler EE, Alkan C, Abdullaev Z, Pack SD, Dutra A, Pak E, Hardy J, Singleton A, Williams NM, Heutink P, Pickering-Brown S, Morris HR, Tienari PJ, Traynor BJ (2011) A hexanucleotide repeat expansion in C9ORF72 is the cause of chromosome 9p21-linked ALS-FTD. Neuron 72:257–268

Richter JD (2007) CPEB: a life in translation. Trends Biochem Sci 32:279–285

Robberecht W, Philips T (2013) The changing scene of amyotrophic lateral sclerosis. Nat Rev Neurosci 14:248–264

Robinson DO, Wills AJ, Hammans SR, Read SP, Sillibourne J (2006) Oculopharyngeal muscular dystrophy: a point mutation which mimics the effect of the PABPN1 gene triplet repeat expansion mutation. J Med Genet 43:e23

Rohrer JD, Guerreiro R, Vandrovcova J, Uphill J, Reiman D, Beck J, Isaacs AM, Authier A, Ferrari R, Fox NC, Mackenzie IR, Warren JD, de Silva R, Holton J, Revesz T, Hardy J, Mead S, Rossor MN (2009) The heritability and genetics of frontotemporal lobar degeneration. Neurology 73:1451–1456

Rub U, Schols L, Paulson H, Auburger G, Kermer P, Jen JC, Seidel K, Korf HW, Deller T (2013) Clinical features, neurogenetics and neuropathology of the polyglutamine spinocerebellar ataxias type 1, 2, 3, 6 and 7. Prog Neurobiol 104:38–66

Rudnicki DD, Holmes SE, Lin MW, Thornton CA, Ross CA, Margolis RL (2007) Huntington's disease–like 2 is associated with CUG repeat-containing RNA foci. Ann Neurol 61:272–282

Rudnicki DD, Pletnikova O, Vonsattel JP, Ross CA, Margolis RL (2008) A comparison of huntington disease and huntington disease-like 2 neuropathology. J Neuropathol Exp Neurol 67:366–374

Satterfield TF, Pallanck LJ (2006) Ataxin-2 and its Drosophila homolog, ATX2, physically assemble with polyribosomes. Hum Mol Genet 15:2523–2532

Schneider-Gold C, Timchenko LT (2010) CCUG repeats reduce the rate of global protein synthesis in myotonic dystrophy type 2. Rev Neurosci 21:19–28

Seixas AI, Holmes SE, Takeshima H, Pavlovich A, Sachs N, Pruitt JL, Silveira I, Ross CA, Margolis RL, Rudnicki DD (2012) Loss of junctophilin-3 contributes to Huntington disease-like 2 pathogenesis. Ann Neurol 71:245–257

Sellier C, Freyermuth F, Tabet R, Tran T, He F, Ruffenach F, Alunni V, Moine H, Thibault C, Page A, Tassone F, Willemsen R, Disney MD, Hagerman PJ, Todd PK, Charlet-Berguerand N (2013) Sequestration of DROSHA and DGCR8 by expanded CGG RNA repeats alters microRNA processing in fragile X-associated tremor/ataxia syndrome. Cell Rep 3:869–880

Sellier C, Rau F, Liu Y, Tassone F, Hukema RK, Gattoni R, Schneider A, Richard S, Willemsen R, Elliott DJ, Hagerman PJ, Charlet-Berguerand N (2010) Sam68 sequestration and partial loss of function are associated with splicing alterations in FXTAS patients. EMBO J 29:1248–1261

Serrano-Munuera C, Corral-Juan M, Stevanin G, San Nicolas H, Roig C, Corral J, Campos B, de Jorge L, Morcillo-Suarez C, Navarro A, Forlani S, Durr A, Kulisevsky J, Brice A, Sanchez I, Volpini V, Matilla-Duenas A (2013) New subtype of spinocerebellar ataxia with altered vertical eye movements mapping to chromosome 1p32. JAMA Neurol 70:764–771

Shibata H, Huynh DP, Pulst SM (2000) A novel protein with RNA-binding motifs interacts with ataxin-2. Hum Mol Genet 9:1303–1313

Shieh SY, Bonini NM (2011) Genes and pathways affected by CAG-repeat RNA-based toxicity in Drosophila. Hum Mol Genet 20:4810–4821

Sicot G, Gomes-Pereira M (2013) RNA toxicity in human disease and animal models: from the uncovering of a new mechanism to the development of promising therapies. Biochim Biophys Acta 1832:1390–1409

Simon-Sanchez J, Hanson M, Singleton A, Hernandez D, McInerney A, Nussbaum R, Werner J, Gallardo M, Weiser R, Gwinn-Hardy K, Singleton AB, Clarimon J (2005) Analysis of SCA-2 and SCA-3 repeats in Parkinsonism: evidence of SCA-2 expansion in a family with autosomal dominant Parkinson's disease. Neurosci Lett 382:191–194

Siomi H, Siomi MC, Nussbaum RL, Dreyfuss G (1993) The protein product of the fragile X gene, FMR1, has characteristics of an RNA-binding protein. Cell 74:291–298

Sofola OA, Jin P, Qin Y, Duan R, Liu H, de Haro M, Nelson DL, Botas J (2007) RNA-binding proteins hnRNP A2/B1 and CUGBP1 suppress fragile X CGG premutation repeat-induced neurodegeneration in a Drosophila model of FXTAS. Neuron 55:565–571

Sun S, Zhang Z, Fregoso O, Krainer AR (2012) Mechanisms of activation and repression by the alternative splicing factors RBFOX1/2. RNA 18:274–283

Sutcliffe JS, Nelson DL, Zhang F, Pieretti M, Caskey CT, Saxe D, Warren ST (1992) DNA methylation represses FMR-1 transcription in fragile X syndrome. Hum Mol Genet 1:397–400

Taneja KL, McCurrach M, Schalling M, Housman D, Singer RH (1995) Foci of trinucleotide repeat transcripts in nuclei of myotonic dystrophy cells and tissues. J Cell Biol 128:995–1002

Tassone F, Hagerman R (2012) The fragile X-associated tremor ataxia syndrome. Results Probl Cell Differ 54:337–357

Tassone F, Hagerman RJ, Garcia-Arocena D, Khandjian EW, Greco CM, Hagerman PJ (2004a) Intranuclear inclusions in neural cells with premutation alleles in fragile X associated tremor/ataxia syndrome. J Med Genet 41:e43

Tassone F, Hagerman RJ, Loesch DZ, Lachiewicz A, Taylor AK, Hagerman PJ (2000a) Fragile X males with unmethylated, full mutation trinucleotide repeat expansions have elevated levels of FMR1 messenger RNA. Am J Med Genet 94:232–236

Tassone F, Hagerman RJ, Taylor AK, Gane LW, Godfrey TE, Hagerman PJ (2000b) Elevated levels of FMR1 mRNA in carrier males: a new mechanism of involvement in the fragile-X syndrome. Am J Hum Genet 66:6–15

Tassone F, Hagerman RJ, Taylor AK, Mills JB, Harris SW, Gane LW, Hagerman PJ (2000c) Clinical involvement and protein expression in individuals with the FMR1 premutation. Am J Med Genet 91:144–152

Tassone F, Iwahashi C, Hagerman PJ (2004b) FMR1 RNA within the intranuclear inclusions of fragile X-associated tremor/ataxia syndrome (FXTAS). RNA Biol 1:103–105

Teive HA, Munhoz RP, Arruda WO, Raskin S, Werneck LC, Ashizawa T (2011) Spinocerebellar ataxia type 10—a review. Parkinsonism Relat Disord 17:655–661

Teplova M, Patel DJ (2008) Structural insights into RNA recognition by the alternative-splicing regulator muscleblind-like MBNL1. Nat Struct Mol Biol 15:1343–1351

Timchenko LT, Miller JW, Timchenko NA, DeVore DR, Datar KV, Lin L, Roberts R, Caskey CT, Swanson MS (1996) Identification of a (CUG)n triplet repeat RNA-binding protein and its expression in myotonic dystrophy. Nucleic Acids Res 24:4407–4414

Timchenko NA, Patel R, Iakova P, Cai ZJ, Quan L, Timchenko LT (2004) Overexpression of CUG triplet repeat-binding protein, CUGBP1, in mice inhibits myogenesis. J Biol Chem 279:13129–13139

Todd PK, Oh SY, Krans A, He F, Sellier C, Frazer M, Renoux AJ, Chen KC, Scaglione KM, Basrur V, Elenitoba-Johnson K, Vonsattel JP, Louis ED, Sutton MA, Taylor JP, Mills RE, Charlet-Berguerand N, Paulson HL (2013) CGG repeat-associated translation mediates neurodegeneration in fragile X tremor ataxia syndrome. Neuron 78:440–455

Tome FM, Chateau D, Helbling-Leclerc A, Fardeau M (1997) Morphological changes in muscle fibers in oculopharyngeal muscular dystrophy. Neuromuscul Disord 7(Suppl 1):S63–S69

Tome FM, Fardeau M (1980) Nuclear inclusions in oculopharyngeal dystrophy. Acta Neuropathol 49:85–87

Udd B, Krahe R (2012) The myotonic dystrophies: molecular, clinical, and therapeutic challenges. Lancet Neurol 11:891–905

Underwood JG, Boutz PL, Dougherty JD, Stoilov P, Black DL (2005) Homologues of the Caenorhabditis elegans Fox-1 protein are neuronal splicing regulators in mammals. Mol Cell Biol 25:10005–10016

Van Damme P, Robberecht W (2013) Clinical implications of recent breakthroughs in amyotrophic lateral sclerosis. Curr Opin Neurol 26(5):466–472

Vance C, Rogelj B, Hortobagyi T, De Vos KJ, Nishimura AL, Sreedharan J, Hu X, Smith B, Ruddy D, Wright P, Ganesalingam J, Williams KL, Tripathi V, Al-Saraj S, Al-Chalabi A, Leigh PN, Blair IP, Nicholson G, de Belleroche J, Gallo JM, Miller CC, Shaw CE (2009) Mutations in FUS, an RNA processing protein, cause familial amyotrophic lateral sclerosis type 6. Science 323:1208–1211

Verkerk AJ, Pieretti M, Sutcliffe JS, Fu YH, Kuhl DP, Pizzuti A, Reiner O, Richards S, Victoria MF, Zhang FP et al (1991) Identification of a gene (FMR-1) containing a CGG repeat coincident with a breakpoint cluster region exhibiting length variation in fragile X syndrome. Cell 65:905–914

Wakamiya M, Matsuura T, Liu Y, Schuster GC, Gao R, Xu W, Sarkar PS, Lin X, Ashizawa T (2006) The role of ataxin 10 in the pathogenesis of spinocerebellar ataxia type 10. Neurology 67:607–613

Walker RH, Morgello S, Davidoff-Feldman B, Melnick A, Walsh MJ, Shashidharan P, Brin MF (2002) Autosomal dominant chorea-acanthocytosis with polyglutamine-containing neuronal inclusions. Neurology 58:1031–1037

Wang ET, Cody NA, Jog S, Biancolella M, Wang TT, Treacy DJ, Luo S, Schroth GP, Housman DE, Reddy S, Lecuyer E, Burge CB (2012a) Transcriptome-wide regulation of pre-mRNA splicing and mRNA localization by muscleblind proteins. Cell 150:710–724

Wang LC, Chen KY, Pan H, Wu CC, Chen PH, Liao YT, Li C, Huang ML, Hsiao KM (2011) Muscleblind participates in RNA toxicity of expanded CAG and CUG repeats in Caenorhabditis elegans. Cell Mol Life Sci 68:1255–1267

Wang T, Bray SM, Warren ST (2012b) New perspectives on the biology of fragile X syndrome. Curr Opin Genet Dev 22:256–263

Warrick JM, Paulson HL, Gray-Board GL, Bui QT, Fischbeck KH, Pittman RN, Bonini NM (1998) Expanded polyglutamine protein forms nuclear inclusions and causes neural degeneration in Drosophila. Cell 93:939–949

Wellington CL, Ellerby LM, Hackam AS, Margolis RL, Trifiro MA, Singaraja R, McCutcheon K, Salvesen GS, Propp SS, Bromm M, Rowland KJ, Zhang T, Rasper D, Roy S, Thornberry N, Pinsky L, Kakizuka A, Ross CA, Nicholson DW, Bredesen DE, Hayden MR (1998) Caspase cleavage of gene products associated with triplet expansion disorders generates truncated fragments containing the polyglutamine tract. J Biol Chem 273:9158–9167

White M, Xia G, Gao R, Wakamiya M, Sarkar PS, McFarland K, Ashizawa T (2012) Transgenic mice with SCA10 pentanucleotide repeats show motor phenotype and susceptibility to seizure: a toxic RNA gain-of-function model. J Neurosci Res 90:706–714

White MC, Gao R, Xu W, Mandal SM, Lim JG, Hazra TK, Wakamiya M, Edwards SF, Raskin S, Teive HA, Zoghbi HY, Sarkar PS, Ashizawa T (2010) Inactivation of hnRNP K by expanded intronic AUUCU repeat induces apoptosis via translocation of PKCdelta to mitochondria in spinocerebellar ataxia 10. PLoS Genet 6:e1000984

Wilburn B, Rudnicki DD, Zhao J, Weitz TM, Cheng Y, Gu X, Greiner E, Park CS, Wang N, Sopher BL, La Spada AR, Osmand A, Margolis RL, Sun YE, Yang XW (2011) An antisense CAG repeat transcript at JPH3 locus mediates expanded polyglutamine protein toxicity in Huntington's disease-like 2 mice. Neuron 70:427–440

Winter R, Kuhn U, Hause G, Schwarz E (2012) Polyalanine-independent conformational conversion of nuclear poly(A)-binding protein 1 (PABPN1). J Biol Chem 287:22662–22671

Wojciechowska M, Krzyzosiak WJ (2011) Cellular toxicity of expanded RNA repeats: focus on RNA foci. Hum Mol Genet 20:3811–3821

Xu Z, Poidevin M, Li X, Li Y, Shu L, Nelson DL, Li H, Hales CM, Gearing M, Wingo TS, Jin P (2013) Expanded GGGGCC repeat RNA associated with amyotrophic lateral sclerosis and frontotemporal dementia causes neurodegeneration. Proc Natl Acad Sci U S A 110:7778–7783

Zoghbi HY, Orr HT (2009) Pathogenic mechanisms of a polyglutamine-mediated neurodegenerative disease, spinocerebellar ataxia type 1. J Biol Chem 284:7425–7429

Zu T, Gibbens B, Doty NS, Gomes-Pereira M, Huguet A, Stone MD, Margolis J, Peterson M, Markowski TW, Ingram MA, Nan Z, Forster C, Low WC, Schoser B, Somia NV, Clark HB, Schmechel S, Bitterman PB, Gourdon G, Swanson MS, Moseley M, Ranum LP (2011) Non-ATG-initiated translation directed by microsatellite expansions. Proc Natl Acad Sci U S A 108:260–265

Chapter 11
RNA-Binding Proteins in Heart Development

Jimena Giudice and Thomas A. Cooper

Abstract RNA-binding proteins (RBPs) are key players of posttranscriptional regulation occurring during normal tissue development. All tissues examined thus far have revealed the importance of RBPs in the regulation of complex networks involved in organ morphogenesis, maturation, and function. They are responsible for controlling tissue-specific gene expression by regulating alternative splicing, mRNA stability, translation, and poly-adenylation. The heart is the first organ form during embryonic development and is also the first to acquire functionality. Numerous remodeling processes take place during late cardiac development since fetal heart first adapts to birth and then undergoes a transition to adult functionality. This physiological remodeling involves transcriptional and posttranscriptional networks that are regulated by RBPs. Disruption of the normal regulatory networks has been shown to cause cardiomyopathy in humans and animal models. Here we review the complexity of late heart development and the current information regarding how RBPs control aspects of postnatal heart development. We also review how activities of RBPs are modulated adding complexity to the regulation of developmental networks.

Keywords RNA-binding proteins • Heart development • Alternative splicing • Cardiac disorders • mRNA stability • CELF • MBNL • RBFOX • SR proteins • Sarcomere

J. Giudice (✉)
Department of Pathology and Immunology, Baylor College of Medicine, Houston, TX 77030, USA
e-mail: giudice@bcm.edu; jimena.giudice@gmail.com

T.A. Cooper (✉)
Department of Pathology and Immunology, Baylor College of Medicine, Houston, TX 77030, USA

Department of Molecular and Cellular Biology, Baylor College of Medicine, Houston, TX 77030, USA

Department of Molecular Physiology and Biophysics, Baylor College of Medicine, Houston, TX 77030, USA
e-mail: tcooper@bcm.edu

G.W. Yeo (ed.), *Systems Biology of RNA Binding Proteins*, Advances in Experimental Medicine and Biology 825, DOI 10.1007/978-1-4939-1221-6_11

Abbreviations

AS	Alternative splicing
CELF1	CUGBP, ELAV-like family 1
CF	Cardiac fibroblasts
CLIP-seq	Cross-linking/immunoprecipitation/sequencing
CM	Cardiomyocytes
cycD1	Cyclin D1
cycD3-cdk4	Cyclin D3-cdk4
DCM	Dilated cardiomyopathy
DM	Myotonic distrophy
DM1	Myotonic dystrophy type 1
DM2	Myotonic dystrophy type 2
ECC	Excitation-contraction coupling
eIF2α	Eukaryotic translation initiation factor 2α
eIF4E	Eukaryotic translation initiation factor 4E
HDAC	Histone deacetylase
Hsp32	Heat-shock protein 32
MBNL	Muscleblind-like proteins
miRNA	MicroRNA
nPTB	Neuronal homolog of PTB
NXF1	Nuclear export factor 1
PN	Postnatal day
PTB	Polypyrimidine tract-binding protein
PEVK	Titin region rich in proline, glutamate, valine, and lysine
PKC	Protein kinase C
PKR	RNA protein kinase
QKI	Quaking RBP
RBM	RNA-binding motif protein
RBPs	RNA-binding proteins
RISC	RNA-induced silencing complex
RNA-seq	RNA sequencing
RRM	RNA recognition motif region
S	Serine
Sam68	Src-associated substrate in mitosis of 68 kDa
SR	Serine/arginine-rich protein
SRSF	Serine/arginine-rich splicing factor
T	Threonine
UTR	Untranslated region

1 Introduction

RNA-binding proteins (RBPs) play important roles in transcription and posttranscriptional processing of RNA molecules. RBPs are involved in multiple regulated processes including alternative splicing (AS), alternative poly-adenylation, nuclear export, subcellular localization, translation control, and stabilization and degradation of mRNAs (Gesteland et al. 1999; Glisovic et al. 2008). In many cases, RBPs are differentially expressed in tissues or even in the same tissue they show specific expression patterns within different cell types.

Over the years, key roles of RBPs were investigated and more deeply understood by RBP-associated human diseases. For example, CUGBP, ELAV-like family 1 (CELF1) and muscleblind-like 1 (MBNL1) proteins are mis-regulated in myotonic dystrophy (DM) type 1 (DM1) (Timchenko et al. 1996, 2001a, b; Miller et al. 2000; Lin et al. 2006; Kuyumcu-Martinez et al. 2007) generating aberrant AS, associated with disease features such as insulin resistance, myotonia, muscle wasting, transverse tubule alterations, and cardiac dysfunction (Philips et al. 1998; Savkur et al. 2001; Charlet et al. 2002; Mankodi et al. 2002; Fugier et al. 2011). Another example is the vast number of neurodegenerative disorders. One of the most dramatic causes of neurodegeneration, amyotrophic lateral sclerosis, is associated with alterations in RBP levels and defects in RNA metabolism (Lukong et al. 2008; Buratti and Baralle 2012; Hanson et al. 2012; Mills and Janitz 2012). Brain is one of the human tissues where AS is more highly prevalent (Gabut et al. 2008). The neurodegenerative disorder paraneoplastic opsoclonus-myoclonus ataxia involves the Nova RBPs and AS of target mRNAs is mis-regulated affecting synaptic transmission (Jensen et al. 2000; Ule et al. 2003; Licatalosi and Darnell 2006). Spinal muscular atrophy is associated with mutations or deletions of the survival motor neuron (*SMN*) gene affecting ribonucleoprotein complex assembly and localization. Finally, a growing number of RBPs are known to be involved in cancer affecting regulation of cell growth, proliferation, and migration. Many chromosomal translocations observed in cancer implicate altered expression of RBPs (Riggi et al. 2007; Torchia et al. 2007), different tumor types show up-regulation of RBPs such as the eukaryotic translation initiation factor 4E (eIF4E) (Wendel et al. 2007), the Src-associated substrate in mitosis of 68 kDa (Sam68) (Busà et al. 2007), and the serine/arginine-rich protein (SR) splicing factor 1 (SRSF1 also known as SF2/ASF, for nomenclature of SR proteins see Manley and Krainer 2010) (Karni et al. 2007). Gliomas express lower levels of the Quaking RBP (QKI) than normal cells (Chénard and Richard 2008).

Various forms of heart disease have been linked with failures in mRNA processing (Kong et al. 2010), and alterations in several RBPs are associated with abnormal cardiac development. Much of the current knowledge about the regulatory roles of RBPs during heart development came from investigating the mechanisms of cardiomyopathy in human diseases or in mouse models. Therefore, complementary information is (and was) achieved by efforts in understanding the physiology of both normal heart development and pathology.

2 The Heart Is the First Organ to Form and Function During Embryogenesis

In mice, heart is morphologically complete by embryonic day 14.5. Multiple transcription factors (NKX, TBX, GATA), epigenetic regulators (chromatin-remodeling protein SMYD1, histone deacetylases HDACs, SWI/SNF complex), signaling pathways (Wnt and bone morphogenic protein, BMP, cascades), and microRNAs (miRNAs) (miR-1, miR-133a) are involved in this process in which mesodermal cells became cardiomyocytes (CM) and the cardiac morphology is acquired (Olson 2001; Srivastava 2006). After morphological development, the heart undergoes extensive remodeling to respond to increased workload demands in the developing organism (Olson 2006). By day 7 after birth (PN7), CM lose the proliferative capacity and exit cell cycle. Therefore, after PN7 postnatal heart growth is mainly due to CM hypertrophy (Porrello et al. 2011a; Mahmoud et al. 2013).

CM make up 30 % of the cell population within the human heart, cardiac fibroblasts (CF) make up 66 %, and 4 % are endothelial and vascular smooth muscle cells (Camelliti et al. 2005; Baudino et al. 2006). CM are responsible for contraction, while CF act as mechanical scaffolds for effective pumping (Miragoli et al. 2006). However, interactions between the two cell types are crucial for the three-dimensional structure of the heart. CM-CF communication involves paracrine signaling, direct cell-cell contact through gap junctions, and pathways triggered by extracellular matrix proteins secreted by CF (reviewed in Kakkar and Lee 2010; Tirziu et al. 2010).

Heart development, and specifically postnatal development, is an outstanding biological system to study transcriptome changes for two main reasons. First, it gives the opportunity to study coordinated networks regulating gene expression in a complete physiological context. Second, due to the low cell turnover (CM stop dividing and constitute the majority of heart volume, ~75 %), the transcriptome changes observed during development reflect transitions within a cell population that remains unchanged over time.

3 The Roles of Regulated Gene Expression in Normal Heart Development and Cardiomyopathy

Embryonic and postnatal heart development is characterized by extensive physiological remodeling. The functional conversion from fetal to adult heart involves many transcriptional and posttranscriptional mechanisms, including coordinated networks of AS and other regulatory pathways (Xu et al. 2005; Olson 2006; Kalsotra et al. 2008, 2010). More than 95 % of the human genes are alternative spliced (Wang et al. 2008; Merkin et al. 2012), revealing that splicing is a key mechanism for gene expression regulation and for generating large proteomic diversity from a limited number of genes. Tissue-specific AS provides specialized properties to individual organs such as the heart. Additional support of the key role of AS in normal development and conditions is given by the evidence that many human pathological mutations alter mRNA splicing (Blencowe 2006; Srebrow and Kornblihtt 2006; Singh and Cooper 2012).

Heart disease is one of the leading causes of mortality and mRNA splicing has been found to be broadly altered in human cardiac diseases. In particular, AS of multiple sarcomere genes has been shown to be affected in cardiac pathologies (Kong et al. 2010). Individuals with ischemic cardiomyopathy, dilated cardiomyopathy (DCM), or aortic stenosis show strong changes in AS of key sarcomeric genes such as cardiac troponin T (*TNNT2*), cardiac troponin I (*TNNI3*), myosin heavy chain 7 (*MYH7*), filamin C, gamma (*FLNC*) (Kong et al. 2010), and titin (Neagoe et al. 2002). Another important mutated gene is the serum response factor (*SRF*), involved in cardiac cell growth and muscle gene regulation (Davis et al. 2002). Genes encoding ion channels (*Serca2, Ryr, Camk2d, Scn5a*) and signaling molecules (*Igf1, Vegfa, Tbx3, Tbx5, Tbx20*) are also regulated by AS during development and are affected in some cardiac defects (nicely reviewed by Lara-Pezzi et al. 2013). All together, these data strongly support the hypothesis that, similar to other diseases, mRNA splicing is broadly mis-regulated in human heart failure.

More recently, deep RNA sequencing (RNA-seq) was applied to study the transcriptome changes induced by pathological or physiological hypertrophy in mice. The study revealed different signatures for each type of hypertrophy at the gene expression and AS levels (Song et al. 2012). These results are interesting because the enlargement of the heart in both hypertrophies appears similar at the morphological level; however, the final functional consequences are different. While physiological hypertrophy leads to an improvement in life quality by increasing oxygenation capacity and maximal cardiac output, pathological hypertrophy leads to a progressive deterioration of cardiac function ending with fatal heart failure (Song et al. 2012). RNA-seq was also recently applied to show that the developmental AS program is recapitulated during pressure-overload cardiac hypertrophy in rats (Ames et al. 2013). The comparison of AS events in hypertrophy and during heart development revealed a significant overlap, suggesting that hypertrophic hearts reuse some of the regulatory mechanisms involved in normal development (Ames et al. 2013).

During heart development a complex transcriptional and posttranscriptional program takes place. This network includes *cis*-acting elements, located within the genes, and *trans*-acting factors including RBPs, splicing factors, and also miRNAs (Fig. 11.1, lower panel). Both types of regulatory elements act together and cooperatively to drive the normal switches within the appropriate cells and at the proper time. Developmental switches include AS events, gene expression changes by mRNA stability, and translation regulation through binding of *trans*-acting factors to 3′ and 5′ untranslated regions (UTRs) of target molecules. However, one more level of regulation exists because the *trans*-acting factors are also subject to regulation by AS, proteolytic cleavage, mRNA stability, phosphorylation, and other posttranslational modifications.

The scenario is complex and reveals a tight regulation during embryonic formation of the heart, acute adaptation to birth, and then the postnatal transition from fetal to adult function. Alterations in any of these steps are likely to lead to defects in cardiac function and normal development. In this review we focus on the roles of RBPs in normal heart development as well as in pathological conditions. In order to help to schematically summarize RBP roles in heart development and function, we organize the information available in two tables. Table 11.1 contains available information about RBPs particularly relevant to heart development, SR, RBFOX,

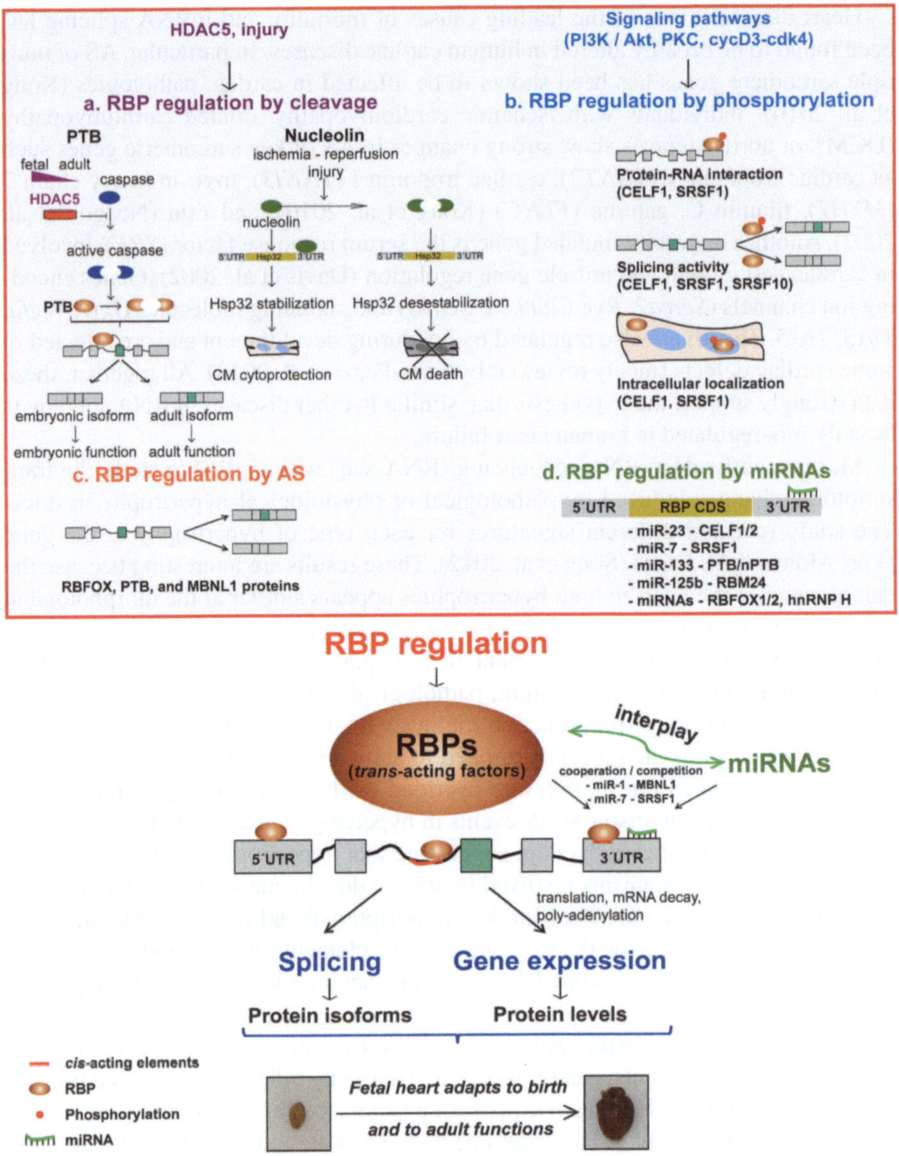

Fig. 11.1 RBP roles in heart development and regulation of RBPs. RBPs are regulated at the level of their expression, localization, and activity by cleavage (**a**), phosphorylation (**b**), AS (**c**), and/or miRNAs (**d**). References in (**a**): Ye et al. 2013 (PTB) and Wang et al. 2012b; Jiang et al. 2013 (Nucleolin). References in (**b**): Kuyumcu-Martinez et al. 2007; Wang et al. 2007; Salisbury et al. 2008; Wang et al. 2008; Huichalaf et al. 2010 (CELF1), Huang et al. 2007 (SRSF1); Shin and Manley 2002; Shin et al. 2004; Feng et al. 2008 (SRSF10). References in (**c**): Wollerton et al. 2001 (PTB), Kuroyanagi 2009 (RBFOX), Tran et al. 2011 (MBNL1). References in (**d**): Kalsotra et al. 2010 (CELF1/2, RBFOX1/2, PTB, hnRNPH); Boutz et al. 2007a, b (PTB/nPTB); Ge et al. 2011; Vacchi-Suzzi et al. 2013 (RBM24); Wu et al. 2010 (SRSF1)

Table 11.1 SR, RBFOX, RBM, PTB, and Hermes protein animal models and associated disease

RBP	Animal model or disease	Cardiac and other features	Altered molecular processes evidenced in the study	References
SRSF2	Conditioned KO (CM) mouse	Normal heart development and life-span DCM, ECC defects, chamber dilatation, hypertrophy	Ryanodin receptor 2 downregulation (no global gene expression alteration)	Ding et al. (2004)
	Human heart failure (aortic stenosis)	SC25 up-regulation. Elevated left ventricular end-diastolic pressure	Not studied	Hein et al. (2003)
	Viral infections	DCM and myocarditis	SRSF2 loss of function (proposed) SRSF2 mis-localization affects spliceosome assembly leading to splicing inhibition	Vercellotti (2001) Mason (2002)
SRSF1	Conditioned KO (CM) mouse	Hyper-contraction ECC defects	Altered AS: Camk2d (mis-targeting to sarcolemmal membranes), Tnnt (neonatal pattern), Cypher (from cardiac to skeletal muscle isoform)	Xu et al. (2005)
SRSF10	Constitutive KO mouse	Embryonic lethality Cardiac defects (atrial and ventricular septal defects, thin, undifferentiated and disorganized myocardium)	Altered AS and downregulation of triadin and calsequestrin 2 Ca^{2+} release defects (more frequent Ca^{2+} sparks and lower amplitudes)	Feng et al. (2009)
RBFOX1	Human tetralogy of Fallot (congenital heart defect)	Low blood oxygenation, right ventricular hypertrophy, ventricular septal defect, pulmonary infundibular stenosis	Mis-regulation of gene expression (Rbfox1 up-regulation)	Kaynak et al. (2003)
RBFOXl and RBFOX2	Morpholinos in zebrafish (*rbfox1l* and *rbfox2* knockdown)	Cardiac morphology defects, reduced heart rate Skeletal muscle defects Paralysis and no normal coiling and swimming Fiber disorganization	AS mis-regulation (Syne1a, Sivil, Gabrg2, Scn8aa, Fxr1)	Gallagher et al. (2011)

(continued)

Table 11.1 (continued)

RBP	Animal model or disease	Cardiac and other features	Altered molecular processes evidenced in the study	References
RBM24	Morpholinos in embryonic zebrafish HL-1 cell knockdown assays	Early heart development: normal. Then, heart morphology, contractility, and blood circulation deficiencies Vasculature defects	Contractile, sarcomere, and vasculature related genes downregulation (Tnnt2, Tpm, Meromyosin, Actn2) Nppa up-regulation (linked to cardiomyopathy)	Poon et al. (2012) Maragh et al. (2011)
RBM38	C2C12 cells knock down/ over-expression experiments	Knocked down: differentiation inhibition Over-expression: myogenesis promotion	Cell cycle arrest regulation through binding to the 3′ UTR of p23 transcript	Miyamoto et al. (2009)
RBM20	Spontaneous rat strain: titin AS deficiency, ~complete *Rbm20* gene deletion	Phenotype similar to pathology of human patients with DCM	Titin AS mis-regulation RBM20 mediates intron retention, exon skipping, and exon shuffling of titin mRNA RBM20 aggregates on titin pre-mRNA repressing splicing of certain regions	Greaser et al. (2008) Guo et al. (2012) Li et al. 2013.
	Human mutations in *RBM20* gene	DCM		Brauch et al. (2009) Li et al. (2010) Refaat et al. (2012)
PTB	Primary CM over-expressing PTB	Not applicable	Induction of pro-apoptotic protein expression (translation regulation) Ischemia: caspase-dependent DNA fragmentation	Zhang et al. (2009)
	HDAC5 KO mouse	Cardiac hypertrophy in response to stress Propensity to lethal cardiac defects	Downregulation of PTB protein (no mRNA changes) Caspase induction AS mis-regulation (α- and β-tropomyosin, Mef2a, Mef2d)	Chang et al. (2004) Ye et al. (2013)
Hermes	Hermes over-expression in *Xenopus* embryos	Abolishment of myocardium differentiation and morphogenesis	Hermes localized in the cytoplasm interacting with polyA-mRNAs (roles in mRNA localization, stability, and/or translation regulation?)	Gerber et al. (2002)

KO refers to knocked out animals

Table 11.2 CELF and MBNL protein animal model or associated diseases

RBP	Animal model or disease	Cardiac and other features	Altered molecular processes evidenced in the study	References
CELF1 *CELF2*	Celf nuclear dominant negative specifically expressed in heart (mouse)	DCM and sever cardiac dysfunction	Mis-regulation of the AS developmentally regulated Gene expression mis-regulation (contraction, Ca^{2+} signaling, serum response factor pathway)	Ladd et al. (2005) Terenzi et al. (2009) Dasgupta et al. (2013)
CELF1	CM-specific CELF1 over-expressing mouse (inducible)	Lethality DCM and similar defects as those observed in hearts of DM1 patients	Developmental AS program mis-regulation Downregulation of mitochondria metabolism-related genes	Kalsotra et al. (2008) Koshelev et al. (2010) Unpublished data
CELF2	CM-specific CELF2 over-expressing mouse (inducible)	No lethality No detected cardiac problems	Mis-regulation of some AS events regulated during development	Unpublished data
CELF1 *MBNL1*	Heart-specific over-expression of 960 CUG interrupted repeats (*DMPK* gene context) (inducible)	Cardiomyopathy, diastolic and systolic dysfunction, arrhythmias, mortality within 2 weeks of induction Cardiac features similar to DM1 defects	RNA foci formation co-localizing with MBNL1 protein, elevated CELF1 protein expression, and mis-regulation of AS	Wang et al. (2007)
MBNL1	Constitutive KO (*Mbnl1*$\Delta^{E3/}\Delta^{E3}$)	Myotonia, delayed muscle relaxation	AS mis-regulation reverting back to neonatal pattern 3′ UTR binding and mRNA localization	Kanadia et al. (2003b) Du et al. (2010) Wang et al. (2012a)
	Constitutive MBNL1 over-expressing mouse	Reduced body mass Increased mortality (often sudden death)	Neonates: premature AS transitions to adult pattern of MBNL1 targets	Chamberlain and Ranum (2012)
MBNL2	Zebrafish knocked down *Mbnl2*	DM features Cardiac morphology defects and slow beating Skeletal/cardiac muscle: abnormal myofibril alignment, Z-bands, and sarcoplasm organization Brain and eye morphology defects, unrestricted and uncoordinated movements	AS mis-regulation of specific targets (tnnt2, clcn1)	Machuca-Tzili, et al. (2011)

KO refers to knocked out animals

Table 11.3 RBPs involved in heart development, expression pattern, binding motif, and functions

RBP	Expression protein changes during heart development (mouse/rat)		Binding motif	References	RBP function
	Protein	References			
SRSF1	Not studied[a]		RGAAGAAC	Reviewed by Chen and Manley (2009)	AS, mRNA nuclear export, mRNA stability, protein translation Transcription pause release (SC35)
SRSF2	Not studied[a]		UGCUGUU		
SRSF10	Not studied[a]		AAAGACAAA		
RBFOX1	Up-regulated (~10-fold)	Kalsotra et al. (2008)	UGCAUG	Jin et al. (2003) Auweter et al. (2006) Ponthier et al. (2006) Yeo et al. (2009)	AS
RBFOX2	Downregulated (~3-fold)	Kalsotra et al. (2008)			
RBM24	Downregulated (~2-fold)	Poon et al. (2012)	Not known[a]		Translation regulation, mRNA stability
RBM38	Not studied[a]		UGUGU	Kuroyanagi et al. (2007)	
RBM20	Not studied[a]		Not studied[a]		AS, mRNA localization
PTB	Not changed/ downregulated (2.8-fold)	Kalsotra et al. (2008) Zhang et al. (2009) Ye et al. 2013.	UCUU and UCUCU	• Chan and Black 1995. Singh et al. (1995)	AS and translation
Hermes	Not studied[a]		Not studied[a]		mRNA stability, localization, and translation regulation (proposed)
CELF1	Downregulated (~10-fold)	Kalsotra et al. (2008)	GU-rich motifs	Kalsotra et al. (2008) Vlasova et al. (2008) Rattenbacher et al. (2010) Masuda et al. (2012)	AS, mRNA stability, and translation
CELF2	Downregulated (~18-fold)	Kalsotra et al. (2008)			
MBNL1	Up-regulated (~4-fold)	Kalsotra et al. (2008)	YGCY	Masuda et al. (2012) Wang et al. (2012a)	AS, mRNA stability, localization, protein translation, and secretion

[a]Refers not known to our knowledge

RNA-binding motif proteins (RBM), polypyrimidine tract-binding proteins (PTB), and Hermes. We present animal models and/or diseases from where the data were generated, as well as the cardiac features and the known molecular roles of the correspondent RBP from the studies. All references are detailed in the last column. Similarly, Table 11.2 includes available information about CELF and MBNL proteins. Table 11.3 summarizes known developmental expression patterns, binding motifs, and main functions for each RBP relevant for heart function.

Finally, we present an overview of the mechanisms by which the activities of RBPs in heart are regulated and the impact on the cardiac developmental program.

4 RBP Roles During Heart Development and Cardiac Disorders

4.1 SR Proteins

SR proteins are phylogenetically conserved and structurally related. They contain a serine- and arginine-rich domain and one or two RNA recognition motif regions (RRM). The RRMs are responsible for sequence-specific binding to the pre-mRNA and the RS domain is involved in protein-protein and protein-RNA interactions. SR proteins are essential for constitutive and alternative pre-mRNA splicing (Graveley 2000; Lin and Fu 2007; Long and Caceres 2009), and they participate in mRNA nuclear export (Huang and Steitz 2001; Huang et al. 2003), mRNA stability (Lemaire et al. 2002), as well as protein translation (Sanford et al. 2004). In addition, SRSF2 (originally known as SC35) was just recently shown to be involved in the regulation of transcription pause release (Ji et al. 2013).

Knockout studies in mice have shown that SR proteins have non-redundant roles during development: deletion of *Srsf3* (also known as SRp20) (Jumaa et al. 1999), *Srsf2* (Wang et al. 2001; Ding et al. 2004), and *Srsf1* (Xu et al. 2005) genes in mice leads to early embryonic phenotype.

Human heart failure was associated with up-regulation of SRSF2 (Hein et al. 2003). Viral infections are known to cause myocarditis and DCM and different virus were shown to regulate SR proteins leading to its loss of function and splicing inhibition (Vercellotti 2001; Mason 2002; Ding et al. 2004). Using knocked out mice, the tissue-specific ablation of *Srsf2* gene causes DCM and excitation-contraction coupling (ECC) defects associated with the up-regulation of the cardiac-specific ryanodine receptor 2 (Ryr2) (Ding et al. 2004).

SRSF1 is involved in postnatal heart remodeling and a heart-specific knockout mouse showed DCM and ECC defects mainly mediated by AS alteration in a restricted set of transcripts, including Tnnt2, the Z-line protein Cypher (LIM domain binding protein 3, Ldb3), and the Ca^{2+}/calmodulin-dependent kinase IIδ (Camk2d). For Tnnt2 and Camk2d, Srsf1 coordinates the AS switch that occurs during heart development. Cypher is not regulated during cardiac development; it is alternatively spliced in different tissues by mutually exclusive usage of exon 4 in cardiac muscle and exons 5–7 in skeletal muscle (Xu et al. 2005). Srsf1 is required to maintain the tissue splicing isoform since when *Srsf1* gene is deleted, the cardiac isoform switches to the skeletal muscle isoform (Xu et al. 2005). *Srsf1* gene ablation in CM generates the retention of the CaMKIIδA isoform in adult stages without switching it off as normally occurs after birth. CaMKIIδ isoforms localize in different cellular compartments (Bayer and Schulman 2001). The CaMKIIδA isoform exhibits a striated pattern of localization, which is different from that of the CaMKIIδB isoform in the nucleus and CaMKIIδC isoform in the cytoplasm. The striated localization of the CaMKIIδA isoform indicates its association with transverse tubules explaining the defects in Ca^{2+} handling detected in *Srsf1* knocked out mice (Xu et al. 2005).

SRSF10 (known as SRp38) is an unusual SR protein since it functions as a splicing repressor when it is desphosphorylated (Shin and Manley 2002; Shin et al. 2004) or as a sequence-dependent splicing activator when it is phosphorylated (Feng et al. 2008). *Srsf10* deletion in mice is embryonically lethal (Feng et al. 2009) and embryos exhibit cardiac alterations including atrial and ventricular septal defects and thin, undifferentiated, and disorganized myocardium. Gene expression showed only limited differences in hearts from knocked out animals comparing with wild-type mice. However, two cardiac proteins, triadin and the interacting protein calsequestrin 2, exhibit different splicing patterns. Since both proteins play critical roles in ECC regulating Ca^{2+} release from the sarcoplasmic reticulum, AS mis-regulation would be involved in the phenotypes (Feng et al. 2009).

4.2 RBFOX Proteins

In mammals, the RBFOX family includes three paralogs: RBFOX1 (Fox-1 or A2BP1), RBFOX2 (Fox-2, RBM9, or FXH), and RBFOX3 (Fox-3, HRNBP3, or NeuN). The three mouse paralogs show high sequence conservation, especially within the RRM, which is identical in Rbfox1 and Rbfox2 and 94 % amino acid identity in Rbfox3. While RBFOX1 protein is expressed in neurons, heart, and skeletal muscle (Kiehl et al. 2001; Underwood et al. 2005), RBFOX2 is more ubiquitously expressed (Jin et al. 2003; Underwood et al. 2005; Ponthier et al. 2006; Yeo et al. 2007) and RBFOX3 is exclusively expressed in neurons (McKee et al. 2005; Kim et al. 2009). RBFOX proteins contain a single RRM and regulate splicing by binding the motif (U)GCAUG in the introns flanking the alternative exon. In contrast to other splicing factors for which binding motifs are degenerate, binding of RBFOX to (U)GCAUG element is highly specific (Auweter et al. 2006). RBFOX is an important contributor to the "tissue splicing code." The effect of RBFOX proteins (as well as other RBPs) depends on the location of its binding site such that binding downstream of the alternative exon generally leads to exon inclusion while upstream binding promotes exon skipping (Jin et al. 2003; Yeo et al. 2009). Various studies using different technologies and biological systems demonstrated that the (U)GCAUG motifs are enriched in introns downstream of exons that are specifically alternatively spliced in muscle tissues (Das et al. 2007; Zhang et al. 2008; Bland et al. 2010) or in brain (Brudno et al. 2001). In addition, the first genome-wide expression compendium of human AS events found by microarray identified that 9,516 AS events were differentially expressed in at least one tissue and the UGCAUG motif was enriched in introns downstream of alternative cassette exons in skeletal muscle and heart, with limited enrichment in brain, adipose, and colon (Castle et al. 2008). Deep sequencing analysis revealed also that UGCAUG motifs are substantially enriched within downstream introns of exons with increased inclusion in heart and brain, and within upstream introns of exons with increased skipping in skeletal muscle (Wang et al. 2008).

Mutations and abnormal expression of the *Rbfox1* gene were found in human neurological disorders (Bhalla et al. 2004; Martin et al. 2007) as well as in heart diseases (Kaynak et al. 2003). Many predicted RBFOX1/2 targets encode neuromuscular proteins, which consistently were reported to be disrupted in muscular dystrophy and neurological, sensory, and cardiac diseases (McKusick 1998; Zhang et al. 2009). Zhang et al. found that those targets are more likely to be disease-associated genes than what would be expected by chance (Zhang et al. 2009). This evidence supports the idea that RBFOX proteins are important regulators of heart functions.

During murine postnatal heart development, RBFOX1 is up-regulated ~10-fold between PN1 and adult stages, and RBFOX2 is downregulated ~3-fold (Kalsotra et al. 2008). Analysis of developmentally regulated AS revealed enrichment in GCAUG motifs in the immediate downstream intron in agreement with the other high-throughput studies. This motif is conserved in eight mammalian species and enriched within this region in both mouse and chicken, supporting a relevant role for RBFOX proteins in normal heart development (Kalsotra et al. 2008).

Analysis in zebrafish revealed that Rbfox-regulated AS is critical for cardiac and muscle functions (Gallagher et al. 2011). Knocked down of *Rbfox* (*Rbfox1l* and *Rbfox2*) animals by morpholinos develop severe defects in skeletal and cardiac muscle functions, complete paralysis, and absence of normal spontaneous coiling. Cardiac morphology is altered, heart rate is reduced, and muscle fibers are disorganized in knocked down animals. Mis-regulation of several AS events was detected in transcripts such as Syne1a (spectrin repeat containing, nuclear envelope 1a), a protein involved in human Emery-Dreifuss muscular dystrophy (Zhang et al. 2007); Svil (supervillin), a membrane protein for which the muscle-specific isoform interacts with F-actin (Oh et al. 2003); Gabrg2 (GABA receptor subunit γ2) and Scn8aa (Na^+ channel, voltage-gated, type VIII, α-a) membrane transporters (Baulac et al. 2001; Chen et al. 2008a); and Fxr1 (fragile X mental retardation, autosomal homolog 1), an RBP involved in somite formation regulation (Huot et al. 2005).

4.3 *RBM24 and RBM38 Proteins*

In mouse the Rbm24 and its paralog Rbm38 are preferentially expressed in heart and skeletal muscle tissues (Miyamoto et al. 2009). Deletion of Rbm38 blocks differentiation of the mouse C2C12 myoblast cell line and its over-expression promotes binding to the 3′ UTR of p21 transcripts (Shu et al. 2006) inducing cell cycle arrest and promoting myogenesis (Miyamoto et al. 2009).

Human RBM24 is a 236 amino acid protein that shows 99 % homology to mouse and rat RBM24, 89 % homology to *Xenopus* Seb4, and 83 % homology to zebrafish Rbm24 (Poon et al. 2012). RBM24 is enriched in human embryonic stem cell-derived CM, suggesting a role in cardiogenesis (Xu et al. 2009). As found for RBM38, RBM24 regulates myogenic differentiation (Miyamoto et al. 2009); however RBM24 exerts this control through regulation of myogenin expression (Jin et al. 2010). In cardiac progenitors, the mouse Rbm24 transcript is up-regulated and expressed from the earliest stages, in the cardiac crescent, and then within the heart tube and looping heart (Miller et al. 2008).

Rbm24 deficiency in mice provokes the downregulation of cardiac sarcomeric proteins leading to diverse abnormalities in heart morphology, contractility, blood circulation, and vasculature development (Maragh et al. 2011; Poon et al. 2012). Those results suggest that RBM24 acts early during myocardial differentiation playing key roles in sarcomere assembly, contractility, and vasculature development.

4.4 RBM20 Protein

The titin gene contains the largest number of exons (363 in humans) and encodes the largest known protein, 3.7 MDa in neonates and 2.97 MDa in adults (Labeit and Kolmerer 1995; Bang et al. 2001). The reduction in protein size during development is a consequence of an increased skipping of exons 50–219 (Krüger and Linke 2011). This splicing transition is functionally relevant due to the structure and characteristics of titin molecule and its localization within the CM. Titin is an elastic molecule which spans half of the sarcomere with the N-terminus located at the Z-line and the C-terminus in the M-line (Labeit and Kolmerer 1995; Obermann et al. 1997; Gregorio et al. 1998). Titin elasticity is primarily due to the polymeric immunoglobulin regions and the PEVK region (rich in proline, glutamate, valine, and lysine) (Labeit and Kolmerer 1995). Titin mechanically maintains sarcomere length and integrity and also acts as a scaffold involved in myofibrillar assembly (Krüger and Linke 2011). Exons 50–96 encode part of the immunoglobulin region and exons 115–225 are within the PEVK zone. In heart, the ratios of the isoform are crucial since the titin-based passive tension determines the stiffness of the myocardial wall during ventricular filling (Li et al. 2013). In fact, mis-regulation of titin AS has been associated with cardiac diseases (Neagoe et al. 2002; Warren et al. 2003; Makarenko et al. 2004; Williams et al. 2009).

In 2008, Greaser et al. showed that a mutant rat strain deficient in titin AS suffered alterations in CM passive tension. Four years later, it was reported that those animals carry on a nearly complete deletion of the *Rbm20* gene (Guo et al. 2012) exhibiting a phenotype similar to the pathology seen in human patients with DCM caused by *Rbm20* mutations (Brauch et al. 2009; Li et al. 2010; Refaat et al. 2012). Those animals exhibit a mis-regulation in titin AS as well as in other 31 genes. Four of those genes, Ttn, Camk2d, Ldb3, and Cacna1c, revealed conserved regulation by AS between rats and humans, supporting an important cardiac role for Rbm20 protein as well as its splicing targets.

Recently, the molecular mechanism linking RBM20 and titin AS was nicely demonstrated. RBM20 binds to certain regions of the newly transcribed titin pre-mRNA before splicing starts in those regions. In consequence, RBM20 binding blocks intron removal in those regions. However, the rest of titin pre-mRNA is normally spliced and the partially spliced pre-mRNAs are retained in the nucleus. The final step consists in the splicing between the 5′ and 3′ splice sites of the exons flanking the RBM20-bound region. Therefore, the exons/introns where the RBM20 was bound are skipped (Li et al. 2013). This two-step sequential splicing mechanism was also recently demonstrated in a different model (Kameyama et al. 2012).

Taken together, these reports give insights into the mechanisms underlying RBM20-related pathology suggesting that RBM20 regulates multiple aspects of cardiac function through effects on AS of genes involved in biomechanics (*Ttn* and *Tpm1*), ion homeostasis and electrical activity (*Camk2d* and C*acna1c*), and signal transduction (*Camk2d* and *Spen*). Similar to titin and Camk2d transcripts, mis-regulation of Ldb3 and Cacna1c AS was linked with DCM and arrhythmia, respectively (Cheng et al. 2011; Tang et al. 2011).

4.5 PTB Proteins

PTB protein (also known as heterogeneous nuclear ribonucleoprotein I, hnRNPI) is an ubiquitous RBP highly conserved in metazoans. Three alternatively spliced isoforms, PTBP1, PTBP2, and PTBP4, and a neuronal homolog of PTB (nPTB) have been described in mammals/humans. PTB contains four RRMs, binds to pyrimidine-rich sequences such as UCUUC and CUCUCU (Singh et al. 1995), and was originally characterized as a protein involved in AS regulation (reviewed by Keppetipola et al. 2012). Regulated transcripts include those encoding the sarcomeric proteins β-tropomyosin (Mulligan et al. 1992), α-tropomyosin (Patton et al. 1991), α-actinin (Southby et al. 1999), and cardiac troponin T (Charlet et al. 2002). However, it is now known that PTB also regulates other aspects of RNA metabolism including poly-adenylation, mRNA stability, and translation initiation (reviewed by Sawicka et al. 2008). PTB can regulate translation by binding to the internal ribosome entry site in the 5′ UTR (Mitchell et al. 2001; Schepens et al. 2005) and cap-independent protein translation. Specificity of PTB function is given by its cellular localization (nucleus/cytoplasm transport is tightly controlled) and its interaction with other proteins (*trans*-acting factors) (Sawicka et al. 2008).

PTB protein expression during heart development is controversial. While Kalsotra et al. (2008) reported that their levels do not change from PN1 to adult stages, Zhang et al. (2009) and Ye et al. (2013) demonstrated its downregulation in rat and mouse. A correlation between the PTB developmental downregulation and the developmental downregulation of apoptotic genes was demonstrated in mouse (Zhang et al. 2009). Consistently, over-expression of PTB in adult CM induces the expression of pro-apoptotic proteins without affecting their transcript levels suggesting that PTB regulates translation of apoptotic genes. When PTB is re-expressed in CM, caspase activity is enhanced and ischemia triggers caspase-dependent DNA fragmentation. By contrast, in the absence of PTB differentiated CM show caspase-independent DNA fragmentation (Zhang et al. 2009). These results suggest that PTB proteins play important roles in CM maturation similarly to what has been shown in C2C12 myoblast and neuron differentiation (Boutz et al. 2007a, b; Makeyev et al. 2007).

4.6 Hermes

The RNA-binding protein with multiple splicing (RBPMS) is commonly known as Hermes. It contains a single RRM and was the first RBP directly implicated in regulation of cardiogenesis through binding to mature RNAs in *Xenopus* embryos (Gerber et al. 2002). Hermes is prominently expressed in differentiating cardiac myocardium (Gerber et al. 1999) and its appearance correlates with the expression of markers of complete differentiation (Gerber et al. 1999). This correlation suggests that Hermes plays a key role in the differentiation process. Hermes is mainly localized in the cytoplasm and its over-expression in developing *Xenopus* embryos inhibits heart development, myocardial differentiation, and overall morphogenesis. While Gata4 expression is normal, the levels of the early cardiogenesis marker Nkx2.5 mRNA are strongly reduced. Three putative mRNA targets negatively regulated by Hermes (RINGO/Spy, Mos, and Xcat2) encode proteins involved in meiotic maturation, early cleavage, and germline development (Song et al. 2007).

4.7 CELF Proteins

The CELF protein family, also called Bruno-like (BRUNOL), contains six paralogs (CELF1-6) (Good et al. 2000; Ladd et al. 2001). Six CELF paralogs also exist in mouse and five in chicken due to a partial deletion of *Celf3* gene. *Xenopus* and zebrafish have paralogs of *Celf1-5*, but not *Celf6*.

CELF proteins contain three RRMs, two at the N-terminus and one at the C-terminus. A linker domain called divergent domain separates the RRM2 and RRM3. CELF proteins were shown to exert roles both in the nucleus (AS and RNA editing) and in the cytoplasm (mRNA stability, translation, and deadenylation regulation) (reviewed by Dasgupta and Ladd 2012). Similarities in the divergent domain divide CELF proteins into two subfamilies, CELF1-2 and CELF3-6 (Dasgupta and Ladd 2012). CELF1 and CELF2 are expressed in skeletal muscle, heart, and brain (Good, et al. 2000; Ladd et al. 2001), while CELF3 and CELF5 proteins are primarily expressed in the nervous system, and CELF6 was detected in kidney, testis, and the nervous system (Dasgupta and Ladd 2012). While the entire family is involved in nervous system functions, CELF1-2 subfamily plays key roles in function of striated muscle.

In mouse hearts, CELF1 and CELF2 protein levels are developmentally down-regulated ~10- and ~18-fold, respectively (Kalsotra et al. 2008). However, mRNA expression does not show this level of regulation suggesting that CELF1 and CELF2 protein levels are posttranscriptionally regulated (see below).

4.7.1 Nuclear CELF Roles in Heart

In heart, a nuclear role of CELF was demonstrated at the level of AS. Downregulation of CELF proteins during heart development correlates with AS transitions occurring from fetal to adult stages (Kalsotra et al. 2008). Nucleotide motifs (GUGUG) were found to be enriched in downstream flanking introns of exons undergoing developmental splicing transitions. Motifs were associated with increased inclusion or skipping of the variable exon when located, respectively, in the beginning or in the end of the downstream intron (Kalsotra et al. 2008). Therefore, CELF proteins repress or activate AS depending on the location of the binding sites in the target mRNAs. Confirmation of CELF1 roles in AS came from experiments using mouse models where CELF1 was over-expressed in adults (where endogenous protein levels are low) or where CELF was blocked by a dominant negative (Ladd et al. 2005; Koshelev et al. 2010) (see below).

4.7.2 Cytoplasmic CELF Roles in Heart

Many transcripts are down- or up-regulated during heart development, and this level of regulation plays a key role in the adaptation of the fetal heart to the adult functions. As many other RBPs, CELF proteins do not have a complete consensus binding sequence. It was shown that CELF1 and CELF2 bind to $(CUG)_8$ repeats (Timchenko et al. 1996; Lu et al. 1999); however Takahashi et al. (2000) demonstrated that CELF1 binds specifically to UG dinucleotide repeats by yeast three-hybrid assays. GU-rich sequences were also detected as binding sites for CELF1 and CELF2 using systematic evolution of ligands by exponential enrichment, SELEX (Faustino and Cooper 2005; Marquis et al. 2006). CELF1 has been shown to destabilize mRNAs by binding to GU-rich motifs within their 3′ UTRs in human primary T cells, in HeLa and muscle C2C12 cells (Vlasova et al. 2008; Lee et al. 2010; Rattenbacher et al. 2010). High-throughput sequencing of RNA isolated by cross-linking immunoprecipitation (HITS-CLIP) experiments showed that CELF1 preferentially bind to 3′ UTRs facilitating mRNA decay in C2C12 cells. CLIP tags were mainly located in 3′ UTRs (~50 %) and motif analysis demonstrated enrichment of GU-rich sequences for CELF1 (Masuda et al. 2012). Given the dramatic postnatal downregulation of CELF1 and CELF2, it is expected that in addition to their roles in AS, these RBPs would also be involved in mRNA stability and translation regulation during heart development.

Although no reports were published yet showing CELF1 acting through binding to 5′ UTRs in heart, many studies have demonstrated that this is the case in other biological systems. For example, CELF1 was shown to bind to a GC-rich sequence in the 5′ UTR of the p21 transcript, inducing translation and therefore inhibiting the cell cycle in senescent human fibroblasts (Iakova et al. 2004). Similar results were obtained during differentiation of cultured skeletal muscle. While CELF1 is induced in the cytoplasm during differentiation of cells from normal patients, cells from DM patients do not accumulate cytoplasmic CELF1. Failure in the cytoplasmic accumulation of CELF1 reduces its binding to p21 mRNA, and therefore cells do not increase p21 protein levels during differentiation as normal cells (Timchenko et al. 2001b). Given the role demonstrated in skeletal muscle cells, it is expectable that some proteins regulated at the expression level during heart development are being regulated at the translation level by CELF1 binding to 5′ UTRs or also at the 3′ UTR inducing deadenylation (Moraes et al. 2006; Graindorge et al. 2008).

4.7.3 Insights into Cardiac CELF Roles from Diseases and Animal Models

Celf1 constitutive knocked out mice have growth retardation and severe viability and fertility defects (Kress et al. 2007). Unfortunately, there are no reports regarding cardiac function in these animals. However, a mouse model with the cardiac-specific expression of a nuclear dominant negative CELF protein under the myosin heavy-chain promoter was reported allowing the study of the nuclear CELF role in AS regulation (Ladd et al. 2005). Mis-regulation of the AS networks developmentally regulated by CELF1 and CELF2 proteins was detected and transgenic animals developed DCM and other cardiac pathologies (Ladd et al. 2005; Terenzi et al. 2009). Microarray experiments of heart samples from these animals and gene ontology analysis of differential gene expression have shown that contraction, calcium signaling, and serum response factor network were the most affected processes. Serum response factor controls transcriptional activation of many cardiac genes involved in contractile apparatus structure and function (Zhang et al. 2001; Nelson et al. 2005; Balza and Misra 2006). Since the dominant negative is exclusively acting in the nucleus, thus affecting only CELF nuclear roles (mainly AS), the authors proposed that the gene expression mis-regulation observed is a consequence of the AS alteration by depletion of nuclear CELF proteins. However, other nuclear roles of CELF proteins such as RNA editing cannot be ruled out (Dasgupta and Ladd 2012).

Another experimental approach was the cardiac-specific over-expression of CELF1. This was achieved by the generation of a transgenic mouse model where human CELF1 is inducible specifically in CM and in a tetracycline-dependent manner (Koshelev et al. 2010). Inducible CELF1 re-expression in adult hearts resulted in echo-cardiographic, electrocardiographic, and histological defects similar to those observed in DM patients (Koshelev et al. 2010). CELF1 over-expression alters AS such that a subset of developmentally regulated events reverts back to neonatal patterns (Kalsotra et al. 2008; Koshelev et al. 2010). In addition, ~40 % of the developmentally up-regulated genes are downregulated when CELF1 is over-expressed in adults suggesting a role for CELF1 in the regulation of gene expression levels, probably by mRNA stability (our unpublished data). By contrast, CELF2 over-expressing animals do not show cardiac abnormalities and a lethal phenotype as CELF1 animals, suggesting that the CELF1-regulated transitions are likely to be the most physiologically relevant for cardiac function.

Arrythmogenic right ventricular dysplasia is a genetic disorder characterized by fibrosis and loss of myocardial cells (Li et al. 2001). One familial form maps to the locus 10p12-14 containing the human *CELF2* locus. Although no mutation was detected to change CELF2 protein sequence, variations in the DNA sequence were found co-segregating with the disease (Li et al. 2001). On the other hand, the partial monosomy 10p is a chromosomal aberration with similar symptoms to the DiGeorge syndrome. The phenotype is the result of haploinsufficiency within two regions on 10p; the HDR1 region associated with hypo-parathyroidism, sensorineural deafness, and renal defects; and the DGCR2 region is responsible for heart defects and thymus hypoplasia/aplasia. While *GATA3* was identified as the gene causing the HDR1 syndrome, *CELF2* was the only known gene found within the critical 300 kb region of the DGCR2 (Lichtner et al. 2002).

DM, one of the most variable human hereditary diseases, is an autosomal dominant syndrome caused by microsatellite expansions in two different genes (Harper 2001). DM1 is caused by the expression of expanded CUG repeats (37–3,500) in the 3′ UTR of the *DMPK1* gene, while DM2 results from the CCTG repeats (75–11,000) in the first intron of the *ZNF9* gene (Ranum and Cooper 2006; Cooper et al. 2009). In both cases, symptoms include skeletal muscle wasting, myotonia, insulin resistance, cataracts, and cardiac defects (Harper 2001). More than 80 % of the patients with DM1 develop cardiac abnormalities such as DCM, conduction defects, and atrioventricular block (Pelargonio et al. 2002; Bhakta et al. 2004; Groh et al. 2008). The RNAs containing the repeats form double-stranded hairpins and accumulate in the nucleus where they disrupt RNA processing by at least two mechanisms (Cooper et al. 2009; Schoser and Timchenko 2010). First, the CUG/CCUG RNA hairpins bind the MBNL family of proteins leading to their sequestration and loss of function in the nucleus (Miller et al. 2000; Lin et al. 2006). Second, CUG repeats induce stabilization and up-regulation of CELF1 ending with its gain of function (Timchenko et al. 2001a; Kuyumcu-Martinez et al. 2007). As a consequence, the expression of the repeats generates the reversion of the gain of MBNL1 and the loss of CELF1 during normal postnatal development. Therefore, DM1 is characterized by the reversion to a neonatal-embryonic AS pattern for many events resulting in the main disease symptoms (Cooper et al. 2009). Representative targets of CELF1 splicing regulation that are mis-regulated in DM1 are cardiac troponin T (TNNT2) (Philips et al. 1998; Ho et al. 2005), chloride channel I (CLCN1) (Charlet et al. 2002; Ho et al. 2005), and insulin receptor (INSR) (Savkur et al. 2001). Consistently, a mouse model over-expressing 960 interrupted CUG repeats in the *DMPK* context specifically in heart reproduces DM1 cardiac abnormalities, mis-regulated AS, RNA foci formation co-localizing with MBNL1, and CELF1 up-regulation (Wang et al. 2007).

4.8 MBNL Proteins

MBNL proteins are highly conserved and they are regulated during development. MBNL roles have been shown in muscle and eye development, and in DM (Artero et al. 1998; Begemann et al. 1997; Miller et al. 2000). Mammals possess three *Mbnl* genes (Fardaei et al. 2002). In mouse and human, MBNL1 and MBNL2 proteins are expressed across many tissues (brain, heart, muscle), whereas MBNL3 is mainly expressed in placenta (Squillace et al. 2002; Kanadia et al. 2003a), suggesting functional specialization.

MBNL proteins contain two pairs of highly conserved zinc finger domains, which bind to pre-mRNA to regulate AS (Pascual et al. 2006). In DM1 and DM2, MBNL proteins (MBNL1, MBNL2, and MBNL3) have been shown to bind to the expanded repeats in nuclear foci (Mankodi et al. 2001; Fardaei et al. 2002) being sequestered from their normal target mRNAs (Miller et al. 2000). Therefore, a reversion to neonatal AS patterns is present for many developmentally regulated transitions. As described above for CELF proteins, MBNL proteins are also expressed in the cytoplasm and, from there, they regulate mRNA stability (Osborne et al. 2009; Du et al. 2010; Masuda et al. 2012) or localization (Adereth et al. 2005; Wang et al. 2012a).

MBNL1 is the major *MBNL* gene expressed in skeletal muscle. Some isoforms (containing exon 3) bind to CUG repeat expansion and others (not containing exon 3) do not bind the repeats (Miller et al. 2000; Fardaei et al. 2002). A *Mbnl1* knocked out mouse model was generated by targeting exon 3, *Mbnl1*$\Delta^{E3/}\Delta^{E3}$, (Kanadia et al. 2003b). Knocked out mice exhibited overt myotonia and delayed muscle relaxation, similarly to DM human patients (Kanadia et al. 2003b). Although cardiac functions have not been yet characterized in these animals, heart (and other tissues) transcriptome was recently analyzed by RNA-seq in combination with ultraviolet cross-linking/immunoprecipitation/sequencing (CLIP-seq) in order to identify binding targets (Wang et al. 2012a).

4.8.1 Nuclear MBNL1 Roles in Heart

Hundreds of Mbnl1-responsive AS cassette exons, alternative 3′ and 5′splice sites, and other types of AS events were detected in heart from *Mbnl1*$\Delta^{E3/}\Delta^{E3}$ mice. Several individual Mbnl1-regulated exons were previously described and shown to be activated or repressed by Mbnl1 depending on the location of the binding sites upstream or downstream the exons, respectively (Warf and Berglund 2007; Du et al. 2010; Goers et al. 2010). CLIP-seq confirmed many previously discovered binding sites and revealed thousands of new locations for Mbnl1-binding regions associated with AS activity (Wang et al. 2012a). As found for several splicing regulators, splicing repression is generally associated with Mbnl1 binding in the intron upstream of the alternative exon, and splicing activation is linked with Mbnl1 binding in the downstream intron (Wang et al. 2012a).

4.8.2 Cytoplasmic MBNL1 Roles in Heart

Many Mbnl1-dependent alternative 3′ UTRs were identified in C2C12 cells (Masuda et al. 2012) and in different tissues including heart (Wang et al. 2012a) by high-throughput technologies. A substantial proportion of the Mbnl1 CLIP tags were located within 3′ UTRs (~50 %) in two independent studies (Masuda et al. 2012; Wang et al. 2012a). In fact, Mbnl1 destabilizes endogenous mRNAs by binding to 3′ UTRs and the level of destabilization correlates with the number of CLIP tags in the 3′ UTR (Masuda et al. 2012). Targeted 3′ UTRs include mRNA-encoding proteins involved in synaptic vesicle fusion (Snap25, Vamp1), signaling proteins (Camk2a), and extracellular matrix proteins such as β1 integrin and collagens. Those results suggest a functional association between Mbnl1 and mRNAs encoding proteins with specific localizations, and cell culture experiments using C2C12 myoblasts demonstrated that Mbnl1 depletion alters localization of the bound mRNAs. Interestingly, the degree of redistribution correlates with the CLIP-tag densities. Taken together, the evidence strongly supports a model in which Mbnl1 directly affects mRNA trafficking and localization and regulates translation (Wang et al. 2012a). Mechanistically of interest is the finding that nuclear and cytoplasmic Mbnl1 roles would be linked since genes regulated by Mbnl1 at the AS level were twice likely to have CLIP-tag clusters within their 3′ UTRs (Wang et al. 2012a).

4.8.3 MBNL and Diseases and Other Animal Models

Besides CELF1, MBNL proteins are involved in the DM1 pathogenesis. The zinc finger domains bind to the CUG repeats in the 3′ UTR of the *DMPK* gene (Begemann et al. 1997; Ho et al. 2004; Teplova and Patel 2008) leading to MBNL1 sequestration in nuclear foci and thus its loss of function. The *Mbnl1* Δ^{E3}/Δ^{E3} provides a mouse model of DM1 (Miller et al. 2000; Lin et al. 2006) and many mis-regulated AS events found in the *Mbnl1* Δ^{E3}/Δ^{E3} mice were found in human DM1 patients (Du et al. 2010). In addition, ~80–90 % of the altered splicing phenotype in CUG expansion expressing mice in skeletal muscle (Mankodi et al. 2000) is present in *Mbnl1* Δ^{E3}/Δ^{E3} mice (Du et al. 2010). This evidence supports the critical role of MBNL1 protein in DM1 molecular mechanism at the level of AS.

The complementary approach to the loss of function of a protein is its gain of function. MBNL1 over-expressing mice were recently generated specifically in skeletal muscle or a ubiquitously expressing line that included expression in heart (Chamberlain and Ranum 2012). The ubiquitous line exhibits reduced body mass, increased mortality at 76 weeks of age, and often sudden death. Neonates over-expressing MBNL1 showed a premature AS transition to adult patterns of MBNL1 mRNA targets in heart (Tnnt2, Sorbs2, Ryr2, Cacna1d) as well as in brain and skeletal muscle (Chamberlain and Ranum 2012). However, more studies are needed to understand the role of MBNL1 over-expression in the heart.

The role of MBNL2 in heart (and other tissues) was addressed in zebrafish using morpholinos to knock down the *mbnl2* gene, without up-regulation of the other Mbnl proteins by compensatory mechanisms. Mbnl2 knocked down animals showed disruption of heart functions and morphology. The effects included slow heart rate as well as abnormal Z-bands and myofibril alignment, and sarcoplasmic reticulum disorganization in skeletal and cardiac muscle. At the molecular level, AS is disrupted in specific targets, tnnt2 and clcn1, explaining muscle and cardiac abnormalities (Machuca-Tzili et al. 2011).

Although heart functions were not studied, it is useful to mention here that two mouse models were recently reported in which *Mbnl2* or *Mbnl3* genes were consti-

Table 11.4 Available *Mbnl* knocked out (KO) mouse models

RBPs	Expression	KO mouse model	Recapitulation of DM features	Recapitulated molecular DM features	References
MBNL1	High in heart and muscle	*Mbnl1*$^{\Delta E3/\Delta E3}$	*Muscle:* Myotonia, delayed muscle relaxation, muscle pathology (central nuclei, fibrosis, split fibers)	Strong AS mis-regulation in heart and skeletal muscle Modest AS mis-regulation in brain	Kanadia et al. (2003b) Kalsotra et al. (2008) Du et al. (2010) Suenaga et al. (2012) Wang et al. (2012a)
MBNL2	High in brain, heart, and lungs. Low in skeletal muscle	*Mbnl2*$^{\Delta E2/\Delta E2}$	*Central nervous system:* Abnormal REM sleep propensity, spatial memory and learning deficits, altered GABA sensitivity Skeletal muscle: No pathology no motor deficits	AS mis-regulation in brain and not in skeletal muscle	Charizanis et al. (2012)
MBNL3	High in placenta. In adults, MBNL3 expression picks after injury	*Mbnl3*$^{\Delta E2/\Delta E2}$	*Muscle regeneration:* Age-dependent impairment of adult muscle regeneration	Only modest changes in AS and gene expression	Poulos et al. (2013)

tutively knocked out, *Mbnl2*$\Delta^{E2/}\Delta^{E2}$ and *Mbnl3*$\Delta^{E2/}\Delta^{E2}$, respectively (Charizanis et al. 2012; Poulos et al. 2013). *Mbnl2*$\Delta^{E2/}\Delta^{E2}$ mice develop several DM-associated central nervous system features and show splicing mis-regulation of hundreds of exons without altering transcript levels. The majority of those exons are also mis-regulated in DM supporting a role for MBNL2 in neuronal features of DM caused by the disruption of the MBNL2-mediated developmental AS program (Charizanis et al. 2012). *Mbnl3*$\Delta^{E2/}\Delta^{E2}$ mice show progressive impairment of muscle regeneration such that regeneration is normal in young animals, but delayed in older animals (Poulos et al. 2013). Those results suggest that this mouse model would provide insights into the age-associated muscular atrophy in the disease. For the three Mbnl1 proteins a high proportion (51–57 %) of the CLIP tags are located within 3′ UTRs (Charizanis et al. 2012; Masuda et al. 2012; Wang et al. 2012a; Poulos et al. 2013) supporting Mbnl1 roles in mRNA stability and localization as well as translation regulation that may also be involved in pathological conditions. The summary information about the *Mbnl1* knocked out mouse models is summarized in Table 11.4.

5 RBPs Are Regulated

Although RBPs play key roles in the regulation of gene expression during heart development as well as in cardiac pathologies, they are also regulated adding one more level of regulation for their cardiac roles. The activities of RBPs can be regulated through posttranscriptional or posttranslational modification. Cleavage and phosphorylation are two key posttranslational mechanisms of RBP regulation (Fig. 11.1a, b). Several RBPs exist as multiple isoforms due to AS mechanisms and the different isoforms show different activities or localization (Fig. 11.1c). This is the case for RBFOX1/2 proteins (reviewed by Kuroyanagi 2009), PTB which splicing variants have different splicing activities (Wollerton et al. 2001), and MBNL1 where AS of exons 3, 5, and 7 controls its localization, interaction with mRNA targets, dimerization, and splicing activity (Tran et al. 2011). Finally (but not excluding other modifications), miRNAs can regulate expression of RBPs (Fig. 11.1d) and also compete or cooperate with RBPs by binding to the same mRNA target (Fig. 11.1, lower panel).

5.1 Phosphorylation

One of the best characterized RBP modifications is phosphorylation. Phosphorylation controls local concentration of splicing factors that are adjacent to pre-mRNA targets, by altering their intracellular localization (Habelhah et al. 2001; Daoud et al. 2002; Huang et al. 2004) and stability (Kuyumcu-Martinez et al. 2007), protein-protein interactions (Xiao and Manley 1997), protein-RNA interactions (Huang et al. 2007), and splicing activity (Shin and Manley 2002; Feng et al. 2008).

The SR domains of the SR proteins are responsible for their phosphorylation (Misteli et al. 1998) and the degree of phosphorylation is important for their functions. SR proteins regulate mRNA nuclear export by interacting with the export receptor TAP/nuclear export factor 1 (NXF1). The SR proteins SRSF7 (previously known as 9G8) and SRSF1 interact with higher affinity with hypo-phosphorylated TAP/NXF1 and also splicing is accompanied by their own dephosphorylation (Huang et al. 2007) (Fig. 11.2a). Phosphorylation status can also determine whether an RBP acts as a splicing repressor or activator as is the case for SRSF10. While dephosphorylated SRSF10 acts as a global splicing repressor (Shin and Manley 2002; Shin et al. 2004), it becomes a sequence-specific activator when phosphorylated (Feng et al. 2008).

Several groups have shown that the phosphorylation state of CELF1 regulates its localization and its activity in AS, mRNA decay, and translation (Kuyumcu-Martinez et al. 2007; Timchenko et al. 2006; Salisbury et al. 2008; Wang et al. 2008; Huichalaf et al. 2010; Kalsotra et al. 2010) (Fig. 11.2b).

CELF1 can be phosphorylated in serine (S) 28 by Akt controlling its intracellular localization (CELF1 nuclear splicing regulation/CELF1 cytoplasmic role in mRNA stability and translation) and its interaction with cyclin D1 (cycD1) mRNA (Salisbury et al. 2008; Huichalaf et al. 2010). CUG repeat expression induces Akt

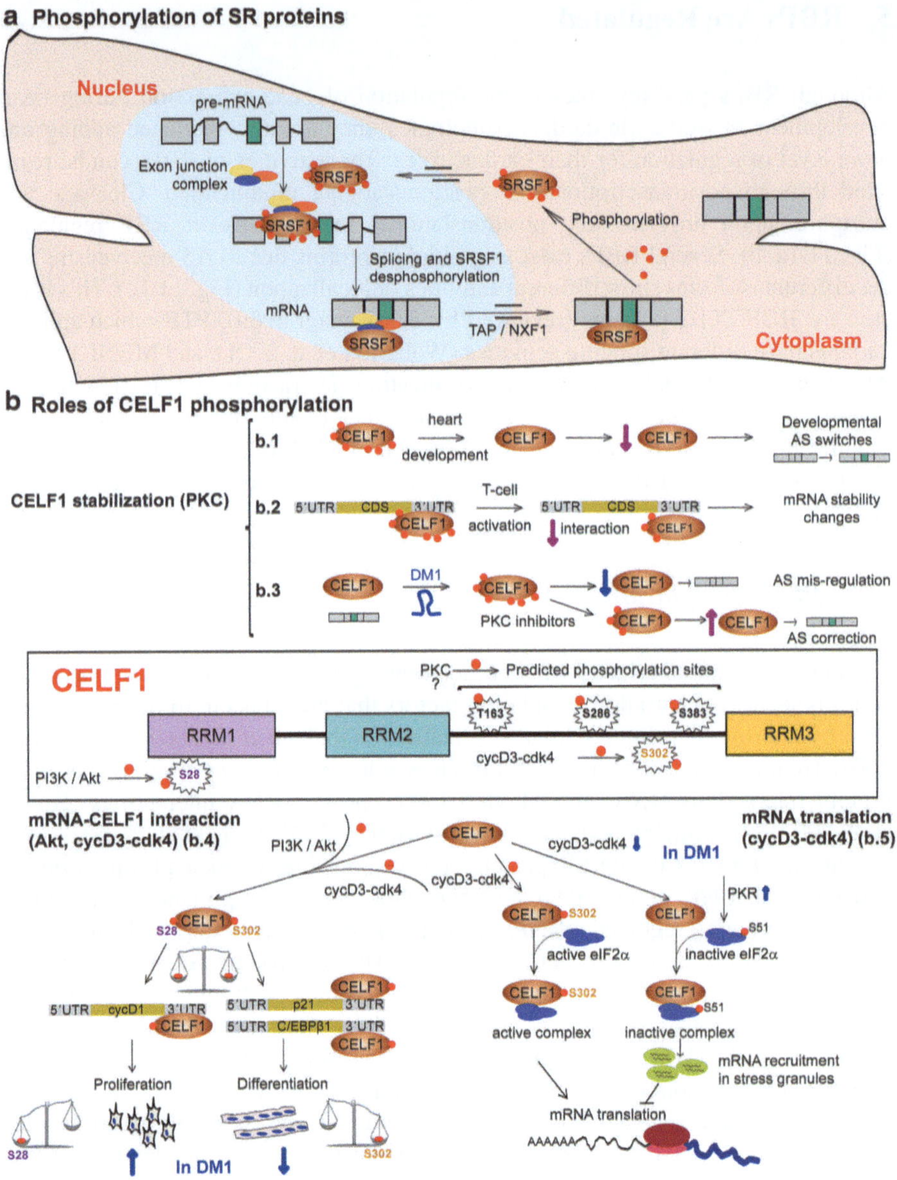

Fig. 11.2 Regulation of RBPs by phosphorylation. (**a**) Activity of SRSF1 and other SR proteins is regulated by phosphorylation (Huang et al. 2007; Feng et al. 2008). (**b**) Phosphorylation regulates CELF1 activities by its stabilization. References: Kalsotra et al. 2008, 2010 (**b1**); Beisang et al. 2012 (**b2**); Kuyumcu-Martinez et al. 2007; Wang et al. 2007 (**b3**). Balance between phosphorylation of S28 by PI3K/Akt pathway and S302 by cycD3-cdk4 affects CELF1 interaction with different mRNA targets by their 3′ UTRs (**b4**). References: Huichalaf et al. 2010. Phosphorylation of CELF1 and eIF2**α** regulates mRNA translation in control and DM1 conditions (**b5**). References: Tian et al. 2000; Holcik and Sonenberg 2005; Ikezoe et al. 2007; Huichalaf et al. 2010

phosphorylation, increasing its interaction with CELF1 and thus the interaction between CELF1 and cycD1 mRNA. S302 is a residue which is phosphorylated by cyclin D3/cdk4 (cycD3-cdk4), controlling interaction with C/EBPβ1 and p21 mRNAs promoting myoblast differentiation (Salisbury et al. 2008) and with the active (phosphorylated) translation initiation factor 2α (eIF2α) activating translation (Timchenko et al. 2006). In DM1, CUG repeats activate the RNA protein kinase (PKR) which phosphorylates S51 of eIF2α (stress response) leading to its inactivation. On the other hand, CUG repeats downregulate cycD3-cdk4, reducing S302 phosphorylation in CELF1 molecules (Huichalaf et al. 2010). In consequence, the complex formed between un-phosphorylated CELF1 and phosphorylated eIF2α in S51 is inactive and translation is blocked by targeting mRNAs into stress granules (Tian et al. 2000; Holcik and Sonenberg 2005; Ikezoe et al. 2007; Huichalaf et al. 2010).

Finally, CELF1 phosphorylation stabilizes the protein in DM1 cells affecting its activity in AS regulation, contributing to the pathogenic effects of repeat expression (Kuyumcu-Martinez et al. 2007). During heart development, Celf1 phosphorylation is reduced potentially contributing to its postnatal downregulation at the protein level (Kalsotra et al. 2010). In addition, protein kinase C (PKC) inhibition in the heart-specific DM1 mouse model mentioned above (Wang et al. 2007) increases animal survival with a reduced phosphorylation and steady-state levels of Celf1 protein (Wang et al. 2009). Cardiac conduction and contraction defects are ameliorated after PKC inhibition and AS events regulated by Celf1 are partially corrected (Wang et al. 2009). Although the PKC-phosphorylated residues of Celf1 have not been experimentally identified, threonine (T) 63, S286, and S383 might be involved due to their high scores as PKC predicted phosphorylation sites (Huichalaf et al. 2010). All together this evidence supports the model that phosphorylation plays a key role in CELF1 nuclear and cytoplasmic functions (AS, mRNA decay, translation) during cardiac development and in DM1 features.

5.2 Cleavage

As we previously described, PTB is regulated by an upstream pathway which involves HDAC5 (downregulated during heart development) and a caspase inhibitor (downregulated during heart development) (Fig. 11.1a). Therefore, the active caspase is up-regulated during development, leading to PTB cleavage and thus reducing its activity. Since PTB generally acts as a repressor of AS, this mechanism promotes adult isoforms including alternative exons (Ye et al. 2013). The heteroribonuclear proteins hnRNP A, C, and K, components of the spliceosome, are also target of caspase cleavage reducing RNA processing activity. SRSF1 has been shown to be cleaved by caspases; however consequences have not been yet studied (reviewed by Fischer et al. 2003).

Nucleolin, the most abundant RBP in the nucleolus and involved in RNA transport, is also regulated by cleavage (Fig. 11.1a). Nucleolin is stable in proliferating cells, but undergoes self-cleavage in quiescent cells or in response to oxidative stimulus in C2C12 cells (Chen et al. 1991; Wang et al. 2012b). Recently, rat hearts showed that after ischemia-reperfusion injury nucleolin expression is reduced because protein cleavage is increased. Nucleolin over-expression shows cytoprotective effects in cell culture as well as in transgenic mice, while nucleolin deletion results in hypoxia-induced CM death. This protective effect of nucleolin is due to binding to the 3′ UTR of the heat-shock protein 32 (Hsp32) slowing its degradation and thus leading to its up-regulation (Jiang et al. 2013).

5.3 miRNAs/RBP Interplay

miRNAs bind to target mRNAs by partial sequence complementarity after incorporation into the RNA-induced silencing complex (RISC), leading to initiation of mRNA decay and/or translation inhibition (Eulalio et al. 2008; Bartel 2009; Fabian and Sonenberg 2012). They are key players in the gene expression regulation networks during differentiation and tissue development as well as in homeostasis (van Rooij et al. 2007; Bartel 2009). Many recent studies have established critical functions for miRNAs during cardiac development and in cardiac diseases (reviewed by Porrello 2013). Different mouse models have revealed that miRNAs are important in different stages of heart development: Dicer knocked out specifically in heart using Nkx2.5-Cre revealed defects in embryonic heart development and embryonic lethality (Zhao et al. 2007); conditional Dicer knocked out using α-myosin heavy-chain promoter-driven Cre resulted in severe DCM and heart failure shortly after birth with numerous contractile protein expression patterns mis-regulated (Chen et al. 2008b); Dicer deletion induction in younger and older mice specifically in CM has also shown serious cardiac defects, DCM, and reactivation of fetal cardiac genes (da Costa Martins et al. 2008). Additionally, Dicer is up-regulated in human patients with DCM and heart failure (Chen et al. 2008b) and large miRNA expression alteration is present in cardiac disease (van Rooij et al. 2006; Sayed et al. 2007).

miRNAs target numerous mRNAs such as cell cycle and extracellular matrix genes being responsible for their regulation during heart development and in cardiac pathologies (van Rooij et al. 2008; Ikeda and Pu 2010; Porrello et al. 2011b; Cao et al. 2013). However, miRNAs can also bind mRNA-encoding RBPs, regulating RBP expression (Fig. 11.1d). Therefore, important regulation of numerous target genes of those RBPs would be exerted by miRNAs. During heart development, a large number of miRNAs is up- and downregulated suggesting that this level of regulation would significantly impact in the developmental transitions of the transcriptome.

CELF1 and CELF2 are regulated by miRNAs in heart. Both of them are downregulated during heart development and are dramatically up-regulated after two days of Dicer knocked out specifically in heart of adult mice (Kalsotra et al. 2010). In particular, miR-23a/b plays a direct role in the downregulation of the two CELF proteins by binding to their 3′ UTRs, and therefore in the cardiac splicing program involved during postnatal development (Kalsotra et al. 2010). Several other RBPs are also up-regulated after Dicer deletion including RBFOX1, RBFOX2, PTP, TIA1, and hnRNP H. Protein expression of other splicing regulators is not affected (MBNL1,

SRSF1, hnRNPC, and hnRNP L), suggesting that only a specific set of splicing factors are under the regulation of miRNAs in the murine heart (Kalsotra et al. 2010).

As we have already mentioned, PTP/nPTB expression is controlled by miRNAs in a cell- and tissue-specific manner (Boutz et al. 2007a; Makeyev et al. 2007). While miR-124 controls nPTB expression in neurons (Makeyev et al. 2007), miR-133 controls its protein expression in C2C12 myoblasts without altering mRNA levels (Boutz et al. 2007a). Rbm24 is a target of miR-125b, and while Rbm24 is up-regulated during myoblast differentiation, miR-125b is downregulated (Ge et al. 2011). The sequence of the predicted miR-125b-binding site in the 3′ UTR of Rbm24 transcript is highly conserved across species supporting that this miRNA-mRNA interaction can be relevant for cardiac function (Vacchi-Suzzi et al. 2013). Recent experiments in culture HPASM and HEK293 cells revealed that expression of RBM24 is directly inhibited by miR-125b and mutation of the predicted binding site rescues luciferase activity (Vacchi-Suzzi et al. 2013).

On the other hand, some RBPs are able to regulate miRNA Drosha and Dicer processing. SRSF1 and miR-7 can form a negative feedback loop where SRSF1 promotes miR-7 maturation, which negatively regulates SRSF1 expression at the translational level (Fig. 11.3a). In addition, SRSF1 is also involved in miR-221, miR-222, and miR-29b-1 maturation, suggesting that SRSF1 would regulate AS- and miRNA-mediated gene repression (Wu et al. 2010).

miR-1/MBNL1 is another example of miRNAs/RBP interplay where one player regulates the other one and vice versa (Fig. 11.3b). miR-1 family is involved in heart diseases in humans as well as in mouse models (Yang et al. 2007; Zhao et al. 2007). Pre-miR-1 can be bound by LIN28 promoting uridylation by ZCCHC11 (TUT4) and thus Dicer processing is blocked. In DM1 and DM2 patients mature miR-1 is significantly downregulated and miR-1 targets are up-regulated. In normal conditions, MBNL1 protein binds to the terminal loop of pre-miR-1 competing with LIN28 and, thus, Dicer processing takes place. By contrast, in DM, since MBNL1 is sequestered by the CUG/CCUG repeats, LIN28 binds to pre-miR-1 blocking Dicer processing. Therefore, expression of the mature miR-1 is lost, leading to an up-regulation of miR-1 targets such as cardiac L-type calcium channel (CACNA1C), the main Ca^{2+} channel in heart, connexin 43 (GJA1), responsible of intra-CM conductance, contributing to the pathology of the disease (Rau et al. 2011). Other pre-miRNAs possess conserved UGC motifs in their loops suggesting possible roles of MBNL1 in the regulation of the processing of other miRNAs in heart (Rau et al. 2011).

The interplay between RBPs and miRNAs is also implicated in different normal and pathological processes including angiogenesis (reviewed by Chang and Hla 2011) and cancer (reviewed by van Kouwenhove et al. 2011). RBPs are involved in miRNA function determination by controlling different stages of miRNA biogenesis, localization, degradation, and activity. Additionally, regulation relies also on the miRNA- and RBP-binding ability to common mRNA targets probably under a very tight spatiotemporal control (Kouwenhove et al. 2011). miRNA action might occur in conjunction with RBP-mRNA interactions cooperating in translation repression (let7/HuR/cMyc 3′UTR; miR-200/HuR/Vegfa 3′UTR) or activation (miR-369/FXR1/TNFα 3′UTR), cooperating in mRNA decay (miR-16/TTP/TNFα 3′UTR),

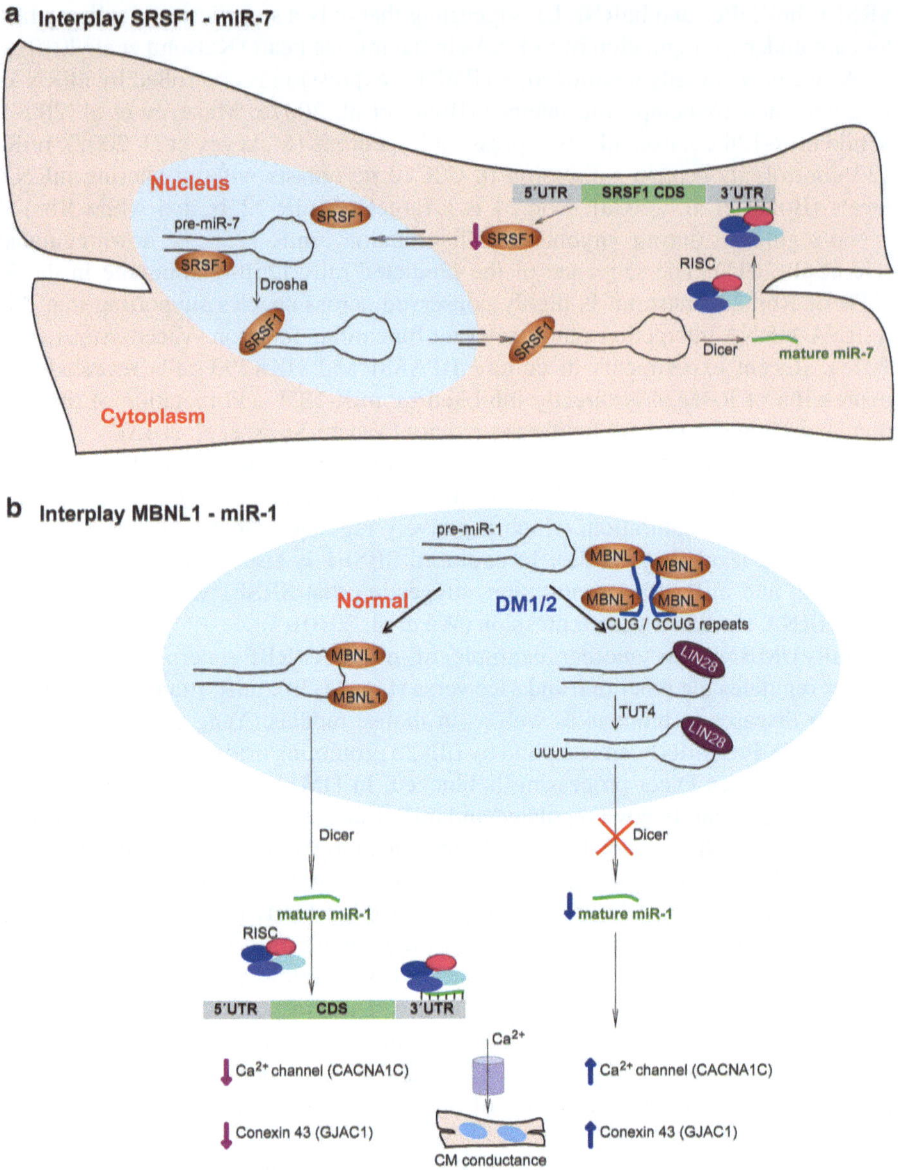

Fig. 11.3 Interplay between miRNAs and RBPs. (**a**) SRSF1 regulates pre-miR-7 processing and miR-7 downregulates SRSF1 by binding to its 3′ UTR generating a negative feedback (Wu et al. 2010). (**b**) MBNL1 regulates miR-1 processing and this pathway is mis-regulated in DM1 leading to CM conductance problems and Ca^{2+} defects (Rau et al. 2011)

or competing (miR-122/HuR/CAT1 3′UTR) (Chang and Hla 2011; Chang et al. 2013). The miRNA-RBP cooperation and competition add an additional level of gene expression regulation. This type of interplay can also be involved in transcription regulation during cardiac development as well as in heart failure due to the extensive changes in miRNA (Cao et al. 2013) and RBP expression levels (Kalsotra et al. 2008; and our unpublished data) in both processes.

6 Conclusion and Perspectives

Throughout this chapter we have described the diverse roles of RBPs in heart remodeling normally occurring during development. We have presented an overview of RBP actions on transcriptome changes occurring in physiological conditions as well as when pathological defects alter the normal regulation. We hope to have transmitted to the reader the complexity, the diverse levels of regulation, and the interconnections that should be tightly and spatiotemporally controlled to be able to have a complete heart pumping and beating.

Two interesting questions remain open in the field related with the topic of our chapter and it is appropriate to mention here. Since the majority of the available data comes from cell culture experiments using CM or myoblast cultures, mouse models where deletions or over-expression were performed specifically in CM, or complete heart tissues (remember that CM are ~75 % of the cardiac volume) our present knowledge about RBP roles is mainly focused in one cardiac cell type. This cell type is important in the cardiac functionality, but other cells exist in the heart. Therefore, this raises the first open question: How is the regulation exerted by RBPs in other cell types in the heart? And the second one is the following: How similar or different are the transcriptional and posttranscriptional programs in CM, CF, endothelial, and vascular smooth cells during development and which is the contribution of one regulatory network to another one?

Insights into these two questions would provide a more complete understanding of physiological heart development as well as diseases. This potential knowledge would be also valuable in investigating and directing therapeutic approaches in adult heart failure as well as in congenital cardiac disorders.

Acknowledgments J.G. is a Pew Latin American Fellow in the Biomedical Sciences supported by the Pew Charitable Trusts. T.A.C. is funded by the National Institutes of Health (R01HL045565, R01AR060733, and R01AR045653) and Muscular Dystrophy Association grants.

References

Adereth Y, Dammai V, Kose N, Li R, Hsu T (2005) RNA-dependent integrin alpha3 protein localization regulated by the Muscleblind-like protein MLP1. Nat Cell Biol 7:1240–1247

Ames EG, Lawson MJ, Mackey AJ, Holmes JW (2013) Sequencing of mRNA identifies re-expression of fetal splice variants in cardiac hypertrophy. J Mol Cell Cardiol 62:99–107

Artero R, Prokop A, Paricio N, Begemann G, Pueyo I, Mlodzik M, Perez-Alonso M, Baylies MK (1998) The muscleblind gene participates in the organization of Z-bands and epidermal attachments of Drosophila muscles and is regulated by Dmef2. Dev Biol 195:131–143

Auweter SD, Fasan R, Reymond L, Underwood JG, Black DL, Pitsch S, Allain FH (2006) Molecular basis of RNA recognition by the human alternative splicing factor Fox-1. EMBO J 25:163–173

Balza RO Jr, Misra RP (2006) Role of the serum response factor in regulating contractile apparatus gene expression and sarcomeric integrity in cardiomyocytes. J Biol Chem 281:6498–6510

Bang ML, Centner T, Fornoff F, Geach AJ, Gotthardt M, McNabb M, Witt CC, Labeit D, Gregorio CC, Granzier H, Labeit S (2001) The complete gene sequence of titin, expression of an unusual approximately 700-kDa titin isoform, and its interaction with obscurin identify a novel Z-line to I-band linking system. Circ Res 89:1065–1072

Bartel DP (2009) MicroRNAs: target recognition and regulatory functions. Cell 136:215–233

Baudino TA, Carver W, Giles W, Borg TK (2006) Cardiac fibroblasts: friend or foe? Am J Physiol Heart Circ Physiol 291:H1015–H1026

Baulac S, Huberfeld G, Gourfinkel-An I, Mitropoulou G, Beranger A, Prud'homme JF, Baulac M, Brice A, Bruzzone R, LeGuern E (2001) First genetic evidence of GABA(A) receptor dysfunction in epilepsy: a mutation in the gamma2-subunit gene. Nat Genet 28:46–48

Bayer KU, Schulman H (2001) Regulation of signal transduction by protein targeting: the case for CaMKII. Biochem Biophys Res Commun 289:917–923

Begemann G, Paricio N, Artero R, Kiss I, Pérez-Alonso M, Mlodzik M (1997) Muscleblind, a gene required for photoreceptor differentiation in Drosophila, encodes novel nuclear Cys3His-type zinc-finger-containing proteins. Development 124:4321–4331

Beisang D, Rattenbacher B, Vlasova-St Louis IA, Bohjanen PR (2012) Regulation of CUG-binding protein 1 (CUGBP1) binding to target transcripts upon T cell activation. J Biol Chem 287:950–960

Bhalla K, Phillips HA, Crawford J, McKenzie OL, Mulley JC, Eyre H, Gardner AE, Kremmidiotis G, Callen DF (2004) The de novo chromosome 16 translocations of two patients with abnormal phenotypes (mental retardation and epilepsy) disrupt the A2BP1 gene. J Hum Genet 49:308–311

Bhakta D, Lowe MR, Groh WJ (2004) Prevalence of structural cardiac abnormalities in patients with myotonic dystrophy type I. Am Heart J 147:224–227

Bland CS, Wang ET, Vu A, David MP, Castle JC, Johnson JM, Burge CB, Cooper TA (2010) Global regulation of alternative splicing during myogenic differentiation. Nucleic Acids Res 38:7651–7664

Blencowe BJ (2006) Alternative splicing: new insights from global analyses. Cell 126:37–47

Boutz PL, Chawla G, Stoilov P, Black DL (2007a) MicroRNAs regulate the expression of the alternative splicing factor nPTB during muscle development. Genes Dev 21:71–84

Boutz PL, Stoilov P, Li Q, Lin CH, Chawla G, Ostrow K, Shiue L, Ares M Jr, Black DL (2007b) A post-transcriptional regulatory switch in polypyrimidine tract-binding proteins reprograms alternative splicing in developing neurons. Genes Dev 21:1636–1652

Brauch KM, Karst ML, Herron KJ, de Andrade M, Pellikka PA, Rodeheffer RJ, Michels VV, Olson TM (2009) Mutations in ribonucleic acid binding protein gene cause familial dilated cardiomyopathy. J Am Coll Cardiol 54:930–941

Brudno M, Gelfand MS, Spengler S, Zorn M, Dubchak I, Conboy JG (2001) Computational analysis of candidate intron regulatory elements for tissue-specific alternative pre-mRNA splicing. Nucleic Acids Res 29:2338–2348

Buratti E, Baralle FE (2012) TDP-43: gumming up neurons through protein-protein and protein-RNA interactions. Trends Biochem Sci 37:237–247

Busà R, Paronetto MP, Farini D, Pierantozzi E, Botti F, Angelini DF, Attisani F, Vespasiani G, Sette C (2007) The RNA-binding protein Sam68 contributes to proliferation and survival of human prostate cancer cells. Oncogene 26:4372–4382

Camelliti P, Borg TK, Kohl P (2005) Structural and functional characterisation of cardiac fibroblasts. Cardiovasc Res 65:40–51

Cao X, Wang J, Wang Z, Du J, Yuan X, Huang W, Meng J, Gu H, Nie Y, Ji B, Hu S, Zheng Z (2013) MicroRNA profiling during rat ventricular maturation: A role for miR-29a in regulating cardiomyocyte cell cycle re-entry. FEBS Lett 587:1548–1555

Castle JC, Zhang C, Shah JK, Kulkarni AV, Kalsotra A, Cooper TA, Johnson JM (2008) Expression of 24,426 human alternative splicing events and predicted cis regulation in 48 tissues and cell lines. Nat Genet 40:1416–1425

Chamberlain CM, Ranum LP (2012) Mouse model of muscleblind-like 1 overexpression: skeletal muscle effects and therapeutic promise. Hum Mol Genet 21:4645–4654

Chan RC, Black DL (1995) Conserved intron elements repress splicing of a neuron-specific c-src exon in vitro. Mol Cell Biol 15:6377–6385

Chang S, McKinsey TA, Zhang CL, Richardson JA, Hill JA, Olson EN (2004) Histone deacetylases 5 and 9 govern responsiveness of the heart to a subset of stress signals and play redundant roles in heart development. Mol Cell Biol 24:8467–8476

Chang SH, Hla T (2011) Gene regulation by RNA binding proteins and microRNAs in angiogenesis. Trends Mol Med 17:650–658

Chang SH, Lu YC, Li X, Hsieh WY, Xiong Y, Ghosh M, Evans T, Elemento O, Hla T (2013) Antagonistic function of the RNA-binding protein HuR and miR-200b in post-transcriptional regulation of vascular endothelial growth factor-A expression and angiogenesis. J Biol Chem 288(7):4908–4921

Charizanis K, Lee KY, Batra R, Goodwin M, Zhang C, Yuan Y, Shiue L, Cline M, Scotti MM, Xia G et al (2012) Muscleblind-like 2-mediated alternative splicing in the developing brain and dysregulation in myotonic dystrophy. Neuron 75:437–450

Charlet BN, Savkur RS, Singh G, Philips AV, Grice EA, Cooper TA (2002) Loss of the muscle-specific chloride channel in type 1 myotonic dystrophy due to misregulated alternative splicing. Mol Cell 10:45–53

Chen CM, Chiang SY, Yeh NH (1991) Increased stability of nucleolin in proliferating cells by inhibition of its self-cleaving activity. J Biol Chem 266:7754–7758

Chen YH, Huang FL, Cheng YC, Wu CJ, Yang CN, Tsay HJ (2008a) Knockdown of zebrafish Nav1.6 sodium channel impairs embryonic locomotor activities. J Biomed Sci 15:69–78

Chen JF, Murchison EP, Tang R, Callis TE, Tatsuguchi M, Deng Z, Rojas M, Hammond SM, Schneider MD, Selzman CH et al (2008b) Targeted deletion of Dicer in the heart leads to dilated cardiomyopathy and heart failure. Proc Natl Acad Sci U S A 105:2111–2116

Chen M, Manley JL (2009) Mechanisms of alternative splicing regulation: insights from molecular and genomics approaches. Nat Rev Mol Cell Biol 10:741–754

Chénard CA, Richard S (2008) New implications for the QUAKING RNA binding protein in human disease. J Neurosci Res 86:233–242

Cheng H, Zheng M, Peter AK, Kimura K, Li X, Ouyang K, Shen T, Cui L, Frank D, Dalton ND et al (2011) Selective deletion of long but not short Cypher isoforms leads to late-onset dilated cardiomyopathy. Hum Mol Genet 20:1751–1762

Cooper TA, Wan L, Dreyfuss G (2009) RNA and disease. Cell 136(4):777–793

da Costa Martins PA, Bourajjaj M, Gladka M, Kortland M, van Oort RJ, Pinto YM, Molkentin JD, De Windt LJ (2008) Conditional dicer gene deletion in the postnatal myocardium provokes spontaneous cardiac remodeling. Circulation 118:1567–1576

Daoud R, Mies G, Smialowska A, Oláh L, Hossmann KA, Stamm S (2002) Ischemia induces a translocation of the splicing factor tra2-beta 1 and changes alternative splicing patterns in the brain. J Neurosci 22:5889–5899

Das D, Clark TA, Schweitzer A, Yamamoto M, Marr H, Arribere J, Minovitsky S, Poliakov A, Dubchak I, Blume JE, Conboy JG (2007) A correlation with exon expression approach to identify cis-regulatory elements for tissue-specific alternative splicing. Nucleic Acids Res 35:4845–4857

Dasgupta T, Ladd AN (2012) The importance of CELF control: molecular and biological roles of the CUG-BP, Elav-like family of RNA-binding proteins. Wiley Interdiscip Rev RNA 3:104–121

Dasgupta T, Stillwagon SJ, Ladd AN (2013) Gene expression analyses implicate an alternative splicing program in regulating contractile gene expression and serum response factor activity in mice. PLoS One 8:e56590. doi:10.1371/journal.pone.0056590

Davis FJ, Gupta M, Pogwizd SM, Bacha E, Jeevanandam V, Gupta MP (2002) Increased expression of alternatively spliced dominant-negative isoform of SRF in human failing hearts. Am J Physiol Heart Circ Physiol 282:H1521–H1533

Ding JH, Xu X, Yang D, Chu PH, Dalton ND, Ye Z, Yeakley JM, Cheng H, Xiao RP, Ross J et al (2004) Dilated cardiomyopathy caused by tissue-specific ablation of SC35 in the heart. EMBO J 23:885–896

Du H, Cline MS, Osborne RJ, Tuttle DL, Clark TA, Donohue JP, Hall MP, Shiue L, Swanson MS, Thornton CA, Ares M Jr (2010) Aberrant alternative splicing and extracellular matrix gene expression in mouse models of myotonic dystrophy. Nat Struct Mol Biol 17:187–193

Eulalio A, Huntzinger E, Izaurralde E (2008) Getting to the root of miRNA-mediated gene silencing. Cell 132:9–14

Fabian MR, Sonenberg N (2012) The mechanics of miRNA-mediated gene silencing: a look under the hood of miRISC. Nat Struct Mol Biol 19:586–593

Fardaei M, Rogers MT, Thorpe HM, Larkin K, Hamshere MG, Harper PS, Brook JD (2002) Three proteins, MBNL, MBLL and MBXL, co-localize in vivo with nuclear foci of expanded-repeat transcripts in DM1 and DM2 cells. Hum Mol Genet 11:805–814

Faustino N, Cooper T (2005) Identification of putative new splicing targets for ETR-3 using its SELEX sequences. Mol Cell Biol 25:879–887

Feng Y, Chen M, Manley JL (2008) Phosphorylation switches the general splicing repressor SRp38 to a sequence-specific activator. Nat Struct Mol Biol 15:1040–1048

Feng Y, Valley MT, Lazar J, Yang AL, Bronson RT, Firestein S, Coetzee WA, Manley JL (2009) SRp38 regulates alternative splicing and is required for Ca(2+) handling in the embryonic heart. Dev Cell 16:528–538

Fischer U, Jänicke RU, Schulze-Osthoff K (2003) Many cuts to ruin: a comprehensive update of caspase substrates. Cell Death Differ 10:76–100

Fugier C, Klein AF, Hammer C, Vassilopoulos S, Ivarsson Y, Toussaint A, Tosch V, Vignaud A, Ferry A, Messaddeq N, Kokunai Y, Tsuburaya R, de la Grange P, Dembele D, Francois V, Precigout G, Boulade-Ladame C, Hummel MC, Lopez de Munain A, Sergeant N, Laquerrière A, Thibault C, Deryckere F, Auboeuf D, Garcia L, Zimmermann P, Udd B, Schoser B, Takahashi MP, Nishino I, Bassez G, Laporte J, Furling D, Charlet-Berguerand N (2011) Misregulated alternative splicing of BIN1 is associated with T tubule alterations and muscle weakness in myotonic dystrophy. Nat Med 17:720–725

Gabut M, Chaudhry S, Blencowe BJ (2008) SnapShot: the splicing regulatory machinery. Cell 133:192, e1

Gallagher TL, Arribere JA, Geurts PA, Exner CR, McDonald KL, Dill KK, Marr HL, Adkar SS, Garnett AT, Amacher SL, Conboy JG (2011) Rbfox-regulated alternative splicing is critical for zebrafish cardiac and skeletal muscle functions. Dev Biol 359:251–261

Ge Y, Sun Y, Chen J (2011) IGF-II is regulated by microRNA-125b in skeletal myogenesis. J Cell Biol 192:69–81

Gerber WV, Yatskievych TA, Antin PB, Correia KM, Conlon RA, Krieg PA (1999) The RNA-binding protein gene, hermes, is expressed at high levels in the developing heart. Mech Dev 80:77–86

Gerber WV, Vokes SA, Zearfoss NR, Krieg PA (2002) A role for the RNA-binding protein, hermes, in the regulation of heart development. Dev Biol 247:116–126

Gesteland RF, Cech TR, Atkins JF (eds) (1999) The RNA World, 2nd edn. Cold Spring Harbor Laboratory Press, Cold Spring Harbor, New York

Glisovic T, Bachorik JL, Yong J, Dreyfuss G (2008) RNA-binding proteins and post-transcriptional gene regulation. FEBS Lett 582:1977–1986

Goers ES, Purcell J, Voelker RB, Gates DP, Berglund JA (2010) MBNL1 binds GC motifs embedded in pyrimidines to regulate alternative splicing. Nucleic Acids Res 38:2467–2484

Good PJ, Chen Q, Warner SJ, Herring DC (2000) A family of human RNA-binding proteins related to the Drosophila Bruno translational regulator. J Biol Chem 275:28583–28592

Graindorge A, Le Tonquèze O, Thuret R, Pollet N, Osborne HB, Audic Y (2008) Identification of CUG-BP1/EDEN-BP target mRNAs in Xenopus tropicalis. Nucleic Acids Res 36:1861–1870

Graveley BR (2000) Sorting out the complexity of SR protein functions. RNA 6:1197–1211

Greaser ML, Warren CM, Esbona K, Guo W, Duan Y, Parrish AM, Krzesinski PR, Norman HS, Dunning S, Fitzsimons DP, Moss RL (2008) Mutation that dramatically alters rat titin isoform expression and cardiomyocyte passive tension. J Mol Cell Cardiol 44:983–991

Gregorio CC, Trombitás K, Centner T, Kolmerer B, Stier G, Kunke K, Suzuki K, Obermayr F, Herrmann B, Granzier H et al (1998) The NH2 terminus of titin spans the Z-disc: its interaction with a novel 19-kD ligand (T-cap) is required for sarcomeric integrity. J Cell Biol 143:1013–1027

Groh WJ, Groh MR, Saha C, Kincaid JC, Simmons Z, Ciafaloni E, Pourmand R, Otten RF, Bhakta D, Nair GV et al (2008) Electrocardiographic abnormalities and sudden death in myotonic dystrophy type 1. N Engl J Med 358:2688–2697

Guo W, Schafer S, Greaser ML, Radke MH, Liss M, Govindarajan T, Maatz H, Schulz H, Li S, Parrish AM et al (2012) RBM20, a gene for hereditary cardiomyopathy, regulates titin splicing. Nat Med 18:766–773

Habelhah H, Shah K, Huang L, Ostareck-Lederer A, Burlingame AL, Shokat KM, Hentze MW, Ronai Z (2001) ERK phosphorylation drives cytoplasmic accumulation of hnRNP-K and inhibition of mRNA translation. Nat Cell Biol 3:325–330

Hanson KA, Kim SH, Tibbetts RS (2012) RNA-binding proteins in neurodegenerative disease: TDP-43 and beyond. Wiley Interdiscip Rev RNA 3:265–285

Harper PS (2001) Myotonic dystrophy. WB Saunders, London

Hein S, Arnon E, Kostin S, Schönburg M, Elsässer A, Polyakova V, Bauer EP, Klövekorn WP, Schaper J (2003) Progression from compensated hypertrophy to failure in the pressure-overloaded human heart: structural deterioration and compensatory mechanisms. Circulation 107:984–991

Ho TH, Charlet-B N, Poulos MG, Singh G, Swanson MS, Cooper TA (2004) Muscleblind proteins regulate alternative splicing. EMBO J 23:3103–3112

Ho TH, Bundman D, Armstrong DL, Cooper TA (2005) Transgenic mice expressing CUG-BP1 reproduce splicing mis-regulation observed in myotonic dystrophy. Hum Mol Genet 14:1539–1547

Holcik M, Sonenberg N (2005) Translational control in stress and apoptosis. Nat Rev Mol Cell Biol 6:318–327

Huang C, Zhou Q, Liang P, Hollander MS, Sheikh F, Li X, Greaser M, Shelton GD, Evans S, Chen J (2003) Characterization and in vivo functional analysis of splice variants of cypher. J Biol Chem 278:7360–7365

Huang Y, Steitz JA (2001) Splicing factors SRp20 and 9G8 promote the nucleocytoplasmic export of mRNA. Mol Cell 7:899–905

Huang Y, Yario TA, Steitz JA (2004) A molecular link between SR protein dephosphorylation and mRNA export. Proc Natl Acad Sci U S A 101:9666–9670

Huang CJ, Tang Z, Lin RJ, Tucker PW (2007) Phosphorylation by SR kinases regulates the binding of PTB-associated splicing factor (PSF) to the pre-mRNA polypyrimidine tract. FEBS Lett 581:223–232

Huichalaf C, Sakai K, Jin B, Jones K, Wang GL, Schoser B, Schneider-Gold C, Sarkar P, Pereira-Smith OM, Timchenko N, Timchenko L (2010) Expansion of CUG RNA repeats causes stress and inhibition of translation in myotonic dystrophy 1 (DM1) cells. FASEB J 24:3706–3719

Huot ME, Bisson N, Davidovic L, Mazroui R, Labelle Y, Moss T, Khandjian EW (2005) The RNA-binding protein fragile X-related 1 regulates somite formation in Xenopus laevis. Mol Biol Cell 16:4350–4361

Iakova P, Wang GL, Timchenko L, Michalak M, Pereira-Smith OM, Smith JR, Timchenko NA (2004) Competition of CUGBP1 and calreticulin for the regulation of p21 translation determines cell fate. EMBO J 23:406–417

Ikeda S, Pu WT (2010) Expression and function of microRNAs in heart disease. Curr Drug Targets 11:913–925

Ikezoe K, Nakamori M, Furuya H, Arahata H, Kanemoto S, Kimura T, Imaizumi K, Takahashi MP, Sakoda S, Fujii N, Kira J (2007) Endoplasmic reticulum stress in myotonic dystrophy type 1 muscle. Acta Neuropathol 114:527–535

Jensen KB, Dredge BK, Stefani G, Zhong R, Buckanovich RJ, Okano HJ, Yang YY, Darnell RB (2000) Nova-1 regulates neuron-specific alternative splicing and is essential for neuronal viability. Neuron 25:359–371

Ji X, Zhou Y, Pandit S, Huang J, Li H, Lin CY, Xiao R, Burge CB, Fu XD (2013) SR proteins collaborate with 7SK and promoter-associated nascent RNA to release paused polymerase. Cell 153:855–868

Jiang B, Zhang B, Liang P, Chen G, Zhou B, Lv C, Tu Z, Xiao X (2013) Nucleolin protects the heart from ischaemia-reperfusion injury by up-regulating heat shock protein 32. Cardiovasc Res 99:92–101

Jin Y, Suzuki H, Maegawa S, Endo H, Sugano S, Hashimoto K, Yasuda K, Inoue K (2003) A vertebrate RNA-binding protein Fox-1 regulates tissue-specific splicing via the pentanucleotide GCAUG. EMBO J 22:905–912

Jin D, Hidaka K, Shirai M, Morisaki T (2010) RNA-binding motif protein 24 regulates myogenin expression and promotes myogenic differentiation. Genes Cells 15:1158–1167

Jumaa H, Wei G, Nielsen PJ (1999) Blastocyst formation is blocked in mouse embryos lacking the splicing factor SRp20. Curr Biol 9:899–902

Kakkar R, Lee RT (2010) Intramyocardial fibroblast myocyte communication. Circ Res 106:47–57

Kalsotra A, Xiao X, Ward AJ, Castle JC, Johnson JM, Burge CB, Cooper TA (2008) A postnatal switch of CELF and MBNL proteins reprograms alternative splicing in the developing heart. Proc Natl Acad Sci U S A 105:20333–20338

Kalsotra A, Wang K, Li PF, Cooper TA (2010) MicroRNAs coordinate an alternative splicing network during mouse postnatal heart development. Genes Dev 24:653–658

Kameyama T, Suzuki H, Mayeda A (2012) Re-splicing of mature mRNA in cancer cells promotes activation of distant weak alternative splice sites. Nucleic Acids Res 40:7896–7906

Kanadia RN, Urbinati CR, Crusselle VJ, Luo D, Lee YJ, Harrison JK, Oh SP, Swanson MS (2003a) Developmental expression of mouse muscleblind genes Mbnl1, Mbnl2 and Mbnl3. Gene Expr Patterns 3:459–462

Kanadia RN, Johnstone KA, Mankodi A, Lungu C, Thornton CA, Esson D, Timmers AM, Hauswirth WW, Swanson MS (2003b) A muscleblind knockout model for myotonic dystrophy. Science 302:1978–1980

Karni R, de Stanchina E, Lowe SW, Sinha R, Mu D, Krainer AR (2007) The gene encoding the splicing factor SF2/ASF is a proto-oncogene. Nat Struct Mol Biol 14:185–193

Kaynak B, von Heydebreck A, Mebus S, Seelow D, Hennig S, Vogel J, Sperling HP, Pregla R, Alexi-Meskishvili V, Hetzer R et al (2003) Genome-wide array analysis of normal and malformed human hearts. Circulation 107:2467–2474

Keppetipola N, Sharma S, Li Q, Black DL (2012) Neuronal regulation of pre-mRNA splicing by polypyrimidine tract binding proteins, PTBP1 and PTBP2. Crit Rev Biochem Mol Biol 47:360–378

Kiehl TR, Shibata H, Vo T, Huynh DP, Pulst SM (2001) Identification and expression of a mouse ortholog of A2BP1. Mamm Genome 12:595–601

Kim KK, Adelstein RS, Kawamoto S (2009) Identification of neuronal nuclei (NeuN) as Fox-3, a new member of the Fox-1 gene family of splicing factors. J Biol Chem 284:31052–31061

Kong SW, Hu YW, Ho JW, Ikeda S, Polster S, John R, Hall JL, Bisping E, Pieske B, dos Remedios CG, Pu WT (2010) Heart failure-associated changes in RNA splicing of sarcomere genes. Circ Cardiovasc Genet 3:138–146

Koshelev M, Sarma S, Price RE, Wehrens XH, Cooper TA (2010) Heart-specific overexpression of CUGBP1 reproduces functional and molecular abnormalities of myotonic dystrophy type 1. Hum Mol Genet 19:1066–1075

Kress C, Gautier-Courteille C, Osborne HB, Babinet C, Paillard L (2007) Inactivation of CUG-BP1/CELF1 causes growth, viability, and spermatogenesis defects in mice. Mol Cell Biol 27:1146–1157

Krüger M, Linke WA (2011) The giant protein titin: a regulatory node that integrates myocyte signaling pathways. J Biol Chem 286:9905–9912

Kuroyanagi H, Ohno G, Mitani S, Hagiwara M (2007) The Fox-1 family and SUP-12 coordinately regulate tissue-specific alternative splicing in vivo. Mol Cell Biol 27:8612–8621

Kuroyanagi H (2009) Fox-1 family of RNA-binding proteins. Cell Mol Life Sci 66:3895–3907
Kuyumcu-Martinez NM, Wang GS, Cooper TA (2007) Increased steady-state levels of CUGBP1 in myotonic dystrophy 1 are due to PKC-mediated hyperphosphorylation. Mol Cell 28:68–78
Labeit S, Kolmerer B (1995) Titins: giant proteins in charge of muscle ultrastructure and elasticity. Science 270:293–296
Ladd AN, Charlet N, Cooper TA (2001) The CELF family of RNA binding proteins is implicated in cell-specific and developmentally regulated alternative splicing. Mol Cell Biol 21:1285–1296
Ladd AN, Taffet G, Hartley C, Kearney DL, Cooper TA (2005) Cardiac tissue-specific repression of CELF activity disrupts alternative splicing and causes cardiomyopathy. Mol Cell Biol 25:6267–6278
Lara-Pezzi E, Gómez-Salinero J, Gatto A, García-Pavía P (2013) The alternative heart: impact of alternative splicing in heart disease. J Cardiovasc Transl Res 6(6):945–955
Lee JE, Lee JY, Wilusz J, Tian B, Wilusz CJ (2010) Systematic analysis of cis-elements in unstable mRNAs demonstrates that CUGBP1 is a key regulator of mRNA decay in muscle cells. PLoS One 5:e11201. doi:10.1371/journal.pone.0011201
Lemaire R, Prasad J, Kashima T, Gustafson J, Manley JL, Lafyatis R (2002) Stability of a PKCI-1-related mRNA is controlled by the splicing factor ASF/SF2: a novel function for SR proteins. Genes Dev 16:594–607
Li D, Bachinski LL, Roberts R (2001) Genomic organization and isoform-specific tissue expression of human NAPOR (CUGBP2) as a candidate gene for familial arrhythmogenic right ventricular dysplasia. Genomics 74:396–401
Li S, Guo W, Dewey CN, Greaser ML (2013) Rbm20 regulates titin alternative splicing as a splicing repressor. Nucleic Acids Res 41:2659–2672
Li D, Morales A, Gonzalez-Quintana J, Norton N, Siegfried JD, Hofmeyer M, Hershberger RE (2010) Identification of novel mutations in RBM20 in patients with dilated cardiomyopathy. Clin Transl Sci 3:90–97
Lin S, Fu XD (2007) SR proteins and related factors in alternative splicing. Adv Exp Med Biol 623:107–122
Licatalosi DD, Darnell RB (2006) Splicing regulation in neurologic disease. Neuron 52:93–101
Lin X, Miller JW, Mankodi A, Kanadia RN, Yuan Y, Moxley RT, Swanson MS, Thornton CA (2006) Failure of MBNL1-dependent post-natal splicing transitions in myotonic dystrophy. Hum Mol Genet 15:2087–2097
Lichtner P, Attié-Bitach T, Schuffenhauer S, Henwood J, Bouvagnet P, Scambler PJ, Meitinger T, Vekemans M (2002) Expression and mutation analysis of BRUNOL3, a candidate gene for heart and thymus developmental defects associated with partial monosomy 10p. J Mol Med (Berl) 80:431–442
Long JC, Caceres JF (2009) The SR protein family of splicing factors: master regulators of gene expression. Biochem J 417:15–27
Lu X, Timchenko NA, Timchenko LT (1999) Cardiac elav-type RNA-binding protein (ETR-3) binds to RNA CUG repeats expanded in myotonic dystrophy. Hum Mol Genet 8:53–60
Lukong KE, Chang KW, Khandjian EW, Richard S (2008) RNA-binding proteins in human genetic disease. Trends Genet 24:416–425
Machuca-Tzili LE, Buxton S, Thorpe A, Timson CM, Wigmore P, Luther PK, Brook JD (2011) Zebrafish deficient for Muscleblind-like 2 exhibit features of myotonic dystrophy. Dis Model Mech 4:381–392
Mahmoud AI, Kocabas F, Muralidhar SA, Kimura W, Koura AS, Thet S, Porrello ER, Sadek HA (2013) Meis1 regulates postnatal cardiomyocyte cell cycle arrest. Nature 497:249–253
Makarenko I, Opitz CA, Leake MC, Neagoe C, Kulke M, Gwathmey JK, del Monte F, Hajjar RJ, Linke WA (2004) Passive stiffness changes caused by upregulation of compliant titin isoforms in human dilated cardiomyopathy hearts. Circ Res 95:708–716
Makeyev EV, Zhang J, Carrasco MA, Maniatis T (2007) The MicroRNA miR-124 promotes neuronal differentiation by triggering brain-specific alternative pre-mRNA splicing. Mol Cell 27:435–448
Mankodi A, Logigian E, Callahan L, McClain C, White R, Henderson D, Krym M, Thornton CA (2000) Myotonic dystrophy in transgenic mice expressing an expanded CUG repeat. Science 289:1769–1773

Mankodi A, Urbinati CR, Yuan QP, Moxley RT, Sansone V, Krym M, Henderson D, Schalling M, Swanson MS, Thornton CA (2001) Muscleblind localizes to nuclear foci of aberrant RNA in myotonic dystrophy types 1 and 2. Hum Mol Genet 10:2165–2170

Mankodi A, Takahashi MP, Jiang H, Beck CL, Bowers WJ, Moxley RT, Cannon SC, Thornton CA (2002) Expanded CUG repeats trigger aberrant splicing of ClC-1 chloride channel pre-mRNA and hyperexcitability of skeletal muscle in myotonic dystrophy. Mol Cell 10:35–44

Manley JL, Krainer AR (2010) A rational nomenclature for serine/arginine-rich protein splicing factors (SR proteins). Genes Dev 24:1073–1074

Maragh S, Miller RA, Bessling SL, McGaughey DM, Wessels MW, de Graaf B, Stone EA, Bertoli-Avella AM, Gearhart JD, Fisher S, McCallion AS (2011) Identification of RNA binding motif proteins essential for cardiovascular development. BMC Dev Biol 11:62. doi:10.1186/1471-213X-11-62

Marquis J, Paillard L, Audic Y, Cosson B, Danos O, Le Bec C, Osborne HB (2006) CUG-BP1/CELF1 requires UGU-rich sequences for high-affinity binding. Biochem J 400:291–301

Martin CL, Duvall JA, Ilkin Y, Simon JS, Arreaza MG, Wilkes K, Alvarez-Retuerto A, Whichello A, Powell CM, Rao K et al (2007) Cytogenetic and molecular characterization of A2BP1/FOX1 as a candidate gene for autism. Am J Med Genet B Neuropsychiatr Genet 144B:869–876

Mason JW (2002) Viral latency: a link between myocarditis and dilated cardiomyopathy? J Mol Cell Cardiol 34:695–698

Masuda A, Andersen HS, Doktor TK, Okamoto T, Ito M, Andresen BS, Ohno K (2012) CUGBP1 and MBNL1 preferentially bind to 3' UTRs and facilitate mRNA decay. Sci Rep 2:209. doi:10.1038/srep00209

McKee AE, Minet E, Stern C, Riahi S, Stiles CD, Silver PA (2005) A genome-wide in situ hybridization map of RNA-binding proteins reveals anatomically restricted expression in the developing mouse brain. BMC Dev Biol 5:14

McKusick VA (1998) Mendelian inheritance in man. A catalog of human genes and genetic disorders. Johns Hopkins University Press, Baltimore, MD

Merkin J, Russell C, Chen P, Burge CB (2012) Evolutionary dynamics of gene and isoform regulation in Mammalian tissues. Science 338:1593–1599

Miller JW, Urbinati CR, Teng-Umnuay P, Stenberg MG, Byrne BJ, Thornton CA, Swanson MS (2000) Recruitment of human muscleblind proteins to (CUG)(n) expansions associated with myotonic dystrophy. EMBO J 19:4439–4448

Miller RA, Christoforou N, Pevsner J, McCallion AS, Gearhart JD (2008) Efficient array-based identification of novel cardiac genes through differentiation of mouse ESCs. PLoS One 3:e2176. doi:10.1371/journal.pone.0002176

Mills JD, Janitz M (2012) Alternative splicing of mRNA in the molecular pathology of neurodegenerative diseases. Neurobiol Aging 33(1012):e11–e24

Miragoli M, Gaudesius G, Rohr S (2006) Electrotonic modulation of cardiac impulse conduction by myofibroblasts. Circ Res 98:801–810

Misteli T, Cáceres JF, Clement JQ, Krainer AR, Wilkinson MF, Spector DL (1998) Serine phosphorylation of SR proteins is required for their recruitment to sites of transcription in vivo. J Cell Biol 143:297–307

Mitchell SA, Brown EC, Coldwell MJ, Jackson RJ, Willis AE (2001) Protein factor requirements of the Apaf-1 internal ribosome entry segment: roles of polypyrimidine tract binding protein and upstream of N-ras. Mol Cell Biol 21:3364–3374

Miyamoto S, Hidaka K, Jin D, Morisaki T (2009) RNA-binding proteins Rbm38 and Rbm24 regulate myogenic differentiation via p21-dependent and -independent regulatory pathways. Genes Cells 14:1241–1252

Moraes KC, Wilusz CJ, Wilusz J (2006) CUG-BP binds to RNA substrates and recruits PARN deadenylase. RNA 12:1084–1091

Mulligan GJ, Guo W, Wormsley S, Helfman DM (1992) Polypyrimidine tract binding protein interacts with sequences involved in alternative splicing of beta-tropomyosin pre-mRNA. J Biol Chem 267:25480–25487

Neagoe C, Kulke M, del Monte F, Gwathmey JK, de Tombe PP, Hajjar RJ, Linke WA (2002) Titin isoform switch in ischemic human heart disease. Circulation 106:1333–1341

Nelson TJ, Balza R Jr, Xiao Q, Misra RP (2005) SRF-dependent gene expression in isolated cardiomyocytes: regulation of genes involved in cardiac hypertrophy. J Mol Cell Cardiol 39:479–489

Obermann WM, Gautel M, Weber K, Fürst DO (1997) Molecular structure of the sarcomeric M band: mapping of titin and myosin binding domains in myomesin and the identification of a potential regulatory phosphorylation site in myomesin. EMBO J 16:211–220

Oh SW, Pope RK, Smith KP, Crowley JL, Nebl T, Lawrence JB, Luna EJ (2003) Archvillin, a muscle-specific isoform of supervillin, is an early expressed component of the costameric membrane skeleton. J Cell Sci 116:2261–2275

Olson EN (2001) Development. The path to the heart and the road not taken. Science 291:2327–2328

Olson EN (2006) Gene regulatory networks in the evolution and development of the heart. Science 313:1922–1927

Osborne RJ, Lin X, Welle S, Sobczak K, O'Rourke JR, Swanson MS, Thornton CA (2009) Transcriptional and post-transcriptional impact of toxic RNA in myotonic dystrophy. Hum Mol Genet 18:1471–1481

Pascual M, Vicente M, Monferrer L, Artero R (2006) The Muscleblind family of proteins: an emerging class of regulators of developmentally programmed alternative splicing. Differentiation 74:65–80

Patton JG, Mayer SA, Tempst P, Nadal-Ginard B (1991) Characterization and molecular cloning of polypyrimidine tract-binding protein: a component of a complex necessary for pre-mRNA splicing. Genes Dev 5:1237–1251

Pelargonio G, Dello Russo A, Sanna T, De Martino G, Bellocci F (2002) Myotonic dystrophy and the heart. Heart 88:665–670

Philips AV, Timchenko LT, Cooper TA (1998) Disruption of splicing regulated by a CUG-binding protein in myotonic dystrophy. Science 280:737–741

Ponthier JL, Schluepen C, Chen W, Lersch RA, Gee SL, Hou VC, Lo AJ, Short SA, Chasis JA, Winkelmann JC, Conboy JG (2006) Fox-2 splicing factor binds to a conserved intron motif to promote inclusion of protein 4.1R alternative exon 16. J Biol Chem 281:12468–12474

Poon KL, Tan KT, Wei YY, Ng CP, Colman A, Korzh V, Xu XQ (2012) RNA-binding protein RBM24 is required for sarcomere assembly and heart contractility. Cardiovasc Res 94:418–427

Porrello ER, Mahmoud AI, Simpson E, Hill JA, Richardson JA, Olson EN, Sadek HA (2011a) Transient regenerative potential of the neonatal mouse heart. Science 331:1078–1080

Porrello ER, Johnson BA, Aurora AB, Simpson E, Nam YJ, Matkovich SJ, Dorn GW 2nd, van Rooij E, Olson EN (2011b) MiR-15 family regulates postnatal mitotic arrest of cardiomyocytes. Circ Res 109:670–679

Porrello ER (2013) microRNAs in cardiac development and regeneration. Clin Sci (Lond) 125:151–166

Poulos MG, Batra R, Li M, Yuan Y, Zhang C, Darnell RB, Swanson MS (2013) Progressive impairment of muscle regeneration in muscleblind-like 3 isoform knockout mice. Hum Mol Genet 22(17):3547–3558

Rau F, Freyermuth F, Fugier C, Villemin JP, Fischer MC, Jost B, Dembele D, Gourdon G, Nicole A, Duboc D et al (2011) Misregulation of miR-1 processing is associated with heart defects in myotonic dystrophy. Nat Struct Mol Biol 18:840–845

Ranum LP, Cooper TA (2006) RNA-mediated neuromuscular disorders. Annu Rev Neurosci 29:259–277

Rattenbacher B, Beisang D, Wiesner DL, Jeschke JC, von Hohenberg M, St Louis-Vlasova IA, Bohjanen PR (2010) Analysis of CUGBP1 targets identifies GU-repeat sequences that mediate rapid mRNA decay. Mol Cell Biol 30:3970–3980

Refaat MM, Lubitz SA, Makino S, Islam Z, Frangiskakis JM, Mehdi H, Gutmann R, Zhang ML, Bloom HL, MacRae CA et al (2012) Genetic variation in the alternative splicing regulator RBM20 is associated with dilated cardiomyopathy. Heart Rhythm 9:390–396

Riggi N, Cironi L, Suvà ML, Stamenkovic I (2007) Sarcomas: genetics, signalling, and cellular origins. Part 1: the fellowship of TET. J Pathol 213:4–20

Salisbury E, Sakai K, Schoser B, Huichalaf C, Schneider-Gold C, Nguyen H, Wang GL, Albrecht JH, Timchenko LT (2008) Ectopic expression of cyclin D3 corrects differentiation of DM1 myoblasts through activation of RNA CUG-binding protein, CUGBP1. Exp Cell Res 314: 2266–2278

Sanford JR, Gray NK, Beckmann K, Cáceres JF (2004) A novel role for shuttling SR proteins in mRNA translation. Genes Dev 18:755–768

Savkur RS, Philips AV, Cooper TA (2001) Aberrant regulation of insulin receptor alternative splicing is associated with insulin resistance in myotonic dystrophy. Nat Genet 29:40–47

Sayed D, Hong C, Chen IY, Lypowy J, Abdellatif M (2007) MicroRNAs play an essential role in the development of cardiac hypertrophy. Circ Res 100:416–424

Schepens B, Tinton SA, Bruynooghe Y, Beyaert R, Cornelis S (2005) The polypyrimidine tract-binding protein stimulates HIF-1alpha IRES-mediated translation during hypoxia. Nucleic Acids Res 33:6884–6894

Schoser B, Timchenko L (2010) Myotonic dystrophies 1 and 2: complex diseases with complex mechanisms. Curr Genomics 11:77–90

Shin C, Manley JL (2002) The SR protein SRp38 represses splicing in M phase cells. Cell 111:407–417

Shin C, Feng Y, Manley JL (2004) Dephosphorylated SRp38 acts as a splicing repressor in response to heat shock. Nature 427:553–558

Shu L, Yan W, Chen X (2006) RNPC1, an RNA-binding protein and a target of the p53 family, is required for maintaining the stability of the basal and stress-induced p21 transcript. Genes Dev 20:2961–2972

Singh R, Valcárcel J, Green MR (1995) Distinct binding specificities and functions of higher eukaryotic polypyrimidine tract-binding proteins. Science 268:1173–1176

Singh RK, Cooper TA (2012) Pre-mRNA splicing in disease and therapeutics. Trends Mol Med 18:472–482

Song HW, Cauffman K, Chan AP, Zhou Y, King ML, Etkin LD, Kloc M (2007) Hermes RNA-binding protein targets RNAs-encoding proteins involved in meiotic maturation, early cleavage, and germline development. Differentiation 75:519–528

Song HK, Hong SE, Kim T, Kim do H (2012) Deep RNA sequencing reveals novel cardiac transcriptomic signatures for physiological and pathological hypertrophy. PLoS One 7:e35552. doi:10.1371/journal.pone.0035552

Southby J, Gooding C, Smith CW (1999) Polypyrimidine tract binding protein functions as a repressor to regulate alternative splicing of alpha-actinin mutually exclusive exons. Mol Cell Biol 19:2699–2711

Squillace RM, Chenault DM, Wang EH (2002) Inhibition of muscle differentiation by the novel muscleblind-related protein CHCR. Dev Biol 250:218–230

Srebrow A, Kornblihtt AR (2006) The connection between splicing and cancer. J Cell Sci 119:2635–2641

Srivastava D (2006) Making or breaking the heart: from lineage determination to morphogenesis. Cell 126:1037–1048

Suenaga K, Lee KY, Nakamori M, Tatsumi Y, Takahashi MP, Fujimura H, Jinnai K, Yoshikawa H, Du H, Ares M Jr et al (2012) Muscleblind-like 1 knockout mice reveal novel splicing defects in the myotonic dystrophy brain. PLoS One 7:e33218. doi:10.1371/journal.pone.0033218

Sawicka K, Bushell M, Spriggs KA, Willis AE (2008) Polypyrimidine-tract-binding protein: a multifunctional RNA-binding protein. Biochem Soc Trans 36:641–647

Takahashi N, Sasagawa N, Suzuki K, Ishiura S (2000) The CUG-binding protein binds specifically to UG dinucleotide repeats in a yeast three-hybrid system. Biochem Biophys Res Commun 277:518–523

Tang ZZ, Sharma S, Zheng S, Chawla G, Nikolic J, Black DL (2011) Regulation of the mutually exclusive exons 8a and 8 in the CaV1.2 calcium channel transcript by polypyrimidine tract-binding protein. J Biol Chem 286:10007–10016

Teplova M, Patel DJ (2008) Structural insights into RNA recognition by the alternative-splicing regulator muscleblind-like MBNL1. Nat Struct Mol Biol 15:1343–1351

Terenzi F, Brimacombe KR, Penn MS, Ladd AN (2009) CELF-mediated alternative splicing is required for cardiac function during early, but not later, postnatal life. J Mol Cell Cardiol 46:395–404

Tian B, White RJ, Xia T, Welle S, Turner DH, Mathews MB, Thornton CA (2000) Expanded CUG repeat RNAs form hairpins that activate the double-stranded RNA-dependent protein kinase PKR. RNA 6:79–87

Timchenko LT, Miller JW, Timchenko NA, DeVore DR, Datar KV, Lin L, Roberts R, Caskey CT, Swanson MS (1996) Identification of a (CUG)n triplet repeat RNA-binding protein and its expression in myotonic dystrophy. Nucleic Acids Res 24:4407–4414

Timchenko NA, Cai ZJ, Welm AL, Reddy S, Ashizawa T, Timchenko LT (2001a) RNA CUG repeats sequester CUGBP1 and alter protein levels and activity of CUGBP1. J Biol Chem 276:7820–7826

Timchenko NA, Iakova P, Cai ZJ, Smith JR, Timchenko LT (2001b) Molecular basis for impaired muscle differentiation in myotonic dystrophy. Mol Cell Biol 21:6927–6938

Timchenko LT, Salisbury E, Wang GL, Nguyen H, Albrecht JH, Hershey JW, Timchenko NA (2006) Age-specific CUGBP1-eIF2 complex increases translation of CCAAT/enhancer-binding protein beta in old liver. J Biol Chem 281:32806–32819

Tirziu D, Giordano FJ, Simons M (2010) Cell communications in the heart. Circulation 122:928–937

Tran H, Gourrier N, Lemercier-Neuillet C, Dhaenens CM, Vautrin A, Fernandez-Gomez FJ, Arandel L, Carpentier C, Obriot H, Eddarkaoui S et al (2011) Analysis of exonic regions involved in nuclear localization, splicing activity, and dimerization of Muscleblind-like-1 isoforms. J Biol Chem 286:16435–16446

Torchia EC, Boyd K, Rehg JE, Qu C, Baker SJ (2007) EWS/FLI-1 induces rapid onset of myeloid/erythroid leukemia in mice. Mol Cell Biol 27:7918–7934

Ule J, Jensen KB, Ruggiu M, Mele A, Ule A, Darnell RB (2003) CLIP identifies Nova-regulated RNA networks in the brain. Science 302:1212–1215

Underwood JG, Boutz PL, Dougherty JD, Stoilov P, Black DL (2005) Homologues of the Caenorhabditis elegans Fox-1 protein are neuronal splicing regulators in mammals. Mol Cell Biol 25:10005–10016

Vacchi-Suzzi C, Hahne F, Scheubel P, Marcellin M, Dubost V, Westphal M, Boeglen C, Büchmann-Møller S, Cheung MS, Cordier A et al (2013) Heart structure-specific transcriptomic atlas reveals conserved microRNA-mRNA interactions. PLoS One 8:e52442. doi:10.1371/journal.pone.0052442

van Kouwenhove M, Kedde M, Agami R (2011) MicroRNA regulation by RNA-binding proteins and its implications for cancer. Nat Rev Cancer 11:644–656

van Rooij E, Sutherland LB, Liu N, Williams AH, McAnally J, Gerard RD, Richardson JA, Olson EN (2006) A signature pattern of stress-responsive microRNAs that can evoke cardiac hypertrophy and heart failure. Proc Natl Acad Sci U S A 103:18255–18260

van Rooij E, Sutherland LB, Qi X, Richardson JA, Hill J, Olson EN (2007) Control of stress-dependent cardiac growth and gene expression by a microRNA. Science 316:575–579

van Rooij E, Sutherland LB, Thatcher JE, DiMaio JM, Naseem RH, Marshall WS, Hill JA, Olson EN (2008) Dysregulation of microRNAs after myocardial infarction reveals a role of miR-29 in cardiac fibrosis. Proc Natl Acad Sci U S A 105:13027–13032

Vlasova IA, Tahoe NM, Fan D, Larsson O, Rattenbacher B, Sternjohn JR, Vasdewani J, Karypis G, Reilly CS, Bitterman PB, Bohjanen PR (2008) Conserved GU-rich elements mediate mRNA decay by binding to CUG-binding protein 1. Mol Cell 29:263–270

Vercellotti GM (2001) Overview of infections and cardiovascular diseases. J Allergy Clin Immunol 108:S117–S120

Wang HY, Xu X, Ding JH, Bermingham JR Jr, Fu XD (2001) SC35 plays a role in T cell development and alternative splicing of CD45. Mol Cell 7:331–342

Wang GS, Kearney DL, De Biasi M, Taffet G, Cooper TA (2007) Elevation of RNA-binding protein CUGBP1 is an early event in an inducible heart-specific mouse model of myotonic dystrophy. J Clin Invest 117:2802–2811

Wang ET, Sandberg R, Luo S, Khrebtukova I, Zhang L, Mayr C, Kingsmore SF, Schroth GP, Burge CB (2008) Alternative isoform regulation in human tissue transcriptomes. Nature 456: 470–476

Wang GS, Kuyumcu-Martinez MN, Sarma S, Mathur N, Wehrens XH, Cooper TA (2009) PKC inhibition ameliorates the cardiac phenotype in a mouse model of myotonic dystrophy type 1. J Clin Invest 119:3797–3806

Wang ET, Cody NA, Jog S, Biancolella M, Wang TT, Treacy DJ, Luo S, Schroth GP, Housman DE, Reddy S et al (2012a) Transcriptome-wide regulation of pre-mRNA splicing and mRNA localization by muscleblind proteins. Cell 150:710–724

Wang K, Deng G, Chen G, Liu M, Yi Y, Yang T, McMillan DR, Xiao X (2012b) Heat shock protein 70 inhibits hydrogen peroxide-induced nucleolar fragmentation via suppressing cleavage and down-regulation of nucleolin. Cell Stress Chaperones 17:121–130

Warf MB, Berglund JA (2007) MBNL binds similar RNA structures in the CUG repeats of myotonic dystrophy and its pre-mRNA substrate cardiac troponin T. RNA 13:2238–2251

Warren CM, Jordan MC, Roos KP, Krzesinski PR, Greaser ML (2003) Titin isoform expression in normal and hypertensive myocardium. Cardiovasc Res 59:86–94

Wendel HG, Silva RL, Malina A, Mills JR, Zhu H, Ueda T, Watanabe-Fukunaga R, Fukunaga R, Teruya-Feldstein J, Pelletier J, Lowe SW (2007) Dissecting eIF4E action in tumorigenesis. Genes Dev 21:3232–3237

Williams L, Howell N, Pagano D, Andreka P, Vertesaljai M, Pecor T, Frenneaux M, Granzier H (2009) Titin isoform expression in aortic stenosis. Clin Sci (Lond) 117:237–242

Wollerton MC, Gooding C, Robinson F, Brown EC, Jackson RJ, Smith CW (2001) Differential alternative splicing activity of isoforms of polypyrimidine tract binding protein (PTB). RNA 7:819–832

Wu H, Sun S, Tu K, Gao Y, Xie B, Krainer AR, Zhu J (2010) A splicing-independent function of SF2/ASF in microRNA processing. Mol Cell 38:67–77

Xiao SH, Manley JL (1997) Phosphorylation of the ASF/SF2 RS domain affects both protein-protein and protein-RNA interactions and is necessary for splicing. Genes Dev 11: 334–344

Xu X, Yang D, Ding JH, Wang W, Chu PH, Dalton ND, Wang HY, Bermingham JR Jr, Ye Z, Liu F et al (2005) ASF/SF2-regulated CaMKIIdelta alternative splicing temporally reprograms excitation-contraction coupling in cardiac muscle. Cell 120:59–72

Xu XQ, Soo SY, Sun W, Zweigerdt R (2009) Global expression profile of highly enriched cardiomyocytes derived from human embryonic stem cells. Stem Cells 27:2163–2174

Yang B, Lin H, Xiao J, Lu Y, Luo X, Li B, Zhang Y, Xu C, Bai Y, Wang H et al (2007) The muscle-specific microRNA miR-1 regulates cardiac arrhythmogenic potential by targeting GJA1 and KCNJ2. Nat Med 13:486–491

Ye J, Llorian M, Cardona M, Rongvaux A, Moubarak RS, Comella JX, Bassel-Duby R, Flavell RA, Olson EN, Smith CW, Sanchis D (2013) A pathway involving HDAC5, cFLIP and caspases regulates expression of the splicing regulator polypyrimidine tract binding protein in the heart. J Cell Sci 126:1682–1691

Yeo GW, Xu X, Liang TY, Muotri AR, Carson CT, Coufal NG, Gage FH (2007) Alternative splicing events identified in human embryonic stem cells and neural progenitors. PLoS Comput Biol 3:1951–1967

Yeo GW, Coufal NG, Liang TY, Peng GE, Fu XD, Gage FH (2009) An RNA code for the FOX2 splicing regulator revealed by mapping RNA-protein interactions in stem cells. Nat Struct Mol Biol 16:130–137

Zhang X, Azhar G, Chai J, Sheridan P, Nagano K, Brown T, Yang J, Khrapko K, Borras AM, Lawitts J et al (2001) Cardiomyopathy in transgenic mice with cardiac-specific overexpression of serum response factor. Am J Physiol Heart Circ Physiol 280:H1782–H1792

Zhang Q, Bethmann C, Worth NF, Davies JD, Wasner C, Feuer A, Ragnauth CD, Yi Q, Mellad JA, Warren DT et al (2007) Nesprin-1 and –2 are involved in the pathogenesis of Emery Dreifuss muscular dystrophy and are critical for nuclear envelope integrity. Hum Mol Genet 16:2816–2833

Zhang C, Zhang Z, Castle J, Sun S, Johnson J, Krainer AR, Zhang MQ (2008) Defining the regulatory network of the tissue-specific splicing factors Fox-1 and Fox-2. Genes Dev 22:2550–2563
Zhang J, Bahi N, Llovera M, Comella JX, Sanchis D (2009) Polypyrimidine tract binding proteins (PTB) regulate the expression of apoptotic genes and susceptibility to caspase-dependent apoptosis in differentiating cardiomyocytes. Cell Death Differ 16:1460–1468
Zhao Y, Ransom JF, Li A, Vedantham V, von Drehle M, Muth AN, Tsuchihashi T, McManus MT, Schwartz RJ, Srivastava D (2007) Dysregulation of cardiogenesis, cardiac conduction, and cell cycle in mice lacking miRNA-1-2. Cell 129:303–317

Chapter 12
Identification of miRNAs and Their Targets in *C. elegans*

Sarah Azoubel Lima and Amy E. Pasquinelli

Abstract MicroRNAs (miRNAs) are small noncoding RNAs that direct posttranscriptional regulation of specific target genes. Since their discovery in *Caenorhabditis elegans*, they have been associated with the control of virtually all biological processes and are known to play major roles in development and cellular homeostasis. Yet the biological roles of most miRNAs remain to be fully known. Furthermore, the precise rules by which miRNAs recognize their targets and mediate gene silencing are still unclear. Systematic identification of miRNAs and of the RNAs they regulate is essential to close these knowledge gaps. Studies in *C. elegans* have been instrumental not only in the discovery phase of miRNA biology but also in the elucidation of mechanisms regulating miRNA expression, target recognition and regulation. This chapter highlights some of the main challenges still present in the field, while introducing the major studies and methods used to find miRNAs and their targets in the worm.

Keywords microRNAs • miRNAs • Posttranscriptional regulation • Argonaute • C. elegans • Caenorhabditis elegans • RNA-seq • CLIP

1 Introduction

MicroRNAs (miRNAs) are small, approximately 22 nucleotide (nt) RNAs that mediate posttranscriptional regulation of gene expression throughout the plant and animal kingdoms (Aalto and Pasquinelli 2012). These small RNAs regulate a large portion of the transcriptome and are involved in virtually all biological processes (Friedman et al. 2009). A single miRNA has the potential to modulate the expression of hundreds or even thousands of targets, producing both specific and cumulative changes in gene expression that are enough to trigger developmental transitions and maintain the balance between homeostasis and disease (Alvarez-Garcia and Miska 2005; Esquela-Kerscher and Slack 2006; Kloosterman and Plasterk 2006). A conservative, and likely underestimated, prediction suggests that at least 27 % of *Caenorhabditis elegans* 3′UTRs are under selective pressure to preserve miRNA target sites (Jan et al. 2011).

S.A. Lima • A.E. Pasquinelli (✉)
Division of Biology, University of California, San Diego, La Jolla, CA 92093-0349, USA
e-mail: apasquinelli@ucsd.edu

G.W. Yeo (ed.), *Systems Biology of RNA Binding Proteins*, Advances in Experimental Medicine and Biology 825, DOI 10.1007/978-1-4939-1221-6_12

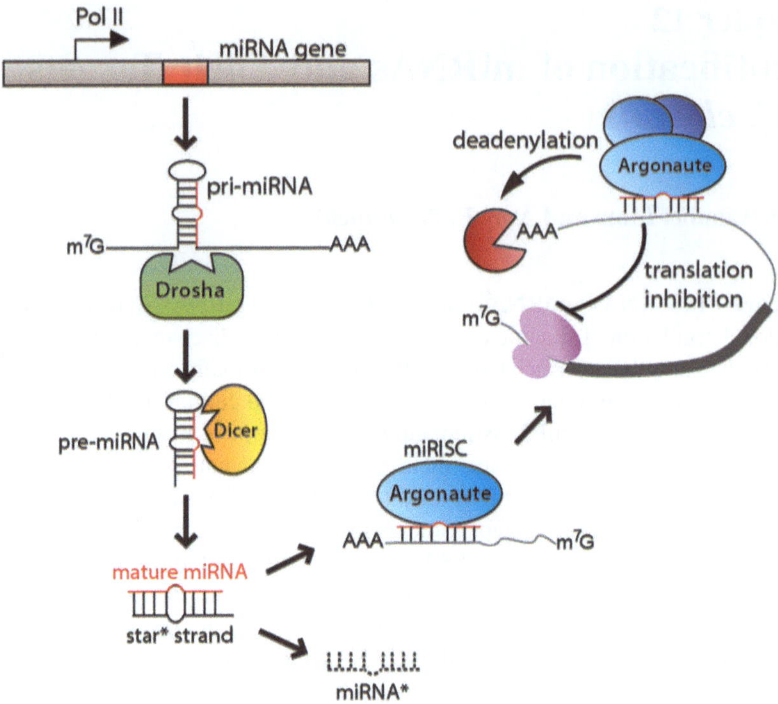

Fig. 12.1 microRNA biogenesis. miRNAs are transcribed into long primary transcripts (pri-miRNAs) and processed by Drosha and Dicer (DRSH-1 and DCR-1 in *C. elegans*). Drosha cleaves the precursor hairpin (pre-miRNA) and Dicer generates a double stranded heteroduplex, formed by the mature miRNA and the miRNA* (star strand). The mature miRNA is loaded onto Argonaute proteins (ALG-1 and ALG-2 in *C. elegans*) forming miRNA-mediated silencing complexes (miRISCs) and the miRNA* is usually degraded. Targets are recognized by partial base pairing of the miRNA. The miRISC promotes silencing by recruiting factors that promote target degradation and/or translational inhibition

The biogenesis of miRNAs is a complex and highly regulated process that has been the focus of extensive reviews (Finnegan and Pasquinelli 2013; Krol et al. 2010) and is summarized in Fig. 12.1. Briefly, miRNAs originate from stem-loops embedded in long primary transcripts (pri-miRNAs) or introns (mirtrons). The endonucleases Drosha and Dicer (DRSH-1 and DCR-1 in *C. elegans*) are responsible for processing most miRNAs into their mature forms. Drosha cleaves the miRNA stem-loop from the pri-miRNA, generating the precursor hairpin (pre-miRNA). Dicer then processes the pre-miRNA into a small double stranded heteroduplex, formed by the mature miRNA paired with the passenger (star) strand. The duplex is separated and the mature miRNA is loaded onto Argonaute proteins (ALG-1 and ALG-2 in *C. elegans*), forming miRNA-mediated silencing complexes (miRISCs). The miRNAs act as guides for miRISC by partially base pairing to specific target RNA sequences. Perfect pairing of nucleotides 2–7 of the miRNA, called the seed region, is involved in many but not all target interactions (Bartel 2009).

Once a target is recognized, silencing is achieved by recruiting factors that promote its degradation and/or translational inhibition (Fabian et al. 2010; Pasquinelli 2012).

Ever since the discovery of miRNAs in *C. elegans* in 1993, many researchers have concentrated their efforts on identifying these small RNAs and their targets across species (Kozomara and Griffiths-Jones 2011). *C. elegans* has many advantages for the study of miRNAs, such as an extensive library of strains, which includes mutants in many miRNA and target genes (Harris et al. 2010; Miska et al. 2007). There is also an abundance of data on spatial/temporal expression of small RNAs and protein-coding genes from several RNA-seq, microarray, and reporter studies (Gerstein et al. 2010). Ultimately, the broad spectrum of information on small RNA biology that is available in *C. elegans* makes it an ideal model organism for traditional and more sophisticated studies to elucidate general features of the miRNA pathway. This chapter aims to facilitate navigation through the major studies and methods involved in the identification of miRNAs and their targets in *C. elegans*.

2 miRNA Identification and Expression Profiling in *C. elegans*

2.1 Finding miRNAs in the Worm

Following the original discovery in 1993 of the first miRNA (*lin-4*) in *C. elegans* (Lee et al. 1993; Wightman et al. 1993), interest in small regulatory RNAs picked up steam 7 years later with the revelation that another miRNA, *let-7*, was in fact conserved across many organisms, including humans (Pasquinelli et al. 2000; Reinhart et al. 2000). Anticipating the existence of other genes like *lin-4* and *let-7*, several groups entered the race to systematically identify more members of this curious class in worms, mammalian cells, Drosophila, and Arabidopsis (Lagos-Quintana et al. 2001; Lau et al. 2001; Lee and Ambros 2001; Lim et al. 2003; Llave et al. 2002; Reinhart et al. 2002). Soon, thousands of novel small RNAs were described and annotated as miRNAs (Ambros 2003; Griffiths-Jones et al. 2006; Kozomara and Griffiths-Jones 2011).

Forward genetics enabled not only the early identification of miRNAs in *C. elegans* but also the first insights into their functions (Fig. 12.3). Notably, *lin-4* (Lee et al. 1993; Wightman et al. 1993) and *let-7* (Reinhart et al. 2000) were characterized as essential regulators of developmental timing, whereas *lsy-6* was found to control neuronal patterning (Johnston and Hobert 2003). However, the amount of time, labor, and luck needed to identify miRNAs through classical genetics renders this approach impractical for executing large-scale searches for novel miRNAs. The understanding that miRNAs are derived from precursor stem-loops of ~65 nt that undergo Dicer processing to produce ~22 nt RNAs with 5′ terminal monophosphates and 3′ terminal hydroxyl groups enabled more effective methods for discovering miRNAs (Bernstein et al. 2001; Elbashir et al. 2001; Grishok et al. 2001; Hutvágner et al. 2001). These features were used to search for new miRNAs by experimental and computational methods (Fig. 12.2) by developing protocols for

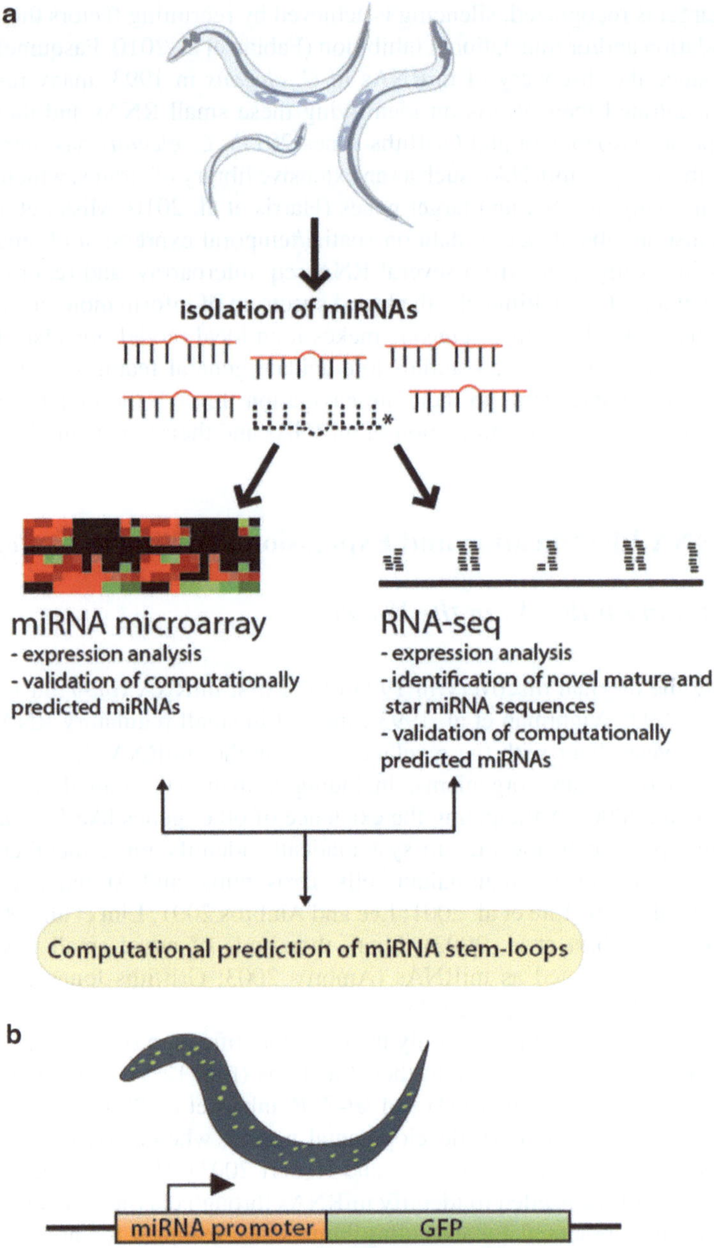

Fig. 12.2 miRNA identification and expression profiling. (**a**) Identification and expression analysis of miRNAs through microarrays and RNA-seq. Global methods to detect miRNAs are used to analyze expression patterns and confirm predicted miRNAs, especially when the rare miRNA* strands are also identified. (**b**) Spatiotemporal analysis of miRNA expression through reporter constructs

size exclusion and cloning of small RNAs as well as algorithms to analyze the genome for potential miRNA stem-loops (Grad et al. 2003; Lagos-Quintana et al. 2001; Lau et al. 2001; Lee and Ambros 2001; Lim et al. 2003; Reinhart et al. 2002). Several of the candidate miRNAs (sequenced ~22 nt RNAs embedded in predicted hairpin structures) were confirmed by northern blotting and, on occasion, by showing that they accumulated as precursors upon downregulation of Dicer activity. By 2003, when almost 100 worm miRNAs had been identified and experimentally validated, estimates of the total number of *C. elegans* miRNAs varied from 120 (Lim et al. 2003) to over 300 (Grad et al. 2003). In 2006 the Bartel group performed high-throughput pyrosequencing of worm miRNA libraries, raising the number of experimentally detected worm miRNAs to 112 (Ruby et al. 2006). Although subsequent studies helped confirm some previously predicted miRNAs and identify a few new candidates (Gu et al. 2007; Zhang et al. 2007), the biggest jump in the discovery of new miRNAs came with advancements in sequencing technology. Using the Solexa deep sequencing platform, 66 new miRNAs were identified in *C. elegans* (Kato et al. 2009). This was also the first study to report high levels of star strand sequences, which are a good indicator of a hairpin having undergone bona fide miRNA processing. Previous efforts to globally sequence miRNAs in the worm had reported only 1 % of star strand sequences among their libraries (Ruby et al. 2006), whereas in Kato et al. (2009) 35 % of the newly found miRNA candidates also had reads for the opposite strand. The latest version of miRBase, a database that compiles miRNA sequences and annotations across organisms, counts 223 precursors and 368 mature miRNAs and miRNAs* in *C. elegans* (Kozomara and Griffiths-Jones 2011). Recently, there has been a push to improve the quality of miRBase miRNA annotations by utilizing even more sensitive RNA-seq technology. The greatly increased number of detected reads allows for more reliable identification of mature miRNAs and the alternative products generated during their processing (star and loop sequences) (Warf et al. 2011).

2.2 *Profiling miRNA Expression Patterns in the Worm*

Determining where and when miRNAs are expressed provides important clues into their functions. From the first studies on miRNAs, there was evidence that these small regulators could exhibit distinct expression patterns during development (Grad et al. 2003; Lagos-Quintana et al. 2001; Lau et al. 2001; Lee et al. 1993; Lee and Ambros 2001; Lim et al. 2003; Reinhart et al. 2000; Wightman et al. 1993). So far, the efforts to produce global profiles of miRNA expression in *C. elegans* have relied on deep sequencing (Kato et al. 2009), microarrays (Gu et al. 2007; Ibáñez-Ventoso et al. 2006) and promoter:reporter constructs (Isik et al. 2010; Martinez et al. 2008b) (Fig. 12.2). These approaches are complementary as the first two provide quantitative data on the timing of miRNA expression and the last can be used to assay tissue specificity.

Studies that followed miRNA expression from embryogenesis to young adulthood in *C. elegans* (Gu et al. 2007; Kato et al. 2009) revealed that temporal changes in abundance during development are a common feature of many worm miRNAs. Microarray analysis on miRNAs isolated from miRISC demonstrated that the levels of over 60 % of the detected miRNAs exhibit changes greater then fourfold across development (Gu et al. 2007) and small RNA deep sequencing indicated that the abundance of 16 % of the miRNAs change by more than tenfold (Kato et al. 2009). Interestingly, when miRNA expression was evaluated throughout the worm lifespan, there was a trend of global downregulation in miRNA levels after adulthood, with only a few miRNAs upregulated during aging (among them miR-34 and miR-231) (Ibáñez-Ventoso et al. 2006; Kato et al. 2011). However, a recent study shows that even if miRNAs are generally downregulated after development, they are still needed to maintain normal physiology and are important regulators of aging in *C. elegans* (Lehrbach et al. 2012).

Several studies report that members of the same miRNA family (comprised of miRNAs that share a seed sequence but not necessarily a genomic locus) tend to have similar expression patterns and, presumably, act on similar targets (Gu et al. 2007; Ibáñez-Ventoso et al. 2006; Kato et al. 2009; Martinez et al. 2008a). MiRNAs in a genomic cluster, which sometimes contain identical seed sequences, are also frequently expressed at the same time, in agreement with the assumption that most of them are processed from a single transcript (Gu et al. 2007; Ibáñez-Ventoso et al. 2006; Kato et al. 2009). However, a global assessment of worm miRNA promoter:GFP fusions across all developmental stages (Martinez et al. 2008b), showed that some exceptions to overlapping profiles of related and/or clustered miRNAs do occur. Examples include the *miR-75*, *miR-232*, and *miR-251* families, whose members exhibit distinct spatial expression patterns. Extreme differences in miRNA levels have also been observed for some co-expressed miRNAs. The seven members of the *miR-35-41* cluster are expressed in the embryo and seem to derive from a common transcript (Lau et al. 2001). However, they vary greatly in expression levels: deep sequencing data (Kato et al. 2009) uncovered over 100,000 reads for miR-35 but only 40 for miR-41. These striking differences point to posttranscriptional regulation with modulation of miRNA processing or decay.

Analysis of transgenic worms expressing GFP driven by predicted miRNA promoters is a powerful approach for determining spatiotemporal expression in vivo (Isik et al. 2010; Martinez et al. 2008b) (Fig. 12.2b). However, this strategy is limited by the risk of excluding important elements in the cloned promoter regions and the inability to account for posttranscriptional events that might regulate production of the functional mature miRNA. Despite these caveats, expression data from northern blots mostly correlated with the temporal manifestation of the GFP reporters (Martinez et al. 2008b). Interestingly, it was observed that although functional miRNA promoters were found in almost all worm tissues, 50 % of individual promoters seemed to be active only in three or fewer tissues. The promoters of *let-7*, *lin-4*, and *miR-53* were among the few that were able to confer ubiquitous somatic tissue expression (Martinez et al. 2008b). Additionally, both studies (Isik et al. 2010; Martinez et al. 2008b) characterized several intronic miRNAs with expression patterns distinct from that of their host genes, indicating that these embedded miRNAs are transcribed independently.

3 Identifying miRNA Targets

3.1 *Finding Targets by Genetic Suppression*

The first miRNA targets were identified through phenotypic suppression (Fig. 12.3a). This was possible because mutations in the *lin-4* and *let-7* miRNA genes cause obvious loss-of-function phenotypes. Both *lin-4* and *let-7* participate in the developmental timing pathway (Kaufman and Miska 2010; Lee et al. 1993; Reinhart et al. 2000) and disruption of either miRNA gene results in failure to develop adult

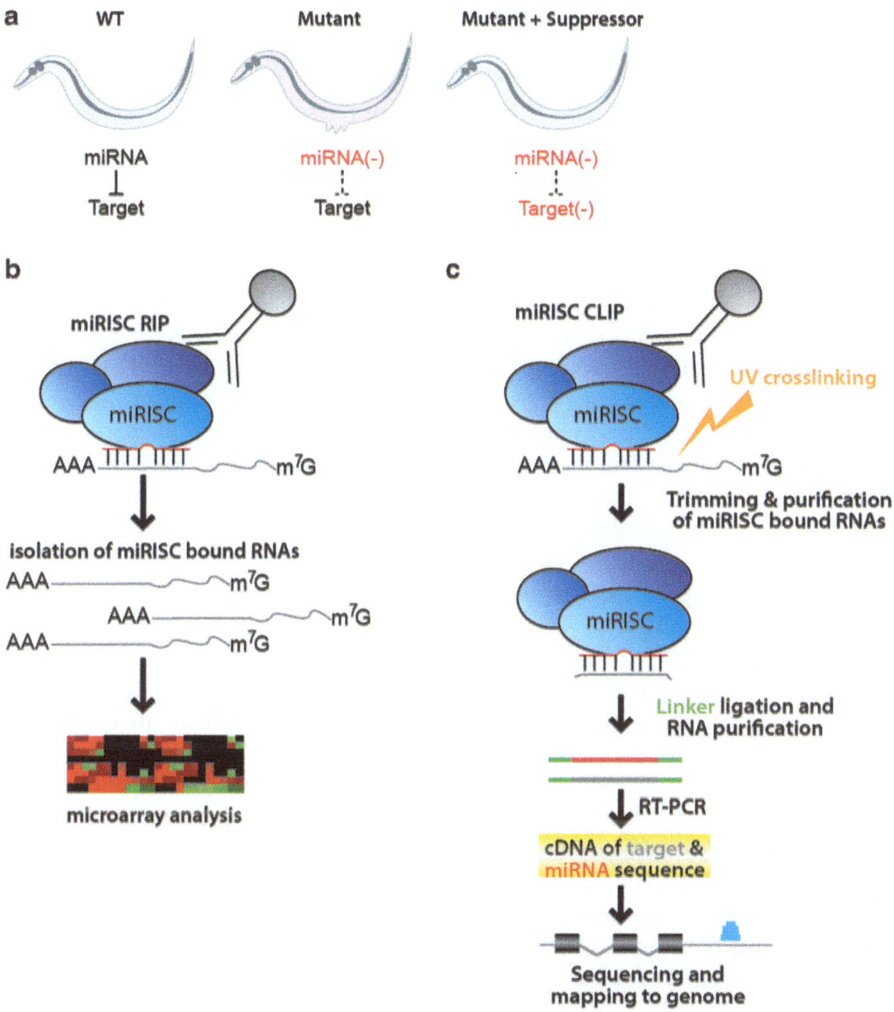

Fig. 12.3 Experimental strategies for finding miRNA targets in *C. elegans*. (**a**) Suppressor genetics, (**b**) miRISC RIP-Chip, (**c**) miRISC CLIP-seq. See text for details

structures and inability to lay eggs (Ambros 2008; Mondol and Pasquinelli 2012). Additionally, *let-7* mutants often die through vulval bursting (Mondol and Pasquinelli 2012). Since the best characterized miRNA–target interactions result in negative regulation, decreased expression of target genes can sometimes compensate for the absence of miRNA activity. Thus, suppression of the miRNA mutant phenotype by a secondary mutation or by RNAi depletion can be used to reveal potential targets.

Multiple let-7 miRNA targets have been identified through forward and reverse genetic screens in *C. elegans*. An advantage of *let-7* is that there are several mutant alleles providing a range of phenotypes that can be scored in different experimental designs. These phenotypes vary from non-conditional lethality, where mature let-7 is completely absent, to temperature sensitive lethality, where mature let-7 contains a point mutation and is expressed at very low levels, to more mild defects in seam cell and alae development, where ~2-fold less mature let-7 is produced (Mondol and Pasquinelli 2012; Reinhart et al. 2000). The *lin-41* gene was uncovered as a potent suppressor of the bursting and lethal phenotypes associated with *let-7* (Reinhart et al. 2000; Slack et al. 2000). The recognition of two closely spaced sequences in the *lin-41* 3′UTR with partial complementarity to let-7 miRNA helped establish the model of imperfect base-pairing between miRNAs and targets. Remarkably, regulation of *lin-41* by let-7 seems to be broadly conserved across species (Lancman et al. 2005; O'Farrell et al. 2008; Pasquinelli et al. 2000; Schulman et al. 2005). Several other let-7 targets in *C. elegans*, such as *hbl-1* (Abrahante et al. 2003; Lin et al. 2003), *daf-12* (Grosshans et al. 2005) and *let-60/RAS* (Johnson et al. 2005) have been identified through phenotypic suppression using mutagenesis or RNAi screens. Combinatorial approaches that consider misregulation of mRNA or protein levels in *let-7* mutants, predicted let-7 binding sites and suppression of *let-7* phenotypes have yielded an assortment of targets important for different aspects of the *let-7* pathway (Ding et al. 2008; Grosshans et al. 2005; Hunter et al. 2013; Jovanovic et al. 2010; Lall et al. 2006). It is perhaps surprising that reduced activity of a single target can suppress the loss of a miRNA, which has many targets. This phenomenon suggests that there are complex cross-regulatory and feedback loops connecting the pathways regulated by miRNAs.

Suppression genetics is a valuable approach for finding miRNA targets that are biologically relevant. It also has the advantage of not relying on preconceived notions of target recognition rules. However, it can be difficult to distinguish between direct and indirect targets and further authentication of the target is necessary. Also, miRNAs can regulate the expression of many different targets and it is likely that some phenotypes are very complex and do not depend on a single candidate gene. Furthermore, it seems that *C. elegans* miRNAs are highly redundant and that most miRNA mutants do not have detectable phenotypes under standard lab conditions (Miska et al. 2007). These limitations point to the importance of developing sensitive biochemical methods and accurate prediction algorithms to facilitate the identification of miRNA targets.

3.2 Computational Prediction of Targets

miRNAs act as guides for selective gene regulation by tethering the effector proteins of the RISC complex to their targets. The majority of known functional miRNA/target interactions is mediated by base pairing of the miRNA seed sequence (nucleotides 2–7) with a complementary site located in the 3′UTR of the target mRNA (Bartel 2009). The significance of seed pairing is also reinforced by the fact that, to date, many studies involving high-throughput identification of miRNA targets report enrichment of seed sequences in their experimental datasets (Hafner et al. 2010; Hunter et al. 2013; Jovanovic et al. 2012; Leung et al. 2011; Lim et al. 2005; Zhang et al. 2007, 2009; Zisoulis et al. 2010). Although perfect seed pairing is clearly an important motif for recognizing miRNA targets, it is not the only one. Some of the best studied *let-7* targets in *C. elegans*, *lin-41* (Reinhart et al. 2000) and *hbl-1* (Abrahante et al. 2003; Lin et al. 2003), contain functional target sites with seed mismatches. Imperfections in seed pairing can sometimes be compensated by a high degree of complementarity in the 3′end of the miRNA with its target. Nevertheless, there are several examples of miRNA target sites that rely on more centered or distributed interactions and the precise set of rules governing target recognition is still largely unknown (Pasquinelli 2012).

Currently, computational prediction of miRNA targets mostly relies on perfect or nearly perfect complementarity, allowing for occasional mismatches or G:U wobbles, of the seed region with its 3′UTR target site (Bartel 2009). However, seed-pairing predictions in animals generate hundreds to thousands of possible targets for a given miRNA, including many false positives (Thomas et al. 2010). To increase the reliability of predictions, targets with sufficient seed complementarity are usually scored according to secondary features of their sites, such as conservation, structural accessibility of the site in the 3′UTR (measured by the amount of AU rich regions surrounding the site), overall binding energy of the heteroduplex and number of sites for a given miRNA (Bartel 2009; Pasquinelli 2012; Thomas et al. 2010).

Some of the more commonly used algorithms for target prediction in *C. elegans* are: TargetScan (Jan et al. 2011; Lewis et al. 2005), PicTar (Lall et al. 2006), Miranda (Betel et al. 2008), PITA (Kertesz et al. 2007), rna22 (Miranda et al. 2006), and mirWIP (Hammell et al. 2008) (Table 12.1). TargetScan and PicTar both rely on stringent seed pairing. However, while TargetScan looks for perfect seed complementarity, PicTar allows for imperfect pairing as long as the binding energy is within the cutoff. Miranda, PITA, mirWIP and rna22 have more moderate seed pairing requirements, being more tolerant towards mismatches or wobbles. TargetScan, PicTar, Miranda and mirWIP consider conservation of the target sites. PITA relies heavily on the stability of the heteroduplex and on the structural accessibility of the site within the 3′UTR. The rna22 algorithm uses a distinct system that evaluates shared miRNA sequence patterns and heteroduplex stability. MirWIP is also unique in the fact that it was based and trained on an experimental dataset of targets in *C. elegans*. Table 12.1 provides a summarized comparison among these prediction algorithms.

Table 12.1 Comparison between algorithms for miRNA target site prediction in *C. elegans*

Target prediction tool	Seed-pairing[a]	Conservation	Site accessibility	Heteroduplex stability	Site number	Reference
TargetScan	S	✓			✓	Jan et al. (2011)
PicTar	S	✓		✓	✓	Lall et al. (2006)
Miranda	M	✓		✓	✓	Betel et al. (2008)
mirWIP	M	✓	✓	✓	✓	Hammell et al. (2008)
PITA	M		✓	✓	✓	Kertesz et al. (2007)
rna22	M			✓		Miranda et al. (2006)

[a]Stringent (S), Moderate (M)

The application of bioinformatics for prediction of miRNA targets has been very useful for enabling the identification and experimental analysis of many new target sites (Maragkakis et al. 2009). Yet these prediction algorithms walk a fine line between sensitivity of target identification and reliability of their predictions (e.g., emphasizing conservation of target sites can increase their reliability, but it can also prevent the identification of many non-conserved functional targets). Furthermore, the use of different 3′UTR and miRNA databases along with substantial biases in prediction parameters yields an unsatisfying level of overlap among candidate targets identified by current algorithms (Hammell et al. 2008; Hua et al. 2009; Ritchie et al. 2009). Also, noncanonical miRNA pairing and the presence of regulatory sites outside of the 3′UTR still represent major challenges in target prediction, creating a high demand for experimental approaches that can help define broader and more accurate target recognition rules.

3.3 *Biochemical Detection of miRNA Targets*

3.3.1 RNA Immunoprecipitation and Microarray (RIP-Chip) Assays

The ability to capture miRNA targets through biochemical methods has offered many insights into the dynamics of target selection (Fig. 12.3b, c). These approaches provide a snapshot of mRNAs bound by miRISC in vivo that does not rely on our limited knowledge of miRNA target recognition rules. Results from these types of experiments can be used to optimize target prediction algorithms and reveal new features of miRNA targeting. In *C. elegans*, the first attempt to isolate potential miRNA targets used DNA microarrays to analyze transcripts that co-immunoprecipitated with proteins related to GW182 (RIP-chip) (Fig. 12.3b) (Zhang et al. 2007). The GW182 family proteins associate with Argonaute and recruit factors that mediate the deadenylation and translational repression of miRNA targets (Braun et al. 2013). In *C. elegans* the GW182 family members, AIN-1 and AIN-2, are thought to have largely overlapping roles in binding the miRISC Argonautes ALG-1 and ALG-2 and directing target mRNA repression (Ding et al. 2005; Zhang et al. 2007). Over 3,500 mRNAs were found to be enriched in AIN-1/2 IPs, including most of the previously validated worm miRNA targets such as *lin-14, lin-28, hbl-1*, and *let-60/ras* (Zhang et al. 2007). This strategy has also been used to detect mRNAs associated with miRISC at different worm stages, which supported a broad role for miRNAs in regulating development and signaling processes (Zhang et al. 2009). More recently, IPs of tagged AIN-2 expressed in specific tissues were used to identify targets in the intestine and muscle (Kudlow et al. 2012). This strategy revealed a strong bias for miRISC regulation of pathogen responsive genes in the gut and pointed to new roles for the miRNA pathway in host–pathogen interactions (Kudlow et al. 2012). One of the biggest caveats of RIP-chip approaches is that they are dependent on computer predictions to locate potential miRNA target sites within the isolated transcripts. Since these predictions can be biased towards particular pairing requirements and often only search for sites in 3′UTRs, critical features of endogenous miRNA target recognition can be missed. Also, the use of microarrays limits the pool of genes analyzed and prevents the identification of targets outside the scope of the chip, such as noncoding RNAs.

3.3.2 Cross-Linking Immunoprecipitation of miRISC Coupled with RNA Sequencing (CLIP-seq)

Some of the sensitivity problems and biases in standard RIP approaches are solved when cross-linking is used to stabilize miRISC interactions for IP and deep sequencing is used to identify mRNA regions in direct contact with the complex (CLIP-seq) (Darnell 2010) (Fig. 12.3c). Cross-linking assures the fidelity of endogenous interactions and the use of RNA deep sequencing technology allows for a more global and unbiased identification of the interacting RNAs. The downside of this approach is that the ability to detect weaker interactions could lead to higher false-positive rates, as it is possible that Argonaute proteins "scan" sites that are not functional. The key innovation and advantage of CLIP-seq is that it reveals direct sites of interaction between miRISC and its targets. This simplifies prediction of the actual miRNA binding site and enables identification of noncanonical miRNA binding patterns and locations. CLIP-seq of Argonaute proteins has uncovered thousands of targets sites in mouse brain (Chi et al. 2009), human cells (Hafner et al. 2010; Leung et al. 2011), and *C. elegans* (Zisoulis et al. 2010).

C. elegans offers several unique advantages for using CLIP-seq to identify endogenous miRNA target sites (Zisoulis et al. 2011). A strain with a mutation in *alg-1* that removes the epitope for the anti-ALG-1 antibody serves as a powerful negative control for distinguishing authentic interactions from background. A list of well-established miRNA targets provides a set of positive controls that should be identified in successful CLIP-seq experiments. Furthermore, a large collection of miRNA mutants offers a valuable resource for using CLIP-seq to determine which ALG-1 sites are dependent on specific miRNAs (Miska et al. 2007). By performing ALG-1 CLIP-seq in wild-type versus *alg-1* mutant worms at the L4 stage, almost 5,000 specific sites in over 3,000 genes were identified, including 10 out of 13 previously validated targets (Zisoulis et al. 2010). Compared to control, unbound sequences, ALG-1 interaction sites showed greater sequence conservation and accessibility and were associated with CU-rich motifs (Zisoulis et al. 2010). Moreover, this study provided the first global map of miRISC binding in a whole animal.

Recently, the more sensitive iCLIP (individual-nucleotide resolution CLIP) protocol has been optimized for *C. elegans* ALG-1 (Broughton and Pasquinelli 2013). iCLIP follows the same basic principles of original CLIP-seq protocols with the addition of steps that improve the precision of mapping binding sites and greatly increase the recovery of target reads (König et al. 2010; Sugimoto et al. 2012). Efforts have also been made in developing techniques that will allow the recovery of the miRNAs paired with their target sites. In *C. elegans* one study attempted to use miRNAs as reverse transcription primers in order to clone their target sites (Andachi 2008). Although this approach identified a known target site, the broad applicability of this technique is yet to be established. In human cells, a method similar to CLIP-seq called CLASH (cross-linking, ligation and sequencing of hybrids) was modified to include a ligation step between the Argonaute bound miRNAs and their target RNA fragments (Helwak et al. 2013). Despite the inefficiency of this step (less then 2 % of the reads were chimeras between miRNAs and targets), the method shows promise and provides further confirmation that many miRNA–target interactions are noncanonical and reside outside of 3′UTRs.

Currently, the IP-based methods for identifying miRNA targets provide no intrinsic information about the functional consequences of this interaction. Further validation by monitoring changes in RNA and protein levels is necessary to confirm an output of miRISC binding. Intriguingly, the results of ALG-1 CLIP-seq in *C. elegans* suggested that target recognition and regulation differ based on the position of the site. Consistent with CLIP in other species, the majority of the ALG-1 sites localized to 3′UTRs and coding exons. Target sites in 3′UTRs were highly enriched for the seed pairing motif and correlated with mRNA destabilization (Zisoulis et al. 2010). In contrast, neither of those features described targets with sites in coding exons. Thus, further investigations are needed to establish that those sites mediate regulation, perhaps at the translational level.

3.4 Support and Validation for Experimentally Identified Targets

The expression of biologically relevant targets is expected to change upon regulation by miRISC. Since target mRNA levels are often downregulated in response to miRNA regulation (Pasquinelli 2012), microarrays and RNA-seq are useful methods for globally analyzing changes in mRNA expression in the presence or absence of miRNA activity (Fig. 12.4). A more quantitative assessment of select target mRNAs can be achieved through northern blotting and qRT-PCR (Fig. 12.4). Since loss of a specific miRNA is expected to produce direct and indirect effects, further validation of potential miRNA targets is required. Reporter genes fused to the 3′UTR of the predicted target should be assayed for the requirement of the complementary site as well as the miRNA for regulation. In these types of assays, care should be taken to test the reporter under physiological conditions. Transgenic overexpression of the reporter or miRNA or forced expression in non-native tissues can result in false conclusions about endogenous miRNA function.

There are two general methods, ribosome profiling and quantitative mass spectrometry, for analyzing miRISC dependent changes in protein levels on a genome wide scale (Fig. 12.4). Ribosome profiling (RP) offers an indirect approach to assess protein expression by monitoring the association of transcripts with translating ribosomes (Ingolia et al. 2009). Results from RP assays in mammalian cell culture led to the conclusion that mRNA degradation, as opposed to translational repression,

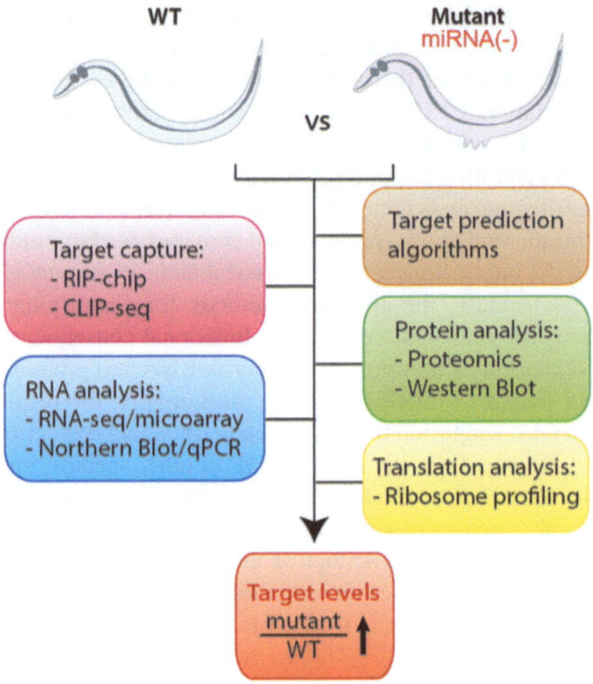

Fig. 12.4 Combinatorial approaches for identifying and validating miRNA targets. Potential miRNA targets from computational predictions or experimental approaches are tested for effects on protein or mRNA expression in wild-type versus mutant backgrounds to confirm regulation by a given miRNA

is the predominant outcome of miRNA target regulation (Guo et al. 2010). This method was used to assess the contributions of mRNA destabilization and translational inhibition to the regulation of known miRNA targets important for development in *C. elegans* (Stadler et al. 2012). Curiously, this high throughput approach led to conclusions about mRNA levels and ribosome association for some of these targets that seem to contradict previous more detailed studies (Bagga et al. 2005; Ding and Grosshans 2009). Thus, multiple validation efforts should be utilized to assess the functional outcome of miRNA targeting.

The global analysis of miRNA dependent changes in protein levels through proteomics methods was first developed in mammalian cell culture (Fig. 12.4) (Baek et al. 2008; Selbach et al. 2008). A more targeted proteomics approach has been used to detect miRNA targets in *C. elegans*. Selected reaction monitoring (SRM) was employed to quantify expression differences in proteins of interest from wild-type versus strains with mutations in specific miRNAs (Jovanovic et al. 2010, 2012). This method was combined with ALG-1 RIP assays to detect targets regulated by miR-58 at the mRNA and protein level (RIP-chip-SRM) (Jovanovic et al. 2012). Candidates dependent on miR-58 for association with ALG-1 were screened by quantitative proteomics and microarrays for differential expression in extracts from wild-type versus *miR-58* mutants. Based on these global assays, the authors concluded that miR-58 targets are largely resistant to destabilization and, instead, regulated primarily at the translational level (Jovanovic et al. 2012).

4 Integrating Datasets and Building Functional Networks

The biggest challenge faced by researchers sifting through the massive amount of data that has been generated in just over a decade of miRNA research is managing and integrating it. In order to facilitate this task, there have been efforts to validate and combine material from different sources. miRBase compiles a great amount of information on miRNAs, including the known miRNA sequences of several organisms as well as information on their expression, genomic location, stem-loop, family, cluster, and links to platforms that provide predicted or validated targets (Griffiths-Jones 2006; Kozomara and Griffiths-Jones 2011). One of these platforms is TarBase, a compilation of experimentally validated miRNA targets that summarizes the amount of evidence on the targets and provides links to the associated publications (Papadopoulos et al. 2009). Likewise, the 3′UTRome project compiles extensive data on *C. elegans* 3′UTRs, including the developmental stage(s) in which a certain 3′UTR was found and the method used to find it (Mangone et al. 2008, 2010). Another very useful approach for visualization of miRNA data is provided by starBase, which integrates several types of information in a genome browser format (Yang et al. 2011). This makes it possible for one to view numerous information tracks (CLIP-seq and Degradome-seq data, target predictions, 3′UTR annotations, etc.) in a combined and intuitive manner.

The modENCODE (model organism Encyclopedia of DNA Elements) project has aimed to systematically annotate functional genomic elements in flies and worms (Gerstein et al. 2010). As part of this project, Martinez et al. (2008a, b) analyzed the interplay between miRNAs and transcription factors (TFs). Interestingly, by merging networks of TFs that bind miRNA promoters and TFs that could be regulated by miRNAs, the study proposed a very complex mesh of interactions and several types of regulatory motifs. Most striking among those were the 23 composite miRNA-TF feedback loops in which the TF that binds the promoter of a certain miRNA is also a predicted target of that same miRNA (Martinez et al. 2008a). Additionally, modENCODE has provided a wealth of data on the *C. elegans* transcriptome, including annotation of novel and known ncRNAs and assignment of transcription start sites (TSS). Also, several worm deep sequencing datasets have been generated to determine changes in expression of small RNAs upon development, aging, and stress (modencode.org). Analyzing the modulation of miRNA levels and potential targets during those conditions could provide important clues to the functions of the many worm miRNAs whose biological roles are still unknown.

5 Conclusions

Our knowledge of miRNAs has evolved greatly considering the relatively short period of research since their discovery, and *C. elegans* has been instrumental in this progress. Numerous high quality studies have generated an enormous volume of data, calling for databases capable of integrating information from different experimental and computational sources. Despite the thousands of miRNAs now annotated in different organisms, the biological role of the majority of miRNAs is still largely unknown. Considering that the function of miRNAs is directly associated with their influence on gene expression, establishing reliable rules for target recognition remains a paramount challenge in the field. Complementary experimental and computational methods coupled with rigorous validation studies are essential for advancing the field and deciphering the complex miRNA–target interactions that shape most biological pathways.

Acknowledgements We thank members of the Pasquinelli lab for critical review of this manuscript. This work was supported by funding from the NIH (GM071654), Keck, and Peter Gruber Foundations.

References

Aalto AP, Pasquinelli AE (2012) Small non-coding RNAs mount a silent revolution in gene expression. Curr Opin Cell Biol 24:333–340. doi:10.1016/j.ceb.2012.03.006

Abrahante JE, Daul AL, Li M et al (2003) The *Caenorhabditis elegans* hunchback-like gene lin-57/hbl-1 controls developmental time and is regulated by microRNAs. Dev Cell 4:625–637

Alvarez-Garcia I, Miska EA (2005) MicroRNA functions in animal development and human disease. Development 132:4653–4662. doi:10.1242/dev.02073

Ambros V (2003) A uniform system for microRNA annotation. RNA 9:277–279. doi:10.1261/rna.2183803

Ambros V (2008) The evolution of our thinking about microRNAs. Nat Med 14:1036–1040. doi:10.1038/nm1008-1036

Andachi Y (2008) A novel biochemical method to identify target genes of individual microRNAs: identification of a new *Caenorhabditis elegans* let-7 target. RNA 14:2440–2451. doi:10.1261/rna.1139508

Baek D, Villén J, Shin C et al (2008) The impact of microRNAs on protein output. Nature 455:64–71. doi:10.1038/nature07242

Bagga S, Bracht J, Hunter S et al (2005) Regulation by let-7 and lin-4 miRNAs results in target mRNA degradation. Cell 122:553–563. doi:10.1016/j.cell.2005.07.031

Bartel DP (2009) MicroRNAs: target recognition and regulatory functions. Cell 136:215–233. doi:10.1016/j.cell.2009.01.002

Bernstein E, Caudy AA, Hammond SM, Hannon GJ (2001) Role for a bidentate ribonuclease in the initiation step of RNA interference. Nature 409:363–366. doi:10.1038/35053110

Betel D, Wilson M, Gabow A et al (2008) The microRNA.org resource: targets and expression. Nucleic Acids Res 36:D149–D153. doi:10.1093/nar/gkm995

Braun JE, Huntzinger E, Izaurralde E (2013) The role of GW182 proteins in miRNA-mediated gene silencing. Adv Exp Med Biol 768:147–163. doi:10.1007/978-1-4614-5107-5_9

Broughton JP, Pasquinelli AE (2013) Identifying Argonaute binding sites in *Caenorhabditis elegans* using iCLIP. Methods. doi:10.1016/j.ymeth.2013.03.033

Chi SW, Zang JB, Mele A, Darnell RB (2009) Argonaute HITS-CLIP decodes microRNA-mRNA interaction maps. Nature 460:479–486. doi: 10.1038/nature08170

Darnell RB (2010) HITS-CLIP: panoramic views of protein-RNA regulation in living cells. Wiley Interdiscip Rev RNA 1:266–286. doi:10.1002/wrna.31

Ding XC, Grosshans H (2009) Repression of *C. elegans* microRNA targets at the initiation level of translation requires GW182 proteins. EMBO J 28:213–222. doi:10.1038/emboj.2008.275

Ding L, Spencer A, Morita K, Han M (2005) The developmental timing regulator AIN-1 interacts with miRISCs and may target the argonaute protein ALG-1 to cytoplasmic P bodies in *C. elegans*. Mol Cell 19:437–447. doi:10.1016/j.molcel.2005.07.013

Ding XC, Slack FJ, Grosshans H (2008) The let-7 microRNA interfaces extensively with the translation machinery to regulate cell differentiation. Cell Cycle 7:3083–3090

Elbashir SM, Lendeckel W, Tuschl T (2001) RNA interference is mediated by 21- and 22-nucleotide RNAs. Genes Dev 15:188–200

Esquela-Kerscher A, Slack FJ (2006) Oncomirs—microRNAs with a role in cancer. Nat Rev Cancer 6:259–269. doi:10.1038/nrc1840

Fabian MR, Sonenberg N, Filipowicz W (2010) Regulation of mRNA translation and stability by microRNAs. Annu Rev Biochem 79:351–379. doi:10.1146/annurev-biochem-060308-103103

Finnegan EF, Pasquinelli AE (2013) MicroRNA biogenesis: regulating the regulators. Crit Rev Biochem Mol Biol 48:51–68. doi:10.3109/10409238.2012.738643

Friedman RC, Farh KK-H, Burge CB, Bartel DP (2009) Most mammalian mRNAs are conserved targets of microRNAs. Genome Res 19:92–105. doi:10.1101/gr.082701.108

Gerstein MB, Lu ZJ, Van Nostrand EL et al (2010) Integrative analysis of the *Caenorhabditis elegans* genome by the modENCODE project. Science 330:1775–1787. doi:10.1126/science.1196914

Grad Y, Aach J, Hayes GD et al (2003) Computational and experimental identification of *C. elegans* microRNAs. Mol Cell 11:1253–1263

Griffiths-Jones S (2006) miRBase: the microRNA sequence database. Methods Mol Biol 342:129–138. doi:10.1385/1-59745-123-1:129

Griffiths-Jones S, Grocock RJ, van Dongen S et al (2006) miRBase: microRNA sequences, targets and gene nomenclature. Nucleic Acids Res 34:D140–D144. doi:10.1093/nar/gkj112

Grishok A, Pasquinelli AE, Conte D et al (2001) Genes and mechanisms related to RNA interference regulate expression of the small temporal RNAs that control *C. elegans* developmental timing. Cell 106:23–34

Grosshans H, Johnson T, Reinert KL et al (2005) The temporal patterning microRNA let-7 regulates several transcription factors at the larval to adult transition in *C. elegans*. Dev Cell 8:321–330. doi:10.1016/j.devcel.2004.12.019

Gu SG, Pak J, Barberan-Soler S et al (2007) Distinct ribonucleoprotein reservoirs for microRNA and siRNA populations in *C. elegans*. RNA 13:1492–1504. doi:10.1261/rna.581907

Guo H, Ingolia NT, Weissman JS, Bartel DP (2010) Mammalian microRNAs predominantly act to decrease target mRNA levels. Nature 466:835–840. doi:10.1038/nature09267

Hafner M, Landthaler M, Burger L et al (2010) Transcriptome-wide identification of RNA-binding protein and MicroRNA target sites by PAR-CLIP. Cell 141:129–141. doi:10.1016/j.cell.2010.03.009

Hammell M, Long D, Zhang L et al (2008) mirWIP: microRNA target prediction based on microRNA-containing ribonucleoprotein-enriched transcripts. Nat Methods 5:813–819. doi:10.1038/nmeth.1247

Harris TW, Antoshechkin I, Bieri T et al (2010) WormBase: a comprehensive resource for nematode research. Nucleic Acids Res 38:D463–D467. doi:10.1093/nar/gkp952

Helwak A, Kudla G, Dudnakova T, Tollervey D (2013) Mapping the human miRNA interactome by CLASH reveals frequent noncanonical binding. Cell 153:654–665. doi:10.1016/j.cell.2013.03.043

Hua Y-J, Tang Z-Y, Tu K et al (2009) Identification and target prediction of miRNAs specifically expressed in rat neural tissue. BMC Genomics 10:214. doi:10.1186/1471-2164-10-214

Hunter SE, Finnegan EF, Zisoulis DG et al (2013) Functional genomic analysis of the let-7 regulatory network in *Caenorhabditis elegans*. PLoS Genet 9:e1003353. doi:10.1371/journal.pgen.1003353

Hutvágner G, McLachlan J, Pasquinelli AE et al (2001) A cellular function for the RNA-interference enzyme Dicer in the maturation of the let-7 small temporal RNA. Science 293:834–838. doi:10.1126/science.1062961

Ibáñez-Ventoso C, Yang M, Guo S et al (2006) Modulated microRNA expression during adult lifespan in *Caenorhabditis elegans*. Aging Cell 5:235–246. doi:10.1111/j.1474-9726.2006.00210.x

Ingolia NT, Ghaemmaghami S, Newman JRS, Weissman JS (2009) Genome-wide analysis in vivo of translation with nucleotide resolution using ribosome profiling. Science 324:218–223. doi:10.1126/science.1168978

Isik M, Korswagen HC, Berezikov E (2010) Expression patterns of intronic microRNAs in *Caenorhabditis elegans*. Silence 1:5. doi:10.1186/1758-907X-1-5

Jan CH, Friedman RC, Ruby JG, Bartel DP (2011) Formation, regulation and evolution of *Caenorhabditis elegans* 3'UTRs. Nature 469:97–101. doi:10.1038/nature09616

Johnson SM, Grosshans H, Shingara J et al (2005) RAS is regulated by the let-7 microRNA family. Cell 120:635–647. doi:10.1016/j.cell.2005.01.014

Johnston RJ, Hobert O (2003) A microRNA controlling left/right neuronal asymmetry in *Caenorhabditis elegans*. Nature 426:845–849. doi:10.1038/nature02255

Jovanovic M, Reiter L, Picotti P et al (2010) A quantitative targeted proteomics approach to validate predicted microRNA targets in *C. elegans*. Nat Methods 7:837–842. doi:10.1038/nmeth.1504

Jovanovic M, Reiter L, Clark A et al (2012) RIP-chip-SRM—a new combinatorial large-scale approach identifies a set of translationally regulated bantam/miR-58 targets in *C. elegans*. Genome Res 22:1360–1371. doi:10.1101/gr.133330.111

Kato M, de Lencastre A, Pincus Z, Slack FJ (2009) Dynamic expression of small non-coding RNAs, including novel microRNAs and piRNAs/21U-RNAs, during *Caenorhabditis elegans* development. Genome Biol 10:R54. doi:10.1186/gb-2009-10-5-r54

Kato M, Chen X, Inukai S et al (2011) Age-associated changes in expression of small, noncoding RNAs, including microRNAs, in *C. elegans*. RNA 17:1804–1820. doi:10.1261/rna.2714411

Kaufman EJ, Miska EA (2010) The microRNAs of *Caenorhabditis elegans*. Semin Cell Dev Biol 21:728–737. doi:10.1016/j.semcdb.2010.07.001

Kertesz M, Iovino N, Unnerstall U et al (2007) The role of site accessibility in microRNA target recognition. Nat Genet 39:1278–1284. doi:10.1038/ng2135

Kloosterman WP, Plasterk RHA (2006) The diverse functions of microRNAs in animal development and disease. Dev Cell 11:441–450. doi:10.1016/j.devcel.2006.09.009

König J, Zarnack K, Rot G et al (2010) iCLIP reveals the function of hnRNP particles in splicing at individual nucleotide resolution. Nat Struct Mol Biol 17:909–915. doi:10.1038/nsmb.1838

Kozomara A, Griffiths-Jones S (2011) miRBase: integrating microRNA annotation and deep-sequencing data. Nucleic Acids Res 39:D152–D157. doi:10.1093/nar/gkq1027

Krol J, Loedige I, Filipowicz W (2010) The widespread regulation of microRNA biogenesis, function and decay. Nat Rev Genet 11:597–610. doi:10.1038/nrg2843

Kudlow BA, Zhang L, Han M (2012) Systematic analysis of tissue-restricted miRISCs reveals a broad role for microRNAs in suppressing basal activity of the *C. elegans* pathogen response. Mol Cell 46:530–541. doi:10.1016/j.molcel.2012.03.011

Lagos-Quintana M, Rauhut R, Lendeckel W, Tuschl T (2001) Identification of novel genes coding for small expressed RNAs. Science 294:853–858. doi:10.1126/science.1064921

Lall S, Grün D, Krek A et al (2006) A genome-wide map of conserved microRNA targets in *C. elegans*. Curr Biol 16:460–471. doi:10.1016/j.cub.2006.01.050

Lancman JJ, Caruccio NC, Harfe BD et al (2005) Analysis of the regulation of lin-41 during chick and mouse limb development. Dev Dyn 234:948–960. doi:10.1002/dvdy.20591

Lau NC, Lim LP, Weinstein EG, Bartel DP (2001) An abundant class of tiny RNAs with probable regulatory roles in *Caenorhabditis elegans*. Science 294:858–862. doi:10.1126/science.1065062

Lee RC, Ambros V (2001) An extensive class of small RNAs in *Caenorhabditis elegans*. Science 294:862–864. doi:10.1126/science.1065329

Lee RC, Feinbaum RL, Ambros V (1993) The *C. elegans* heterochronic gene lin-4 encodes small RNAs with antisense complementarity to lin-14. Cell 75:843–854

Lehrbach NJ, Castro C, Murfitt KJ et al (2012) Post-developmental microRNA expression is required for normal physiology, and regulates aging in parallel to insulin/IGF-1 signaling in *C. elegans*. RNA 18:2220–2235. doi:10.1261/rna.035402.112

Leung AKL, Young AG, Bhutkar A et al (2011) Genome-wide identification of Ago2 binding sites from mouse embryonic stem cells with and without mature microRNAs. Nat Struct Mol Biol 18:237–244. doi:10.1038/nsmb.1991

Lewis BP, Burge CB, Bartel DP (2005) Conserved seed pairing, often flanked by adenosines, indicates that thousands of human genes are microRNA targets. Cell 120:15–20. doi:10.1016/j.cell.2004.12.035

Lim LP, Lau NC, Weinstein EG et al (2003) The microRNAs of *Caenorhabditis elegans*. Genes Dev 17:991–1008. doi:10.1101/gad.1074403

Lim LP, Lau NC, Garrett-Engele P et al (2005) Microarray analysis shows that some microRNAs downregulate large numbers of target mRNAs. Nature 433:769–773. doi:10.1038/nature03315

Lin S-Y, Johnson SM, Abraham M et al (2003) The *C. elegans* hunchback homolog, hbl-1, controls temporal patterning and is a probable microRNA target. Dev Cell 4:639–650

Llave C, Xie Z, Kasschau KD, Carrington JC (2002) Cleavage of Scarecrow-like mRNA targets directed by a class of Arabidopsis miRNA. Science 297:2053–2056. doi:10.1126/science.1076311

Mangone M, MacMenamin P, Zegar C et al (2008) UTRome.org: a platform for 3'UTR biology in *C. elegans*. Nucleic Acids Res 36:D57–D62. doi:10.1093/nar/gkm946

Mangone M, Manoharan AP, Thierry-Mieg D et al (2010) The landscape of *C. elegans* 3'UTRs. Science 329:432–435. doi:10.1126/science.1191244

Maragkakis M, Reczko M, Simossis VA et al (2009) DIANA-microT web server: elucidating microRNA functions through target prediction. Nucleic Acids Res 37:W273–W276. doi:10.1093/nar/gkp292

Martinez NJ, Ow MC, Barrasa MI et al (2008a) A *C. elegans* genome-scale microRNA network contains composite feedback motifs with high flux capacity. Genes Dev 22:2535–2549. doi:10.1101/gad.1678608

Martinez NJ, Ow MC, Reece-Hoyes JS et al (2008b) Genome-scale spatiotemporal analysis of *Caenorhabditis elegans* microRNA promoter activity. Genome Res 18:2005–2015. doi:10.1101/gr.083055.108

Miranda KC, Huynh T, Tay Y et al (2006) A pattern-based method for the identification of MicroRNA binding sites and their corresponding heteroduplexes. Cell 126:1203–1217. doi:10.1016/j.cell.2006.07.031

Miska EA, Alvarez-Saavedra E, Abbott AL et al (2007) Most *Caenorhabditis elegans* microRNAs are individually not essential for development or viability. PLoS Genet 3:e215. doi:10.1371/journal.pgen.0030215

Mondol V, Pasquinelli AE (2012) Let's make it happen: the role of let-7 microRNA in development. Curr Top Dev Biol 99:1–30. doi:10.1016/B978-0-12-387038-4.00001-X

O'Farrell F, Esfahani SS, Engström Y, Kylsten P (2008) Regulation of the Drosophila lin-41 homologue dappled by let-7 reveals conservation of a regulatory mechanism within the LIN-41 subclade. Dev Dyn 237:196–208. doi:10.1002/dvdy.21396

Papadopoulos GL, Reczko M, Simossis VA et al (2009) The database of experimentally supported targets: a functional update of TarBase. Nucleic Acids Res 37:D155–D158. doi:10.1093/nar/gkn809

Pasquinelli AE (2012) MicroRNAs and their targets: recognition, regulation and an emerging reciprocal relationship. Nat Rev Genet 13:271–282. doi:10.1038/nrg3162

Pasquinelli AE, Reinhart BJ, Slack F et al (2000) Conservation of the sequence and temporal expression of let-7 heterochronic regulatory RNA. Nature 408:86–89. doi:10.1038/35040556

Reinhart BJ, Slack FJ, Basson M et al (2000) The 21-nucleotide let-7 RNA regulates developmental timing in *Caenorhabditis elegans*. Nature 403:901–906. doi:10.1038/35002607

Reinhart BJ, Weinstein EG, Rhoades MW et al (2002) MicroRNAs in plants. Genes Dev 16:1616–1626. doi:10.1101/gad.1004402

Ritchie W, Flamant S, Rasko JEJ (2009) Predicting microRNA targets and functions: traps for the unwary. Nat Methods 6:397–398. doi:10.1038/nmeth0609-397

Ruby JG, Jan C, Player C et al (2006) Large-scale sequencing reveals 21U-RNAs and additional microRNAs and endogenous siRNAs in *C. elegans*. Cell 127:1193–1207. doi:10.1016/j.cell.2006.10.040

Schulman BRM, Esquela-Kerscher A, Slack FJ (2005) Reciprocal expression of lin-41 and the microRNAs let-7 and mir-125 during mouse embryogenesis. Dev Dyn 234:1046–1054. doi:10.1002/dvdy.20599

Selbach M, Schwanhäusser B, Thierfelder N et al (2008) Widespread changes in protein synthesis induced by microRNAs. Nature 455:58–63. doi:10.1038/nature07228

Slack FJ, Basson M, Liu Z et al (2000) The lin-41 RBCC gene acts in the *C. elegans* heterochronic pathway between the let-7 regulatory RNA and the LIN-29 transcription factor. Mol Cell 5:659–669

Stadler M, Artiles K, Pak J, Fire A (2012) Contributions of mRNA abundance, ribosome loading, and post- or peri-translational effects to temporal repression of *C. elegans* heterochronic miRNA targets. Genome Res 22:2418–2426. doi:10.1101/gr.136515.111

Sugimoto Y, König J, Hussain S et al (2012) Analysis of CLIP and iCLIP methods for nucleotide-resolution studies of protein-RNA interactions. Genome Biol 13:R67. doi:10.1186/gb-2012-13-8-r67

Thomas M, Lieberman J, Lal A (2010) Desperately seeking microRNA targets. Nat Struct Mol Biol 17:1169–1174. doi:10.1038/nsmb.1921

Warf MB, Johnson WE, Bass BL (2011) Improved annotation of *C. elegans* microRNAs by deep sequencing reveals structures associated with processing by Drosha and Dicer. RNA 17:563–577. doi:10.1261/rna.2432311

Wightman B, Ha I, Ruvkun G (1993) Posttranscriptional regulation of the heterochronic gene lin-14 by lin-4 mediates temporal pattern formation in *C. elegans*. Cell 75:855–862

Yang J-H, Li J-H, Shao P et al (2011) starBase: a database for exploring microRNA-mRNA interaction maps from Argonaute CLIP-Seq and Degradome-Seq data. Nucleic Acids Res 39:D202–D209. doi:10.1093/nar/gkq1056

Zhang L, Ding L, Cheung TH et al (2007) Systematic identification of *C. elegans* miRISC proteins, miRNAs, and mRNA targets by their interactions with GW182 proteins AIN-1 and AIN-2. Mol Cell 28:598–613. doi:10.1016/j.molcel.2007.09.014

Zhang L, Hammell M, Kudlow BA et al (2009) Systematic analysis of dynamic miRNA-target interactions during *C. elegans* development. Development 136:3043–3055. doi:10.1242/dev.039008

Zisoulis DG, Lovci MT, Wilbert ML et al (2010) Comprehensive discovery of endogenous Argonaute binding sites in *Caenorhabditis elegans*. Nat Struct Mol Biol 17:173–179. doi:10.1038/nsmb.1745

Zisoulis DG, Yeo GW, Pasquinelli AE (2011) Comprehensive identification of miRNA target sites in live animals. Methods Mol Biol 732:169–185. doi:10.1007/978-1-61779-083-6_13

Chapter 13
Splicing Code Modeling

Yoseph Barash and Jorge Vaquero-Garcia

Abstract How do *cis* and *trans* elements involved in pre-mRNA splicing come together to form a splicing "code"? This question has been a driver of much of the research involving RNA biogenesis. The variability of splicing outcome across developmental stages and between tissues coupled with association of splicing defects with numerous diseases highlights the importance of such a code. However, the sheer number of elements involved in splicing regulation and the context-specific manner of their operation have made the derivation of such a code challenging. Recently, machine learning-based methods have been developed to infer computational models for a splicing code. These methods use high-throughput experiments measuring mRNA expression at exonic resolution and binding locations of RNA-binding proteins (RBPs) to infer what the regulatory elements that control the inclusion of a given pre-mRNA segment are. The inferred regulatory models can then be applied to genomic sequences or experimental conditions that have not been measured to predict splicing outcome. Moreover, the models themselves can be interrogated to identify new regulatory mechanisms, which can be subsequently tested experimentally. In this chapter, we survey the current state of this technology, and illustrate how it can be applied by non-computational or RNA splicing experts to study regulation of specific exons by using the AVISPA web tool.

Keywords Splicing code • Posttranscriptional regulation • Alternative splicing • Machine learning • Computational biology

---

Y. Barash (✉) • J. Vaquero-Garcia
Department of Genetics, University of Pennsylvania, University Park, PA, USA

Department of Computer and Information Science, University of Pennsylvania, University Park, PA, USA
e-mail: yosephb@mail.med.upenn.edu

G.W. Yeo (ed.), *Systems Biology of RNA Binding Proteins*, Advances in Experimental Medicine and Biology 825, DOI 10.1007/978-1-4939-1221-6_13

1 Background: Splicing Regulation

Approximately 90 % of human genes are composed of multiple exons, and RNA-Seq studies across multiple tissues indicate that nearly 95 % of human multi-exon genes exhibit alternative splicing (AS) (Pan et al. 2008; Wang et al. 2008). The process of splicing is inherently stochastic yet far from random. The stochastic nature is exemplified by the employment of the nonsense-mediated decay (NMD) mechanism to degrade defective transcripts, while the nonrandom nature of splicing is reflected by thousands of conserved splicing changes between developmental stages and tissues (Barbosa-Morais et al. 2012a; Merkin et al. 2012) (see also Chaps. 12 and 14).

The chief executioner of splicing is the spliceosome, a massive complex comprising five small nuclear ribonucleoprotein particles (snRNPs) and a large number of auxiliary proteins. Through a series of biochemical reactions these components recognize the core splicing signals, namely the 5′ and 3′ splice site (5′ss, 3′ss), the branch point sequence (BPS), and the polypyrimidine tract (cf Matlin et al. 2005).

In addition to the spliceosomal component recognition of pre-mRNA, in human there are over 270 RNA-binding proteins (RBPs) with an RNA recognition motif (RRM) or K homology (KH) domain (Ray et al. 2013). Many of those proteins have been shown to operate as splice factors (SFs), influencing the splicing process and consequently exon inclusion. SFs bind splicing regulatory elements (SREs) that are commonly classified based on their effect and location into four categories: exonic splicing enhancers (ESEs), exonic splicing silencers (ESSs), intronic splicing enhancers (ISEs), and intronic splicing silencers (ISSs). Many splice factors are ubiquitous in most cell types. Such ubiquitous factors include members of the SR (Ser–Arg) protein family and the family of heterogeneous nuclear RNPs (hnRNPs). Other factors, such as FOX1/2 or the members of the muscleblind-like (MBNL) RBP family, change their expression and protein levels across developmental stages and tissue types (cf Chen and Manley 2009).

Structural elements have also been shown to affect splicing outcome. For example, local RNA structures can serve to expose or obscure the recognition of sequence motifs by the core spliceosome machinery or auxiliary splicing factors (Hiller et al. 2007). A recent study found 22 RBPs display a significant preference in vitro for or against predicted hairpin loops (Ray et al. 2013), while other studies focusing on 3′ splice site processing reported a relation between the 3′ splice site strength, the polypyrimidine tract, the branch point distance, and its composition (Corvelo et al. 2010).

Overall, the picture that has emerged from years of elaborate research efforts with model systems is that of a complex, stochastic, yet highly regulated and context-specific process. Consequently, deducing a splicing "code" in the form of predictive rules for splicing given primary sequence proved challenging. Fortunately, the advances in experimental and computational methods have offered new hope for deriving such predictive splicing code models. RNA-Seq experiments now offer unprecedented quantification accuracy of expression level at exonic resolution. Methods involving cross-linking immunoprecipitation (CLIP) followed by sequencing allow researchers to identify thousands of in vivo

RBP-binding sites (Kishore et al. 2011), and methods such as RNAcompete (Chap. 2) allow systematic in vitro analysis of the RNA sequence preferences of RBPs. These data need to be matched with adequate computational analysis, which is the subject of the following section.

2 Computational Analysis of Splicing Regulation

The computational analysis of splicing regulation is tightly coupled to what data are available and what information one may hope to retrieve from these data. The first large datasets used for splicing analysis were derived from EST/cDNA libraries. Researchers used these datasets to build models for the core splicing signals (Lim and Burge 2001; Yeo and Burge 2004), to identify enriched k-mers in exons vs. introns, near strong vs. weak splice sites (Fairbrother et al. 2002; Zhang and Chasin 2004), and in highly conserved intronic regions (Yeo et al. 2007). In parallel, functional SELEX (systematic evolution of ligands by exponential enrichment) enabled researchers to characterize position-specific scoring matrices (PSSMs) representing the binding affinity of several RBPs in vitro (Cartegni et al. 2003). Researchers then used these motif occurrences as features to discriminate between alternative and constitutively spliced exons (Dror et al. 2005).

The next advance in computational analysis of splicing regulation occurred when high-throughput experiments enabled gene expression measurements at exonic resolution. Using dedicated microarrays (Pan et al. 2004; Sugnet et al. 2006) and later high-throughput sequencing (RNA-Seq), researchers could detect relative changes in exon inclusion across experimental conditions. Researchers used these new technologies to monitor global splicing changes between tissues, developmental stages, or following a splice factor knockdown (Fagnani et al. 2007; Castle et al. 2008; Kalsotra et al. 2008). The computational analysis that followed these experiments concentrated mostly on detecting motif enrichment in exons that exhibited an inclusion change (Sugnet et al. 2006; Fagnani et al. 2007; Castle et al. 2008).

Several new technologies have enabled researchers to characterize RBP-binding elements at unprecedented detail and resolution. UV cross-linking immunoprecipitation (CLIP) followed by microarray or sequencing enabled researcher to identify thousands of in vivo binding locations of RBPs (Kishore et al. 2011; Ule et al. 2006; Licatalosi et al. 2008). These experiments led to computational analysis focused on characterizing the binding site motifs and relative locations of RBPs (aka "Motif Maps" (Ule et al. 2006)). Combining CLIP with the matching SF knockdown followed by RNA-Seq also enabled researchers to derive high-confidence sets of splice factor's target genes (Wang et al. 2012) and to compare targets of many different splice factors (Huelga et al. 2012). Recently, researchers used RNACompete to systematically characterize the in vitro binding preference of 207 RBPs. This large-scale study enabled the researchers to analyze binding motif similarities across species and by amino acid sequence identity in their respective RBDs (Ray et al. 2013).

The high-throughput data described above and the matching computational analysis gave immensely useful genome-wide "slices" of the overall complex process of splicing and its regulation. Returning to the original challenge of creating a splicing "code" model, it therefore seems natural to integrate these sources of data into a unified and predictive framework.

3 Predictive Splicing Code Models

The high amount of diverse and noisy measurements produced by sequencing experiments combined with the stochastic nature of the splicing process motivated the application of machine learning techniques to construct predictive splicing code models. Applying machine learning algorithms can be thought of as a process by which we take a template for a certain model type and set or "train" it for a specific task. Such a template can be, for example, that for simple linear models that weigh splice factor-binding site occurrences around an exon to predict the exon inclusion level. The training in this case would involve feeding the algorithm many examples of exons, the binding sites around each of them, and their respective inclusion. During training the algorithm would then try to find what weights to associate with each motif, or "feature," so as to minimize some predefined error function. Various definitions for the error function may be used (e.g., average squared loss or average absolute deviation) but intuitively they will all measure how "far" the predictions, based on the motif occurrences and weights, are from the observed exon inclusion levels. In what follows we review some of the key issues pertaining to the development of such algorithms for deriving predictive splicing models.

3.1 Splicing Quantification

Deriving a predictive splicing code involves fitting a model to some experimental measurements of splicing outcome, as done using RNA-Seq. The raw output of high-throughput experiments requires considerable preprocessing before these can be fed into any splicing learning algorithm. Many works have addressed the processing of raw sequence reads and reviewing those is outside the scope of this chapter. Importantly, quantifying high-throughput mRNA measurements can be divided into three main categories: quantifying general gene expression, quantifying transcript abundance, and quantifying local exon inclusion levels. The available data, based on short sequence reads or microarray probes, combined with the locality of many splicing determinants, have led to a focus on the last category with respect to splicing prediction algorithms.

The first splicing code was derived from custom junction probe arrays. In that work, the percent spliced in (PSI) of exons was quantified from microarray probes using the GenASAP method (Shai et al. 2006). After quantifying the inclusion of

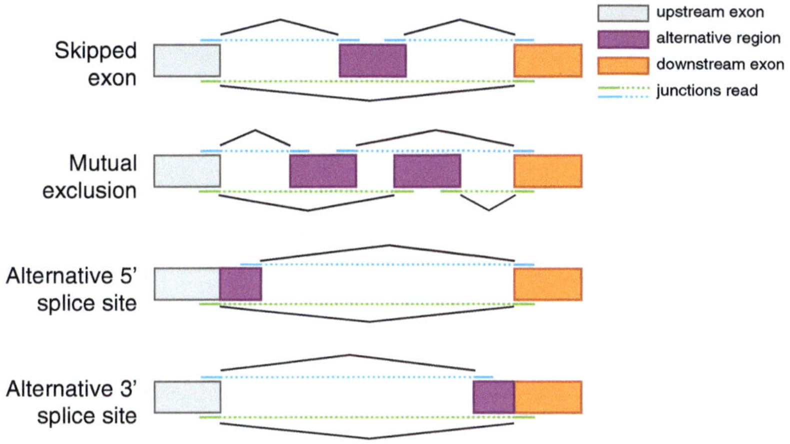

Fig. 13.1 RNA-Seq reads mapped to alternatively spliced exons. Junction reads are mapped across splice junctions to exonic segments in the inclusion (*blue*) or the exclusion isoforms (*green*), capturing different types of alternative splicing

3,707 exons in 27 mouse tissues and cell lines, a second algorithm based on sparse factor analysis was applied to the estimated PSI values (Barash et al. 2010a). This second algorithm helped overcome much of the biological and experimental noise in the estimated PSI values by identifying four main tissue groups where splicing changes commonly occurred: central nervous system, muscle, digestive, and embryonic. A change in exon inclusion level in those tissue groups was termed a splicing "pattern." Every exon was assigned a probability of exhibiting such a pattern given the data, i.e., the chance that it was differentially included, excluded, or did not change in each of these four tissue groups.

With the advance of RNA-Seq, many methods were developed specifically for the task of exon inclusion quantification. The short RNA-Seq reads can be mapped to exon bodies or across exon junctions in order to quantify relative abundance of locally defined alternative splicing types such as cassette exons and alternative 3′ and 5′ splice sites (Fig. 13.1). Some methods, such as RSEM (Li and Dewey 2011) and MISO (Katz et al. 2010), can be used for the entire transcript quantification or be applied to short segments, such as those involving cassette exons. MISO, for example, gives its user a posterior distribution over exon inclusion levels given the observed reads, and uses that to report confidence in cases where a significant change in exon inclusion is detected between two experiments. The method MATS (Shen et al. 2012) shares a similar probabilistic framework for read assignment to transcripts, but focuses on identifying differential splicing between experiments. MATS does not report exon inclusion levels in each experiment. Instead, pairs of experiments are directly compared against each other, with a shared uniform prior, and the posterior confidence in a splicing change is reported (e.g., a 90 % confidence in exon inclusion increase of at least 15 %). DESEQx is also geared towards detecting exon inclusion changes across many samples, including replicates (Anders et al. 2012). Unlike MISO and MATS, it uses a generalized linear model to identify cases where the exon

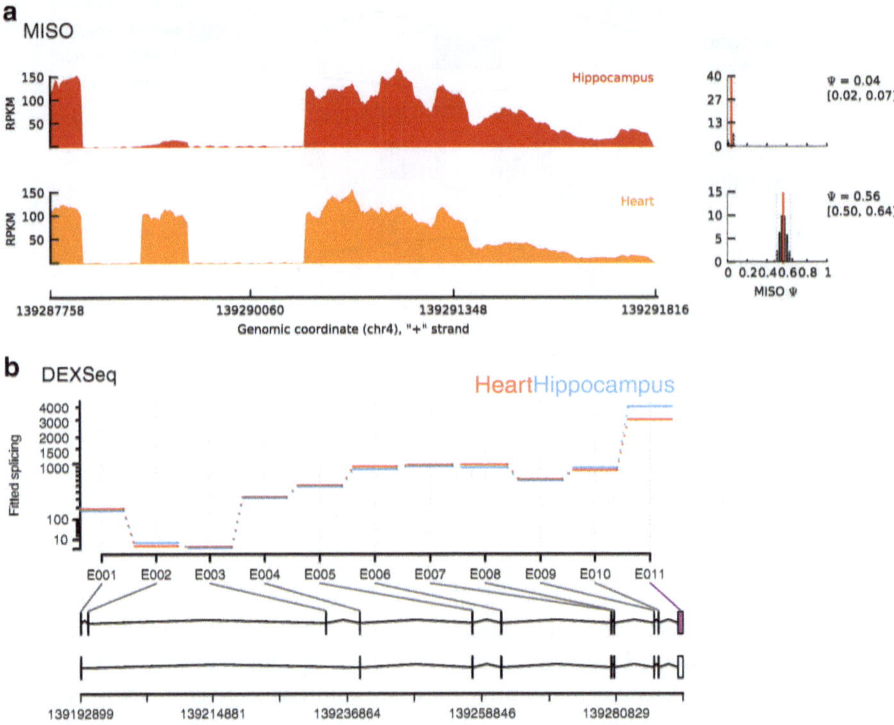

Fig. 13.2 Methods for quantifying exon inclusion levels and for detecting differential splicing. (**a**) RNA-Seq reads for cassette exon 11 in the *Capzb* gene from two experiments analyzed using MISO (Katz et al. 2010). MISO's output includes (*left*) the reads mapped to the alternative exon and two flanking exons in the two conditions (hippocampus in *red* and heart in *orange*). On the *right*, the output includes visualization of the posterior distribution over Ψ (PSI, or "percent spliced in"), with the maximum a posteriori value (*red*) and a confidence interval (*grey lines*). (**b**) Analysis of the same dataset using DEXSeq (Anders et al. 2012). DEXSeq requires a set of known transcript structures, shown at the *bottom*. The *x*-axis represents the different exons of the gene, and the *y*-axis is the fitted expression under the DEXSeq model in hippocampus (*blue*) and heart (*red*). *Capzb* exon 11 was detected as differentially expressed in the two conditions and marked in *purple* (*bottom right*). Note that the flanking downstream exon 12 is annotated as a 3′ UTR region and is therefore not included in the DEXSeq diagram

inclusion level changes significantly compared to the overall gene expression. Examples from the output of the different methods can be seen in Fig. 13.2.

The choice of how splicing measurements are quantified determines much of the downstream analysis. For example, the output of methods such as MISO can be used with regression techniques to fit the quantified exon inclusion levels. The output of DESeqx is easy to parse into binary labels of exons that change between two conditions and those that do not. More generally, the output may be binary, discrete, or continuous. It may represent absolute inclusion levels (e.g., MISO), or relative inclusion (e.g., MATS, DESeqx).

Finally, it is important to note that using a learning algorithm to train a splicing code model does not necessitate exhaustive annotation of the genome. It is important though to strike a good balance between quantity and quality of the output generated by the splicing quantification algorithms. A set of high-quality cases which is too small to deduce a model from and a large set with many erroneously labeled exons will both make the downstream learning task much harder. Using several methods or the procedures described in Sect. 3.4 may help to define a good set with which to train and test a code model.

3.2 *Defining a Feature Set*

The feature set is what a machine-learning algorithm uses to predict the observed quantifications of splicing levels (Fig. 13.3). Genomic features used for splicing prediction can be roughly divided into the following categories: primary sequence motifs, structural features, and sequence conservation (Barash et al. 2010b).

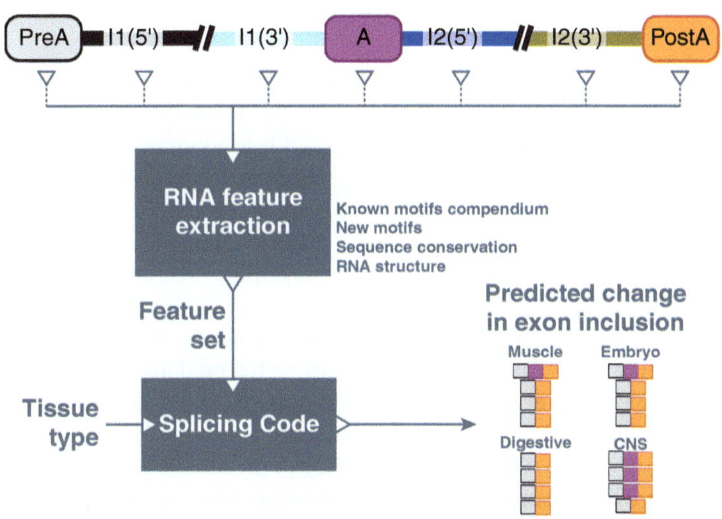

Fig. 13.3 A splicing code analysis scheme. RNA-related features such as occurrences of known splice factor-binding sites, local secondary structure, and conservation levels are extracted for several regions around a putative alternative exon. These regions include the alternative exon ("A"), the flanking up ("PreA") and downstream ("PostA") exons, and proximal intronic regions (5′ and 3′ ends of the upstream intron "I1," downstream intron "I2"). The splicing code model then uses this feature set to output tissue-dependent predictions for the relative inclusion of the alternative exon, shown schematically on the *right*. The splicing code is first trained so that its predictions fit quantified exon inclusion levels from RNA-Seq experiments (see Fig. 13.2 and main text). Once trained, the code can then be used for various tasks such as detecting novel regulatory elements (Barash et al. 2010b; Gazzara et al. 2014), predicting exon inclusion in related species (Barbosa-Morais et al. 2012b), and predicting novel tissue-dependent splicing (Barash et al. 2010b, 2013; Gazzara et al. 2014)

Primary sequence motifs include short sequence kmers, IUPAC consensus motifs, probabilistic models such as position-specific scoring matrices (PSSM), and nucleotide frequencies. These motifs are typically collected from experimentally validated *cis* regulatory elements or computational motif enrichment analysis (Yeo et al. 2007; Cartegni et al. 2003). Structural features include relative length of exonic and intronic regions and marginal probabilities for secondary structure-free regions. In addition, CLIP-Seq data can be parsed to represent splice factor-binding sites (Ule et al. 2006; Licatalosi et al. 2008).

For general splicing prediction the feature set may grow to be quite large. For example, the first tissue-dependent splicing code derivation used more than a thousand features as input (Barash et al. 2010b). In contrast, specific modeling tasks may include a much smaller feature set. For example, modeling Nova-regulated alternative exons involved a Bayesian network with 17 nodes representing entities such as YCAY Nova motif clusters, sequence conservation, reading frame, and AS conservation (Zhang et al. 2010).

3.3 Choice of Model, Target Function, and Optimization Technique

The computational derivation of a predictive splicing code requires specifying the type of model to learn, the loss function, and the technique used to optimize it. Naturally, the appropriate choice of these should lead to increased prediction accuracy. It is also desirable to construct models that can be interpreted in terms of putative regulatory mechanisms and assign a measure of confidence to those. In general, many different machine-learning algorithms may be applied to achieve splicing prediction. However, several elements should be taken into account irrespective of the particular algorithm chosen.

One key element is whether the dependency structure between the features is known. In specific cases, a small set of features with a predefined dependency structure may suffice (Zhang et al. 2010). Typically though, given the partial understanding of splicing regulation, many putative regulatory features will be collected and the relations between those will not be known. Part of the learning process would therefore be to infer which features are relevant, in what combinations, and in what context. The expanded set of putative regulatory features and the limited amount of experimental data require a learning technique that controls for model complexity in order to avoid over-fitting the model to the training data. In the first derivation of a splicing code model the authors used a variant of the boosting learning technique (Barash et al. 2010b). In a later work, a spike and slab prior was used to encourage the learned model to use only a small subset of features and a Markov Chain Monte Carlo (MCMC) technique was used to sample from the space of possible model structures (Xiong et al. 2011). Similarly, the type of dependency structure allowed by the model should be a consideration. While nonlinear models give improved accuracy controlling the complexity of the inferred interactions is crucial (Xiong et al. 2011).

Another important aspect to consider is the ability to incorporate prior biological knowledge and interpret the resulting models. Some methods may exhibit significant improvement in splicing prediction accuracy when coupled with feature preprocessing steps such as PCA, but interpretation of the learned models becomes challenging (Xiong et al. 2011). Models previously used for splicing prediction include a Bayesian network, mixtures of decision trees, and a Bayesian neural network (Barash et al. 2010b; Zhang et al. 2010; Xiong et al. 2011). In the case involving a Bayesian network, the model was used to represent a fixed dependency structure reflecting prior expert knowledge over 17 variables. The mixtures of decision trees were used for the original derivation of a tissue-specific splicing code model as it offered a relatively simple and intuitive representation of stochastic "rules," elevated the need to scale the various features, yet allowed for nonlinear dependency structures. The Bayesian neural network model (Fig. 13.4a) may not be as easily interpretable as the decision trees but exhibited improved prediction accuracy while still maintaining the ability to interrogate the learned model structure (Xiong et al. 2011).

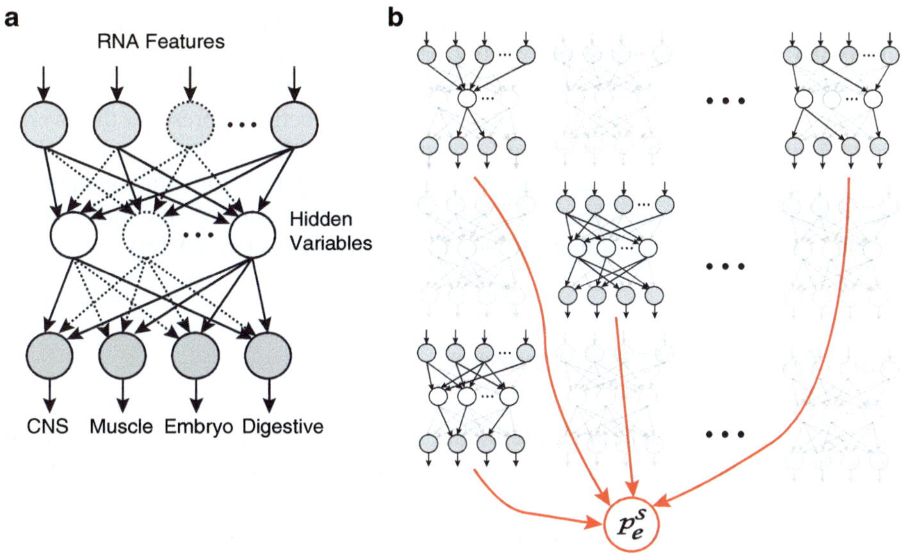

Fig. 13.4 Using a Bayesian neural network for splicing code modeling. (**a**) The Bayesian neural network model. Variables are represented as nodes and nonzero weight dependencies are represented as *arrows*. The model's input is RNA features (*top*) and the output is inclusion probabilities in the different tissues (*bottom*). Hidden units (*middle*) connect the input and output units via sigmoid "activation" functions. *Dashed lines* represent additional nodes and edges that can be added during the learning process. Specifically, the number of hidden units, the connections between the three layers of the networks, and the nonzero weights associated with the sigmoid functions are part of the learning process when trying to fit observed exon inclusion levels (Fig. 13.2) using the exons' RNA features. (**b**) Instead of assuming a single structure for the splicing code model, a set of model structures are sampled from the posterior over possible model structures given the data. The sampled models (*highlighted structures*) are then combined (*red arrows*) to give splicing predictions (p_e^s, *bottom node*) for unseen test exons

The choice of what function to optimize depends in large part on the quantification procedure chosen (see Sect. 3.1). For example, for binary classification SVM algorithms may optimize the hinge loss while probabilistic algorithms such as logistic regression will minimize the conditional log likelihood using a logit function (Scholkopf and Smola 2001; Bishop 2007). Given the nature of the data, it is important to choose a function that will take into account both the uncertainty and the possible noise in the splicing quantifications. One approach to achieve this is to use discretized labels representing either relative changes in inclusion levels between conditions (e.g., "up," "down," and "no change") or absolute levels of inclusion (e.g., "low," "mid," "high"). Each such label for each exon is then assigned a probability or a "confidence." The optimization function can then be defined as

$$H = \sum_e \sum_s q_e^s \log\left(\frac{p_e^s}{\overline{q}^s}\right),$$

where "*e*" sums over exons and "*s*" sums over the three types of splicing changes described above. Here p^s is the probability for the type of change *s* as predicted by the model, q^s is the confidence that that change occurred based on the experimental data, and $\overline{q}^s$ is the marginal probability or what a Naïve predictor would guess. Note that the sum over q^s for any given exon may be less than 1, representing overall less confidence in the exon's labeling. The above function represents the overall *code quality*. It has the nice property of minimizing the Kullback–Leibler (KL) divergence between the two distributions or maximizing the information gain in bits per instance, and is equivalent to optimizing the conditional log likelihood of the labels given the data in a probabilistic framework (Barash et al. 2010b).

3.4 Evaluating Model Accuracy

The usefulness of a splicing code model is largely determined by its accuracy. A model's accuracy refers to its ability to predict splicing outcome for yet unseen "cases," or exons. If structure learning is part of the task, we may also evaluate how well the structure captures correct regulatory interactions.

In machine learning, it is common to evaluate prediction accuracy by using the K-fold cross validation (CV) procedure. In this procedure, the data is randomly divided into K equally sized subsets. Iterating K times, each subset is then held out as test data while the other K-1 subsets are used to train the model. Finally, the overall accuracy is estimated using the average prediction accuracy over the K test sets.

If the learned model is more of a "black box" that uses complex feature transformations the model structure is hard to interrogate in terms of inferred regulatory interactions. Such a model can still be probed for regulatory effects such as "does removing this FOX-binding site affect predictions for the adjacent exon?" but general "rules" of regulation are hard to decipher.

Since the learned model serves as a proxy to the underlying regulation, it is also important to take into account that there is not a single combination of regulatory elements that determines a specific splicing outcome. Consequently, algorithms that agglomerate many possible models can offer improved performance and robust tests of regulatory structures. For example, a bootstrap-like method was used for learning many mixtures of decision trees from subsamples of the data in Barash et al. (2010b). All the features selected during this process were evaluated in terms of the number of times they were selected and their overall weight in the trees. Selected features or combinations of those were then tested for enrichment in exons that exhibited specific splicing patterns (e.g., increased inclusion in the brain). In another work, a Bayesian neural network was used with features connected via a sparse prior to a varying number of hidden units. These hidden "activation" units were in turn connected to the observed inclusion probabilities. A Markov Chain Monte Carlo (MCMC) technique was used to agglomerate thousands of possible structures from the posterior distribution over network structures (Fig. 13.4b), which were then used for predicting splicing outcome for test exons. In this model the number of hidden units and the connections in the network were not fixed and were part of the learning process. Consequently, the hidden units were not identifiable but the ensemble of models was interrogated to detect combination of features that were commonly selected and wired together. In general, identified features can be compared against known regulatory connections from the literature, and tested for their predicted affect on tissue-dependent exons using mini-gene reporter assays (Barash et al. 2010b). If a feature corresponds to a known splice factor-binding site, the factor can be knocked down to test a regulatory effect (Gazzara et al. 2014). For example, a recent study used CLIP-Seq and siRNA knockdown to identify several hundred Mbnl/2-dependent exons in mouse brain, muscle, and heart. Splicing code predictions for those were then compared against predictions for nontissue-dependent exons, achieving an AUC of ~94 % (Barash et al. 2013).

Finally, it is important to note the need to validate putative causal effects. In principle, modeling causal relations using probabilistic models is plausible but both the data and the models need to comply with certain constraints (Pearl 2000). For splicing predictions, some features such as sequence conservation may be highly useful yet are obviously not causal. Careful choice of the represented features and the allowed connections between those with follow-up experimental validations are therefore crucial. For example, in mammals (U)GCAUG-binding sites for Fox-1 (A2bp1) and its paralog Fox-2 (Rbm9) are enriched in the introns proximal to differentially spliced exons in brain and muscle tissues. In general, these binding sites were found more frequently in introns upstream of exons that were differentially excluded in those tissues. However, in the case of the mouse Daam1 exon 16 the exon was predicted to be differentially included in brain tissues, even though a Fox1/2-binding site was found upstream of the exon. Mutating the binding site was found to further increase exon inclusion, pointing to a possible balancing act between the various regulators across different tissues (Barash et al. 2010b).

4 Using AVISPA for Exon *In Silico* Splicing Analysis

The previous sections reviewed how splicing code models can be derived and how they can help us increase our understanding of splicing regulation beyond single case or single regulatory element analysis. However, these models are of limited use if they remain a theoretical proof of concept applied in specific studies. To address the need for a "practical" splicing code, AVISPA was recently developed as a web tool for *in silico* tissue-dependent splicing analysis (Barash et al. 2013). Unlike previous works, which focused on a predetermined set of alternative exons, AVISPA allows users to plug in an exon from any gene of interest and perform a whole array of splicing predictions and regulatory element analysis. It is designed to make the splicing code models accessible to researchers working in diverse areas such as RNA biogenesis, development, and disease studies, who are not necessarily splicing or computational experts. Here, we review how AVISPA can be used for such a task by focusing on exon 11 of the mouse *Capzb* gene as a case study.

AVISPA is integrated into the Galaxy framework (Giardine et al. 2005). First, a user uploads a query exon by specifying either the genomic sequence or coordinates. AVISPA then maps it to internal databases of known cases of alternative splicing, known transcripts, or directly to the genome. If the query cannot be mapped, an error is reported. Next, the uploaded data is used to submit an AVISPA splicing analysis query. The query processing comprises two main stages. First, the exon is evaluated for whether it is likely to be alternatively or constitutively (i.e., always included) spliced. This is in contrast to previous works, which concentrated on predefined sets of alternative exons and thus precluded genome-wide analysis. Next, the exon is scored for how likely it is to exhibit a tissue-dependent splicing profile. For each stage of this analysis, the exon's score is compared to a large pool of pre-annotated exons in the AVISPA database, allowing the user to control the statistical significance of the results reported. Finally, every putative regulatory feature is evaluated for enrichment in the query, and every putative regulatory motif is removed *in silico*. The effect of feature removal on splicing predictions is quantified and compared using a normalized feature effect (NFE) score (Barash et al. 2013).

The AVISPA analysis pipeline is illustrated here for exon 11 of the mouse *Capzb* gene. The C terminus of Capzb, an actin filament-capping protein, is modified by alternative inclusion of exon 11 in heart and skeletal muscle (Schafer et al. 1994). AVISPA predicts the exon to be alternative with an estimated false-positive rate (FPR) of 0.028. As for tissue-dependent splicing, AVISPA predicts that it is most likely differentially included in muscle (relative rank 0.014 compared to a known set of regulated exons) and excluded in digestive and embryonic tissues. It also predicts differential splicing in brain, but is not confident whether it would be differentially included or excluded in that tissue. Figure 13.5 shows a histogram of the normalized feature effect (NFE) for the motifs that affect the splicing predictions in muscle. The most dominant effectors are the CU-rich elements in the up- and downstream introns, known to bind members of the polypyrimidine tract-binding protein family (Ptb), and ACUAAY elements, known to bind Quaking (QkI) of the STAR (signal transduction and activation of RNA) family of proteins. This result is in line with a recent study showing Ptb- and QkI-regulated exon 11 inclusion in C2C12 myoblasts (Hall et al. 2013).

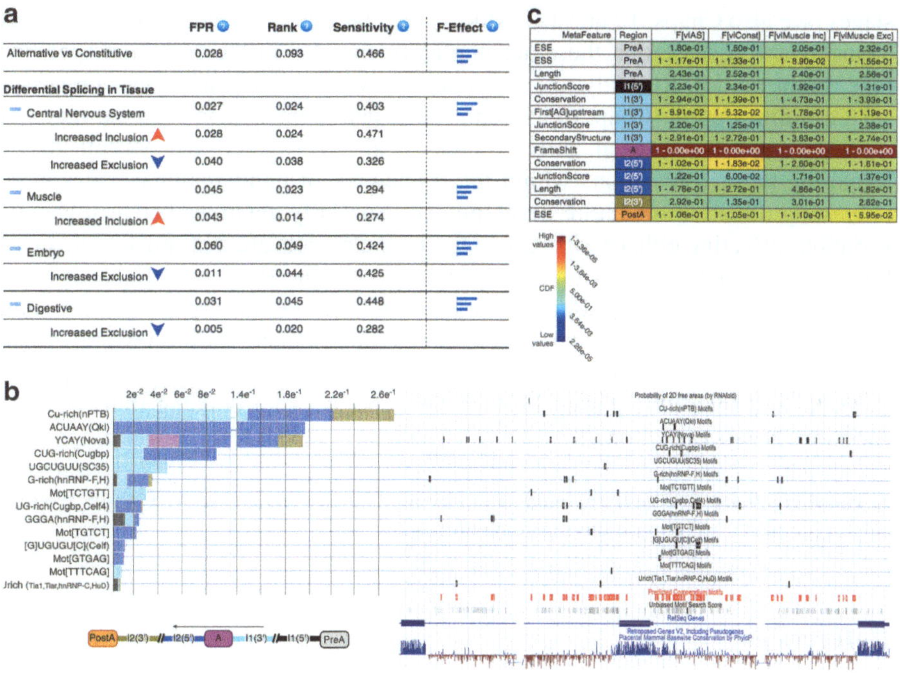

Fig. 13.5 AVISPA's splicing analysis. (**a**) Exon 11 of *Capzb* gene, shown in Fig. 13.2, is predicted by AVISPA to be an alternative exon with a false-positive rate (FPR) of 0.028. It is also predicted to be differentially included in muscle tissues (rank 0.014), and excluded in embryonic and digestive tissues. AVISPA predictions are not conclusive regarding the direction of change in CNS (see main text). (**b**) Enrichment of non-motif features. Each feature (*row*) in the query is compared against several reference sets (*columns*) indicated by the header (*AS* alternative set, *Const* constitutive set, *Muscle Inc* muscle inclusion set, and *Muscle Exc* muscle exclusion set). The table entries correspond to the relative rank of the query's feature value compared to the reference set of each column. Enrichment is visually represented as a heat map (*below*) where *red* is associated with high values, *green* is associated with values closer to the median, and *blue* is associated with low values. The region containing each feature is indicated in the second column, with colors matching the regional legend above. (**c**) The putative regulatory motifs with the highest normalized feature effect (NFE) for muscle-dependent splicing are listed on the *left* along with name of the RBP known to bind that motif. The matching stacked bars represent each feature's NFE score while the *colors* of the bars correspond to the region in which the motif was found (see legend below). A custom track maps these motif occurrences to the genome browser (*right*), where they are combined with other information such as conservation (*bottom right*)

5 Future Directions in Splicing Code Development

Current splicing code models can be viewed as capturing some rough outlines of a complex biological process (splicing) with a crude computational camera. Many additions and refinements are still necessary. First, the models described focus only on the most common type of alternative splicing in mammals: cassette exons. Finer resolution is also needed in terms of tissue specificity and developmental stages (see also Chaps. 12 and 14). Models thus far do not include changes to *trans* elements and assume that these are implicitly encoded in the condition specified (e.g., brain tissue). Recently, the cross talk between splicing and other processes such as transcription (Chap. 6) and epigenetic determinants has been the focus of much research (Luco et al. 2010; Shukla et al. 2011; Luco and Misteli 2011). Exploring these connections within the framework of computational codes is an interesting direction for future developments. Similarly, relations between genetic variations affecting either *trans* (Chap. 6) or *cis* elements are a promising direction for splicing model applications. Finally, much work is needed to develop code models across different species. Recent work showed that a mouse splicing code could be applied successfully to other species, including human (Barbosa-Morais et al. 2012b). Nonetheless, higher accuracy models are likely to combine species-specific elements and should help shed light on the process of introducing transcriptome variations across evolution.

Acknowledgments The authors would like to thank Matthew Gazzara and Alex Amlie-Wolf for helpful comments and suggestions regarding the manuscript.

References

Anders S, Reyes A, Huber W (2012) Detecting differential usage of exons from RNA-seq data. Genome Res 22:2008–2017. doi:10.1101/gr.133744.111

Barash Y, Blencowe BJ, Frey BJ (2010a) Model-based detection of alternative splicing signals. Bioinformatics 26:i325–i333

Barash Y, Calarco JA, Gao W et al (2010b) Deciphering the splicing code. Nature 465:53–59. doi:10.1038/nature09000

Barash Y, Vaquero-Garcia J, Gonzalez-Vallinas J et al (2013) AVISPA: a web tool for the prediction and analysis of alternative splicing. Genome Biol 14:R114. doi:10.1186/gb-2013-14-10-r114

Barbosa-Morais NL, Irimia M, Pan Q et al (2012) The evolutionary landscape of alternative splicing in vertebrate species. Science 338:1587–1593. doi:10.1126/science.1230612

Bishop CM (2007) Pattern recognition and machine learning, Information science and statistics. Springer, New York, 1st ed. 2006. Corr. 2nd printing 2011

Cartegni L, Wang J, Zhu Z et al (2003) ESEfinder: a web resource to identify exonic splicing enhancers. Nucleic Acids Res 31:3568–3571

Castle JC, Zhang C, Shah JK et al (2008) Expression of 24,426 human alternative splicing events and predicted cis regulation in 48 tissues and cell lines. Nat Genet 40(12):1416–1425

Chen M, Manley JL (2009) Mechanisms of alternative splicing regulation: insights from molecular and genomics approaches. Nat Rev Mol Cell Biol 10:741–754. doi:10.1038/nrm2777

Corvelo A, Hallegger M, Smith CW, Eyras E (2010) Genome-wide association between branch point properties and alternative splicing. PLoS Comput Biol 6:e1001016

Dror G, Sorek R, Shamir R (2005) Accurate identification of alternatively spliced exons using support vector machine. Bioinformatics 21:897–901

Fagnani M, Barash Y, Ip J et al (2007) Functional coordination of alternative splicing in the mammalian central nervous system. Genome Biol 8:R108

Fairbrother WG, Yeh RF, Sharp PA, Burge CB (2002) Predictive identification of exonic splicing enhancers in human genes. Science 297:1007–1013

Gazzara M, Vaquero-Garcia J, Lynch KW, Barash Y (2014) In silico to in vivo splicing analysis using splicing code models. Methods 67(1):3–12

Giardine B, Riemer C, Hardison RC et al (2005) Galaxy: a platform for interactive large-scale genome analysis. Genome Res 15:1451–1455
Hall MP, Nagel RJ, Fagg WS et al (2013) Quaking and PTB control overlapping splicing regulatory networks during muscle cell differentiation. RNA 19:627–638. doi:10.1261/rna.038422.113
Hiller M, Zhang Z, Backofen R, Stamm S (2007) Pre-mRNA secondary structures influence exon recognition. PLoS Genet 3:e204
Huelga SC, Vu AQ, Arnold JD et al (2012) Integrative genome-wide analysis reveals cooperative regulation of alternative splicing by hnRNP proteins. Cell Rep 1:167–178. doi:10.1016/j.celrep.2012.02.001
Kalsotra A, Xiao X, Ward AJ et al (2008) A postnatal switch of CELF and MBNL proteins reprograms alternative splicing in the developing heart. Proc Natl Acad Sci 105:20333–20338
Katz Y, Wang ET, Airoldi EM, Burge CB (2010) Analysis and design of RNA sequencing experiments for identifying isoform regulation. Nat Methods 7:1009–1015. doi:10.1038/nmeth.1528
Kishore S, Jaskiewicz L, Burger L et al (2011) A quantitative analysis of CLIP methods for identifying binding sites of RNA-binding proteins. Nat Methods 8:559–564. doi:10.1038/nmeth.1608
Li B, Dewey C (2011) RSEM: accurate transcript quantification from RNA-Seq data with or without a reference genome. BMC Bioinformatics 12:323
Licatalosi DD, Mele A, Fak JJ et al (2008) HITS-CLIP yields genome-wide insights into brain alternative RNA processing. Nature 456(7221):464–469
Lim LP, Burge CB (2001) A computational analysis of sequence features involved in recognition of short introns. Proc Natl Acad Sci 98:11193–11198
Luco RF, Misteli T (2011) More than a splicing code: integrating the role of RNA, chromatin and non-coding RNA in alternative splicing regulation. Curr Opin Genet Dev. doi:10.1016/j.gde.2011.03.004
Luco RF, Pan Q, Tominaga K et al (2010) Regulation of alternative splicing by histone modifications. Science 327:996–1000. doi:10.1126/science.1184208
Matlin AJ, Clark F, Smith CWJ (2005) Understanding alternative splicing: towards a cellular code. Nat Rev Mol Cell Biol 6:386–398. doi:10.1038/nrm1645
Merkin J, Russell C, Chen P, Burge CB (2012) Evolutionary dynamics of gene and isoform regulation in mammalian tissues. Science 338:1593–1599. doi:10.1126/science.1228186
Pan Q, Shai O, Misquitta C et al (2004) Revealing global regulatory features of mammalian alternative splicing using a quantitative microarray platform. Mol Cell 16:929–941
Pan Q, Shai O, Lee LJ et al (2008) Deep surveying of alternative splicing complexity in the human transcriptome by high-throughput sequencing. Nat Genet 40:1413–1415
Pearl J (2000) Causality: models, reasoning, and inference. Cambridge University Press, Cambridge
Ray D, Kazan H, Cook KB et al (2013) A compendium of RNA-binding motifs for decoding gene regulation. Nature 499:172–177. doi:10.1038/nature12311
Schafer DA, Korshunova YO, Schroer TA, Cooper JA (1994) Differential localization and sequence analysis of capping protein beta-subunit isoforms of vertebrates. J Cell Biol 127:453–465. doi:10.1083/jcb.127.2.453
Scholkopf B, Smola AJ (2001) Learning with Kernels: support vector machines, regularization, optimization, and beyond. MIT Press, Cambridge, MA
Shai O, Morris QD, Blencowe BJ, Frey BJ (2006) Inferring global levels of alternative splicing isoforms using a generative model of microarray data. Bioinformatics 22:606
Shen S, Park JW, Huang J et al (2012) MATS: a Bayesian framework for flexible detection of differential alternative splicing from RNA-Seq data. Nucleic Acids Res 40:e61. doi:10.1093/nar/gkr1291
Shukla S, Kavak E, Gregory M et al (2011) CTCF-promoted RNA polymerase II pausing links DNA methylation to splicing. Nature 479:74–79. doi:10.1038/nature10442
Sugnet CW, Srinivasan K, Clark TA et al (2006) Unusual intron conservation near tissue-regulated exons found by splicing microarrays. PLoS Comput Biol 2:e4
Ule J, Stefani G, Mele A et al (2006) An RNA map predicting Nova-dependent splicing regulation. Nature 444:580–586

Wang ET, Sandberg R, Luo S et al (2008) Alternative isoform regulation in human tissue transcriptomes. Nature 456:470–476. doi:10.1038/nature07509

Wang ET, Cody NAL, Jog S et al (2012) Transcriptome-wide regulation of Pre-mRNA splicing and mRNA localization by muscleblind proteins. Cell 150:710–724

Xiong HY, Barash Y, Frey BJ (2011) Bayesian prediction of tissue-regulated splicing using RNA sequence and cellular context. Bioinformatics 27:2554–2562

Yeo G, Burge CB (2004) Maximum entropy modeling of short sequence motifs with applications to RNA splicing signals. J Comput Biol 11:377–394

Yeo GW, Nostrand EL, Liang TY (2007) Discovery and analysis of evolutionarily conserved intronic splicing regulatory elements. PLoS Genet 3:e85

Zhang XH, Chasin LA (2004) Computational definition of sequence motifs governing constitutive exon splicing. Genes Dev 18:1241–1250

Zhang C, Frias MA, Mele A et al (2010) Integrative modeling defines the nova splicing-regulatory network and its combinatorial controls. Science 329:439–443. doi:10.1126/science.1191150

Index

G.W. Yeo (ed.), *Systems Biology of RNA Binding Proteins*, Advances in Experimental Medicine and Biology 825, DOI 10.1007/978-1-4939-1221-6